Engineering Circuit Analysis

McGRAW-HILL ELECTRICAL AND ELECTRONIC ENGINEERING SERIES

Frederick Emmons Terman, *Consulting Editor*
W. W. Harman, J. G. Truxal, and R. A. Rohrer,
Associate Consulting Editors

Angelakos and Everhart • Microwave Communications
Angelo • Electronic Circuits
Angelo • Electronics: BJTs, FETs, and Microcircuits
Aseltine • Transform Method in Linear System Analysis
Atwater • Introduction to Microwave Theory
Bennett • Introduction to Signal Transmission
Beranek • Acoustics
Bracewell • The Fourier Transform and Its Application
Brenner and Javid • Analysis of Electric Circuits
Brown • Analysis of Linear Time-invariant Systems
Bruns and Saunders • Analysis of Feedback Control Systems
Carlson • Communication Systems: An Introduction to Signals and Noise in Electrical Communication
Chen • The Analysis of Linear Systems
Chen • Linear Network Design and Synthesis
Chirlian • Analysis and Design of Electronic Circuits
Chirlian • Basic Network Theory
Chirlian • Electronic Circuits: Physical Principles, Analysis, and Design
Chirlian and Zemanian • Electronics
Clement and Johnson • Electrical Engineering Science
Cunningham • Introduction to Nonlinear Analysis
D'Azzo and Houpis • Feedback Control System Analysis and Synthesis
Elgerd • Control Systems Theory
Elgerd • Electric Energy Systems Theory: An Introduction
Eveleigh • Adaptive Control and Optimization Techniques
Feinstein • Foundations of Information Theory
Fitzgerald, Higginbotham, and Grabel • Basic Electrical Engineering
Fitzgerald, Kingsley, and Kusko • Electric Machinery
Frank • Electrical Measurement Analysis
Friedland, Wing, and Ash • Principles of Linear Networks
Gehmlich and Hammond • Electromechanical Systems
Ghausi • Principles and Design of Linear Active Circuits
Ghose • Microwave Circuit Theory and Analysis
Glasford • Fundamentals of Television Engineering
Greiner • Semiconductor Devices and Applications
Hammond and Gehmlich • Electrical Engineering
Hancock • An Introduction to the Principles of Communication Theory
Harman • Fundamentals of Electronic Motion
Harman • Principles of the Statistical Theory of Communication
Harman and Lytle • Electrical and Mechanical Networks
Hayashi • Nonlinear Oscillations in Physical Systems
Hayt • Engineering Electromagnetics
Hayt and Kemmerly • Engineering Circuit Analysis
Hill • Electronics in Engineering
Javid and Brown • Field Analysis and Electromagnetics
Johnson • Transmission Lines and Networks
Koenig, Tokad, and Kesavan • Analysis of Discrete Physical Systems
Kraus • Antennas
Kraus • Electromagnetics
Kuh and Pederson • Principles of Circuit Synthesis
Kuo • Linear Networks and Systems

LEDLEY • Digital Computer and Control Engineering
LEPAGE • Complex Variables and the Laplace Transform for Engineering
LEPAGE AND SEELY • General Network Analysis
LEVI AND PANZER • Electromechanical Power Conversion
LEY, LUTZ, AND REHBERG • Linear Circuit Analysis
LINVILL AND GIBBONS • Transistors and Active Circuits
LITTAUER • Pulse Electronics
LYNCH AND TRUXAL • Introductory System Analysis
LYNCH AND TRUXAL • Principles of Electronic Instrumentation
LYNCH AND TRUXAL • Signals and Systems in Electrical Engineering
MCCLUSKEY • Introduction to the Theory of Switching Circuits
MANNING • Electrical Circuits
MEISEL • Principles of Electromechanical-energy Conversion
MILLMAN AND HALKIAS • Electronic Devices and Circuits
MILLMAN AND TAUB • Pulse, Digital, and Switching Waveforms
MINORSKY • Theory of Nonlinear Control Systems
MISHKIN AND BRAUN • Adaptive Control Systems
MOORE • Traveling-wave Engineering
MURDOCH • Network Theory
NANAVATI • An Introduction to Semiconductor Electronics
OBERMAN • Disciplines in Combinational and Sequential Circuit Design
PETTIT AND MCWHORTER • Electronic Switching, Timing, and Pulse Circuits
PETTIT AND MCWHORTER • Electronic Amplifier Circuits
PFEIFFER • Concepts of Probability Theory
REZA • An Introduction to Information Theory
REZA AND SEELY • Modern Network Analysis
RUSTON AND BORDOGNA • Electric Networks: Functions, Filters, Analysis
RYDER • Engineering Electronics
SCHILLING AND BELOVE • Electronic Circuits: Discrete and Integrated
SCHWARTZ • Information Transmission, Modulation, and Noise
SCHWARTZ AND FRIEDLAND • Linear Systems
SEELY • Electromechanical Energy Conversion
SEIFERT AND STEEG • Control Systems Engineering
SHOOMAN • Probabilistic Reliability: An Engineering Approach
SISKIND • Direct-current Machinery
SKILLING • Electric Transmission Lines
STEVENSON • Elements of Power System Analysis
STEWART • Fundamentals of Signal Theory
STRAUSS • Wave Generation and Shaping
SU • Active Network Synthesis
TAUB AND SCHILLING • Principles of Communication Systems
TERMAN • Electronic and Radio Engineering
TERMAN AND PETTIT • Electronic Measurements
TOU • Digital and Sampled-data Control Systems
TOU • Modern Control Theory
TRUXAL • Automatic Feedback Control System Synthesis
TUTTLE • Electric Networks: Analysis and Synthesis
VALDES • The Physical Theory of Transistors
VAN BLADEL • Electromagnetic Fields
WEEKS • Antenna Engineering
WEINBERG • Network Analysis and Synthesis

Engineering Circuit Analysis

Second Edition

William H. Hayt, Jr.
Professor of Electrical Engineering
Purdue University

Jack E. Kemmerly
Professor of Engineering
California State University, Fullerton

McGraw-Hill Book Company

New York
St. Louis
San Francisco
Düsseldorf
Johannesburg
Kuala Lumpur
London
Mexico
Montreal
New Delhi
Panama
Rio de Janeiro
Singapore
Sydney
Toronto

Engineering Circuit Analysis

Library of Congress Catalog Card Number 70-141920

07-027382-0

101112131415 HDHD 798765

This book was set in Laurel by York Graphic Services, Inc., printed on
permanent paper by Halliday Lithograph Corporation, and bound by
The Book Press, Inc. The designer was Merrill Haber; the drawings
were done by John Cordes, J. & R. Technical Services, Inc. The
editors were Michael Elia and Madelaine Eichberg. Robert R.
Laffler supervised production.

Contents

PART TWO: The Transient Circuit

PART THREE: Sinusoidal Analysis

PART FOUR: Complex Frequency

PART FIVE: Two-port Networks

PART SIX: Network Analysis

PART SEVEN: Appendixes

Preface

This book is intended for use with a first course in electrical engineering. In many colleges and universities such a course will be preceded or accompanied by an introductory physics course in which the basic concepts of electricity and magnetism are introduced, most often from the field aspect. Such a background is not a prerequisite, however. Instead, several of the requisite basic concepts of electricity and magnetism are discussed (or reviewed) in the first chapter. Only a basic course in the differential and integral calculus need be considered as a prerequisite, or possibly a corequisite, to the reading of the book. Circuit elements are introduced and defined here in terms of their circuit equations; only incidental comments are offered about the pertinent field relationships.

It is the authors' intention that this text be one from which a student may teach himself; it is written to the student and not to the instructor. If at all possible, each new term is clearly defined when it is first introduced. The basic material appears toward the beginning of each chapter and is explained carefully and in detail; numerical examples are usually used to introduce and suggest general results. Drill problems appear at the end of most sections; they are generally simple, and the answers to the three parts are given in random order. The more difficult problems which appear at the end of the chapters are in the general order of presentation of the text material. These problems are occasionally used to introduce less important or more advanced topics through a guided step-by-step procedure, as well as to introduce topics which will appear in the following chapter. The introduction and resulting repetition are both important to the learning process. In all, there are 234 drill problems, each consisting of three parts, and 447 additional problems at the ends of the chapters. These problems are all new in this edition.

The general order of the material has been selected so that the student may learn as many of the techniques of circuit analysis as possible in the simplest context, namely, the resistive circuit which is the subject of the first part of this text. Basic laws, a few theorems, and some elementary network topology enable most of the basic analytical techniques to be developed. Numerous examples and problems are possible since the solutions are not mathematically complicated. The extension of these techniques to more advanced circuits in subsequent parts of the text affords the opportunity both for review and generalization. This first part of the text may be covered in three to six weeks, depending on the students' background and ability, and on the course intensity.

The second part of the text is devoted to the natural response and the complete response to dc excitation of the simpler *RL*, *RC*, and *RLC* circuits. Differential and integral calculus are necessary, but a background in differential equations is not required. The unit step is introduced as an important singularity function in this part, but the introduction of the unit impulse is withheld until the sixth part when our need for it is greater.

The third part of the text introduces the frequency domain and initiates operations with complex numbers by concentrating on sinusoidal analysis in the steady state. Part three also includes a discussion of average power, rms values, and polyphase circuits, all of which are associated with the sinusoidal steady state.

In the fourth part of the book the complex-frequency concept is introduced and its use in relating the forced response and the natural response is emphasized. The determination of the complete response of sinusoidally excited circuits begins to tie together the material of the first three parts.

The fifth part begins with a consideration of magnetic coupling, which is basically a two-port phenomenon, and logically leads into a consideration of two-port network analysis and the linear modeling of various electronic devices.

Part six of the text describes the use of transform methods in circuit analysis. Starting with the Fourier series description of periodic waveforms, the treatment is extended to nonperiodic forcing and response functions, using the Fourier transform. The final chapter covers the more important Laplace transform techniques and their use in obtaining the complete response of more complicated circuits.

It is felt that the material in this book is more than adequate for a two-semester course, but some selection may be made from the last four or five chapters. No material is included in the text which will not be of some value in the following term; thus, signal-flow graphs, the relationship of circuit theory to field theory, advanced topological concepts, and the details of computer-aided analysis are among those subjects which are relegated to subsequent courses. The goal is the student's ability to write correct circuit equations for any circuit and his ability to solve the less complex cases and to understand the solutions.

A number of changes have been made in this revision. The generally accepted International System of Units has been adopted in this edition, and a notational summary is provided inside the back cover. The material on topology is presented earlier in the text, where it can be utilized to greater advantage. The problems, which have undergone the scrutiny of thousands of students and teachers, have all been replaced. The last two chapters on Fourier and Laplace transforms are new and provide more powerful analytical tools with which to attack problems that defy the more elementary approaches. Finally, the authors wished to place a greater emphasis on dependent voltage and current sources, thus permitting models for electronic devices to be studied from the outset.

Throughout the book there is a logical trail leading from definition, through explanation, description, illustration, and numerical example, to problem-solving ability and to the obvious expectation of the authors that the student continually ask himself, "Why does this happen? How is it related to last week's work? Where do we logically go next?" There is a tremendous amount of enthusiastic momentum in a beginning engineering student, and this may be preserved by providing frequent drill problems whose successful solution confirms the student's progress in his own mind, by integrating the various sections into a coherent whole, by pointing out future applications and more advanced techniques, and by maintaining an interested, inquisitive attitude in the student.

If the book occasionally appears to be informal, or even lighthearted, it is because the authors feel that it is not necessary to be dry or pompous to be educational. An amused smile on the face of a student is seldom an obstacle to his absorbing information. If the writing of the text had its entertaining moments, then why not the reading too?

Much of the material in the text is based on courses taught at California State College, Fullerton, and at Purdue University.

William H. Hayt, Jr.
Jack E. Kemmerly

Engineering Circuit Analysis

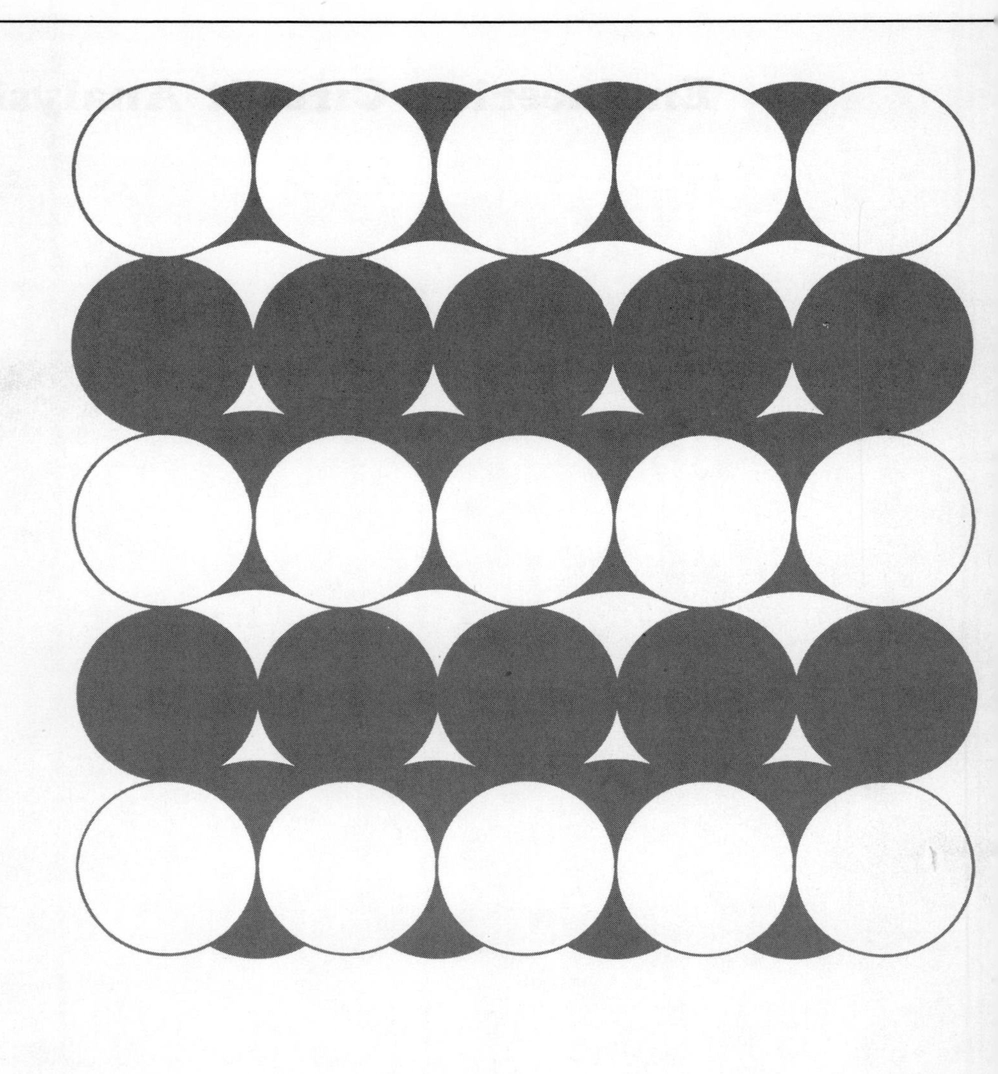

Part One
THE RESISTIVE CIRCUIT

Chapter One
Definitions and Units

1-1 INTRODUCTION

Twenty years ago, an introductory textbook on electric-circuit analysis would have begun with a description of the electrical engineer's place in science and industry, his glorious achievements of the past, and the lucrative and challenging life ahead. Now, however, it is becoming increasingly common for students of mechanical engineering, civil engineering, chemical engineering, and other engineering curricula, as well as an occasional student of mathematics or physics, to study introductory circuit analysis with the electrical engineer. As a matter of fact, courses based largely on beginning circuit analysis are now being taken by engineering students before they have become identified with a particular branch of engineering.

If we have already entered or intend to enter an electrical engineering program, then circuit analysis simply may represent the introductory course in our chosen field. If we are associated with some other branch of engineering, then circuit analysis may represent a large fraction of our total study of electrical engineering, but it also enables us to continue our electrical work in electronics, instrumentation, and other areas. Most important, however, is the possibility given to us to broaden our educational base and become more informed members of a team which may be primarily concerned with the development of some electrical device or system. Effective communication within such a team can be achieved only if the language and definitions used are familiar to all.

Few of the engineering accomplishments of the recent past can be credited to a single individual. The era of the Edison-type inventor has passed, and a graduating engineer should expect to become part of a group consisting of many kinds of engineers, applied mathematicians, and physicists. The group effort will be coordinated by technically trained managers, and the technical products will be produced, sold, and often operated and maintained by men having scientific or engineering training. The engineering graduates of today are not all employed solely to work on the technical design aspects of engineering problems.

In order to contribute to the solution of these engineering problems, an engineer must acquire many skills, one of which is a knowledge of electric-circuit analysis.

We shall begin this study by considering systems of units and several basic definitions and conventions. For those who have no knowledge of basic electricity and magnetism, the elementary subject material is included in this chapter. It should be rapid reading for anyone having an adequate background in introductory physics, but it should be read carefully by all. After these introductory topics have been mastered, we can then turn our attention to a simple electric circuit.

1-2 SYSTEMS OF UNITS

We must first establish a common language. Engineers cannot communicate with one another in a meaningful way unless each term used is clear and definite. It is also true that little learning can be achieved from a textbook which does not define carefully each new quantity as it is introduced. If we speak in the vague generalities of a television commercial—"gets clothes up to 40 per cent whiter"—and do not bother to define whiteness or to provide units by which it may be measured, then we shall certainly not succeed in engineering, although we might sell a lot of soap.

In order to state the value of some measurable quantity, we must give both a *number* and a *unit,* such as "3 inches." Fortunately we all use the same number system and know it well. This is not as true for the units, and some time must be spent in becoming familiar with a suitable system of units. We must agree on a standard unit and be assured of its permanence and its general acceptability. The standard unit of length should not be defined in terms of the distance between two marks on a certain rubber band; this is not permanent, and furthermore everybody else is using another standard.

We shall also need to define each technical term at the time it is introduced, stating the definition in terms of previously defined units and quantities. Here the definition cannot always be as general as the more theoretically minded might wish. For instance, it will soon be necessary to define "voltage." We must either accept a very complete and general definition, which we can neither appreciate nor understand now, or else adopt a less general but simpler definition which will satisfy our purposes for the present. By the time a more general definition is needed, our familiarity with the simpler concepts will help our understanding at that time.

It will also become evident that many quantities are so closely related to each other that the first one defined needs a few subsequent definitions before it can be thoroughly understood. As an example, when the "circuit element" is defined it is most convenient to define it in terms of *current* and *voltage,* and when current and voltage are defined, it is helpful to do so with reference to a circuit element. None of these three definitions can be well understood until all have been stated. Therefore, our first definition of the circuit element may be somewhat inadequate, but then we shall define current and voltage in terms of a circuit element and, finally, go back and define a circuit element more carefully. A later study of electromagnetic theory should provide us with a more general definition of both current and voltage.

We have very little choice open to us with regard to a system of units. The one we shall use was adopted by the National Bureau of Standards in 1964; it is used by all the major professional engineering societies and is the language in which today's textbooks are written. This is the *International System of Units* (abbreviated SI in all languages), adopted by the General Conference on Weights and Measures in 1960. The SI is built upon six basic units: the meter, kilogram, second, ampere, degree Kelvin, and candela. We shall look at the definitions of the first four of these basic units below. Standard abbreviations for them and other SI units are listed inside the back cover of this text and will be used throughout our discussions.

In the late 1700s the meter was defined to be exactly one ten-millionth of the distance from the earth's pole to its equator. This distance was marked off by two fine lines on a platinum-iridium bar which had been cooled to zero degrees Celsius (°C) (formerly centigrade). Although more accurate surveys have shown since that the marks on the bar do not represent this fraction of the earth's meridian exactly, the distance between the marks was nonetheless accepted internationally as the definition of the standard meter until 1960. In that year the General Conference agreed to define the meter (m) as 1,650,763.73 times the wavelength of radiation of the orange line of krypton 86. The definitions are equivalent, but the newer definition is more permanent and reproducible.

The basic unit of mass, the kilogram (kg), was defined in 1901 as the mass of a platinum block kept with the standard meter bar at the International Bureau of Weights and Measures in Sèvres, France. This definition was reaffirmed in 1960. The mass of this block is approximately equal to 1000 times the mass of 1 cm^3 of pure water at 4°C.

The third basic unit, the second (s), was defined prior to 1956 as $1/_{86,400}$ of a mean solar day. At that time it was defined as $1/_{31,556,925.9747}$ of the tropical year 1900. In 1964 it was defined more carefully as 9,192,631,770 periods of the transition frequency between the hyperfine levels $F = 4$, $m_F = 0$ and $F = 3$, $m_F = 0$ of the ground state $^2S_{1/2}$ of the atom of cesium 133, unperturbed by external fields. This latter definition is permanent and more reproducible than the former; it is also comprehensible only to atomic physicists. However, any of these definitions adequately describes the second with which we are all familiar.

The definition of the fourth basic unit, the ampere (A), will appear later in this chapter after we are more familiar with the basic properties of electricity. The remaining two basic units, the degree Kelvin (°K) and the candela (cd), are not of immediate concern to circuit analysts.[1]

The SI incorporates the decimal system to relate larger and smaller units to the basic unit and uses standard prefixes to signify the various powers of ten. These are:

atto-	(a-, 10^{-18})	deci-	(d-, 10^{-1})
femto-	(f-, 10^{-15})	deka-	(da-, 10^{1})
pico-	(p-, 10^{-12})	hecto-	(h-, 10^{2})
nano-	(n-, 10^{-9})	kilo-	(k-, 10^{3})
micro-	(μ-, 10^{-6})	mega-	(M-, 10^{6})
milli-	(m-, 10^{-3})	giga-	(G-, 10^{9})
centi-	(c-, 10^{-2})	tera-	(T-, 10^{12})

[1]Complete definitions of all the basic units and a further discussion of the International System of Units may be found in R. D. Huntoon, Status of National Standards for Physical Measurements, *Science*, vol. 150, no. 3693, pp. 169–178, Oct. 8, 1965, or in C. H. Page et al., IEEE Recommended Practice for Units in Published Scientific and Technical Work, *IEEE Spectrum*, vol. 3, no. 3, pp. 169–173, March, 1966.

These prefixes are worth memorizing, for they will appear often, both in this text and in other scientific work. Thus, a millisecond (ms) is $\frac{1}{1000}$ of a second, and a kilometer (km) is 1000 m. It is apparent now that the gram (g) was originally established as the basic unit of mass, and the kilogram then represented merely 1000 g. Now the kilogram is our basic unit, and we could describe the gram as a millikilogram if we wished to be confusing. Combinations of several prefixes, such as the millimicrosecond, are unacceptable; the term *nanosecond* should be used. Also officially frowned on is the use of micron for 10^{-6} m; the correct term is the micrometer (μm). The angstrom (Å), however, may be used for 10^{-10} m.

This power-of-10 relationship is not present in the so-called *British System of Units*, which is in common use in this country. There are many occasions when the results of an engineering analysis must be transformed into the British System of Units for use in the shop or for clarity in discussions with others.

Most of us have a better mental picture of 2 in. than we do of 5 cm, although this great dependence on the older system is gradually changing.

The fundamental British units are defined in terms of the SI units as follows: 1 in. is exactly 0.0254 m, 1 pound-mass (lbm) is exactly 0.453 592 37 kg, and the second is common to both systems.

As a last item in our discussion of units, we consider the three derived units used to measure force, work or energy, and power. The newton (N) is the fundamental unit of force,[2] and it is the force required to accelerate a 1-kg mass by one meter per second per second (1 m/s^2). A force of 1 N is equivalent to 0.22481 pound of force (lbf), and the average nineteen-year-old male, having a mass of 68 kg, exerts a force of 670 N on the scales.

The fundamental unit of work or energy is the joule (J), defined as one newton-meter (N-m). The application of a constant 1-N force through a 1-m distance requires an energy expenditure of 1 J. The same amount of energy is required to lift this book, weighing about 10 N, a distance of approximately 10 cm. The joule is equivalent to 0.73756 foot pound-force (ft-lbf).

The last derived quantity with which we shall concern ourselves is power, the rate at which work is done or energy is expended. The fundamental unit of power is the watt (W), defined as 1 J/s. One watt is equivalent to 0.73756 ft-lbf/s. It is also equivalent to $\frac{1}{745.7}$ horsepower (hp), a unit which is now being phased out of engineering terminology.

NOTE: Throughout the text, drill problems appear following sections in which a new principle is introduced, in order to allow the student to test his under-

[2]It is worth noting that all units named after famous scientists have abbreviations beginning with capital letters.

standing of the basic fact itself. The problems are useful in gaining familiarization with new terms and ideas and should all be worked. More general problems appear at the ends of the chapters. The answers to the drill problems are given in random order. For example, in Drill Prob. 1-1, the answers are (*a*) 1.09, (*b*) 2.00, and (*c*) 1.67.

Drill Problems

1-1 Fill in each of the following blanks: (*a*) 0.24 lbm = ________ hg; (*b*) 2×10^4 dm = ________ km; (*c*) 10^5 ms = ________ min.

Ans. 1.09; 1.67; 2.00

1-2 Determine the average power required: (*a*) in watts, for a 150-lbf man to ascend a 10-ft ladder in 4 s; (*b*) in kilowatts, to accelerate a 2-ton-mass automobile from a standstill to 60 mi/h in 7 s; (*c*) in picowatts, to accelerate an electron (mass = 9.11×10^{-31} kg) from rest to 10^6 m/s in 1.5 ns.

Ans. 93; 304; 508

1-3 THE UNIT OF CHARGE

Before beginning a discussion of electricity and electric circuits, we may define in terms of an analogy the class of electrical phenomena which we are going to consider. When we hold a baseball out at arm's length and release it, we know that it falls toward the earth because of the gravitational force exerted on it. We can also describe precisely *how* it accelerates, *what* its velocity is at any given instant, *when* it reaches a given point, and *where* it will be at a given instant. Few of us understand, however, *why* it falls. Although we understand very well what gravitational forces do, we do not know what they are.

In an analogous way, an electrical engineer is very familiar with the forces, meter deflections, heating effects, and other measurable responses caused by electricity, but he is only rarely concerned with the theoretical (and philosophical) nature of electricity itself. Therefore, our goal is a competence in observing electrical phenomena, describing them mathematically, and putting them to a practical use. We shall be only incidentally concerned with their cause.

Suppose that we take a small piece of some light material such as pith and suspend it by a fine thread. If we now rub a hard rubber comb with a woolen cloth and then touch the pith ball with the comb, we find that the pith ball tends to swing away from it; a force of repulsion exists between the comb and the pith ball. After laying down the comb and then approaching the pith ball with the woolen cloth, we can see that there is a force of attraction present between the pith ball and the woolen cloth.

We explain both of these forces on the pith ball by saying that they are *electrical* forces caused by the presence of *electrical charges* on the pith ball, the comb, and the woolen cloth. In an analogous way, we attribute the force on the baseball to a *gravitational* force caused by the presence of gravitational masses in the baseball and the earth.

Our experiment shows clearly that the electrical force may be one of either attraction or repulsion, and in this respect the gravitational analogy breaks down. As far as we know at the present time, a gravitational repulsive force does not exist.

We explain the existence of an electrical force of both attraction and repulsion by the hypothesis that there are two kinds of charge and that like charges repel and unlike charges attract. The two kinds of charge are called positive and negative, although we might have called them gold and black or vitreous and resinous (as they were termed many years ago). Arbitrarily, the type of charge originally present on the comb was called negative by Benjamin Franklin, and that on the woolen cloth, positive.

We may now describe our experiment in these new terms. By rubbing the comb with the cloth, a negative charge is produced on the comb and a positive charge on the cloth. Touching the pith ball with the comb transferred some of its negative charge to the pith ball, and the force of repulsion between the like kinds of charge on the pith ball and comb caused the ball to move away. As we brought the positively charged woolen cloth near the negatively charged pith ball, a force of attraction between the two different kinds of charge was evident.

We also know now that all matter is made up of fundamental building blocks called atoms and that the atoms, in turn, are composed of different kinds of fundamental particles. The three most important particles are the electron, the proton, and the neutron. The electron possesses a negative charge, the proton possesses an equal-magnitude positive charge, and the neutron is neutral, or has no charge at all. As we rubbed the rubber comb with the woolen cloth, the comb acquired its negative charge because some of the electrons on the wool were rubbed off onto the comb; the cloth then had an insufficient number of electrons to maintain its electrical neutrality and thus behaved as a positive charge.

The mass of each of the three particles named above has been determined experimentally and is 9.10908×10^{-31} kg for the electron and about 1840 times as large for the proton and the neutron.

Now we are ready to define the fundamental unit of charge, called the coulomb after Charles Coulomb, the first man to make careful quantitative measurements of the force between two charges. The coulomb can, of course, be defined in any way we wish as long as the definition is convenient, universally accepted, permanent, and does not contradict any previous definition. Again, this leaves us no freedom at all because the definition which is already universally accepted is as follows: two small, identically charged particles which are separated one meter in a vacuum and repel each other with a force of $10^{-7}c^2$ newtons possess an identical charge of either plus or minus one coulomb (C). The symbol c represents the velocity of light, 2.997925×10^8 m/s. In terms of this unit, the charge of an electron is a negative 1.60210×10^{-19} C, and 1 C (negative) therefore represents the combined charge of about 6.24×10^{18} electrons.

We shall symbolize charge by Q or q; the capital letter is reserved for a charge which does not change with time, or is a constant, and the lowercase letter represents the general case of a time-varying charge. We often call this the *instantaneous* value of the charge and may emphasize its time dependence by writing it as $q(t)$. This same use of capital and lowercase letters will be carried over to all other electrical quantities as well.

Drill Problem

1-3 Find the charge in aC represented by: (*a*) 7 electrons; (*b*) 15 protons; (*c*) the combination of 15 protons and 7 electrons.

Ans. -1.12; 1.28; 2.40 aC

1-4 CURRENT, VOLTAGE, AND POWER

The electrical phenomena discussed above belong to the field of electrostatics, which is concerned with the behavior of electric charges at rest. This is of interest to us only because it is a beginning and serves as a useful device to define charge.

One part of the experiment, however, departed from electrostatics, the process of transferring charge from the wool cloth to the comb or from the comb to the pith ball. This idea of "transfer of charge" or "charge in motion" is of vital importance to us in studying electric circuits, because, in moving a charge from place to place, we may also transfer energy from one point to another. The familiar cross-country power transmission line is a practical example.

Of equal importance is the possibility of varying the rate at which the charge is transferred in order to communicate or transfer intelligence. This process is the basis of communication systems such as radio, television, and telemetry.

Charge in motion represents a *current,* which we shall define more carefully below. The current present in a discrete path, such as a metallic wire, has both a magnitude and a direction associated with it; it is a measure of the rate at which charge is moving past a given reference point in a specified direction. We now consider a rather arbitrary example which, however, will lead us to the general definition of current as the time rate of change of charge, dq/dt.

Let us consider a discrete path along which charge can move and ask a number of questions about the manner in which charge is traveling along this lead or conductor. As a first-hand observer, we shall place a very small student at point A on the path and ask him to record the total amount of charge which has passed him since some reference time $t = 0$. We ask that he take data every second[3] and then give him these detailed instructions:

1. The positive direction is to your right.
2. If positive charge moves past you in the positive direction, add the magnitude of the charge.
3. If positive charge moves in the negative direction, subtract the charge magnitude.
4. If negative charge moves in the positive direction, also subtract the charge magnitude.
5. If negative charge moves in the negative direction, add the charge magnitude.

The observer watches for 8 s, records his data, and then hands us the graph, Fig. 1-1, explaining that q is the total charge which has moved past him since $t = 0$.

[3]He is a small, *quick* student.

Fig. 1-1 A graph of the total charge q that has passed a given reference point since $t = 0$. The charge is measured over 1-s intervals.

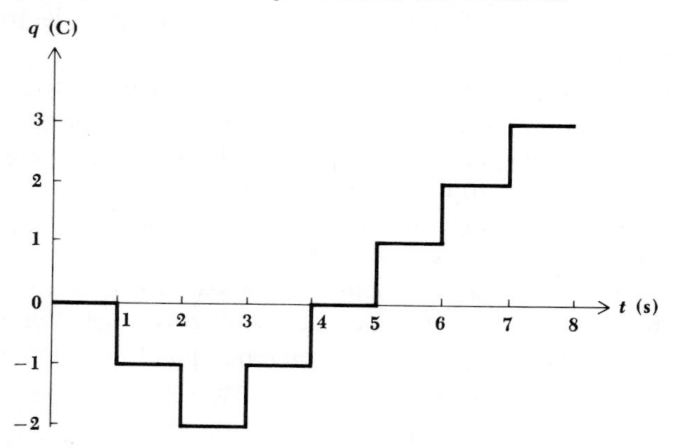

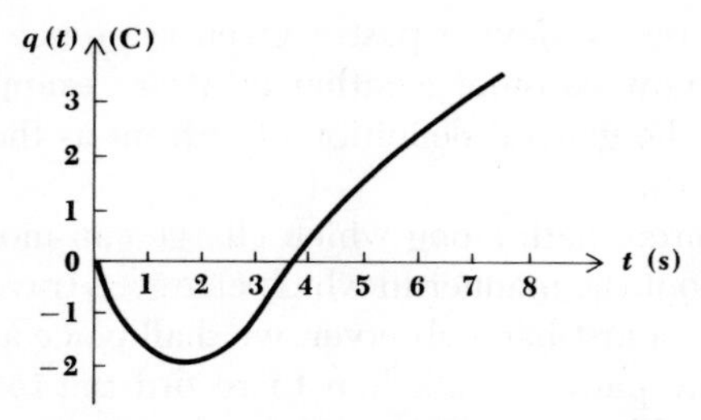

Fig. 1-2 A graph of the instantaneous value of the total charge $q(t)$ which has passed a given reference point since $t = 0$.

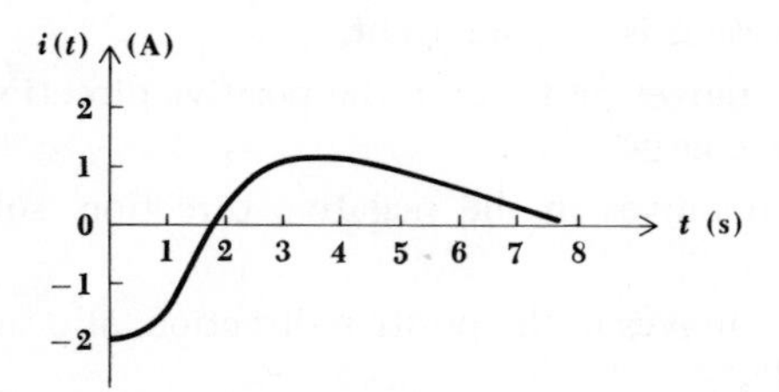

Fig. 1-3 The instantaneous current, $i = dq/dt$, where q is given in Fig. 1-2.

Now we see that there are many ways in which this record might be interpreted. For instance, in the first second either one unit positive charge moved by to the left or one negative charge moved past to the right. The same alternative is present for the second 1-s interval. As a matter of fact, in either of these intervals the observer might have had to count 100 unit positive charges moving to the right and 101 moving to the left. Perhaps positive and negative charges were in motion in both directions.

It is fortunate that we do not need to know which specific one of this infinite number of possibilities actually occurred; the electrical effects produced by each will be the same.

We now refine the data by making measurements much more often, and this requires that smaller and smaller elements of charge be counted. The limit is the amount of charge carried by a single electron. The graphical record now appears as a smooth curve, Fig. 1-2.

We are now ready to consider the *rate* at which charge is being transferred. In the time interval extending from t to $(t + \Delta t)$, the charge transferred past the reference point has increased from q to $(q + \Delta q)$. If the graph is decreasing at this instant, then Δq is a negative value. The rate at which charge is passing the reference point at time t is therefore very closely equal to $\Delta q/\Delta t$, and as the interval Δt decreases, the exact value of the rate is given by the derivative

$$\frac{dq}{dt} = \lim_{\Delta t \to 0} \frac{\Delta q}{\Delta t}$$

We define the current at a specific point and flowing in a specified direction as the instantaneous rate at which net positive charge is moving past that point in the specified direction. Current is symbolized by I or i, and thus

$$i = \frac{dq}{dt}$$

The unit of current is the ampere (A), which corresponds to charge moving at the rate of 1 C/s. The ampere was named after A. M. Ampère, a French physicist of the early nineteenth century. It is often called an "amp," but this is informal and unofficial. The use of the lowercase letter i is again to be associated with an instantaneous value. Using the data of Fig. 1-2, the instantaneous current is given by the slope of the curve at every point. This current is plotted in Fig. 1-3.

The total charge transferred between time t_0 and t may be expressed as a definite integral,

$$q = \int_{t_0}^{t} i \, dt$$

Several different types of current are illustrated in Fig. 1-4. A current which is constant is termed a direct current, or simply dc, and is shown by Fig. 1-4a. We shall find many practical examples of currents which vary sinusoidally with time, Fig. 1-4b; currents of this form are present

Fig. 1-4 Several types of current: (a) Direct current, or dc. (b) Sinusoidal current, or ac. (c) Exponential current. (d) Damped sinusoidal current.

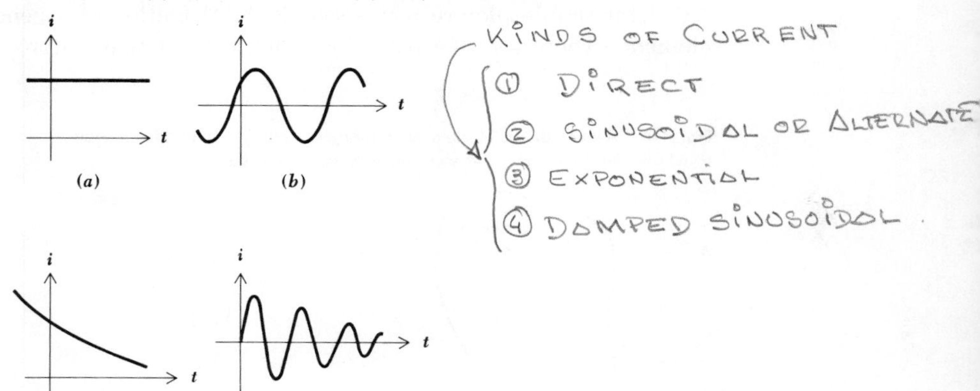

Fig. 1-5 Two methods of representation for the same current.

in the normal household circuits. Such a current is often referred to as alternating current, or ac. Exponential currents and damped sinusoidal currents, sketched in Fig. 1-4c and d, will also be encountered later.

We establish a graphical symbol for current by placing an arrow next to the conductor. Thus, in Fig. 1-5a the direction of the arrow and the value "3 A" indicate *either* that a net positive charge of 3 C/s is moving to the right or that a net negative charge of -3 C/s is moving to the left each second. In Fig. 1-5b there are again two possibilities: either -3 C/s is flowing to the left or $+3$ C/s is flowing to the right. All four of these statements and both figures represent currents which are equivalent in their electrical effects, and we say that they are equal.

It is convenient to think of current as the motion of positive charge even though it is known that current flow in metallic conductors results from electron motion. In ionized gases, in electrolytic solutions, and in some semiconductor materials, positively charged elements in motion constitute part or all of the current. Thus, any definition of current can agree only with the physical nature of conduction part of the time. The definition and symbolism we have adopted are standard.

We must next define a circuit element. Such electrical devices as fuses, light bulbs, resistors, batteries, capacitors, generators, and spark coils can be represented by combinations of simple circuit elements. We shall begin by showing a very general circuit element as a shapeless object possessing two terminals at which connections to other elements may be made, Fig. 1-6. This simple picture may serve as the definition of a general circuit element. There are two paths by which current may enter or leave the

Fig. 1-6 A general circuit element is characterized by a pair of terminals to which other general circuit elements may be connected.

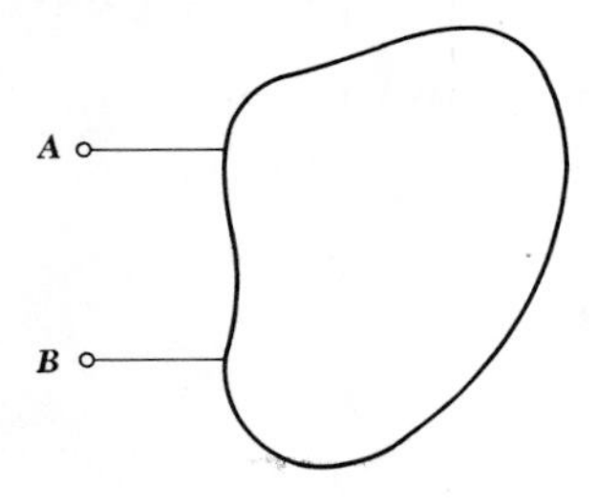

element. Later we shall define particular circuit elements by describing the electrical characteristics which may be observed at their pairs of terminals.

Let us suppose that direct current is directed into terminal A of Fig. 1-6, through the general element, and out of B. Let us also assume that the passage of this charge through the element requires an expenditure of energy. We then say that an electrical *voltage* or a *potential difference* exists between the two terminals, or that there is a voltage or potential difference "across" the element. Thus, the voltage across a terminal pair is a measure of the work required to move charge through the element. Specifically, we shall define the voltage across the element as the work required to move a positive charge of 1 C from one terminal through the device to the other terminal. The sign of the voltage will be discussed below. The unit of voltage is the *volt* (V), which is 1 J/C and voltage is represented by V or v. We are indeed fortunate that the full name of the eighteenth-century Italian physicist, Alessandro Giuseppe Antonio Anastasio Volta, is not used for our unit of potential difference.

The energy which is expended in forcing the charges through the element must appear somewhere else by the principle of conservation of energy. When we later meet specific circuit elements, we should note whether the energy is stored in some form which is readily available or whether it changes irreversibly into heat, acoustic energy, and so forth.

We must now establish a convention by which we can distinguish between energy supplied to the element by some external source and energy which may be supplied by the element itself to some external device. We do this by our choice of a sign for the voltage of terminal A with respect to terminal B. If a positive current is entering terminal A of the element and if an external source must expend energy to establish this current, then terminal A is positive with respect to terminal B. Alternatively, we may say also that terminal B is negative with respect to terminal A.

The sense of the voltage is indicated by a plus-minus pair of algebraic signs. In Fig. 1-7*a*, for example, the placement of the plus sign at terminal A indicates that terminal A is v volts positive with respect to terminal B. If we later find that v happens to have a numerical value of -5 V, then we may say either that A is -5 V positive with respect to B or that B

Fig. 1-7 In (*a*) and (*b*), terminal B is 5 V positive with respect to terminal A; in (*c*) and (*d*), terminal A is 5 V positive with respect to terminal B.

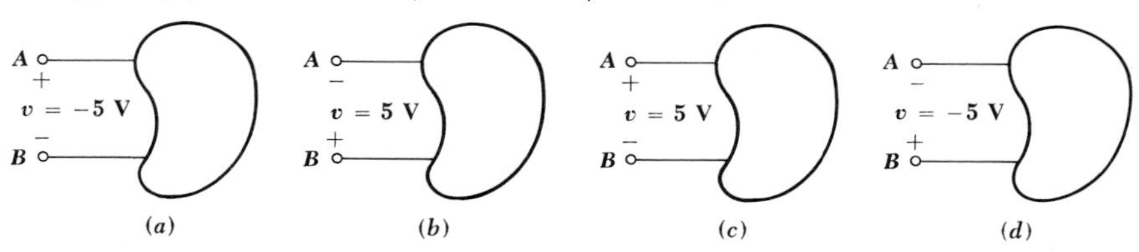

is 5 V positive with respect to A. Other cases are shown and described in Fig. 1-7b, c, and d.

No statement can be made concerning energy transfer in any of these four cases until the direction of the current is specified. Let us assume that a current arrow is placed by each upper lead, directed to the right, and labeled "+2 A"; then, since in both cases c and d terminal A is 5 V positive with respect to terminal B and since a positive current is entering terminal A, energy is being supplied to the element. In the remaining two cases, the element is delivering energy to some external device.

We have already defined power, and we shall represent it by P or p. If one joule of energy is expended in transferring one coulomb of charge through the device, then the rate of energy expenditure in transferring one coulomb of charge per second through the device is one watt. The power must be proportional both to the number of coulombs transferred per second, or current, and to the energy needed to transfer one coulomb through the element, or voltage. Thus,

$$p = vi$$

Dimensionally, the right side of this equation is the product of joules per coulomb and coulombs per second, which produces the expected dimension of joules per second, or watts.

With a current arrow placed by each upper lead of Fig. 1-7, directed to the right and labeled "2 A," 10 W is absorbed by the element in c and d and -10 W is absorbed (or 10 W is generated) in a and b.

The conventions for current, voltage, and power are summarized in Fig. 1-8. The sketch shows that if one terminal of the element is v volts positive with respect to the other terminal, and if a current i is entering the element through the first terminal, then a power $p = vi$ is being absorbed by or delivered to the element. This convention should be studied carefully, understood, and memorized. In other words, it says that if the current arrow and the voltage polarity signs are placed at the terminals of the element such that the current enters that end of the element marked with the positive sign, and if both the arrow and the sign pair are labeled

Fig. 1-8 The power absorbed by the element is given by the product, $p = vi$.

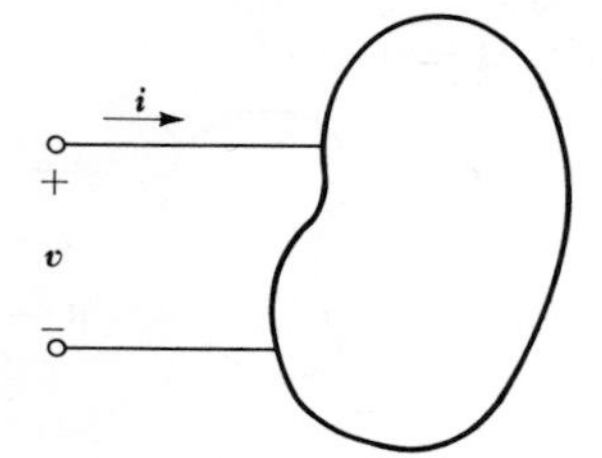

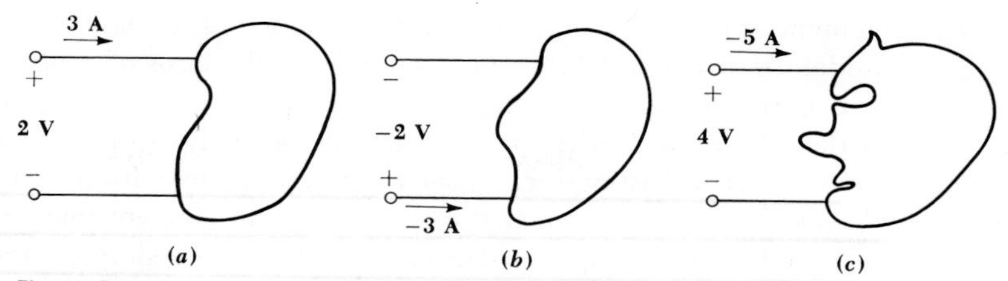

Fig. 1-9 (*a*) A power, $p = (2)(3) = 6$ W, is absorbed by the element. (*b*) A power, $p = (-2)(-3) = 6$ W, is absorbed by the element. (*c*) A power, $p = (4)(-5) = -20$ W, is absorbed by the element, or 20 W is delivered by the element.

with the appropriate algebraic quantities, then the power *absorbed* by the element can be expressed by the algebraic product of these two quantities. If the numerical value of the product is negative, then the element is absorbing negative power or delivering power to some external element. The three examples of Fig. 1-9 further illustrate this convention.

Drill Problems

1-4 The total charge $q(t)$ entering the upper terminal of the circuit element in Fig. 1-8 is given by $20\ e^{-0.2t} \sin(\pi t/4)$ C. Find i at $t =$: (*a*) 0; (*b*) 2 s; (*c*) −1 s.

Ans. −2.68; 15.7; 17.0 A

1-5 The current entering the upper terminal of the circuit element of Fig. 1-8 is given by $600t^2 - 8t$ A. If the total charge that has entered that terminal of the element is 60 μC at $t = -10$ ms, find q at $t =$: (*a*) −5 ms; (*b*) 0; (*c*) 10 ms.

Ans. 460; 535; 660 μC

1-6 (*a*) If the power absorbed by the element of Fig. 1-7*b* is 30 W, find the current entering terminal *B*. (*b*) If the power absorbed by the element of Fig. 1-7*a* is 30 W, find the current entering terminal *A*. (*c*) If the power delivered by the element of Fig. 1-7*d* is 30 W, find the current entering terminal *A*.

Ans. −6; −6; 6 A

1-5 TYPES OF CIRCUITS AND CIRCUIT ELEMENTS

Using the concepts of current and voltage, it is now possible to be more specific in defining a circuit element.

It is important to differentiate between the physical device itself and the mathematical model of this device which we shall use to analyze its behavior in a circuit. Let us agree that we will use the expression "circuit element" to refer to the mathematical model. The choice of a particular model for any real device must be made on the basis of experimental data or experience; we shall usually assume that this choice has already been made. We must first learn the methods of analysis of idealized circuits.

Now let us distinguish a *general circuit element* from a *simple circuit element* by the statement that a general circuit element may be composed of more than one simple circuit element, but that a simple circuit element cannot be further subdivided into other simple circuit elements. For brevity, we shall agree that the term circuit element generally refers to a simple circuit element.

All the simple circuit elements that will be considered in the work that follows can be classified according to the relationship of the current through the element to the voltage across the element. For instance, if the voltage across the element is directly proportional to the current through it, or $v = ki$, we shall call the element a resistor. Other types of simple circuit elements have a terminal voltage which is proportional to the time derivative or the integral with respect to time of the current. There are also elements in which the voltage is completely independent of the current or the current is completely independent of the voltage; these are the independent sources. Furthermore, we shall need to define special kinds of sources in which the source voltage or current depends upon a current or voltage elsewhere in the circuit; such sources will be termed dependent sources or controlled sources.

By definition, a simple circuit element is the mathematical model of a two-terminal electrical device, and it can be completely characterized by its voltage-current relationship but cannot be subdivided into other two-terminal devices.

The first element which we shall need is an *independent voltage source.* It is characterized by a terminal voltage which is completely independent of the current through it. Thus, if we are given an independent voltage source and are notified that the terminal voltage is $50t^2$ V, we can be sure that at $t = 1$ s the voltage will be 50 V, regardless of the current that was flowing, is flowing, or is going to flow. The representation of an independent voltage source is shown in Fig. 1-10. The subscript s merely identifies the voltage as a "source" voltage.

A point worth repeating here is that the presence of the plus sign at

Fig. 1-10 The circuit symbol of an independent voltage source. The circuit symbol of a dependent or controlled voltage source is shown in Fig. 1-13*a*.

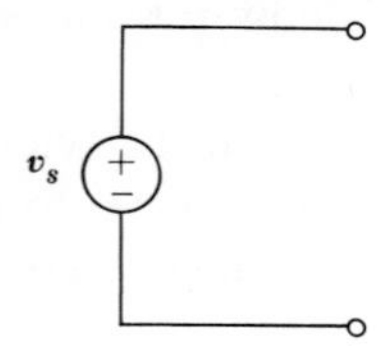

the upper end of the symbol for the independent voltage source in Fig. 1-10 does not necessarily mean that the upper terminal is always positive with respect to the lower terminal. Instead, it means that the upper terminal is v_s volts positive with respect to the lower. If, at some instant, v_s happens to be negative, then the upper terminal is actually negative with respect to the lower at that instant.

If a current arrow is placed adjacent to the upper conductor of this source and directed to the left, then the current i is entering the terminal at which the positive sign is located and the source thus absorbs a power $p = v_s i$. More often than not, a source is expected to deliver power to a network and not to absorb it. Consequently, we might choose to direct the arrow to the right in order that $v_s i$ will represent the power delivered by the source. Either direction may be used.

The independent voltage source is an *ideal* source and does not represent exactly any real physical device, because the ideal source could theoretically deliver an infinite amount of energy from its terminals. Each coulomb passing through it receives an energy of v_s joules, and the number of coulombs per second is unlimited. This idealized voltage source does, however, furnish a reasonable approximation to several practical voltage sources. An automobile storage battery, for example, has a terminal voltage of 12 V that remains essentially constant as long as the current through it does not exceed a few amperes. The small current may have direction either through the battery, corresponding to power furnished by the battery to the headlights while it is discharging, or to power absorbed by the battery from the generator or a battery charger while it is charging. An ordinary household electrical outlet also approximates an independent voltage source provided the voltage $v_s = 115\sqrt{2} \cos 2\pi 60t$ V; the representation is valid for currents less than perhaps 20 A.

An independent voltage source which has a constant terminal voltage is often termed an independent dc voltage source and is represented by either symbol shown in Fig. 1-11. Note in Fig. 1-11*b* that, when the physical plate structure of the battery is suggested, the longer plate is placed at the positive terminal; the plus and minus signs then represent redundant notation, but they are usually included anyway.

Fig. 1-11 Alternative representations of a constant, or dc, independent voltage source. In (*a*) the source is delivering 12 W and in (*b*) the battery is absorbing 12 W.

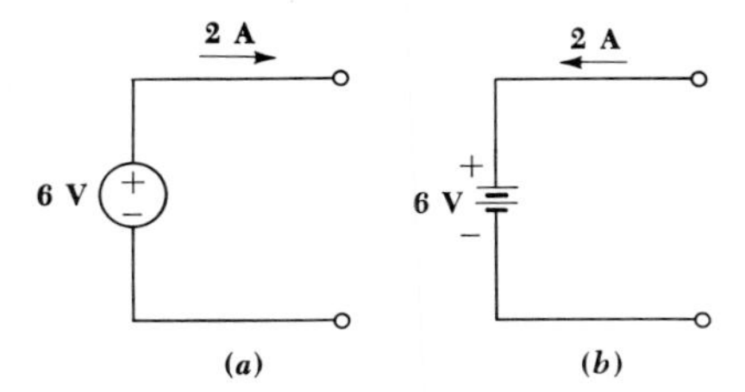

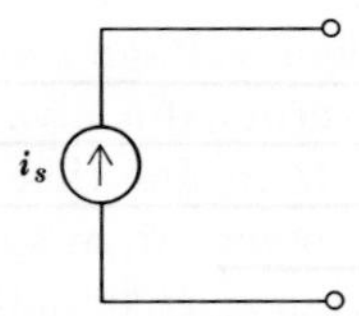

Fig. 1-12 The circuit symbol of an independent current source.

Another ideal source which we will need is the *independent current source*. Here, the current through the element is completely independent of the voltage across it. The symbol for an independent current source is shown in Fig. 1-12. If i_s is constant, we call the source an independent dc source.

Like the independent voltage source, the independent current source is at best a reasonable approximation for a physical element. In theory it can deliver infinite power from its terminals, because it produces the same finite current for any voltage across it, no matter how large that voltage may be. It is, however, a good approximation for many practical sources. For example, the independent dc source represents very closely the electron beam of a synchrotron which is operating at a constant beam current of perhaps 1 μA and will continue to deliver 1 μA to almost any device placed across its "terminals" (the beam and the earth).

The two types of ideal sources that we have discussed up to now are called *independent* sources because the value of the source quantity is not affected in any way by activities in the remainder of the circuit. This is in contrast with yet another kind of ideal source, the *dependent* or *controlled source*, in which the source quantity is determined by a voltage or current existing at some other location in the electrical system under examination. To distinguish between independent and dependent sources, we introduce the additional symbols shown in Fig. 1-13. Sources such as these will appear in the equivalent electrical models for many electronic devices, such as transistors, vacuum tubes, and integrated circuits. We shall see all of these in the following chapters.

Dependent and independent voltage and current sources are *active* elements; they are capable of delivering power to some external device. For the present we shall think of a *passive* element as one which is capable

Fig. 1-13 The diamond shape characterizes the circuit symbols for (*a*) the dependent voltage source and (*b*) the dependent current source.

(*a*) (*b*)

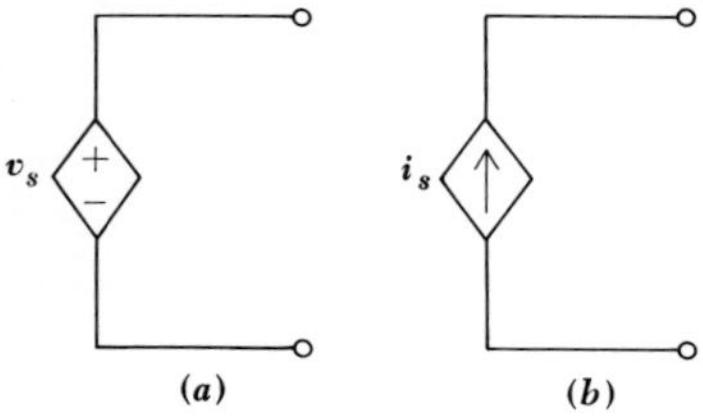

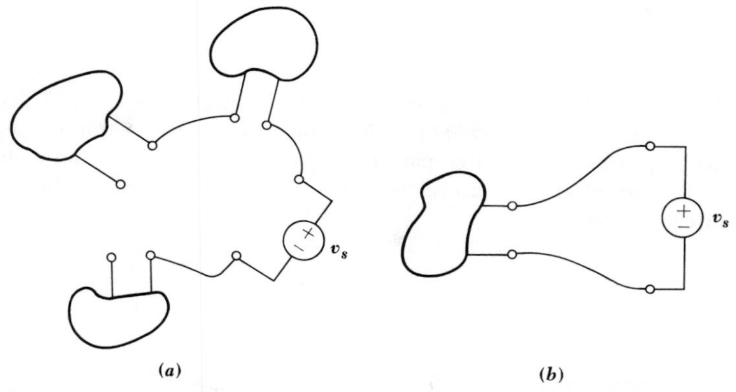

Fig. 1-14 (*a*) An electrical network which is not a circuit. (*b*) A network which is a circuit.

only of receiving power. However, we shall later see that several passive elements are able to store a finite amount of energy and then return it later to an external element, and since we shall still wish to call such an element passive, it will be necessary to improve upon our two definitions then.

The interconnection of two or more simple circuit elements is called an electrical *network*. If the network contains at least one closed path, we shall call it an electric *circuit*. Every circuit is a network, but not all networks are circuits. Figure 1-14*a* shows a network which is not a circuit, and Fig. 1-14*b* shows a network which is a circuit.

A network which contains at least one active element, such as an independent voltage or current source, is an *active network*. A network which does not contain any active elements is a *passive network*.

Drill Problem

1-7 Determine the power being supplied by the sources shown in Fig. 1-15*a*, *b*, and *c*.

Ans. -21; -20; 18 W

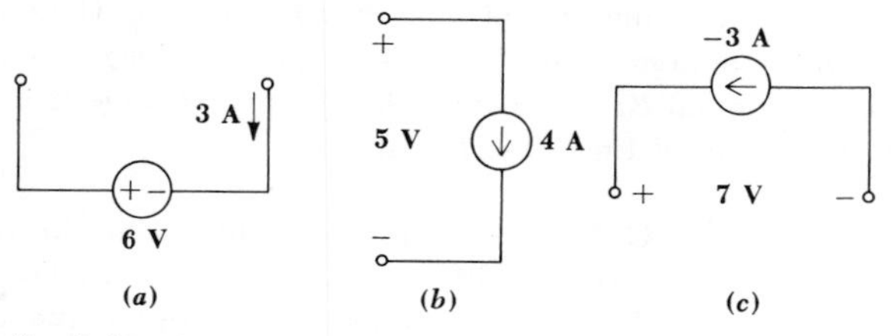

Fig. 1-15 See Drill Prob. 1-7.

PROBLEMS

☒ **1** Curve (*a*) of Fig. 1-16 represents the total charge ($q = y$, in μC) that has passed a specified reference point in a given direction as a function of time ($x = t$, in ms). (*a*) How much charge passed the point between $t = 3$ and 6 ms? (*b*) What is the current at $t = 4.2$ ms? (*c*) What is the largest value of the current in the interval shown?

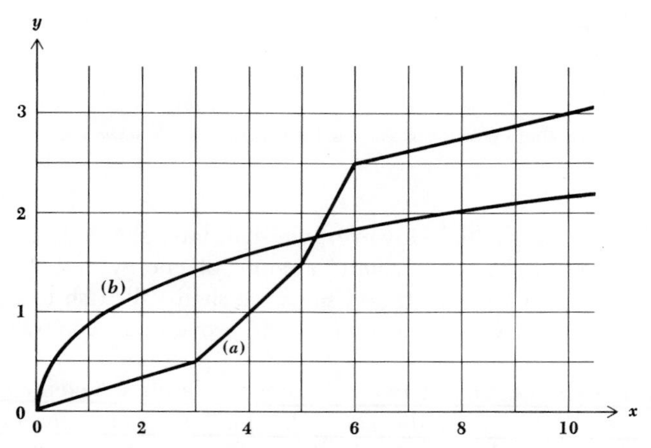

Fig. 1-16 Graphical relationships used in various problems.

☐ **2** Curve (*b*) of Fig. 1-16 represents the total charge ($q = y$, in nC) that has entered terminal A of a circuit element as a function of time ($x = t$, in μs). (*a*) How long a time is required for 2 nC to enter terminal A? (*b*) What is the value of the current entering terminal A at $t = 2$ μs? (*c*) Make a sketch of the current as a function of time, $0 < t < 10$ μs.

☐ **3** Curve (*a*) of Fig. 1-16 represents the current ($i = y$, in A) at a certain point in a specified direction as a function of time ($x = t$, in ms). (*a*) How much charge passed the point between $t = 0$ and 3 ms? (*b*) Between $t = 3$ and 6 ms? (*c*) At what rate is charge passing the point at $t = 4$ ms?

☒ **4** Curve (*b*) of Fig. 1-16 represents the current ($i = y$, in mA) entering terminal B of a circuit element as a function of time ($x = t$, in μs). (*a*) How much charge enters terminal B in the interval, $6 < t < 10$ μs? (*b*) How long a time is required for 2 nC to enter terminal B? (*c*) Make a sketch of the charge that has entered terminal B as a function of time, $0 < t < 10$ μs.

☒ **5** (*a*) If $q = -10^{-7}e^{-10^5 t}$ C, find the average current during the time interval from $t = -5$ μs to $+5$ μs. Given $i = 12 \cos (1000t + \pi/6)$ A; (*b*) find the average current during the interval, $0 < t < \pi/3$ ms; (*c*) determine the total charge transferred between $t = 0$ and $t = 2\pi/3$ ms.

☐ **6** The charge (in C) that has entered a circuit element since $t = -\infty$ is numerically equal to 50 times the current (in A) at that point for every instant of time. (*a*) If the current is 4 A at $t = 10$ s, find $q(t)$. (*b*) If the total charge is 5 C at $t = -20$ s, find $i(t)$.

☒ **7** A 12-V storage battery is charged by supplying a current entering its positive terminal that is a constant 3 A for 2 h and then decreases linearly to zero during the next hour. Assuming the battery voltage is constant: (*a*) What is the total charge delivered to the battery? (*b*) At what time is the power being delivered to the battery 24 W? (*c*) What is the average power delivered to the battery over the 3-h interval? (*d*) How much energy is supplied to the battery?

☐ **8** For the independent current source shown in Fig. 1-12, $i_s = 10 \sin 1000t$ mA. Let the voltage of the upper terminal with respect to the lower be v. (*a*) If $v = 25 \sin 1000t$ V, find the power being absorbed by the source at $t = 1.5\pi$ ms. (*b*) If $v = 25 \cos 1000t$ V, find the power absorbed at $t = 0.25\pi$ and 0.75π ms.

☐ **9** The output of a certain independent current source is a function of temperature $T(°\mathrm{K})$, $i_s = 10^{-6}T^2 - 6 \times 10^{-4}T + 0.1$ A. (*a*) At what temperature in the range $250 \leq T \leq 320°\mathrm{K}$ is i_s a maximum? (*b*) At a constant temperature of 320°K, how much charge can the source deliver at a voltage of 3722 V in 10 min?

☐ **10** Let the current i_1 enter the plus-marked terminal of a circuit element across which the voltage v_1 is present. Find the power being absorbed by the element at $t = 10$ s if $i_1 = 2\,e^{-0.1t}$ A and: (*a*) $v_1 = 6\,di_1/dt$; (*b*) $v_1 = (1/6)\int_0^t i_1\,dt + 2$ V.

☐ **11** An independent current source i_s and an independent voltage source v_s are joined together by connecting the terminal with the arrowhead to the plus-marked terminal and then connecting the remaining two terminals together. The current leaving the current source is thus constrained to enter the voltage source. If $v_s = 2 \cos 10^6\pi t$ V and $i_s = 0.03$ A, find the power absorbed by each source at: (*a*) $t = 0$; (*b*) $t = 1\ \mu$s.

☐ **12** The charge leaving the positive reference terminal of a voltage source is $q = 6 \sin 120\pi t - 8 \cos 120\pi t$ mC. If $v_s = 160 \cos 120\pi t$ V: (*a*) find the power being supplied by the voltage source at $t = 0$; (*b*) find the average power supplied over the interval, $0 < t < 1/60$ s; (*c*) find the energy supplied by the source over the same time interval.

Chapter Two Experimental Laws and Simple Circuits

2-1 INTRODUCTION

In the last chapter we became familiar with both independent and dependent voltage and current sources and were cautioned that they were idealized elements which could only be approximated in a real circuit. Another idealized element, the linear resistor, will be introduced in this chapter.

We must next accept two fundamental laws as being axiomatic. With these laws, with the five simple circuit elements, and with the few definitions we already have, we may then begin to study simple electric circuits. This study will be restricted almost completely to *analysis,* which is the process by which the voltage and current associated with each element in a given circuit are determined. Fortunately, a complete analysis is not usually necessary, for often only a specific current, voltage, or perhaps power is needed.

After a proficiency in analysis has been achieved in this and other early courses, problems in *synthesis* may be considered. Here we are given a mathematical description of the desired behavior of a circuit and must determine the necessary elements and their interconnection in order to obtain the desired response. Synthesis problems may often have more than one solution.

The final type of circuit problem, and the one for which engineering salaries are most often paid, is that of *design*. A real, physical, manufacturable, salable, economical, reliable device is the desired end product. Sometimes size, weight, temperature characteristics, and even eye appeal must be considered in the design. It is obvious that experience is a prerequisite for design proficiency; it is also evident that analysis and synthesis must come first.

This chapter and the following one are restricted to the analysis of simple circuits containing only current sources, voltage sources, and resistors; the sources may be independent or dependent. In analyzing these circuits we shall use several network transformations, network theorems, and mathematical methods which we shall later be able to apply, with only slight modifications, to circuits containing other types of passive elements excited by time-varying sources. We shall learn the methods useful in circuit analysis by applying them to the simplest possible case, the resistive circuit.

2-2 OHM'S LAW

The simplest passive element, the resistor, may be introduced by considering the work of an obscure German physicist, George Simon Ohm, who published a pamphlet in 1827 entitled "Die galvanische Kette mathematisch bearbeitet."[1] In it were contained the results of one of the first efforts to measure currents and voltages and to describe and relate them mathematically. One result was a statement of the fundamental relationship we now call Ohm's law, even though it has since been shown that this result was discovered 46 years earlier in England by Henry Cavendish, a brilliant semirecluse. However, no one, including Ohm, we shall hope, knew of the work done by Cavendish because it was not uncovered and published until long after both were dead.

Ohm's pamphlet received much undeserved criticism and ridicule for

[1]"The Galvanic Circuit Investigated Mathematically."

several years after its first publication, but it was later accepted and served to remove the obscurity associated with his name.

Ohm's law states that the voltage across many types of conducting materials is directly proportional to the current flowing through the material,

$$v = Ri$$

where the constant of proportionality R is called the *resistance*. The unit of resistance is the *ohm*, which is 1 V/A and customarily abbreviated by a capital omega, Ω.

When this equation is plotted on v versus i axes, it is a straight line passing through the origin. The equation is a linear equation, and we shall consider it as the definition of a *linear resistor*. Hence, if the ratio of the current and voltage associated with any simple circuit element is a constant, then the element is a linear resistor and has a resistance equal to the voltage-current ratio.

Again, it must be emphasized that the linear resistor is an idealized circuit element; it is a mathematical model of a physical device. "Resistors" may be easily purchased or manufactured, but it is soon found that the voltage-current ratio of this physical device is reasonably constant only within certain ranges of current, voltage, or power and depends also on temperature and other environmental factors. We shall usually refer to a linear resistor as simply a resistor, using the longer term only when the linear nature of the element needs emphasis. Any resistor which is nonlinear will always be described as such. Nonlinear resistors should not necessarily be considered as undesirable elements. Although it is true that their presence complicates an analysis, the performance of the device may depend on or be greatly improved by the nonlinearity. Zener diodes, tunnel diodes, and fuses are such elements.

Figure 2-1 shows the most common circuit symbol used for a resistor. In accordance with the voltage, current, and power conventions adopted in the last chapter, the product of v and i gives the power absorbed by the resistor. This absorbed power appears physically as heat and is always positive; a resistor is a passive element that cannot deliver power or store energy. Alternative expressions for the absorbed power are

$$p = vi = i^2R = \frac{v^2}{R}$$

Fig. 2-1 The circuit symbol for a resistor; $R = v/i$ and $p = vi = i^2R = v^2/R$.

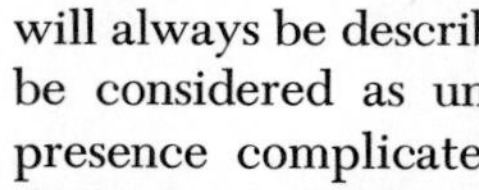

i $+ \; v \; -$

R

The ratio of current to voltage is also a constant,

$$\frac{i}{v} = G$$

where G is called the *conductance*. The unit of conductance is the *mho*, 1 A/V, and is abbreviated by an inverted omega, ℧. The same circuit symbol is used to represent both resistance and conductance. The absorbed power is again necessarily positive and may be expressed in terms of the conductance by

$$p = vi = v^2G = \frac{i^2}{G}$$

Thus a 2-Ω resistor has a conductance of ½ ℧, and if a current of 5 A is flowing through it, a voltage of 10 V is present across the terminals and a power of 50 W is being absorbed.

All the expressions above have been written in terms of instantaneous current, voltage, and power, such as $v = Ri$ and $p = vi$. It is apparent that the current through and voltage across a resistor must both vary with time in the same manner. Thus, if $R = 10\ \Omega$ and $v = 2 \sin 100t$ V, then $i = 0.2 \sin 100t$ A; the power, however, is $0.4 \sin^2 100t$ W, and a simple sketch will illustrate the different nature of its variation with time. Although the current and voltage are each negative during certain time intervals, the absorbed power is never negative.

Drill Problems

2-1 With reference to Fig. 2-1, find v if: (*a*) $G = 10^{-2}$ ℧ and $i = -2.5$ A; (*b*) $R = 40\ \Omega$ and the resistor absorbs 250 W; (*c*) $i = 2.5$ A and the resistor absorbs 500 W.

Ans. −250; ±100; 200 V

2-2 If the current through a 10-kΩ resistor is $2 \cos 10^6\pi t$ mA, and the voltage across it is positive at $t = 0$, find the voltage at $t =$: (*a*) 0.2 μs; (*b*) 0.5 μs; (*c*) 1 μs.

Ans. −20; 0; 16.2 V

2-3 KIRCHHOFF'S LAWS

We are now ready to consider current and voltage relations in simple networks resulting from the interconnection of two or more simple circuit elements. The elements will be connected by electrical conductors, or

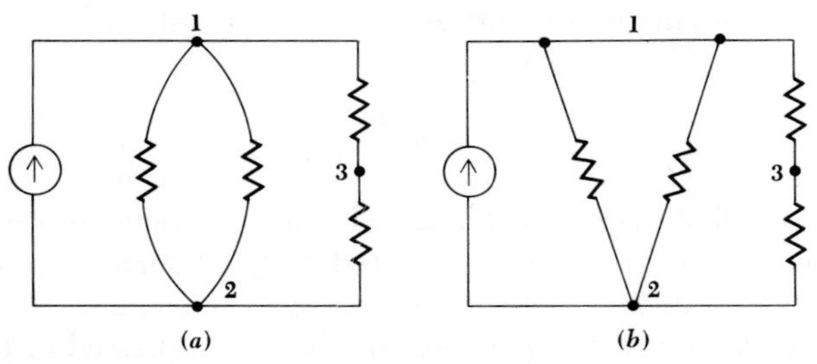

Fig. 2-2 (*a*) A circuit containing three nodes. (*b*) Node 1 is redrawn to look like two nodes; it is still one node.

leads, which have zero resistance, or are *perfectly conducting*. Since the network then appears as a number of simple elements and a set of connecting leads, it is called a *lumped-constant* network. A more difficult analysis problem arises when we are faced with a *distributed-constant* network, which essentially contains an infinite number of vanishingly small elements. This latter type of network is considered in later courses.

A point at which two or more elements have a common connection is called a *node*. Figure 2-2*a* shows a circuit containing three nodes. Sometimes networks are drawn so as to trap an unwary student into believing that there are more nodes present than is actually the case. This occurs when a node, such as node 1 in Fig. 2-2*a*, is shown as two separate junctions connected by a (zero-resistance) conductor. However, all that has been done is to spread the common point out into a common line. Node 1 has been redrawn in this fashion in Fig. 2-2*b*.

Another term whose use will prove convenient is a branch. We may define a *branch* as a single path containing one simple element which connects one node to any other node. The circuit shown in Fig. 2-2*a* and *b* contains five branches.

We are now ready to consider the first of the two laws named for Gustav Robert Kirchhoff, a German university professor who was born about the time Ohm was doing his experimental work. This axiomatic law is called Kirchhoff's *current law*, and it states that the algebraic sum of all the currents entering any node is zero.

Although we cannot rigorously prove the law at this time,[2] we should at least agree that it seems plausible. Suppose we consider the node shown in Fig. 2-3 at which three elements are joined together. Surrounding this node is a closed surface containing a volume V, within which we shall let the total charge be q. The total current entering V is therefore the algebraic sum of all the currents entering V, $i_A + i_B + i_C$. This is equivalent to stating

[2]The proof is based on the hypothesis that charge is conserved, or that it can be neither created nor destroyed. We might just as well accept Kirchhoff's current law as being axiomatic instead of accepting the conservation of charge.

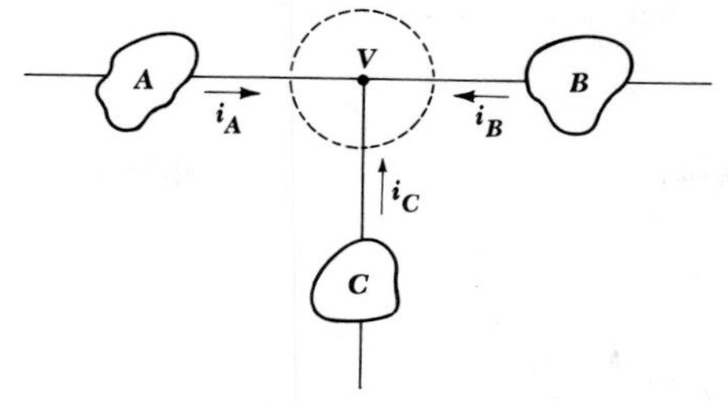

Fig. 2-3 A closed surface surrounds a volume V and encloses a node; the total current entering the closed surface is zero: $i_A + i_B + i_C = 0$.

that charge is entering V at the rate of $i_A + i_B + i_C$ C/s. Hence, the charge q within V must be increasing at the rate of $i_A + i_B + i_C$ C/s, and

$$i_A + i_B + i_C = \frac{dq}{dt}$$

Now we let V become vanishingly small and find that no electrical device other than the node itself is within V. A node, however, cannot store, destroy, or generate charge. Therefore, dq/dt must be zero, and

$$i_A + i_B + i_C = 0$$

This argument may be extended to any number of branches joined at a node.

It is evident that we may also state Kirchhoff's current law in other ways. For instance, the algebraic sum of all the currents *leaving* a node is zero, or the algebraic sum of all the currents entering a node must equal the algebraic sum of all the currents leaving a node. These three forms lead directly to the three equivalent equations written below for the node shown in Fig. 2-4,

$$i_A + i_B - i_C - i_D = 0$$

$$i_C + i_D - i_A - i_B = 0$$

$$i_A + i_B = i_C + i_D$$

Fig. 2-4 Kirchhoff's current law enables us to write $i_A + i_B - i_C - i_D = 0$, $i_C + i_D - i_A - i_B = 0$, or $i_A + i_B = i_C + i_D$.

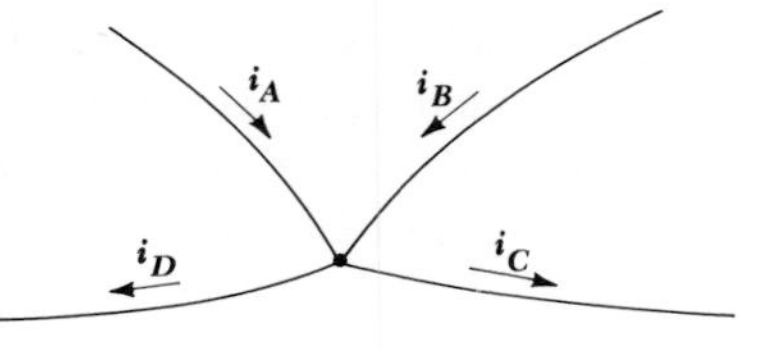

A compact expression for Kirchhoff's current law is

$$\sum_{n=1}^{N} i_n = 0$$

and this is just a shorthand statement for

$$i_1 + i_2 + i_3 + \cdots + i_N = 0$$

When this form is used, it is understood that the N current arrows either are all directed toward the node in question or are all directed away from it.

It is sometimes helpful to interpret Kirchhoff's current law in terms of a hydraulic analogy. Water, like charge, cannot be stored at a point, and thus if we identify a junction of several pipes as a node, it is evident that the number of gallons of water entering the node every second must equal the number of gallons leaving the node each second.

We now turn to Kirchhoff's *voltage law*. This law states that the algebraic sum of the voltages around any closed path in a circuit is zero. Again, we must accept this law as an axiom, even though it is developed in introductory electromagnetic theory.

In view of our definition of the voltage across an element as the energy expended in moving a unit positive charge through the element, an equivalent statement of Kirchhoff's voltage law would seem to say that no energy is expended in moving the unit positive charge about any closed path. It turns out that this statement is correct for a circuit, or at least for its mathematical model; it is *not* true for a general path in a region of space containing time-varying magnetic fields. This statement may be clarified by considering a gravitational analogy. To raise a mass from a point at one elevation to a higher point and then to lower it to the original point requires no net expenditure of energy, regardless of the path taken. This interpretation corresponds to the application of Kirchhoff's voltage law to a circuit. However, let us suppose that there is a strong narrow jet of wind blowing upward at one point on the earth's surface. If we lift our mass outside the jet and then lower it within the jet, we must obviously do more work than if we had reversed the procedure, that is, raising it within the jet and lowering it outside. The work we do depends on the path. The presence of a time-varying magnetic field has much the same effect; the energy we expend in moving the charge depends on the particular path we take through the magnetic field. Now, time-varying magnetic fields are intimately associated with inductors, as we shall see in Chap. 4, but Kirchhoff's voltage law may still be applied to circuits containing inductors. The reason is that, as soon as an inductance value is specified, a path is inferred.

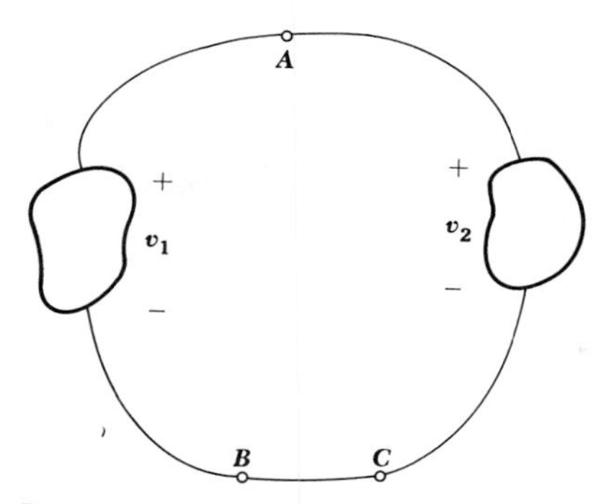

Fig. 2-5 A simple circuit used to emphasize the fact that $v_1 = v_2$.

In summary, then, Kirchhoff's voltage law is applicable to circuits because physical devices are reduced to mathematical models in such a way that the law will be applicable. We should not be surprised to find that the work expended in carrying a unit positive charge around some more general closed path in space is not always zero.

In a *circuit,* then, there is a single, definite value of energy associated with the two terminals of the element. Let us see how this statement leads to Kirchhoff's voltage law. Everybody who performs the coulomb-carrying experiment arrives at the same answer, regardless of the path taken in moving the charge about the circuit. Suppose that we have a choice of two paths along which we may move our charge from one point to another in a circuit, such as that available when two elements are connected as shown in Fig. 2-5. We may proceed from A to B either through the element on the left or through that on the right. We should expect the energy expended to be the same along either path. Thus the voltages v_1 and v_2 are equal,

$$v_1 = v_2 \qquad \text{or} \qquad v_2 - v_1 = 0$$

Since $v_2 - v_1$ is the algebraic sum of the voltages around this closed path, then Kirchhoff's voltage law is satisfied.

Let us consider this same circuit in a slightly different way by determining the voltage between points B and C. We first carry our coulomb from B to A through the left element, supplying v_1 joules of electrical energy to our coulomb, and then proceed to C through the right element, losing v_2 joules of electrical energy from our coulomb. The voltage between B and C along this path is thus

$$-v_1 + v_2 = v_2 - v_1$$

However, we might also choose the short path along the perfect conductor

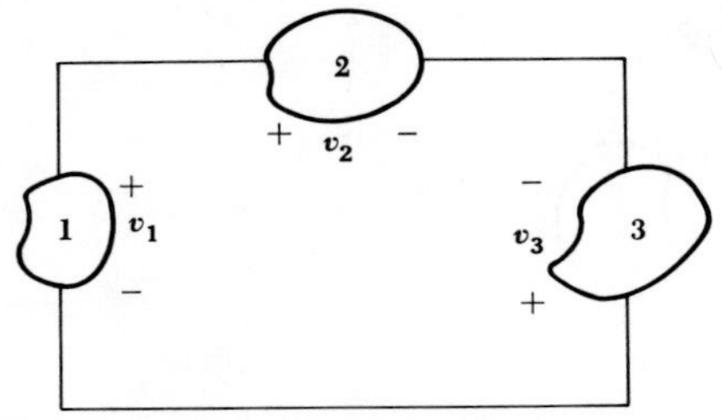

Fig. 2-6 Kirchhoff's voltage law leads to the equation $-v_1 + v_2 - v_3 = 0$.

connecting B directly to C. The voltage here must be zero, for any nonzero voltage across a zero resistance will give rise to an infinite current which we reject as a physical impossibility. If we again assume that the voltage or energy obtained along different paths is the same, then, once more,

$$v_2 - v_1 = 0$$

This latter point of view enables us to consider circuits composed of any number of elements connected in a single loop. Since the voltage across a short piece of one of the perfect conductors must be zero, then the voltage around the remainder of the circuit must be zero. Thus, we may write

$$\sum_{n=1}^{N} v_n = 0 \qquad \text{or} \qquad v_1 + v_2 + v_3 + \cdots + v_N = 0$$

where we agree that every element is entered at its positive voltage reference in a clockwise (or counterclockwise) traversal of the circuit.

We may apply Kirchhoff's voltage law in several different ways. For instance, in the circuit of Fig. 2-6, a clockwise trip around the circuit unfortunately meets one + sign and two (−) signs. We might choose to reverse the signs across elements 1 and 3, letting those voltages be $-v_1$ and $-v_3$. It follows then that

$$(-v_1) + v_2 + (-v_3) = 0$$

We could instead reverse the voltage reference signs on element 2 and change the sign of v_2; counterclockwise travel then leads to

$$v_3 + (-v_2) + v_1 = 0$$

Usually, it is much easier to move around the circuit mentally, writing down directly the voltage of each element whose + terminal is entered and writing down the negative of every voltage first met at the minus sign.

Clockwise travel around the above circuit then gives

$$-v_1 + v_2 - v_3 = 0$$

whereas counterclockwise travel yields

$$v_3 - v_2 + v_1 = 0$$

These last four results are of course identical.

Drill Problems

2-3 Determine the number of branches and nodes present in each of the circuits of Fig. 2-7.

Ans. 2, 2; 5, 4; 6, 4

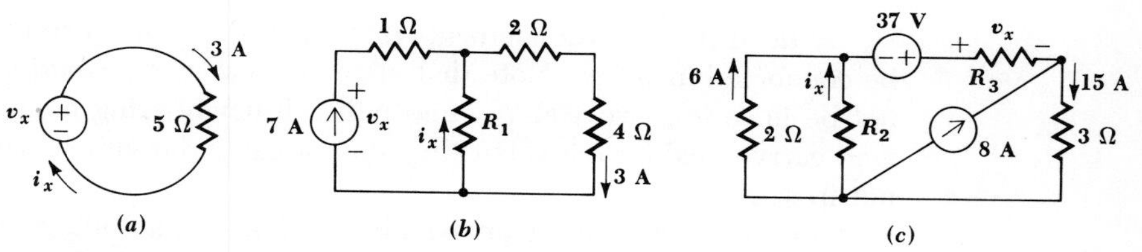

Fig. 2-7 See Drill Probs. 2-3, 2-4, and 2-5.

2-4 Determine i_x in each of the circuits of Fig. 2-7.

Ans. −4; 1; 3 A

2-5 Determine v_x in each of the circuits of Fig. 2-7.

Ans. −20; 15; 25 V

2-4 ANALYSIS OF A SINGLE-LOOP CIRCUIT

Having established Ohm's and Kirchhoff's laws, we may flex our analytical muscles by applying these tools in the analysis of a simple resistive circuit, such as the one shown in Fig. 2-8*a*. We shall assume that the resistance values and the source voltages are known and attempt to determine the current through each element, the voltage across each element, and the power delivered to or absorbed by each element.

Our first step in the analysis is the assumption of reference directions for the unknown currents since we do not know a priori what these

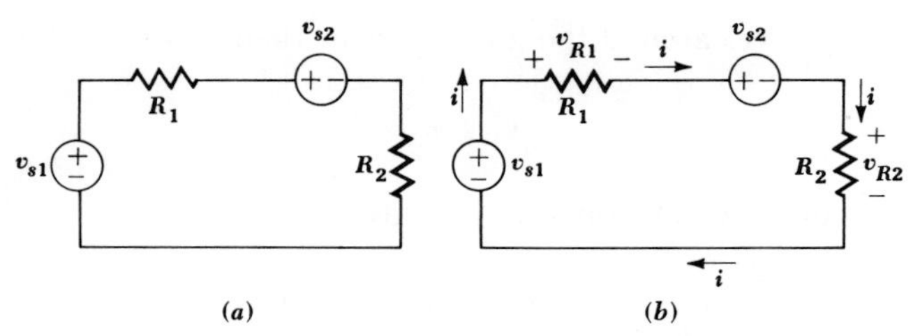

Fig. 2-8 (*a*) A single-loop circuit in which the source voltages and resistances are given. (*b*) Current and voltage reference signs have been added to the circuit.

directions are. Arbitrarily, let us select an unknown current i which flows out of the upper terminal of the left voltage source. This choice is indicated by an arrow at that point in the circuit, as shown in Fig. 2-8*b*. A trivial application of Kirchhoff's current law assures us that this same current must also flow through every other element in the circuit. We may emphasize this fact this one time by placing several other current symbols about the circuit.

By definition, all the elements that carry the *same* current are said to be connected in *series*. Note that elements may carry *equal* currents and not be in series; two 100-W lamp bulbs in neighboring houses may very well carry equal currents, but they do not carry the same current and are not in series.

Our second step in the analysis is the choice of a voltage reference for each of the two resistors. We have already found that the application of Ohm's law, $v = Ri$, demands that the sense of the current and voltage be selected so that the current enters the terminal at which the positive voltage reference is located. If the choice of the current direction is arbitrary, then the selection of the voltage sense is fixed if we intend to use Ohm's law in the form $v = Ri$. The voltages v_{R1} and v_{R2} are shown in Fig. 2-8*b*.

The third step is the application of Kirchhoff's voltage law to the single closed path present. Let us decide to move around the circuit in the clockwise direction, beginning at the lower left corner, and write down directly every voltage first met at its positive reference and write down the negative of every voltage encountered at the negative terminal. Thus,

$$-v_{s1} + v_{R1} + v_{s2} + v_{R2} = 0$$

Finally, we apply Ohm's law to the resistive elements,

$$v_{R1} = R_1 i \qquad \text{and} \qquad v_{R2} = R_2 i$$

and obtain

$$-v_{s1} + R_1 i + v_{s2} + R_2 i = 0$$

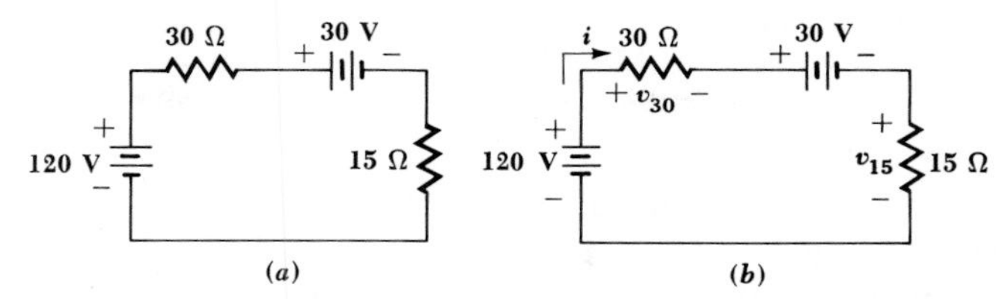

Fig. 2-9 (*a*) A given series circuit. (*b*) The circuit with current and voltage references assigned.

This equation is solved for i, and thus

$$i = \frac{v_{s1} - v_{s2}}{R_1 + R_2}$$

where all the quantities on the right side are known and enable us to determine i. The voltage or power associated with any element may now be obtained in one step by applying $v = Ri$, $p = vi$, or $p = i^2R$.

Let us consider the numerical example illustrated in Fig. 2-9*a*. Two batteries and two resistors are connected in a series circuit. Currents and voltages are assigned to the circuit and indicated in Fig. 2-9*b*, and Kirchhoff's voltage law yields

$$-120 + v_{30} + 30 + v_{15} = 0$$

An application of Ohm's law to each resistor permits us to write

$$-120 + 30i + 30 + 15i = 0$$

from which

$$i = \frac{120 - 30}{30 + 15} = 2 \quad \text{A}$$

Thus, the voltage across each resistor is

$$v_{30} = 2(30) = 60 \quad \text{V} \qquad v_{15} = 2(15) = 30 \quad \text{V}$$

The power absorbed by each element has been shown to be given by the product of the voltage across the element and the current flowing into the element terminal at which the positive voltage reference is located. For the 120-V battery, then, the power absorbed is

$$p_{120\text{V}} = 120(-2) = -240 \quad \text{W}$$

and thus 240 W is *delivered* to other elements in the circuit by this source.

In a similar manner,

$$p_{30V} = 30(2) = 60 \quad \text{W}$$

and we find that this nominally active element is actually absorbing power (or being charged), delivered to it by the other battery.

The power absorbed by each resistor is necessarily positive and may be calculated by

$$p_{30} = v_{30}i = 60(2) = 120 \quad \text{W}$$

or by

$$p_{30} = i^2R = 2^2(30) = 120 \quad \text{W}$$

and

$$p_{15} = v_{15}i = i^2R = 60 \quad \text{W}$$

The results check because the total power absorbed must be zero, or in other words, the power delivered by the 120-V battery is exactly equal to the sum of the powers absorbed by the three other elements. A power balance is often a useful method of checking for careless mistakes.

Before leaving this example, it is important that we be convinced that our initial assumption of a direction for current flow had nothing to do with the answers obtained. Let us suppose that we assumed the current i to be directed in a counterclockwise direction. Both resistor voltages must then be assigned opposite directions also, and we should have obtained

$$-120 - 30i + 30 - 15i = 0$$

and $i = -2$ A, $v_{30} = -60$ V, and $v_{15} = -30$ V. Since each voltage reference is now reversed and each quantity is the negative of the previously obtained value, it is evident that the results are the same. Each absorbed power will be the same.

Any random or convenient choice of current direction may be made. Those who insist on positive answers may always go back and reverse the direction of the current arrow and rework the problem.

Now let us complicate the analysis slightly by letting one of the voltage sources be a dependent source, as exemplified by Fig. 2-10. We again assign a reference direction for the current i and the voltage v_{30}. There is no need to assign a voltage to the 15-Ω resistor, since the controlling voltage v_A for the dependent source is already available. It is worth noting, however, that the reference signs for v_A are reversed from those we would have assigned, and that Ohm's law for this element must thus be expressed as $v_A = -15i$. We apply Kirchhoff's voltage law around the loop,

$$-120 + v_{30} + 2v_A - v_A = 0$$

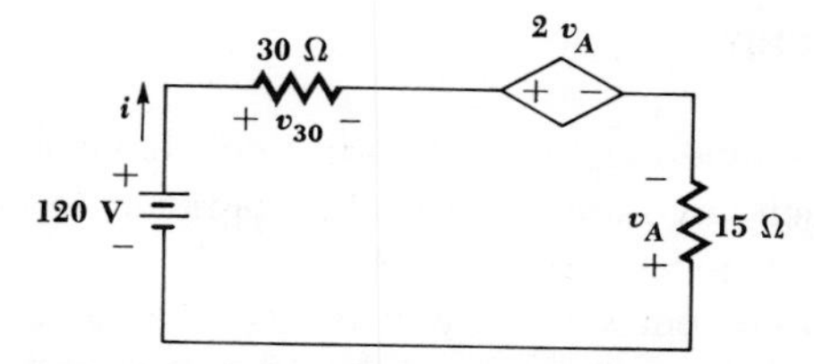

Fig. 2-10 A current i and voltage v_{30} are assigned in a single-loop circuit containing a dependent source.

utilize Ohm's law twice,

$$v_{30} = 30i$$
$$v_A = -15i$$

and obtain

$$-120 + 30i - 30i + 15i = 0$$
$$i = 8 \quad \text{A}$$

The power relationships show that the 120-V battery supplies 960 W, the dependent source supplies 1920 W, and the two resistors together dissipate 2880 W.

More practical applications of the dependent source, such as transistor and vacuum-tube equivalent circuits, will begin to appear in the following chapter.

Drill Problems

2-6 For the circuit shown in Fig. 2-11*a*, find the power: (*a*) delivered to the 5-kΩ resistor; (*b*) supplied by the −25-V source; (*c*) supplied by the 10-V source.

Ans. −20; 20; 50 mW

2-7 With reference to the circuit of Fig. 2-11*b*, find the power absorbed by: (*a*) the 100-Ω resistor; (*b*) the 300-V source; (*c*) the dependent source.

Ans. −1200; −640; 1600 W

Fig. 2-11 See Drill Probs. 2-6 and 2-7.

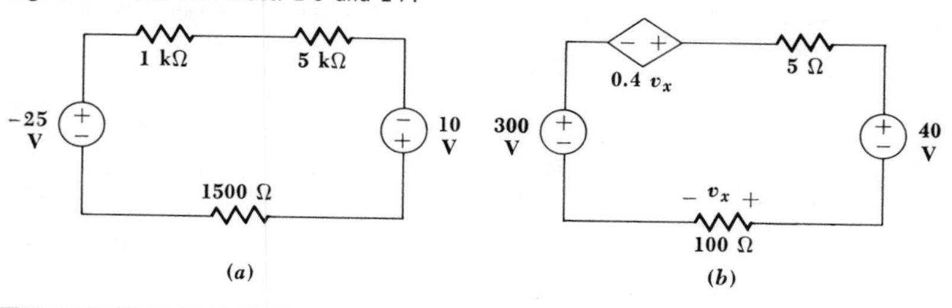

2-5 THE SINGLE NODE-PAIR CIRCUIT

The companion of the single-loop circuit discussed above is the single node-pair circuit in which any number of simple elements are connected between the same pair of nodes. An example of such a circuit is shown in Fig. 2-12*a*. The two current sources and the conductance values are known, and we are to find the voltage, current, and power associated with each element once more.

Our first step is now to assume a voltage across any element, assigning an arbitrary reference polarity. Then Kirchhoff's voltage law forces us to recognize that the voltage across each branch is the same because a closed path proceeds through any branch from one node to the other and then is completed through any other branch. A total voltage of zero requires an identical voltage across every element. We shall say that elements having a common voltage across them are connected in *parallel*. Let us call this voltage v and arbitrarily select it as shown in Fig. 2-12*b*.

Two currents, flowing in the resistors, are then selected in conformance with the convention established with Ohm's law. These currents are also shown in Fig. 2-12*b*.

Our third step in the analysis of the single node-pair circuit is the application of Kirchhoff's current law to either of the two nodes in the circuit. It is usually clearer to apply it to the node at which the positive voltage reference is located, and thus we shall equate the algebraic sum of the currents leaving the upper node to zero,

$$-120 + i_{30} + 30 + i_{15} = 0$$

Finally, the current in each resistor is expressed in terms of v and the conductance of the resistor by Ohm's law,

$$i_{30} = 30v \qquad \text{and} \qquad i_{15} = 15v$$

and we obtain

$$-120 + 30v + 30 + 15v = 0$$

Fig. 2-12 (*a*) A single node-pair circuit. (*b*) A voltage and two currents are assigned.

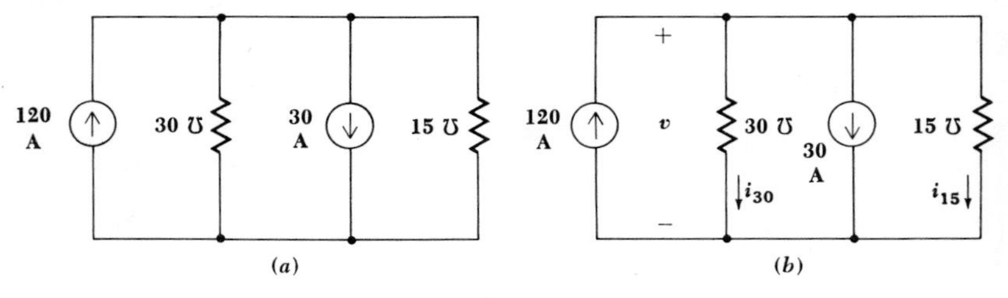

Thus,

$$v = 2 \quad \text{V}$$

$$i_{30} = 60 \quad \text{A} \qquad \text{and} \qquad i_{15} = 30 \quad \text{A}$$

The several values of absorbed power are now easily obtained. In the two resistors,

$$p_{30} = 30(2)^2 = 120 \quad \text{W} \qquad p_{15} = 15(2)^2 = 60 \quad \text{W}$$

and for the two sources,

$$p_{120\text{a}} = 120(-2) = -240 \quad \text{W} \qquad p_{30\text{a}} = 30(2) = 60 \quad \text{W}$$

Thus, the larger current source delivers 240 W to the other three elements in the circuit, and the conservation of energy is verified again.

The similarity of this example to the one previously completed, illustrating the solution of the series circuit with independent sources only, should not have gone unnoticed. The numbers are all the same, but currents and voltages, resistances and conductances, and "series" and "parallel" are interchanged. This is an example of *duality,* and the two circuits are said to be *exact duals* of each other. If the element values or source values were changed in either circuit, without changing the configuration of the network, the two circuits would be *duals,* although not exact duals. We shall study and use duality later, and at this time should only suspect that any result we obtain in terms of current, voltage, and resistance in a series circuit will have its counterpart in terms of voltage, current, and conductance for a parallel circuit.

With this introduction to the duality principle, let us consider the dual of Fig. 2-10, the series circuit containing a dependent source. The single node-pair circuit with a dependent current source is illustrated in Fig. 2-13. A voltage v is assigned as shown, as is a current i_{30} through the 30-℧ conductance. Since the current in the 15-℧ element is already defined as i_A, we need assign no redundant current there.

Fig. 2-13 A voltage v and current i_{30} are assigned in a single node-pair circuit containing a dependent source.

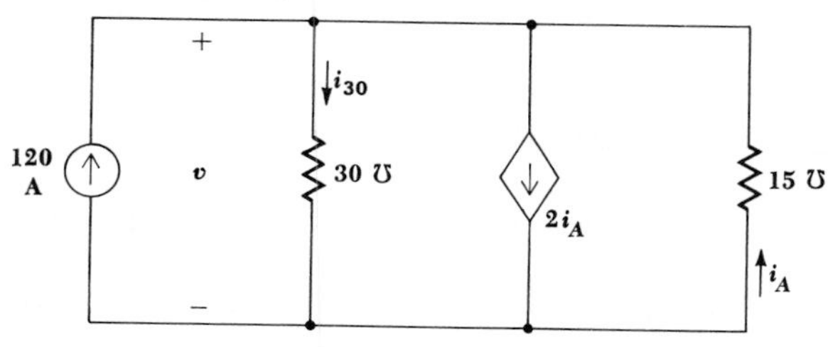

Applying Kirchhoff's current law once,

$$-120 + i_{30} + 2i_A - i_A = 0$$

and Ohm's law twice,

$$i_{30} = 30v$$
$$-i_A = 15v$$

we have

$$-120 + 30v - 30v + 15v = 0$$

and

$$v = 8 \quad \text{V}$$

This corresponds to our previous result of 8 A, and duality triumphs again.

Drill Problems

2-8 In the circuit of Fig. 2-14*a*, find the energy absorbed in 1 h by the: (*a*) 1-mA source; (*b*) -2-mA source; (*c*) 40-μ℧ conductance.

Ans. -108; *129.6; 216* J

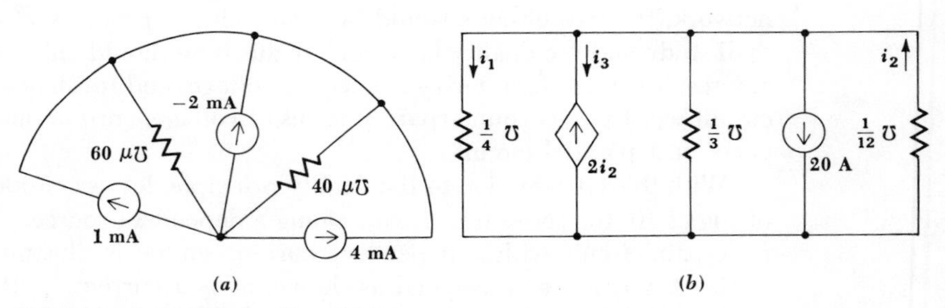

Fig. 2-14 See Drill Probs. 2-8 and 2-9.

2-9 For the circuit of Fig. 2-14*b*, find: (*a*) i_1; (*b*) i_2; (*c*) i_3.

Ans. -6; -4; *2* A

2-6 RESISTANCE AND SOURCE COMBINATION

Some of the equation writing that we have been doing for the simple series and parallel circuits can be avoided. This is achieved by replacing relatively complicated resistor combinations by a single equivalent resistor whenever

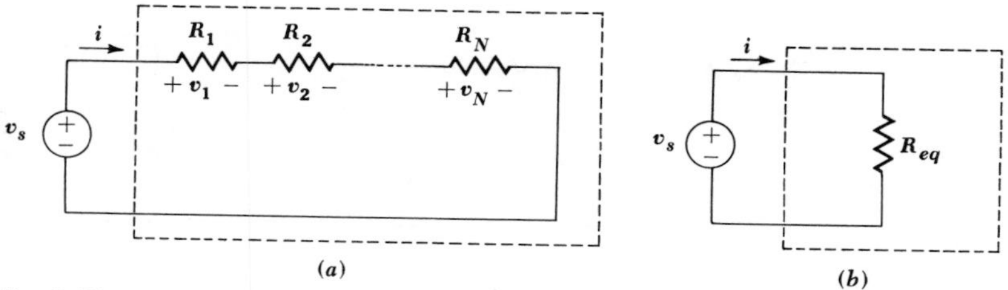

Fig. 2-15 (*a*) A circuit containing a series combination of N resistors. (*b*) A simpler equivalent circuit: $R_{eq} = R_1 + R_2 + \cdots + R_N$.

we are not specifically interested in the current, voltage, or power associated with any of the individual resistors in the combinations. All the current, voltage, and power relationships in the remainder of the circuit will be the same.

We first consider the series combination of N resistors, shown schematically in Fig. 2-15*a*. The broken line surrounding the resistors is intended to suggest that they are enclosed in a "black box," or perhaps in another room, and we wish to replace the N resistors by a single resistor with resistance R_{eq} so that the remainder of the circuit, in this case only the voltage source, does not realize that any change has been made. The source current, power, and, of course, the voltage will be the same before and after.

We apply Kirchhoff's voltage law

$$v_s = v_1 + v_2 + \cdots + v_N$$

and Ohm's law

$$v_s = R_1 i + R_2 i + \cdots + R_N i = (R_1 + R_2 + \cdots + R_N)i$$

and then compare this result with the simple equation applying to the equivalent circuit shown in Fig. 2-15*b*,

$$v_s = R_{eq} i$$

Thus, the value of the equivalent resistance for N series resistances is

$$R_{eq} = R_1 + R_2 + \cdots + R_N$$

It should be emphasized again that we might be particularly interested in the current, voltage, or power of one of the original elements, as would be the case when the voltage of a dependent voltage source depends upon, say, the voltage across R_3. Once R_3 is combined with several series resistors to form an equivalent resistance, then it is gone and the voltage across

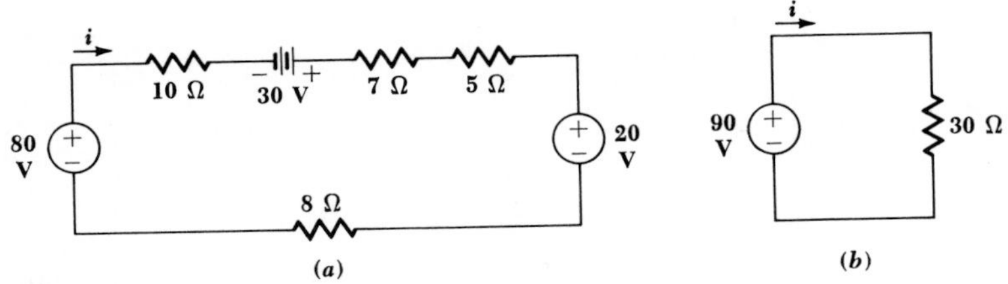

Fig. 2-16 (*a*) A given series circuit. (*b*) A simpler equivalent circuit.

it cannot be determined until R_3 is identified by removing it from the combination. It would have been better to have looked ahead and not made R_3 a part of the combination initially.

An inspection of the Kirchhoff voltage equation for a series circuit also shows two other possible simplifications. The order in which elements are placed in a series circuit makes no difference, and several voltage sources in series may be replaced by an equivalent voltage source having a voltage equal to the algebraic sum of the individual sources. There is usually little advantage in including a dependent voltage source in a series combination.

These simplifications may be illustrated by considering the circuit shown in Fig. 2-16*a*. We first interchange the element positions in the circuit, being careful to preserve the proper sense of the sources, and then combine the three voltage sources into an equivalent 90-V source and the four resistors into an equivalent 30-Ω resistance, as shown in Fig. 2-16*b*. Thus, instead of writing

$$-80 + 10i - 30 + 7i + 5i + 20 + 8i = 0$$

we have simply

$$-90 + 30i = 0$$

and

$$i = 3 \quad \text{A}$$

In order to calculate the power delivered to the circuit by the 80-V source appearing in the given circuit, it is necessary to return to that circuit

Fig. 2-17 (*a*) A circuit containing N parallel resistors having conductances $G_1, G_2, \ldots, G_N$. (*b*) A simpler equivalent circuit: $G_{eq} = G_1 + G_2 + \cdots + G_N$.

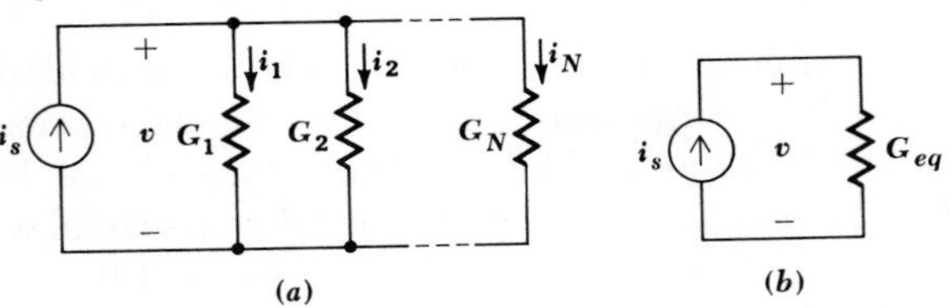

with the knowledge that the current is 3 A. The desired power is 240 W.

It is interesting to note that no element of the original circuit remains in the equivalent circuit, unless we are willing to count the interconnecting wires as elements.

Similar simplifications can be applied to parallel circuits.[3] A circuit containing N conductances in parallel, as in Fig. 2-17a, leads to the Kirchhoff current law equation,

$$i_s = i_1 + i_2 + \cdots + i_N$$

or

$$i_s = G_1 v + G_2 v + \cdots + G_N v = (G_1 + G_2 + \cdots + G_N)v$$

whereas its equivalent in Fig. 2-17b gives

$$i_s = G_{eq} v$$

and thus

$$G_{eq} = G_1 + G_2 + \cdots + G_N$$

In terms of resistance instead of conductance,

$$\frac{1}{R_{eq}} = \frac{1}{R_1} + \frac{1}{R_2} + \cdots + \frac{1}{R_N}$$

or

$$R_{eq} = \frac{1}{1/R_1 + 1/R_2 + \cdots + 1/R_N}$$

This last equation is probably the most often used means of combining parallel resistive elements.

The special case of only two parallel resistors

$$R_{eq} = \frac{1}{1/R_1 + 1/R_2} \qquad \text{or} \qquad R_{eq} = \frac{R_1 R_2}{R_1 + R_2}$$

is needed very often. The last form is worth memorizing.

Parallel current sources may also be combined by algebraically adding the individual currents, and the order of the parallel elements may be rearranged as desired.

The various combinations described in this section are used to simplify the circuit of Fig. 2-18a. Let us suppose that we wish to know the power and voltage of the dependent source. We may just as well leave it alone, then, and combine the remaining two sources into one 2-A source. The resistances are combined by beginning with the parallel combination of the two 6-Ω resistors into a 3-Ω resistance, followed by the series combi-

[3]Duality.

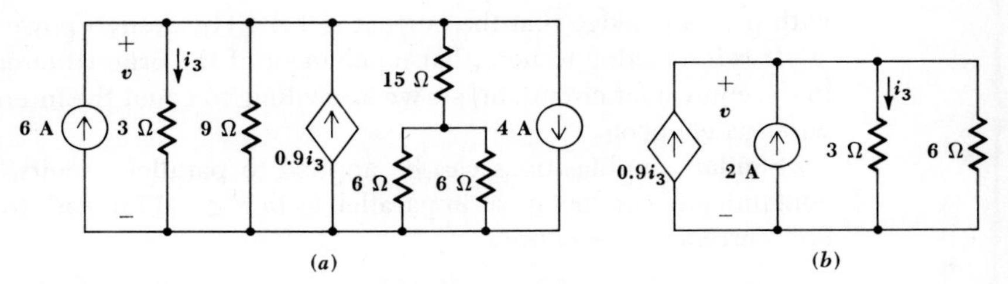

Fig. 2-18 (*a*) A given circuit. (*b*) A simplified equivalent circuit.

nation of 3 Ω and 15 Ω. The 18-Ω and 9-Ω elements combine in parallel to produce 6 Ω, and this is as far as we can proceed profitably. Certainly 6 Ω in parallel with 3 Ω is 2 Ω, but the current i_3 on which the source depends then disappears.[4]

From the equivalent circuit in Fig. 2-18*b*, we have

$$-0.9i_3 - 2 + i_3 + \frac{v}{6} = 0$$

and

$$v = 3i_3$$

yielding

$$i_3 = 10/3 \quad \text{A}$$

$$v = 10 \quad \text{V}$$

Thus, the dependent source furnishes $(0.9)(^{10}\!/_3)(10) = 30$ W to the remainder of the circuit.

Now if we are belatedly asked for the power dissipated in the 15-Ω resistor, we must return to the original circuit. This resistor is in series with an equivalent 3-Ω resistance; a voltage of 10 V is across the 18-Ω total; thus, a current of $^5\!/_9$ A flows through the 15-Ω resistor and the power absorbed by this element is $(^5\!/_9)^2(15)$, or 4.63 W.

To conclude the discussion of parallel and series element combinations, we should consider the parallel combination of two voltage sources and the series combination of two current sources. For instance, what is the equivalent of a 5-V source in parallel with a 10-V source? By the definition of a voltage source, the voltage across the source cannot change; by Kirchhoff's voltage law, then, 5 equals 10 and we have hypothesized a physical impossibility. Thus, voltage sources in parallel are permissible only when each has the same terminal voltage at every instant. Later, we shall see that *practical* voltage sources may be combined in parallel without any theoretical difficulty.

[4]Of course, we could have preserved it by using the given circuit to write $i_3 = v/3$, thus expressing i_3 in terms of variables appearing in the final circuit.

In a similar way, two current sources may not be placed in series unless each has the same current, including sign, for every instant of time.

A voltage source in parallel or series with a current source presents an interesting little intellectual diversion. The two possible cases are illustrated by Prob. 29 at the end of the chapter.

Drill Problems

2-10 Find R_{eq} for the network of Fig. 2-19*a*: (*a*) as it is shown; (*b*) with the 5-Ω resistor replaced by a short circuit (0 Ω); (*c*) with the 5-Ω resistor replaced by an open circuit (∞ Ω).

Ans. *9.933; 10; 10.2* Ω

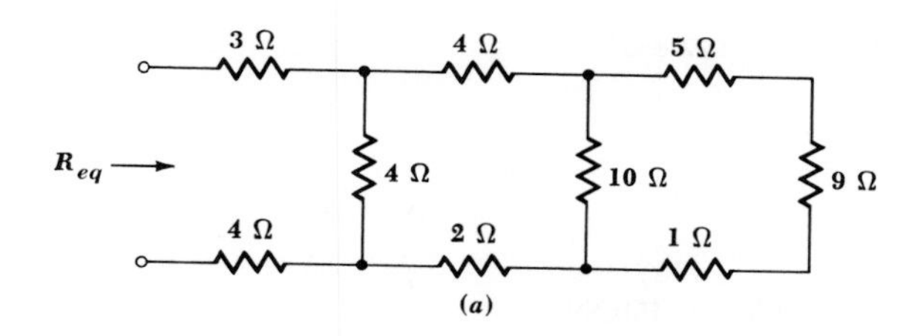

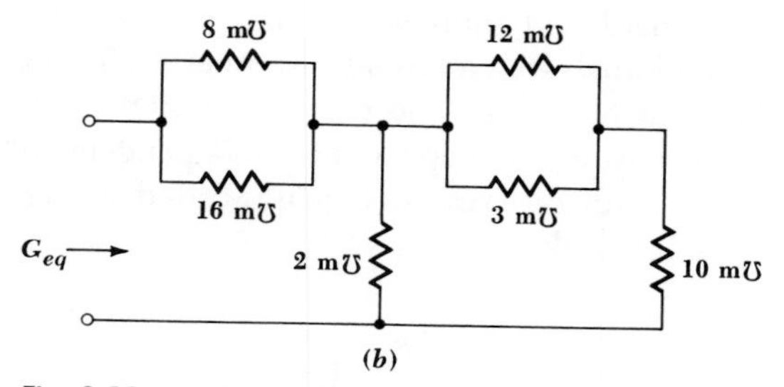

Fig. 2-19 See Drill Probs. 2-10 and 2-11.

2-11 Find G_{eq} for the network of Fig. 2-19*b*: (*a*) as it is shown; (*b*) if it is disconnected and the six elements reconnected in parallel; (*c*) if it is disconnected and the six elements reconnected in series.

Ans. *0.830; 6; 51* m℧

2-12 Find v_x in the circuits of Fig. 2-20*a*, *b*, and *c*.

Ans. *−60; 6; 48* V

2-13 Find i_y in the circuits of Fig. 2-20*a*, *b*, and *c*.

Ans. *−2; 0.5; 5* A

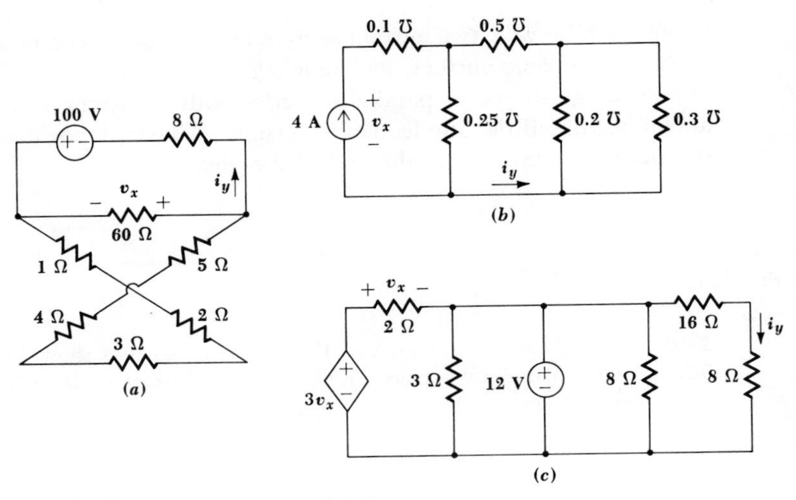

Fig. 2-20 See Drill Probs. 2-12 and 2-13.

2-7 VOLTAGE AND CURRENT DIVISION

By combining resistances and sources, we have found one method of shortening the work of analyzing a circuit. Another useful shortcut is the application of the ideas of voltage and current division.

Voltage division occurs when a dependent or independent voltage source is connected in series with two resistors, as illustrated in Fig. 2-21. The voltage across R_2 is obviously

$$v_2 = R_2 i = R_2 \frac{v}{R_1 + R_2}$$

or

$$v_2 = \frac{R_2}{R_1 + R_2} v$$

Fig. 2-21 An illustration of voltage division, $v_2 = \dfrac{R_2}{R_1 + R_2} v$.

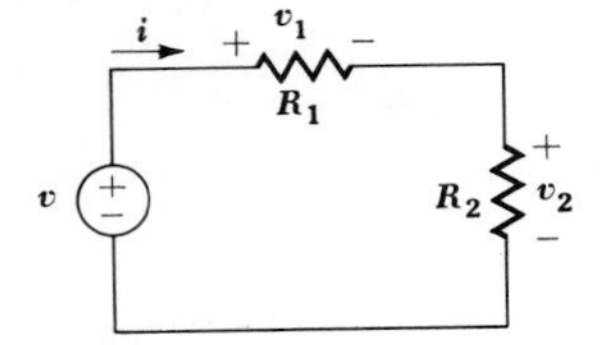

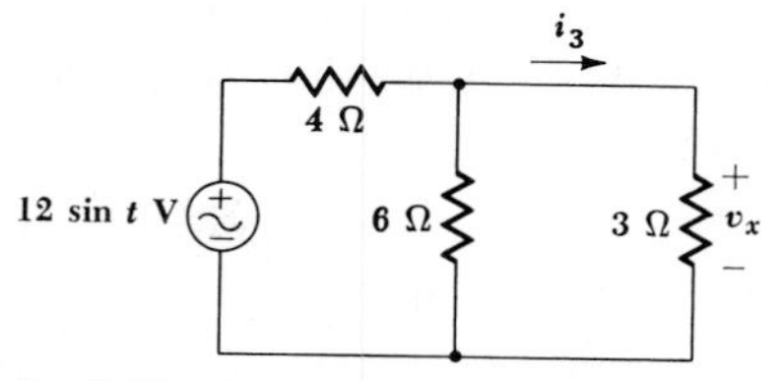

Fig. 2-22 A numerical example illustrating resistance combination and voltage division. The wavy line within the source symbol indicates a sinusoidal variation with time.

and the voltage across R_1 is, similarly,

$$v_1 = \frac{R_1}{R_1 + R_2} v$$

The voltage appearing across either of the series resistors is the applied voltage times the ratio of that resistance to the total resistance. Voltage division and resistance combination may both be applied, as in the circuit shown in Fig. 2-22. We mentally combine the 3- and 6-Ω resistances, obtaining 2 Ω, and thus find that v_x is $2/6$ of 12 sin t, or 4 sin t V.

The dual of voltage division is current division. We now are given a current source in parallel with two conductances, as exemplified by the circuit of Fig. 2-23. The current flowing through G_2 is

$$i_2 = G_2 v = G_2 \frac{i}{G_1 + G_2}$$

or

$$i_2 = \frac{G_2}{G_1 + G_2} i$$

and, similarly,

$$i_1 = \frac{G_1}{G_1 + G_2} i$$

Thus the current flowing through either of the parallel conductances is the applied current times the ratio of that conductance to the total conductance.

Fig. 2-23 An illustration of current division.

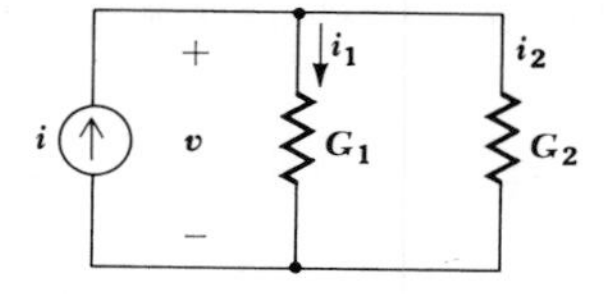

Since we are given the value of the resistance more often than the conductance, a more important form of the last result is obtained by replacing G_1 by $1/R_1$ and G_2 by $1/R_2$,

$$i_2 = \frac{R_1}{R_1 + R_2} i \qquad \text{and} \qquad i_1 = \frac{R_2}{R_1 + R_2} i$$

Nature has not smiled on us here, for these last two equations have a factor which differs subtly from the factor used with voltage division, and some effort is going to be needed to avoid errors. Many students look on the expression for voltage division as "obvious" and that for current division as being "different." It also helps to realize that the larger resistor always carries the smaller current.

As an example of the use of both current division and resistance combination, let us return to the example of Fig. 2-22 and write an expression for the current through the 3-Ω resistor. The total current flowing into the 3- and 6-Ω combination is

$$i = \frac{12 \sin t}{4 + (6)(3)/(6 + 3)}$$

and thus the desired current is

$$i_3 = \frac{12 \sin t}{4 + (6)(3)/(6 + 3)} \frac{6}{6 + 3} = \tfrac{4}{3} \sin t$$

PROBLEMS

☐ **1** For a section of conducting material of uniform cross-sectional area A, length l, and conductivity σ (℧/m) or resistivity ρ ($\rho = 1/\sigma$, in Ω-m), the resistance of the section is $R = l/\sigma A = \rho l/A$. The resistivity of copper varies as a function of the absolute temperature T ($100° < T < 500°$K) as: $\rho = 1.724 \times 10^{-8}[1 + 0.00393(T - 273)]$ Ω-m. (*a*) Find the resistance of a 1000-turn coil of wire, average radius 4 cm, wire radius 1.4×10^{-2} cm, at 273°K. (*b*) By what percentage does the resistance increase as the temperature increases 100°C? (*c*) If the total power dissipated in this coil is limited to 2 W, what is the maximum current it can carry?

☐ **2** If the resistance of a copper wire is proportional to its length and inversely proportional to its area, by what factor is the resistance increased when the same mass of material is used to produce a wire of: (*a*) twice the length? (*b*) half the diameter?

☐ **3** For the network shown in Fig. 2-24*a*, find i, v_s, R, and the power supplied by the voltage source.

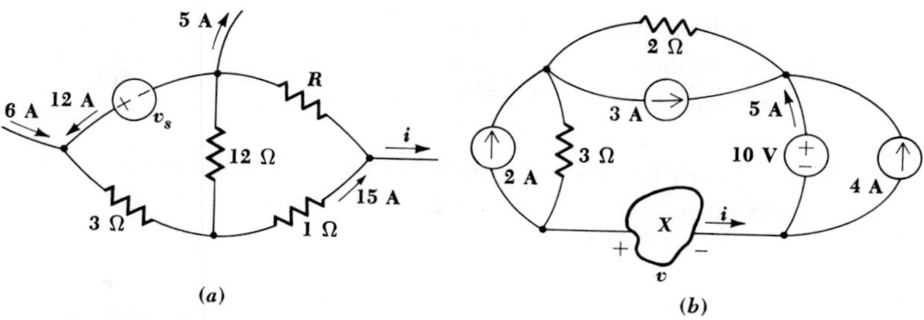

Fig. 2-24 (*a*) See Prob. 3. (*b*) See Prob. 4.

☐ **4** For the circuit shown in Fig. 2-24*b*, find v, i, and the power absorbed by element X.

☐ **5** If $i_x = 4$ A and $R = 3\ \Omega$ in the circuit of Fig. 2-25*a*, find the power absorbed by each of the five circuit elements.

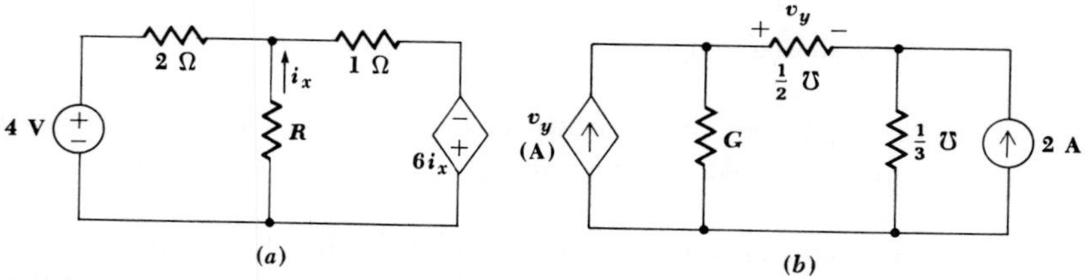

Fig. 2-25 (*a*) See Prob. 5. (*b*) See Prob. 6.

☐ **6** If $v_y = 12$ V and $G = \frac{1}{6}$ ℧ in the circuit of Fig. 2-25*b*, find the power absorbed by each of the five circuit elements.

☐ **7** A circuit contains four nodes lettered A, B, C, and D. There are six branches, one between each pair of nodes. Let i_{AB} be the current in branch AB directed from node A to node B through the element. Then, given $i_{AB} = 16$ mA, and $i_{DA} = 39$ mA, find i_{AC}, i_{BC}, and i_{BD} if $i_{CD} =$: (*a*) 23 mA; (*b*) -23 mA.

☐ **8** A circuit contains six nodes lettered A, B, C, D, E, and F. Let v_{AB} be the voltage between nodes A and B with its positive reference at the first-named node, here A. Find v_{AC}, v_{AD}, v_{AE}, and v_{AF} if $v_{AB} = 6$ V, $v_{BD} = -3$ V, $v_{CF} = -8$ V, $v_{EC} = 4$ V, and: (*a*) $v_{DE} = 1$ V; (*b*) $v_{CD} = 1$ V; (*c*) $v_{FE} = 4$ V.

☐ **9** A series loop contains the following circuit elements in order: a 6-V source, a 2-kΩ resistor, a 3-kΩ resistor, an 18-V source, and a 7-kΩ resistor. The voltage sources aid. (*a*) Find the magnitude of the voltage across each resistor. (*b*) Determine the power absorbed by each element.

☐ **10** A 12-V battery is connected to a 5.5-Ω load through wiring having a total resistance of 0.5 Ω. Find: (*a*) p_{load}; (*b*) p_{wiring}; (*c*) the total power supplied by the source; (*d*) efficiency $= p_{load}/(p_{load} + p_{wiring})$.

☐ **11** Specify i, v, and the power absorbed by the unknown circuit element in Fig. 2-26*a* if the 100-V source supplies: (*a*) 100 W; (*b*) 300 W.

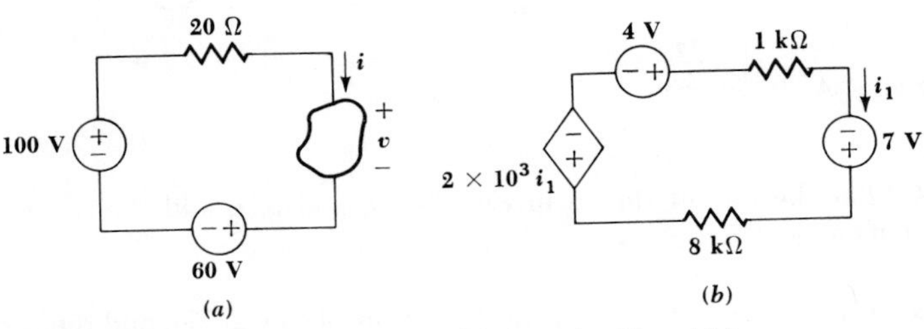

Fig. 2-26 (*a*) See Probs. 11 and 12. (*b*) See Probs. 13 and 14.

☐ **12** In the circuit of Fig. 2-26*a*, consider the unknown element as a voltage source. What value should v be so that a charge of 600 C is delivered to the 100-V source in 1 min?

☐ **13** (*a*) In the circuit shown in Fig. 2-26*b*, find i_1. (*b*) Replace the dependent source by a 2-kΩ resistor and again find i_1.

☐ **14** In the circuit shown in Fig. 2-26*b*, relabel the dependent source $13v_1$, where v_1 is the voltage across the 1-kΩ resistor, positive reference at the left terminal of the resistor. Find i_1.

☐ **15** The following circuit elements are connected between a pair of nodes: a 2-m℧ conductance, a 6-mA source, a 3-m℧ conductance, an 18-mA source, and a 7-m℧ conductance. The current sources are both directed into the upper node. (*a*) Find the magnitude of the current through each conductance. (*b*) Determine the power absorbed by each element.

☐ **16** Specify v, i, and the power absorbed by the unknown circuit element in Fig. 2-27*a* if the 0.1-A source supplies: (*a*) 1 W; (*b*) 3 W.

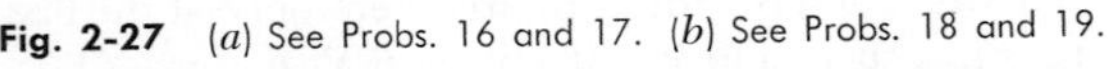

Fig. 2-27 (*a*) See Probs. 16 and 17. (*b*) See Probs. 18 and 19.

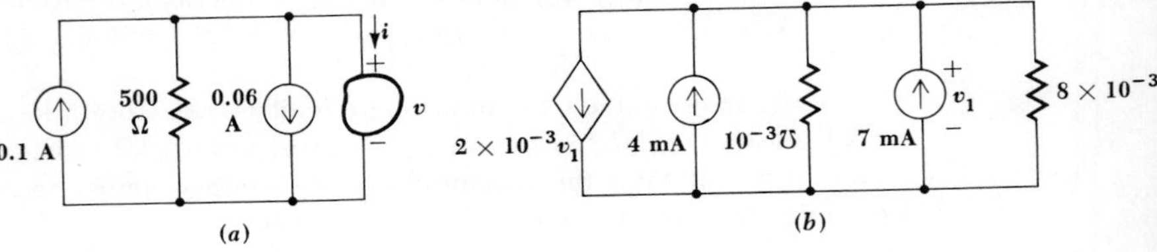

☐ **17** In the circuit of Fig. 2-27a, consider the unknown as a current source. What would be the value of i if an energy of 120 J is delivered to the 0.1-A source in 0.5 min?

☐ **18** (a) In the circuit of Fig. 2-27b, find v_1. (b) Replace the dependent source by a 500-Ω resistor and again find v_1.

☐ **19** In the circuit shown in Fig. 2-27b, relabel the dependent source $13i_1$, where i_1 is the downward current through the 10^{-3}-℧ conductance. Find v_1.

☐ **20** Find the power delivered to each of the 10-Ω resistors in the circuit of Fig. 2-28a.

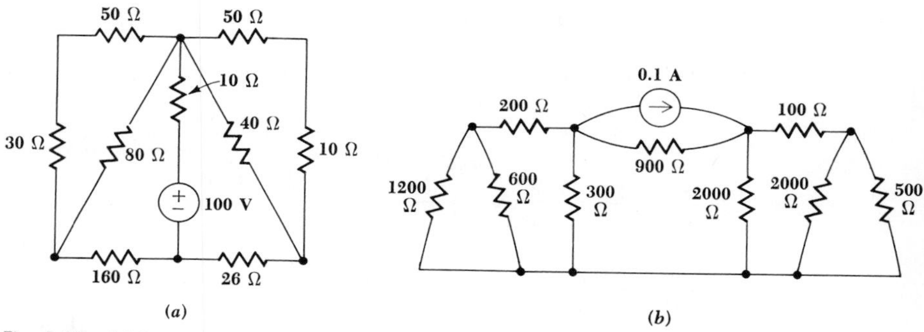

Fig. 2-28 (a) See Prob. 20. (b) See Probs. 21 and 22.

☐ **21** By combining resistances in the circuit of Fig. 2-28b, find the power supplied by the source and also the power absorbed by the 900-Ω resistor.

☐ **22** The destruction (by open circuit) of which resistor in the circuit of Fig. 2-28b would cause the greatest increase in the power supplied by the current source? What is this power?

☐ **23** In the circuit shown in Fig. 2-29a, it is known that $v_1 = 6$ V. Find i_s and v.

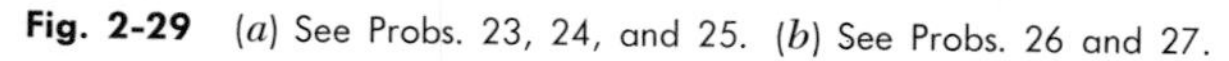

Fig. 2-29 (a) See Probs. 23, 24, and 25. (b) See Probs. 26 and 27.

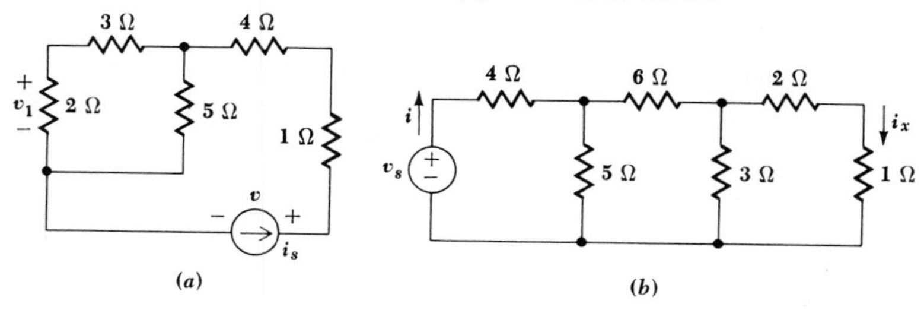

□ **24** In parallel with the independent current source in Fig. 2-29*a*, install a dependent current source, directed to the right, and labeled $0.6v_1$. Find v_1 and the voltage v across both sources if i_s is given as 12 A.

□ **25** If $i_s = 4$ A in the circuit of Fig. 2-29*a*, calculate in a single step (each) the current, voltage, and power associated with the 3-Ω resistor.

□ **26** (*a*) Given $i_x = 1$ A in the circuit of Fig. 2-29*b*, find i and v_s. (*b*) Use the results of (*a*) to determine i_x if $v_s = 70$ V; (*c*) if $v_s = 100$ V.

□ **27** Let $v_s = 50$ V in the circuit of Fig. 2-29*b*. Using the concepts of voltage division, current division, and resistance combination, write a single expression that will yield the current i_x.

□ **28** What must be the resistance of R in the circuit of Fig. 2-30*a*?

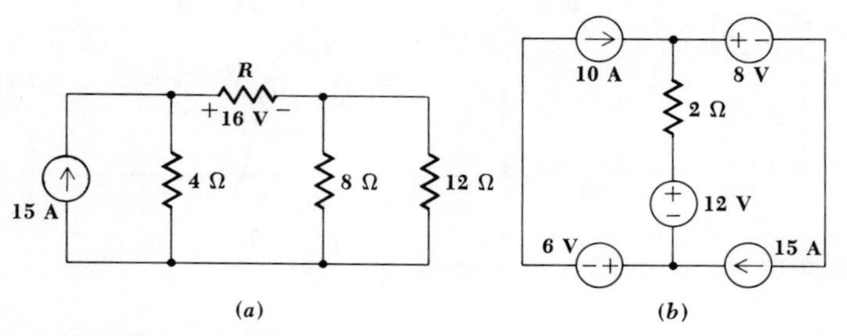

Fig. 2-30 (*a*) See Prob. 28. (*b*) See Prob. 29.

□ **29** (*a*) Find the power absorbed by each element in the circuit of Fig. 2-30*b*. (*b*) How would the results change if the 2-Ω resistor were reduced to 0?

□ **30** Assume that the heating element for an electric hotplate consists of three concentric circles of resistance wire having radii of 2, 4, and 6 cm. Assume further that an ideal 230-V (dc) source is connected to diametrically opposite points on each ring. It is desired to provide a power of 10 W/cm length on every ring. Find the total source current and the resistance per unit length for each ring.

Chapter Three
Some Useful Techniques of Circuit Analysis

3-1 INTRODUCTION

We should now be familiar with Ohm's law and Kirchhoff's laws and their application in the analysis of simple series and parallel resistive circuits. When they will produce results more easily we should be able to combine resistances or sources in series or parallel and be able to use the principles of voltage and current division. The circuits on which we have been practicing are simple and of questionable practical importance; they are useful in helping us learn to apply the fundamental laws. Now we must begin analyzing more complicated circuits.

Physical systems which we shall want to analyze and design in the coming years will include electric and electronic control circuits, communication systems, energy converters such as motors and generators, power distribution systems, and entertainment or other devices which are now unknown. Many of us will be confronted with allied problems involving heat flow, fluid flow, and the behavior of various mechanical systems. In the analysis of any of these cases it is often helpful to replace the system with an equivalent electric circuit. As an example, we might consider a transistor amplifier, an electronic device which is a part of many communication systems and control circuits. The transistor, along with several resistors and other passive circuit elements, is used to amplify or magnify an electrical signal and to direct the amplified signal to a desired load. It is possible to replace the transistor, the resistors, the other passive circuit elements, the signal source, and the load by combinations of simple circuit elements, such as current sources, voltage sources, and resistors. The solution of the problem is then achieved by circuit methods and techniques which we either know already or will meet in this chapter.

When we are able to describe mathematically the behavior of fluid-flow and heat-flow systems, the dynamic response of aircraft control surfaces, and other nonelectrical phenomena, we shall see that the resultant equations are often precisely analogous to those describing current and voltage relationships in electric circuits. We may decide, then, that it is much easier and cheaper to construct the analogous electric circuit than it is to build a prototype of the actual physical system. The electric circuit may then be used to predict the performance of the other system as various elements are changed and may help achieve a better final design. This is the basis on which the electronic analog computer operates.

It is evident that one of the primary goals of this chapter must be learning methods of simplifying the analysis of more complicated circuits. Among these methods will be superposition, loop, mesh, and nodal analysis. We shall also try to develop the ability to select the most convenient analysis method. Most often we are interested only in the detailed performance of an isolated portion of a complex circuit; a method of replacing the remainder of the circuit by a greatly simplified equivalent is then very desirable. The equivalent is often a single resistor in series or parallel with an ideal source. Thévenin's and Norton's theorems will enable us to do this.

We shall begin studying methods of simplifying circuit analysis by considering a powerful general method, that of nodal analysis.

3-2 NODAL ANALYSIS

In the previous chapter we considered the analysis of a simple circuit containing only two nodes. We found then that the major step of the analysis was taken as we obtained a single equation in terms of a single unknown quantity, the voltage between the pair of nodes. We shall now let the number of nodes increase, and correspondingly provide one additional unknown quantity and one additional equation for each added node. Thus, a three-node circuit should have two unknown voltages and two equations; a ten-node circuit will have nine unknown voltages and nine equations; and an N-node circuit will need $(N - 1)$ voltages and $(N - 1)$ equations.

We consider the mechanics of nodal analysis in this section, but the justification for our methods will be developed toward the end of this chapter.

As an example, let us consider the three-node circuit shown in Fig. 3-1*a*. We may emphasize the locations of the 3 nodes by redrawing the circuit, as shown in Fig. 3-1*b*, in which each node is identified by a number. We would now like to associate a voltage with each node, but we must remember that a voltage must be defined as existing between *two nodes* in a network. We thus select one node as a reference node, and then define a voltage between each remaining node and the reference node. Hence, we note again that there will be only $(N - 1)$ voltages defined in an N-node circuit.

We choose node 3 as the reference node. Either of the other nodes could have been selected, but a little simplification in the resultant equa-

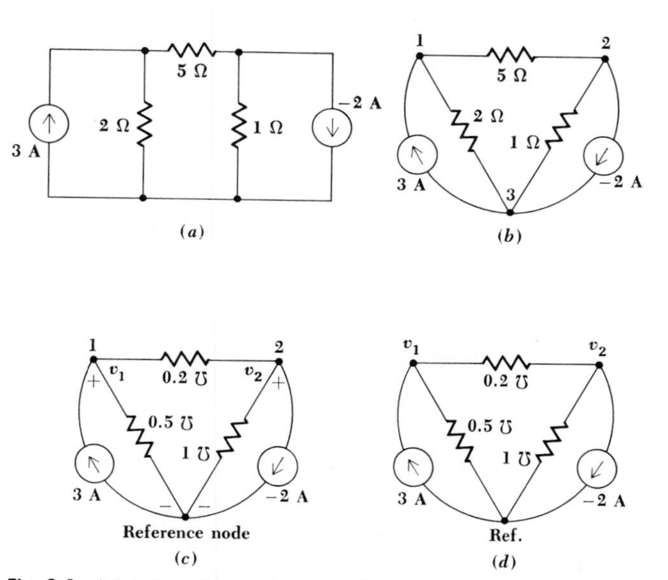

Fig. 3-1 (*a*) A given three-node circuit. (*b*) The circuit is redrawn to emphasize the three nodes, and each node is numbered. (*c*) A voltage, including polarity reference, is assigned between each node and the reference node. (*d*) The voltage assignment is simplified by eliminating the polarity references; it is understood that each voltage is sensed positive relative to the reference node.

tions is obtained if the node to which the greatest number of branches is connected is identified as the reference node. In some practical circuits there may be a large number of elements connected to a metallic case or chassis on which the circuit is built; the chassis is often connected through a good conductor to the earth. Thus, the metallic case may be called "ground," and this ground node becomes the most convenient reference node. More often than not, it appears as a common lead across the bottom of a circuit diagram.

The voltage between node 1 and the reference node 3 is identified as v_1, and v_2 is defined between node 2 and the reference. These two voltages are sufficient, and the voltage between any other pair of nodes may be found in terms of them. For example, the voltage of node 1 with respect to node 2 is $(v_1 - v_2)$. The voltages v_1 and v_2 and their reference signs are shown in Fig. 3-1*c*. In this figure the resistance values have also been replaced with conductance values.

The circuit diagram is finally simplified in Fig. 3-1*d* by eliminating all voltage reference symbols. A reference node is plainly marked, and the voltages placed at each remaining node are *understood* to be the voltage of that node with respect to the reference node.

We must now apply Kirchhoff's current law to nodes 1 and 2. We do this by equating the total current leaving the node through the several conductances to the total source current entering the node. Thus,

$$0.5v_1 + 0.2(v_1 - v_2) = 3$$

or

$$0.7v_1 - 0.2v_2 = 3 \tag{1}$$

At node 2 we obtain

$$v_2 + 0.2(v_2 - v_1) = 2$$

or

$$-0.2v_1 + 1.2v_2 = 2 \tag{2}$$

Equations (1) and (2) are the desired two equations in two unknowns, and they may be solved easily. The result is:

$$v_1 = 5 \quad \text{V}$$

$$v_2 = 2.5 \quad \text{V}$$

Also, the voltage of node 1 relative to node 2 is $(v_1 - v_2)$, or 2.5 V, and any current or power in the circuit may now be found in one step. For example, the current directed downward through the 0.5-℧ conductance is $0.5v_1$, or 2.5 A.

Now let us increase the number of nodes by one. A new circuit is shown in Fig. 3-2*a*, and it is redrawn in Fig. 3-2*b*, with the nodes identified, a con-

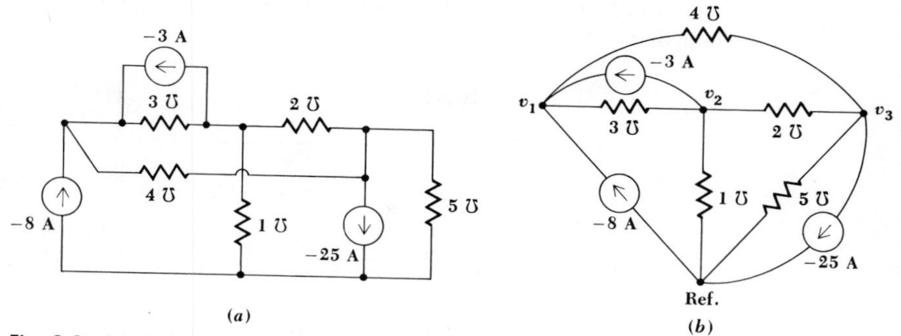

Fig. 3-2 (a) A circuit containing four nodes and eight branches. (b) The same circuit redrawn with node voltages assigned.

venient reference node chosen, and the node voltages specified. We next sum the currents leaving node 1:

$$3(v_1 - v_2) + 4(v_1 - v_3) - (-8) - (-3) = 0$$

or

$$7v_1 - 3v_2 - 4v_3 = -11 \tag{3}$$

At node 2:

$$3(v_2 - v_1) + 1v_2 + 2(v_2 - v_3) - 3 = 0$$

or

$$-3v_1 + 6v_2 - 2v_3 = 3 \tag{4}$$

and at node 3:

$$4(v_3 - v_1) + 2(v_3 - v_2) + 5v_3 - 25 = 0$$

or

$$-4v_1 - 2v_2 + 11v_3 = 25 \tag{5}$$

Equations (3) through (5) may be solved by a simple process of elimination of variables, or by Cramer's rule and determinants.[1] Using the latter method, we have

$$v_1 = \frac{\begin{vmatrix} -11 & -3 & -4 \\ 3 & 6 & -2 \\ 25 & -2 & 11 \end{vmatrix}}{\begin{vmatrix} 7 & -3 & -4 \\ -3 & 6 & -2 \\ -4 & -2 & 11 \end{vmatrix}}$$

[1]Appendix 1 provides a short review of determinants and the solution of a system of simultaneous linear equations by Cramer's rule.

Expanding the numerator and denominator determinants by minors along their first columns leads to

$$v_1 = \frac{-11\begin{vmatrix} 6 & -2 \\ -2 & 11 \end{vmatrix} - 3\begin{vmatrix} -3 & -4 \\ -2 & 11 \end{vmatrix} + 25\begin{vmatrix} -3 & -4 \\ 6 & -2 \end{vmatrix}}{7\begin{vmatrix} 6 & -2 \\ -2 & 11 \end{vmatrix} - (-3)\begin{vmatrix} -3 & -4 \\ -2 & 11 \end{vmatrix} + (-4)\begin{vmatrix} -3 & -4 \\ 6 & -2 \end{vmatrix}}$$

$$= \frac{-11(62) - 3(-41) + 25(30)}{7(62) + 3(-41) - 4(30)} = \frac{-682 + 123 + 750}{434 - 123 - 120}$$

$$= \frac{191}{191} = 1 \quad \text{V}$$

Similarly,

$$v_2 = \frac{\begin{vmatrix} 7 & -11 & -4 \\ -3 & 3 & -2 \\ -4 & 25 & 11 \end{vmatrix}}{191} = 2 \quad \text{V}$$

and

$$v_3 = \frac{\begin{vmatrix} 7 & -3 & -11 \\ -3 & 6 & 3 \\ -4 & -2 & 25 \end{vmatrix}}{191} = 3 \quad \text{V}$$

The denominator determinant is common to each of the three evaluations above. For circuits that do not contain either voltage sources or dependent sources (i.e., circuits containing only independent current sources), this denominator determinant may be written as a matrix[2] and defined as the *conductance matrix* of the circuit:

$$\mathbf{G} = \begin{bmatrix} 7 & -3 & -4 \\ -3 & 6 & -2 \\ -4 & -2 & 11 \end{bmatrix}$$

It should be noted that the nine elements of the matrix are the ordered array of the coefficients of (3), (4), and (5), each of which is a conductance value. The first row is composed of the coefficients of the Kirchhoff current law equation at the first node, the coefficients being given in the order of v_1, v_2, and v_3. The second row applies to the second node, and so on.

[2]There is no appendix on matrices because we shall not rely on matrix equations or matrix operations in this book. These topics, however, will certainly appear later in every engineer's professional courses as well as somewhere in his mathematics sequence. We shall identify the important matrices of circuit theory as we meet them.

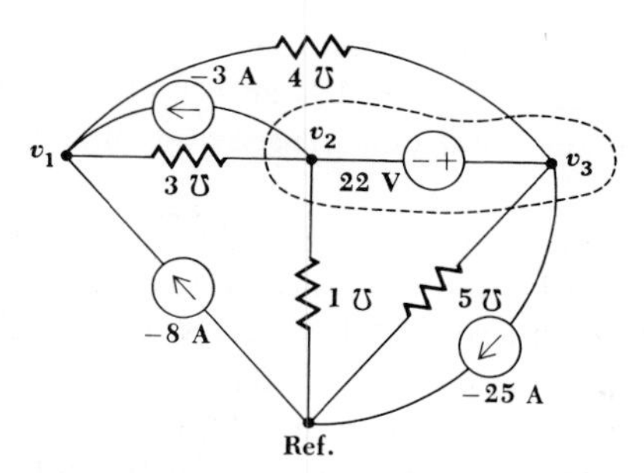

Fig. 3-3 The 2-℧ conductance in the circuit of Fig. 3-2 is replaced by an independent voltage source. Kirchhoff's current law is used on the supernode enclosed by the broken line, and the source voltage is set equal to $v_3 - v_2$.

The conductance matrix is symmetrical about the major diagonal (upper left to lower right). This is a general consequence of the systematic way in which we assigned variables, applied Kirchhoff's current law, and ordered the equations, as well as of the reciprocity theorem, which we shall discuss in Chap. 16. For the present, we merely acknowledge the symmetry in these circuits that have only independent current sources and accept the check that it provides us in discovering errors we may have committed in writing circuit equations.

We still must see how voltage sources and dependent sources affect the strategy of nodal analysis. We now investigate the consequences of including a voltage source.

As a typical example, consider the circuit shown in Fig. 3-3. Our previous four-node circuit has been changed by replacing the 2-℧ conductance between nodes 2 and 3 by a 22-V voltage source. We still assign the same node-to-reference voltages, v_1, v_2, and v_3. Previously, the next step was the application of Kirchhoff's current law at each of the three nonreference nodes. If we try to do that once again, we see that we shall run into some difficulty at both nodes 2 and 3, for we do not know what the *current* is in the branch with the *voltage* source. There is no way by which we can express the current as a function of the voltage, for the definition of a voltage source is exactly that the voltage is independent of the current.

There are two ways out of these difficulties. The more difficult is to assign an unknown current to the branch with the voltage source, proceed to apply Kirchhoff's current law three times, and then apply Kirchhoff's voltage law once between nodes 2 and 3; the result is four equations in four unknowns for this example.

The easier method is to agree that we are primarily interested in the node voltages, so that we may avoid utilizing the voltage-source branch that is causing our problems. We do this by treating node 2, node 3, and the voltage source together as a sort of supernode and applying Kirchhoff's current law to both nodes at the same time. This is certainly possible, because, if the total current leaving node 2 is zero and the total current leaving node 3 is zero, then the total current leaving the totality of the two nodes is zero.

A simple way of considering the two nodes at the same time is to think of the voltage source as a short circuit physically joining the two nodes together. Thus, each voltage source will effectively reduce the number of nonreference nodes at which we must apply Kirchhoff's current law by one, regardless of whether the voltage source extends between two nonreference nodes or is connected between a node and the reference.

So, let us visualize the 22-V source as a short circuit joining nodes 2 and 3 together. We find six branches connected to this supernode (suggested by a broken line in Fig. 3-3). Beginning with the 3-℧ conductance branch and working clockwise, we sum the six currents leaving this supernode:

$$3(v_2 - v_1) - 3 + 4(v_3 - v_1) - 25 + 5v_3 + 1v_2 = 0$$

or

$$-7v_1 + 4v_2 + 9v_3 = 28$$

The Kirchhoff current law equation at node 1 is unchanged from (3):

$$7v_1 - 3v_2 - 4v_3 = -11$$

We need one additional equation since we have three unknowns, and it must utilize the fact that there is a 22-V voltage source between nodes 2 and 3,

$$v_3 - v_2 = 22$$

Rewriting these last three equations,

$$\begin{aligned} 7v_1 - 3v_2 - 4v_3 &= -11 \\ -7v_1 + 4v_2 + 9v_3 &= 28 \\ -v_2 + v_3 &= 22 \end{aligned}$$

the determinant solution for v_1 is

$$v_1 = \frac{\begin{vmatrix} -11 & -3 & -4 \\ 28 & 4 & 9 \\ 22 & -1 & 1 \end{vmatrix}}{\begin{vmatrix} 7 & -3 & -4 \\ -7 & 4 & 9 \\ 0 & -1 & 1 \end{vmatrix}} = \frac{-189}{42} = -4.5 \quad \text{V}$$

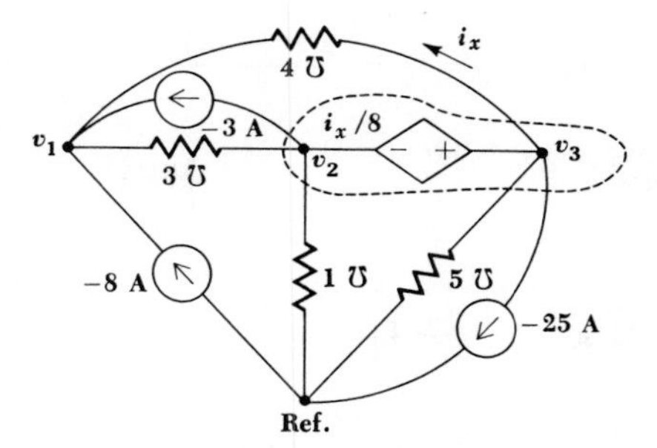

Fig. 3-4 The 2-℧ conductance in the circuit of Fig. 3-2 is replaced by a dependent voltage source. The region within the broken line is treated as a supernode and the voltage $i_x/8$ is expressed as $4(v_3 - v_1)/8$ and equated to $v_3 - v_2$.

Note the lack of symmetry in the denominator determinant. This is a result of the presence of the voltage source. Note also that it would not make much sense to call the denominator determinant the conductance matrix of the circuit, for the bottom row comes from the equation $-v_2 + v_3 = 22$, and this equation does not depend on any conductances in any way.

Now let us include a dependent source. We again replace the 2-℧ conductance in Fig. 3-2, this time by a dependent voltage source, as indicated in Fig. 3-4. We select a dependent voltage source rather than a dependent current source, as we have found that voltage sources offer us more of a challenge than do current sources in nodal analysis. Furthermore, we let the dependent source be controlled by a current rather than by a voltage, because this also is the less easy of the two cases to analyze.

In Fig. 3-4 then, the dependent voltage source is $i_x/8$, where i_x is the current toward the left in the 4-℧ conductance. The same three node-to-reference voltages are selected, and, since we again do not know the current in the dependent source branch, we let nodes 2 and 3 and the source be a supernode for which Kirchhoff's current law again yields

$$3(v_2 - v_1) - 3 + 4(v_3 - v_1) - 25 + 5v_3 + 1v_2 = 0$$

or

$$-7v_1 + 4v_2 + 9v_3 = 28$$

At node 1

$$7v_1 - 3v_2 - 4v_3 = -11$$

which is also unchanged. Turning our attention finally to the dependent source, we have

$$v_3 - v_2 = \frac{i_x}{8} = \frac{4(v_3 - v_1)}{8}$$

or

$$-0.5v_1 + v_2 - 0.5v_3 = 0$$

With determinants, v_1 is

$$v_1 = \frac{\begin{vmatrix} -11 & -3 & -4 \\ 28 & 4 & 9 \\ 0 & 1 & -0.5 \end{vmatrix}}{\begin{vmatrix} 7 & -3 & -4 \\ -7 & 4 & 9 \\ -0.5 & 1 & -0.5 \end{vmatrix}} = \frac{-33}{-33} = 1 \text{ V}$$

This result happens to be the same as that for the original four-node circuit (Fig. 3-2), as are the values for v_2 and v_3.

Let us summarize the method by which we may obtain a set of nodal equations for any resistive circuit:

1 Make a neat, simple, circuit diagram. Indicate all element and source values. Conductance values are preferable to resistance values. Each source should have its reference symbol.

2 Assuming that the circuit has N nodes, choose one of these nodes as a reference node. Then write the node voltages $v_1, v_2, \ldots, v_{N-1}$ at their respective nodes, remembering that each node voltage is understood to be measured with respect to the chosen reference.

3 If the circuit contains only current sources, apply Kirchhoff's current law at each nonreference node. To obtain the conductance matrix if a circuit has only independent current sources, equate the total current leaving each node through all conductances to the total source current entering that node, and order the terms from v_1 to v_{N-1}. For each dependent current source present, relate the source current and the controlling quantity to the variables $v_1, v_2, \ldots, v_{N-1}$, if they are not already in that form.

4 If the circuit contains voltage sources, mentally replace each such source by a short circuit, thus reducing the number of nodes by one, and apply Kirchhoff's current law at each of this reduced number of nodes. Relate the source voltage to the variables $v_1, v_2, \ldots, v_{N-1}$, if it is not already in that form.

With these suggestions in mind, let us consider the circuit displayed in Fig. 3-5, one which contains all four types of sources and has five nodes.

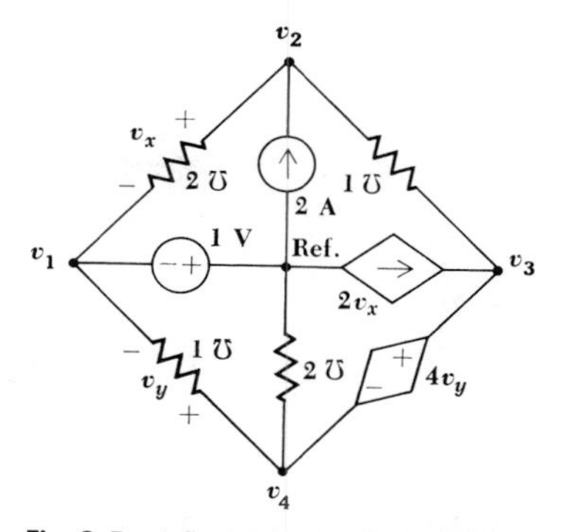

Fig. 3-5 A five-node circuit containing all of the four different types of sources.

We select the central node as the reference, and assign v_1 to v_4 in a clockwise direction starting from the left node.

After thinking of each voltage source as a short circuit, we see that we need to write Kirchhoff's current law equations only at node 2 and at the supernode containing both nodes 3 and 4 and the dependent voltage source.

At node 2,

$$2(v_2 - v_1) + 1(v_2 - v_3) = 2$$

while at the supernode,

$$1(v_3 - v_2) - 2v_x + 2v_4 + 1(v_4 - v_1) = 0$$

We next relate the source voltages to the node voltages:

$$v_1 = -1$$

$$v_3 - v_4 = 4v_y = 4(v_4 - v_1)$$

And finally we express the dependent current source in terms of the assigned variables:

$$2v_x = 2(v_2 - v_1)$$

Thus, we obtain four equations in the four node voltages:

$$-2v_1 + 3v_2 - v_3 = 2$$

$$v_1 = -1$$

$$v_1 - 3v_2 + v_3 + 3v_4 = 0$$

$$4v_1 + v_3 - 5v_4 = 0$$

to which the solutions are

$$v_1 = -1 \quad \text{V}$$
$$v_2 = 17/9 \quad \text{V}$$
$$v_3 = 17/3 \quad \text{V}$$
$$v_4 = 1/3 \quad \text{V}$$

Drill Problems

3-1 In the circuit of Fig. 3-6, use nodal analysis to find: (*a*) v_A; (*b*) v_B; (*c*) v_C.
Ans. -2.5; 4.5; 7 V

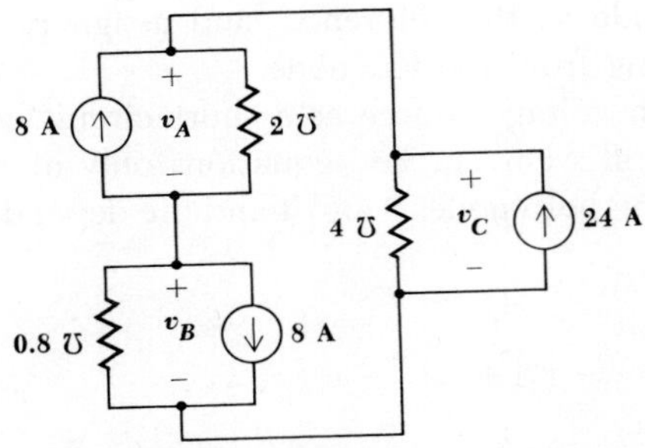

Fig. 3-6 See Drill Probs. 3-1, 3-2, 3-3, and 3-4.

3-2 Replace the 24-A source in Fig. 3-6 with a dependent current source, with the upward-directed arrow labeled $-6i_B$, where i_B is the downward current in the 0.8-℧ conductance. Find: (*a*) v_A; (*b*) v_B; (*c*) v_C.
Ans. $-25/7$; $46/7$; 3 V

3-3 Replace the 24-A source in Fig. 3-6 with a 22-V voltage source, positive reference at the top, and find: (*a*) v_A; (*b*) v_B; (*c*) v_C.
Ans. 10; 12; 22 V

3-4 Replace the 24-A source in Fig. 3-6 with a dependent voltage source, positive reference at the bottom, labeled $0.6v_B$. Find: (*a*) v_A; (*b*) v_B; (*c*) v_C.
Ans. -4; 2.4; 6.4 V

3-3 MESH ANALYSIS

The technique of nodal analysis described in the preceding section is completely general and can always be applied to any electrical network. This is not the only method for which a similar claim can be made,

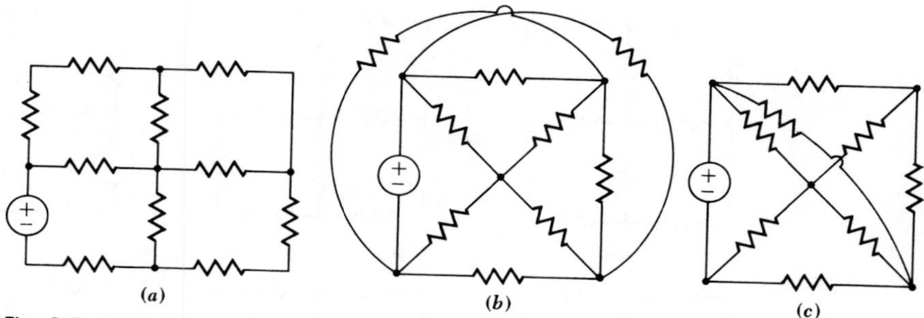

Fig. 3-7 (*a*) A planar network can be drawn on a plane surface without crossovers. (*b*) A nonplanar network cannot be drawn on a plane surface without at least one crossover. (*c*) A planar network can be drawn so that it may look nonplanar.

however. In particular, we shall meet a generalized nodal analysis method and a technique known as *loop analysis* in the concluding sections of this chapter.

First, however, let us consider a method known as *mesh analysis*. Even though this technique is not applicable to every network, it can be applied to most of the networks we shall need to analyze, and it is widely used. Mesh analysis is applicable only to those networks which are planar, a term we now proceed to define.

If it is possible to draw the diagram of a circuit on a plane surface in such a way that no branch passes over or under any other branch, then that circuit is said to be a *planar* circuit. Thus, Fig. 3-7*a* shows a planar network, Fig. 3-7*b* shows a nonplanar network, and Fig. 3-7*c* shows a planar network, although it is drawn in such a way as to make it appear nonplanar at first glance.

In the first chapter, a circuit was defined as a network containing at least one *closed path* about which current might flow. The official name for this closed path is a loop. Thus, if we begin at a certain node and trace a continuous closed path through the network, passing through no node or element more than once and ending at the same node, that path is a *loop*. Figure 3-8 shows several examples. The particular paths considered are those which are drawn with heavy lines, and it is evident that the first two paths cannot be loops since the path either is not closed or passes through a node twice. The remaining four paths are all loops. The circuit contains 11 branches.

The mesh is a property of a planar circuit and is not defined for a nonplanar circuit. We define a *mesh* as a loop which does not contain any other loops within it. Thus, the loops indicated in Fig. 3-8*c* and *d* are not meshes, whereas those of *e* and *f* are meshes. This circuit contains four meshes.

If a network is planar, mesh analysis can be used to accomplish its

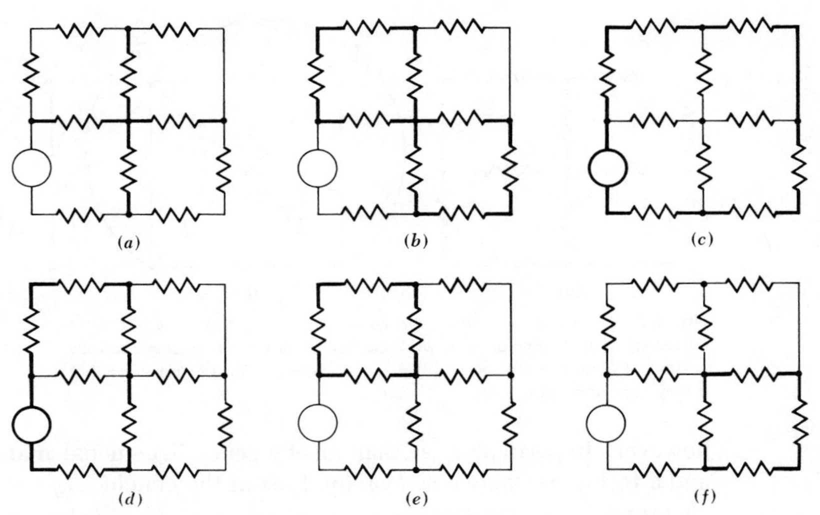

Fig. 3-8 (*a*) The path identified by the heavy line is not a loop since it is not closed. (*b*) The path here is not a loop since it passes through the central node twice. (*c*) This path is a loop but not a mesh since it encloses other loops. (*d*) This path is also a loop but not a mesh. (*e*) and (*f*) Each of these paths is both a loop and a mesh.

analysis. This technique involves the concept of a *mesh current,* which we shall introduce by considering the analysis of the two-mesh circuit of Fig. 3-9.

As we did in the single-loop circuit, we shall begin by assuming a current through one of the branches. Let us call the current flowing to the right through the 6-Ω resistor i_1. We intend to apply Kirchhoff's voltage law around each of the two meshes, and the resulting two equations are sufficient to determine two unknown currents. Therefore we select a second current i_2 flowing to the right in the 4-Ω resistor. We might also choose to call the current flowing downward through the central branch i_3, but it is evident from Kirchhoff's current law that i_3 may be expressed in terms of the two previously assumed currents as $(i_1 - i_2)$. The assumed currents are shown in Fig. 3-9.

Fig. 3-9 Two currents, i_1 and i_2, are assumed in a two-mesh circuit.

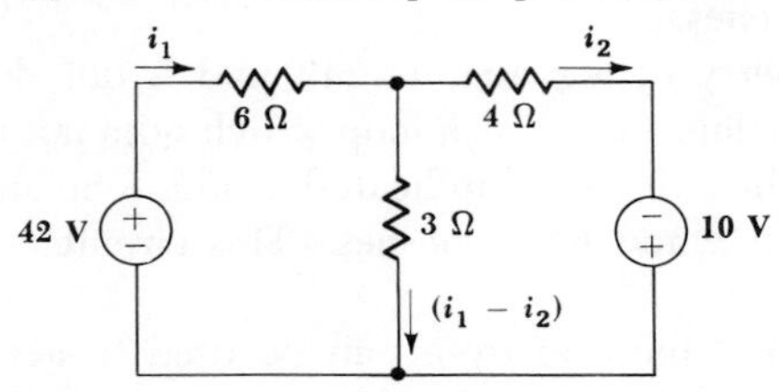

Following the method of solution for the single-loop circuit, we now apply Kirchhoff's voltage law to the left-hand mesh,

$$-42 + 6i_1 + 3(i_1 - i_2) = 0$$

or

$$9i_1 - 3i_2 = 42 \tag{6}$$

and then to the right-hand mesh,

$$-3(i_1 - i_2) + 4i_2 - 10 = 0$$

or

$$-3i_1 + 7i_2 = 10 \tag{7}$$

Equations (6) and (7) are independent equations; one cannot be derived from the other.[3] There are two equations and two unknowns, and the solution is easily obtained: i_1 is 6 A, i_2 is 4 A, and $(i_1 - i_2)$ is therefore 2 A. The voltage and power relationships may be quickly obtained if desired.

If our circuit had contained M meshes, then we should have had to assume M branch currents and write M independent equations.[4] The solution in general may be systematically obtained through the use of determinants.

Now let us consider this same problem in a slightly different manner by using mesh currents. We define a *mesh current* as a current which flows only around the perimeter of a mesh. If we label the left-hand mesh of our problem as mesh 1, then we may establish a mesh current i_1 flowing in a clockwise direction about this mesh. A mesh current is indicated by a curved arrow that almost closes on itself and is drawn inside the appropriate mesh, as shown in Fig. 3-10. The mesh current i_2 is established in the remaining mesh, again in a clockwise direction. Although the direction is arbitrary, we shall always choose clockwise mesh currents because a certain error-minimizing symmetry then results in the equations.

We no longer have a current or current arrow shown directly on each

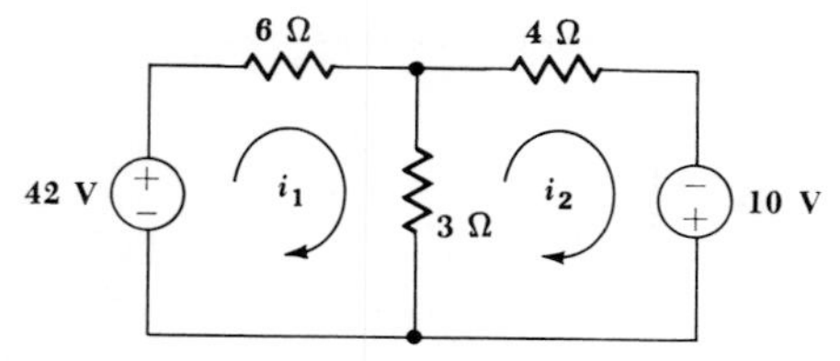

Fig. 3-10 A clockwise mesh current is assigned to each mesh of a planar circuit.

[3] It will be shown in Sec. 3-8 that mesh equations are always independent.
[4] The proof of this statement will be found in Sec. 3-8.

branch in the circuit. The current through any branch must be determined by considering the mesh currents flowing in every mesh in which that branch appears. This is not difficult because it is obvious that no branch can appear in more than two meshes. For example, the 3-Ω resistor appears in both meshes, and the current flowing downward through it is $(i_1 - i_2)$. The 6-Ω resistor appears only in mesh 1, and the current flowing to the right in that branch is equal to the mesh current i_1.

A mesh current may often be identified as a branch current, as i_1 and i_2 are identified above. This is not always true, however, for consideration of a square nine-mesh network soon shows that the central mesh current cannot be identified as the current in any branch.

One of the greatest advantages in the use of mesh currents is the fact that Kirchhoff's current law is automatically satisfied. If a mesh current flows into a given node, it obviously flows out of it also.

We therefore may turn our attention to the application of Kirchhoff's voltage law to each mesh. For the left-hand mesh,

$$-42 + 6i_1 + 3(i_1 - i_2) = 0$$

while for the right-hand mesh,

$$3(i_2 - i_1) + 4i_2 - 10 = 0$$

and these two equations are the same as (6) and (7).

Let us next consider the five-node, seven-branch, three-mesh circuit shown in Fig. 3-11. The three required mesh currents are assigned as indicated, and we methodically apply Kirchhoff's voltage law about each mesh:

$$-7 + 1(i_1 - i_2) + 6 + 2(i_1 - i_3) = 0$$

$$1(i_2 - i_1) + 2i_2 + 3(i_2 - i_3) = 0$$

$$2(i_3 - i_1) - 6 + 3(i_3 - i_2) + 1i_3 = 0$$

Fig. 3-11 Mesh currents i_1, i_2, and i_3 are assumed in a five-node seven-branch three-mesh circuit.

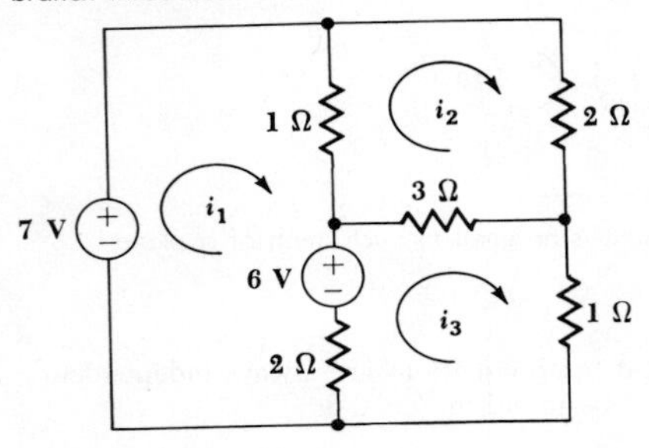

Simplifying,

$$3i_1 - i_2 - 2i_3 = 1$$
$$-i_1 + 6i_2 - 3i_3 = 0$$
$$-2i_1 - 3i_2 + 6i_3 = 6$$

and Cramer's rule leads to the formulation for i_3:

$$i_3 = \frac{\begin{vmatrix} 3 & -1 & 1 \\ -1 & 6 & 0 \\ -2 & -3 & 6 \end{vmatrix}}{\begin{vmatrix} 3 & -1 & -2 \\ -1 & 6 & -3 \\ -2 & -3 & 6 \end{vmatrix}} = \frac{117}{39} = 3 \quad \text{A}$$

The other mesh currents are $i_1 = 3$ A, and $i_2 = 2$ A.

Again we notice that we have a denominator determinant that is symmetrical about the major diagonal. This occurs for circuits that contain only independent voltage sources when clockwise mesh currents are assigned, where the elements appearing in the first row of the determinant are the ordered coefficients of $i_1, i_2, \ldots, i_M$ in the Kirchhoff voltage law equation about the first mesh, where the second row corresponds to the second mesh, and so on. This symmetrical array appearing in the denominator determinant is termed the *resistance matrix* of the network,

$$\mathbf{R} = \begin{bmatrix} 3 & -1 & -2 \\ -1 & 6 & -3 \\ -2 & -3 & 6 \end{bmatrix}$$

How must we modify this straightforward procedure when a current source is present in the network? Taking our lead from nodal analysis (and duality), we should feel that there are two possible methods. First, we could assign an unknown voltage across the current source, apply Kirchhoff's voltage law around each mesh as before, and then relate the source current to the assigned mesh currents. This is generally the more difficult approach.

A better technique is the dual of the supernode approach of nodal analysis. Instead of short-circuiting voltage sources to create supernodes, thus reducing the total number of nodes, we now open-circuit or remove current sources to create loops ("supermeshes"), thereby reducing the total number of meshes. We apply Kirchhoff's voltage law only to those meshes in the modified network.

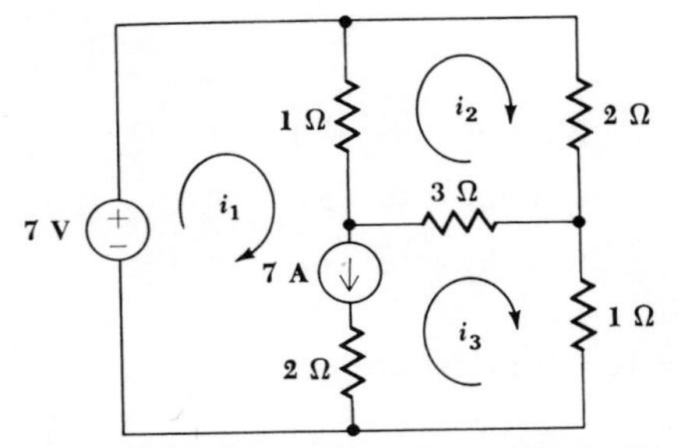

Fig. 3-12 Mesh analysis is applied to this circuit containing a current source by writing the Kirchhoff-voltage-law equation about the loop: 7 V, 1 Ω, 3 Ω, 1 Ω.

As an example of this procedure, consider the network shown in Fig. 3-12, in which a 7-A independent current source is in the common boundary of two meshes. Mesh currents i_1, i_2, and i_3 are assigned, and the current source is mentally open-circuited, thus creating a new mesh whose interior is that of meshes 1 and 3. Applying Kirchhoff's voltage law about this loop,

$$-7 + 1(i_1 - i_2) + 3(i_3 - i_2) + 1i_3 = 0$$

or

$$i_1 - 4i_2 + 4i_3 = 7 \tag{8}$$

and around mesh 2,

$$1(i_2 - i_1) + 2i_2 + 3(i_2 - i_3) = 0$$

or

$$-i_1 + 6i_2 - 3i_3 = 0 \tag{9}$$

Finally, the source current is related to the assumed mesh currents,

$$i_1 - i_3 = 7 \tag{10}$$

Solving (8) through (10), we have

$$i_3 = \frac{\begin{vmatrix} -1 & 6 & 0 \\ 1 & -4 & 7 \\ 1 & 0 & 7 \end{vmatrix}}{\begin{vmatrix} -1 & 6 & -3 \\ 1 & -4 & 4 \\ 1 & 0 & -1 \end{vmatrix}} = \frac{28}{14} = 2 \quad \text{A}$$

We may also find that $i_1 = 9$ A and $i_2 = 2.5$ A.

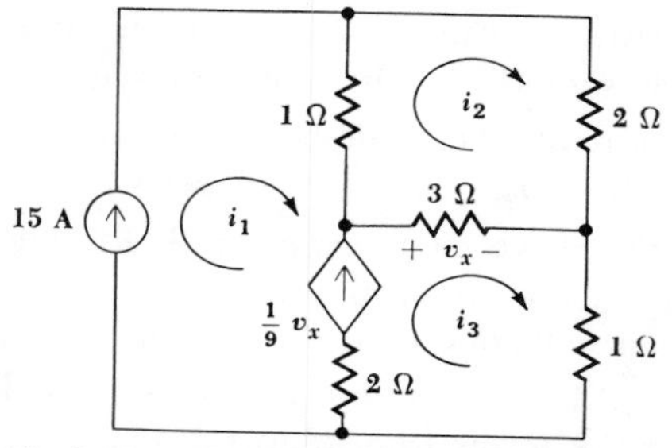

Fig. 3-13 The presence of two current sources in this three-mesh circuit makes it necessary to apply Kirchhoff's voltage law only once, around mesh 2.

The presence of one or more dependent sources merely requires each of these source quantities and the variable on which it depends to be expressed in terms of the assigned mesh currents. In Fig, 3-13, for example, we note that both a dependent and an independent current source are included in the network. Three mesh currents are assigned and Kirchhoff's voltage law is applied to mesh 2:

$$1(i_2 - i_1) + 2i_2 + 3(i_2 - i_3) = 0$$

The current sources appear in meshes 1 and 3; when they are open-circuited, only mesh 2 remains, and we already have an equation for it. We therefore turn our attention to the source quantities, obtaining

$$i_1 = 15$$

and

$$\tfrac{1}{9}v_x = i_3 - i_1 = \tfrac{1}{9}[3(i_3 - i_2)]$$

Thus,

$$-i_1 + 6i_2 - 3i_3 = 0$$
$$i_1 = 15$$
$$-i_1 + \tfrac{1}{3}i_2 + \tfrac{2}{3}i_3 = 0$$

from which we have $i_1 = 15$, $i_2 = 11$, and $i_3 = 17$ A. We might note that we wasted a little time in assigning a mesh current i_1 to the left mesh; we should simply have indicated a mesh current and labeled it 15 A.

Let us summarize the method by which we may obtain a set of mesh equations for a resistive circuit:

1. Make certain that the network is a planar network. If it is nonplanar, mesh analysis is not applicable.
2. Make a neat, simple, circuit diagram. Indicate all element and

source values. Resistance values are preferable to conductance values. Each source should have its reference symbol.

3 Assuming that the circuit has M meshes, assign a clockwise mesh current in each mesh, $i_1, i_2, \ldots, i_M$.

4 If the circuit contains only voltage sources, apply Kirchhoff's voltage law around each mesh. To obtain the resistance matrix if a circuit has only independent voltage sources, equate the clockwise sum of all the resistor voltages to the counterclockwise sum of all the source voltages, and order the terms from i_1 to i_M. For each dependent voltage source present, relate the source voltage and the controlling quantity to the variables $i_1, i_2, \ldots, i_M$, if they are not already in that form.

5 If the circuit contains current sources, mentally replace each such source by an open circuit, thus reducing the number of meshes by one, and apply Kirchhoff's voltage law around the meshes of the resultant network. Relate the source current to the variables $i_1, i_2, \ldots, i_M$, if it is not already in that form.

Drill Problems

3-5 In the circuit of Fig. 3-14, use mesh analysis to find: (*a*) i_A; (*b*) i_B; (*c*) i_C.

Ans. $-8/3$; 1; $5/3$ mA

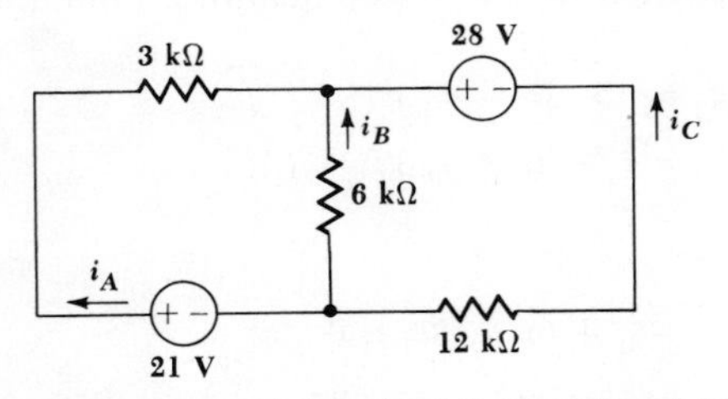

Fig. 3-14 See Drill Probs. 3-5, 3-6, 3-7, and 3-8.

3-6 Replace the 28-V source in Fig. 3-14 by a dependent voltage source, positive reference on the left, labeled $21{,}000i_B$. Find: (*a*) i_A; (*b*) i_B; (*c*) i_C.

Ans. -3; $-4/3$; $13/3$ mA

3-7 Replace the 6-kΩ resistor of Fig. 3-14 with a 4-mA current source, reference arrow directed upward, and find: (*a*) i_A; (*b*) i_B; (*c*) i_C.

Ans. $-11/3$; $-1/3$; 4 mA

3-8 Replace the 3-kΩ resistor of Fig. 3-14 with a dependent current source, arrow directed to the right, labeled $0.5i_c$. Find: (*a*) i_A; (*b*) i_B; (*c*) i_C.

Ans. -2; $2/3$; $4/3$ mA

3-4 SOURCE TRANSFORMATIONS

In all our previous work we have been making continual use of *ideal* voltage and current sources; it is now time to take a step closer to reality by considering *practical* sources. These sources will enable us to make more realistic representations of physical devices. Having defined the practical sources, we shall then study methods whereby practical current and voltage sources may be interchanged without affecting the remainder of the circuit. Such sources will be called *equivalent* sources. Our methods will be applicable for both independent and dependent sources.

The ideal voltage source was defined as a device whose terminal voltage is independent of the current through it. A 1-V dc source produces a current of 1 A through a 1-Ω resistor and a current of 1,000,000 A through a 1-μΩ resistor; it may provide an unlimited amount of power. No such device exists practically, of course, and we agreed that a real physical source might be represented by an ideal source only as long as relatively small currents, or powers, were drawn from it. For example, an automobile storage battery may be approximated by an ideal dc voltage source if its current is limited to a few amperes. However, anyone who has ever tried to start an automobile with the headlights on must have observed that the lights dimmed perceptibly when the battery was asked to deliver the heavy starter current, 100 A or more, in addition to the headlight current. Under these conditions, an ideal voltage source may be a very poor representation of the storage battery.

The ideal voltage source must be modified to account for the lowering of its terminal voltage when large currents are drawn from it. Let us suppose that we observe experimentally that a storage battery has a terminal voltage of 12 V when no current is flowing through it and a reduced voltage of 11 V when 100 A is flowing. Thus, a more accurate representation might be an ideal voltage source of 12 V in series with a resistor across which 1 V appears when 100 A flows through it. The resistor must be 0.01 Ω, and the ideal voltage source and this series resistance comprise a *practical voltage source*. This particular practical voltage source is shown connected to a general load resistor R_L in Fig. 3-15*a*; the terminal voltage of the practical source is the same as the voltage across R_L and is marked v_L. In Fig. 3-15*b* a plot of the terminal voltage as a function of the load resistance points out the fact that the output voltage approaches that of the ideal source only for large values of the load resistance where relatively small currents are drawn. The terminal voltage is only one-half the voltage of the ideal source when the load resistance is equal to the internal resistance of the practical source.

Let us now consider a general practical voltage source, as shown in Fig. 3-16. The voltage of the ideal source is v_s, and a resistance R_{sv}, called an *internal resistance*, is placed in series with it. The resistor is not one

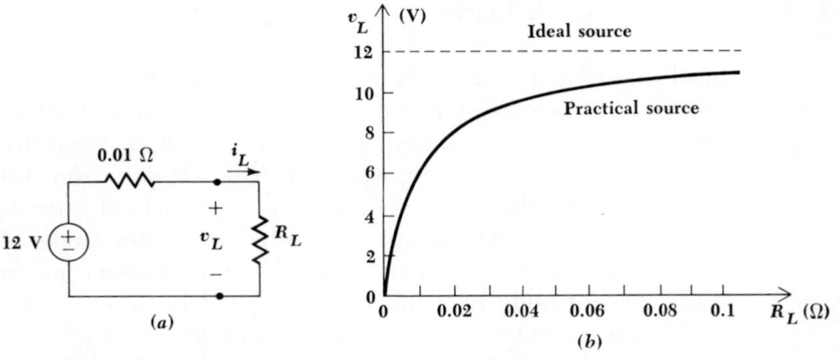

Fig. 3-15 (*a*) A practical source which approximates the behavior of a certain 12-V storage battery is shown connected to a load resistor R_L. (*b*) The terminal voltage decreases as R_L decreases and i_L increases.

which we would wire or solder into a circuit, but merely serves to account for a terminal voltage which decreases as the load current increases, and thus its presence enables us to represent a physical voltage source more closely. The voltage v_L across the load resistor R_L is thus

$$v_L = \frac{v_s}{R_{sv} + R_L} R_L \tag{11}$$

and the load current i_L is

$$i_L = \frac{v_s}{R_{sv} + R_L} \tag{12}$$

An ideal current source is also nonexistent in the real world; there is no physical device which will deliver a constant current, regardless of the load resistance to which it is connected or the voltage across its terminals. A pentode vacuum-tube amplifier and certain transistor circuits will deliver a constant current to a wide range of load resistances, but the load resistance can always be made sufficiently large that the current through it becomes very small. Infinite power is simply never available.

Fig. 3-16 A general practical voltage source connected to a load resistor R_L.

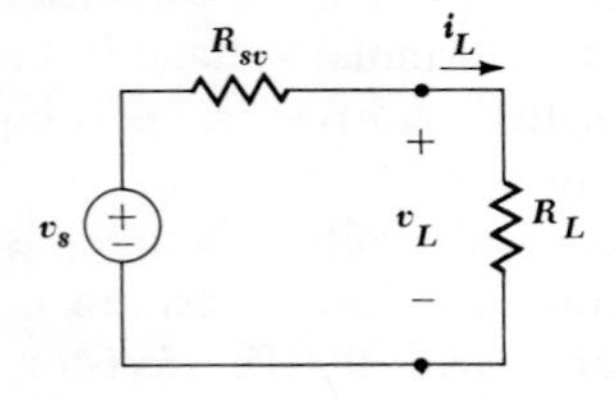

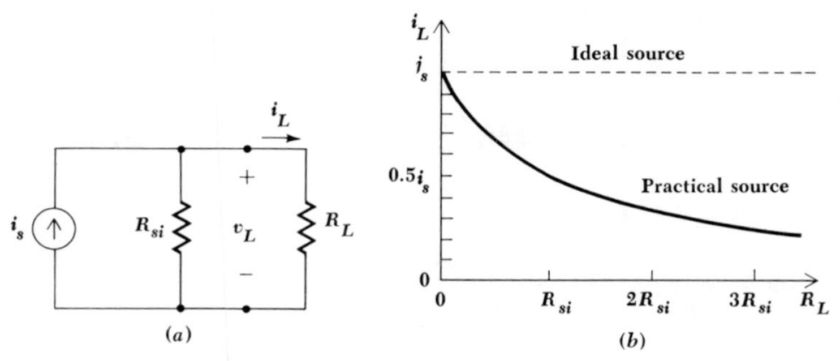

Fig. 3-17 (*a*) A general practical current source connected to a load resistor R_L. (*b*) The load current provided by the practical current source is shown as a function of the load resistance.

A practical current source is defined as an ideal current source in parallel with an internal resistance R_{si}. Such a source is shown in Fig. 3-17*a*, and the current i_L and voltage v_L produced across a load resistor R_L are indicated. It is apparent that

$$v_L = \frac{R_{si}R_L}{R_{si} + R_L} i_s \tag{13}$$

and

$$i_L = \frac{R_{si}}{R_{si} + R_L} i_s \tag{14}$$

The variation of load current with changing load resistance is shown in Fig. 3-17*b*, and it is evident that the load current and ideal source current are approximately equal only for values of load resistance which are small compared with R_{si}.

Having defined both practical sources, we are now ready to discuss their equivalence. We shall define two sources as being *equivalent* if each produces identical current and identical voltage in *any* load which is placed across its terminals. In other words, if we are confronted with two black boxes, each having a single pair of terminals on it, then there is no way in which we can differentiate between the boxes by measuring current or voltage in a resistive load.

It should be noted carefully that although equivalent sources will deliver the same current, voltage, and power to identical resistive loads, the power which the two *ideal* sources supply and the power absorbed in R_{sv} and R_{si} may be quite different.

The conditions for equivalence are now quickly established. Since the load currents are to be identical, from (12) and (14) we have

$$i_L = \frac{v_s}{R_{sv} + R_L} = \frac{R_{si}i_s}{R_{si} + R_L}$$

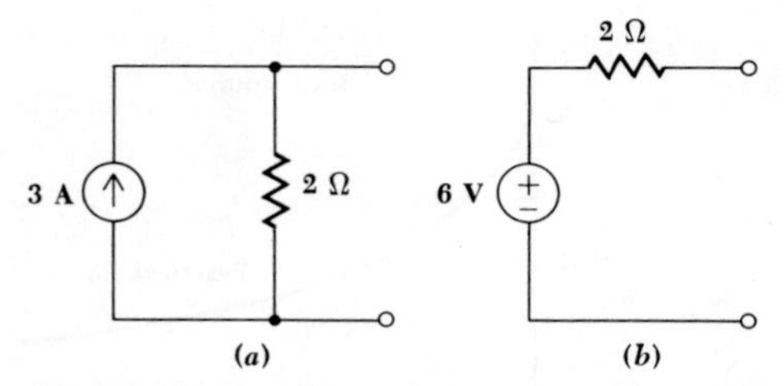

Fig. 3-18 *(a)* A given practical current source. *(b)* The equivalent practical voltage source: $R_{sv} = R_{si} = R_s$; $v_s = R_s i_s$.

and if these two expressions are to be the same for any R_L, then it follows that

$$R_{sv} = R_{si} = R_s \tag{15}$$

and

$$v_s = R_s i_s \tag{16}$$

where we shall now let R_s represent the internal resistance of either practical source. It is easily shown that these same two conditions may be obtained from (11) and (13).

As an example of the use of these ideas, consider the practical current source shown in Fig. 3-18*a*. Since its internal resistance is 2 Ω, the internal resistance of the equivalent practical voltage source is also 2 Ω; the voltage of the ideal voltage source contained within the practical voltage source is (2)(3) = 6 V. The equivalent practical voltage source is shown in Fig. 3-18*b*.

To check the equivalence, let us visualize a 4-Ω resistor connected to each source. In both cases, a current of 1 A, a voltage of 4 V, and a power of 4 W are associated with the 4-Ω load. However, the ideal current source is delivering a total power of 12 W, while the ideal voltage source is delivering only 6 W. Correspondingly different powers are dissipated in the two internal resistances.

It is enlightening to develop the equivalence conditions by a slightly different method. Suppose that we are given a practical current source and are asked to determine the equivalent voltage source. There are two unknowns, v_s and R_{sv}. We therefore need two facts, or equations, and we may easily obtain these by forcing the current that each source delivers to a short circuit and the voltage that each delivers to an open circuit to be identical. We first equate the two short-circuit currents

$$i_{Lsc} = i_s = \frac{v_s}{R_{sv}}$$

and, hence,

$$v_s = R_{sv} i_s$$

and then the two open-circuit voltages

$$v_{Loc} = v_s = R_{si}i_s \qquad \text{or} \qquad v_s = R_{si}i_s$$

Finally,

$$R_{sv} = R_{si} = R_s \qquad \text{and} \qquad v_s = R_s i_s$$

as before.

A very useful power theorem may be developed with reference to a practical voltage or current source. For the practical voltage source (Fig. 3-16 with $R_{sv} = R_s$), the power delivered to the load R_L is

$$p_L = i_L^2 R_L = \frac{v_s^2 R_L}{(R_s + R_L)^2}$$

To find the value of R_L that absorbs a maximum power from the given practical source, we differentiate with respect to R_L:

$$\frac{dp_L}{dR_L} = \frac{(R_s + R_L)^2 v_s^2 - v_s^2 R_L(2)(R_s + R_L)}{(R_s + R_L)^4}$$

and equate the derivative to zero, obtaining,

$$2R_L(R_s + R_L) = (R_s + R_L)^2$$

or

$$R_s = R_L$$

Since the values, $R_L = 0$ and $R_L = \infty$, both give a minimum ($p_L = 0$), and since we have already developed the equivalence between practical voltage and current sources, we have therefore proved the following *maximum power transfer* theorem:

> An independent voltage source in series with a resistance R_s or an independent current source in parallel with a resistance R_s delivers a maximum power to that load resistance R_L for which $R_L = R_s$.

Drill Problems

3-9 For the circuit shown in Fig. 3-19*a*: (*a*) use nodal analysis to find the power supplied by the 3-A source; (*b*) transform the practical voltage source into a practical current source and use nodal analysis to find the power supplied by the new ideal current source; (*c*) transform the practical current source into a practical

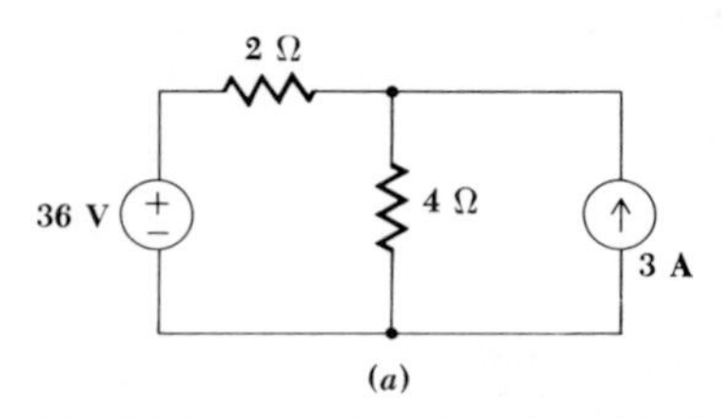

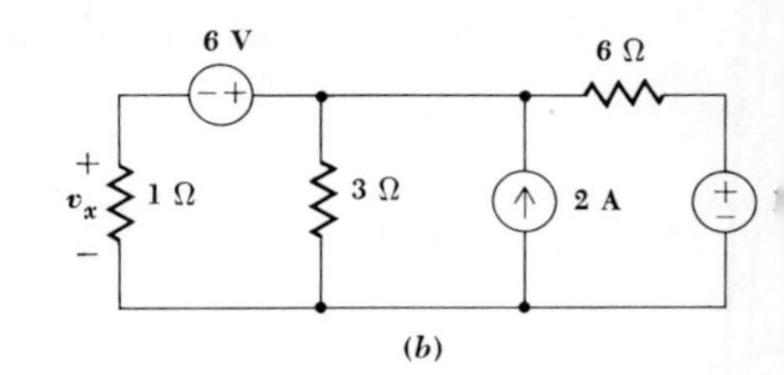

Fig. 3-19 See Drill Probs. 3-9 and 3-10.

voltage source and use mesh analysis to find the power supplied by the new ideal voltage source.

Ans. -48; 84; 504 W

3-10 Find v_x in the circuit shown in Fig. 3-19*b* by: (*a*) nodal analysis; (*b*) mesh analysis; (*c*) beginning on the right side of the circuit and alternating source transformations and source and resistance combinations until only a single loop circuit remains.

Ans. $2/3$; $2/3$; $2/3$ V

3-5 LINEARITY AND SUPERPOSITION

All the circuits which we have analyzed up to now (and which we shall analyze later) are linear circuits. At this time we must be more specific in defining a linear circuit. Having done this, we can then consider the most important consequence of linearity, the principle of superposition. This principle is very basic and will appear repeatedly in our study of linear circuit analysis. As a matter of fact, the nonapplicability of superposition to nonlinear circuits is the reason they are so difficult to analyze.

The principle of superposition states that the response (a desired current or voltage) at any point in a linear circuit having more than one independent source can be obtained as the sum of the responses caused by each independent source acting alone. In the following discussion, we shall investigate the meaning of "linear" and "acting alone." We shall also take note of a slightly broader form of the theorem.

The difference between linearity and nonlinearity may first be illustrated by a nonelectrical example. Let us place one loaded truck (the forcing function) in the middle of a suspension bridge and measure the deflection (response) of the bridge. It turns out to be 3 cm. A second truck may produce another 3 cm, indicating a linear system, but if we try to extend the results too far we may extend the bridge too far. The system becomes nonlinear, and response is no longer proportional to force or excitation.

Let us first define a *linear element* as a passive element that has a linear voltage-current relationship. By a "linear voltage-current relationship" we shall mean simply that multiplication of the time-varying current through the element by a constant K results in the multiplication of the time-varying voltage across the element by the same constant K. At this time, only one passive element has been defined, the resistor, and its voltage-current relationship

$$v(t) = Ri(t)$$

is obviously linear. As a matter of fact, if $v(t)$ is plotted as a function of $i(t)$, the result is a straight *line*. We shall see in Chap. 4 that the defining voltage-current equations for inductance and capacitance are also linear relationships, as is the defining equation for mutual inductance presented in Chap. 15.

We must also define a *linear* dependent source as a dependent current or voltage source whose output current or voltage is proportional only to the first power of some current or voltage variable in the circuit or to the sum of such quantities. That is, a dependent voltage source, $v_s = 0.6i_1 - 14v_2$, is linear, but $v_s = 0.6i_1^2$ and $v_s = 0.6i_1v_2$ are not.

We may now define a *linear circuit* as a circuit composed entirely of independent sources, linear dependent sources, and linear elements. From this definition, it is possible to show[5] that "the response is proportional to the source," or that multiplication of all independent source voltages and currents by a constant K increases all the current and voltage responses by the same factor K (including the dependent source voltage or current outputs).

The most important consequence of linearity is superposition. Let us prove the superposition principle by considering first the circuit of Fig. 3-20, which contains two independent sources, the current generators which force the currents i_a and i_b into the circuit. Sources are often called *forcing functions* for this reason, and the voltages which they produce between node 1 or 2 and the reference node may be termed *response functions,* or simply *responses*. Both the forcing functions and the responses may be functions of time.

The two nodal equations for this circuit are

$$0.7v_1 - 0.2v_2 = i_a \qquad (17)$$

$$-0.2v_1 + 1.2v_2 = i_b \qquad (18)$$

[5]The proof involves first showing that the use of nodal analysis on the linear circuit can produce only linear equations of the form:

$$a_1v_1 + a_2v_2 + \cdots + a_Nv_N = b$$

where the a_i are constants (combinations of resistance or conductance values, constants appearing in dependent source expressions, 0, or ± 1), the v_i are the unknown node voltages (responses), and b is an independent source value or a sum of independent source values. Given a set of such equations, if we multiply all the b's by K, then it is evident that the solution of this new set of equations will be the node voltages $Kv_1, Kv_2, \ldots, Kv_N$.

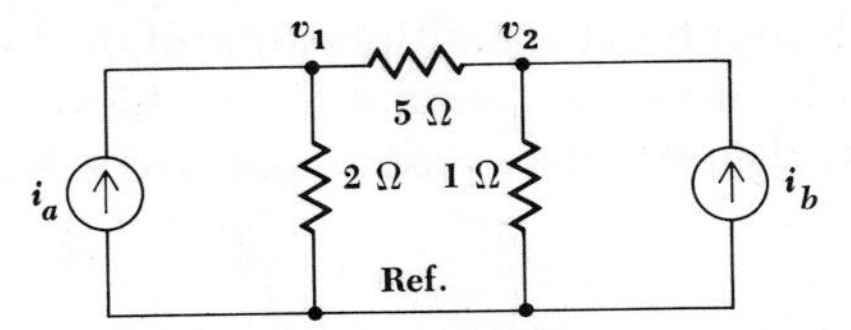

Fig. 3-20 A three-node circuit containing two forcing functions used to illustrate the superposition principle.

Now let us perform experiment x. We change the two forcing functions to i_{ax} and i_{bx}; the two unknown voltages will now be different, and we shall let them be v_{1x} and v_{2x}. Thus,

$$0.7v_{1x} - 0.2v_{2x} = i_{ax} \tag{19}$$

$$-0.2v_{1x} + 1.2v_{2x} = i_{bx} \tag{20}$$

We next perform experiment y by changing the source currents to i_{ay} and i_{by} and by letting the responses be v_{1y} and v_{2y},

$$0.7v_{1y} - 0.2v_{2y} = i_{ay} \tag{21}$$

$$-0.2v_{1y} + 1.2v_{2y} = i_{by} \tag{22}$$

These three sets of equations describe the same circuit with different source currents. Let us *add* or *superpose* the last two sets of equations. Adding (19) to (21),

$$(0.7v_{1x} + 0.7v_{1y}) - (0.2v_{2x} + 0.2v_{2y}) = i_{ax} + i_{ay} \tag{23}$$

$$0.7v_1 \quad - \quad 0.2v_2 \quad = \quad i_a \tag{17}$$

and adding (20) to (22),

$$-(0.2v_{1x} + 0.2v_{1y}) + (1.2v_{2x} + 1.2v_{2y}) = i_{bx} + i_{by} \tag{24}$$

$$-0.2v_1 \quad + \quad 1.2v_2 \quad = \quad i_b \tag{18}$$

where (17) has been written immediately below (23), and (18) below (24) for easy comparison.

The linearity of all these equations allows us to compare (23) with (17) and (24) with (18) and draw an interesting conclusion. If we select i_{ax} and i_{ay} such that their sum is i_a, select i_{bx} and i_{by} such that their sum is i_b, then the desired responses v_1 and v_2 may be found by *adding* v_{1x} to v_{1y} and v_{2x} to v_{2y}, respectively. In other words, we may perform experiment x and note the responses, perform experiment y and note the responses, and finally add the corresponding responses. These are the responses of the original circuit to independent sources which are the sums of the independent sources used in experiments x and y. This is the fundamental concept involved in the superposition principle.

It is evident that we may extend these results by breaking up either source current into as many pieces as we wish; there is no reason why we cannot perform experiments z and q also. It is only necessary that the algebraic sum of the pieces be equal to the original current.

The *superposition theorem* usually appears in a form similar to the following:

> In any linear resistive network containing several sources, the voltage across or the current through any resistor or source may be calculated by adding algebraically all the individual voltages or currents caused by each independent source acting alone, with all other independent voltage sources replaced by short circuits and all other independent current sources replaced by open circuits.

Thus if there are N independent sources, we perform N experiments. Each independent source is active in only one experiment, and only one independent source is active in each experiment. An inactive independent voltage source is identical with a short circuit, and an inactive independent current source is an open circuit. Note that *dependent* sources are in general active in *every* experiment.

Our proof of superposition, however, should indicate that a much stronger theorem might be written; a group of independent sources may be made active and inactive collectively, if we wish. For example, suppose there are three independent sources. The theorem above states that we may find a given response by considering each of the three sources acting alone and adding the three results. Alternatively, we may find the response due to the first and second sources operating with the third inactive, and then add to this the response caused by the third source acting alone. This amounts to treating several sources collectively as a sort of supersource.

There is also no reason that an independent source must assume only its given value or a zero value in the several experiments; it is only necessary for the sum of the several values to be equal to the original value. An inactive source almost always leads to the simplest circuit, however.

Let us illustrate the application of the superposition principle by considering an example in which both types of independent source are present. It is not necessary to transform either source. For the circuit of Fig. 3-21, let us use superposition to write an expression for the unknown branch current i_x. We may first kill the current source and obtain the portion of i_x due to the voltage source as 0.2 A. Then if we kill the voltage source and apply current division, the remaining portion of i_x is seen to be 0.8 A. We might write the answer in detail as

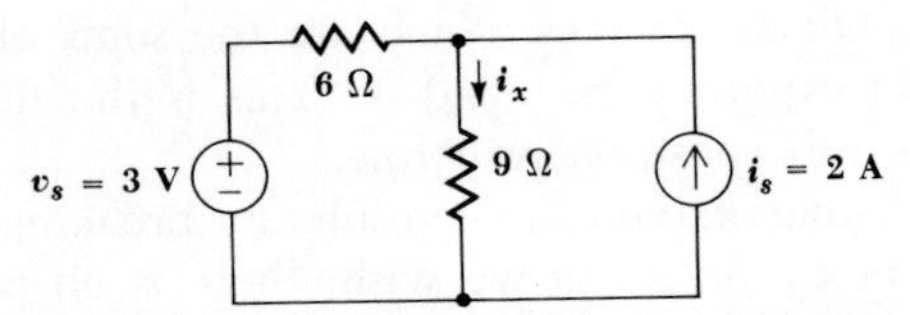

Fig. 3-21 A circuit containing both an independent current and a voltage source which is easily analyzed by the superposition principle.

$$i_x = i_x|_{i_s=0} + i_x|_{v_s=0} = \frac{3}{6+9} + 2\frac{6}{6+9} = 0.2 + 0.8 = 1.0 \quad \text{A}$$

As an example of the application of the superposition principle to a circuit containing a dependent source, consider Fig. 3-22. We seek i_x, and we first open-circuit the 3-A source. The single mesh equation is

$$-10 + 2i'_x + 1i'_x + 2i'_x = 0$$

so that

$$i'_x = 2$$

Next, we short-circuit the 10-V source and write the single-node equation,

$$\frac{v''}{2} + \frac{v'' - 2i''_x}{1} = 3$$

and relate the dependent-source-controlling quantity to v'',

$$v'' = -2i''_x$$

We find

$$i''_x = -0.6$$

and, thus,

$$i_x = 2 - 0.6 = 1.4$$

Fig. 3-22 Superposition may be used to analyze this circuit by first replacing the 3-A source by an open circuit and then replacing the 10-V source by a short circuit. The dependent voltage source is always active (unless $i_x = 0$).

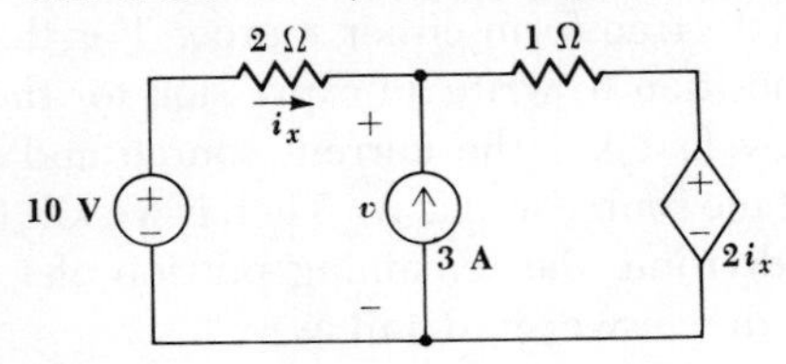

It usually turns out that little if any time is saved in analyzing a circuit containing one or more dependent sources by use of the superposition principle, for there must always be at least two sources in operation: one independent source and all the dependent sources.

We must constantly be aware of the limitations of superposition. It is applicable only to linear responses, and thus the most common nonlinear response—power—is not subject to superposition. For example, consider two 1-V batteries in series with a 1-Ω resistance. The power delivered to the resistor is obviously 4 W, but if we mistakenly try to apply superposition we might say that each battery alone furnished 1 W and thus the total power is 2 W. This is incorrect.

Drill Problems

3-11 Use superposition to find v in each of the circuits shown in Fig. 3-23.

Ans. -3; -1.6; 6 V

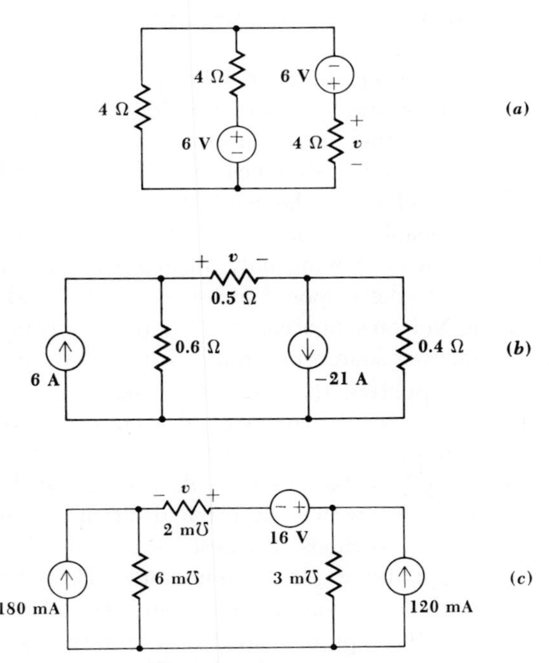

Fig. 3-23 See Drill Prob. 3-11.

3-12 First find v_x in the circuit of Fig. 3-24 by using the superposition principle and then determine the power generated by: (*a*) the 5-A source; (*b*) the 6-V source; (*c*) the dependent source.

Ans. -12.32; 14; 21.6 W

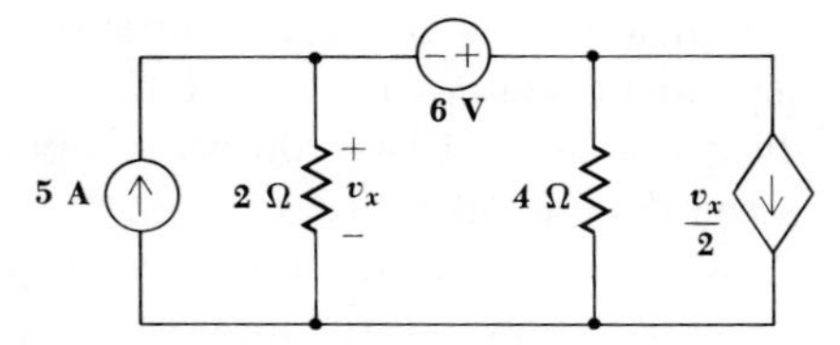

Fig. 3-24 See Drill Prob. 3-12.

3-6 THÉVENIN'S AND NORTON'S THEOREMS

Now that we have the superposition principle, it is possible to develop two more theorems which will greatly simplify the analysis of many linear circuits. The first of these theorems is named after M. L. Thévenin, a French engineer working in telegraphy, who first published a statement of the theorem in 1883; the second may be considered a corollary of the first and is credited to E. L. Norton, a scientist with the Bell Telephone Laboratories.

Let us suppose that we need to make only a partial analysis of a circuit; perhaps we wish to determine the current, voltage, and power delivered to a single load resistor by the remainder of the circuit, which may consist of any number of sources and resistances; or perhaps we wish to find the response for different values of the load resistor. Thévenin's theorem then tells us that it is possible to replace everything except the load resistor by an equivalent circuit containing only an independent voltage source in series with a resistor; the response measured at the load resistor will be unchanged. Using Norton's theorem, we obtain an equivalent composed of an independent current source in parallel with a resistor.

It should thus be apparent that one of the main uses of Thévenin's and Norton's theorems is the replacement of a large part of a network, often a complicated and uninteresting part, by a very simple equivalent. The new, simpler circuit enables us to make rapid calculations of the voltage, current, and power which the original circuit is able to deliver to a load. It also helps us to choose the best value of this load resistance. In a transistor power amplifier, for example, the Thévenin or Norton equivalent enables us to determine the maximum power that can be taken from the amplifier and the type of load that is required to accomplish a maximum transfer of power or to obtain maximum practical voltage or current amplification.

Consider the circuit shown in Fig. 3-25. The broken lines separate the circuit into networks A and B; we shall assume that our main interest is in network B, which consists only of the load resistor R_L. Network A may be simplified by making repeated source transformations. We first treat the

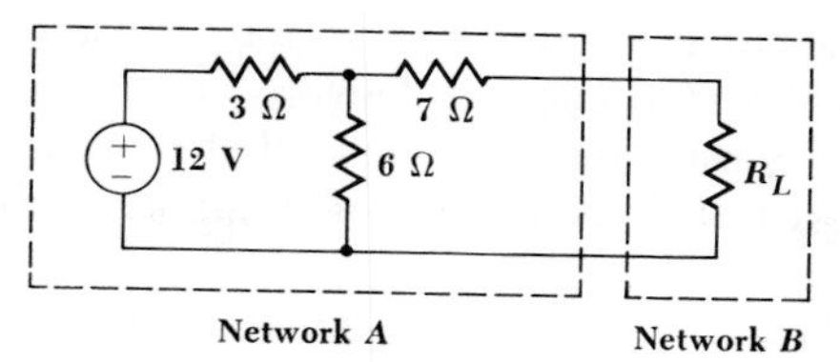

Fig. 3-25 A simple resistive circuit is divided into network A, in which we have no detailed interest, and network B, a load resistor with which we are fascinated.

12-V source and the 3-Ω resistor as a practical voltage source and replace it with a practical current source consisting of a 4-A source in parallel with 3 Ω. The parallel resistances are then combined into 2 Ω, and the practical current source which results is transformed back into a practical voltage source. The steps are indicated in Fig. 3-26, the final result appearing in Fig. 3-26*d*. From the viewpoint of the load resistor R_L, this circuit (the Thévenin equivalent) is equivalent to the original circuit; from our viewpoint, the circuit is much simpler and we can now easily compute the power delivered to the load. It is

$$p_L = \left(\frac{8}{9 + R_L}\right)^2 R_L$$

Furthermore we can see from the equivalent circuit that the maximum voltage which can be obtained across R_L is 8 V; a quick transformation

Fig. 3-26 The source transformations and resistance combinations involved in simplifying network A are shown in order. The result, given in (*d*), is the Thévenin equivalent.

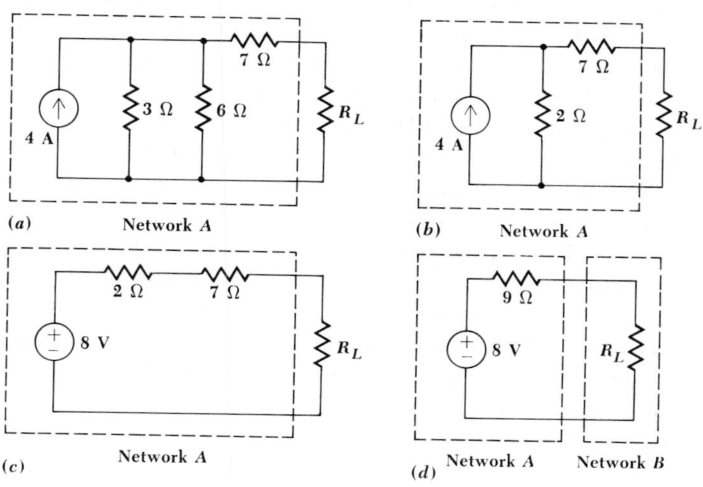

of network A to a practical current source (the Norton equivalent) indicates that the maximum current which may be delivered to the load is $8/9$ A; and the maximum power transfer theorem shows that a maximum power is delivered to R_L when R_L is 9 Ω. None of these facts is readily apparent from the original circuit.

If network A had been more complicated, the number of source transformations and resistance combinations necessary to obtain the Thévenin or Norton equivalent could easily become prohibitive. Thévenin's and Norton's theorems allow us to find the equivalent circuit much more quickly and easily, even in more complicated circuits.

Let us now state *Thévenin's theorem* formally:

> Given any linear circuit, rearrange it in the form of two networks A and B that are connected together by two resistanceless conductors. Define a voltage v_{oc} as the open-circuit voltage which would appear across the terminals of A if B were disconnected so that no current is drawn from A. Then all the currents and voltages in B will remain unchanged if A is killed (all independent voltage sources and independent current sources in A replaced by short circuits and open circuits, respectively) and an independent voltage source v_{oc} is connected, with proper polarity, in series with the dead (inactive) A network.

The terms *killed* and *dead* are a little bloodthirsty, but they are descriptive and concise, and we shall use them in a friendly way. Moreover, it is possible that network A may only be sleeping, for it may still contain *dependent* sources which come to life whenever their controlling currents or voltages are nonzero.

Let us see if we can apply Thévenin's theorem successfully to the circuit we considered in Fig. 3-25. Disconnecting R_L, voltage division enables us to determine that v_{oc} is 8 V. Killing the A network, that is, replacing the 12-V source by a short circuit, we see looking back into the dead A network a 7-Ω resistor connected in series with the parallel combination of 6 Ω and 3 Ω. Thus the dead A network can be represented here by simply a 9-Ω resistor. This agrees with the previous result.

The equivalent circuit we have obtained is completely independent of the B network, because we have been instructed first to remove the B network and measure the open-circuit voltage produced by the A network, an operation which certainly does not depend on the B network in any way, and then to place the inactive A network in series with a voltage source v_{oc}. The B network is mentioned in the theorem and proof only to indicate that an equivalent for A may be obtained *no matter what arrangement of elements is connected to the* A *network;* the B network represents this general network.

A proof of Thévenin's theorem in the form in which we have stated it is rather lengthy, and therefore it has been placed in Appendix 2 where the curious or rigorous may peruse it.

There are several points about the theorem which deserve emphasis. First, it is not necessary to impose any restrictions on A or B, other than requiring that the original circuit composed of A and B be a linear circuit. No restrictions were imposed on the complexity of A or B; either one may contain any combination of independent voltage or current sources, linear dependent voltage or current sources, resistors, or any other circuit elements which are linear. The general nature of the theorem (and its proof) will enable it to be applied to networks containing inductors and capacitors, which are linear passive circuit elements to be defined in the following chapter. At this time, however, resistors are the only passive circuit elements which have been defined, and the application of Thévenin's theorem to resistive networks is a particularly simple special case. The dead A network can be represented by a single equivalent resistance, R_{th}.

Norton's theorem bears a close resemblance to Thévenin's theorem, another consequence of duality. As a matter of fact, the two statements will be used as an example of dual language when the duality principle is discussed in the following chapter.

Norton's theorem may be stated as follows:

> Given any linear circuit, rearrange it in the form of two networks A and B that are connected together by two resistanceless conductors. Define a current i_{sc} as the short-circuit current which would appear at the terminals of A if B were short-circuited so that no voltage is provided by A. Then all the voltages and currents in B will remain unchanged if A is killed (all independent current sources and independent voltage sources in A replaced by open circuits and short circuits, respectively) and an independent current source i_{sc} is connected, with proper polarity, in parallel with the dead (inactive) A network.

If A is an active *resistive* network, then it is obvious that the inactive A network may be replaced by a single equivalent resistance, which we shall also call the Thévenin resistance, since it is once again the resistance viewed at the terminals of the inactive A network. The Norton equivalent of an active resistive network is therefore the Norton current source i_{sc} in parallel with the Thévenin resistance R_{th}.

There is an important relationship between the Thévenin and Norton equivalents of an active *resistive* network. The relationship may be obtained by applying a source transformation to either equivalent network. For example, if we transform the Norton equivalent, we obtain a voltage source $R_{th}i_{sc}$ in series with the resistance R_{th}; this network is in the form

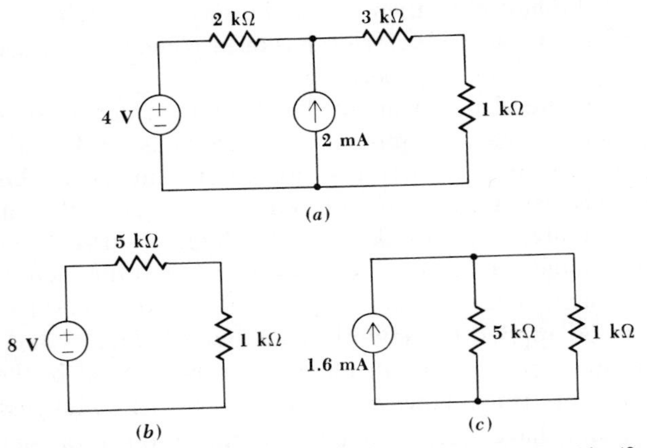

Fig. 3-27 (*a*) A given circuit in which the 1-kΩ resistor is identified as network *B*. (*b*) The Thévenin equivalent is shown for network *A*. (*c*) The Norton equivalent is shown for network *A*.

of the Thévenin equivalent, and thus

$$v_{oc} = R_{th} i_{sc} \tag{25}$$

In resistive circuits containing *dependent* sources as well as independent sources, we shall often find it more convenient to determine either the Thévenin or Norton equivalent by finding both the open-circuit voltage and the short-circuit current and then determining the value of R_{th} as their quotient. It is therefore advisable to become adept at finding both open-circuit voltages and short-circuit currents, even in the simple problems which follow. If the Thévenin and Norton equivalents are determined independently, (25) may serve as a useful check.

Let us consider three examples of the determination of a Thévenin or Norton equivalent circuit. The first is shown in Fig. 3-27*a*; the Thévenin and Norton equivalents are desired for the network faced by the 1-kΩ resistor. That is, network *B* is this resistor, and network *A* is the remainder of the given circuit.

We first kill both independent sources to determine the form of the dead *A* network. With the 4-V source short-circuited and the 2-mA source open-circuited, the result is the series combination of a 2-kΩ and 3-kΩ resistor, or the equivalent, a 5-kΩ resistor. The open-circuit voltage is easily determined by superposition. With only the 4-V source operating, the open-circuit voltage is 4 V; when only the 2-mA source is on, the open-circuit voltage is also 4 V; with both independent sources operating, we see that $v_{oc} = 4 + 4 = 8$ V. This determines the Thévenin equivalent, shown in Fig. 3-27*b*, and from it the Norton equivalent of Fig. 3-27*c* can be drawn quickly. As a check, let us determine i_{sc} for the given circuit.

We use superposition and a little current division:

$$i_{sc} = i_{sc}|_{4\ V} + i_{sc}|_{2\ mA} = \frac{4}{2+3} + 2\frac{2}{2+3} = 0.8 + 0.8 = 1.6 \quad \text{mA}$$

which completes the check.

As the second example, we consider the network A shown in Fig. 3-28, which contains a dependent source. We desire the Thévenin equivalent. To find v_{oc} we note that $v_x = v_{oc}$, and that the dependent source current must pass through the 2-kΩ resistor since there is an open circuit to the right. Summing voltages around the outer loop:

$$-4 + 2 \times 10^3\left(\frac{-v_x}{4000}\right) + 3 \times 10^3(0) + v_x = 0$$

and

$$v_x = 8 = v_{oc}$$

By Thévenin's theorem, then, the equivalent could be formed with the dead A network in series with an 8-V source, as shown in Fig. 3-28*b*. This is correct, but is not very simple and not very helpful; in the case of linear resistive networks, we should certainly show a much simpler equivalent for the inactive A network, namely R_{th}. We therefore seek i_{sc}. Upon short-circuiting the output terminals in Fig. 3-28*a*, it is apparent that $v_x = 0$ and the dependent current source is zero. Hence, $i_{sc} = 4/(5 \times 10^3) =$

Fig. 3-28 (*a*) A given network whose Thévenin equivalent is desired. (*b*) A possible, but rather useless, form of the Thévenin equivalent. (*c*) The best form of the Thévenin equivalent for this linear resistive network.

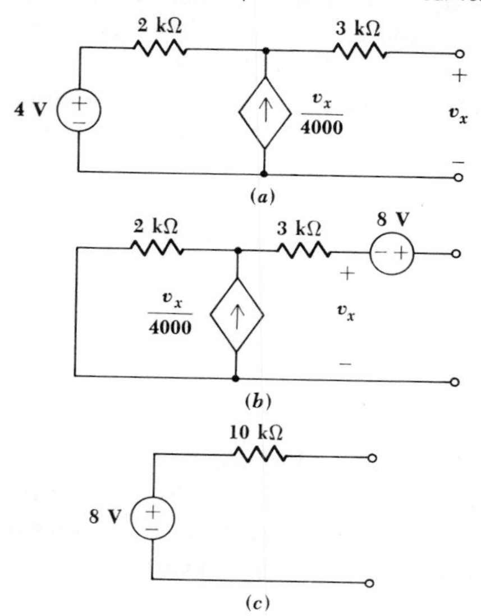

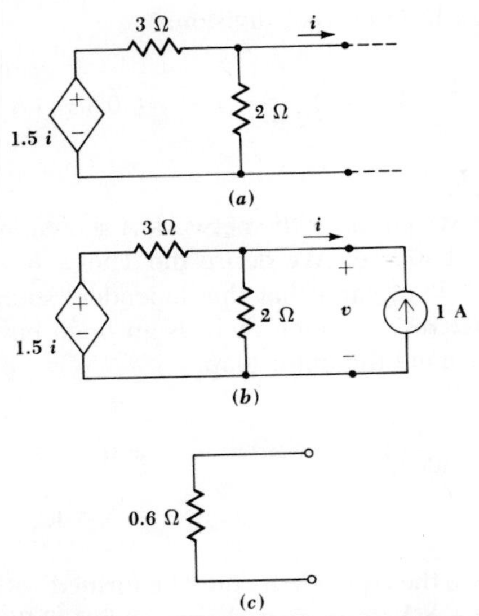

Fig. 3-29 (*a*) A network, containing no independent sources, whose Thévenin equivalent is desired. (*b*) R_{th} is numerically equal to v. (*c*) The Thévenin equivalent of (*a*).

0.8 mA. Thus, $R_{th} = v_{oc}/i_{sc} = 8/(0.8 \times 10^{-3}) = 10\ \text{k}\Omega$, and the accepted Thévenin equivalent of Fig. 3-28*c* is obtained.

As our final example, let us consider a network having a dependent source but no independent source, such as that shown in Fig. 3-29*a*. The network therefore qualifies already as the dead *A* network, and $v_{oc} = 0$. We thus seek the value of R_{th} represented by this two-terminal network. However, we cannot find v_{oc} and i_{sc} and take their quotient, for there is no independent source in the network and both v_{oc} and i_{sc} are zero. Let us, therefore, be a little tricky. We apply a 1-A source *externally*, measure the resultant voltage, and then set $R_{th} = v/1$. Referring to Fig. 3-29*b*, we see that $i = -1$ and

$$\frac{v - 1.5(-1)}{3} + \frac{v}{2} = 1$$

so that

$$v = 0.6 \quad \text{V}$$

and

$$R_{th} = 0.6 \quad \Omega$$

The Thévenin equivalent is shown in Fig. 3-29*c*.

Although we are devoting our attention almost entirely to the analysis of *linear* circuits, it is enlightening to know that Thévenin's and Norton's theorems are both valid if network *B* is nonlinear; only network *A* must be linear.

Drill Problems

3-13 Find the Thévenin equivalent of the network shown in Fig. 3-30*a* if: (*a*) $C_1 = 0$, $C_2 = 12$ V; (*b*) $C_1 = 0.2$ A/V, $C_2 = 12$ V; (*c*) $C_1 = 0.2$ A/V, $C_2 = 0$.

Ans. 0 V, 12.5 Ω; 6 V, 5 Ω; 15 V, 12.5 Ω

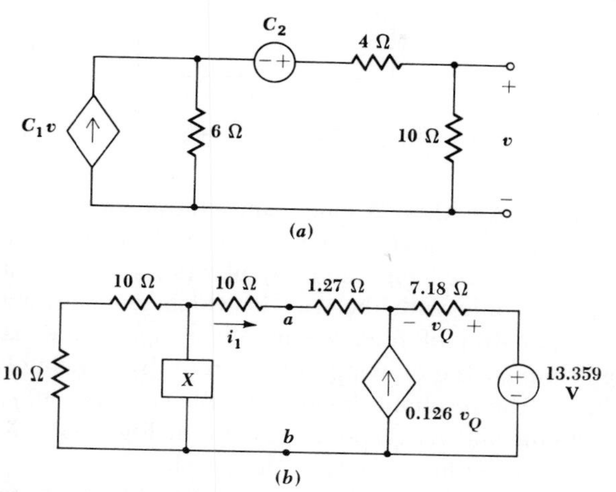

Fig. 3-30 See Drill Probs. 3-13 and 3-14.

3-14 Find the Norton equivalent of the network to the left of terminals *ab* in Fig. 3-30*b* if element *X* is: (*a*) a 5-Ω resistor; (*b*) a 3-A independent current source, arrow directed upward; (*c*) a dependent voltage source, labeled $2i_1$, positive reference at top.

Ans. 0 A, 8 Ω; 0 A, 14 Ω; 2 A, 30 Ω

3-7 TREES AND GENERAL NODAL ANALYSIS

In this section we shall generalize the method of nodal analysis that we have come to know and love. Since this method is applicable to any general network, we cannot promise that we shall be able to solve a wider class of circuit problems. We can, however, look forward to being able to select a nodal analysis method for any particular problem that may result in fewer equations and less work.

We must first extend our list of definitions relating to network topology. We begin by defining *topology* itself as a branch of geometry which is concerned with those properties of a geometrical figure which are unchanged when the figure is twisted, bent, folded, stretched, squeezed, or tied in knots, with the provision that no parts of the figure are to be cut

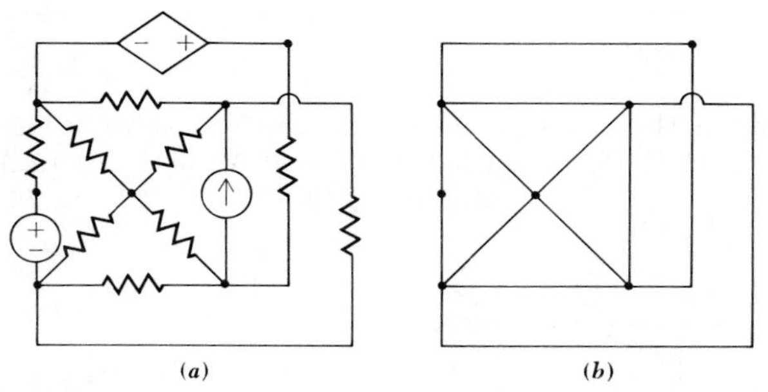

Fig. 3-31 (*a*) A given circuit. (*b*) The linear graph of this circuit.

apart or to be joined together. A sphere and a tetrahedron are topologically identical, as are a square and a circle. In terms of electric circuits, then, we are not now concerned with the particular types of elements appearing in the circuit, but only with the way in which branches and nodes are arranged. As a matter of fact, we usually suppress the nature of the elements and simplify the drawing of the circuit by showing the elements as straight lines. The resultant drawing is called a *linear graph,* or simply a *graph.* A circuit and its graph are shown in Fig. 3-31. Note that all nodes are identified by heavy dots in the graph.

Since the topological properties of the circuit or its graph are unchanged when it is distorted, the three graphs shown in Fig. 3-32 are all topologically identical with the circuit and graph of Fig. 3-31.

Topological terms which we already know and have been using correctly are:

node: a point at which two or more elements have a common connection.

branch: a single path, containing one simple element, which connects one node to any other node.

Fig. 3-32 The three graphs shown are topologically identical to each other and to the graph of Fig. 3-31*b*, and each is a graph of the circuit shown in Fig. 3-31*a*.

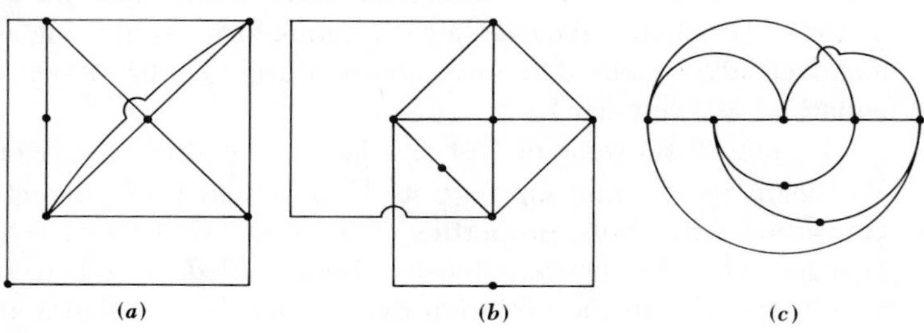

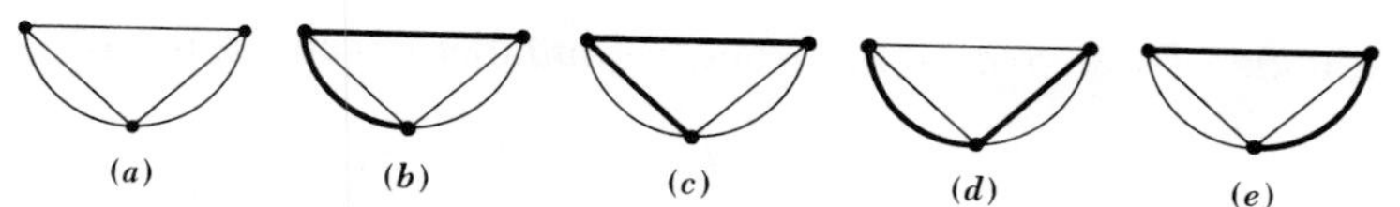

Fig. 3-33 (*a*) The linear graph of a three-node network. (*b*), (*c*), (*d*), and (*e*) Four of the eight different trees which may be drawn for this graph are shown by the heavy lines.

loop: a set of branches forming a closed path which passes through no node more than once.

mesh: a loop which does not contain any other loops within it.

planar circuit: a circuit which may be drawn on a plane surface in such a way that no branch passes over or under any other branch.

nonplanar circuit: any circuit which is not planar.

The graphs of Fig. 3-32 each contain 12 branches and 7 nodes.

Two new properties of a linear graph must now be defined, a tree and a link. We may define a *tree* as any set of branches which does not contain any loops but which connects every node to every other node, not necessarily directly. There are usually a number of different trees which may be drawn for a network, and the number increases rapidly as the complexity of the network increases. The simple graph shown in Fig. 3-33*a* has eight possible trees, four of which are shown by heavy lines in Fig. 3-33*b*, *c*, *d*, and *e*.

In Fig. 3-34*a* a more complex graph is shown. Figure 3-34*b* shows one possible tree, and Fig. 3-34*c* and *d* shows sets of branches which are *not* trees because neither set satisfies the definition above.

Once we understand the construction of a tree, the concept of the link is very simple, for a *link* is any branch in a linear graph which is not a branch of the tree. It is evident that any particular branch may or may not be a link, depending on the particular tree which is selected.

The number of links in a graph may be related to the number of branches and nodes very simply. If the graph has N nodes, then exactly $(N - 1)$ branches are required to construct a tree because the first branch chosen connects two nodes and each additional branch includes one more

Fig. 3-34 (*a*) A linear graph. (*b*) A possible tree for this graph. (*c*) and (*d*) These sets of branches do not satisfy the definition of a tree.

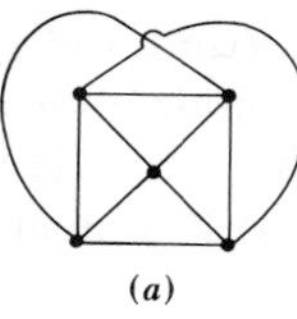

(*a*)

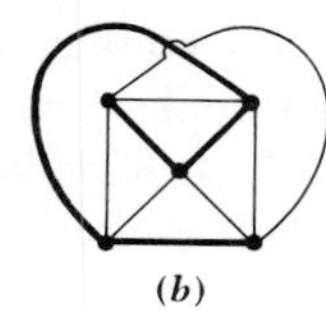

(*b*)

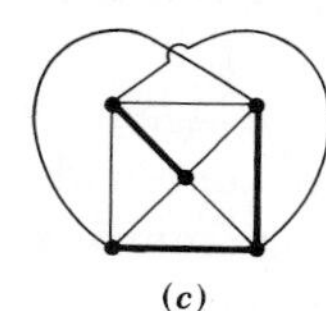

(*c*)

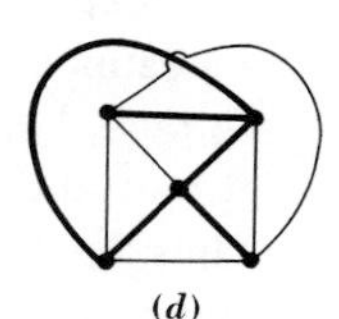

(*d*)

node. Thus, given B branches, the number of links L must be

$$L = B - (N - 1)$$

or

$$L = B - N + 1 \tag{26}$$

In any of the graphs shown in Fig. 3-33, we note that $3 = 5 - 3 + 1$, and in the graph of Fig. 3-34*b*, $6 = 10 - 5 + 1$. A network may be in several disconnected parts, and (26) may be made more general by replacing $+1$ with $+S$, where S is the number of separate parts. However, it is also possible to connect two separate parts by a *single* conductor, thus causing two nodes to form one node; no current can flow through this single conductor. This process may be used to join any number of separate parts, and thus we shall not suffer any loss of generality if we restrict our attention to circuits for which $S = 1$.

We are now ready to discuss a method whereby a set of nodal equations may be written for a network that are independent and sufficient. The method will enable us to obtain many different sets of equations for the same network, and all the sets will be valid. However, the method does not provide us with *every* possible set of equations. Let us first describe the procedure, illustrate it by an example, and then point out the reason that the equations are independent and sufficient.

Given a network, we first draw its graph and then construct a tree. Any tree will do, and we shall see shortly how to choose that tree which is most convenient for our purposes. We next focus our attention on the branches in the tree. Each tree branch is assigned a voltage, extending from the node at one end of the tree branch to the other. Thus, there are $(N - 1)$ voltages chosen since the tree contains $(N - 1)$ branches. Kirchhoff's current law is then applied to any $(N - 1)$ nodes, and the remaining node may be ignored. This corresponds to the presence of the reference node in our previous attack on nodal analysis. The equation written at the Nth node is not independent; it may be obtained from the other $(N - 1)$ equations.

If there are voltage sources in the network, they should be placed in the tree and the source voltage should be assigned as that tree-branch voltage. Any dependent source that is voltage controlled should have that control voltage placed in a tree branch if possible. Current sources should be placed in links, and currents that control dependent sources should also be placed in links whenever possible.

In applying Kirchhoff's current law at $(N - 1)$ nodes, it is evident that we cannot express the current of any voltage source in terms of the source voltage; hence, we again mentally replace each voltage source by a short circuit and form supernodes at which Kirchhoff's current law is applied. Thus, each voltage source will result in the $(N - 1)$ Kirchhoff current law equations being one less in number.

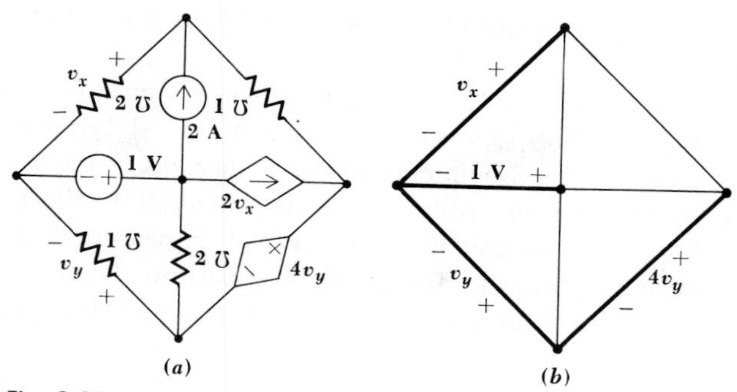

Fig. 3-35 (*a*) The circuit of Fig. 3-5 repeated. (*b*) A tree is chosen such that both voltage sources and both control voltages are tree branches.

The procedure may be illustrated by considering an example we analyzed earlier by defining all node voltages with respect to a reference node. The circuit is that of Fig. 3-5, repeated as Fig. 3-35*a*. We draw a tree so that both voltage sources and both control voltages appear as tree-branch voltages and, hence, as assigned variables. As it happens, these four branches constitute a tree, Fig. 3-35*b*, and tree-branch voltages v_x, 1, v_y, and $4v_y$ are chosen, as shown.

Considering both voltage sources as short circuits, we apply Kirchhoff's current law twice, once to the top node,

$$2v_x + 1(v_x - v_y - 4v_y) = 2$$

and once to the supernode consisting of the right node, the bottom node, and the dependent voltage source,

$$1v_y + 2(v_y - 1) + 1(4v_y + v_y - v_x) = 2v_x$$

Instead of the four equations we had previously, we have only two, and we find easily that $v_x = 26/9$ V and $v_y = 4/3$ V, both values agreeing with the earlier solution.

Now let us discuss the sufficiency of the assumed voltages and the independence of the nodal equations. If these tree-branch voltages are *sufficient*, then every branch voltage must be obtainable from a knowledge of the values of all the tree-branch voltages. Once every branch voltage is known, it is apparent that every branch current or power can be readily obtained and thus the tree-branch voltages will be "sufficient" information to enable a complete solution to be obtained if desired. But we certainly can express every branch voltage in terms of the tree-branch voltages. This is already accomplished for those branches in the tree. For the links, we know that each link extends between two nodes, and, by definition, the tree must also connect those two nodes. Hence, every link voltage may also be established in terms of the tree-branch voltages. Thus, sufficiency is demonstrated.

To show that the $(N - 1)$ nodal equations are *independent,* visualize the application of Kirchhoff's current law to the $(N - 1)$ different nodes. Each time we write the Kirchhoff current law equation, there is a new tree branch involved—the one which connects that node to the remainder of the tree. Since that circuit element has not appeared in any previous equation, we must obtain an independent equation. This is true for each of the $(N - 1)$ nodes in turn, and hence we have $(N - 1)$ independent equations.

Drill Problem

3-15 By constructing a suitable tree, rework part (*b*) of Drill Prob.: (*a*) 3-1, using only two equations; (*b*) 3-3, using only one equation; (*c*) 3-4, using only one equation.

Ans. -4; -2.5; 10 V

3-8 LINKS AND LOOP ANALYSIS

Now we shall consider the use of a tree to obtain a suitable set of loop equations. In some respects this is the dual of the method of writing nodal equations. Again it should be pointed out that, although we are able to guarantee that any set of equations we write will be both sufficient and independent, we should not expect that the method will lead directly to every possible set of equations.

We begin by constructing any convenient tree for the given network and then focus our attention on any arbitrary link. If that link is added to the tree, a closed loop must be formed; to this loop we assign our first loop current. It flows through certain branches in the tree and through this particular link. Now let us temporarily ignore that first link and the first loop current. We select another link, imagine its addition to the tree, note that once again a loop is formed, and assign our second loop current to this loop. The process is repeated for each link and, since there are $(B - N + 1)$ links, we end up with $(B - N + 1)$ loop currents; no link has more than one loop current flowing in it. We thus may describe the current interchangeably as a link current or a loop current. The procedure is now familiar; Kirchhoff's voltage law is applied to each loop, as defined by its loop current, and the resultant equations may be solved by determinants, substitution, inspection, or digital computer.

If there are current sources in the network, they should be placed in the links and the source current should be assigned as that link current. Any dependent source that is current controlled should have that control current placed in a link, if possible. Voltage sources should be placed in tree branches, and voltages that control dependent sources should also be placed in tree branches whenever possible. Note that the general rule either for nodal analysis or loop analysis is to place voltages in tree branches and currents in links, a mandatory rule for sources and a desirable rule for controlling quantities.

In applying Kirchhoff's voltage law around $(B - N + 1)$ loops, it is evident that we cannot express the voltage of any current source in terms of the source current; hence, we again replace each current source mentally by an open circuit, thus discarding the loop that would be formed by the closure of that link. Each current source therefore results in the $(B - N + 1)$ Kirchhoff voltage law equations being one less in number.

Let us illustrate this procedure by reworking an earlier example solved by mesh currents. The circuit is shown in Fig. 3-12 and redrawn in Fig. 3-36*a*. The tree is selected so that the voltage source is in a tree branch while the current source is in a link, Fig. 3-36*b*. Upon closing the short, central, vertical link, a loop (also mesh) current is established and labeled 7 A. This link is opened and the short, central, horizontal link is closed; the resultant loop current is also a mesh current, and it is called i_A. This link is opened and the last link is closed, permitting a loop current to be present around the perimeter of the circuit, i_B. Note that in each case the closure of the link defines only one loop current that is present in tree branches and that link only.

The equation about the i_A loop is

$$1(i_A - 7) + 2(i_A + i_B) + 3i_A = 0$$

and for the i_B loop,

$$-7 + 2(i_A + i_B) + 1i_B = 0$$

Fig. 3-36 (*a*) The circuit of Fig. 3-12 is shown again. (*b*) A tree is chosen such that the current source is in a link and the voltage source is in a tree branch.

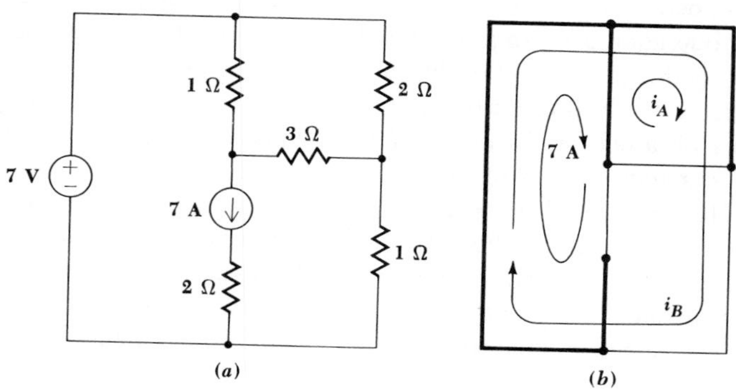

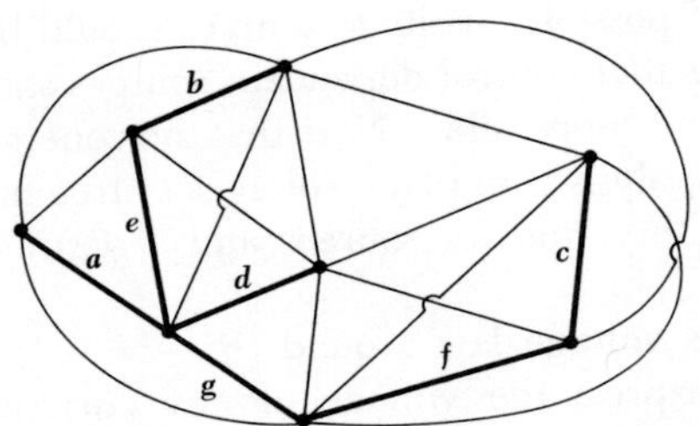

Fig. 3-37 A tree which is used as an example to illustrate the sufficiency of the link currents.

and we have $i_A = 0.5$ A, $i_B = 2$ A, as before. The solution was achieved with one less equation than before.

How may we demonstrate sufficiency? Let us visualize a tree. It contains no loops and therefore contains at least two nodes to each of which only one tree branch is connected. The current in each of these two branches is easily found from the known link currents by applying Kirchhoff's current law. If there are other nodes at which only one tree branch is connected, these tree-branch currents may also be immediately obtained. In the tree shown in Fig. 3-37, we thus have found the currents in branches *a*, *b*, *c*, and *d*. Now we move along the branches of the tree, finding the currents in the tree branches *e* and *f*; the process may be continued until all the branch currents are determined. The link currents are therefore sufficient to determine all branch currents. It is helpful to look at the situation where an incorrect "tree" has been drawn which contains a loop. Even if all the link currents were zero, a current might still circulate about this "tree loop." Hence, the link currents could not determine this current, and they would not represent a sufficient set. Such a "tree" is by definition impossible.

In order to demonstrate that the $(B - N + 1)$ loop equations are independent, it is only necessary to point out that each represents the application of Kirchhoff's voltage law around a loop which contains one link not appearing in any other equation. We might visualize a different resistance $R_1, R_2, \ldots, R_{B-N+1}$ in each of these links, and it is then apparent that one equation can never be obtained from the others since it contains one coefficient not appearing in any other equation.

Hence, the link currents are sufficient to enable a complete solution to be obtained, and the set of loop equations which we use to find the link currents is a set of independent equations.

Having looked at both general nodal analysis and loop analysis, we should now consider the advantages and disadvantages of each method so that an intelligent choice of a plan of attack can be made.

The nodal method in general requires $(N - 1)$ equations, but this number is reduced by one for each independent or dependent voltage source

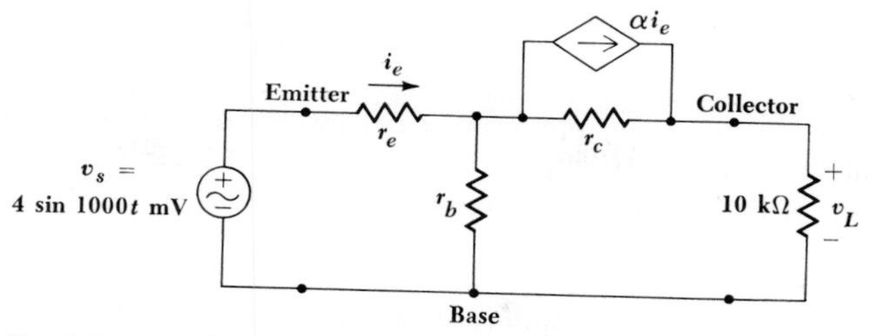

Fig. 3-38 A sinusoidal voltage source and a 10-kΩ load are connected to the T-equivalent circuit of a transistor. The common connection between the input and output is at the base terminal of the transistor and the arrangement is called the common-base configuration.

in a tree branch, and increased by one for each dependent source that is voltage controlled by a link voltage or current controlled.

The loop method basically involves $(B - N + 1)$ equations. However, each independent or dependent current source in a link reduces this number by one, while each dependent source that is current controlled by a tree-branch current or is voltage controlled increases the number by one.

As a grand finale for this discussion, let us inspect the T-equivalent-circuit model for a transistor, shown in Fig. 3-38, to which are connected a sinusoidal source, 4 sin 1000t mV, and a 10-kΩ load. We select typical values for the emitter resistance, $r_e = 50\ \Omega$; for the base resistance, $r_b = 500\ \Omega$; for the collector resistance, $r_c = 20$ kΩ; and for the common-base forward-current-transfer ratio, $\alpha = 0.98$. Suppose that we wish to find the input (emitter) current i_e and the load voltage v_L.

Although the details are requested in Drill Probs. 3-17 and 3-18 below, we should see readily that the analysis of this circuit might be accomplished by drawing trees requiring three general nodal equations $(N - 1 - 1 + 1)$ or two loop equations $(B - N + 1 - 1)$. We might also note that three equations are required in terms of node-to-reference voltages, as are three mesh equations.

No matter which method we choose, these results are obtained for this specific circuit:

$$i_e = 18.1 \sin 1000t \quad \mu\text{A}$$

$$v_L = 120 \sin 1000t \quad \text{mV}$$

and we therefore find that this transistor circuit provides a voltage gain (v_L/v_s) of 30, a current gain $(v_L/10{,}000i_e)$ of 0.659, and a power gain equal to the product, 30(0.659) = 19.8. Higher gains could be secured by operating this transistor in a common-emitter configuration, as illustrated by Prob. 30.

Drill Problems

3-16 By constructing an appropriate tree, rework: (*a*) part (*b*) of Drill Prob. 3-5, using only two equations; (*b*) part (*c*) of Drill Prob. 3-7, using only one equation; (*c*) part (*c*) of Drill Prob. 3-8, using only one equation.

Ans. $-8/3$; $-1/3$; $4/3$ mA

3-17 For the transistor amplifier equivalent circuit shown in Fig. 3-38, let $r_e = 50\ \Omega$, $r_b = 500\ \Omega$, $r_c = 20\ \text{k}\Omega$, $\alpha = 0.98$, and find both i_e and v_L by drawing a suitable tree and using: (*a*) two loop equations; (*b*) three nodal equations with a common reference node for the voltages; (*c*) three nodal equations without a common reference node.

Ans. 18.1 sin 1000*t* μA and 120 sin 1000*t* mV

3-18 Determine the Thévenin and Norton equivalent circuits presented to the 10-kΩ load in Fig. 3-38 by finding: (*a*) the open-circuit value of v_L; (*b*) the (downward) short-circuit output current; (*c*) the Thévenin equivalent resistance. All circuit values are given in Drill Prob. 3-17.

Ans. 146 sin 1000*t* mV; 65.6 sin 1000*t* μA; 2230 Ω

PROBLEMS

☐ **1** In Fig. 3-39, let circuit element *X* be a 2-℧ conductance. Find the power supplied by the 10-A source.

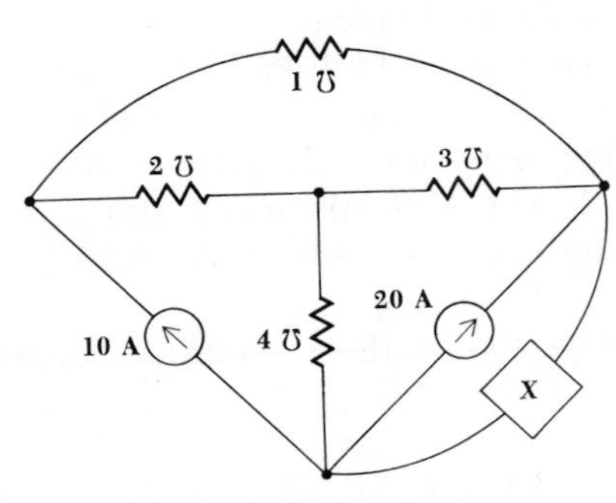

Fig. 3-39 See Probs. 1 to 4.

☐ **2** In Fig. 3-39, let circuit element *X* be a 10-V source with its positive voltage reference at the upper right node. Find the power supplied by the 10-A current source.

☐ **3** In Fig. 3-39, let circuit element *X* be a dependent voltage source with its positive voltage reference at the right node and having a voltage of $v_1/3$, where the 10-A source supplies a power $10v_1$. Find v_1.

☐ **4** In Fig. 3-39, circuit element X is a dependent current source, $4v_1/3$, with the reference arrow directed toward the upper right node. Let the power supplied by the 10-A source be $10v_1$; find v_1.

☐ **5** For the circuit shown in Fig. 3-40a: (a) let $R = 4\ \Omega$ and use nodal methods to find the power supplied by the 7-A source; (b) repeat if R is infinite.

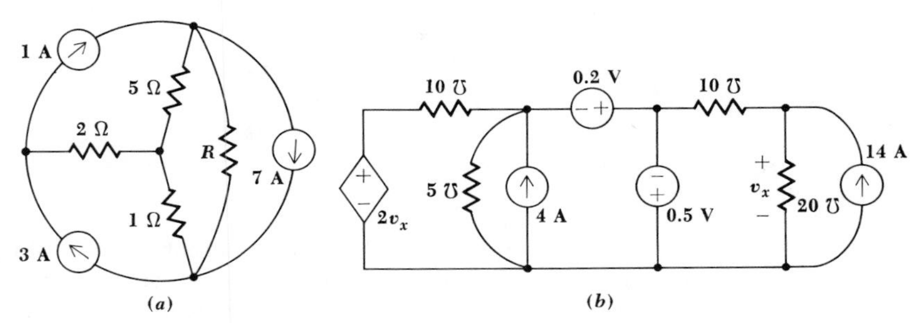

Fig. 3-40 See Probs. 5 and 6.

☐ **6** With reference to Fig. 3-40b, let the lower node be the reference, and write the single nodal equation required to find v_x. How much power is delivered by the dependent source?

☐ **7** In Fig. 3-41, let circuit element X be a 2-Ω resistance. Find the power supplied by the 6-V source.

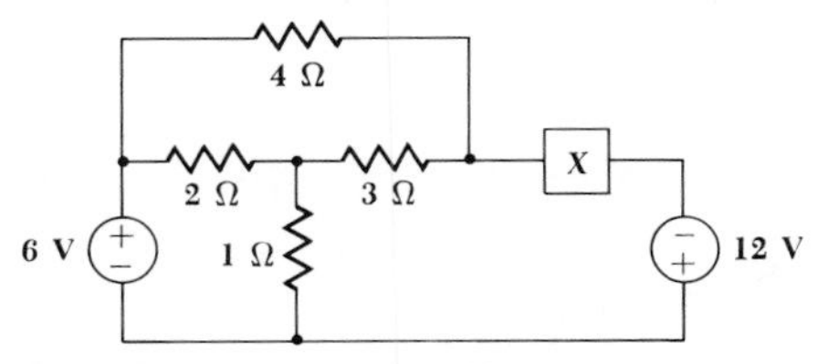

Fig. 3-41 See Probs. 7 to 10.

☐ **8** In Fig. 3-41, let circuit element X be a 6-A source with its arrow directed to the right. Find the power supplied by the 6-V voltage source.

☐ **9** In Fig. 3-41, let circuit element X be a dependent current source with its arrow directed to the right and having a current of $i_1/3$, where i_1 is the upward current through the 6-V source. Find i_1.

☐ **10** In Fig. 3-41, circuit element X is a dependent voltage source, $4i_1/3$, with the positive reference on the right side of the element. Let the power supplied by the 6-V source be $6i_1$; find i_1.

☐ **11** In Fig. 3-42a, let S_1, S_2, and S_3 be 30-, 14-, and -2-V voltage sources, respectively, all with positive reference symbols on top. Use mesh analysis to find the power supplied by the 30-V source.

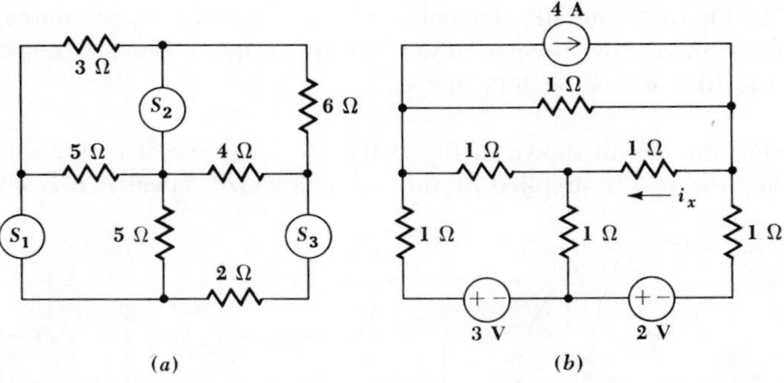

Fig. 3-42 See Probs. 11 to 13 and 27.

☐ **12** In Fig. 3-42*a*, let S_1, S_2, and S_3 be three current sources with currents of -9, -1.5, and 6 A, respectively, flowing in an upward direction. Use mesh analysis to find the power supplied by the 6-A source.

☐ **13** Find the current i_x in Fig. 3-42*b* by: (*a*) nodal analysis; (*b*) mesh analysis; (*c*) changing the two practical voltage sources to practical current sources and then using nodal analysis; (*d*) changing the practical current source to a practical voltage source and again using nodal analysis.

☐ **14** (*a*) By beginning with the practical current source at the right in Fig. 3-43*a*, make repeated source transformations and resistance combinations in order to find the power supplied by the 24-V source. (*b*) What power does the 18-A source provide?

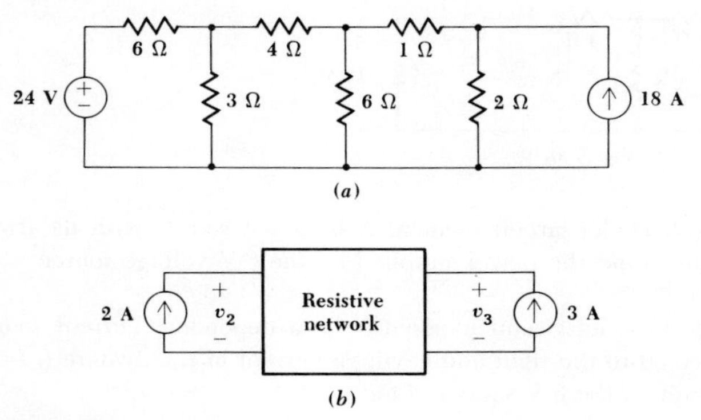

Fig. 3-43 See Probs. 14 to 16.

☐ **15** (*a*) Use superposition to determine the magnitude and direction of the current in the 4-Ω resistor of Fig. 3-43*a*. (*b*) How would the above result change if there were also a 40-V source, positive reference to the right, in series with the 4-Ω resistor?

☐ **16** Two current sources are connected to a resistive network as shown in Fig. 3-43*b*. When the 3-A source is disconnected, it is found that the 2-A source delivers 28 W to the network and v_3 is 8 V. However, when the 2-A source is disconnected, the 3-A source supplies 54 W and v_2 is 12 V. Find the power supplied by each when both are operating.

☐ **17** An automotive battery is found to be capable of supplying a load with 20 A at 12.3 V and 50 A at 12.0 V. (*a*) Represent the battery as a practical voltage source. (*b*) To what load resistance does the maximum power transfer theorem dictate a maximum load power is supplied? (*c*) What is the load power? (*d*) Under these conditions, what power would be dissipated within the battery?

☐ **18** Two terminals, *a* and *b*, extend from a black box containing an active resistive network. An ideal voltage source is connected across the terminals and it is found that when the voltage v_{ab} is adjusted to 7.2 V no current flows from the source. Then an ideal current source is connected to the terminals, and it is found that $v_{ab} = 0$ when the current into the *b* terminal of the black box is 40 mA. Determine the Thévenin and Norton equivalents of the black box.

☐ **19** Find the Thévenin and Norton equivalents of the circuits shown in: (*a*) Fig. 3-44*a*; (*b*) Fig. 3-44*b* with $R = \infty$; (*c*) Fig. 3-44*b* with $R = 1.6$ kΩ.

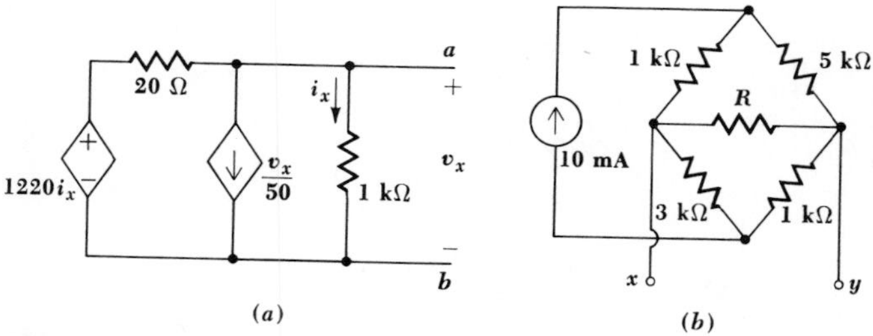

Fig. 3-44 See Prob. 19.

☐ **20** For the two-terminal network shown in Fig. 3-45: (*a*) obtain a single equation relating *v* and *i*; (*b*) plot a curve of *i* versus *v*.

Fig. 3-45 See Probs. 20 and 21.

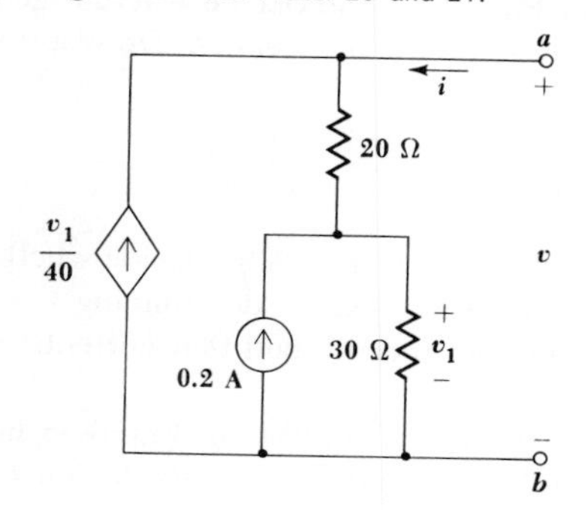

☐ **21** What is the maximum power that can be delivered to an external load resistor R_L by the network of Fig. 3-45?

☐ **22** Two Thévenin equivalent circuits (v_{s1} and R_1, v_{s2} and R_2) are connected in parallel. Find the Thévenin equivalent of the resultant two-terminal network by using a method based on: (*a*) superposition; (*b*) source transformation.

☐ **23** Refer to the circuit shown in Fig. 3-35*a* and change the dependent voltage source $4v_y$ to a 2-℧ conductance. Draw a suitable tree and use general node voltages to determine v_x and v_y.

☐ **24** Draw a suitable tree and analyze the circuit of Fig. 3-36*a* by using general node voltages to determine the power supplied by the 7-A source.

☐ **25** Draw an appropriate tree for the circuit of Fig. 3-35*a*, assign link currents, and use loop analysis to determine the power supplied by the dependent voltage source. (Note that the two control voltages can be expressed easily as currents and placed in links.)

☐ **26** Use loop analysis on the circuit shown in Fig. 3-46 to determine the magnitude and direction of the current in the upper 2-Ω resistor.

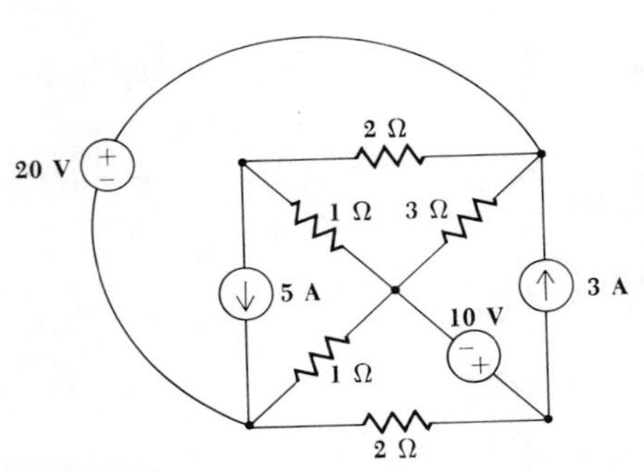

Fig. 3-46 See Probs. 26 and 29.

☐ **27** Refer to the circuit shown in Fig. 3-42*b* and decide whether general node or loop analysis will lead to the solution for i_x most easily. Draw a suitable tree and find i_x.

☐ **28** Show that the network drawn in Fig. 3-47 is planar.

☐ **29** After inspection of the network shown in Fig. 3-46, decide whether general node or loop analysis would be most expeditious in determining the current in the upper 2-Ω resistor. Draw a suitable tree and find that current.

☐ **30** Reconnect the common-base transistor amplifier of Fig. 3-38 in common-emitter configuration by interchanging r_e and r_b, by interchanging the labels

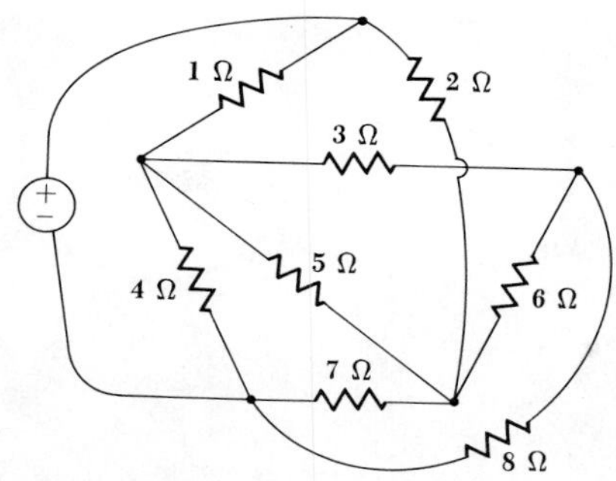

Fig. 3-47 See Prob. 28.

Emitter and *Base*, and by noting that i_e is now directed vertically upward in the center leg. Find: (*a*) the ratio of the load voltage to the source voltage; (*b*) the ratio of the load current to the source current; (*c*) the ratio of the power delivered to the load to that supplied by the source.

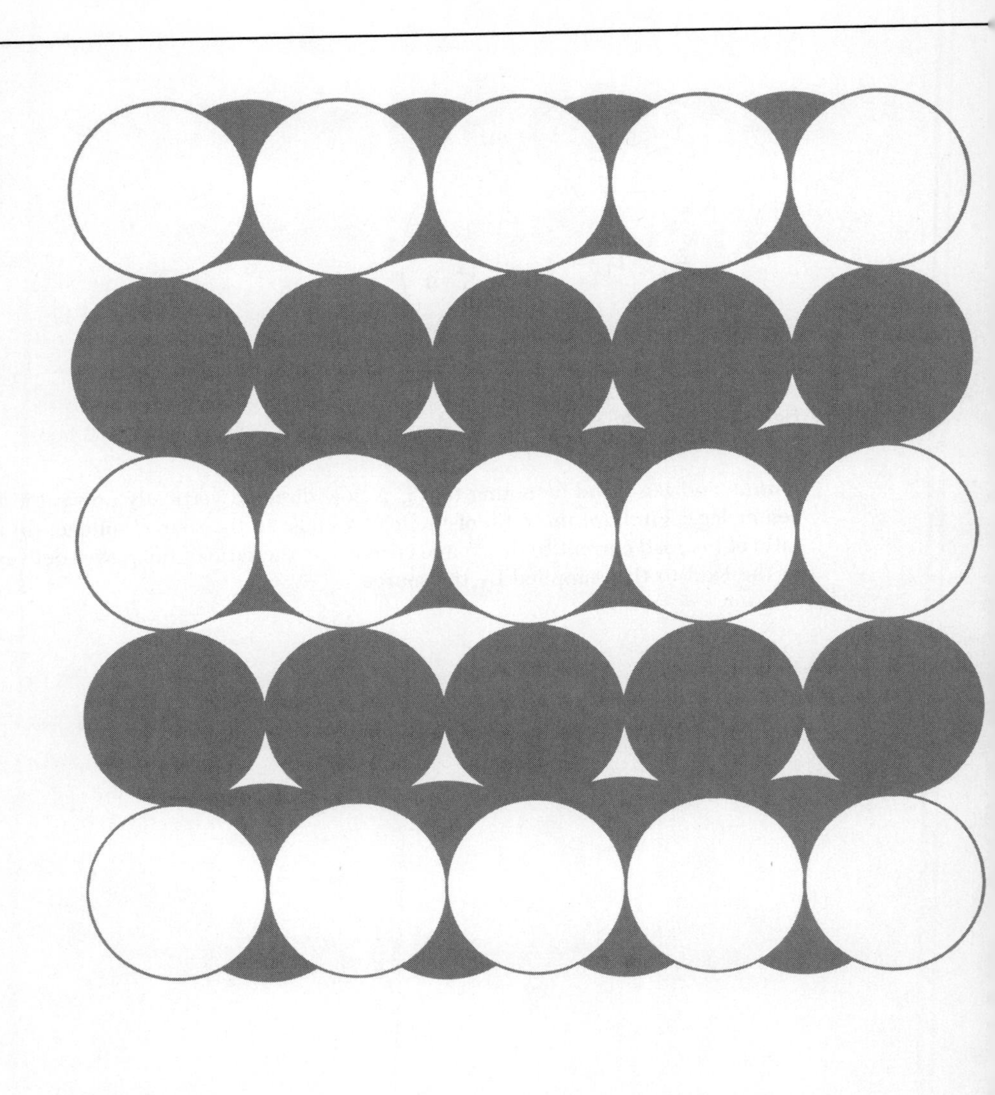

Part Two
THE TRANSIENT CIRCUIT

Chapter Four Inductance and Capacitance

4-1 INTRODUCTION

We are now ready to begin the second major portion of our study of circuits. In this chapter we shall introduce two new simple circuit elements whose voltage-current relationships involve the rate of change of a voltage or current. Before we begin this new study, it will be worthwhile to pause for a moment and look back upon our study of the analysis of resistive circuits. A little bit of philosophical review will aid in our understanding of the coming work.

After setting up a satisfactory system of units, we began our discussion of electric circuits by defining current, voltage, and five simple circuit elements. The independent and dependent voltage and current sources were called active elements and the linear resistor was termed a passive element, although our definitions of "active" and "passive" are still slightly fuzzy and need to be brought into sharper focus. We think of an active element as one which is capable of delivering net power to some external device, and the ideal sources were classified as such; a passive element, however, is capable only of receiving power and the resistor falls into this category. The energy it receives is usually transformed into heat.

Each of these elements was defined in terms of the restrictions placed on its voltage-current relationship. In the case of the independent voltage source, for example, the terminal voltage must be completely independent of the current drawn from its terminals. We then considered circuits composed of the different building blocks. In general, we used only constant voltages and currents, but now that we have gained a familiarity with the basic analytical techniques by treating only the resistive circuit, we may begin to consider the much more interesting and practical circuits in which inductance and capacitance may be present and in which both the forcing functions and the responses usually vary with time.

4-2 THE INDUCTOR

Both the inductor, which is the subject of this and the following section, and the capacitor, which is discussed later in the chapter, are passive elements which are capable of storing and delivering finite amounts of energy. Unlike an ideal source, they cannot provide an unlimited amount of energy or a finite average power. This concept may be used to improve upon our definition of an *active element* as being one which is capable of delivering an infinite amount of energy and of a *passive element* as one which is not capable of delivering infinite energy.

Although we shall define an inductor and inductance strictly from a circuit point of view, that is, by a voltage-current equation, a few comments about the historic development of the magnetic field may provide a better understanding of the definition. In the early 1800s the Danish scientist Oersted showed that a current-carrying conductor produced a magnetic field, or that compass needles were affected in the presence of a current-carrying conductor. In France, shortly thereafter, Ampère made some careful measurements which demonstrated that this magnetic field was linearly related to the current which produced it. The next step occurred some twenty years later when the English experimentalist Michael Faraday and the American inventor Joseph Henry discovered almost simultaneously[1] that a changing magnetic field could produce a voltage in a

[1]Faraday won.

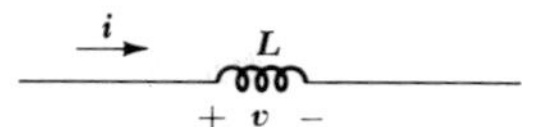

Fig. 4-1 The reference signs for voltage and current are shown on the circuit symbol for an inductor: $v = L\, di/dt$.

neighboring circuit. They showed that this voltage was proportional to the time rate of change of the current which produced the magnetic field. The constant of proportionality we now call the *inductance,* symbolized by L, and therefore

$$v = L\frac{di}{dt} \tag{1}$$

where we must realize that v and i are both functions of time. When we wish to emphasize this, we may do so by using the symbols $v(t)$ and $i(t)$.

The circuit symbol for the inductor is shown in Fig. 4-1, and it should be noted that the passive sign convention is used, just as it was with the resistor. The unit in which inductance is measured is the *henry*[2] (H), and the defining equation shows that the henry is just a shorter expression for a volt-second per ampere.

The inductor whose inductance is defined by (1) is a mathematical model; it is an ideal element which we may use to approximate the behavior of a real device. A physical inductor may be constructed by winding a length of wire into a coil. This serves effectively to increase the current which is causing the magnetic field and also to increase the "number" of neighboring circuits into which Faraday's voltage may be induced. The result of this twofold effect is that the inductance of a coil is approximately proportional to the square of the number of complete turns made by the conductor out of which it is formed. An inductor, or "coil," which has the form of a long helix of very small pitch is found to have an inductance of $\mu N^2 A/s$, where A is the cross-sectional area, s is the axial length of the helix, N is the number of complete turns of wire, and μ (mu) is a constant of the material inside the helix, called the *permeability*. For air, $\mu = \mu_0 = 4\pi \times 10^{-7}$ H/m.

Physical inductors should be on view in an accompanying laboratory course, and the topics concerned with the magnetic flux, permeability, and the methods of using the characteristics of the physical coil to calculate a suitable inductance for the mathematical model are treated in both physics courses and courses in electromagnetic field theory.

Let us now scrutinize (1) to determine some of the electrical characteristics of this mathematical model. This equation shows that the voltage across an inductor is proportional to the time rate of change of the current

[2] An empty victory.

through it. In particular, it shows that there is no voltage across an inductor carrying a constant current, regardless of the magnitude of this current. Accordingly, we may view an inductor as a "short circuit to dc." Another fact which is evidenced by this equation is related to an infinite rate of change of the inductor current, such as that caused by an abrupt change in current from one finite value to some other finite value. This sudden or discontinuous change in the current must be associated with an infinite voltage across the inductor. In other words, if we wish to produce an abrupt change in an inductor current, we must apply an infinite voltage. Although an infinite-voltage forcing function might be acceptable theoretically, it can never be a part of the phenomena displayed by a real physical device. As we shall see shortly, an abrupt change in the inductor current also requires an abrupt change in the energy stored in the inductor, and this sudden change in energy requires infinite power at that instant; infinite power is again not a part of the real physical world. In order to avoid infinite voltage and infinite power, an inductor current must not be allowed to jump instantaneously from one value to another. If an attempt is made to open-circuit a physical inductor through which a finite current is flowing, an arc may appear across the switch. The stored energy is dissipated in ionizing the air in the path of the arc. This is useful in the ignition system of an automobile, where the current through the spark coil is interrupted by the distributor, and the arc appears across the spark plug.

We shall not consider any circuits at the present time in which an inductor is suddenly open-circuited. It should be pointed out, however, that we shall remove this restriction later when we hypothesize the exist-

Fig. 4-2 (*a*) The current waveform in a 3-H inductor. (*b*) The corresponding voltage waveform, $v = 3\ di/dt$.

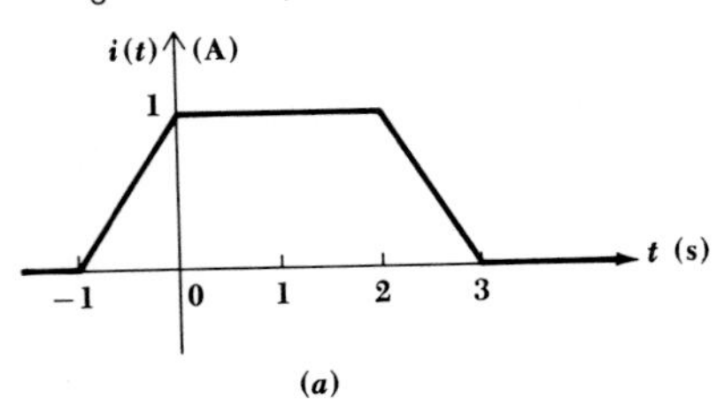

(*a*)

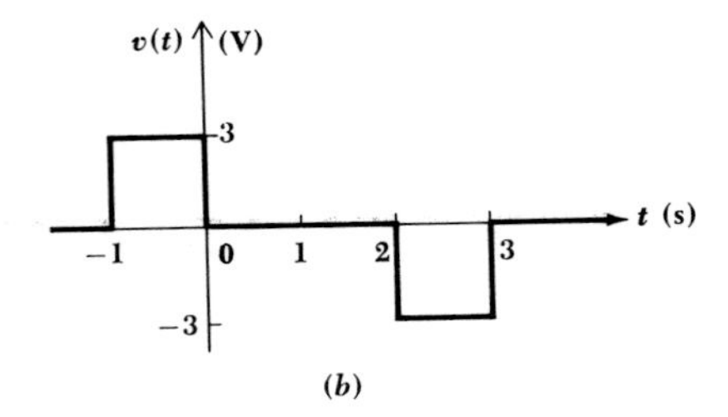

(*b*)

ence of a voltage forcing function or response which does become infinite instantaneously.

Equation (1) may also be interpreted (and solved, if necessary) by graphical methods. Let us assume a current which is zero prior to $t = -1$ s, increases linearly to 1 A in the next second, remains at 1 A for 2 s, and then decreases to zero in the next second, remaining zero thereafter. The current waveform is sketched as a function of time in Fig. 4-2*a*. If this current is present in a 3-H inductor, and if the voltage and current senses are assigned to satisfy the passive sign convention, then we may use (1) to obtain the voltage waveform. Since the current is zero and constant for $t < -1$, the voltage is zero in this interval. The current then begins to increase at the linear rate of 1 A/s, and thus a constant voltage of 3 V is produced. During the following 2-s interval, the current is constant and the voltage is therefore zero. The final decrease of the current causes a negative 3 V and no response thereafter. The voltage waveform is sketched in Fig. 4-2*b* on the same time scale.

Let us now investigate the effect of a more rapid rise and decay of the current between the zero and 1-A values. If the intervals required for the rise and fall are decreased to 0.1 s, then the derivative must be ten times as great in magnitude. This condition is shown in the current and voltage sketches of Fig. 4-3*a* and *b*. In the voltage waveforms of Figs.

Fig. 4-3 (*a*) The time required for the current of Fig. 4-2*a* to change from 0 to 1 and from 1 to 0 is decreased by a factor of 10. (*b*) The resultant voltage waveform. Note that the pulse widths are exaggerated slightly for clarity.

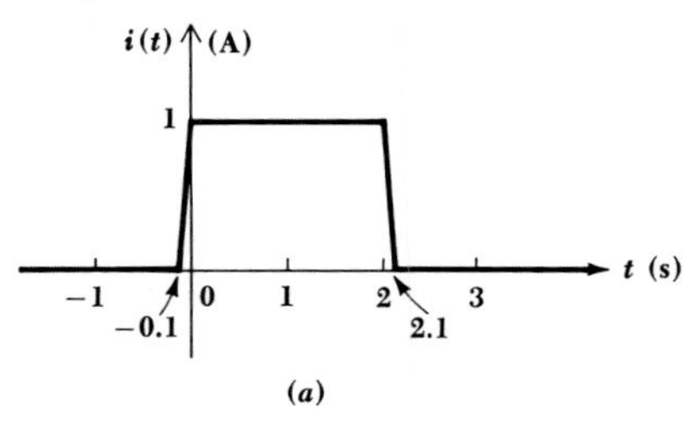

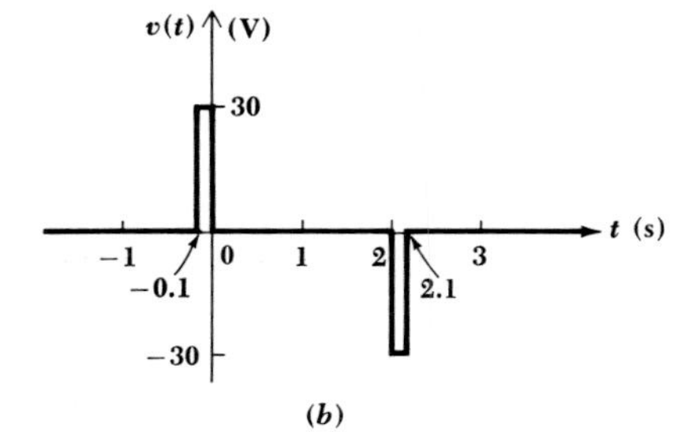

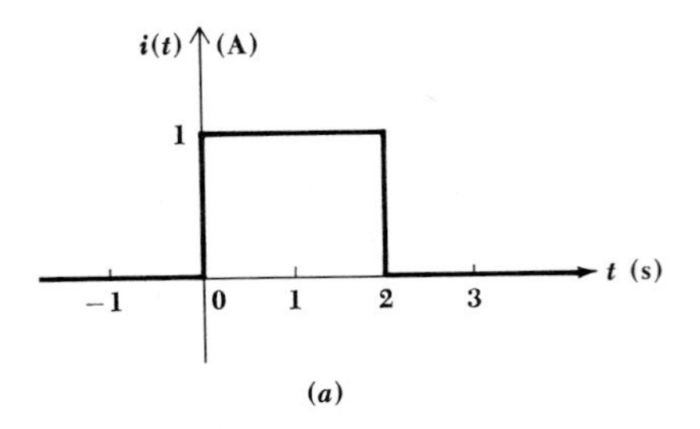

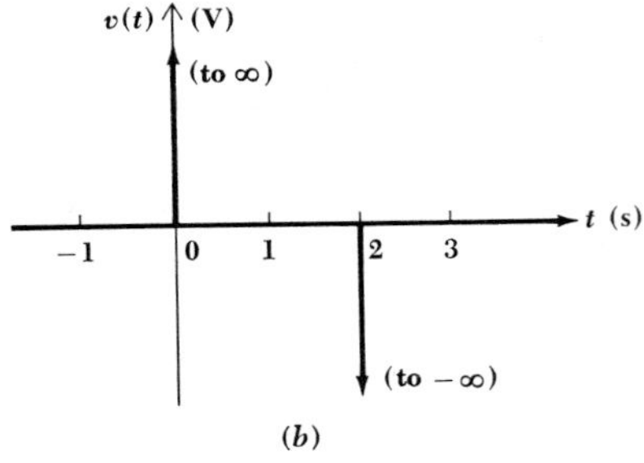

Fig. 4-4 (*a*) The time required for the current of Fig. 4-2*a* to change from 0 to 1 and from 1 to 0 is decreased to zero; the rise and fall are abrupt. (*b*) The associated voltage across the 3-H inductor consists of a positive and negative infinite spike.

4-2*b* and 4-3*b*, it is interesting to note that the area under each voltage pulse is 3 V-s.

A further decrease in the length of these two intervals will produce a proportionally larger voltage magnitude, but only within the interval in which the current is increasing or decreasing. An abrupt change in the current will cause the infinite voltage "spikes" (each having an area of 3 V-s) that are suggested by the waveforms of Fig. 4-4*a* and *b*; or, from the equally valid but opposite point of view, these infinite voltage spikes are required to produce the abrupt changes in the current. It will be convenient later to provide such infinite voltages (and currents), and we shall then call them "impulses"; for the present, however, we shall stay closer to physical reality by not permitting infinite voltage, current, or power. An abrupt change in the inductor current is therefore temporarily outlawed.

Drill Problems

4-1 For the circuit of Fig. 4-5*a*, find: (*a*) v_1; (*b*) v_2; (*c*) v_3.

Ans. -0.45; 0; 1.2 V

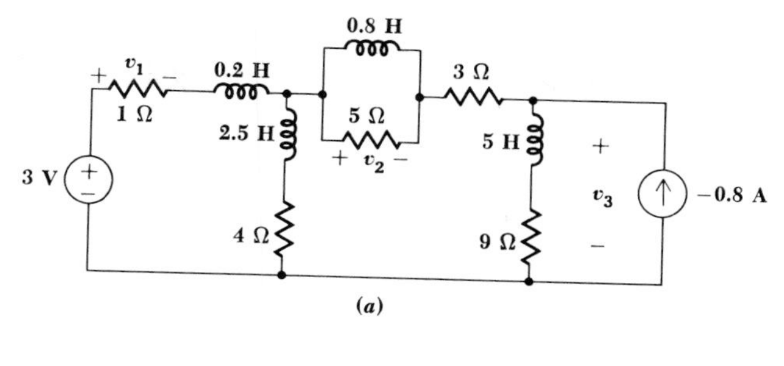

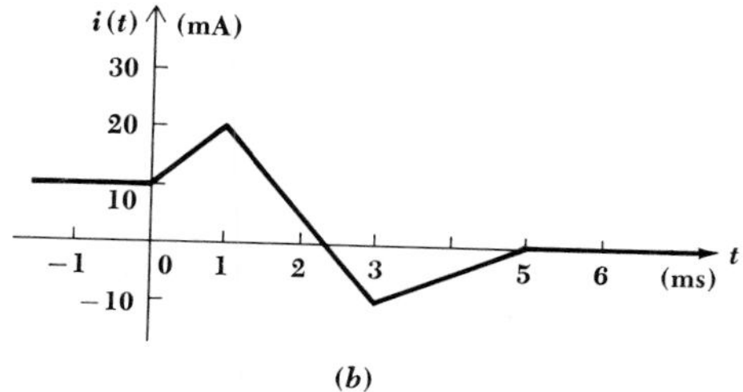

Fig. 4-5 See Drill Probs. 4-1 and 4-2.

4-2 The variation of current through a 20-mH inductor as a function of time is shown in Fig. 4-5*b*. Find the inductor voltage at $t =$: (*a*) 0.5; (*b*) 2.7; (*c*) 4.01 ms.

Ans. −*0.3; 0.1; 0.2* V

4-3 INTEGRAL RELATIONSHIPS FOR THE INDUCTOR

We have defined inductance by a simple differential equation

$$v = L\frac{di}{dt} \tag{2}$$

and we have been able to draw several conclusions about the characteristics of an inductor from this relationship. For example, we have found that we may consider an inductor as a short circuit to direct current, and we have agreed that we cannot permit an inductor current to change abruptly from one value to another because this would require that an infinite voltage and power be associated with the inductor. The defining equation for an inductance still contains more information, however. Rewritten in a slightly different form,

$$di = \frac{1}{L}v\,dt$$

it invites integration. Let us first consider the limits to be placed on the two integrals. We desire the current i at time t, and this pair of quantities therefore provides the upper limits on the integrals appearing on the left and right side of the equation, respectively; the lower limits may also be kept general by merely assuming that the current is $i(t_0)$ at time t_0. Thus

$$\int_{i(t_0)}^{i(t)} di = \frac{1}{L}\int_{t_0}^{t} v\,dt$$

or

$$i(t) - i(t_0) = \frac{1}{L}\int_{t_0}^{t} v\,dt$$

and

$$i(t) = \frac{1}{L}\int_{t_0}^{t} v\,dt + i(t_0) \qquad (3)$$

Equation (2) expresses the inductor voltage in terms of the current, whereas (3) gives the current in terms of the voltage. Other forms are also possible for this latter equation. We may write the integral as an indefinite integral and include a constant of integration k,

$$i(t) = \frac{1}{L}\int v\,dt + k \qquad (4)$$

We may assume that we are solving a realistic problem in which the selection of t_0 as $-\infty$ ensures no current or energy in the inductor. Thus, if $i(t_0) = i(-\infty) = 0$, then

$$i(t) = \frac{1}{L}\int_{-\infty}^{t} v\,dt \qquad (5)$$

Let us consider the use of these several integrals by working a simple example. Suppose that the voltage across a 2-H inductor is known to be $6 \cos 5t$ V; what information is then available about the inductor current? From Eq. (3),

$$i(t) = \frac{1}{2}\int_{t_0}^{t} 6 \cos 5t\,dt + i(t_0)$$

or $\quad i(t) = \frac{1}{2}\frac{6}{5} \sin 5t - \frac{1}{2}\frac{6}{5} \sin 5t_0 + i(t_0) = 0.6 \sin 5t - 0.6 \sin 5t_0 + i(t_0)$

The first term indicates that the inductor current varies sinusoidally; the second and third terms together merely represent a constant which becomes known when the current is numerically specified at some instant of time. Let us assume that the statement of our example problem also shows us that the current is 1 A at $t = -\pi/2$ s. We thus identify t_0 as $-\pi/2$, $i(t_0)$ as 1, and find that

$$i(t) = 0.6 \sin 5t - 0.6 \sin(-2.5\pi) + 1$$

or

$$i(t) = 0.6 \sin 5t + 1.6$$

We may obtain the same result from (4). We have

$$i(t) = 0.6 \sin 5t + k$$

and we establish the numerical value of k by forcing the current to be 1 A at $t = -\pi/2$,

$$1 = 0.6 \sin(-2.5\pi) + k$$

or

$$k = 1 + 0.6 = 1.6$$

and

$$i(t) = 0.6 \sin 5t + 1.6$$

once more.

Equation (5) is going to cause trouble; it is based on the assumption that the current is zero when $t = -\infty$. To be sure, this must be true in the real physical world, but we are working in the land of the mathematical model; our elements and forcing functions are all idealized. The difficulty arises after we integrate,

$$i(t) = 0.6 \sin 5t \Big|_{-\infty}^{t}$$

and attempt to evaluate the integral at the lower limit,

$$i(t) = 0.6 \sin 5t - 0.6 \sin(-\infty)$$

The sine of $\pm\infty$ is indeterminate; we might just as well represent it by an unknown constant,

$$i(t) = 0.6 \sin 5t + k$$

and we see that this result is identical with that which we obtained when we assumed an arbitrary constant of integration in (4).

We should not make any snap judgments, based on this example, as to which single form we are going to use forever; each has its advantages, depending on the problem and the application. Equation (3) represents a long, general method, but it shows clearly that the constant of integration is a current. Equation (4) is a somewhat more concise expression of (3), but the nature of the integration constant is suppressed. Finally, (5) is an excellent expression since no constant is necessary; however, it applies only when the current is zero at $t = -\infty$ and when the analytical expression for the current is not indeterminate there.

Let us now turn our attention to power and energy. The absorbed power is given by the current-voltage product

$$p = vi = Li\frac{di}{dt} \qquad \text{W}$$

The energy w_L accepted by the inductance is stored in the magnetic field around the coil and is expressed by the integral of the power over the desired time interval,

$$\int_{t_0}^{t} p\,dt = L\int_{t_0}^{t} i\frac{di}{dt}\,dt = L\int_{i(t_0)}^{i(t)} i\,di = \tfrac{1}{2}L\{[i(t)]^2 - [i(t_0)]^2\}$$

and thus

$$w_L(t) - w_L(t_0) = \tfrac{1}{2}L\{[i(t)]^2 - [i(t_0)]^2\} \qquad \text{J} \tag{6}$$

where we have again assumed that the current is $i(t_0)$ at time t_0. In using the energy expression, it is customary to assume that a value of t_0 is selected at which the current is zero; it is also customary to assume that the energy is zero at this time. We then have simply

$$w_L(t) = \tfrac{1}{2}Li^2 \tag{7}$$

where we now understand that our reference for zero energy is any time at which the inductor current is zero. At any subsequent time at which the current is zero, we also find no energy stored in the coil. Whenever the current is not zero, and regardless of its direction or sign, energy is stored in the inductance. It follows, therefore, that power must be delivered to the inductor for a part of the time and recovered from the inductor later. All the stored energy may be recovered from an ideal inductor; there are no storage charges or agent's commissions in the mathematical model. A physical coil, however, must be constructed out of real wire and thus will always have a resistance associated with it. Energy can no longer be stored and recovered without loss.

These ideas may be illustrated by a simple example. In Fig. 4-6 an inductance of 3 H is shown in series with a resistance of 0.1 Ω and a sinusoidal current source. The resistor may be interpreted, if we wish, as the resistance of the wire which must be associated with the physical coil.

The voltage across the resistor is given by

$$v_R = Ri = 1.2\sin\frac{\pi}{6}t$$

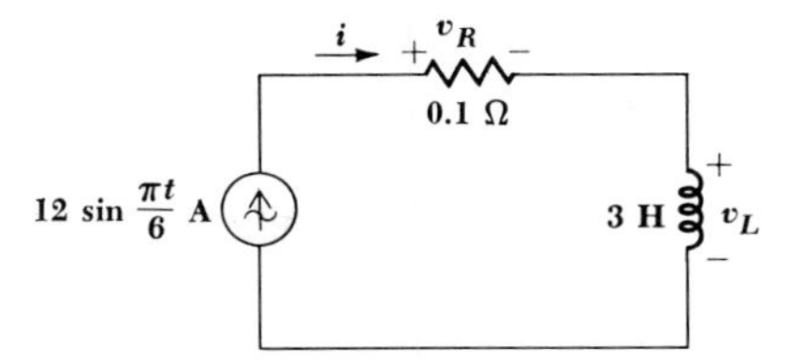

Fig. 4-6 A sinusoidal current is applied as a forcing function to a series *RL* circuit.

and the voltage across the inductance is found by applying the defining equation for an inductance,

$$v_L = L\frac{di}{dt} = 3\frac{d}{dt}\left(12 \sin \frac{\pi}{6}t\right) = 6\pi \cos \frac{\pi}{6}t$$

The energy stored in the inductor is

$$w_L = \tfrac{1}{2}Li^2 = 216 \sin^2 \frac{\pi}{6}t$$

and it is apparent that this energy increases from zero at $t = 0$ to 216 J at $t = 3$ s. During the next 3 s, the energy leaves the inductor completely. Let us see what price we have paid in this coil for the privilege of storing and removing 216 J in these few seconds. The power dissipated in the resistor is easily found as

$$p_R = i^2R = 14.4 \sin^2 \frac{\pi}{6}t$$

and the energy converted into heat in the resistor within this 6-s interval is therefore

$$w_R = \int_0^6 p_R\, dt = \int_0^6 14.4 \sin^2 \frac{\pi}{6}t\, dt$$

or

$$w_R = \int_0^6 14.4(\tfrac{1}{2})\left(1 - \cos \frac{\pi}{3}t\right) dt = 43.2 \quad \text{J}$$

This represents 20 per cent of the maximum stored energy and is a reasonable value for many coils having this large an inductance. For coils having an inductance of about 100 μH, we should expect a figure closer to 3 per cent. In Chap. 14 we shall formalize this concept by defining a quality factor *Q* that is proportional to the ratio of the maximum energy stored to the energy lost per period.

Let us now recapitulate by listing several characteristics of an inductor which result from its defining equation:

1. There is no voltage across an inductor if the current through it is not changing with time. An inductance is therefore a short circuit to dc.
2. A finite amount of energy can be stored in an inductor even if the voltage across the inductance is zero, such as when the current through it is constant.
3. It is impossible to change the current through an inductor by a finite amount in zero time, for this requires an infinite voltage across the inductor. It will be advantageous later to *hypothesize* that such a voltage may be generated and applied to an inductor, but for the present we shall avoid such a forcing function or response. An inductor resists an abrupt change in the current through it in a manner analogous to the way a mass resists an abrupt change in its velocity.
4. The inductor never dissipates energy, but only stores it. Although this is true for the mathematical model, it is not true for a physical inductor.

Drill Problems

4-3 Find the magnitude of the voltage across an inductor: (*a*) of 30 mH if the current through it is increasing at the rate of 20 mA/ms; (*b*) of 0.4 mH at $t = 0$ if the current in it is $50e^{-10^4 t}$ mA; (*c*) at $t = 0$ if the power entering it is given by $12 \cos 100\pi t$ mW and the inductor current is 150 mA at $t = 0$.

Ans. *0.2; 0.08; 0.6* V

4-4 Assuming the passive sign convention, find the current through a 0.1-H inductor at $t = 2$ s if: (*a*) $v_L = 0.5t$ V for $0 \leq t \leq 3$ s, $v_L = 0$ for $t < 0$ and $t > 3$, and $i_L = 0$ at $t = -3$; (*b*) $v_L = 0.5(t + 2)$ V for $-2 \leq t \leq 1$ s, $v_L = 0$ for $t < -2$ and $t > 1$, and $i_L = 0$ at $t = -3$; (*c*) $v_L = 2/(t^2 + 4)$ V for all t, $i_L(-\infty) = 0$.

Ans. *10.0; 22.5; 23.6* A

4-4 THE CAPACITOR

Our next passive circuit element is the capacitor. We shall define *capacitance C* by the voltage-current relationship

$$i = C\frac{dv}{dt} \tag{8}$$

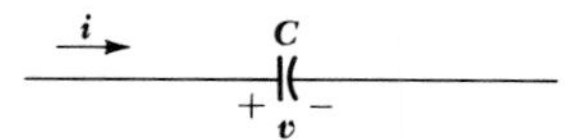

Fig. 4-7 The current and voltage reference marks are shown on the circuit symbol for a capacitor so that $i = C\,dv/dt$.

where v and i satisfy the conventions for a passive element, as shown in Fig. 4-7. From (8), we may determine the unit of capacitance as an ampere-second per volt, or coulomb per volt, but we shall now define the *farad* (F) as one coulomb per volt.

The capacitor whose capacitance is defined by (8) is again a mathematical model of a real device. The construction of the physical device is suggested by the circuit symbol shown in Fig. 4-7, in much the same way as the helical symbol used for the inductor represents the coiled wire in that physical element. A capacitor, physically, consists of two conducting surfaces on which charge may be stored, separated by a thin insulating layer which has a very large resistance. If we assume that this resistance is sufficiently large that it may be considered infinite, then equal and opposite charges placed on the capacitor "plates" can never recombine, at least by any path *within* the element. Let us visualize some external device, such as a current source, connected to this capacitor and causing a positive current to flow into one plate of the capacitor and out of the other plate. Equal currents are entering and leaving the two terminals of the element, and this is no more than we expect for any circuit element. Now let us examine the interior of the capacitor. The positive current entering one plate represents positive charge moving toward that plate through its terminal lead; this charge cannot pass through the interior of the capacitor, and it therefore accumulates on the plate. As a matter of fact, the current and the increasing charge are related by the familiar equation

$$i = \frac{dq}{dt}$$

Now let us pose ourselves a troublesome problem by considering this plate as an overgrown node and applying Kirchhoff's current law. It apparently does not hold; current is approaching the plate from the external circuit, but it is not flowing out of the plate into the "internal circuit." This dilemma bothered a famous Scottish scientist about a century ago, and the unified electromagnetic theory which James Clerk Maxwell then developed hypothesizes a "displacement current" which is present wherever an electric field or voltage is varying with time. The displacement current flowing internally between the capacitor plates is exactly equal to the conduction current flowing in the capacitor leads; Kirchhoff's current law is therefore satisfied if we include both conduction and displacement currents. However, circuit analysis is not concerned with this internal displacement current, and since it is fortunately equal to the conduction current, we may consider Maxwell's hypothesis as relating the conduction current to the changing voltage across the capacitor. The relationship is linear, and the constant of proportionality is obviously the capacitance C,

$$i_{\text{disp}} = i = C\frac{dv}{dt}$$

A capacitor constructed of two parallel conducting plates of area A, separated a distance d, has a capacitance $C = \epsilon A/d$, where ϵ is the permittivity, a constant of the insulating material between the plates, and where the linear dimensions of the conducting plates are all very much greater than d. For air or vacuum, $\epsilon = \epsilon_0 = 8.854$ pF/m $\doteq (1/36\pi)$ nF/m.

The concepts of the electric field, displacement current, and the generalized form of Kirchhoff's current law are more appropriate subjects for courses in physics and electromagnetic field theory, as is the determination of a suitable mathematical model to represent a specific physical capacitor.

Several important characteristics of our new mathematical model can be discovered from the defining equation (8). A constant voltage across a capacitor requires zero current passing through it; a capacitor is thus an "open circuit to dc." This fact is certainly represented by the capacitor symbol. It is also apparent that a sudden jump in the voltage requires an infinite current. Just as we outlawed abrupt changes in inductor currents and the associated infinite voltages on physical grounds, we shall not permit abrupt changes in capacitor voltage; the infinite current (and infinite power) which results is nonphysical. We shall remove this restriction at the time we assume the existence of the current impulse.

The capacitor voltage may be expressed in terms of the current by integrating (8). We first obtain

$$dv = \frac{1}{C} i \, dt$$

and then integrate between the times t_0 and t and between the corresponding voltages $v(t_0)$ and $v(t)$,

$$v(t) = \frac{1}{C}\int_{t_0}^{t} i \, dt + v(t_0) \tag{9}$$

Equation (9) may also be written as an indefinite integral plus a constant of integration,

$$v(t) = \frac{1}{C}\int i \, dt + k \tag{10}$$

Finally, in many real problems, t_0 may be selected as $-\infty$ and $v(-\infty)$ as zero,

$$v(t) = \frac{1}{C}\int_{-\infty}^{t} i\,dt \tag{11}$$

Since the integral of the current over any time interval is the charge accumulated in that period on the capacitor plate into which the current is flowing, it is apparent that capacitance might have been defined as

$$q = Cv$$

The similarity between the several integral equations introduced in this section and those appearing in our discussion of inductance is striking and suggests that the duality we observed between mesh and nodal equations in resistive networks may be extended to include inductance and capacitance as well. The principle of duality will be presented and discussed later in this chapter.

As an illustration of the use of the several integral equations displayed above, let us find the capacitor voltage which is associated with the current shown graphically in Fig. 4-8*a*. We shall assume that the single 20 mA rectangular pulse of 2 ms duration is applied to a 5-μF capacitor. Interpreting (9) graphically, we know that the difference between the values of the voltage at t and t_0 is proportional to the area under the current curve between these same two values of time. The proportionality constant is $1/C$. The area can be obtained from Fig. 4-8*a* by inspection for desired

Fig. 4-8 (*a*) The current waveform applied to a 5-μF capacitor. (*b*) The resultant voltage waveform, easily obtained by integrating graphically.

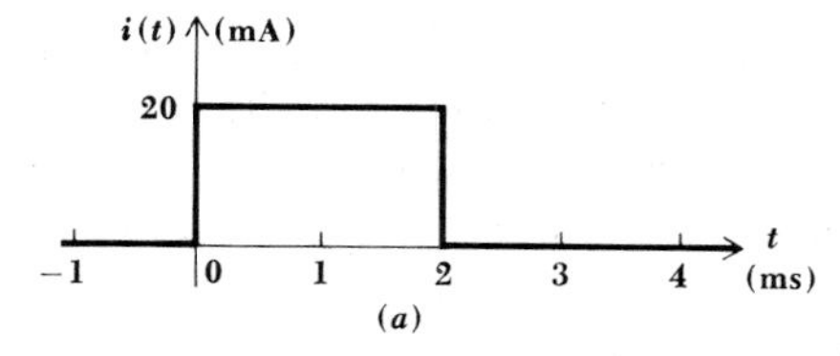

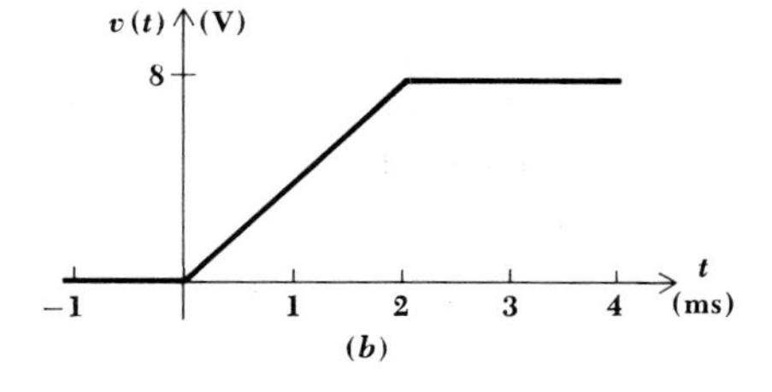

values of t_0 and t. Thus, if $t_0 = -0.5$ and $t = 0.5$ (in ms),

$$v(0.5) = 2 + v(-0.5)$$

or, if $t_0 = 0$ and $t = 3$,

$$v(3) = 8 + v(0)$$

We may express our results in more general terms by dividing the interesting range of time into several intervals. Let us select our starting point t_0 prior to zero time. Then the first interval of t is selected between t_0 and zero,

$$v(t) = 0 + v(t_0) \qquad t_0 \leq t \leq 0$$

and since our waveform implies that no current has ever been applied to this capacitor since the Creation,

$$v(t_0) = 0$$

and, thus,

$$v(t) = 0 \qquad t \leq 0$$

If we now consider the time interval represented by the rectangular pulse, we obtain

$$v(t) = 4000t \qquad 0 \leq t \leq 2 \quad \text{ms}$$

For the semi-infinite interval following the pulse, we have

$$v(t) = 8 \qquad t \geq 2 \quad \text{ms}$$

The results for these three intervals therefore provide us with analytical expressions for the capacitor voltage at any time after $t = t_0$; the time t_0, however, may be selected as early as we wish. The results are expressed much more simply in a sketch than by these analytical expressions, as shown in Fig. 4-8*b*.

The power delivered to a capacitor is

$$p = vi = Cv\frac{dv}{dt}$$

and the energy stored in its electric field is therefore

$$\int_{t_0}^{t} p\,dt = C\int_{t_0}^{t} v\frac{dv}{dt}dt = C\int_{v(t_0)}^{v(t)} v\,dv = \tfrac{1}{2}C\{[v(t)]^2 - [v(t_0)]^2\}$$

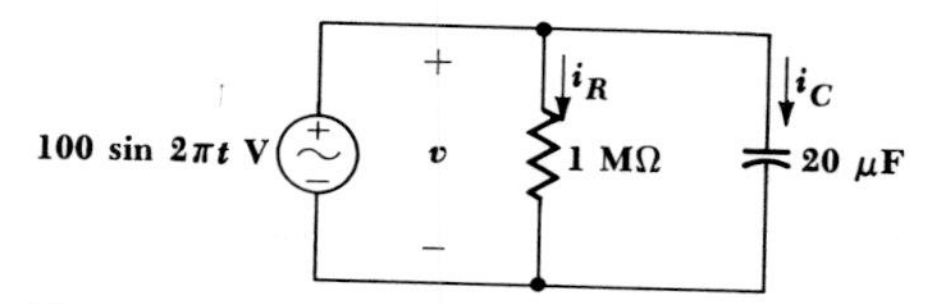

Fig. 4-9 A sinusoidal voltage source is applied to a parallel RC network.

and thus

$$w_C(t) - w_C(t_0) = \tfrac{1}{2}C\{[v(t)]^2 - [v(t_0)]^2\} \tag{12}$$

where the stored energy is $w_C(t_0)$ and the voltage is $v(t_0)$ at t_0. If we select a zero-energy reference at t_0, implying that the capacitor voltage is also zero at that instant, then

$$w_C(t) = \tfrac{1}{2}Cv^2 \tag{13}$$

Let us consider a simple numerical example. As sketched in Fig. 4-9, we shall assume a sinusoidal voltage source in parallel with a 1-MΩ resistor and a 20-μF capacitor. The parallel resistor may be assumed to represent the resistance of the insulator or dielectric between the capacitor plates. The current through the resistor is

$$i_R = \frac{v}{R} = 10^{-4} \sin 2\pi t$$

and the current through the capacitor is

$$i_C = C\frac{dv}{dt} = 20 \times 10^{-6}\frac{d}{dt}(100 \sin 2\pi t) = 4\pi \times 10^{-3} \cos 2\pi t$$

We next obtain the energy stored in the capacitor,

$$w_C = \tfrac{1}{2}Cv^2 = 0.1 \sin^2 2\pi t$$

and see that the energy increases from zero at $t = 0$ to a maximum of 0.1 J at $t = \tfrac{1}{4}$ s and then decreases to zero in another $\tfrac{1}{4}$ s. During this $\tfrac{1}{2}$ s interval, the energy dissipated in the resistor is

$$w_R = \int_0^{0.5} p_R\, dt = \int_0^{0.5} 10^{-2} \sin^2 2\pi t\, dt = 2.5 \quad \text{mJ}$$

Thus, an energy equal to 2.5 per cent of the maximum stored energy is lost in the process of storing and removing the energy in the ideal capacitor.

Much smaller values are possible in "low-loss" capacitors, but these smaller percentages are customarily associated with much smaller capacitors.

Some of the important characteristics of a capacitor are now apparent:

1. The current through a capacitor is zero if the voltage across it is not changing with time. A capacitor is therefore an open circuit to dc.
2. A finite amount of energy can be stored in a capacitor even if the current through the capacitor is zero, such as when the voltage across it is constant.
3. It is impossible to change the voltage across a capacitor by a finite amount in zero time, for this requires an infinite current through the capacitor. It will be advantageous later to *hypothesize* that such a current may be generated and applied to a capacitor, but for the present we shall avoid such a forcing function or response. A capacitor resists an abrupt change in the voltage across it in a manner analogous to the way a spring resists an abrupt change in its displacement.
4. The capacitor never dissipates energy, but only stores it. Although this is true for the mathematical model, it is not true for a physical capacitor.

It is interesting to anticipate our discussion of duality by rereading the previous four statements with certain words replaced by their "duals." If capacitor and inductor, capacitance and inductance, voltage and current, across and through, open circuit and short circuit, spring and mass, and displacement and velocity are interchanged (in either direction), the four statements previously given for inductors are obtained.

Drill Problems

4-5 Find the current through a 0.01-μF capacitor at $t = 0$ if the voltage across it is: (*a*) $2 \sin 2\pi\, 10^6 t$ V; (*b*) $-2e^{-10^7 t}$ V; (*c*) $-2e^{-10^7 t} \sin 2\pi\, 10^6 t$ V.

Ans. -0.1256; *0.1256*; *0.2* A

4-6 Find the voltage across a 0.01-μF capacitor at $t = 0.25\ \mu$s if $v(0) = 0.6$ V and the current through it is: (*a*) $0.2 \sin 2\pi\, 10^6 t$ A; (*b*) $0.2e^{-10^7 t}$ A; (*c*) 0.2 A, $0 \leq t \leq 2\ \mu$s.

Ans. 2.44; 3.78; 5.60 V

4-7 A 1000-pF capacitor, a 1-mH inductor, and a current source of $2 \cos 10^6 t$ mA are in series. The capacitor voltage is zero at $t = 0$. Find the energy stored:

(*a*) in the inductor at $t = 0$; (*b*) in the capacitor at $t = 1.571\ \mu s$; (*c*) in both inductor and capacitor together at $t = 1\ \mu s$.

Ans. 2; 2; 2 nJ

4-5 INDUCTANCE AND CAPACITANCE COMBINATIONS

Now that we have added the inductor and capacitor to our list of passive circuit elements, we need to decide whether or not the methods we have developed in studying resistive circuit analysis are still valid. It will also be convenient to learn how to replace series and parallel combinations of either of these elements with simpler equivalents, just as we did with resistors in Chap. 2.

We look first at Kirchhoff's two laws, both of which are axiomatic. However, when we hypothesized these two laws, we did so with no restrictions as to the types of elements constituting the network. Both, therefore, continue to remain valid.

Now we may extend the procedures we have derived for reducing various combinations of resistors into one equivalent resistor to the analogous cases of inductors and capacitors. We shall first consider an ideal voltage source applied to the series combination of N inductors, as shown in Fig. 4-10*a*. We desire a single equivalent inductor L_{eq} which may replace the series combination such that the source current $i(t)$ is unchanged. The equivalent circuit is sketched in Fig. 4-10*b*. For the original circuit

$$\begin{aligned} v_s &= v_1 + v_2 + \cdots + v_N \\ &= L_1 \frac{di}{dt} + L_2 \frac{di}{dt} + \cdots + L_N \frac{di}{dt} \\ &= (L_1 + L_2 + \cdots + L_N) \frac{di}{dt} \end{aligned}$$

or, written more concisely,

$$v_s = \sum_{n=1}^{N} v_n = \sum_{n=1}^{N} L_n \frac{di}{dt} = \frac{di}{dt} \sum_{n=1}^{N} L_n$$

Fig. 4-10 (*a*) A circuit containing N inductors in series. (*b*) The desired equivalent circuit, in which $L_{eq} = L_1 + L_2 + \cdots + L_N$.

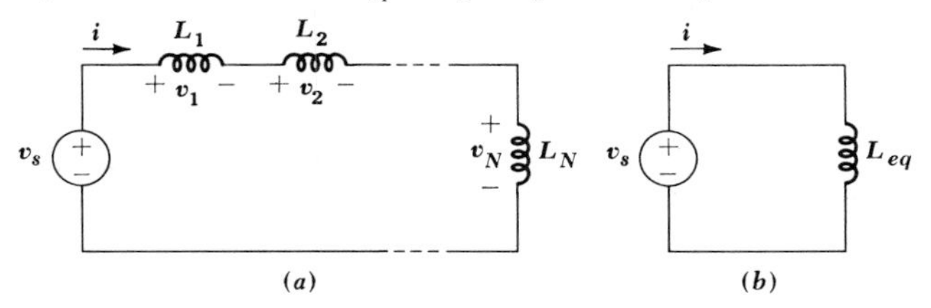

But for the equivalent circuit we have

$$v_s = L_{eq} \frac{di}{dt}$$

and thus the equivalent inductance is

$$L_{eq} = L_1 + L_2 + \cdots + L_N \qquad \text{or} \qquad L_{eq} = \sum_{n=1}^{N} L_n$$

The inductance which is equivalent to several inductances connected in series is simply the sum of the series inductances. This is exactly the same result we obtained for resistors in series.

The combination of a number of parallel inductors is accomplished by writing the single nodal equation for the original circuit, shown in Fig. 4-11*a*,

$$i_s = \sum_{n=1}^{N} i_n = \sum_{n=1}^{N} \left[\frac{1}{L_n} \int_{t_0}^{t} v\, dt + i_n(t_0) \right]$$

$$= \left[\sum_{n=1}^{N} \frac{1}{L_n} \right] \int_{t_0}^{t} v\, dt + \sum_{n=1}^{N} i_n(t_0)$$

and comparing it with the result for the equivalent circuit of Fig. 4-11*b*,

$$i_s = \frac{1}{L_{eq}} \int_{t_0}^{t} v\, dt + i_s(t_0)$$

Since Kirchhoff's current law demands that $i_s(t_0)$ be equal to the sum of the branch currents at t_0, the two integral terms must also be equal; hence,

$$L_{eq} = \frac{1}{1/L_1 + 1/L_2 + \cdots + 1/L_N}$$

For the special case of two inductors in parallel,

$$L_{eq} = \frac{L_1 L_2}{L_1 + L_2}$$

Fig. 4-11 (*a*) The parallel combination of N inductors. (*b*) The equivalent circuit, where $L_{eq} = 1/(1/L_1 + 1/L_2 + \cdots + 1/L_N)$.

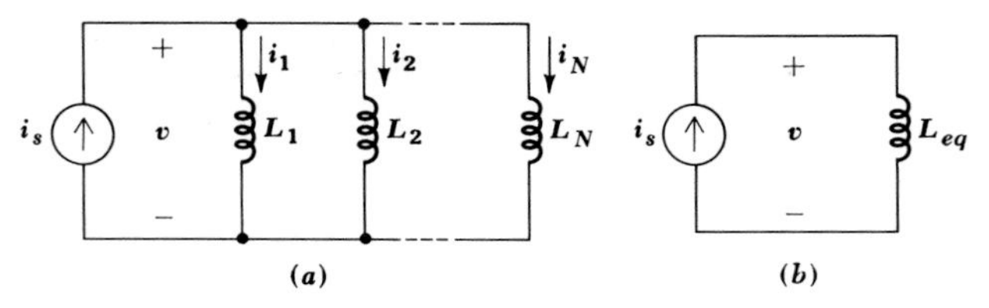

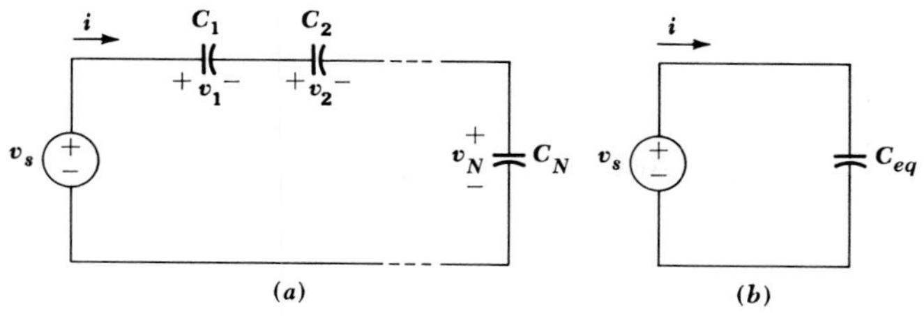

Fig. 4-12 (a) A circuit containing N capacitors in series. (b) The desired equivalent, $C_{eq} = 1/(1/C_1 + 1/C_2 + \cdots + 1/C_N)$.

and we note that inductors in parallel combine exactly as do resistors in parallel.

In order to find a capacitance which is equivalent to N capacitors in series, we use the circuit of Fig. 4-12a and its equivalent Fig. 4-12b to write

$$v_s = \sum_{n=1}^{N} v_n = \sum_{n=1}^{N} \left[\frac{1}{C_n}\int_{t_0}^{t} i\,dt + v_n(t_0)\right]$$

$$= \left[\sum_{n=1}^{N} \frac{1}{C_n}\right]\int_{t_0}^{t} i\,dt + \sum_{n=1}^{N} v_n(t_0)$$

and

$$v_s = \frac{1}{C_{eq}}\int_{t_0}^{t} i\,dt + v_s(t_0)$$

However, Kirchhoff's voltage law establishes the equality of $v_s(t_0)$ and the sum of the capacitor voltages at t_0; thus

$$C_{eq} = \frac{1}{1/C_1 + 1/C_2 + \cdots + 1/C_N}$$

and capacitors in series combine as do resistors in *parallel.*

Finally, the circuits of Fig. 4-13 enable us to establish the value of the capacitance which is equivalent to N parallel capacitors as

$$C_{eq} = C_1 + C_2 + \cdots + C_N$$

Fig. 4-13 (a) The parallel combination of N capacitors. (b) The equivalent circuit, where $C_{eq} = C_1 + C_2 + \cdots + C_N$.

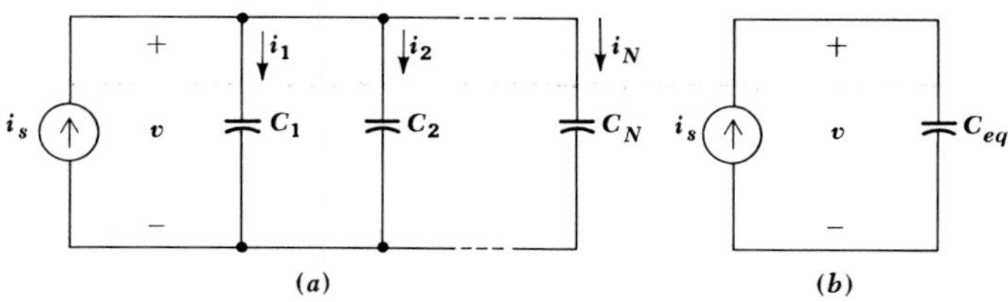

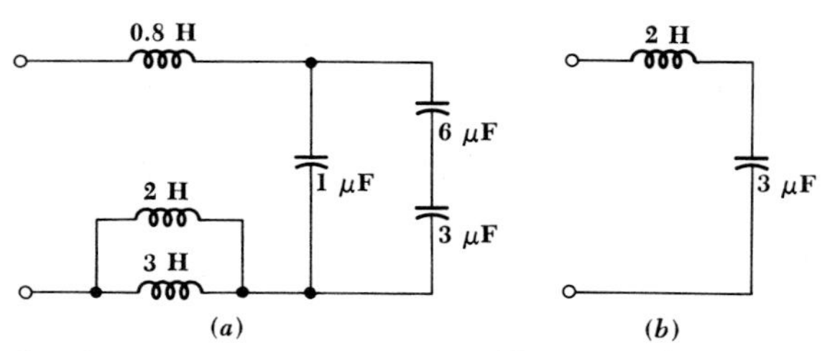

Fig. 4-14 (*a*) A given *LC* network. (*b*) A simpler equivalent circuit.

and it is no great source of amazement to note that capacitors in parallel combine in the same manner in which we combine resistors in series, that is, by simply adding all the individual capacitances.

As an example in which some simplification may be achieved by combining like elements, consider the network of Fig. 4-14*a*. The 6- and 3-μF capacitors are first combined into a 2-μF equivalent, and this capacitor is then combined with the 1-μF element with which it is in parallel to yield an equivalent capacitance of 3 μF. In addition, the 3- and 2-H inductors are replaced by an equivalent 1.2-H inductor which is then added to the 0.8-H element to give a total equivalent inductance of 2 H. The much simpler (and probably less expensive) equivalent network is shown in Fig. 4-14*b*.

The network shown in Fig. 4-15 contains three inductors and three capacitors, but no series or parallel combinations of either the inductors or the capacitors can be achieved. Simplification of this network cannot be accomplished at this time.

Next let us turn to mesh, loop, and nodal analysis. Since we already know that we may safely apply Kirchhoff's laws, we should have little difficulty in writing a set of equations that are both sufficient and independent. They will be constant-coefficient linear integrodifferential equations, however, which are hard enough to pronounce, let alone solve. Consequently, we shall write them now to gain familiarity with the use of Kirchhoff's laws in *RLC* circuits and discuss the solution of the simpler cases in the following chapters.

Fig. 4-15 An *LC* network in which no series or parallel combinations of either the inductors or the capacitors are possible.

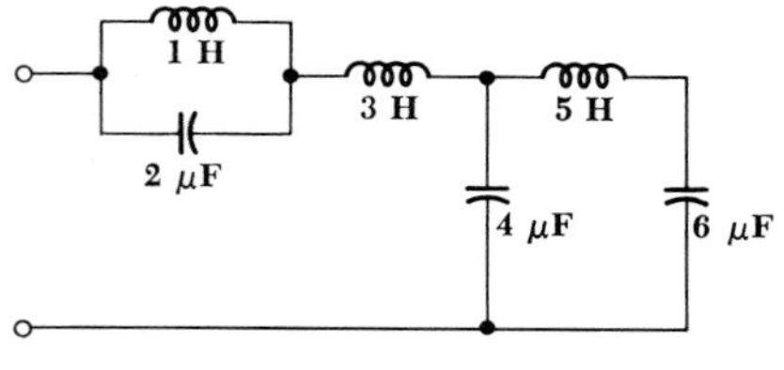

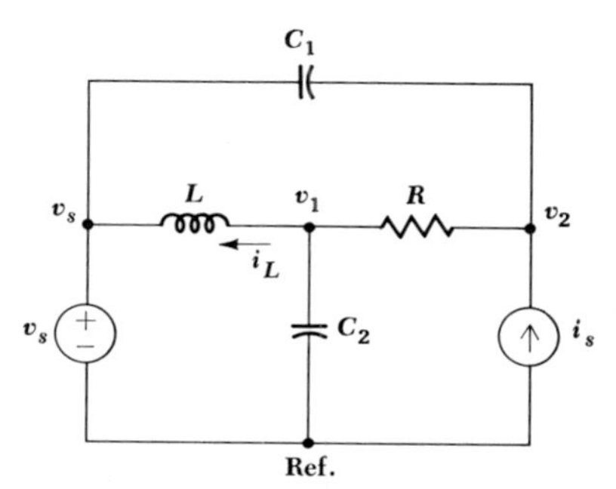

Fig. 4-16 A four-node RLC circuit with node voltages assigned.

Let us attempt to write nodal equations for the circuit of Fig. 4-16. Node voltages are chosen as indicated, and we sum currents leaving the central node,

$$\frac{1}{L}\int_{t_0}^{t}(v_1 - v_s)\,dt + i_L(t_0) + \frac{v_1 - v_2}{R} + C_2\frac{dv_1}{dt} = 0$$

where $i_L(t_0)$ is the value of the inductor current at the time the integration begins, or the initial value. At the right node,

$$C_1\frac{d(v_2 - v_s)}{dt} + \frac{v_2 - v_1}{R} - i_s = 0$$

Rewriting these two equations, we have

$$\frac{v_1}{R} + C_2\frac{dv_1}{dt} + \frac{1}{L}\int_{t_0}^{t} v_1\,dt - \frac{v_2}{R} = \frac{1}{L}\int_{t_0}^{t} v_s\,dt - i_L(t_0)$$

$$-\frac{v_1}{R} + \frac{v_2}{R} + C_1\frac{dv_2}{dt} = C_1\frac{dv_s}{dt} + i_s$$

These are the promised integrodifferential equations, and we may note several interesting points about them. First, the source voltage v_s happens to enter the equations as an integral and as a derivative, but not simply as v_s. Since both sources are specified for all time, we should be able to evaluate the derivative or integral. Secondly, the initial value of the inductor current, $i_L(t_0)$, acts as a (constant) source current at the center node.

We shall not attempt the solution of these equations at this time. It is worthwhile pointing out, however, that when the two voltage forcing functions are sinusoidal functions of time it will be possible to define a voltage-current ratio (called impedance) or a current-voltage ratio (called admittance) for each of the three passive elements. The factors operating on the two node voltages in the equations above will then become simple multiplying factors, and the equations will be linear *algebraic* equations once again. These we may solve by determinants or a simple elimination of variables as before.

Drill Problem

4-8 In the circuit shown in Fig. 4-17, $v_1 = 3e^{-2\times10^4 t}$ V. Find: (*a*) v_2; (*b*) v_3; (*c*) v_s.

Ans. $-2.4e^{-2\times10^4 t}$; $3.6e^{-2\times10^4 t}$; $6e^{-2\times10^4 t}$ V

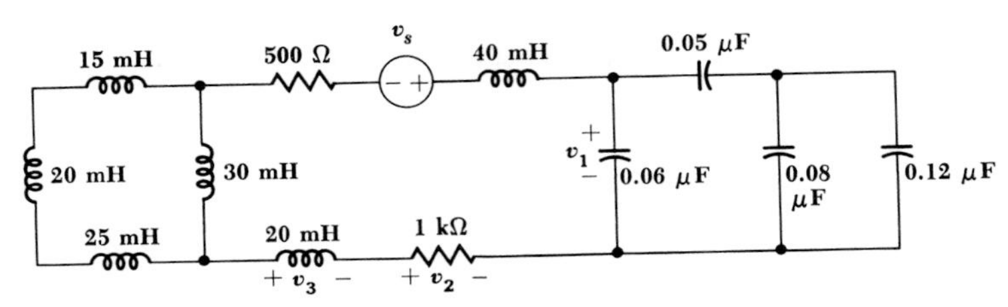

Fig. 4-17 See Drill Prob. 4-8.

4-6 DUALITY

Duality has been mentioned earlier in connection with resistive circuits and more recently in the discussion of inductance and capacitance; the comments made were introductory and a little offhand. Now we may make an exact definition and then use the definition to recognize or construct dual circuits and thus avoid the labor of analyzing both a circuit and its dual.

We shall define duality in terms of the circuit equations. Two circuits are *duals* if the mesh equations that characterize one of them have the same mathematical form as the nodal equations that characterize the other. They are said to be *exact duals* if each mesh equation of the one circuit is numerically identical with the corresponding nodal equation of the other; the current and voltage variables themselves cannot be identical, of course. *Duality* itself merely refers to any of the properties exhibited by dual circuits.

Let us interpret the definition and use it to construct an exact dual circuit by writing the two mesh equations for the circuit shown in Fig. 4-18. Two mesh currents i_1 and i_2 are assigned, and the mesh equations are

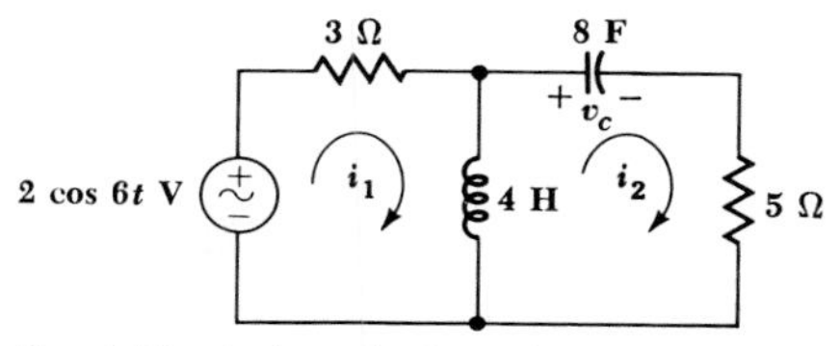

Fig. 4-18 A given circuit to which the definition of duality may be applied to determine the dual circuit.

$$3i_1 + 4\frac{di_1}{dt} - 4\frac{di_2}{dt} = 2\cos 6t \tag{14}$$

$$-4\frac{di_1}{dt} + 4\frac{di_2}{dt} + \frac{1}{8}\int_0^t i_2\,dt + 5i_2 = -10 \tag{15}$$

It should be noted that the capacitor voltage v_C is assumed to be 10 V at $t = 0$.

We may now construct the two equations which are the mathematical exact duals of Eqs. (14) and (15). We wish them to be nodal equations, and we thus begin by replacing the mesh currents i_1 and i_2 by two node-to-reference voltages v_1 and v_2. We obtain

$$3v_1 + 4\frac{dv_1}{dt} - 4\frac{dv_2}{dt} = 2\cos 6t \tag{16}$$

$$-4\frac{dv_1}{dt} + 4\frac{dv_2}{dt} + \frac{1}{8}\int_0^t v_2\,dt + 5v_2 = -10 \tag{17}$$

and we now seek the circuit represented by these two nodal equations.

Let us first draw a line to represent the reference node, and then we may establish two nodes at which the positive references for v_1 and v_2 are located. Equation (16) indicates that a current source $2\cos 6t$ is connected between node 1 and the reference node, oriented to provide a current entering node 1. This equation also shows that a 3-℧ conductance appears between node 1 and the reference node. Turning to (17), we first consider the nonmutual terms, or those terms which do not appear in (16), and they instruct us to connect an 8-H inductor and a 5-℧ conductance (in parallel) between node 2 and the reference. The two similar terms in (16) and (17) represent a 4-F capacitor present mutually at nodes 1 and 2; the circuit is completed by connecting this capacitor between the two nodes. The constant term on the right side of (17) is the value of the inductor current at $t = 0$; thus, $i_L(0) = 10$ A. The dual circuit is shown in Fig. 4-19; since the two sets of equations are numerically identical, the circuits are exact duals.

Dual circuits may be obtained more readily than by the above method, for the equations need not be written. In order to construct the dual of

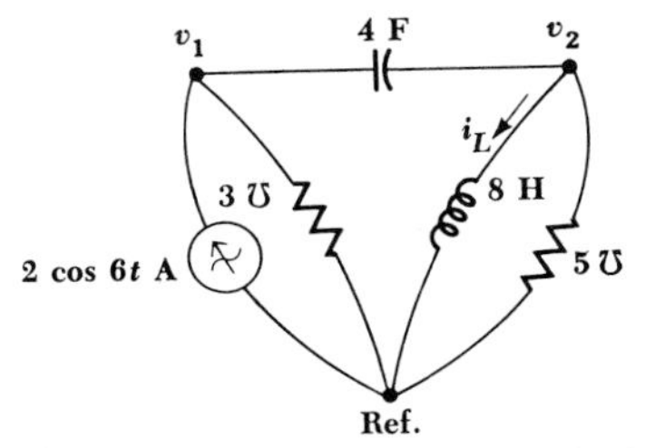

Fig. 4-19 The exact dual of the circuit of Fig. 4-18.

a given circuit, we think of the circuit in terms of its mesh equations. With each mesh we must associate a nonreference node, and, in addition, we must supply the reference node. On a diagram of the given circuit we therefore place a node in the center of each mesh and supply the reference node as a line near the diagram or a loop enclosing the diagram. Each element which appears jointly in two meshes is a *mutual* element and gives rise to identical terms, except for sign, in the two corresponding mesh equations. It must be replaced by an element which supplies the dual term in the two corresponding nodal equations. This dual element must therefore be connected directly between the two nonreference nodes which are within the meshes in which the given mutual element appears. The nature of the dual element itself is easily determined; the mathematical form of the equations will be the same only if inductance is replaced by capacitance, capacitance by inductance, conductance by resistance, and resistance by conductance. Thus, the 4-H inductor which is common to meshes 1 and 2 in the circuit of Fig. 4-18 appears as a 4-F capacitor connected directly between nodes 1 and 2 in the dual circuit.

Elements which appear only in one mesh must have duals which appear between the corresponding node and the reference node. Referring again to Fig. 4-18, the voltage source 2 cos $6t$ V appears only in mesh 1; its dual is a current source 2 cos $6t$ A which is connected only to node 1 and the reference node. Since the voltage source is clockwise sensed, the current source must be into-the-nonreference-node-sensed. Finally, provision must be made for the dual of the initial voltage present across the 8-F capacitor in the given circuit. The equations have shown us that the dual of this initial voltage across the capacitor is an initial current through the inductor in the dual circuit; the numerical values are the same, and the correct sign of the initial current may be determined most readily by considering both the initial voltage in the given circuit and the initial current in the dual circuit as sources. Thus, if v_c in the given circuit is treated as a source, it would appear as $-v_c$ on the right side of the mesh equation; in the dual circuit, treating the current i_L as a source would yield a term $-i_L$ on the right side of the nodal equation. Since each has the same sign when treated as a source, then, if $v_C(0) = 10$ V, $i_L(0)$ must be 10 A.

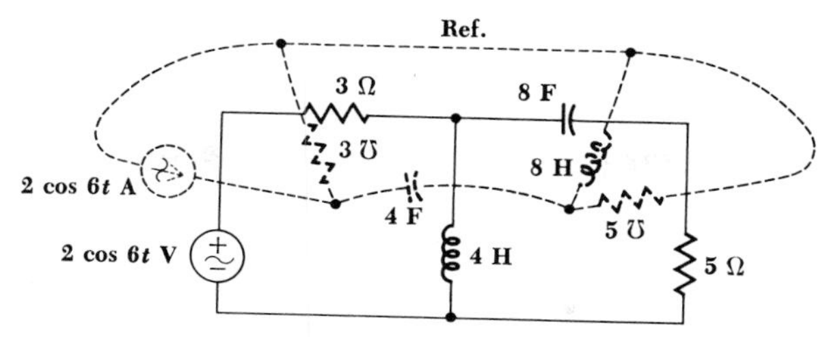

Fig. 4-20 The dual of the circuit of Fig. 4-18 is constructed directly from the circuit diagram.

The circuit of Fig. 4-18 is repeated in Fig. 4-20, and its exact dual is constructed on the circuit diagram itself by merely drawing the dual of each given element between the two nodes which are centered in the two meshes which are common to the given element. A reference node which surrounds the given circuit may be helpful. After the dual circuit is redrawn in more standard form, it appears as shown in Fig. 4-19.

An additional example of the construction of a dual circuit is shown in Fig. 4-21*a* and *b*. Since no particular element values are specified, these two circuits are duals, but not necessarily exact duals. The original circuit may be recovered from the dual by placing a node in the center of each of the five meshes of Fig. 4-21*b* and proceeding as before.

The concept of duality may also be carried over into the language by which we describe circuit analysis or operation. One example of this was discussed previously in Sec. 4-4, and the duals of several words appeared there. Most of these pairs are obvious; whenever there is any question as to the dual of a word or phrase, the dual circuit may always be drawn or visualized and then described in similar language. For example, if we are given a voltage source in series with a capacitor, we might wish to

Fig. 4-21 (*a*) The dual (in light lines) of a given circuit (in heavy lines) is constructed on the given circuit. (*b*) The dual circuit is drawn in more conventional form.

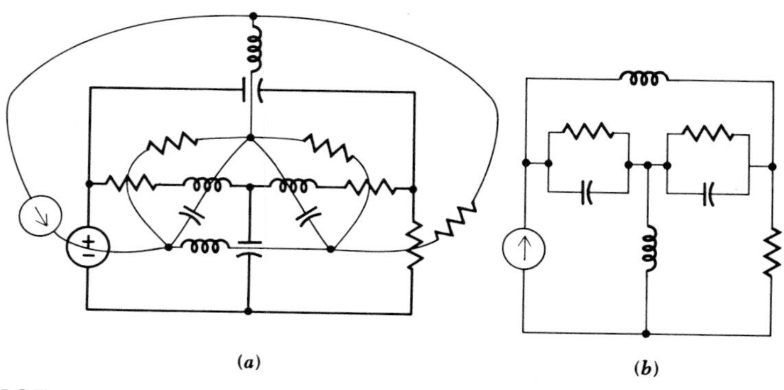

make the important statement, "the voltage source causes a current to flow through the capacitor"; the dual statement is, "the current source causes a voltage to exist across the inductor." The dual of a less carefully worded statement, such as, "the current goes round and round the series circuit," often requires a little inventiveness.[3]

Practice in using dual language can be obtained by reading Thévenin's theorem in this sense; Norton's theorem should result.

We have spoken of dual elements, dual language, and dual circuits. What about a dual *network*? Consider a resistor R and an inductor L in series. The dual of this two-terminal network exists and is most readily obtained by connecting some ideal source to the given network. The dual circuit is then obtained as the dual source in parallel with a conductance G, $G = R$, and a capacitance C, $C = L$. We consider the dual network as the two-terminal network that is connected to the dual source; it is thus a pair of terminals between which G and C are connected in parallel.

Before leaving the definition of duality, it should be pointed out that duality is defined on the basis of mesh and nodal equations. Since nonplanar circuits cannot be described by a system of mesh equations, a circuit which cannot be drawn in planar form does not possess a dual.

We shall use duality principally to reduce the work which we must do to analyze the simple standard circuits. After we have analyzed the series RL circuit, then the parallel RC circuit requires less attention, not because it is less important, but because the analysis of the dual network is already known. Since the analysis of some complicated circuit is not apt to be well known, duality will usually not provide us with any quick solution.

Drill Problem

4-9 Given this set of mesh equations:

$$2\frac{di_1}{dt} + \frac{7}{12}\int_{-\infty}^{t} i_1\,dt - 2\frac{di_2}{dt} - \frac{1}{3}\int_{-\infty}^{t} i_2\,dt - \frac{1}{4}\int_{-\infty}^{t} i_3\,dt = 2\cos 5t$$

$$-2\frac{di_1}{dt} - \frac{1}{3}\int_{-\infty}^{t} i_1\,dt + 6\frac{di_2}{dt} + \frac{5}{6}\int_{-\infty}^{t} i_2\,dt - 3\frac{di_3}{dt} - \frac{1}{2}\int_{-\infty}^{t} i_3\,dt = 0$$

$$-\frac{1}{4}\int_{-\infty}^{t} i_1\,dt - 3\frac{di_2}{dt} - \frac{1}{2}\int_{-\infty}^{t} i_2\,dt + 4i_3 + 3\frac{di_3}{dt} + \frac{3}{4}\int_{-\infty}^{t} i_3\,dt = 0$$

consider the exact dual and let the numbering of the nodes correspond to the

[3]Someone has suggested, "the voltage is across all over the parallel circuit."

numbering of the mesh currents. Then, between what two nodes will there be found a: (*a*) current source? (*b*) 3-F capacitor? (*c*) 4-H inductor?

Ans. 1, Ref.; 1,3; 2,3

4-7 LINEARITY AND ITS CONSEQUENCES AGAIN

In the previous chapter we learned that the principle of superposition is a necessary consequence of the linear nature of the resistive circuits which we were analyzing. The resistive circuits are linear because the voltage-current relationship for the resistor is linear and Kirchhoff's laws are linear.

We now wish to show that the benefits of linearity apply to *RLC* circuits as well. In accordance with our previous definition of a linear circuit, these circuits are also linear because the voltage-current relationships for the inductor and capacitor are linear relationships. For the inductor, we have

$$v = L\frac{di}{dt}$$

and multiplication of the current by some constant K leads to a voltage which is also greater by a factor K. In the integral formulation,

$$i = \frac{1}{L}\int_{t_0}^{t} v\,dt + i_L(t_0)$$

it can be seen that, if each term is to increase by a factor of K, then the initial value of the current must also increase by this same factor. That is, the factor K applies not only to the current and voltage at time t but also to their past values.

A corresponding investigation of the capacitor shows that it too is linear. Thus, a circuit composed of independent sources, linear dependent sources, and linear resistors, inductors, and capacitors is a linear circuit.

In this linear circuit the response is again proportional to the forcing function. The proof of this statement is accomplished by first writing a general system of integrodifferential equations, say, in terms of loop currents. Let us place all the terms having the form of Ri, $L\,di/dt$, and $(1/C)\int i\,dt$ on the left side of each equation and keep the independent source voltages on the right side. As a simple example, one of the equations might have the form

$$Ri + L\frac{di}{dt} + \frac{1}{C}\int_{-t_0}^{t} i\,dt + v_C(t_0) = v_s$$

If every independent source is now increased by a factor K, then the right side of each equation is greater by the factor K. Now each term on the left side is either a linear term involving some loop current or an initial capacitor voltage. In order to cause all the responses (loop currents) to increase by a factor K, it is apparent that we must also increase the initial capacitor voltages by a factor K. That is, we must treat the *initial capacitor voltage as an independent source voltage* and increase it also by a factor K. In a similar manner, initial inductor currents must be treated as independent source currents in nodal analysis.

The principle of proportionality between source and response is thus extensible to the general RLC circuit, and it follows that the principle of superposition is also applicable. It should be emphasized that initial inductor currents and capacitor voltages must be treated as independent sources in applying the superposition principle; each initial value must take its turn in being rendered inactive.

Before we can apply the superposition principle to RLC circuits, however, it is first necessary to develop methods of solving the equations describing these circuits when only one independent source is present. At this time we should feel convinced that a linear circuit will possess a response whose amplitude is proportional to the amplitude of the source. We should be prepared to apply superposition later, considering an inductor current or capacitor voltage specified at $t = t_0$ as a source which must be killed when its turn comes.

Thévenin's and Norton's theorems are based on the linearity of the initial circuit, the applicability of Kirchhoff's laws, and the superposition principle. The general RLC circuit conforms perfectly to these requirements, and it follows, therefore, that all linear circuits which contain any combinations of independent voltage and current sources, linear dependent voltage and current sources, and linear resistors, inductors, and capacitors may be analyzed with the use of these two theorems, if we wish. It is not necessary to repeat the theorems here, for they were previously stated in a manner that is equally applicable to the general RLC circuit.

PROBLEMS

☐ **1** A 2-H inductor is common to meshes 2 and 3 of a circuit. The mesh currents are: $i_2 = 2e^{-0.01t} + 3e^{-0.02t}$ A and $i_3 = -4e^{-0.01t} + 2e^{-0.02t}$ A. At what value of t is the inductor voltage zero?

☐ **2** A current increases linearly from 0 at $t = 0$ to 0.4 A in 2 ms and decreases linearly from 0.4 A to 0 in another 2 ms. The process repeats indefinitely. (*a*) Sketch the current, voltage, power, and energy for a 0.3-H inductor in the interval $0 \leq t \leq 8$ ms. (*b*) At what time is the power leaving the inductor exactly 10 W? (*c*) What energy is stored in the inductor at $t = 6$ ms?

☐ **3** Assume that 10 m of wire is available to construct a solenoid (an inductor in the form of a helix). Let the wire be lossless and of negligible diameter (for simplicity, not for accuracy). The cross section of the solenoid is circular and its length is three times its diameter. The medium is air. Find the dimensions and the number of turns if all the wire is used to make a 0.1-mH coil.

☐ **4** Given a voltage, $v = 36 \cos 200t$ V, across a 3-H inductor, find the inductor current at $t = \pi/400$ s if: (*a*) $i_L(0) = -0.1$ A; (*b*) $i_L(-\pi/600) = 0.02$ A.

☐ **5** The voltage across a 20-μF capacitor is -2 V at $t = -0.1$ s and increases linearly at the rate of 50 V/s thereafter. (*a*) Sketch the capacitor current, voltage, power, and energy as functions of time. (*b*) Find the values of each of these quantities at $t = 0$.

☐ **6** The current through a 4-μF capacitor is an infinite sequence of 100-mA pulses of 1-ms duration and 3-ms separation; that is, the period is 4 ms. The first pulse begins at $t = 0$ when the capacitor has no voltage across it. (*a*) When does the capacitor voltage reach 1000 V? (*b*) What is the capacitor charge (in C) at $t = 12$ ms? (*c*) If the pulse amplitude is reduced to zero after $t = 0.1$ s, what is the energy stored in the capacitor at $t = 0.2$ s?

☐ **7** The voltage across a 10-μF capacitor is 6 V at $t = 0$ and the current is $12 \sin 120\pi t$ mA. (*a*) What is the maximum capacitor voltage and when does it occur? (*b*) What is the maximum power drawn by the capacitor?

☐ **8** Assume that air at standard temperature and pressure between two parallel conducting plates breaks down when the voltage across it is 30,000 V/cm of plate spacing. If each plate has an area of 0.01 m^2 and the separation is 1 mm, find: (*a*) C; (*b*) the maximum voltage possible across the capacitor; (*c*) the maximum energy that can be stored in this capacitor.

☐ **9** Given four nodes, A, B, C, and D, six 5-mH inductors are used to provide one inductor between each node pair. Find the equivalent inductance presented between nodes A and B, if the inductor is open-circuited that joins: (*a*) A to C; (*b*) C to D.

☐ **10** Given an inductor L_1, it is found that when a 1-H coil is placed in parallel with it and another 1-H coil is placed in series with this parallel combination, the equivalent inductance is still L_1. (*a*) Find L_1. (*b*) Repeat if the 1-H coils are connected first in series and then in parallel.

☐ **11** Given a boxful of surplus 1-nF capacitors, how could you provide a: (*a*) 3.5-nF capacitance? (*b*) $\frac{3}{4}$-nF capacitance?

☐ **12** All capacitance values are in μF for the network shown in Fig. 4-22. Find the equivalent capacitance offered at terminals: (*a*) b and e; (*b*) a and b.

☐ **13** (*a*) Write a set of nodal integrodifferential equations for the circuit shown

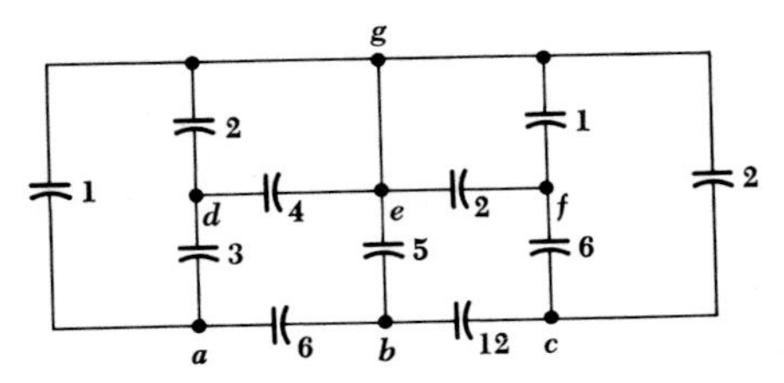

Fig. 4-22 See Prob. 12.

in Fig. 4-23*a*. Assume that the capacitor voltage (+ at top) and inductor current (arrow down) are known to be 0.6 V and 2.5 mA, respectively, at $t = 0$. (*b*) Repeat for loop or mesh equations.

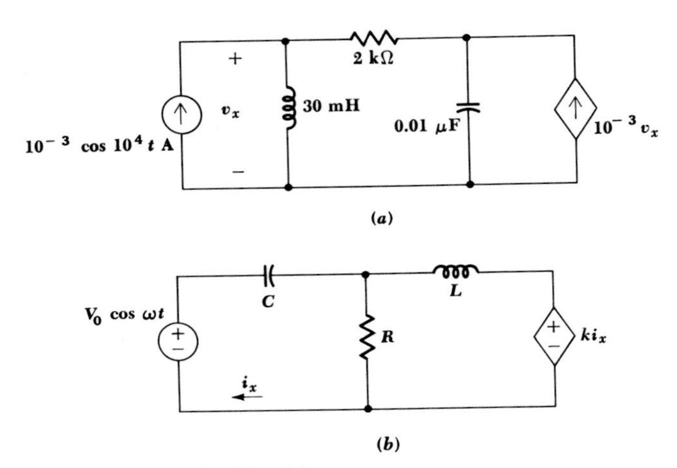

Fig. 4-23 See Probs. 13 to 15.

☐ **14** Write a set of loop or mesh integrodifferential equations for the circuit shown in Fig. 4-23*b*. Assume that the capacitor voltage (+ at left) and inductor current (arrow to right) are v_0 and i_0, respectively, at $t = 0$. (*b*) Repeat for nodal equations.

☐ **15** If the circuit shown in Fig. 4-23*b* is to be the exact dual of that in Fig. 4-23*a*, determine values (including units) for R, L, C, V_0, k, and the initial capacitor voltage (+ to right) and inductor current (arrow to right).

☐ **16** A voltage source, $10e^{-4t}$ V, is connected to the series combination of 8 Ω and 3 H. (*a*) Show that the circuit equations are satisfied by a current, $i = -2.5e^{-4t}$ A. (*b*) Construct the exact dual of the above circuit. (*c*) Find the current through the capacitor in the dual circuit.

☐ **17** Determine the expressions for voltage division across two capacitors in series and for two inductors in series. Initial currents and voltages are assumed to be zero.

☐ **18** Find formulas describing current division between two parallel capacitors and between two parallel inductors. Initial currents and voltages are assumed to be zero.

☐ **19** A capacitor C in series with a voltage source $v_s(t)$ is equivalent to the same capacitor C in parallel with a current source $i_s(t)$. (*a*) If $v_s(t) = 2 \cos 10^6 t$ V and $C = 10^4$ pF, find $i_s(t)$. (*b*) Find $i_s(t)$ in terms of a general $v_s(t)$ and C.

☐ **20** The parallel combination of a current source $i_s(t)$ and an inductor L may be replaced by the series configuration of the inductor and a voltage source $v_s(t)$. Find $v_s(t)$ in terms of $i_s(t)$ and L. If $L = 0.1$ H and $v_s = i_s$, find v_s.

Chapter Five
Source-free *RL* and *RC* Circuits

5-1 INTRODUCTION

In the previous chapter we wrote equations governing the response of several circuits containing both inductance and capacitance, but we did not solve any of them. At this time we are ready to proceed with the solution for the simpler circuits. We shall restrict our attention to certain circuits which contain only resistors and inductors or only resistors and capacitors, and which contain no sources. We shall, however, allow the presence of energy which is stored in the inductors or capacitors, for without such energy every response would be zero.

Although the circuits which we are about to consider have a very elementary appearance, they are also of practical importance. They find use as coupling networks in electronic amplifiers, as compensating networks in automatic control systems, as equalizing networks in communications channels, and in many other ways. A familiarity with these simple circuits will enable us to predict the accuracy with which the output of an amplifier can follow an input which is changing rapidly with time or to predict how quickly the speed of a motor will change in response to a change in its field current. Our knowledge of the performance of the simple *RL* and *RC* circuits will also enable us to suggest modifications to the amplifier or motor in order to obtain a more desirable response.

The analysis of such circuits is dependent upon the formulation and solution of the integrodifferential equations which characterize the circuits. We shall call the special type of equation we obtain a homogeneous linear differential equation, which is simply a differential equation in which every term is of the first degree in the dependent variable and its derivatives. A *solution* is obtained when we have found an expression for the dependent variable, as a function of time, which satisfies the differential equation and also satisfies the prescribed energy distribution in the inductors or capacitors at a prescribed instant of time, usually $t = 0$.

The solution of the differential equation represents a response of the circuit, and it is known by many names. Since this response depends upon the general "nature" of the circuit (the types of elements, their sizes, the interconnection of the elements), it is often called a *natural response*. It is also obvious that any real circuit we construct cannot store energy forever; the resistances necessarily associated with inductors and capacitors will eventually convert all stored energy into heat. The response must eventually die out, and it is therefore referred to as the *transient response*. Finally, we must also be familiar with the mathematician's contribution to the nomenclature; he calls the solution of a homogeneous linear differential equation a *complementary function*. When we consider independent sources acting on a circuit, part of the response will partake of the nature of the particular source used; this part of the response will be "complemented" by the complementary response produced in the source-free circuit, and their sum will be the complete response. The source-free response may be called the natural response, the transient response, the free response, or the complementary function, but because of its more descriptive nature, we shall most often call it the natural response.

We shall consider several different methods of solving these differential equations. This mathematics, however, is not circuit analysis. Our greatest interest lies in the solutions themselves, their meaning, and their interpretation, and we shall try to become sufficiently familiar with the form of the response that we are able to write down answers for new circuits by just plain thinking. Although complicated analytical methods are needed when simpler methods fail, an engineer must always remember that these complex techniques are only tools with which meaningful, informative answers may be obtained; they do not constitute engineering in themselves.

5-2 THE SIMPLE *RL* CIRCUIT

We shall begin our study of transient analysis by considering the simple series RL circuit shown in Fig. 5-1. Let us designate the time-varying current as $i(t)$, and we shall let the value of $i(t)$ at $t = 0$ be prescribed as I_0. We therefore have

$$v_R + v_L = Ri + L\frac{di}{dt} = 0$$

or

$$\frac{di}{dt} + \frac{R}{L}i = 0 \tag{1}$$

and we must determine an expression for $i(t)$ which satisfies this equation and also has the value I_0 at $t = 0$. The solution may be obtained by several different methods.

One very direct method of solving differential equations consists of writing the equation in such a way that the variables are separated and then integrating each side of the equation. The variables in (1) are i and t, and it is apparent that the equation may be multiplied by dt, divided by i, and arranged with the variables separated,

$$\frac{di}{i} = -\frac{R}{L}\,dt \tag{2}$$

Since the current is I_0 at $t = 0$ and $i(t)$ at time t, we may equate the two definite integrals which are obtained by integrating each side between the corresponding limits,

$$\int_{I_0}^{i(t)} \frac{di}{i} = \int_0^t -\frac{R}{L}\,dt$$

and, therefore,

$$\ln i \Big|_{I_0}^{i} = -\frac{R}{L}t \Big|_0^t$$

or

$$\ln i - \ln I_0 = -\frac{R}{L}t$$

Therefore,

$$i(t) = I_0 e^{-Rt/L} \tag{3}$$

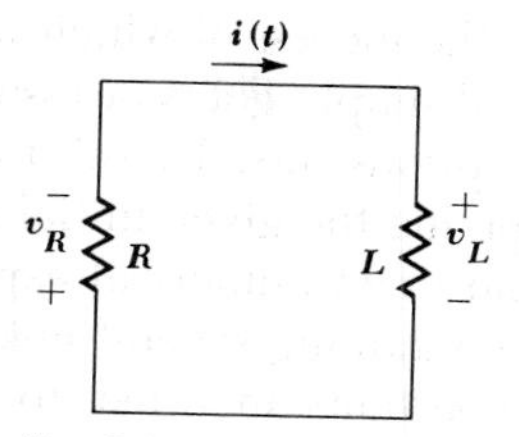

Fig. 5-1 A series RL circuit for which $i(t)$ is to be determined, subject to the initial condition that $i(0) = I_0$.

We check our solution by first showing that substitution of (3) into (1) yields the identity $0 \equiv 0$ and then showing that substitution of $t = 0$ into (3) produces $i(0) = I_0$. Both steps are necessary; the solution must satisfy the differential equation which characterizes the circuit, and it must also satisfy the initial condition, or the response at zero time.

The solution may also be obtained by a slight variation of the method described above. After separating the variables, we may obtain the *indefinite* integral of each side of (2) if we also include a constant of integration. Thus,

$$\int \frac{di}{i} = -\int \frac{R}{L}\,dt + K$$

and integration gives us

$$\ln i = -\frac{R}{L}t + K \tag{4}$$

The constant K cannot be evaluated by substitution of (4) into the original differential equation (1); the identity $0 \equiv 0$ will result, because (4) is a solution of (1) for any value of K. The constant of integration must be selected to satisfy the initial condition $i(0) = I_0$. Thus, at $t = 0$, (4) becomes

$$\ln I_0 = K$$

and we use this value for K in (4) to obtain the desired response

$$\ln i = -\frac{R}{L}t + \ln I_0$$

or

$$i(t) = I_0 e^{-Rt/L}$$

as before.

Either of the methods above can be used when the variables can be separated, but this is only occasionally possible. In the remaining cases we shall rely on a very powerful method, the success of which will depend upon our intuition or experience. We shall simply guess or assume a form for the solution and then test our assumptions, first by substitution into the differential equation and then by applying the given initial conditions. Since we cannot be expected to guess the exact numerical expression for the solution, we shall assume a solution containing several unknown constants and select the values for these constants in order to satisfy the differential equation and the initial conditions. Many of the differential equations encountered in circuit analysis have a solution which may be represented by the exponential function or by the sum of several exponential functions. Let us assume a solution of (1) in exponential form,

$$i(t) = Ae^{s_1 t}$$

where A and s_1 are constants to be determined. After substituting this assumed solution into (1), we have

$$As_1e^{s_1 t} + \frac{R}{L}Ae^{s_1 t} = 0$$

or

$$\left(s_1 + \frac{R}{L}\right)Ae^{s_1 t} = 0$$

In order to satisfy this equation for all values of time, it is necessary that either $A = 0$, or $s_1 = -\infty$, or $s_1 = -R/L$. But if $A = 0$ or $s_1 = -\infty$, then every response is zero; neither can be a solution to our problem. Therefore, we must choose

$$s_1 = -\frac{R}{L}$$

and our assumed solution takes on the form

$$i(t) = Ae^{-Rt/L}$$

The remaining constant must be evaluated by applying the initial condition $i = I_0$ at $t = 0$. Thus,

$$I_0 = A$$

and the final form of the assumed solution is

$$i(t) = I_0e^{-Rt/L} \tag{5}$$

once again.

We shall not consider any other methods for solving (1), although a number of other techniques may be used. Watch for them in a study of differential equations.

Before we turn our attention to the interpretation of the response, let us check the power and energy relationships in this circuit. The power being dissipated in the resistor is

$$p_R = i^2R = I_0{}^2Re^{-2Rt/L}$$

and the total energy turned into heat in the resistor is found by integrating the instantaneous power from zero time to infinite time,

$$W_R = \int_0^\infty p_R\,dt = I_0{}^2R\int_0^\infty e^{-2Rt/L}\,dt$$

$$= I_0{}^2R\left(-\frac{L}{2R}\right)e^{-2Rt/L}\Big|_0^\infty = \tfrac{1}{2}LI_0{}^2$$

This is the result we expect, because the total energy stored initially in the coil is $\frac{1}{2}LI_0{}^2$, and there is no energy stored in the coil at infinite time. All the initial energy is accounted for by dissipation in the resistor.

Drill Problems

5-1 Each circuit shown in Fig. 5-2 has been in the condition shown for an extremely long time. At $t = 0$, the several switches are closed or opened as indicated. The single-pole double-throw switches appearing in b and c are drawn to indicate that they close one circuit before opening the other. They are called *make before break*. After reviewing the characteristics of an inductor, as summarized on page 120, determine $i(0)$ in each circuit.

Ans. -2; 2; 2.5 A

Fig. 5-2 See Drill Probs. 5-1, 5-2, and 5-5.

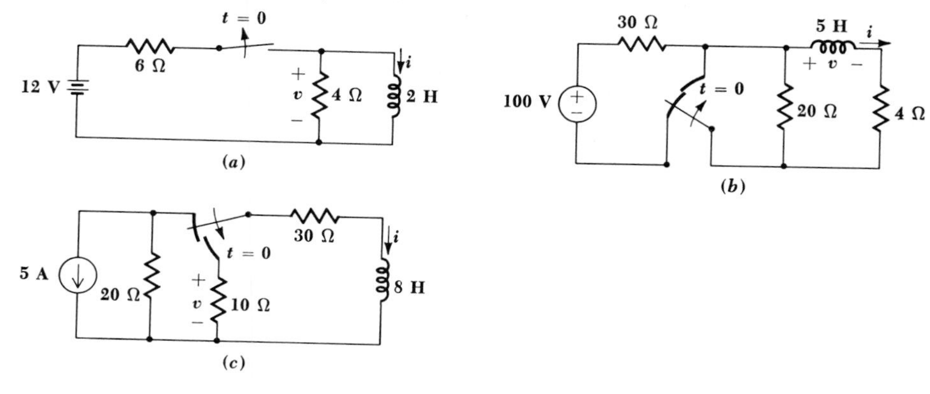

5-2 At the instant just *after* the switches are thrown in the circuits of Fig. 5-2, find v.

Ans. -10; -8; 20 V

5-3 A 30-mH inductor is in series with a 400-Ω resistor. If the energy stored in the coil at $t = 0$ is 0.96 μJ, find the magnitude of the current at $t =$: (*a*) 0; (*b*) 100 μs; (*c*) 300 μs.

Ans. 0.147; 2.11; 8.00 mA

5-3 PROPERTIES OF THE EXPONENTIAL RESPONSE

Let us now consider the nature of the response in the series RL circuit. We found that the current is represented by

$$i(t) = I_0 e^{-Rt/L} \tag{6}$$

At zero time, the current is the assumed value I_0, and as time increases, the current decreases and approaches zero. The shape of this decaying exponential is seen by a plot of $i(t)/I_0$ versus t, as shown in Fig. 5-3. Since the function we are plotting is $e^{-Rt/L}$, the curve will not change if R/L does not change. Thus, the same curve must be obtained for every series RL circuit having the same R/L or L/R ratio. Let us see how this ratio affects the shape of the curve.

If we double the ratio of L to R, then the exponent will be unchanged if t is also doubled. In other words, the original response will occur at a later time, and the new curve is obtained by moving each point on the original curve twice as far to the right. With this larger L/R ratio, the current takes longer to decay to any given fraction of its original value. We might have a tendency to say that the "width" of the curve is doubled,

Fig. 5-3 A plot of $e^{-Rt/L}$ versus t.

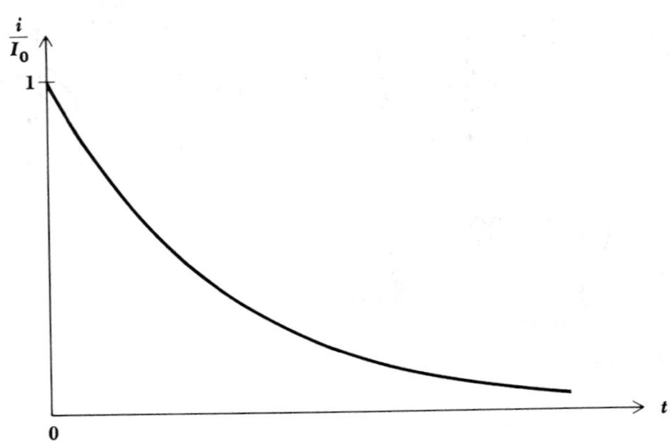

or that the "width" is proportional to L/R. However, we should have to define our term "width," because each curve extends from $t = 0$ to ∞. Instead, let us consider the time that would be required for the current to drop to zero *if it continued to drop at its initial rate.*

The initial rate of decay is found by evaluating the derivative at zero time,

$$\frac{d}{dt}\frac{i}{I_0}\bigg|_{t=0} = -\frac{R}{L}e^{-Rt/L}\bigg|_{t=0} = -\frac{R}{L}$$

Let us designate the value of time it takes for i/I_0 to drop from unity to zero, assuming a constant rate of decay, by the Greek letter τ (*tau*). Thus,

$$\frac{R}{L}\tau = 1$$

or

$$\tau = \frac{L}{R} \tag{7}$$

The ratio L/R has the units of seconds since the exponent $-Rt/L$ must be dimensionless. This value of time τ is called the *time constant;* it is shown in Fig. 5-4. It is apparent that the time constant of a series RL circuit may easily be found graphically from the response curve; it is only necessary to draw the tangent to the curve at $t = 0$ and determine the intercept of this tangent line with the time axis.

An equally important interpretation of the time constant τ is obtained by determining the value of $i(t)/I_0$ at $t = \tau$. We have

$$\frac{i(\tau)}{I_0} = e^{-1} = 0.368 \qquad \text{or} \qquad i(\tau) = 0.368I_0$$

Fig. 5-4 The time constant τ is L/R s for a series RL circuit. It is the time required for the response curve to drop to zero if it decays at a constant rate which is equal to its initial rate of decay.

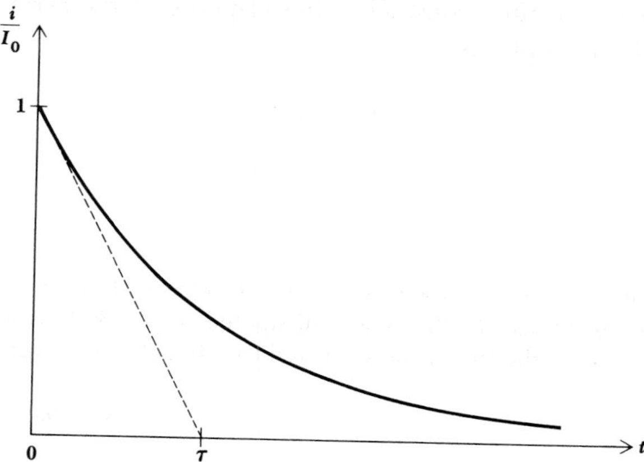

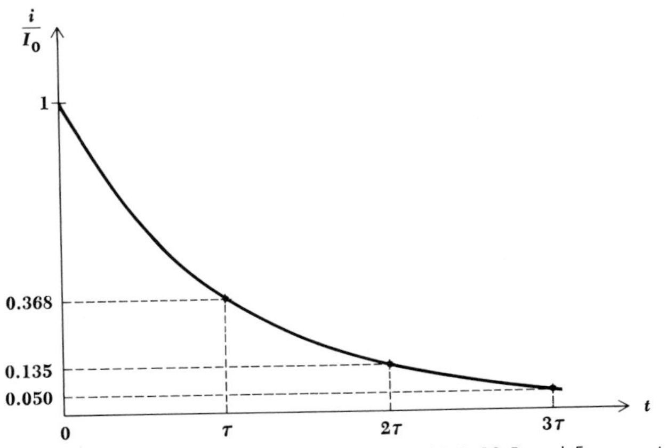

Fig. 5-5 The current in a series RL circuit is 36.8, 13.5, and 5 per cent of its initial value at τ, 2τ, and 3τ, respectively.

Thus, in one time constant the response has dropped to 36.8 per cent of its initial value; the value of τ may also be determined graphically from this fact, as indicated by Fig. 5-5. It is convenient to measure the decay of the current at intervals of one time constant, and recourse to a slide rule or a table of negative exponentials shows that $i(t)/I_0$ is 0.368 at $t = \tau$, 0.135 at $t = 2\tau$, 0.0498 at $t = 3\tau$, 0.0183 at $t = 4\tau$, and 0.0067 at $t = 5\tau$. At some point three to five time constants after zero time, most of us would agree that the current is a negligible fraction of its former self.

Why does a larger value of the time constant L/R produce a response curve which decays more slowly? Let us consider the effect of each element. An increase in L allows a greater energy storage for the same initial current, and this larger energy requires a longer time to be dissipated in the resistor. We may also increase L/R by reducing R. In this case, the power flowing into the resistor is less for the same initial current; again, a greater time is required to dissipate the stored energy.

In terms of the time constant τ, the response of the series RL circuit may be written simply as

$$i(t) = I_0 e^{-t/\tau}$$

Drill Problem

5-4 Find the inductance present in a series RL circuit if: (*a*) R is 40 Ω and the time constant is 30 ms; (*b*) the power dissipated in the 30-Ω resistor is cut in half every 20 ms; (*c*) the time constant is reduced from 16 to 15 ms by increasing the resistance 2 Ω.

Ans. 0.48; 1.2; 1.73 H

5-4 A MORE GENERAL *RL* CIRCUIT

It is not difficult to extend the results obtained for the series RL circuit to a circuit containing any number of resistors and one inductor. We fix our attention on the two terminals of the coil and determine the equivalent resistance across these terminals. The circuit is thus reduced to the simple series case. As an example, consider the circuit shown in Fig. 5-6. The equivalent resistance which the coil faces is

$$R_{eq} = R_3 + R_4 + \frac{R_1 R_2}{R_1 + R_2}$$

and the time constant is therefore

$$\tau = \frac{L}{R_{eq}}$$

The inductor current i_L is

$$i_L = i_L(0)e^{-t/\tau} \tag{8}$$

and (8) represents what we might call the basic solution to the problem. It is quite possible that some current or voltage other than i_L is needed, such as the current i_2 in R_2. We can always apply Kirchhoff's laws and Ohm's law to the resistive portion of the circuit without any difficulty, but current division provides the quickest answer in this circuit,

$$i_2 = -\frac{R_1}{R_1 + R_2} i_L(0)e^{-t/\tau}$$

It may also happen that we know the initial value of some current other than the inductor current. Since the current in a resistor may change instantaneously, we shall indicate the initial value *after* any change that might have occurred at $t = 0$ by use of the symbol 0^+; in more mathematical language, $i_1(0^+)$ is the limit from the right of $i_1(t)$ as t approaches

Fig. 5-6 A source-free circuit containing one inductor and several resistors is analyzed by determining the time constant, $\tau = L/R_{eq}$.

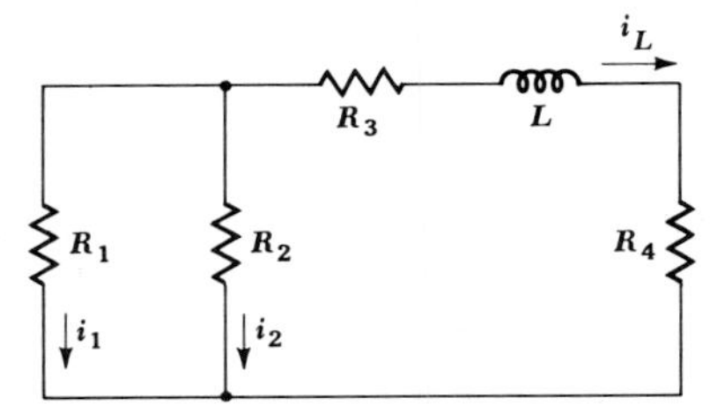

zero. Thus, if we are given the initial value of i_1 as $i_1(0^+)$, then it is apparent that the initial value of i_2 is

$$i_2(0^+) = i_1(0^+)\frac{R_1}{R_2}$$

From these values, we obtain the necessary initial value of $i_L(0)$ [or $i_L(0^-)$ or $i_L(0^+)$],

$$i_L(0) = -[i_1(0^+) + i_2(0^+)] = -\frac{R_1 + R_2}{R_2}i_1(0^+)$$

and the expression for i_2 becomes

$$i_2 = i_1(0^+)\frac{R_1}{R_2}e^{-t/\tau}$$

Let us see if we can obtain this last expression more directly. Since the inductor current decays exponentially as $e^{-t/\tau}$, then every current throughout the circuit must follow the same functional behavior. This is made clear by considering the inductor current as a source current which is being applied to a resistive network. Every current and voltage in the resistive network must have the same time dependence. Using these ideas, we therefore express i_2 as

$$i_2 = Ae^{-t/\tau}$$

where

$$\tau = \frac{L}{R_{eq}}$$

and A must be determined from a knowledge of the initial value of i_2. Since $i_1(0^+)$ is known, then the voltage across R_1 and R_2 is known, and

$$i_2(0^+) = i_1(0^+)\frac{R_1}{R_2}$$

Therefore,

$$i_2 = i_1(0^+)\frac{R_1}{R_2}e^{-t/\tau}$$

A similar sequence of steps will provide a rapid solution to a large number of problems. We first recognize the time dependence of the response as an exponential decay, determine the appropriate time constant by com-

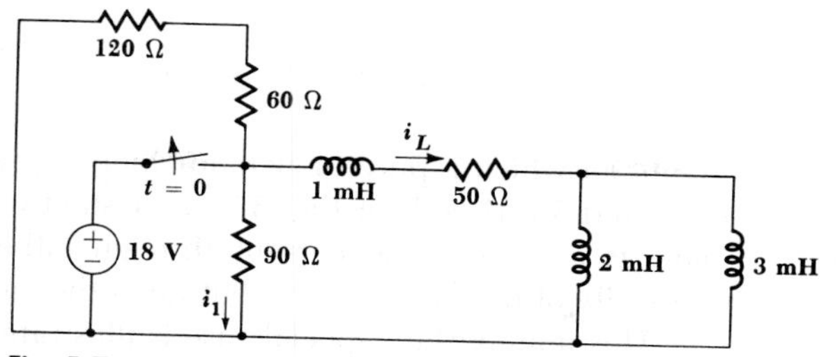

Fig. 5-7 After $t = 0$, this circuit simplifies to an equivalent resistance of 110 Ω in series with $L_{eq} = 2.2$ mH.

bining resistances, write the solution with an unknown amplitude, and then determine the amplitude from a given initial condition.

This same technique is also applicable to a circuit which contains one resistor and any number of inductors, as well as to those special circuits containing two or more inductors and also two or more resistors that may be simplified by resistance or inductance combination until the simplified circuits have only one inductance or one resistance. As an example of such a circuit we may consider Fig. 5-7. After $t = 0$, when the voltage source is disconnected, we easily calculate an equivalent inductance,

$$L_{eq} = \frac{2 \times 3}{2 + 3} + 1 = 2.2 \quad \text{mH}$$

an equivalent resistance,

$$R_{eq} = \frac{90(60 + 120)}{90 + 180} + 50 = 110 \quad \Omega$$

and the time constant,

$$\tau = \frac{L_{eq}}{R_{eq}} = \frac{2.2 \times 10^{-3}}{110} = 20 \quad \mu\text{s}$$

Thus, the form of the natural response is $Ae^{-50{,}000t}$. With the independent source connected ($t < 0$), i_L is $18/50$, or 0.36 A, while i_1 is $18/90$, or 0.2 A. At $t = 0^+$, i_L must still be 0.36 A, but i_1 will jump to a new value determined by $i_L(0^+)$.
Thus,

$$i_1(0^+) = -\, i_L(0^+)\,180/270 = -0.24 \quad \text{A}$$

Hence,

$$\begin{aligned} i_L &= 0.36 \qquad (t < 0) \\ &= 0.36e^{-50{,}000t} \qquad (t > 0) \end{aligned}$$

and
$$i_1 = 0.2 \qquad (t < 0)$$
$$= -0.24e^{-50,000t} \qquad (t > 0)$$

In idealized circuits in which a pure inductance loop is present, such as that through the 2- and 3-mH coils of Fig. 5-7, a constant current may continue to circulate as $t \to \infty$. The current through either of these inductors is not necessarily of the form, $Ae^{-t/\tau}$, but takes the more general form, $A_1 + A_2e^{-t/\tau}$. This unimportant special case is illustrated by Prob. 10 at the end of this chapter.

We have now considered the task of finding the natural response of any circuit which can be represented by an equivalent inductance in series with an equivalent resistance. The most general *RL* circuit will be considered in Sec. 5-7; the analysis is made more complicated because the response is composed of the sum of a number of negative exponentials.

Drill Problems

5-5 After $t = 0$, each of the circuits in Fig. 5-2 is source-free. Find expressions for i and v in each case for $t > 0$.
Ans. $2.5e^{-0.8t}$ A, $-10e^{-0.8t}$ V; $2e^{-2t}$ A, $-8e^{-2t}$ V; $-2e^{-5t}$ A, $20e^{-5t}$ V

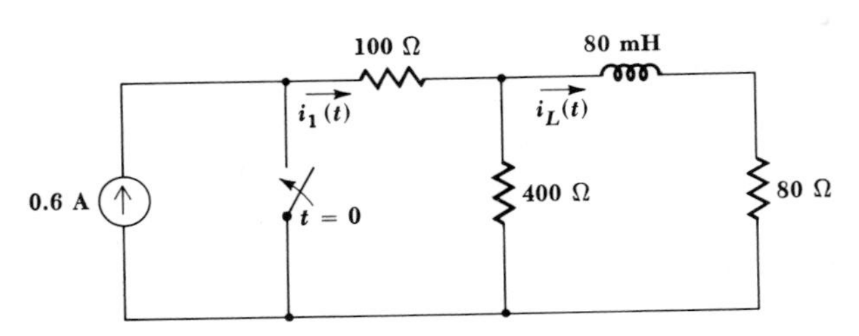

Fig. 5-8 See Drill Prob. 5-6

5-6 For the circuit shown in Fig. 5-8, find: (*a*) $i_L(0^+)$; (*b*) $i_L(10^{-3})$; (*c*) $i_1(10^{-3})$.
Ans. 54.1; 67.7; 500 mA

5-5 THE SIMPLE *RC* CIRCUIT

The series combination of a resistor and a capacitor has a greater practical importance than does the combination of a resistor and an inductor. When an engineer has any freedom of choice between using a capacitor and using an inductor in the coupling network of an electronic amplifier, in the compensation networks of an automatic control system, or in the synthesis

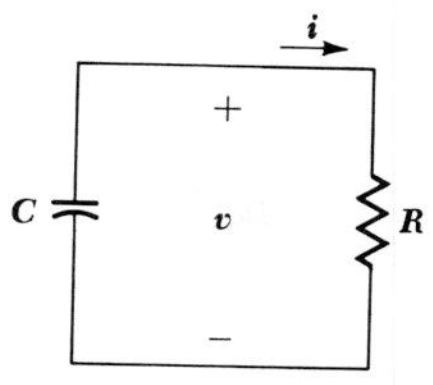

Fig. 5-9 A parallel RC circuit for which $v(t)$ is to be determined, subject to the initial condition that $v(0) = V_0$.

of an equalizing network, for example, he will choose the RC network over the RL network whenever possible. The reasons for this choice are the smaller losses present in a physical capacitor, its lower cost, the better approximation which the mathematical model makes to the physical element it is intended to represent, and the smaller size and lighter weight as exemplified by capacitors in hybrid and integrated circuits.

Let us see how closely the analysis of the parallel (or is it series?) RC circuit corresponds to that of the RL circuit. The RC circuit is shown in Fig. 5-9. We shall assume an initial stored energy in the capacitor by selecting

$$v(0) = V_0$$

The total current leaving the node at the top of the circuit diagram must be zero, and, therefore,

$$C\frac{dv}{dt} + \frac{v}{R} = 0$$

Division by C gives us

$$\frac{dv}{dt} + \frac{v}{RC} = 0 \tag{9}$$

Equation (9) has a familiar form; comparison with (1),

$$\frac{di}{dt} + \frac{R}{L}i = 0 \tag{1}$$

shows that the replacement of i by v and L/R by RC produces the identical equation we considered previously. It should, for the RC circuit we are now analyzing is the dual of the RL circuit we considered first. This duality forces $v(t)$ for the RC circuit and $i(t)$ for the RL circuit to have identical expressions if the resistance of one circuit is equal to the reciprocal of the resistance of the other circuit and if L is numerically equal to C. Thus, the response of the RL circuit,

$$i(t) = i(0)e^{-Rt/L} = I_0 e^{-Rt/L}$$

enables us to write immediately

$$v(t) = v(0)e^{-t/RC} = V_0 e^{-t/RC} \tag{10}$$

for the RC circuit.

Now let us suppose that we had selected the current i as our variable in the RC circuit, rather than the voltage v. Applying Kirchhoff's voltage law,

$$\frac{1}{C}\int_{t_0}^{t} i\,dt - v(t_0) + Ri = 0$$

we obtain an integral equation and not a differential equation. However, if we take the time derivative of both sides of this equation,

$$\frac{i}{C} + R\frac{di}{dt} = 0 \tag{11}$$

and replace i by v/R,

$$\frac{v}{RC} + \frac{dv}{dt} = 0$$

we obtain (9) again. Equation (11) could have been used as our starting point, but duality would not have appeared as naturally.

Let us discuss the physical nature of the voltage response of the RC circuit as expressed by (10). At $t = 0$ we obtain the correct initial condition, and as t becomes infinite the voltage approaches zero. This latter result agrees with our thinking that if there were any voltage remaining across the capacitor, then energy would continue to flow into the resistor and be dissipated as heat. Thus, a final voltage of zero is necessary. The time constant of the RC circuit may be found by using the duality relationships on the expression for the time constant of the RL circuit, or it may be found by simply noting the time at which the response has dropped to 36.8 per cent of its initial value,

$$\frac{\tau}{RC} = 1$$

and

$$\tau = RC \tag{12}$$

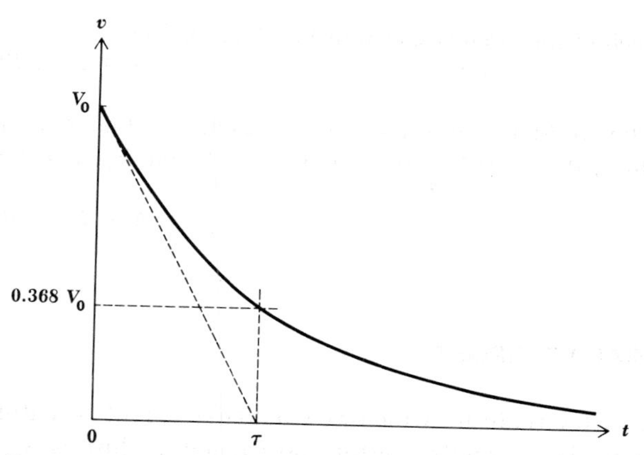

Fig. 5-10 The capacitor voltage $v(t)$ in the parallel RC circuit is plotted as a function of time. The initial value of $v(t)$ is assumed to be V_0.

Our familiarity with the negative exponential and the significance of the time constant τ enables us to sketch the response curve readily, Fig. 5-10. Larger values of R or C provide a larger time constant and a slower dissipation of the stored energy. A larger resistance will dissipate a smaller power[1] with a given voltage across it, thus requiring a greater time to convert the stored energy into heat; a larger capacitance stores a larger energy with a given voltage across it, again requiring a greater time to lose this initial energy.

Drill Problems

5-7 For each of the circuits shown in Fig. 5-11, determine $v(0^+)$.

Ans. 10; 50; 60 V

Fig. 5-11 See Drill Probs. 5-7, 5-8, and 5-10.

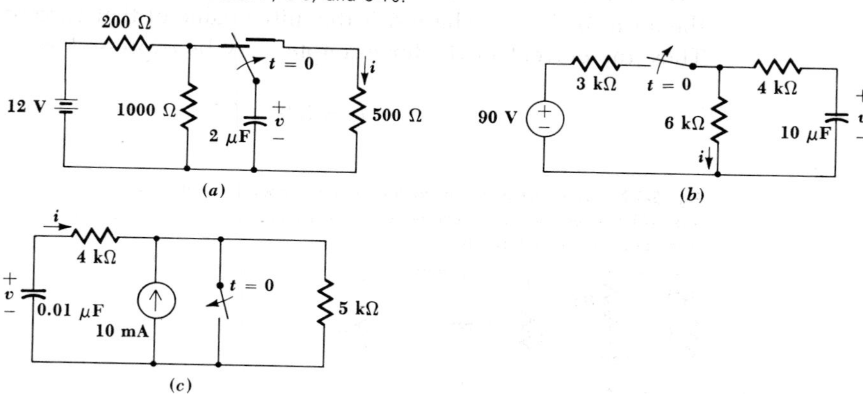

[1]"Greater resistance leads to less dissipation" might be the scholar's motto.

5-8 For each of the circuits shown in Fig. 5-11, find $i(0^+)$.

Ans. 6; 12.5; 20 mA

5-9 A 2500-Ω resistor is in series with a capacitor C. Find C if: (*a*) $v_C(0.1) = 0.1\, v_C(0)$; (*b*) $w_C(0.1) = 0.05\, w_C(0)$; (*c*) $v_C(0) = 20$ V and dv_C/dt at $t = 0$ is -400 V/s.

Ans. 17.4; 20.0; 26.7 μF

5-6 A MORE GENERAL *RC* CIRCUIT

Many of the *RC* circuits for which we would like to find the natural response contain more than a single resistor and capacitor. Just as we did for the *RL* circuits, we shall first consider several special cases and save the general case for the final section in this chapter.

Let us suppose first that we are faced with a circuit containing only one capacitor, but any number of resistors. It is possible to replace the two-terminal resistive network which is across the capacitor terminals by an equivalent resistance, and we may then write down the expression for the capacitor voltage immediately. For example, the circuit shown in Fig. 5-12*a* may be simplified to that of Fig. 5-12*b*, enabling us to write

$$v = V_0 e^{-t/R_{eq}C}$$

where

$$v(0) = V_0 \qquad \text{and} \qquad R_{eq} = R_2 + \frac{R_1 R_3}{R_1 + R_3}$$

Every current and voltage in the resistive portion of the network must have the form $Ae^{-t/R_{eq}C}$, where A is the initial value of that current or voltage. Thus, the current in R_1, for example, may be expressed as

$$i_1 = i_1(0^+)e^{-t/\tau}$$

Fig. 5-12 (*a*) A given circuit containing one capacitor and several resistors. (*b*) The resistors have been replaced by a single equivalent resistor; the time constant is now obvious.

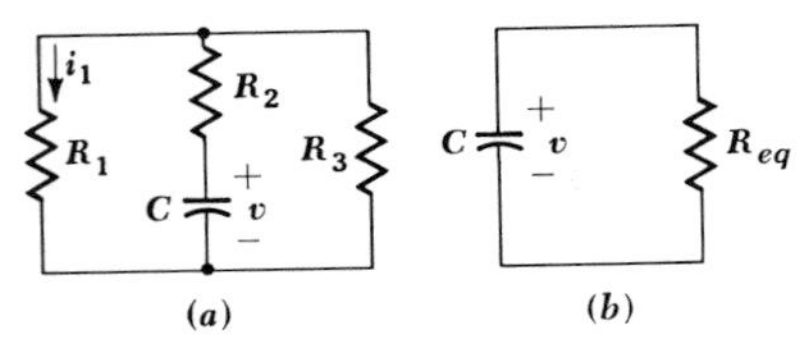

where

$$\tau = \left(R_2 + \frac{R_1R_3}{R_1 + R_3}\right)C$$

and $i_1(0^+)$ remains to be determined from some initial condition. Suppose that $v(0)$ is given. Since v cannot change instantaneously, we may think of the capacitor as being replaced by an independent dc source, $v(0)$. Thus,

$$i_1(0^+) = \frac{v(0)}{R_2 + R_1R_3/(R_1 + R_3)} \frac{R_3}{R_1 + R_3}$$

The solution is obtained by collecting these results.

Another special case includes those circuits containing one resistor and any number of capacitors. The resistor voltage is easily obtained by establishing the value of the equivalent capacitance and determining the time constant. Once again our mathematically perfect elements may lead to phenomena which would not exist in a physical circuit. Here, two capacitors in series may have equal and opposite voltages across each element and yet have zero voltage across the combination. Thus the general form of the voltage across either is $A_1 + A_2e^{-t/\tau}$, while the voltage across the series combination continues to be $Ae^{-t/\tau}$. An example of such a situation is provided by Prob. 20 at the end of the chapter.

Some circuits containing a number of both resistors and capacitors may be replaced by an equivalent circuit containing only one resistor and one capacitor; it is necessary that the original circuit be one which can be broken into two parts, one containing all resistors and the other containing all capacitors, such that the two parts are connected by only two ideal conductors. This is not possible in general.

Drill Problem

5-10 Find $v(t)$ and $i(t)$ for $t > 0$ for each of the circuits of Fig. 5-11.

Ans. $10e^{-1000t}$ V, $20e^{-1000t}$ mA; $50e^{-25{,}000t}$ V, $12.5e^{-25{,}000t}$ mA; $60e^{-10t}$ V, $6e^{-10t}$ mA

5-7 GENERAL *RL* AND *RC* CIRCUITS

A circuit containing several resistances and several inductances does not in general possess a form which allows either the resistances or inductances to be combined into single equivalent elements. There is no single negative exponential term or single time constant associated with the circuit. Rather, there will in general be several negative exponential terms, the number of terms being equal to the number of inductances which remain after all possible inductor combinations have been made. A similar condition holds for a general *RC* circuit, but for simplicity we shall restrict our attention to *RL* circuits.

Let us visualize the problem which now confronts us. We have a circuit which must contain more than one loop, and hence a knowledge of several loop currents is required to describe the response. Each of these loop currents will be expressed as the sum of a number of negative exponentials, and each exponential will have an unknown amplitude and an unknown time constant. This represents a lot of unknown information. Our situation becomes somewhat more cheerful when we realize that each loop current should contain exponential terms with the same time constants as every other loop current. This appears plausible when we stop to consider that the voltages and currents throughout the circuit are related by a constant multiplier (resistors) or by differentiation or integration (inductors). Sums of these operations are also possible. Since none of these operations changes the exponent, we expect that each current or voltage must contain terms which may have every possible exponent. At any rate, we shall make this assumption; if we can satisfy the differential equations and the initial conditions, our assumption will be proved correct.

Before we illustrate this method by an example, it must be realized that there are better and easier ways to find the natural response of these more complex circuits. One of the methods will appear toward the end of Chap. 13; it is based on the concept of complex frequency. The most powerful methods rely on the use of Fourier or Laplace transforms and will arise in Chaps. 18 and 19.

Let us consider a specific circuit containing two inductors and two resistors, shown in Fig. 5-13. Mesh currents are assigned and the appropriate set of mesh equations is written,

$$i_1 + 5\frac{di_1}{dt} - 3\frac{di_2}{dt} = 0 \tag{13}$$

$$-3\frac{di_1}{dt} + 3\frac{di_2}{dt} + 2i_2 = 0 \tag{14}$$

Fig. 5-13 An *RL* circuit having a response which must be described by the sum of two negative exponentials.

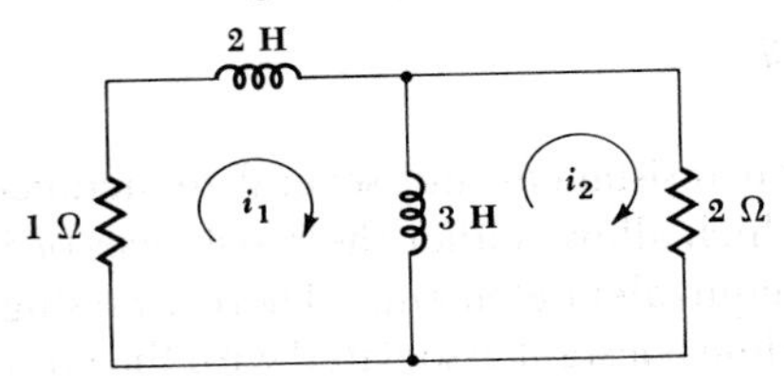

We shall assume that the initial values of i_1 and i_2 are known; let them be $i_1(0)$ and $i_2(0)$. We next assume an appropriate form for i_1 and i_2 and substitute the assumed expressions into the differential equations. We shall select a form for i_1 and i_2 which is the sum of two exponentials,

$$i_1 = Ae^{s_1 t} + Be^{s_2 t} \tag{15}$$

$$i_2 = Ce^{s_1 t} + De^{s_2 t} \tag{16}$$

and then substitute these expressions into the differential equations

$$Ae^{s_1 t} + Be^{s_2 t} + 5As_1e^{s_1 t} + 5Bs_2e^{s_2 t} - 3Cs_1e^{s_1 t} - 3Ds_2e^{s_2 t} = 0$$

$$-3As_1e^{s_1 t} - 3Bs_2e^{s_2 t} + 3Cs_1e^{s_1 t} + 3Ds_2e^{s_2 t} + 2Ce^{s_1 t} + 2De^{s_2 t} = 0$$

After factoring the exponentials,

$$(A + 5As_1 - 3Cs_1)e^{s_1 t} + (B + 5Bs_2 - 3Ds_2)e^{s_2 t} = 0$$

$$(-3As_1 + 3Cs_1 + 2C)e^{s_1 t} + (-3Bs_2 + 3Ds_2 + 2D)e^{s_2 t} = 0$$

we realize that, in order that these expressions be zero for every value of t, each sum in parentheses must be zero. Hence, we obtain the four equations

$$(5s_1 + 1)A - 3s_1C = 0 \tag{17}$$

$$(5s_2 + 1)B - 3s_2D = 0 \tag{18}$$

$$(3s_1 + 2)C - 3s_1A = 0 \tag{19}$$

$$(3s_2 + 2)D - 3s_2B = 0 \tag{20}$$

Eliminating A between (17) and (19), we have

$$\frac{3s_1C}{5s_1 + 1} = \frac{(3s_1 + 2)C}{3s_1}$$

or

$$6s_1^2 + 13s_1 + 2 = 0$$

and

$$s_1 = -1/6, -2$$

A similar manipulation of (18) and (20) yields

$$s_2 = -1/6, -2$$

It makes no difference which answer is called s_1 and which is s_2; so we arbitrarily pick

$$s_1 = -1/6 \qquad s_2 = -2$$

Thus,

$$i_1 = Ae^{-t/6} + Be^{-2t} \tag{21}$$

$$i_2 = Ce^{-t/6} + De^{-2t} \tag{22}$$

The next step is the determination of A, B, C, and D. The given initial conditions must now be applied,

$$i_1(0) = A + B \tag{23}$$

$$i_2(0) = C + D \tag{24}$$

These two equations are not sufficient to determine the four constants, and recourse must be had to (17) to (20). In other words, the differential equations impose certain restrictions among the amplitudes of the several current components. Let us select (17) and (18),

$$(-\tfrac{5}{6} + 1)A + \tfrac{1}{2}C = 0 \qquad (-10 + 1)B + 6D = 0$$

or

$$A + 3C = 0 \tag{25}$$

$$-9B + 6D = 0 \tag{26}$$

The four equations (23) to (26) may be solved by eliminating variables or determinants. The results are easily obtained:

$$A = \tfrac{9}{11}i_1(0) - \tfrac{6}{11}i_2(0)$$
$$B = \tfrac{2}{11}i_1(0) + \tfrac{6}{11}i_2(0)$$
$$C = -\tfrac{3}{11}i_1(0) + \tfrac{2}{11}i_2(0)$$
$$D = \tfrac{3}{11}i_1(0) + \tfrac{9}{11}i_2(0)$$

and the substitution of these results into (21) and (22) completes the mathematical solution. This circuit is analyzed by a much simpler method in Sec. 13-8 in accord with our comments earlier in this section.

Let us select numerical values for the initial conditions and try to interpret the solution physically. Suppose we let $i_1(0) = 11$ A and $i_2(0) = 11$ A. The initial current in the 3-H coil is, therefore,

$$i_1(0) - i_2(0) = 11 - 11 = 0$$

although 11 A is flowing in the 2-H coil at $t = 0$.

After evaluating the four amplitudes, we obtain

$$i_1 = 3e^{-t/6} + 8e^{-2t} \tag{27}$$

$$i_2 = -e^{-t/6} + 12e^{-2t} \tag{28}$$

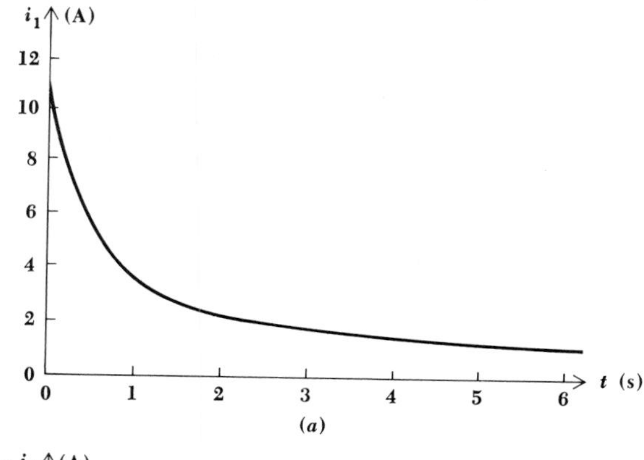

(a)

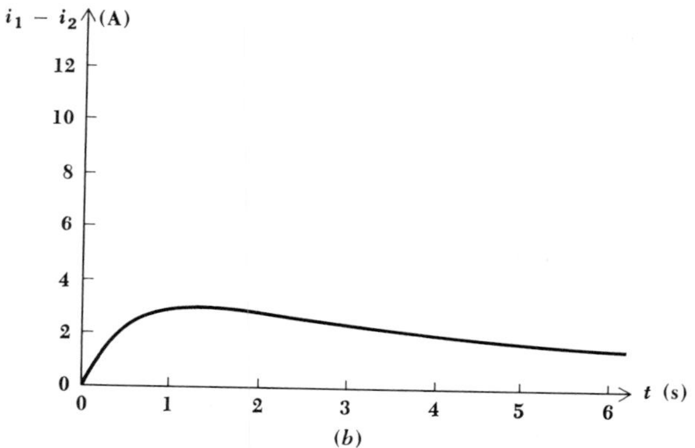

(b)

Fig. 5-14 The current in (a) the 2-H coil and (b) the 3-H coil shown in Fig. 5-13.

and
$$i_1 - i_2 = 4e^{-t/6} - 4e^{-2t}$$

The currents flowing in the two coils, i_1 and $i_1 - i_2$, are sketched in Fig. 5-14*a* and *b* by taking the difference or sum of the two exponential terms graphically. Energy is initially present in the 2-H coil, but its value drops quickly at first as some of the energy is transferred to the 3-H coil. This transfer is required because the initial voltage across the 2-Ω resistor is 22 V, and thus an initial rate of increase of current in the 3-H coil of $^{22}/_3$ A/s is established. Similarly, the current in the 2-H coil decreases initially at the rate of 15.5 A/s. Thus, energy is leaving the 2-H coil, some being transferred to the 3-H coil, but most being dissipated as heat in the two resistors.

The two time constants, 6 s and 0.5 s, are not apparent in the circuit itself. Since they are quite unequal, however, we can at least identify the parts of the circuit which contribute to the longer and shorter time constant. For example, we may think of the circuit initially as being approximately a 2-H coil in series with a 3-Ω resistor. The 3-H coil is drawing a very small current; so we shall temporarily ignore it. The time constant of the early action is thus about $2/(1 + 2)$ or $\frac{2}{3}$ s. At a much later time, the action slows down, a small rate of change of current is evident, and small voltages are present across both coils. Thus, a smaller voltage is across the 2-Ω resistor. If we ignore it completely, the circuit acts as a 5-H coil in series with a 1-Ω resistor. Thus the time constant is about 5 s.

Drill Problems

5-11 For the circuit of Fig. 5-15, for $t > 0$, find: (*a*) the smaller of the two time constants; (*b*) the larger of the two time constants; (*c*) the single time constant that would be present if the larger capacitor were replaced by a short circuit.

Ans. 1; 1.2; 2 s

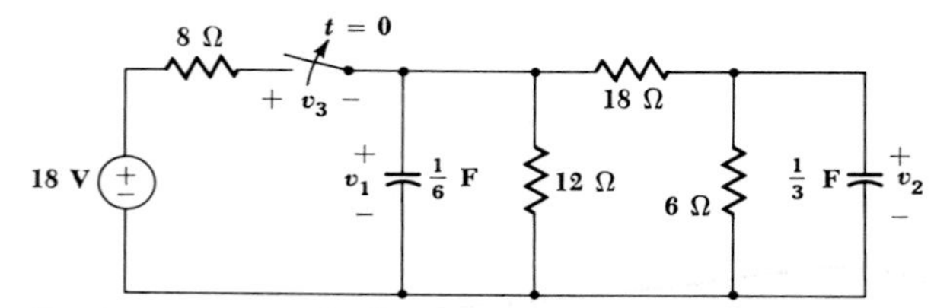

Fig. 5-15 See Drill Probs. 5-11 to 5-13.

5-12 For the circuit of Fig. 5-15, find the value at $t = 0^+$ of: (*a*) v_1; (*b*) v_2; (*c*) v_3.

Ans. 2.25; 9; 9 V

5-13 For the circuit of Fig. 5-15, for $t > 0$, determine: (*a*) $v_1(t)$; (*b*) $v_2(t)$; (*c*) $v_3(t)$.

Ans. $-2.25e^{-t} + 4.5e^{-t/2}$ V; $4.5e^{-t} + 4.5e^{-t/2}$ V; $18 - 4.5e^{-t} - 4.5e^{-t/2}$ V

PROBLEMS

☐ **1** For the circuit shown in Fig. 5-16, find i and v as functions of time for $t > 0$.

Fig. 5-16 See Prob. 1.

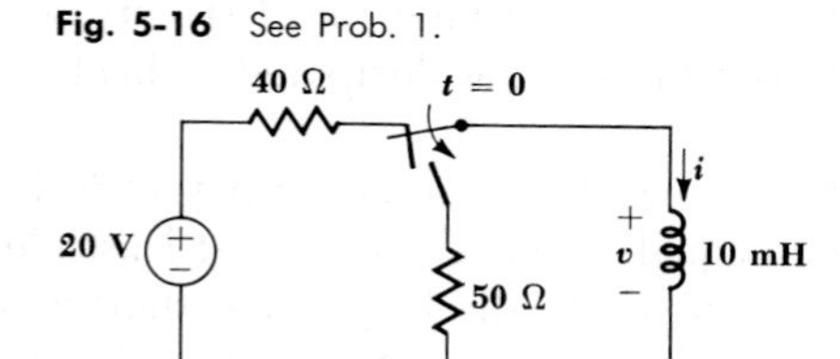

☐ **2** A 200-Ω resistor is in series with an inductor L. The initial value of the inductor current is 5 mA and its value 5 ms later is 3 mA. Find the time constant and the inductance.

☐ **3** Both switches in the circuit of Fig. 5-17 are closed at $t = 0$. Find $i_1(t)$, $i_2(t)$, and $i_3(t)$ for $t > 0$.

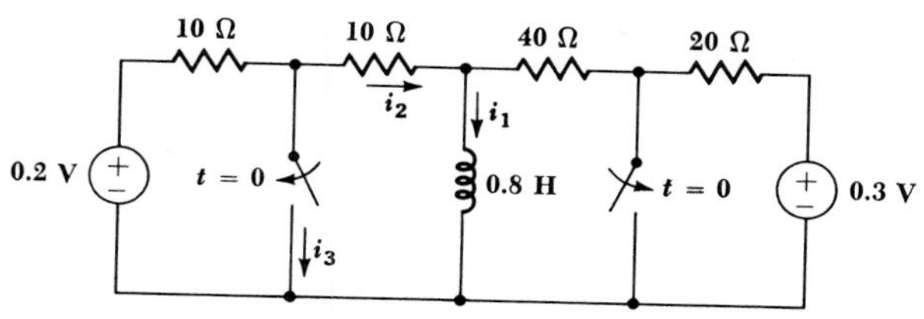

Fig. 5-17 See Prob. 3.

☐ **4** The initial inductor current in the network of Fig. 5-18 is 150 mA. Find $v(t)$ for $t > 0$.

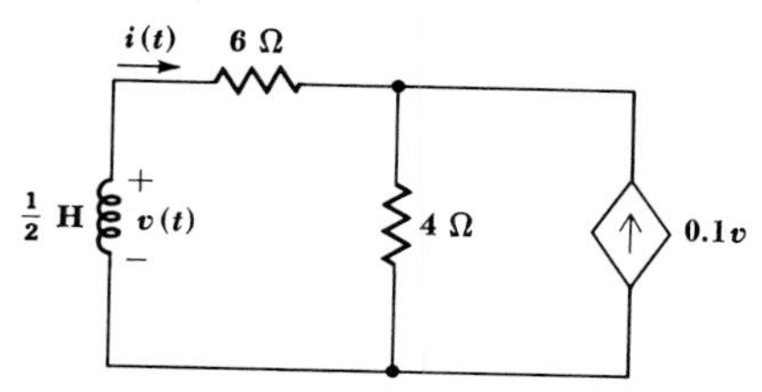

Fig. 5-18 See Prob. 4.

☐ **5** Change the control on the dependent current source in Fig. 5-18 from $0.1v$ to $0.5i$. Let $v(0^+)$ be 2 V and find $i(t)$ for $t > 0$.

☐ **6** The switch in the circuit of Fig. 5-19 is opened at $t = 0$ after having been closed for a very long time. (*a*) What is the voltage v across the switch at $t = 0^+$? (*b*) Find $v(t)$ for $t > 0$.

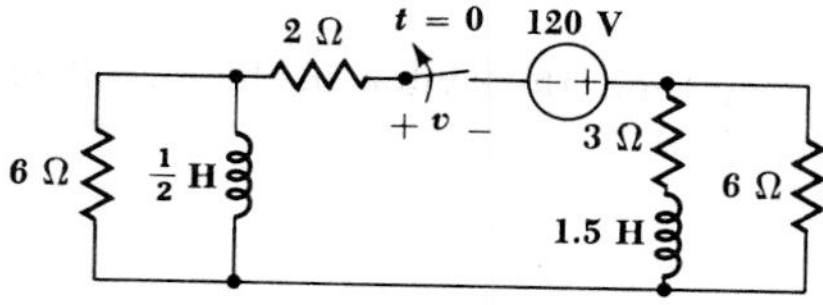

Fig. 5-19 See Prob. 6.

☐ **7** A 1.5-V battery having an internal resistance of 0.5 Ω is in series with a switch and an inductor L. In parallel with the inductor is a resistor R. Assume that the inductor is ideal. Select values for L and R such that the voltage across R

jumps from zero to 1000 V as the switch is opened and does not drop below 100 V until 1 ms later.

☐ **8** Let the initial current in a source-free series RL circuit be I_0. Show that the total charge flowing through R between $t = 0$ and $t = \infty$ is the same as that which would flow if the current remained I_0 for one time constant.

☐ **9** The left switch in the circuit of Fig. 5-20 is closed at $t = 0$. (*a*) Find $i_L(t)$ and $i_1(t)$ for $0 < t < 0.1$ s. (*b*) The right switch is then closed at $t = 0.1$ s. Find $i_L(t)$ and $i_1(t)$ for $t > 0.1$ s.

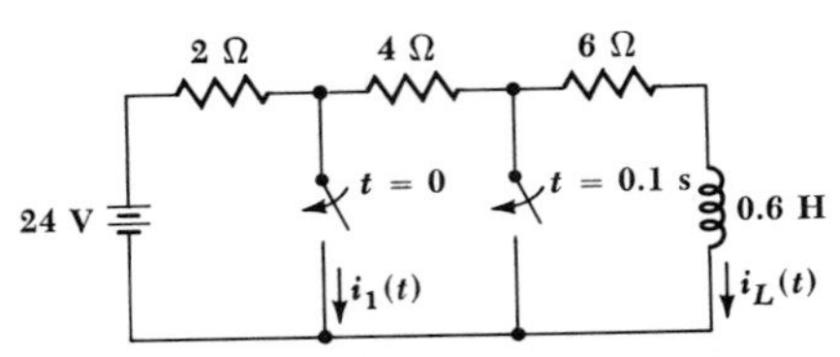

Fig. 5-20 See Prob. 9.

☐ **10** The two switches in the circuit of Fig. 5-21 are thrown simultaneously at $t = 0$. (*a*) Find $i_1(0^+)$, $i_2(0^+)$, and $i(0^+)$. (*b*) find L_{eq} and τ. (*c*) Write $i(t)$ for $t > 0$. (*d*) To find $i_1(t)$ or $i_2(t)$, it is necessary to include a possible constant current present in the inductive loop. This may be done by finding the voltage across the 6-Ω resistor, which is the voltage across each coil, and integrating the inductor voltage to find the current; the known initial value is used as the constant of integration. Find $i_1(t)$ and $i_2(t)$. (*e*) Show that the sum of the energies remaining in the two coils as $t \to \infty$ plus that dissipated since $t = 0$ in the resistor is equal to the sum of the inductor energies at $t = 0$.

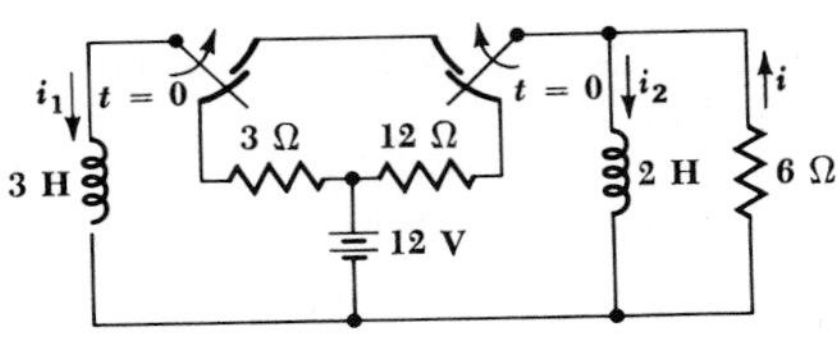

Fig. 5-21 See Prob. 10.

☐ **11** After being closed for a long time, the switch in the circuit of Fig. 5-22 is opened at $t = 0$. Find $v(t)$ for $t > 0$.

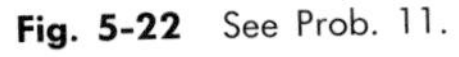

Fig. 5-22 See Prob. 11.

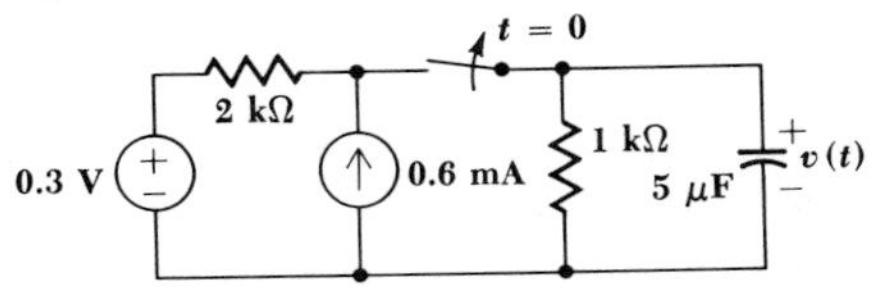

☐ **12** A certain precision 1-μF capacitor has very high resistance material used between its conducting surfaces. The capacitor is charged to 1 V at $t = 0$ and disconnected from the source. It is found that the voltage drops to 0.9 V in 100 hr. Find the insulation resistance.

☐ **13** The independent source in the circuit of Fig. 5-23 is 140 V for $t < 0$ and 0 for $t > 0$. Find $i(t)$ and $v_0(t)$.

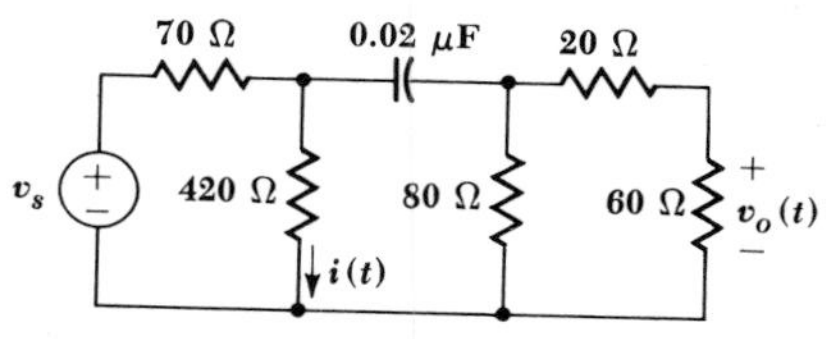

Fig. 5-23 See Prob. 13.

☐ **14** After having been closed for a long time, the switch in the network of Fig. 5-24 is opened at $t = 0$. Find $v_C(t)$ for $t > 0$.

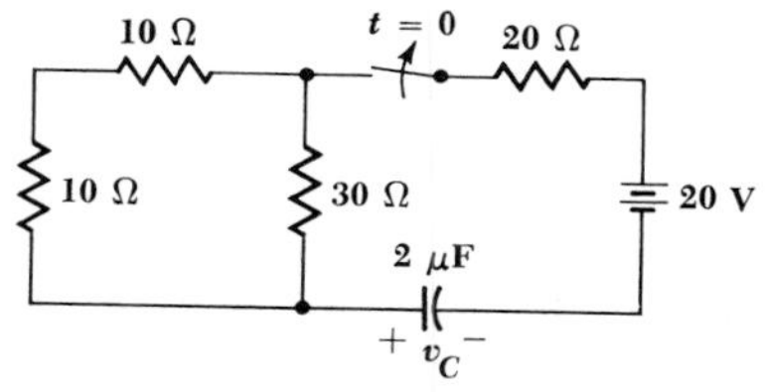

Fig. 5-24 See Prob. 14.

☐ **15** With reference to the circuit shown in Fig. 5-25, let $v(0) = 9$ V. Find $i(t)$ for $t > 0$.

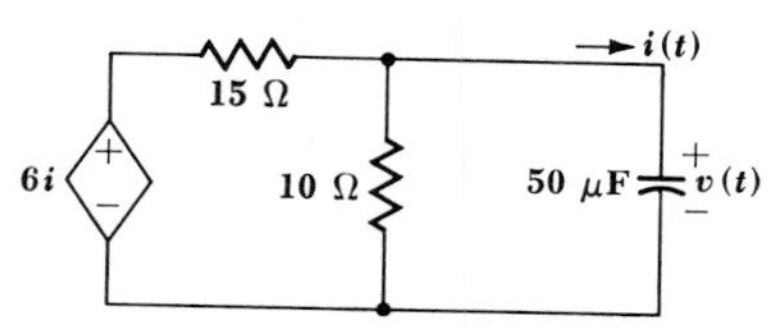

Fig. 5-25 See Prob. 15.

☐ **16** Replace the dependent voltage source in Fig. 5-25 with a dependent current source, arrow directed upward, with control of $-0.1v$. Find $v(t)$ and $i(t)$ if $v(0) = 9$ V.

☐ **17** Two resistive wires, each 1 m long with a resistance of 10 Ω/cm, are parallel and connected together at both ends with a perfect conductor. A 1000-pF capacitor is connected directly between the wires at some unspecified point. (*a*) What is the maximum possible time constant? (*b*) If $\tau = 0.3$ μs, what is the distance from the capacitor to the nearer short circuit?

□ **18** The initial capacitor voltages in the circuit of Fig. 5-26 are $v_1(0) = 10$ V, $v_2(0) = 4$ V. At what value of t is $i = 0$?

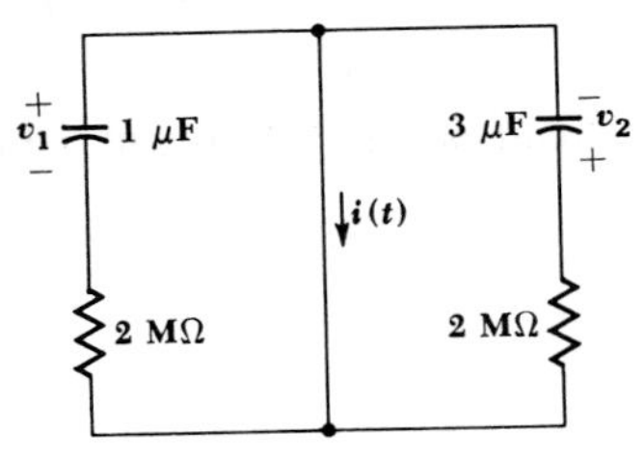

Fig. 5-26 See Prob. 18.

□ **19** The switch in the circuit shown in Fig. 5-27 has been at a for a long time. At $t = 0$ it is moved to b, and at $t = 1$ s it is moved to c. At what value of t is $v = 1$ V?

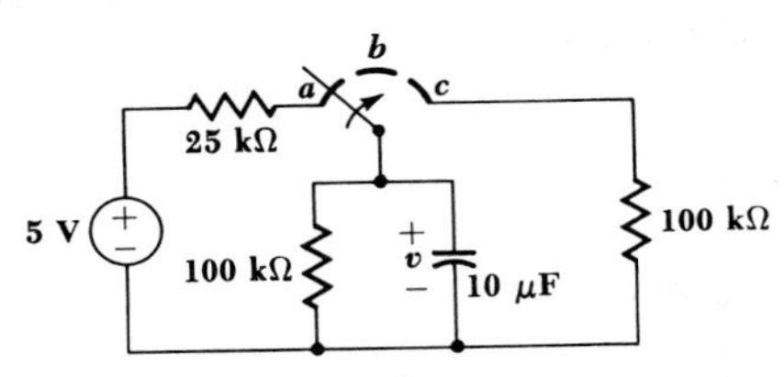

Fig. 5-27 See Prob. 19.

□ **20** Five circuit elements, a 2-, a 3-, and a 6-μF capacitor, a 2-mA current source, and a 0.1-MΩ resistor, have been connected in parallel since a week ago last Tuesday. At $t = 0$, a frenzy of switching removes the current source from the circuit and connects the three capacitors in series with the resistor, such that the initial capacitor voltages are series aiding. (*a*) Find the voltage across the resistor as a function of time. (*b*) Find the current through the capacitors. (*c*) By integrating the current and using the known initial conditions, find the voltage across each capacitor as a function of time. (*d*) What is the voltage across each capacitor as t approaches infinity?

□ **21** After remaining open for a long time, the switch in the circuit of Fig. 5-28 is closed at $t = 0$. At what time after $t = 0$ is $i(t)$ a maximum magnitude?

Fig. 5-28 See Prob. 21.

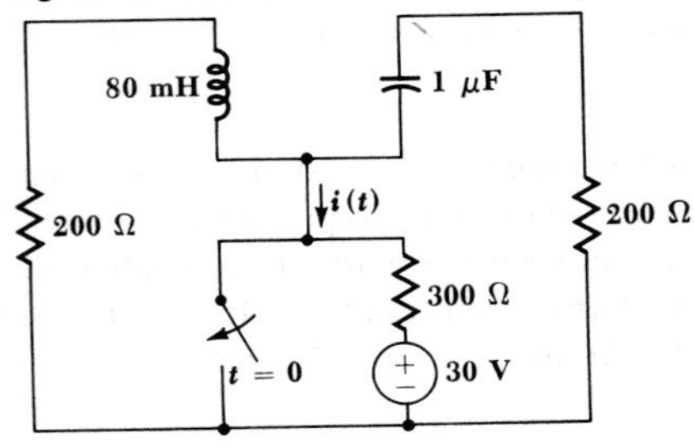

☐ **22** The switch in the circuit of Fig. 5-29 closes at $t = 0$. Find $i_1(t)$ and $i_2(t)$ for $t > 0$.

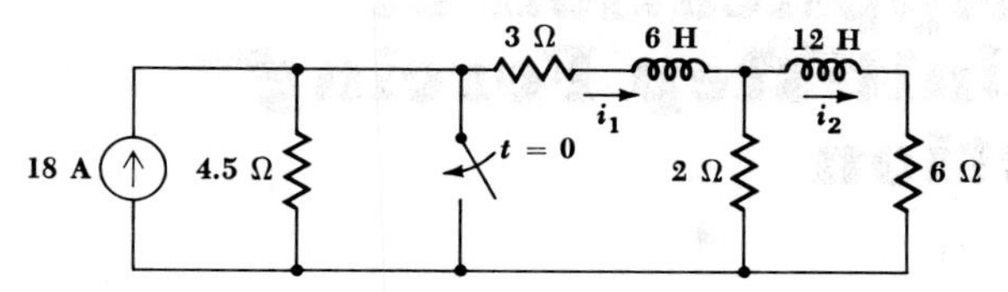

Fig. 5-29 See Prob. 22.

☐ **23** For the circuit shown in Fig. 5-30, find $v_1(t)$ and $v_2(t)$ for $t > 0$.

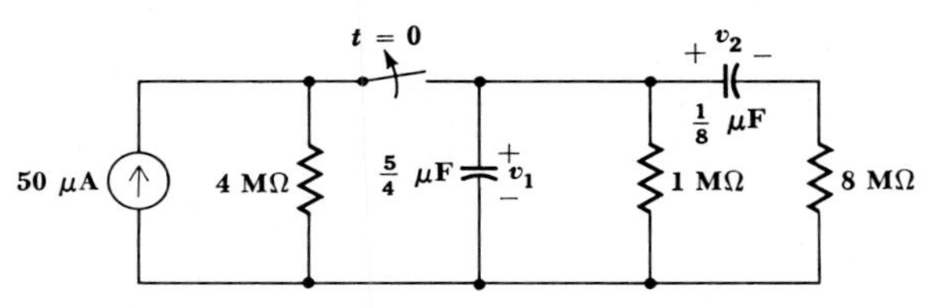

Fig. 5-30 See Prob. 23.

Chapter Six
The Application of the Unit-step Forcing Function

6-1 INTRODUCTION

We have just spent a chapter's worth of our time studying the response of RL and RC circuits when no source or forcing function was present. We termed this response the natural response because its form depends only on the nature of the circuit. The reason that any response at all is obtained arises from the possibility of an initial energy storage within the inductive or capacitive elements in the circuit. In many of the examples and problems, we were confronted with circuits containing sources and switches; we were informed that certain switching operations were performed at $t = 0$ in order to remove all the sources from the circuit, while leaving known amounts of energy stored here and there. In other words, we have been solving problems in which energy sources are suddenly *removed* from the circuit; now we must consider that type of response which results when energy sources are suddenly *applied* to a circuit.

We shall devote this chapter to a study of the response which occurs when the energy sources which are suddenly applied are dc sources. After we have studied sinusoidal and exponential sources, we may then consider the general problem of the sudden application of a more general source. Since every electrical device is intended to be energized at least once, and since most devices are turned on and off many times in the course of their lives, it should be evident that our study will be applicable to many practical cases. Even though we are now restricting ourselves to dc sources, there are still innumerable cases in which these simpler examples correspond to the operation of physical devices. For example, the first circuit we shall analyze may be considered to represent the build-up of the field current when a dc motor is started. The generation and use of the rectangular voltage pulses needed to represent a number or a command in a digital computer provide many examples in the field of electronic or transistor circuitry. Similar circuits are found in the synchronization and sweep circuits of television receivers, in communication systems using pulse modulation, and in radar systems, to name but a few examples. Furthermore, an important part of the analysis of most servomechanisms is the determination of their responses to suddenly applied constant inputs.

6-2 THE UNIT-STEP FORCING FUNCTION

We have been speaking of the "sudden application" of an energy source, and by this phrase we imply its application in zero time. The operation of a switch in series with a battery is thus equivalent to a forcing function which is zero up to the instant that the switch is closed and is equal to the battery voltage thereafter. The forcing function has a break or discontinuity at the instant the switch is closed. Certain special forcing functions which are discontinuous or have discontinuous derivatives are called *singularity functions*, the two most important of these singularity functions being the unit-step function and the unit-impulse function. The unit-step function is the subject of this chapter; the unit impulse is discussed in Chaps. 18 and 19.

We define the *unit-step function* as a function which is zero for all values of its argument which are less than zero and which is unity for all positive values of its argument. If we let x be the argument and represent the unit-step function by u, then $u(x)$ must be zero for all values of x less than zero, and it must be unity for all values of x greater than zero. At $x = 0$, $u(x)$ changes abruptly from 0 to 1. Its value at $x = 0$ is not defined, but

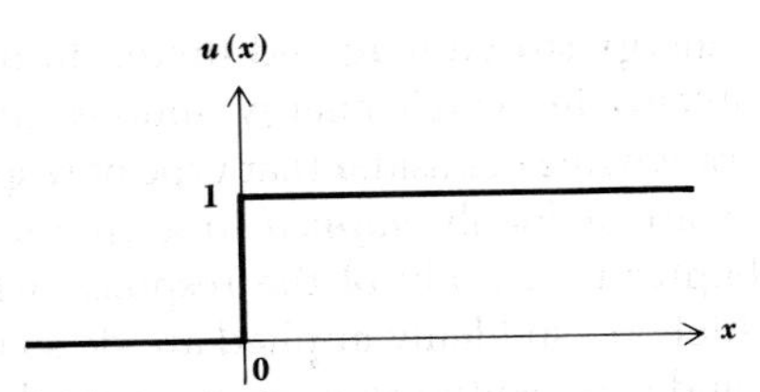

Fig. 6-1 The unit-step function $u(x)$ is shown as a function of x.

its value is known for all points arbitrarily close to $x = 0$. We often indicate this by writing $u(0^-) = 0$ and $u(0^+) = 1$. The concise mathematical definition of the unit-step function is

$$u(x) = \begin{cases} 0 & x < 0 \\ 1 & x > 0 \end{cases}$$

and it is shown graphically in Fig. 6-1.

In order to obtain a unit-step *forcing* function, we must express the unit step as a function of time, and the simplest expression is obtained by replacing x by t,

$$u(t) = \begin{cases} 0 & t < 0 \\ 1 & t > 0 \end{cases}$$

However, we cannot always arrange the operation of a circuit so that every discontinuity occurs at $t = 0$. If two switches are to be thrown in sequence, for example, we might elect to operate one of them at $t = 0$, and it is then necessary to throw the second switch at some later time $t = t_0$. Since the unit-step function provides us with a suitable discontinuity when the value of its argument is zero, then the operation of the second switch may be represented by selecting a unit-step forcing function having an argument $(t - t_0)$. Thus,

$$u(t - t_0) = \begin{cases} 0 & t < t_0 \\ 1 & t > t_0 \end{cases}$$

This function is shown in Fig. 6-2.

Fig. 6-2 The unit-step forcing function, $u(t - t_0)$.

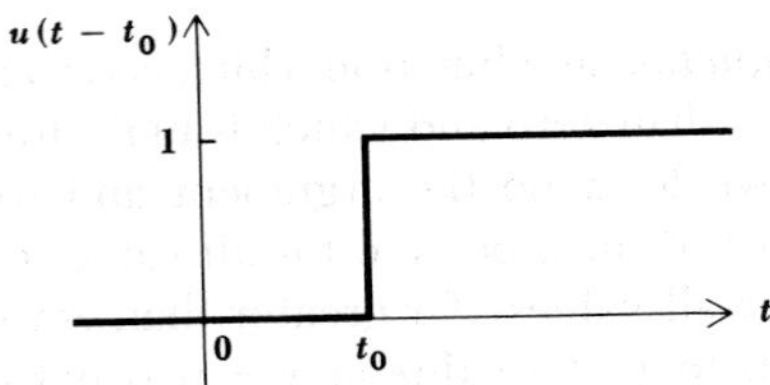

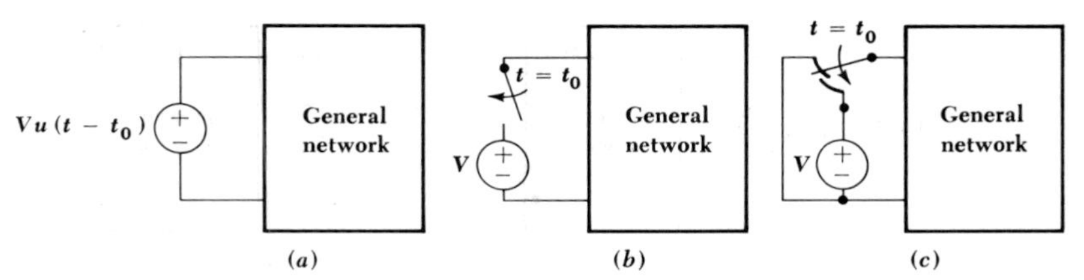

Fig. 6-3 (*a*) A voltage-step forcing function is shown as the source of a general network. (*b*) A simple circuit which, although not the exact equivalent of (*a*), may be used as its equivalent in many cases. (*c*) An exact equivalent of (*a*).

The unit-step forcing function is in itself dimensionless. If we wish it to represent a voltage, it is necessary to multiply $u(t - t_0)$ by some constant voltage, such as V. Thus, $v(t) = Vu(t - t_0)$ is an ideal voltage source which is zero before $t = t_0$ and a constant V after $t = t_0$. This forcing function is shown connected to a general network in Fig. 6-3*a*. We should now logically ask what physical source is the equivalent of this discontinuous forcing function. By equivalent, we mean simply that the voltage-current characteristics of the two networks are identical. For the step-voltage source of Fig. 6-3*a*, the voltage-current characteristic is quite simple; the voltage is zero prior to $t = t_0$, it is V after $t = t_0$, and the current may be any (finite) value in either time interval. Our first thoughts might produce the attempt at an equivalent shown in Fig. 6-3*b*, a dc source V in series with a switch which closes at $t = t_0$. This network is not equivalent for $t < t_0$, however, because the voltage across the battery and switch is completely unspecified in this time interval. The "equivalent" source is an open circuit, and the voltage across it may be anything. After $t = t_0$, the networks are equivalent, and if this is the only time interval in which we are interested, and if the initial currents which flow from the two networks are identical at $t = t_0$, then Fig. 6-3*b* becomes a useful equivalent of Fig. 6-3*a*.

In order to obtain an exact equivalent for the voltage-step forcing function, we may provide a single-pole double-throw switch. Before $t = t_0$, the switch serves to ensure zero voltage across the input terminals of the general network. After $t = t_0$, the switch is thrown to provide a constant input voltage V. At $t = t_0$, the voltage is indeterminate (as is the step function), the battery is momentarily short-circuited, and it is fortunate that we are dealing with mathematical models. This exact equivalent of Fig. 6-3*a* is shown in Fig. 6-3*c*.

Before concluding our discussion of equivalence, it is enlightening to consider the exact equivalent of a battery and a switch. What is the voltage-step forcing function which is equivalent to Fig. 6-3*b*? We are searching for some arrangement which changes suddenly from an open circuit to a constant voltage; a change in resistance is involved, and this

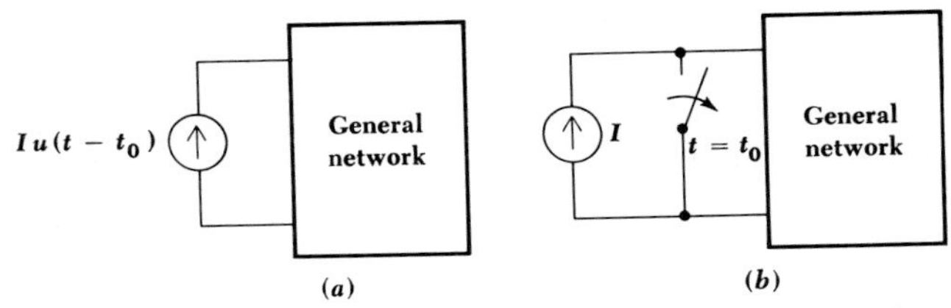

Fig. 6-4 (*a*) A current-step forcing function is applied to a general network. (*b*) A simple circuit which, although not the exact equivalent of (*a*), may be used as its equivalent in many cases.

is the crux of our difficulty. The step function enables us to change a voltage discontinuously (or a current), but here we need a changing resistance as well. The equivalent therefore must contain a resistance or conductance step function, a passive element which is time-varying. Although we might construct such an element with the unit-step function, it should be apparent that the end product is a switch; a switch is merely a resistance which changes instantaneously from zero to infinite ohms, or vice versa. Thus, we conclude that the exact equivalent of a battery and switch in series must be a battery in series with some representation of a time-varying resistance; no arrangement of voltage- and current-step forcing functions is able to provide us with the exact equivalent.[1]

Figure 6-4*a* shows a current-step forcing function driving a general network. If we attempt to replace this circuit by a dc source in parallel with a switch (which *opens at* $t = t_0$), we must realize that the circuits are equivalent after $t = t_0$, but the responses are alike after $t = t_0$ only if the initial conditions are the same. Judiciously, then, we may often use the circuits of Fig. 6-4*a* and *b* interchangeably. The exact equivalent of Fig. 6-4*a* is the dual of the circuit of Fig. 6-3*c*; the exact equivalent of Fig. 6-4*b* cannot be constructed with current- and voltage-step forcing functions alone.[2]

Some very useful forcing functions may be obtained by manipulating the unit-step forcing function. Let us define a rectangular voltage pulse by the following conditions:

$$v(t) = \begin{cases} 0 & t < t_0 \\ V & t_0 < t < t_1 \\ 0 & t_1 < t \end{cases}$$

The pulse is drawn in Fig. 6-5. Can this pulse be represented in terms of the unit-step forcing function? Let us consider the difference of the

[1]An equivalent may always be determined *if some information about the general network is available* (the voltage across the switch for $t < t_0$); we assume no a priori knowledge about the general network.

[2]The equivalent can be drawn if the current through the switch prior to $t = t_0$ is known.

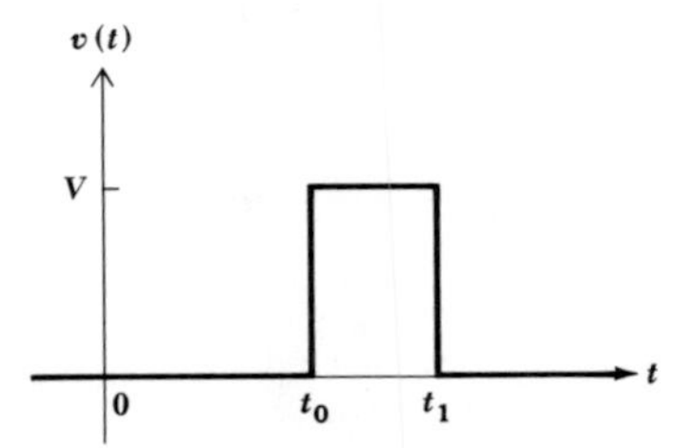

Fig. 6-5 A useful forcing function, the rectangular voltage pulse.

two unit steps $u(t - t_0) - u(t - t_1)$. The two step functions are shown in Fig. 6-6*a*, and their difference is obviously a rectangular pulse. The source $Vu(t - t_0) - Vu(t - t_1)$ which will provide us with the desired voltage is indicated in Fig. 6-6*b*.

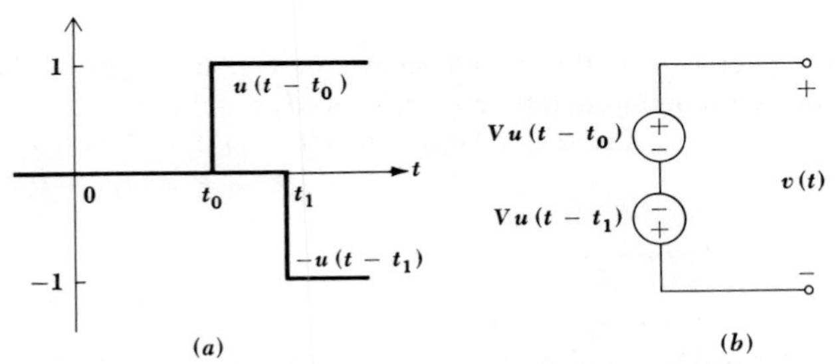

Fig. 6-6 (*a*) The unit steps $u(t - t_0)$ and $-u(t - t_1)$. (*b*) A source which yields the rectangular voltage pulse of Fig. 6-5.

If we have a sinusoidal voltage source $V \sin \omega t$ which is suddenly connected to a network at $t = t_0$, then an appropriate voltage forcing function would be $v(t) = Vu(t - t_0) \sin \omega t$. If we wish to represent one burst of energy from a radar transmitter, we may turn the sinusoidal source off $1/10$ μs later by a second unit-step forcing function. The voltage pulse is thus

$$v(t) = V[u(t - t_0) - u(t - t_0 - 10^{-7})] \sin \omega t$$

This forcing function is sketched in Fig. 6-7.

As a final introductory remark, we should note that the unit-step forcing function must be considered only as the mathematical model of an actual switching operation. No physical resistor, inductor, or capacitor behaves entirely like its idealized circuit element; we also cannot perform a switching operation in zero time. However, switching times less than 1 ns are common in many circuits, and this time is often sufficiently short compared with the time constants in the rest of the circuit that it may be ignored.

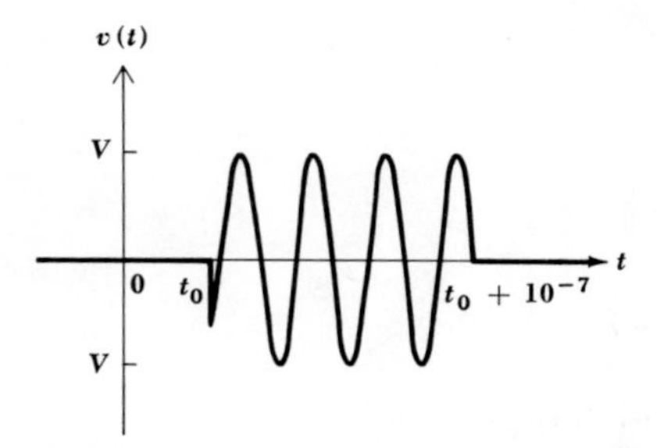

Fig. 6-7 A radio-frequency pulse, described by $v(t) = V[u(t - t_0) - u(t - t_0 - 10^{-7})] \sin \omega t$. The sinusoidal frequency within the pulse shown is about 36 MHz, a value which is too low for radar, but about right for constructing legible drawings.

Drill Problems

6-1 With reference to the circuit shown in Fig. 6-8, express the following currents in step-function notation: (*a*) $i_1(t)$; (*b*) $i_2(t)$; (*c*) $i_3(t)$.

Ans. $2 - 2.8u(t) - 0.2u(t - 1)$; $2 + 1.2u(t) - 0.2u(t - 1)$; $2 + 1.2u(t) + 0.2u(t - 1)$ A

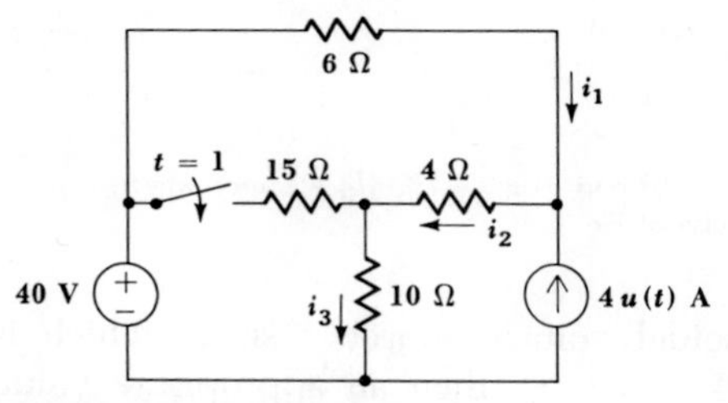

Fig. 6-8 See Drill Prob. 6-1.

6-2 The voltage waveform, $v(t) = 12u(t) - 16u(t - 3)$ V, is applied to a 2-H inductor. Using the passive sign convention, find the inductor current at $t =$: (*a*) 0; (*b*) 2; (*c*) 4 s.

Ans. 0; 12; 16 A

6-3 A FIRST LOOK AT THE DRIVEN *RL* CIRCUIT

We are now ready to subject a simple network to the sudden application of a dc source. The circuit consists of a battery V in series with a switch, a resistance R, and an inductance L. The switch is closed at $t = 0$, as indicated on the circuit diagram of Fig. 6-9*a*. It is evident that the current $i(t)$ is zero before $t = 0$, and we are therefore able to replace the battery

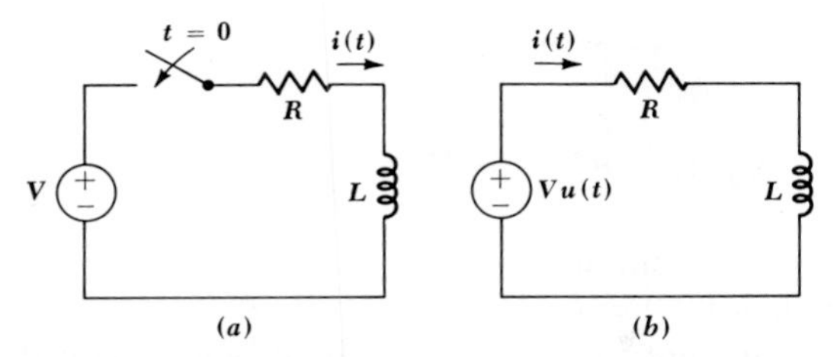

Fig. 6-9 (*a*) The given circuit. (*b*) An equivalent circuit, possessing the same response $i(t)$ for all time.

and switch by a voltage-step forcing function $Vu(t)$, which also produces no response prior to $t = 0$. After $t = 0$, the two circuits are obviously identical. Hence, we seek the current $i(t)$ either in the given circuit of Fig. 6-9*a* or in the equivalent circuit of Fig. 6-9*b*.

We shall find $i(t)$ at this time by writing the appropriate circuit equation and then solving it by separation of the variables and integration. After we obtain the answer and investigate the two parts of which it is composed, we shall next spend some time (the following section) in learning the general significance of these two terms. We can then construct the solution to this problem very easily; moreover, we shall be able to apply the general principles behind this simpler method to produce more rapid and more meaningful solutions to every problem involving the sudden application of any source. Let us now proceed with the more formal method of solution.

Applying Kirchhoff's voltage law to the circuit of Fig. 6-9*b*, we have

$$Ri + L\frac{di}{dt} = Vu(t)$$

Since the unit-step function is discontinuous at $t = 0$, we shall first consider the solution for $t < 0$ and then for $t > 0$. It is evident that the application of zero voltage since $t = -\infty$ has not produced any response, and, therefore,

$$i(t) = 0 \qquad t < 0$$

For positive time, however, $u(t)$ is unity and we must solve the equation

$$Ri + L\frac{di}{dt} = V \qquad t > 0$$

The variables may be separated in several simple algebraic steps, yielding

$$\frac{L\,di}{V - Ri} = dt$$

and each side may be integrated directly,

$$-\frac{L}{R}\ln(V - Ri) = t + k$$

In order to evaluate k, an initial condition must be invoked. Prior to $t = 0$, $i(t)$ is zero, and thus $i(0^-) = 0$; since the current in an inductor cannot change by a finite amount in zero time without being associated with an infinite voltage, we thus have $i(0^+) = 0$. Setting $i = 0$ at $t = 0$, we obtain

$$-\frac{L}{R}\ln V = k$$

and, hence,

$$-\frac{L}{R}[\ln(V - Ri) - \ln V] = t$$

Rearranging,

$$\frac{V - Ri}{V} = e^{-Rt/L}$$

or

$$i = \frac{V}{R} - \frac{V}{R}e^{-Rt/L} \qquad t > 0$$

Thus, an expression for the response valid for all t would be

$$i = \left(\frac{V}{R} - \frac{V}{R}e^{-Rt/L}\right)u(t) \tag{1}$$

This is the desired solution, but it has not been obtained in the simplest manner. In order to establish a more direct procedure, let us try to interpret the two terms appearing in (1). The exponential term has the functional form of the natural response of the RL circuit; it is a negative exponential, it approaches zero as time increases, and it is characterized by the time constant L/R. The *functional form* of this part of the response is thus identical with that which is obtained in the source-free circuit. However, the amplitude of this exponential term depends on V. We might generalize, then, that the response will be the sum of two terms, where one term has a functional form which is identical with that of the source-free response, but has an amplitude which depends on the forcing function. Now let us consider the nature of the second part of the response.

Equation (1) also contains a constant term V/R. Why is it present? The answer is simple: the natural response approaches zero as the energy is gradually dissipated, but the total response must not approach zero. Eventually the circuit behaves as a resistor and inductor in series with a battery, and a direct current V/R flows. This current is a part of the response which is directly attributable to the forcing function, and we call it the *forced response*. It is the response which is present a long time after the switch is closed.

The complete response is composed of two parts, the natural response and the forced response. The natural response is a characteristic of the circuit and not of the sources. Its form may be found by considering the source-free circuit, and it has an amplitude which depends on the initial amplitude of the source and the initial energy storage. The forced response has the characteristics of the forcing function; it is found by pretending that all switches have been thrown a long time ago. Since we are presently concerned only with switches and dc sources, the forced response is merely the solution of a simple dc circuit problem.

The reason for the two responses, forced and natural, may also be seen from physical arguments. We know that our circuit will eventually assume the forced response. However, at the instant the switches are thrown, the initial currents in the coils (or the voltages across the capacitors in other circuits) will have values which depend only on the energy stored in these elements. These currents or voltages cannot be expected to be the same as the currents and voltages demanded by the forced response. Hence, there must be a transient period during which the currents and voltages change from their given initial values to their required final values. The portion of the response which provides the transition from initial to final values is the natural response (often called the transient response, as we found earlier). If we describe the response of the *source-free* simple RL circuit in these terms, then we should say that the forced response is zero and that the natural response serves to connect the initial response produced by stored energy with the zero value of the forced response. This description is appropriate only for those circuits in which the natural response eventually dies out. This always occurs in physical circuits where some resistance is associated with every element, but there are a number of pathologic circuits in which the natural response is nonvanishing as time becomes infinite; those circuits in which trapped currents circulate around inductive loops or voltages are trapped in series strings of capacitors are examples.

Now let us search out the mathematical basis for dividing the response into a natural response and a forced response.

Drill Problem

6-3 A current source of $0.2u(t)$ A, a 100-Ω resistor, and a 0.4-H inductor are in parallel. Find the magnitude of the inductor current: (*a*) as $t \to \infty$; (*b*) at $t = 0^+$; (*c*) at $t = 4$ ms.

Ans. 0; 0.126; 0.200 A

6-4 THE NATURAL AND THE FORCED RESPONSE

There is also an excellent mathematical reason for considering the complete response to be composed of two parts, the forced response and the natural response. The reason is based on the fact that the solution of any linear differential equation may be expressed as the sum of two parts, the complementary solution (natural response) and the particular solution (forced response). Without delving into the general theory of differential equations, let us consider a general equation of the type met in the previous section,

$$\frac{di}{dt} + Pi = Q \tag{2}$$

We may identify Q as a forcing function and express it as $Q(t)$ to emphasize its general time dependence. In all our circuits, P will be a positive constant, but the remarks that follow about the solution of (2) are equally valid for the cases in which P is a general function of time. Let us simplify the discussion by assuming that P is a positive constant. Later, we shall also assume that Q is constant, thus restricting ourselves to dc forcing functions.

In any standard text on elementary differential equations, it is shown that if both sides of this equation are multiplied by a so-called integrating factor, then each side becomes an exact differential which can be integrated directly to obtain the solution. We are not separating the variables, but merely arranging them in such a way that integration is possible. For this equation, the integrating factor is $e^{\int P\,dt}$ or e^{Pt}, since P is a constant. We multiply each side of the equation by this integrating factor and obtain

$$e^{Pt}\frac{di}{dt} + iPe^{Pt} = Qe^{Pt}$$

The form of the left side may now be improved when it is recognized as the exact derivative of ie^{Pt},

$$\frac{d}{dt}(ie^{Pt}) = e^{Pt}\frac{di}{dt} + iPe^{Pt}$$

and, thus,

$$\frac{d}{dt}(ie^{Pt}) = Qe^{Pt}$$

We may now integrate each side with respect to time, finding

$$ie^{Pt} = \int Qe^{Pt}\,dt + A$$

where A is a constant of integration. Since this constant is explicitly shown, we should remember that no integration constant needs to be added to the remaining integral when it is evaluated. Multiplication by e^{-Pt} produces the solution for $i(t)$,

$$i = e^{-Pt} \int Qe^{Pt}\,dt + Ae^{-Pt} \tag{3}$$

If $Q(t)$, the forcing function, is known, then it remains only to evaluate the integral to obtain the exact functional form for $i(t)$. We shall not evaluate such an integral for each problem, however; instead, we are interested in using (3) as an exemplary solution from which we shall draw several very general conclusions.

We should note first that, for a source-free circuit, Q must be zero, and the solution is the natural response

$$i_n = Ae^{-Pt} \tag{4}$$

We shall find that the constant P is never negative; its value depends only on the passive circuit elements[3] and their interconnection in the circuit. The natural response therefore approaches zero as time increases without limit. It must do so, of course, in the simple RL series circuit because the initial energy is gradually dissipated in the resistor. There are also idealized, nonphysical circuits in which P is zero; in these circuits the natural response does not die out, but approaches a constant value, as exemplified by trapped currents or voltages. We therefore find that one of the two terms making up the complete response has the form of the natural response; it has an amplitude which will depend on the initial value of the complete response and thus on the initial value of the forcing function also.

We next observe that the first term of (3) depends on the functional form of $Q(t)$, the forcing function. Whenever we have a circuit in which the natural response dies out as t becomes infinite, then this first term must describe the form of the response completely after the natural response has disappeared. This term we shall call the *forced response;* it is also called the steady-state response, the particular solution, or the particular integral.

For the present, we have elected to consider only those problems involving the sudden application of dc sources, and $Q(t)$ is therefore a constant for all values of time after the switch has been closed. If we wish, we can now evaluate the integral in (3), obtaining the forced response

$$i_f = \frac{Q}{P}$$

[3]If the circuit contains a dependent source or a negative resistance, it is possible for P to be negative.

or the complete response

$$i(t) = \frac{Q}{P} + Ae^{-Pt}$$

For the RL series circuit, Q/P is the constant current V/R and $1/P$ is the time constant τ. We should see that the forced response might have been obtained without evaluating the integral, because it must be the complete response at infinite time; it is merely the source voltage divided by the series resistance. The forced response is thus obtained by inspection.

In the following section we shall attempt to find the complete response for several RL circuits by obtaining the natural and forced responses and then adding them.

Drill Problem

6-4 Let the voltage source $6e^{-2t}u(t)$ V be connected in series with a 2-H inductor and a 10-Ω resistor. Using (3), determine i at $t =$: (*a*) 0^+; (*b*) 0.2 s; (*c*) 0.5 s.

Ans. *0; 0.286; 0.302* A

6-5 *RL* CIRCUITS

Let us use the simple RL series circuit to illustrate how to determine the complete response by the addition of the natural and forced responses. This circuit, shown in Fig. 6-10, has been analyzed earlier, but by a longer method. The desired response is the current $i(t)$, and we first express this current as the sum of the natural and the forced current,

$$i = i_n + i_f$$

The functional form of the natural response must be the same as that obtained without any sources. We therefore replace the step-voltage source

Fig. 6-10 A series RL circuit which is used to illustrate the method by which the complete response is obtained as the sum of the natural and forced response.

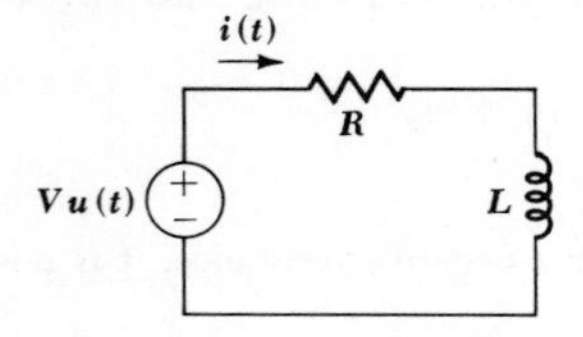

by a short circuit and recognize the old RL series loop. Thus,

$$i_n = Ae^{-Rt/L}$$

where the amplitude A is yet to be determined.

We next consider the forced response, that part of the response which depends upon the nature of the forcing function itself. In this particular problem the forced response must be constant because the source is a constant V for all positive values of time. After the natural response has died out, therefore, there can be no voltage across the inductor; hence, a voltage V appears across R, and the forced response is simply

$$i_f = \frac{V}{R}$$

Note that the forced response is determined completely; there is no unknown amplitude. We next combine the two responses

$$i = Ae^{-Rt/L} + \frac{V}{R}$$

and apply the initial condition to evaluate A. The current is zero prior to $t = 0$, and it cannot change value instantaneously since it is the current flowing through an inductor. Thus, the current is zero immediately after $t = 0$, and

$$0 = A + \frac{V}{R}$$

and

$$i = \frac{V}{R}(1 - e^{-Rt/L}) \tag{5}$$

This response is plotted in Fig. 6-11, and we can see the manner in which the current builds up from its initial value of zero to its final value

Fig. 6-11 The current expressed by (5) is shown graphically. A line extending the initial slope meets the constant forced response at $t = \tau$.

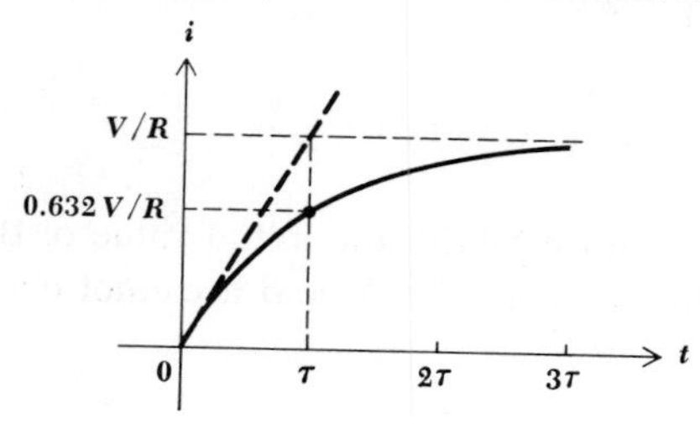

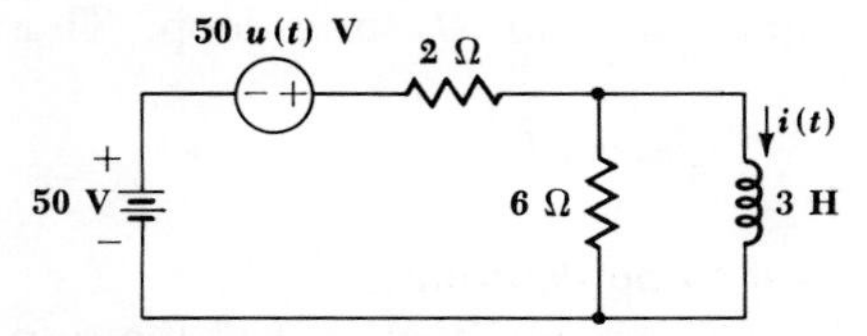

Fig. 6-12 A circuit used as an example.

of V/R. The transition is effectively accomplished in a time 3τ. If our circuit represents the field coil of a large dc motor, we might assign $L = 10$ H and $R = 20\ \Omega$, obtaining $\tau = 0.5$ s. The field current is thus established in about 1.5 s. In one time constant, the current has attained 63.2 per cent of its final value.

Now let us apply this method to a more complicated circuit. In Fig. 6-12, there is shown a circuit containing a dc voltage source as well as a step-voltage source. Let us determine $i(t)$ for all values of time. We might choose to replace everything to the left of the inductor by the Thévenin equivalent, but instead let us merely recognize the form of that equivalent as a resistor in series with some voltage source. The circuit contains only one energy-storage element, the inductor, and the natural response is therefore a negative exponential as before,

$$i = i_f + i_n$$

where

$$i_n = Ae^{-t/2} \qquad t > 0$$

since

$$\tau = \frac{L}{R_{eq}} = \frac{3}{1.5} = 2$$

The forced response must be that produced by a constant voltage of 100 V. The forced response is constant, and no voltage is present across the inductor; it behaves as a short circuit and, therefore,

$$i_f = \frac{100}{2} = 50$$

Thus,

$$i = 50 + Ae^{-0.5t} \qquad t > 0$$

In order to evaluate A, we must establish the initial value of the inductor current. Prior to $t = 0$, this current is 25 A, and it cannot change instantaneously. Thus,

$$25 = 50 + A \qquad \text{or} \qquad A = -25$$

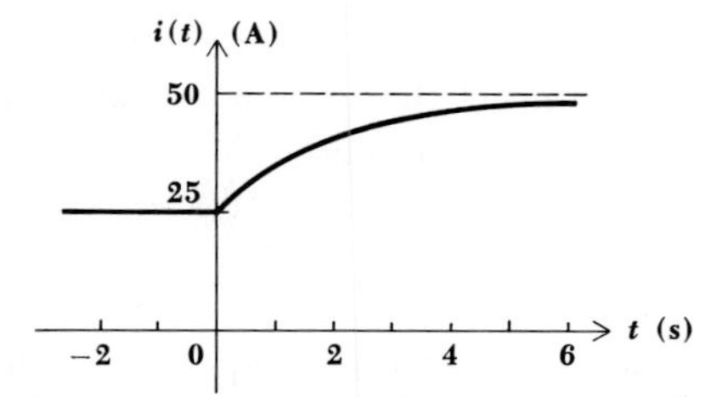

Fig. 6-13 The response $i(t)$ of the circuit shown in Fig. 6-12 is sketched for values of time less and greater than zero.

Hence,

$$i = 50 - 25e^{-0.5t} \qquad t > 0$$

We complete the solution by also stating

$$i = 25 \qquad t < 0$$

or by writing a single expression valid for all t,

$$i = 25 + 25(1 - e^{-0.5t})u(t) \quad \text{A}$$

The complete response is sketched in Fig. 6-13. Note how the natural response serves to connect the response for $t < 0$ with the constant forced response.

As a final example of this method by which the complete response of any circuit subjected to a transient may be written down almost by inspection, let us apply a rectangular voltage pulse of amplitude V and duration t_0 to the simple RL series circuit. We represent the forcing function as the sum of two step-voltage sources $Vu(t)$ and $-Vu(t - t_0)$, as indicated in Fig. 6-14a and b, and plan to obtain the response by using the super-

Fig. 6-14 (a) A rectangular voltage pulse which is to be used as the forcing function in a simple series RL circuit. (b) The series RL circuit, showing the representation of the forcing function by the series combination of two independent voltage-step sources. The current $i(t)$ is desired.

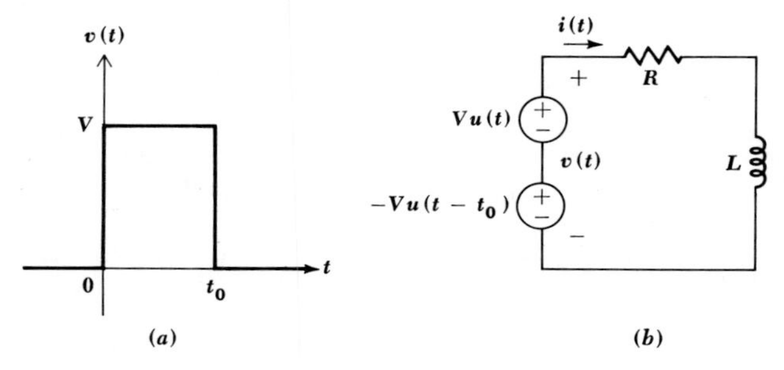

position principle. Suppose we designate that part of $i(t)$ which is due to the upper source $Vu(t)$ acting alone by the symbol $i_1(t)$ and then let $i_2(t)$ represent that part due to $-Vu(t - t_0)$ acting alone. Then,

$$i(t) = i_1(t) + i_2(t)$$

Our object is now to write each of the partial responses i_1 and i_2 as the sum of a natural and a forced response. The response $i_1(t)$ is familiar; this problem was solved two or three pages back,

$$i_1(t) = \frac{V}{R}(1 - e^{-Rt/L}) \qquad t > 0$$

Note that the range of t, $t > 0$, in which this solution is valid, is indicated.

We now turn our attention to the lower source and its response $i_2(t)$. Only the polarity of the source and the time of its application are different. There is thus no need to determine the form of the natural response and the forced response; the solution for $i_1(t)$ enables us to write

$$i_2(t) = -\frac{V}{R}(1 - e^{-R(t-t_0)/L}) \qquad t > t_0$$

where the applicable range of t, $t > t_0$, must again be indicated.

We now add the two solutions, but do so carefully, since each is valid over a different interval of time. Thus,

$$i(t) = \frac{V}{R}(1 - e^{-Rt/L}) \qquad 0 < t < t_0$$

$$i(t) = \frac{V}{R}(1 - e^{-Rt/L}) - \frac{V}{R}(1 - e^{-R(t-t_0)/L}) \qquad t > t_0$$

or

$$i(t) = \frac{V}{R}e^{-Rt/L}(e^{Rt_0/L} - 1) \qquad t > t_0$$

Fig. 6-15 Two possible response curves are shown for the circuit of Fig. 6-14*b*. In (*a*), τ is selected as $t_0/2$, and in (*b*) as $2t_0$.

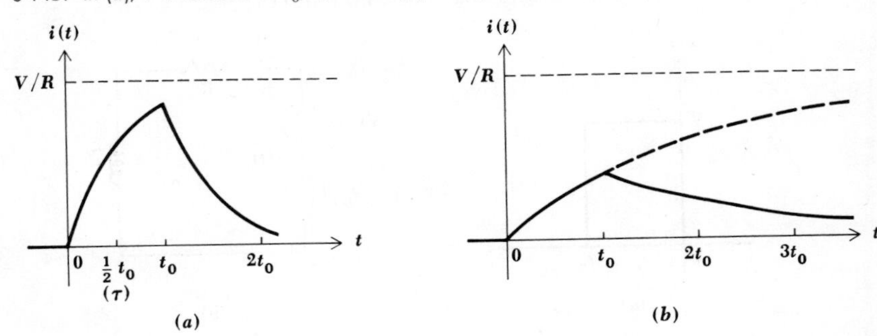

The solution is completed by stating that $i(t)$ is zero for negative t and sketching the response as a function of time. The type of curve obtained depends upon the relative values of t_0 and the time constant τ. Two possible curves are shown in Fig. 6-15. The left curve is drawn for the case where the time constant is only one-half as large as the length of the applied pulse; the rising portion of the exponential has therefore almost reached V/R before the decaying exponential begins. The opposite situation is shown to the right; there, the time constant is twice t_0 and the response never has a chance to reach the larger amplitudes.

Drill Problems

6-5 Find the forced portion of the response $i(t)$ for the circuits of Fig. 6-16*a*, *b*, and *c*.

Ans. 2; 4; 6 A

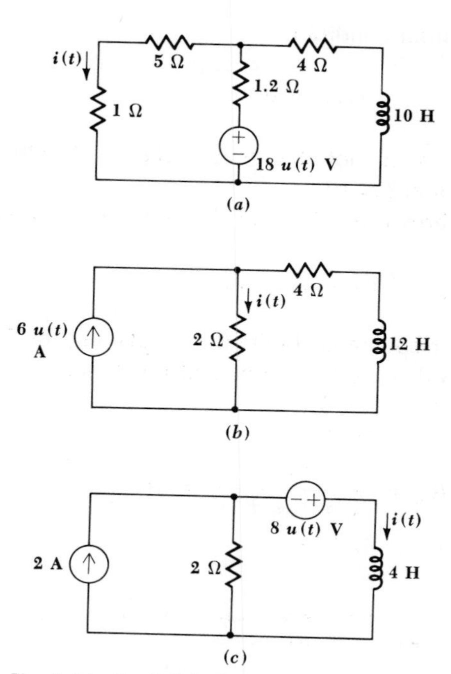

Fig. 6-16 See Drill Probs. 6-5, 6-6, and 6-7.

6-6 Find the value of $i(0^+)$ for the circuits of Fig. 6-16*a*, *b*, and *c*.

Ans. 2; 2.5; 6 A

6-7 Find $i(t)$ for $t > 0$ for the circuits of Fig. 6-16*a*, *b*, and *c*.

Ans. $2 + 0.5e^{-t/2}$; $4 + 2e^{-t/2}$; $6 - 4e^{-t/2}$ A

6-6 RC CIRCUITS

The complete response of any RC circuit may also be obtained as the sum of the natural and the forced response. We shall consider several responses in the circuit shown in Fig. 6-17. The switch is assumed to have been in position "a" for a long time, or, in other words, the natural response which resulted from the original excitation of the circuit has decayed to a negligible amplitude, leaving only a forced response caused by the 120-V source. We are asked for $v_C(t)$, and we thus begin by finding this forced response when the switch is in position "a." The voltages throughout the circuit are all constant, and there is thus no current through the capacitor. Simple voltage division determines the forced response prior to $t = 0$,

$$v_{Cf} = \frac{50}{50 + 10} 120 = 100 \qquad t < 0$$

and we thus have the initial condition

$$v_C(0) = 100$$

Since the capacitor voltage cannot change instantaneously, this voltage is equally valid at $t = 0^-$ and $t = 0^+$.

The switch is now thrown to "b," and the complete response is

$$v_C = v_{Cf} + v_{Cn}$$

The form of the natural response is obtained by replacing the 50-V source by a short circuit and evaluating the equivalent resistance,

$$v_{Cn} = Ae^{-t/R_{eq}C}$$

$$R_{eq} = \frac{1}{1/50 + 1/200 + 1/60} = 24$$

or

$$v_{Cn} = Ae^{-t/1.2}$$

Fig. 6-17 An RC circuit in which the complete responses v_C and i are obtained by adding a forced response and a natural response.

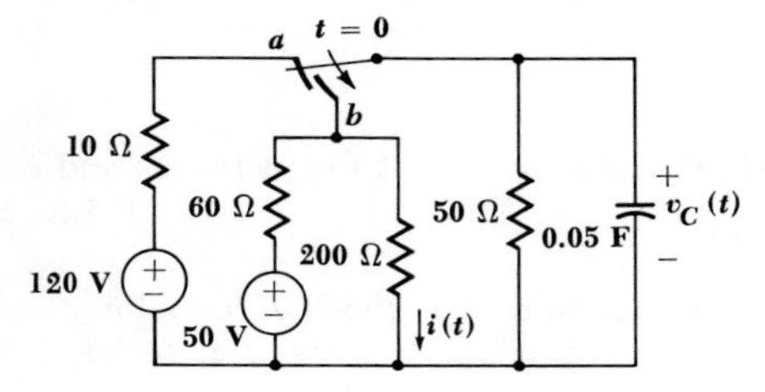

In order to evaluate the forced response with the switch at "b," we wait until all the voltages and currents have stopped changing, thus treating the capacitor as an open circuit, and use voltage division once more,

$$v_{Cf} = \frac{(50)(200)/(50 + 200)}{60 + (50)(200)/(50 + 200)} 50 = 20$$

Thus,

$$v_C = 20 + Ae^{-t/1.2}$$

and from the initial condition already obtained,

$$100 = 20 + A$$

or

$$v_C = 20 + 80e^{-t/1.2} \qquad t > 0$$

This response is sketched in Fig. 6-18a; again the natural response is seen to form a transition from the initial to the final response.

Finally, let us calculate some response that need not remain constant during the instant of switching, such as $i(t)$ in Fig. 6-17. With the contact at a, it is evident that $i = 50/260 = 0.192$ A. When the switch is in position b, the forced response for this current now becomes

$$i_f = \frac{50}{60 + (50)(200)/(50 + 200)} \frac{50}{50 + 200} = 0.1$$

The form of the natural response is the same as that we already determined for the capacitor voltage,

$$i_n = Ae^{-t/1.2}$$

Combining the forced and natural responses, we obtain

$$i = 0.1 + Ae^{-t/1.2}$$

Fig. 6-18 The responses (a) v_C and (b) i are plotted as functions of time for the circuit of Fig. 6-17.

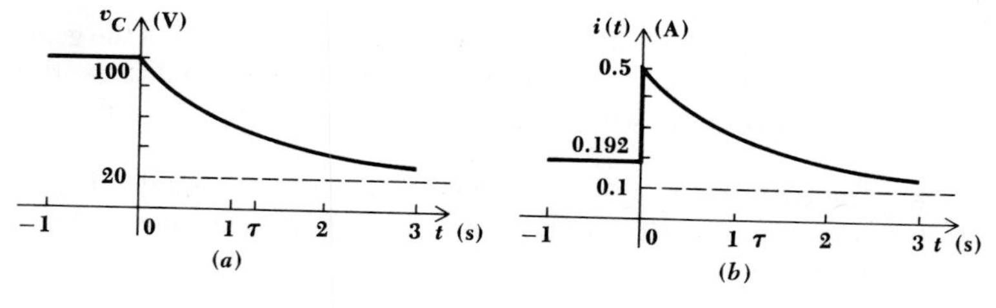

To evaluate A, we need to know $i(0^+)$. This is found by fixing our attention on the energy-storage element, here the capacitor, for the fact that v_C must remain 100 V during the switching interval is the governing condition establishing the other currents and voltages at $t = 0^+$. Since $v_C(0^+) =$ 100 V, and since the capacitor is in parallel with the 200-Ω resistor, we find $i(0^+) = 0.5$ A, $A = 0.4$, and thus,

$$i(t) = 0.192 \qquad t < 0$$

$$i(t) = 0.1 + 0.4e^{-t/1.2} \qquad t > 0$$

or

$$i(t) = 0.192 + (-0.092 + 0.4e^{-t/1.2})u(t) \qquad \text{A}$$

where the latter expression is correct for all t.

The complete response for all t may also be written concisely by using $u(-t)$, which is unity for $t < 0$ and 0 for $t > 0$. Thus,

$$i(t) = 0.192u(-t) + (0.1 + 0.4e^{-t/1.2})u(t) \qquad \text{A}$$

This response is sketched in Fig. 6-18*b*. Note that only four numbers are needed to write the functional form of the response for this single energy-storage-element circuit or to prepare the sketch: the constant value prior to switching (0.192 A), the instantaneous value just after switching (0.5 A), the constant forced response (0.1 A), and the time constant (1.2 s). The appropriate negative exponential function is then easily written or drawn.

PROBLEMS

☐ **1** Sketch each of the following waveforms as functions of time and calculate its value at $t = 2$ s: (*a*) $u(t + 3) - u(t - 3)$; (*b*) $u(3 + t) - u(3 - t)$; (*c*) $tu(t) + u(t - 1.5) - (t - 3)u(t - 3)$; (*d*) $e^{-t}u(t) - e^{-1}u(t - 1)$; (*e*) $u(t - t^2)$.

☐ **2** A 1.5-V battery, a 30-Ω resistor, and an open switch are in series. The switch closes at $t = 0$, opens at $t = 1$ s, closes at $t = 2$ s, and continues in the same pattern thereafter. Express the current as an infinite summation of step functions.

☐ **3** Find the current entering the upper conductor of each of the general networks shown in Fig. 6-3 if $V = 20$ V, $t_0 = 2$ s, and the general network consists of a 5-Ω resistor in series with a 5-V (ideal) battery being charged.

☐ **4** The current, $i(t) = 20u(t) - 20u(t - 10^{-3})$ mA, is applied to a 10-μF capacitor. Find the capacitor voltage as a function of time, and evaluate your result at $t = 0.5$ and 2 ms.

☐ **5** A 2-H inductor, a 50-Ω resistor, and the series combination of a 100-Ω resistor and an independent voltage source $45u(t)$ V are all in parallel. (*a*) Replace everything except the inductor by its Thévenin equivalent and find the inductor current as a function of time. (*b*) Repeat for the Norton equivalent.

☐ **6** A 10-Ω resistor, a 5-H inductor, and an independent voltage source are in series. Find and sketch the current if the source voltage is: (*a*) $50 - 50u(t)$ V; (*b*) $50 + 50u(t)$ V; (*c*) $50 + 50u(t) - 100u(t-1)$ V.

☐ **7** Through the use of an exponential integrating factor, solve the differential equation for the current if a series circuit consists of a 10-Ω resistor, a 1-H inductor, and an independent voltage source: (*a*) $2 \cos 10t\, u(t)$ V; (*b*) $2t\, u(t)$ V.

☐ **8** For the circuit shown in Fig. 6-19, find: (*a*) i_L; (*b*) v_L.

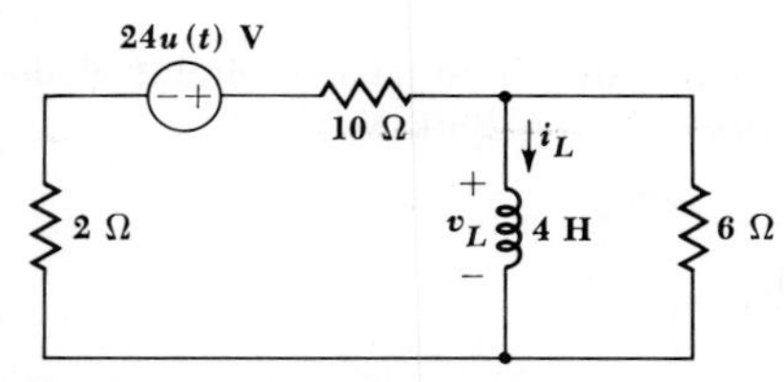

Fig. 6-19 See Probs. 8 and 9.

☐ **9** Change the label on the voltage source in Fig. 6-19 from $24u(t)$ to $12 - 24u(t)$ V and rework Prob. 8.

☐ **10** In the circuit shown in Fig. 6-20, the switch has been in the position shown for a long time. At $t = 0$ it is moved to the left. Find and sketch $i_L(t)$.

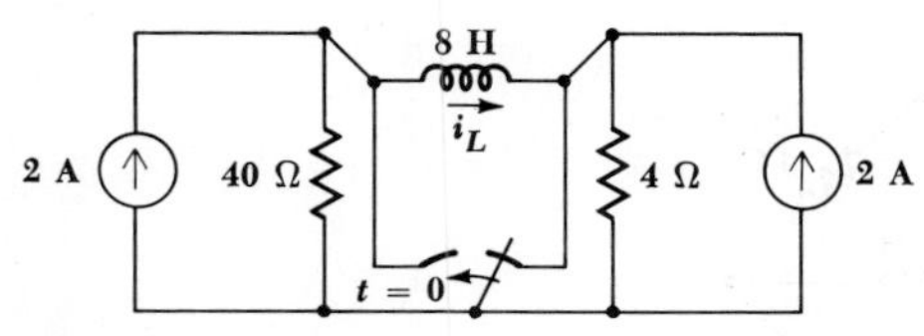

Fig. 6-20 See Prob. 10.

☐ **11** For the circuit of Fig. 6-21, let $i_s = 2u(t)$ A and find: (*a*) i_L; (*b*) i_1.

Fig. 6-21 See Probs. 11 to 13.

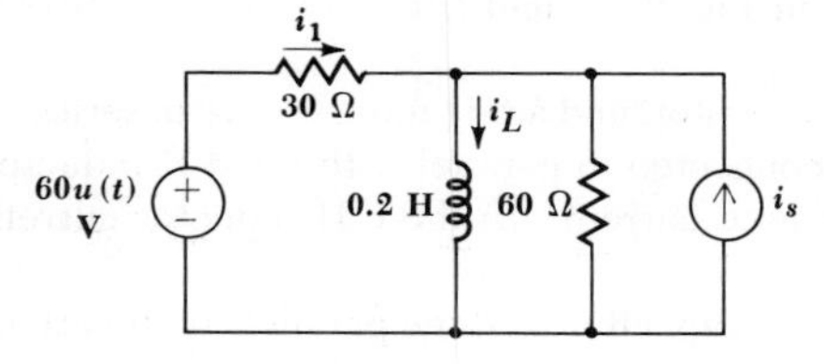

☐ **12** For the circuit of Fig. 6-21, let $i_s = -2u(t)$ A and find: (*a*) i_L; (*b*) i_1.

☐ **13** For the circuit of Fig. 6-21, let $i_s = -2$ A and find: (*a*) i_L; (*b*) i_1.

☐ **14** Find $i_1(t)$, $i_2(t)$, and $v_3(t)$ for the circuit illustrated by Fig. 6-22.

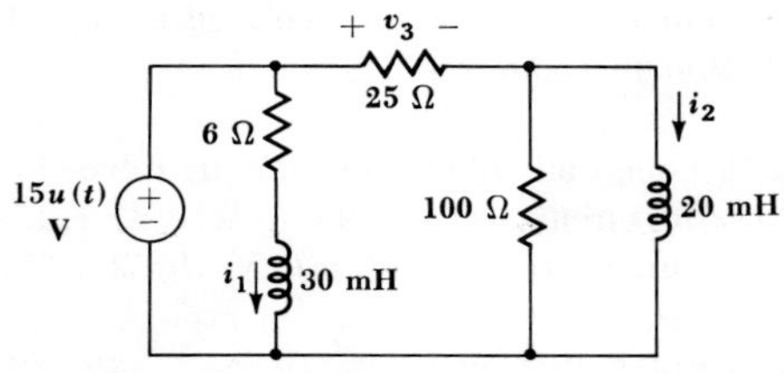

Fig. 6-22 See Prob. 14.

☐ **15** Find the voltage across each of the three passive circuit elements in Fig. 6-23 at $t = 0^+$ if $v_s =$: (*a*) 0; (*b*) 12 V; (*c*) $12u(t)$ V.

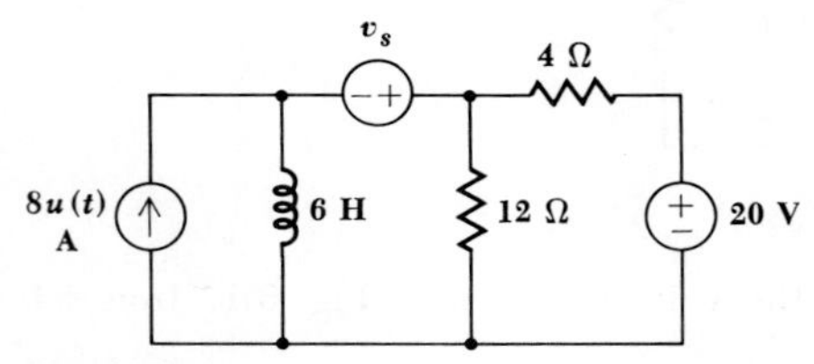

Fig. 6-23 See Prob. 15.

☐ **16** For the circuit shown in Fig. 6-24, find and sketch as functions of time: (*a*) $i_L(t)$; (*b*) $i_R(t)$.

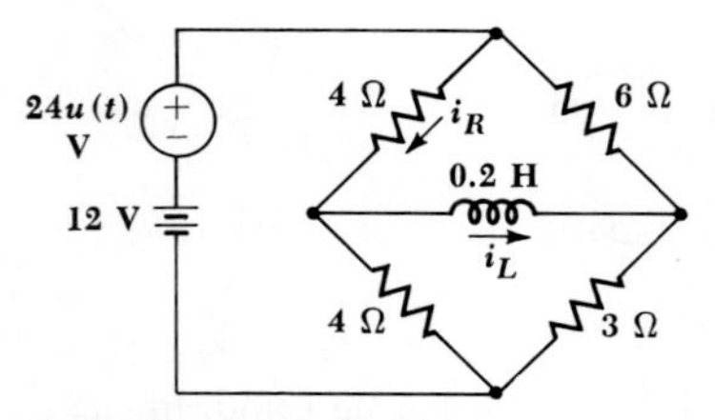

Fig. 6-24 See Prob. 16.

☐ **17** For the circuit shown in Fig. 6-25, find $i_L(t)$.

☐ **18** A $60u(t)$-V source, a 6-Ω resistor, and a 6-H inductor are in series. At $t = 1$ s, a 3-H inductor is suddenly connected in parallel with the 6-H inductor. Find, as functions of time: (*a*) the source current; (*b*) the 6-H inductor current.

☐ **19** If a 20-kΩ resistor, a 10-μF capacitor, and the parallel combination of a 10-kΩ

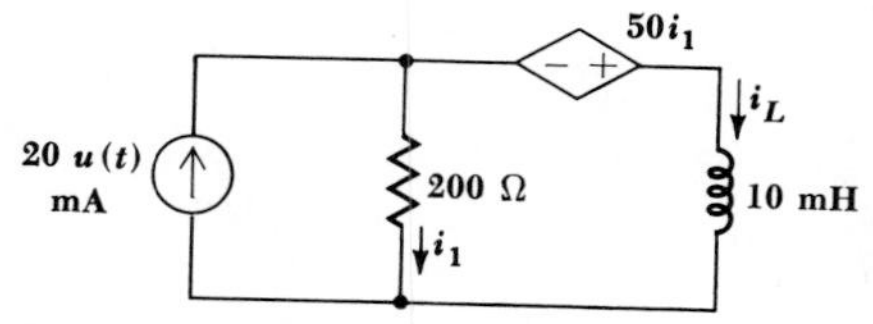

Fig. 6-25 See Prob. 17.

resistor and an independent current source of $6u(t)$ mA are all in series, find: (*a*) the capacitor voltage; (*b*) the current through the 10-kΩ resistor.

☐ **20** A 20-μF capacitor, a 25-kΩ resistor, and an independent current source are in parallel. Find the capacitor voltage $v_C(t)$ if the source current is: (*a*) $4[1 - u(t)]$ mA; (*b*) $4[1 + u(t)]$ mA.

☐ **21** For the circuit shown in Fig. 6-26, find $v_C(t)$ and $i_C(t)$ if $i_s =$: (*a*) $25u(t)$ mA; (*b*) $10 + 15u(t)$ mA.

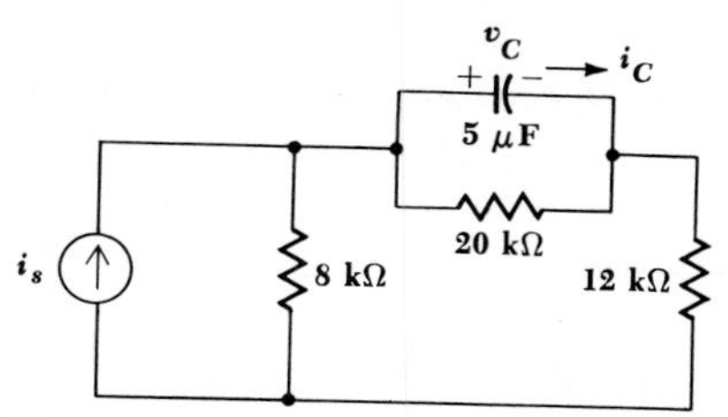

Fig. 6-26 See Prob. 21.

☐ **22** With reference to the circuit shown in Fig. 6-27, let $v_s = 30u(t)$ V and determine: (*a*) $v_C(t)$; (*b*) $v_1(t)$.

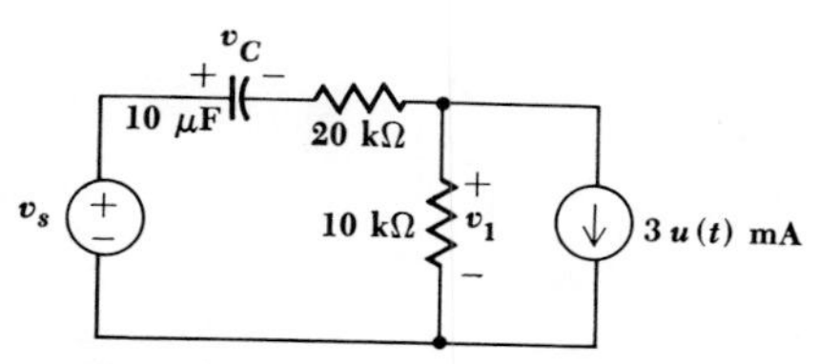

Fig. 6-27 See Probs. 22 and 23.

☐ **23** Let $v_s = -30u(t)$ V in the circuit of Fig. 6-27. Find: (*a*) $v_C(t)$; (*b*) $v_1(t)$.

☐ **24** For the circuit shown in Fig. 6-28, let $v_s = 0$ and find: (*a*) $v_1(t)$; (*b*) $v_2(t)$; (*c*) $i_3(t)$.

☐ **25** For the circuit shown in Fig. 6-28, let $v_s = 4$ V and find: (*a*) $v_1(t)$; (*b*) $v_2(t)$; (*c*) $i_3(t)$.

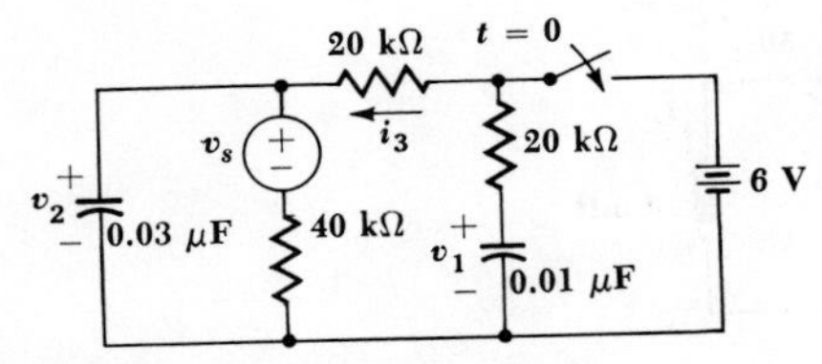

Fig. 6-28 See Probs. 24 and 25.

☐ **26** Find $i_1(0^+)$, $i_2(0^+)$, and $i_3(0^+)$ in the circuit shown in Fig. 6-29 if $i_s =$: (*a*) 0; (*b*) 3 A; (*c*) $3u(t)$ A.

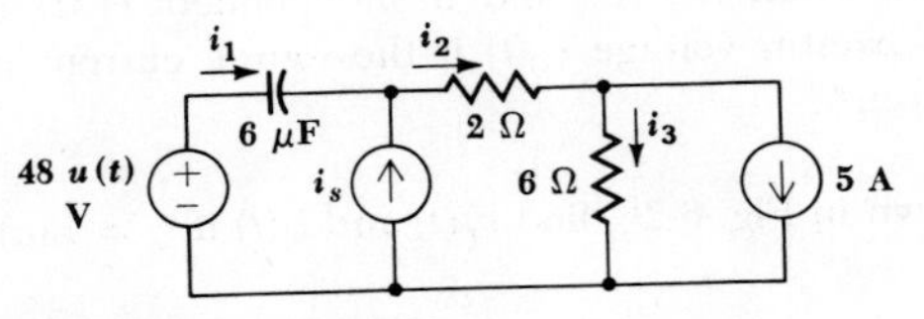

Fig. 6-29 See Prob. 26.

☐ **27** For the circuit shown in Fig. 6-30, find $v_C(t)$ if the dependent source is labeled $v_s =$: (*a*) $500i_1$; (*b*) $0.5v_C$.

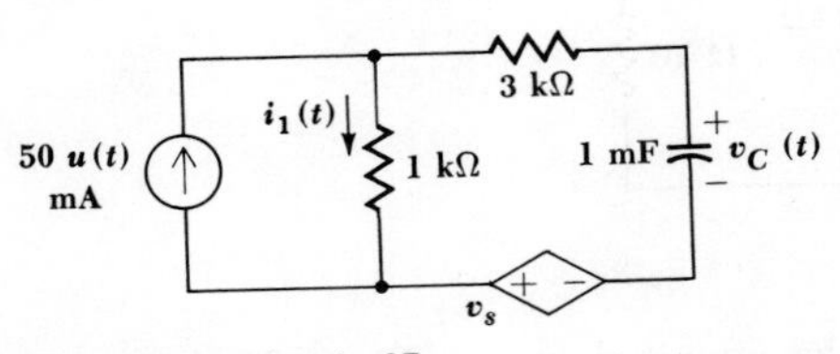

Fig. 6-30 See Prob. 27.

Chapter Seven
The *RLC* Circuit

7-1 INTRODUCTION

It would be very pleasant to learn that the detailed study we have just completed for the *RL* and *RC* circuits will make the analysis of the *RLC* circuit a simple task; unfortunately the analysis remains difficult. The presence of inductance and capacitance in the same circuit produces at least a *second-order system,* that is, one that is characterized by a linear differential equation including a second-order derivative, or by two simultaneous linear first-order differential equations. Although we have considered one second-order system in the final section of Chap. 5, most of our attention has been directed toward the first-order system which is described by a first-order linear differential equation. From the single case we have considered, we should suspect that this increase in order will make it necessary to evaluate two arbitrary constants. Furthermore, it will be necessary to determine initial conditions for derivatives. And finally, we shall see that the presence of inductance and capacitance in the same circuit leads to a response which takes on different functional forms for circuits which have the same configuration but different element values. With this cheerful news, let us quickly review the methods and results we found useful for first-order systems, in order that we may extend this information as intelligently as possible to the second-order system.

We first considered the source-free first-order system. The response was termed the natural response, and it was determined completely by the types of passive elements in the network, by the manner in which they were interconnected, and by the initial conditions which were established by the stored energy. The natural response was invariably an exponentially decreasing function of time, and this response approached a constant value as time became infinite. The constant was usually zero, except in those circuits where paralleled inductors or series-connected capacitors allowed trapped currents or voltages to appear.

The addition of sources to the first-order system resulted in a two-part response, the familiar natural response and an additional term we called the forced response. This latter term was intimately related to the forcing function; its functional form was that of the forcing function itself, plus the integral and first derivative of the forcing function.[1] Since we treated only a constant forcing function, we have not needed to devote much attention to the proper form of the forced response; this problem will not arise until sinusoidal forcing functions are encountered in the following chapter. To the known forced response, we added the correct expression for the natural response, complete except for a multiplicative constant. This constant was evaluated to make the total response fit the prescribed initial conditions.

We now turn to circuits which are characterized by linear second-order differential equations. Our first task is the determination of the natural response. This is most conveniently done by considering initially the source-free circuit. We may then include dc sources, switches, or step sources in the circuit, representing the total response once again as the sum of the natural response and the (usually constant) forced response. The second-order system that we are about to analyze is fundamentally the same as any lumped-constant mechanical second-order system. Our

[1]Higher-order derivatives will appear in higher-order systems, and, strictly speaking, we should say that all derivatives are present, although possibly with zero amplitude. Forcing functions which do not possess a finite number of different derivatives are exceptions which we shall not consider; the singularity functions are exceptions to the exceptions.

results, for example, will be of direct use to a mechanical engineer who is interested in the displacement of a spring-supported mass subjected to viscous damping, or one who is interested in the behavior of a simple pendulum or a torsional pendulum. Our results are still applicable, although less directly, to any distributed-parameter second-order system, such as a short-circuited transmission line, a diving board, a flute, or the ecology of the lemming.

7-2 THE SOURCE-FREE PARALLEL CIRCUIT

Our first goal is the determination of the natural response of a simple circuit formed by connecting R, L, and C in parallel; this modest goal will be reached after completing this and the next three sections. This particular combination of ideal elements is a suitable model for portions of many communications networks. It represents, for example, an important part of some of the electronic amplifiers found in every radio receiver, and it enables the amplifiers to produce a large voltage amplification over a narrow band of signal frequencies and nearly zero amplification outside this band. Frequency selectivity of this kind enables us to listen to the transmission of one station while rejecting the transmission of any other station. Other applications include the use of parallel RLC circuits in multiplexing filters, harmonic suppression filters, and so forth. But even a simple discussion of these principles requires an understanding of such terms as resonance, frequency response, and impedance which we have not yet discussed. Let it suffice to say, therefore, that an understanding of the natural behavior of the parallel RLC circuit is fundamentally important to future studies of communications networks and filter design.

When a physical inductor is connected in parallel with a capacitor, and the inductor has associated with it a nonzero ohmic resistance, the resulting network can be shown to have an equivalent circuit model like that shown in Fig. 7-1. Energy losses in the physical inductor are taken into account by the presence of the ideal resistor whose resistance R is dependent upon (but not equal to) the ohmic resistance of the inductor.

In the following analysis we shall assume that energy may be stored

Fig. 7-1 The source-free parallel RLC circuit.

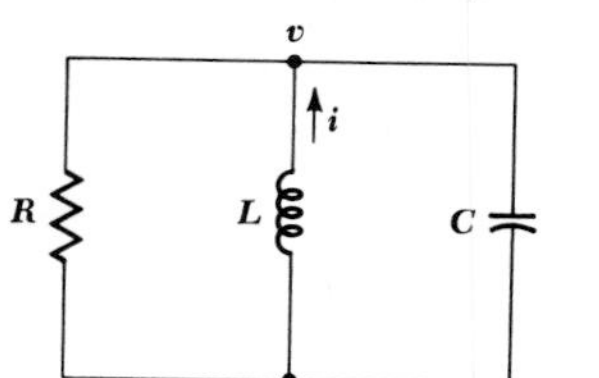

initially in both the inductor and the capacitor, and thus nonzero values of both inductor current and capacitor voltage are initially present. With reference to the circuit of Fig. 7-1, we may then write the single nodal equation

$$\frac{v}{R} + \frac{1}{L}\int_{t_0}^{t} v\,dt - i(t_0) + C\frac{dv}{dt} = 0 \tag{1}$$

Note that the minus sign is a consequence of the assumed direction for i. We must solve (1) subject to the initial conditions

$$i(0^+) = I_0 \tag{2}$$

$$v(0^+) = V_0 \tag{3}$$

When both sides of (1) are differentiated once with respect to time, the result is the linear second-order homogeneous differential equation

$$C\frac{d^2v}{dt^2} + \frac{1}{R}\frac{dv}{dt} + \frac{1}{L}v = 0 \tag{4}$$

whose solution $v(t)$ is the desired natural response.

There are a number of interesting ways to solve (4). These methods we shall leave to a course in differential equations, selecting only the quickest and simplest method to use now. We shall assume a solution, relying upon our intuition and modest experience to select one of the several possible forms which are suitable. Our experience with the first-order equation should suggest that we at least try the exponential form once more. Moreover, the form of (4) indicates that this may work, because we must add three terms, the second derivative, first derivative, and the function itself, each multiplied by a constant factor, and achieve a sum of zero. A function whose derivatives have the same form as the function itself is obviously a sensible choice. With every hope for success, then, we assume

$$v = Ae^{st} \tag{5}$$

where we shall be as general as possible by allowing A and s to be complex numbers if necessary.[2] Substituting (5) into (4), we obtain

$$CAs^2e^{st} + \frac{1}{R}Ase^{st} + \frac{1}{L}Ae^{st} = 0$$

or

$$Ae^{st}\left(Cs^2 + \frac{1}{R}s + \frac{1}{L}\right) = 0$$

[2]There is no cause for panic. Complex numbers will appear in this chapter only in an introductory way in the derivations. Their use as a tool will be necessary in Chap. 9.

In order for this equation to be satisfied for all time, one of the three factors must be zero. If either of the first two factors is put equal to zero, then $v(t) = 0$. This is a trival solution of the differential equation which cannot satisfy our given initial conditions. We therefore equate the remaining factor to zero,

$$Cs^2 + \frac{s}{R} + \frac{1}{L} = 0 \tag{6}$$

This equation is usually called the *auxiliary equation* or the *characteristic equation* by mathematicians. If it can be satisfied, then our assumed solution is correct. Since (6) is a quadratic equation, there are two solutions, identified as s_1 and s_2,

$$s_1 = -\frac{1}{2RC} + \sqrt{\left(\frac{1}{2RC}\right)^2 - \frac{1}{LC}} \tag{7}$$

and

$$s_2 = -\frac{1}{2RC} - \sqrt{\left(\frac{1}{2RC}\right)^2 - \frac{1}{LC}} \tag{8}$$

If either of these two values is used for s in the assumed solution, then that solution satisfies the given differential equation; it thus becomes a valid solution of the differential equation.

Let us assume that we replace s by s_1 in (5), obtaining

$$v_1 = A_1 e^{s_1 t}$$

and, similarly,

$$v_2 = A_2 e^{s_2 t}$$

The former satisfies the differential equation

$$C\frac{d^2v_1}{dt^2} + \frac{1}{R}\frac{dv_1}{dt} + \frac{1}{L}v_1 = 0$$

and the latter satisfies

$$C\frac{d^2v_2}{dt^2} + \frac{1}{R}\frac{dv_2}{dt} + \frac{1}{L}v_2 = 0$$

Adding these two differential equations and combining similar terms,

$$C\frac{d^2(v_1 + v_2)}{dt^2} + \frac{1}{R}\frac{d(v_1 + v_2)}{dt} + \frac{1}{L}(v_1 + v_2) = 0$$

linearity triumphs, and it is seen that the sum of the two solutions is also a solution. We thus have the form of the natural response

$$v = A_1 e^{s_1 t} + A_2 e^{s_2 t} \tag{9}$$

where s_1 and s_2 are given by (7) and (8), and A_1 and A_2 are two arbitrary constants which are to be selected to satisfy the two specified initial conditions.

The form of the natural response as given above can hardly be expected to bring forth any expressions of interested amazement, for, in its present form, it offers little insight into the nature of the curve we might obtain if $v(t)$ were plotted as a function of time. The relative amplitudes of A_1 and A_2, for example, will certainly be important in determining the nature of the response curve. Furthermore, the constants s_1 and s_2 can be real numbers or conjugate complex numbers, depending upon the values of R, L, and C in the given network. These two cases will produce fundamentally different response forms. Therefore it will be helpful to make some simplifying substitutions in (9) for the sake of conceptual clarity.

Since the exponents $s_1 t$ and $s_2 t$ must be dimensionless, s_1 and s_2 must have the unit of some dimensionless quantity "per second." From (7) and (8) it is apparent that the units of $1/2RC$ and $1/\sqrt{LC}$ must also be s^{-1}. Units of this type are called *frequencies*. Although we shall expand this concept in much more detail in Chap. 13, we shall introduce several of the terms now. Let us represent $1/\sqrt{LC}$ by ω_0 (omega),

$$\omega_0 = \frac{1}{\sqrt{LC}} \tag{10}$$

and reserve the term *resonant frequency*[3] for it. On the other hand, we shall call $1/2RC$ the *neper frequency* or the *exponential damping coefficient* and represent it by the symbol α (alpha),

$$\alpha = \frac{1}{2RC} \tag{11}$$

This latter descriptive expression is used because α is a measure of how rapidly the natural response decays or damps out to its steady final value (usually zero). Finally, s, s_1, and s_2, which are quantities that will form the basis for some of our later work, will be called *complex frequencies*.

Let us collect these results. The natural response of the parallel RLC circuit is

$$v(t) = A_1 e^{s_1 t} + A_2 e^{s_2 t} \tag{9}$$

[3]More accurately, the resonant *radian* frequency.

where

$$s_1 = -\alpha + \sqrt{\alpha^2 - \omega_0^2} \tag{12}$$

$$s_2 = -\alpha - \sqrt{\alpha^2 - \omega_0^2} \tag{13}$$

$$\alpha = \frac{1}{2RC} \tag{11}$$

$$\omega_0 = \frac{1}{\sqrt{LC}} \tag{10}$$

and A_1 and A_2 must be found by applying the given initial conditions.

It is now apparent that the nature of the response depends upon the relative magnitudes of α and ω_0. The radical appearing in the expressions for s_1 and s_2 will be real when α is greater than ω_0, imaginary when α is less than ω_0, and zero when α and ω_0 are equal. Each of these cases will be considered separately in the following three sections.

Drill Problem

7-1 Determine values for α, ω_0, and s_1 for a parallel *RLC* circuit in which $C = 0.1\ \mu\text{F}$, $R = 1\ \text{k}\Omega$, and $L =$: (*a*) 0.625 H; (*b*) 0.4 H; (*c*) 0.256 H.

Ans. 5, 4, −2; 5, 5, −5; 5, 6.25, $-5 + j3.75$ $(\text{ms})^{-1}$

7-3 THE OVERDAMPED PARALLEL *RLC* CIRCUIT

It is apparent that α will be greater than ω_0 and α^2 will be greater than ω_0^2 if $LC > 4R^2C^2$. In this case the radical we are concerned with will be real, and both s_1 and s_2 will be real. Moreover, the following inequalities,

$$\sqrt{\alpha^2 - \omega_0^2} < \alpha$$

$$(-\alpha - \sqrt{\alpha^2 - \omega_0^2}) < (-\alpha + \sqrt{\alpha^2 - \omega_0^2}) < 0$$

may be applied to (12) and (13) to show that both s_1 and s_2 are *negative* real numbers. Thus, the response $v(t)$ can be expressed as the (algebraic) sum of two decreasing exponential terms, both of which approach zero as time increases without limit. In fact, since the absolute value of s_2 is larger than that of s_1, the term containing s_2 has the more rapid rate of decrease and, for large values of time, we may write the limiting expression

$$v(t) \to A_1 e^{s_1 t} \to 0 \qquad \text{as } t \to \infty$$

In order to discuss the method by which the arbitrary constants A_1 and

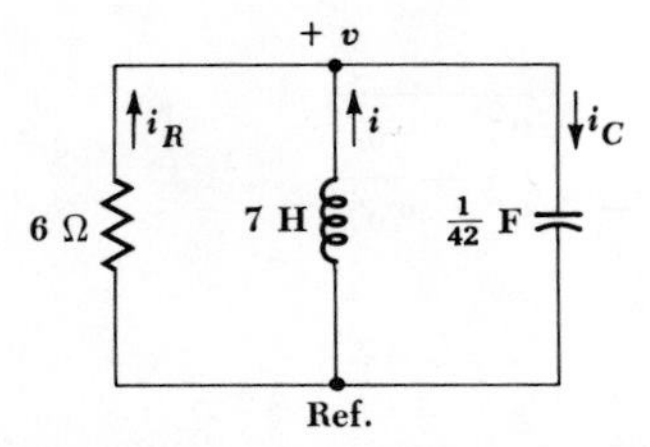

Fig. 7-2 A parallel RLC circuit used as a numerical example. The circuit is overdamped.

A_2 are selected to conform with the initial conditions, and in order to provide a typical example of a response curve, let us turn to a numerical example. We shall select a parallel RLC circuit for which $R = 6\ \Omega$, $L = 7$ H, and, for ease of computation, the impractically large value $C = 1/42$ F; the initial energy storage is specified by choosing an initial voltage across the circuit $v(0) = 0$ and an initial inductor current $i(0) = 10$ A, as shown in Fig. 7-2.

We may easily determine the values of the several parameters

$$\alpha = 3.5 \qquad \omega_0 = \sqrt{6} \qquad s_1 = -1 \qquad s_2 = -6 \qquad \text{(all s}^{-1}\text{)}$$

and immediately write the general form of the natural response

$$v(t) = A_1 e^{-t} + A_2 e^{-6t} \tag{14}$$

Only the evaluation of the two constants A_1 and A_2 remains. If we knew the response $v(t)$ at two different values of time, these two pairs of values could be substituted in (14) and A_1 and A_2 easily found. However, we know only the initial value of $v(t)$,

$$v(0) = 0$$

and, therefore,

$$0 = A_1 + A_2 \tag{15}$$

A second equation relating A_1 and A_2 must be obtained by taking the derivative of $v(t)$ with respect to time in (14), determining the initial value of this derivative through the use of the remaining initial condition $i(0) = 10$, and equating the results. Taking the derivative of both sides of (14),

$$\frac{dv}{dt} = -A_1 e^{-t} - 6A_2 e^{-6t}$$

evaluating the derivative at $t = 0$,

$$\left.\frac{dv}{dt}\right|_{t=0} = -A_1 - 6A_2$$

we next pause to consider how the initial value of the derivative can be found numerically. This next step is always suggested by the derivative itself; dv/dt suggests capacitor current, for

$$i_c = C\frac{dv}{dt}$$

Thus,

$$\left.\frac{dv}{dt}\right|_{t=0} = \frac{i_c(0)}{C} = \frac{i(0) + i_R(0)}{C} = \frac{i(0)}{C}$$
$$= 420 \quad \text{V/s}$$

since zero voltage across the resistor requires zero current through it. We thus have our second equation

$$420 = -A_1 - 6A_2 \tag{16}$$

and simultaneous solution of (15) and (16) provides the two amplitudes $A_1 = 84$ and $A_2 = -84$. Thus, the final numerical solution for the natural response is

$$v(t) = 84(e^{-t} - e^{-6t}) \tag{17}$$

The evaluation of A_1 and A_2 for other conditions of initial energy storage, including initial energy storage in the capacitor, is considered in the first drill problem following this section.

Let us see what information we can glean from (17) without calculating unduly. We note that $v(t)$ is zero at $t = 0$, a comforting check on our original assumption. We may also interpret the first exponential term as having a time constant of 1 s and the other exponential, a time constant of $\frac{1}{6}$ s. Each starts with unity amplitude, but the second decays more rapidly; $v(t)$ is thus always positive. As time becomes infinite, each term approaches zero, and the response itself dies out as it should. We thus have a response curve which is zero at $t = 0$, zero at $t = \infty$, and is always positive; since it is not everywhere zero, it must possess at least one maximum, and this is not a difficult point to determine exactly. We differentiate the response

$$\frac{dv}{dt} = 84(-e^{-t} + 6e^{-6t})$$

set the derivative equal to zero to determine the time t_m at which the voltage becomes maximum,

$$0 = -e^{-t_m} + 6e^{-6t_m}$$

manipulate once,

$$e^{5t_m} = 6$$

and obtain

$$t_m = 0.358 \quad \text{s}$$

and

$$v(t_m) = 48.9 \quad \text{V}$$

A reasonable sketch of the response may be made by plotting the two exponential terms $84e^{-t}$ and $84e^{-6t}$ and then taking their difference. The usefulness of this technique is indicated by the curves of Fig. 7-3; the two exponentials are shown lightly, and their difference, the total response $v(t)$, is drawn with a heavier line. The curves also verify our previous prediction that the functional behavior of $v(t)$ for very large t is $84e^{-t}$, the exponential term containing the smaller magnitude of s_1 and s_2.

Another question that frequently arises in the consideration of network response is concerned with the length of time it takes for the transient part of the response to disappear (or damp out). In practice, it is often desirable to have this transient response approach zero as rapidly as possible, that is, to minimize the *settling time* t_s. Theoretically, of course, t_s is infinite, because $v(t)$ never settles to zero in a finite time. However, a negligible response is present after $v(t)$ has settled to values that are less than 1 per cent of its maximum value v_m. The time which is required for this to occur we define as the settling time. Since v_m is 48.9 V for our example, the settling time is the time required for the response to drop

Fig. 7-3 The response $v(t) = 84(e^{-t} - e^{-6t})$ of the network which is shown in Fig. 7-2.

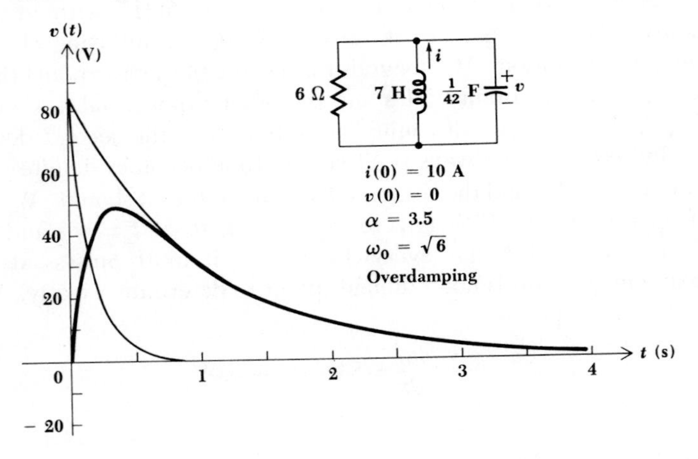

to 0.489 V. Substituting this value for $v(t)$ into (17) and neglecting the second exponential term, known to be negligible here, the settling time is found to be 5.15 s.

In comparison with the responses which we shall obtain in the following two sections, this is a comparatively large settling time; the damping takes overly long, and the response is called overdamped. We shall refer to the case for which α is greater than ω_0 as the *overdamped* case. Now let us see what happens as α is decreased.

Drill Problems

7-2 If the initial voltage across a parallel RLC circuit is not zero, then the resultant value of the initial resistor current must be considered in evaluating the initial value of dv/dt. For the numerical example considered in the above section, determine A_1 and A_2 for the following initial conditions: (*a*) $v(0) = 10$ V, $i(0) = 10$ A; (*b*) $v(0) = 10$ V, $i(0) = -10$ A; (*c*) $v(0) = 24$ V, $i(0) = 2$ A.

Ans. -86, 96; 12, 12; 82, -72 V

7-3 Given a parallel RLC circuit in which $R = 400\ \Omega$, $C = 0.01\ \mu$F, and $L = 10$ mH, determine the voltage across the circuit at the times indicated below if the initial values are $v(0) = 20$ V, $i(0) = 20$ mA: (*a*) $t = 2\ \mu$s; (*b*) $t = 5\ \mu$s; (*c*) $t = 20\ \mu$s.

Ans. 2.70; 10.1; 15.0 V

7-4 CRITICAL DAMPING

The overdamped case is characterized by

$$\alpha > \omega_0$$

or

$$LC > 4R^2C^2$$

and leads to negative real values for s_1 and s_2 and to a response expressed as the algebraic sum of two negative exponentials. Typical forms of the response $v(t)$ are obtained through the numerical example in the last section and in the drill problems following it.

Now let us adjust the element values until α and ω_0 are equal. This is a very special case which is termed *critical damping*. Thus, critical damping is achieved when

$$\left.\begin{aligned} \alpha &= \omega_0 \\ \text{or} \quad LC &= 4R^2C^2 \\ \text{or} \quad L &= 4R^2C \end{aligned}\right\} \text{critical damping}$$

It is obvious that we may produce critical damping by changing the value of any one of the three elements in the numerical example discussed above. We shall select R, increasing its value until critical damping is obtained, and thus leave ω_0 unchanged. The necessary value of R is $7\sqrt{6}/2\ \Omega$; L is still 7 H, and C remains $\frac{1}{42}$ F. We thus find

$$\alpha = \omega_0 = \sqrt{6}$$

$$s_1 = s_2 = -\sqrt{6}$$

and we blithely construct the response as the sum of the two exponentials,

$$v(t) \stackrel{?}{=} A_1 e^{-\sqrt{6}t} + A_2 e^{-\sqrt{6}t}$$

which may be written as

$$v(t) \stackrel{?}{=} A_3 e^{-\sqrt{6}t}$$

At this point, some of us should feel we have lost our way. We have a response which contains only one arbitrary constant, but there are two initial conditions $v(0) = 0$ and $i(0) = 10$ which must be satisfied by this single constant. This is in general impossible. In our case, for example, the first initial condition requires A_3 to be zero, and it is then impossible to satisfy the second initial condition.

Our mathematics and our electricity have been unimpeachable; therefore, if a mistake has not led to our difficulties, we must have begun with an incorrect assumption, and only one assumption has been made. We originally hypothesized that the differential equation could be solved by assuming an exponential solution, and this turns out to be incorrect for this single special case of critical damping. We must therefore return to the differential equation and attempt to solve it by some other means than an assumed solution. Although this detailed solution is carried out below, we should realize that it is the final functional form of the response which is important to us, and not the specific method by which it is obtained. After all, if we were clever enough, we could assume a response of the correct form and then check it by direct substitution in the differential equation.

The original differential equation, (4),

$$C\frac{d^2v}{dt^2} + \frac{1}{R}\frac{dv}{dt} + \frac{1}{L}v = 0$$

may be written in terms of α and ω_0,

$$\frac{d^2v}{dt^2} + 2\alpha\frac{dv}{dt} + \omega_0{}^2 v = 0$$

which becomes, for critical damping,

$$\frac{d^2v}{dt^2} + 2\alpha\frac{dv}{dt} + \alpha^2 v = 0$$

or

$$\frac{d^2v}{dt^2} + \alpha\frac{dv}{dt} + \alpha\frac{dv}{dt} + \alpha^2 v = 0$$

and finally,

$$\frac{d}{dt}\left(\frac{dv}{dt} + \alpha v\right) + \alpha\left(\frac{dv}{dt} + \alpha v\right) = 0$$

If we now let

$$y = \frac{dv}{dt} + \alpha v$$

then

$$\frac{dy}{dt} + \alpha y = 0$$

which is a form we have encountered before. Its solution is

$$y = A_1 e^{-\alpha t}$$

and, therefore,

$$\frac{dv}{dt} + \alpha v = A_1 e^{-\alpha t}$$

This equation has the form of Eq. (6-2), and therefore we should be able to solve it by the use of an integrating factor. The integrating factor is $e^{+\alpha t}$, and we multiply both sides of the equation by it,

$$e^{\alpha t}\frac{dv}{dt} + \alpha e^{\alpha t} v = A_1$$

recognize the derivative of the product and simplify,

$$\frac{d}{dt}(ve^{\alpha t}) = A_1$$

Finally, we integrate each side directly,

$$ve^{\alpha t} = A_1 t + A_2$$

and obtain the desired response form

$$v = e^{-\alpha t}(A_1 t + A_2) \tag{18}$$

It should be noted that the solution may be expressed as the sum of two terms, where one term is the familiar negative exponential but the second is t times a negative exponential. We should also note that the solution contains the *two* expected arbitrary constants.

Let us now complete our numerical example. After we substitute the known value of α into (18),

$$v = A_1te^{-\sqrt{6}t} + A_2e^{-\sqrt{6}t}$$

we establish the values of A_1 and A_2 by first imposing the initial condition on $v(t)$ itself, $v(0) = 0$. Thus, $A_2 = 0$. This simple result occurs because the initial value of the response was selected as zero; the more general case, which leads to an equation determining A_2, may be expected to arise in the drill problems. The second initial condition must be applied to the derivative dv/dt just as in the overdamped case. We therefore differentiate, remembering that $A_2 = 0$,

$$\frac{dv}{dt} = A_1t(-\sqrt{6})e^{-\sqrt{6}t} + A_1e^{-\sqrt{6}t}$$

evaluate at $t = 0$,

$$\left.\frac{dv}{dt}\right|_{t=0} = A_1$$

express the derivative in terms of the initial capacitor current,

$$\left.\frac{dv}{dt}\right|_{t=0} = \frac{i_C(0)}{C} = \frac{i_R(0)}{C} + \frac{i(0)}{C}$$

and thus

$$A_1 = 420$$

The response is, therefore,

$$v(t) = 420te^{-2.45t} \tag{19}$$

Before plotting this response in detail, let us again try to anticipate its form by qualitative reasoning. The specified initial value is zero, and (19) concurs. It is not immediately apparent that the response also approaches zero as t becomes infinitely large because $te^{-2.45t}$ is an indeterminate form. However, this minor obstacle is easily overcome by use of L'Hôpital's rule. Thus,

$$\lim_{t\to\infty} v(t) = 420 \lim_{t\to\infty} \frac{t}{e^{2.45t}} = 420 \lim_{t\to\infty} \frac{1}{2.45e^{2.45t}} = 0$$

and once again we have a response which begins and ends at zero and has positive values at all other times. A maximum value v_m again occurs at time t_m; for our example,

$$t_m = 0.408 \quad \text{s} \qquad \text{and} \qquad v_m = 63.1 \quad \text{V}$$

This maximum is larger than that obtained in the overdamped case and is a result of the smaller losses that occur in the larger resistor; the time of the maximum response is slightly later than it was with overdamping. The settling time may also be determined by solving

$$\frac{v_m}{100} = 420t_s e^{-2.45t_s}$$

for t_s (by trial-and-error methods),

$$t_s = 3.12 \quad \text{s}$$

which is a considerably smaller value than arose in the overdamped case. As a matter of fact, it can be shown that, for given values of L and C, the selection of that value of R which provides critical damping will always give a shorter settling time than any choice of R which produces an overdamped response. However, a slight improvement (reduction) in settling time may be obtained by a further slight increase in resistance;

Fig. 7-4 The response $v(t) = 420te^{-2.45t}$ of the network shown in Fig. 7-2 with R changed to provide critical damping.

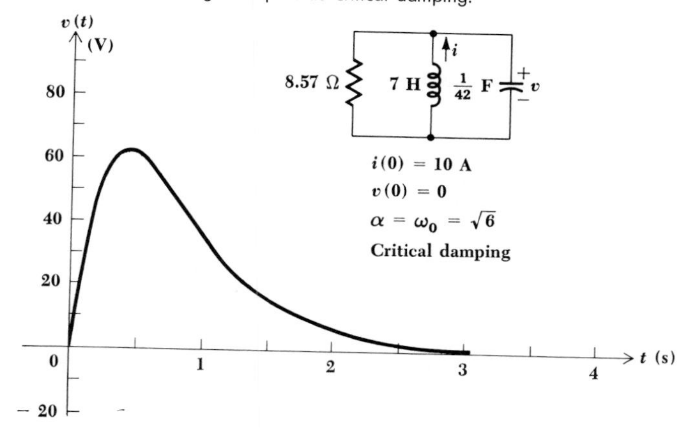

a slightly underdamped response which will undershoot the zero axis before it dies out will yield the shortest settling time.

The response curve for critical damping is drawn in Fig. 7-4; it may be compared with the overdamped (and underdamped) case by reference to Fig. 7-6.

Drill Problem

7-4 Find $v(t)$ for each of the following cases if the damping is critical: (*a*) $R = 1\ \text{k}\Omega$, $L = 2$ H, $v(0) = 10$ V, $i(0) = 10$ mA; (*b*) $L = 0.1$ H, $C = 10\ \mu$F, $v(0) = 10$ V, $i(0) = 10$ mA; (*c*) $R = 20\ \text{k}\Omega$, $C = 0.025\ \mu$F, $v(0) = 10$ V, $i(0) = 10$ mA.

Ans. $10e^{-1000t}(1 - 900t)$; $10e^{-1000t}(1 + 1000t)$; $10e^{-1000t}(1 + 39{,}000t)$ V

7-5 THE UNDERDAMPED PARALLEL *RLC* CIRCUIT

Let us continue the process begun in the last section by increasing R once more. Thus, the damping coefficient α decreases while ω_0 remains constant, α^2 becomes smaller than $\omega_0{}^2$, and the radicand appearing in the expressions for s_1 and s_2 becomes negative. This causes the response to take on quite a different character, but it is fortunately not necessary to return to the basic differential equation again. By using complex numbers, the exponential response turns into a sinusoidal response; this response is composed entirely of real quantities, the complex quantities being necessary only for the derivation.[4]

We therefore begin with the exponential form

$$v(t) = A_1 e^{s_1 t} + A_2 e^{s_2 t}$$

where

$$s_{1,2} = -\alpha \pm \sqrt{\alpha^2 - \omega_0{}^2}$$

and then let

$$\sqrt{\alpha^2 - \omega_0{}^2} = \sqrt{-1}\,\sqrt{\omega_0{}^2 - \alpha^2} = j\sqrt{\omega_0{}^2 - \alpha^2}$$

where

$$j = \sqrt{-1}$$

[4] An introduction to the use of complex numbers appears in Chap. 9 and Appendix 3. At that time we shall emphasize the more general nature of complex quantities by identifying them with boldface type; no special symbolism need be adopted in these few pages.

We now take the new radical, which is real for the underdamped case, and call it ω_d, the *natural resonant frequency,*

$$\omega_d = \sqrt{\omega_0^2 - \alpha^2}$$

Collecting, the response may now be written as

$$v(t) = e^{-\alpha t}(A_1 e^{j\omega_d t} + A_2 e^{-j\omega_d t})$$

or in the longer but equivalent form,

$$v(t) = e^{-\alpha t}\left\{(A_1 + A_2)\left[\frac{e^{j\omega_d t} + e^{-j\omega_d t}}{2}\right] + j(A_1 - A_2)\left[\frac{e^{j\omega_d t} - e^{-j\omega_d t}}{j2}\right]\right\}$$

Two of the most important identities in the field of complex numbers, identities which are later proved in Appendix 3, may now be readily applied. The first bracket in the above equation is identically equal to $\cos \omega_d t$, and the second bracket is identically $\sin \omega_d t$. Hence,

$$v(t) = e^{-\alpha t}[(A_1 + A_2)\cos \omega_d t + j(A_1 - A_2)\sin \omega_d t]$$

and the multiplying factors may be assigned new symbols,

$$v(t) = e^{-\alpha t}(B_1 \cos \omega_d t + B_2 \sin \omega_d t) \tag{20}$$

If we are dealing with the underdamped case, we have now left complex numbers behind. This is true since α, ω_d, and t are real quantities, $v(t)$ itself must be a real quantity (which might be presented on an oscilloscope, a voltmeter, or a sheet of graph paper), and thus B_1 and B_2 are real quantities. Equation (20) is the desired functional form for the underdamped response, and its validity may be checked by direct substitution into the original differential equation; this exercise is left to the doubters. The two real constants B_1 and B_2 are again selected to fit the given initial conditions.

Let us increase the resistance in our example from $7\sqrt{6}/2$ or 8.57 Ω to $10.5\sqrt{2}$ or 14.85 Ω; L and C are unchanged. Thus,

$$\alpha = \frac{1}{2RC} = \sqrt{2}$$

$$\omega_0 = \frac{1}{\sqrt{LC}} = \sqrt{6}$$

and

$$\omega_d = \sqrt{\omega_0^2 - \alpha^2} = 2 \quad \text{(rad/s)}$$

Except for the evaluation of the arbitrary constants, the response is now known,

$$v(t) = e^{-\sqrt{2}t}(B_1 \cos 2t + B_2 \sin 2t)$$

The determination of the two constants proceeds as before. If we again assume that $v(0) = 0$ and $i(0) = 10$, then B_1 must be zero. Hence,

$$v(t) = B_2 e^{-\sqrt{2}t} \sin 2t$$

The derivative is

$$\frac{dv}{dt} = 2B_2 e^{-\sqrt{2}t} \cos 2t - \sqrt{2}B_2 e^{-\sqrt{2}t} \sin 2t$$

and at $t = 0$ it becomes

$$\left.\frac{dv}{dt}\right|_{t=0} = 2B_2 = \frac{i_C(0)}{C} = 420$$

Therefore,

$$v(t) = 210e^{-1.414t} \sin 2t$$

Notice that, as before, this response function has an initial value of zero, because of the initial voltage condition we imposed, and a final value of zero, because the exponential term vanishes for large values of t. As t increases from zero through small positive values, $v(t)$ increases as $210 \sin 2t$ because the exponential term remains essentially equal to unity. But at a time t_m, the exponential function begins to decrease more rapidly than $\sin 2t$ is increasing; so $v(t)$ reaches a maximum v_m and begins to decrease. We should note that t_m is not the value of t for which $\sin 2t$ is a maximum, but must occur somewhat before $\sin 2t$ reaches its maximum value. When $t = \pi/2$, $v(t)$ is zero; for the interval $\pi/2 < t < \pi$ the response is negative, becoming zero again at $t = \pi$. Thus $v(t)$ is an *oscillatory* function of time and crosses the time axis an infinite number of times at $t = n\pi/2$, where n is any positive integer. In our example, however, the response is only slightly underdamped and the exponential term causes the function to die out so rapidly that most of the zero crossings will not be evident in a sketch.

The oscillatory nature of the response becomes more noticeable as α decreases. If α is zero, which corresponds to an infinitely large resistance, then $v(t)$ is an undamped sinusoid which oscillates with constant amplitude. This is not perpetual motion; we have merely assumed an initial energy in the circuit and have not provided any means to dissipate this energy. It is transferred from its initial location in the inductor to the capacitor,

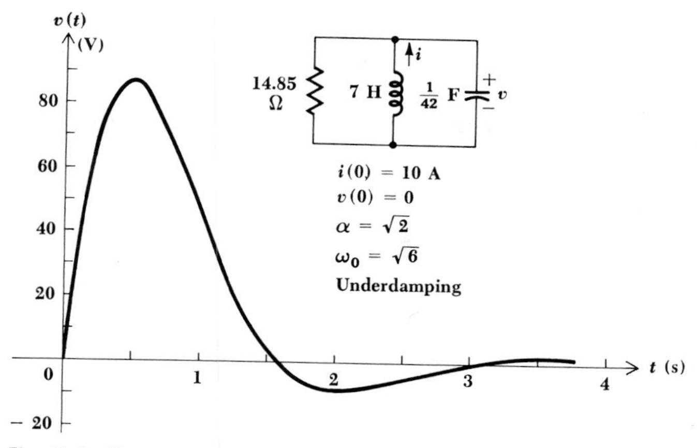

Fig. 7-5 The response $v(t) = 210e^{-1.414t} \sin 2t$ of the network shown in Fig. 7-2 with R increased to produce an underdamped response.

then returns to the inductor, and so on, forever. Actual parallel RLC circuits can be made to have effective values of R so large that a natural undamped sinusoidal response can be maintained for years without supplying any additional energy. We can also build active networks which introduce a sufficient amount of energy during each oscillation of $v(t)$ so that a sinusoidal response which is nearly perfect can be maintained for as long as we wish. This circuit is a sinusoidal oscillator, or signal generator, which is an important laboratory instrument.

Returning to our specific numerical problem, differentiation locates the first maximum of $v(t)$,

$$v_{m1} = 87.3 \quad \text{V} \qquad \text{at } t_{m1} = 0.478 \quad \text{s}$$

the succeeding minimum,

$$v_{m2} = -9.47 \quad \text{V} \qquad \text{at } t_{m2} = 2.05 \quad \text{s}$$

and so on. The response curve is shown in Fig. 7-5.

The settling time may be obtained by a trial-and-error solution, and it turns out to be 3.88 s, a little larger than for critical damping. Problem 15 at the end of this chapter demonstrates that the shortest settling time for this network results for slight underdamping.

The overdamped, critically damped, and underdamped responses for this network are shown on the same graph in Fig. 7-6. A comparison of these three curves makes these general conclusions plausible:

1 When the damping is changed by adjusting the size of the parallel

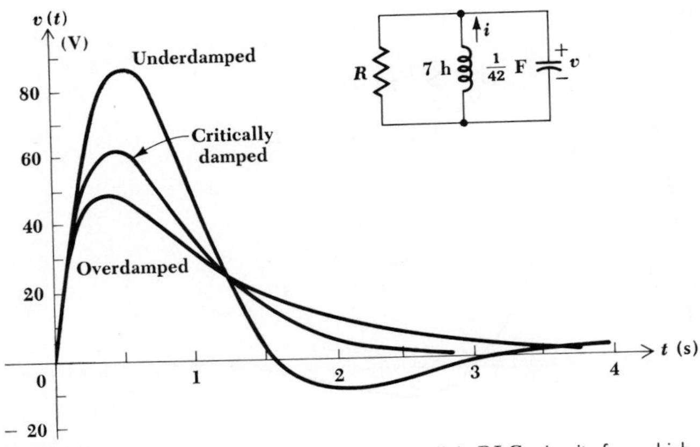

Fig. 7-6 Three response curves for a parallel RLC circuit for which $\omega_0 = \sqrt{6}$, $v(0) = 0$, $i(0) = 10$ A, and α is 3.5 (overdamped), 2.45 (critically damped), and 1.414 (underdamped).

resistance, the maximum magnitude of the response is greater with smaller damping.

2 The minimum settling time occurs approximately for critical damping; actually, the response should be slightly underdamped.

Drill Problems

7-5 The switch in the circuit shown in Fig. 7-7 is closed at $t = 0$. Find v at $t =$: (*a*) 0; (*b*) 2.5 ms; (*c*) $5\pi/3$ ms.

Ans. *0.493; 1.81; 3.00* V

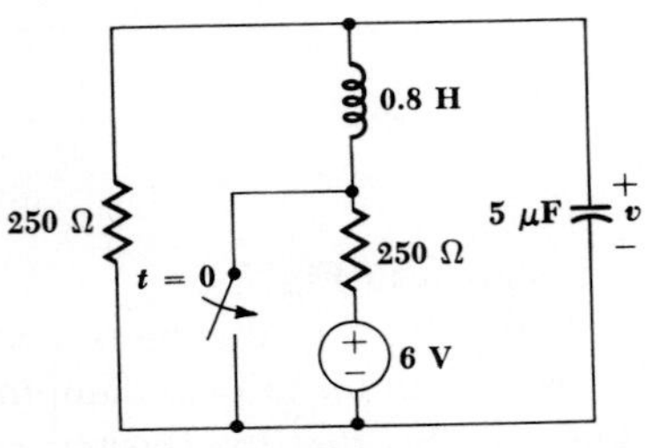

Fig. 7-7 See Drill Prob. 7-5.

7-6 Let the element values in the circuit of Fig. 7-2 be those of the underdamped example in the section above ($R = 10.5\sqrt{2}\ \Omega$, $L = 7$ H, $C = \frac{1}{42}$ F). If the response is: $v(t) = 63e^{-\sqrt{2}t}(\cos 2t - 3 \sin 2t)$ V, find: (*a*) $i_R(0)$; (*b*) $i_C(0)$; (*c*) $i(0)$.

Ans. *−11.12; −6.88; −4.24* A

7-6 THE SOURCE-FREE SERIES *RLC* CIRCUIT

We now wish to determine the natural response of a circuit model composed of an ideal resistor, an ideal inductor, and an ideal capacitor connected in series. The ideal resistor may represent a physical resistor connected into a series *LC* or *RLC* circuit, it may represent the ohmic losses and the losses in the ferromagnetic core of the inductor, or it may be used to represent all these and other energy-absorbing devices. In a special case, the resistance of the ideal resistor may even be exactly equal to the measured resistance of the wire out of which the physical inductor is made.

The series *RLC* circuit is the dual of the parallel *RLC* circuit, and this single fact is sufficient to make its analysis a trivial affair. Figure 7-8*a* shows the series circuit. The fundamental integrodifferential equation is

$$L\frac{di}{dt} + Ri + \frac{1}{C}\int_{t_0}^{t} i\,dt - v_C(t_0) = 0$$

and should be compared with the analogous equation for the parallel *RLC* circuit, drawn again in Fig. 7-8*b*,

$$C\frac{dv}{dt} + \frac{1}{R}v + \frac{1}{L}\int_{t_0}^{t} v\,dt - i_L(t_0) = 0$$

The second-order equations obtained by differentiating each of these equations with respect to time are also duals,

$$L\frac{d^2i}{dt^2} + R\frac{di}{dt} + \frac{i}{C} = 0 \tag{21}$$

$$C\frac{d^2v}{dt^2} + \frac{1}{R}\frac{dv}{dt} + \frac{v}{L} = 0 \tag{22}$$

It is apparent that our complete discussion of the parallel *RLC* circuit is directly applicable to the series *RLC* circuit; the initial conditions on capacitor voltage and inductor current are equivalent to the initial conditions on inductor current and capacitor voltage; the voltage response

Fig. 7-8 (*a*) The series *RLC* circuit which is the dual of (*b*) the parallel *RLC* circuit. Element values are of course not identical in (*a*) and (*b*).

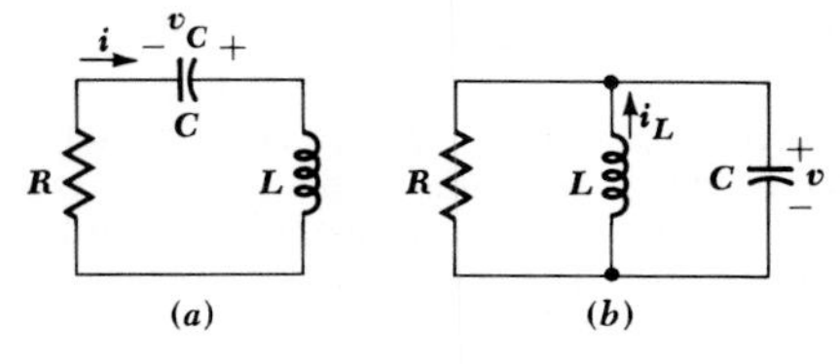

becomes a current response. It is quite possible to reread the previous four sections (including the drill problems) using dual language and thereby obtain a complete description of the series *RLC* circuit.[5] This process, however, is apt to induce a mild neurosis after the first few paragraphs and does not really seem to be necessary.

A brief résumé of the series circuit response is easily collected. In terms of the circuit shown in Fig. 7-8*a*, the overdamped response is

$$i(t) = A_1 e^{s_1 t} + A_2 e^{s_2 t}$$

where

$$\begin{aligned} s_{1,2} &= -\frac{R}{2L} \pm \sqrt{\left(\frac{R}{2L}\right)^2 - \frac{1}{LC}} \\ &= -\alpha \pm \sqrt{\alpha^2 - \omega_0^{\,2}} \\ &= -\alpha \pm j\omega_d \end{aligned}$$

and thus

$$\alpha = \frac{R}{2L}$$

$$\omega_0 = \frac{1}{\sqrt{LC}}$$

$$\omega_d = \sqrt{\omega_0^{\,2} - \alpha^2}$$

The form of the critically damped response is

$$i(t) = e^{-\alpha t}(A_1 t + A_2)$$

and the underdamped case may be written

$$i(t) = e^{-\alpha t}(B_1 \cos \omega_d t + B_2 \sin \omega_d t)$$

It is evident that if we work in terms of the parameters α, ω_0, and ω_d the mathematical forms of the responses for the dual situations are identical. An increase in α in either the series or parallel circuit, while keeping ω_0 constant, tends toward an overdamped response. The only caution that we need exert is in the computation of α, which is $1/2RC$ for the parallel circuit and $R/2L$ for the series circuit; thus, α is increased by increasing the series resistance or decreasing the parallel resistance.

[5]In fact, during the writing of this text, the authors had originally written these first sections to describe the series *RLC* circuit. But, after deciding that it would be better to present the analysis of the more practical parallel *RLC* circuit first, it was easy to go back to the original writing and replace it with its dual. The numerical values of several of the elements were also scaled, a process described later in Chap. 14.

As a numerical example, let us consider a series *RLC* circuit in which $L = 1$ H, $R = 2$ kΩ, $C = \frac{1}{401}$ μF, $i(0) = 2$ mA, and $v_C(0) = 2$ V. We find that α is 1000 and ω_0 is 20.02×10^3, and thus an underdamped response is indicated; we therefore calculate the value of ω_d and obtain 2×10^4. Except for the evaluation of the two arbitrary constants, the response is now known:

$$i(t) = e^{-1000t}(B_1 \cos 2 \times 10^4 t + B_2 \sin 2 \times 10^4 t)$$

By applying the initial value of the current, we find

$$B_1 = 2 \times 10^{-3}$$

and thus

$$i(t) = e^{-1000t}(2 \times 10^{-3} \cos 2 \times 10^4 t + B_2 \sin 2 \times 10^4 t)$$

The remaining initial condition must be applied to the derivative; thus,

$$\frac{di}{dt} = e^{-1000t}(-40 \sin 2 \times 10^4 t + 2 \times 10^4 B_2 \cos 2 \times 10^4 t - 2 \cos 2 \times 10^4 t - 1000 B_2 \sin 2 \times 10^4 t)$$

$$\left.\frac{di}{dt}\right|_{t=0} = 2 \times 10^4 B_2 - 2 = \frac{v_L(0)}{L} = \frac{v_C(0) - Ri(0)}{L}$$

$$= \frac{2 - 2 \times 10^3(2 \times 10^{-3})}{1} = -2$$

or

$$B_2 = 0$$

Fig. 7-9 The current response in an underdamped series *RLC* circuit for which $\alpha = 1000$ s^{-1}, $\omega_d = 2 \times 10^4$ s^{-1}, $i(0) = 2$ mA, $v_C(0) = 2$ V. The graphical construction is simplified by drawing in the envelope, shown as a pair of broken lines.

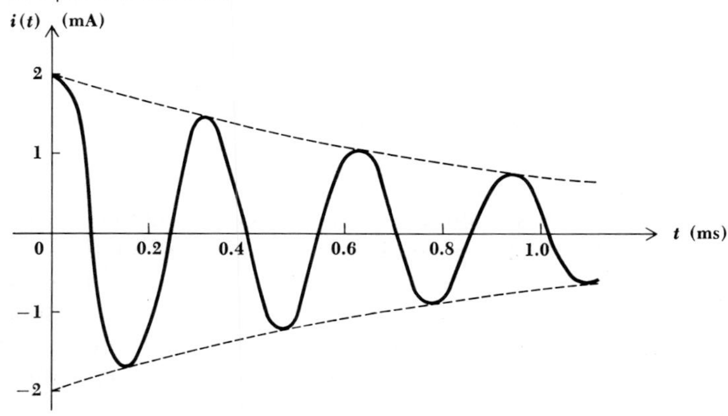

The desired response is, therefore,

$$i(t) = 2 \times 10^{-3} e^{-1000t} \cos 2 \times 10^4 t \qquad \text{A}$$

This response is more oscillatory, or shows less damping, than any we have considered up to this time, and the direct calculation of enough points to graph a smooth response curve is a tedious undertaking. A good sketch may be made by first drawing in the two exponential *envelopes*, $2 \times 10^{-3} e^{-1000t}$ and $-2 \times 10^{-3} e^{-1000t}$, as shown by the broken lines in Fig. 7-9. The location of the quarter-cycle points of the sinusoidal wave at $2 \times 10^4 t = 0, \pi/2, \pi$, etc., or $t = 0.07854k$ ms, $k = 0, 1, 2, \ldots$, by light marks on the time axis then permits the oscillatory curve to be sketched in quickly.

Drill Problems

7-7 Note that the current source in the circuit of Fig. 7-10 goes to zero at $t = 0$, and find $i(0^+)$ and $i'(0^+)$ if $R =$: (*a*) 500 Ω; (*b*) 400 Ω; (*c*) 320 Ω.

Ans. Each is 10 mA, 0 A/s

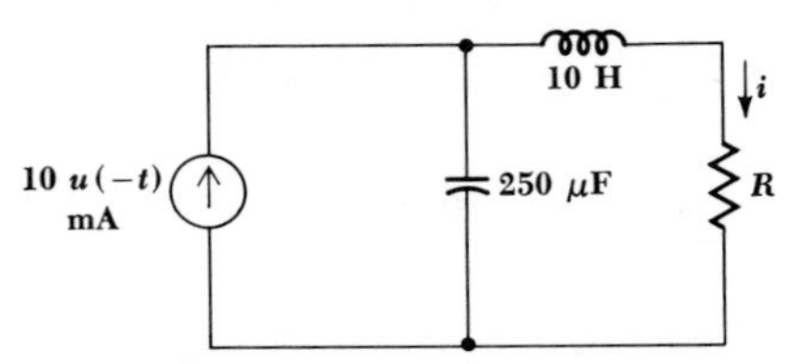

Fig. 7-10 See Drill Probs. 7-7 and 7-8.

7-8 Find $i(t)$ at $t = 50$ ms for the circuit of Fig. 7-10 if $R =$: (*a*) 500 Ω; (*b*) 400 Ω; (*c*) 320 Ω.

Ans. 7.09; 7.36; 7.64 mA

7-7 THE COMPLETE RESPONSE OF THE *RLC* CIRCUIT

We must now consider those *RLC* circuits in which dc sources are switched into the network and produce forced responses that do not vanish as time becomes infinite. The general solution is obtained by the same procedure that was followed in *RL* and *RC* circuits; the forced response is determined completely, the natural response is obtained as a suitable functional form containing the appropriate number of arbitrary constants, the complete response is written as the sum of the forced and the natural responses, and the initial conditions are then determined and applied to the complete response to find the values of the constants. It is this last step which is quite frequently the most troublesome to students. Consequently, although the determination of the initial conditions is basically no different for a circuit containing dc sources than it is for the source-free circuits which we have already covered in some detail, this topic will receive particular emphasis in the examples that follow.

Most of the confusion in determining and applying the initial conditions arises for the simple reason that we do not have laid down for us a rigorous set of rules to follow. At some point in each analysis there usually arises a situation in which some thinking is involved that is more or less unique to that particular problem. This originality and flexibility of thought, as simple as it is to achieve after several problems' worth of practice, is the source of the difficulty.

The complete response (arbitrarily assumed to be a voltage response) of a second-order system consists of a forced response, which is a constant for dc excitation,

$$v_f(t) = V_f$$

and a natural response

$$v_n(t) = Ae^{s_1 t} + Be^{s_2 t}$$

Thus,

$$v(t) = V_f + Ae^{s_1 t} + Be^{s_2 t}$$

We shall now assume that s_1, s_2, and V_f have already been determined from the circuit and the given forcing functions; A and B remain to be found. The last equation shows the functional interdependence of A, B, v, and t, and substitution of the known value of v at $t = 0^+$ thus provides us with a single equation relating A and B. Another relationship between A and B is necessary, and this is normally obtained by taking the derivative of the response,

$$\frac{dv}{dt} = 0 + s_1 Ae^{s_1 t} + s_2 Be^{s_2 t}$$

and inserting in it the known value of dv/dt at $t = 0^+$. There is no reason that this process cannot be continued; a second derivative might be taken, and a third relationship between A and B will then result if the value of d^2v/dt^2 at $t = 0^+$ is used. This value is *not* usually known, however, in a second-order system; as a matter of fact, we are much more likely to use this method to *find* the initial value of the second derivative if we should need it. We thus have only two equations relating A and B, and these may be solved simultaneously to evaluate the two constants.

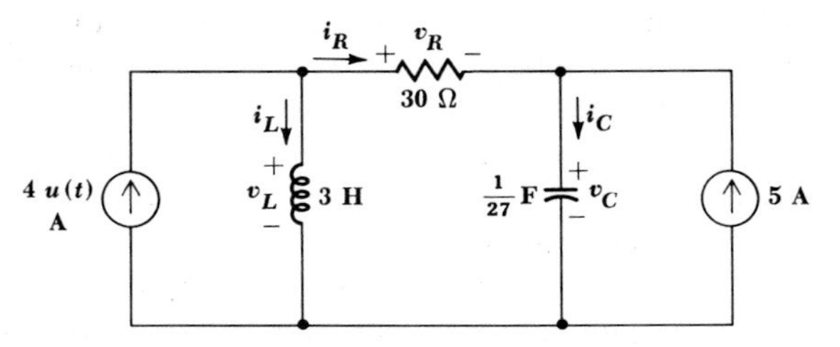

Fig. 7-11 An RLC circuit which is used to illustrate several procedures by which the initial conditions may be obtained. The desired response is nominally taken to be $v_C(t)$.

The only remaining problem is that of determining the values of v and dv/dt at $t = 0^+$. Since $i_C = C\, dv_C/dt$, we should recognize the relationship between the initial value of dv/dt and the initial value of some capacitor current. If we can establish a value for this initial capacitor current, then we shall automatically establish the value of dv/dt. Students are usually able to get $v(0^+)$ very easily, but are inclined to stumble a bit in finding the initial value of dv/dt. If we had selected a current as our response, then the initial value of di/dt must be intimately related to the initial value of some inductor voltage.

Let us illustrate the procedure by the careful analysis of the circuit shown in Fig. 7-11. To simplify the analysis, an unrealistically large capacitor is used again. Our object is to find the value of each current and voltage at both $t = 0^-$ and $t = 0^+$; with these quantities known, the required derivatives may be easily calculated. We shall employ a logical step-by-step method first.

At $t = 0^-$, only the right-hand current source is active. Moreover, the circuit is assumed to have been in this state forever, and all currents and voltages are constant. In other words, a steady-state condition has been reached and the resultant forced response has the form of the forcing function, its integral, and its derivatives. The integral of the forcing function, a linearly increasing function of time, is not present in this circuit, for it can occur only when a constant current is forced through a capacitor or a constant voltage is maintained across an inductor. This situation should not normally be present because the capacitor voltage or inductor current would assume an unrealistic infinite value at $t = 0^-$. Continuing, then, a constant current through the inductor requires zero voltage across it,

$$v_L(0^-) = 0$$

and a constant voltage across the capacitor requires zero current through it,

$$i_C(0^-) = 0$$

We then apply Kirchhoff's current law to the right node to obtain

$$i_R(0^-) = -5 \quad \text{A}$$

which also yields

$$v_R(0^-) = -150 \quad \text{V}$$

We may now use Kirchhoff's voltage law around the central mesh, finding

$$v_C(0^-) = 150 \quad \text{V}$$

while Kirchhoff's current law enables us to find the inductor current,

$$i_L(0^-) = 5 \quad \text{A}$$

Although the derivatives at $t = 0^-$ are of little interest to us, it is evident that they are all zero.

Now let time increase an incremental amount. During the interval from $t = 0^-$ to $t = 0^+$, the left-hand current source becomes active and most of the voltage and current values at $t = 0^-$ will change abruptly. However, we should *begin by focusing our attention on those quantities which cannot change, inductor current and capacitor voltage.* Both of these must remain constant during the switching interval. Thus,

$$i_L(0^+) = 5 \quad \text{A} \qquad \text{and} \qquad v_C(0^+) = 150 \quad \text{V}$$

Since two currents are now known at the left node, we next obtain[6]

$$i_R(0^+) = -1 \quad \text{A} \qquad \text{and} \qquad v_R(0^+) = -30 \quad \text{V}$$

Thus,

$$i_C(0^+) = 4 \quad \text{A} \qquad \text{and} \qquad v_L(0^+) = 120 \quad \text{V}$$

There remain six derivatives which might be evaluated, although not all are needed to evaluate the two arbitrary constants. The procedure must begin with the energy-storage elements by the direct application of their defining equations. For the inductor,

$$v_L = L\frac{di_L}{dt}$$

[6] This current is the only one of the four remaining quantities which can be obtained in one step. In more complicated circuits, it is quite possible that none of the remaining initial values can be obtained with a single step; either circuit equations must then be written or a simpler equivalent resistive circuit must be drawn which can be analyzed by writing simultaneous equations. This latter method will be described shortly.

and, specifically,

$$v_L(0^+) = L\frac{di_L}{dt}\bigg|_{t=0^+}$$

Thus,

$$\frac{di_L}{dt}\bigg|_{t=0^+} = \frac{v_L(0^+)}{L} = 40 \quad \text{A/s}$$

Similarly,

$$\frac{dv_C}{dt}\bigg|_{t=0^+} = \frac{i_C(0^+)}{C} = 108 \quad \text{V/s}$$

The other four derivatives may be determined by realizing that Kirchhoff's current and voltage laws are both satisfied by the derivatives also. For example, at the left node,

$$4 - i_L - i_R = 0 \qquad t > 0$$

and thus

$$0 - \frac{di_L}{dt} - \frac{di_R}{dt} = 0 \qquad t > 0$$

and, therefore,

$$\frac{di_R}{dt}\bigg|_{t=0^+} = -40 \quad \text{A/s}$$

Now let us turn to a slightly different method by which all these currents, voltages, and derivatives may be evaluated at $t = 0^-$ and $t = 0^+$. We shall construct two equivalent circuits, one which is valid for the steady-state condition reached at $t = 0^-$ and a second which is valid during the switching interval. The discussion which follows relies on some of the reasoning we did above and, for that reason, appears shorter than it would be if it were presented first.

Prior to the switching operation, only direct currents and voltages exist in the circuit, and the inductor may therefore be replaced by a short circuit, its dc equivalent, while the capacitor is replaced by an open circuit. Redrawn in this manner, the circuit of Fig. 7-11 appears as shown in Fig. 7-12*a*. The three voltages and three currents at $t = 0^-$ are now easily found by resistive circuit-analysis methods; the numerical values are the same as those found previously.

We now turn to the problem of drawing an equivalent circuit which will assist us in determining the several voltages and currents at $t = 0^+$. Each capacitor voltage and each inductor current must remain constant during the switching interval. These conditions may be ensured by replacing the inductor by a current source and the capacitor by a voltage source.

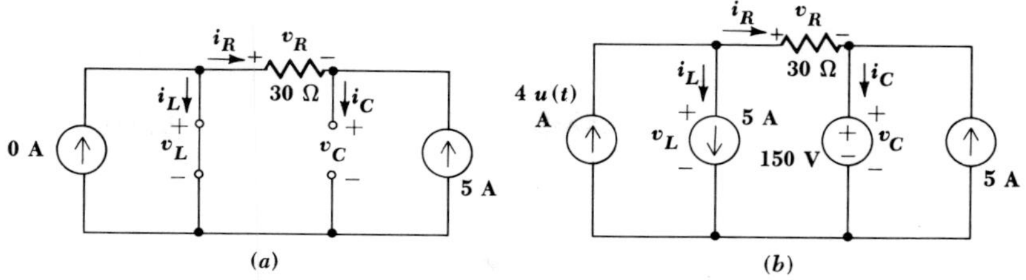

Fig. 7-12 (*a*) A simple circuit which is the equivalent of the circuit of Fig. 7-11 for $t = 0^-$. (*b*) Another equivalent of the circuit of Fig. 7-11, valid during the switching interval, $t = 0^-$ to $t = 0^+$.

Each source serves to maintain the necessary response constant during the discontinuity. The equivalent circuit of Fig. 7-12*b* results; it should be noted that this circuit is a true equivalent at $t = 0^-$ since it possesses the same currents and voltages as the simple equivalent of Fig. 7-12*a*. It is also a true equivalent at $t = 0^+$ since the step-current source appears as a function of time and not merely as 0 A or 4 A.

The voltages and currents at $t = 0^+$ are obtained by letting $4u(t) = 4$ A and solving the dc circuit which results. The solution is not difficult, but the relatively large number of sources present in the network does produce a somewhat strange sight. However, problems of this type were solved in Chap. 3, and nothing new is involved. The six responses at $t = 0^+$ must agree with those found by the previous method.

Before leaving this problem of the determination of the necessary initial values, it should be pointed out that at least one other powerful method of determining them has been omitted; we could have written general nodal or loop equations for the original circuit. Then, the substitution of the known zero values of inductor voltage and capacitor current at $t = 0^-$ would uncover several other response values at $t = 0^-$ and enable the remainder to be found easily. A similar analysis at $t = 0^+$ must then be made. This is an important method, and it becomes a necessary one in more complicated circuits which cannot be analyzed by the simpler step-by-step procedures we have followed. However, we must leave a few topics to be covered at the time operational methods of circuit analysis are introduced later.

Now let us briefly complete the determination of the response $v_C(t)$ for the original circuit of Fig. 7-11. With both sources dead, the circuit appears as a series *RLC* circuit and s_1 and s_2 are easily found to be -1 and -9, respectively. The forced response may be found by inspection or, if necessary, by drawing the dc equivalent, which is similar to Fig. 7-12*a*, with the addition of a 4-A current source. The forced response is 150 V. Thus,

$$v_C(t) = 150 + Ae^{-t} + Be^{-9t}$$

and

$$v_C(0^+) = 150 = 150 + A + B$$

Then,

$$\frac{dv_C}{dt} = -Ae^{-t} - 9Be^{-9t}$$

and

$$\left.\frac{dv_C}{dt}\right|_{t=0^+} = 108 = -A - 9B$$

Finally,

$$A = 13.5 \qquad B = -13.5$$

and

$$v_C(t) = 150 + 13.5(e^{-t} - e^{-9t})$$

Drill Problems

7-9 For the circuit shown in Fig. 7-13, determine values at $t = 0^-$ and $t = 0^+$ for: (*a*) v_1; (*b*) v_2; (*c*) v_3.

Ans. -0.4, -0.4; -0.4, -0.4; 0, 1.2 V

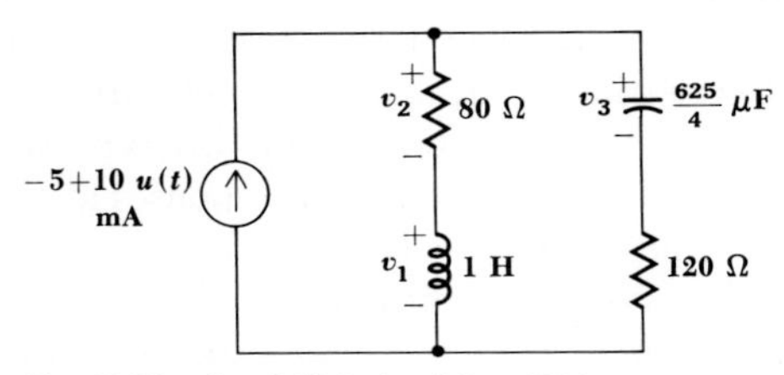

Fig. 7-13 See Drill Probs. 7-9 to 7-11.

7-10 For the circuit of Fig. 7-13, find values at $t = 0^-$ and $t = 0^+$ for: (*a*) dv_1/dt; (*b*) dv_2/dt; (*c*) dv_3/dt.

Ans. 0, -176; 0, 64; 0, 96 V/s

7-11 For the circuit of Fig. 7-13, for $t > 0$, find: (*a*) $v_1(t)$; (*b*) $v_2(t)$; (*c*) $v_3(t)$.

Ans. $0.4 - 0.533e^{-40t} - 0.267e^{-160t}$; $0.133e^{-40t} + 1.067e^{-160t}$; $0.4 - 0.267e^{-40t} - 0.533e^{-160t}$ V

PROBLEMS

☐ **1** A parallel circuit is comprised of the elements, $L = 10$ H, $R = 320\ \Omega$, $C = 125/8\ \mu$F. The initial value of the capacitor voltage is $v_C(0) = -160$ V and the initial value of the capacitor current is $i_C(0^+) = 0.7$ A, where i_C and v_C are related by the passive sign convention. (*a*) Find the initial energy storage in the inductor and capacitor. (*b*) At what value of t is $v_C = 0$? (*c*) At what value of t is v_C a positive maximum?

□ **2** (*a*) The circuit shown on the response curve of Fig. 7-3 is modified by increasing the size of the inductor to $^{1400}/_{23}$ H. If the initial conditions remain unchanged, find and sketch $v(t)$; compare with the response of the original circuit. (*b*) What will the response become as L approaches infinity?

□ **3** The voltage across a source-free parallel RLC circuit is $v_C(t) = 10e^{-1000t} - 5e^{-4000t}$ V for $t > 0$. If the energy stored in the capacitor at $t = 0$ is 100 μJ, find R, L, C, $v_C(0)$, and $i_L(0)$.

□ **4** The circuit shown in Fig. 7-14 is source-free for $t > 0$. Find $v(t)$.

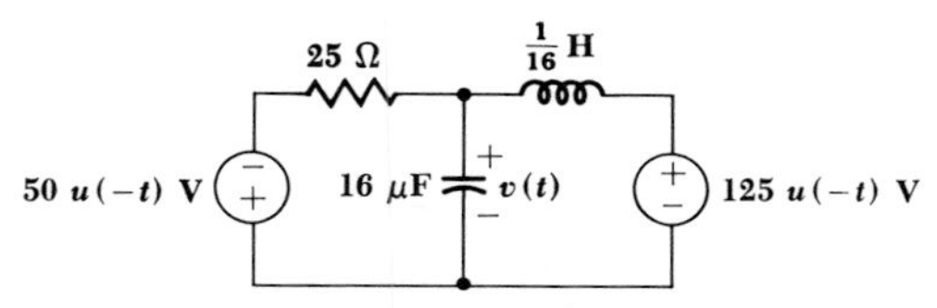

Fig. 7-14 See Probs. 4 and 6.

□ **5** After being open for a long time, the switch in the circuit of Fig. 7-15 is closed at $t = 0$. Find $i(t)$ for $t > 0$.

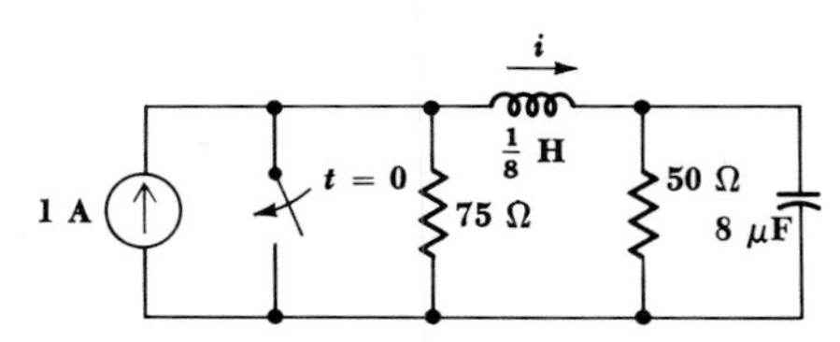

Fig. 7-15 See Probs. 5 and 10.

□ **6** (*a*) What value of resistance must be placed in series with the 25-Ω resistor of Fig. 7-14 to create critical damping? (*b*) With this resistor in place, find $v(t)$ for $t > 0$.

□ **7** A certain parallel RLC circuit has an exponential damping coefficient of 500 s^{-1} and a resonant frequency of 400 rad/s. It is found experimentally that adding a 2-μF capacitor in parallel with the circuit produces critical damping. If the voltage across the critically damped circuit is $v(t)$ and $v(0) = 20$ V while $v'(0^+) = 2000$ V/s, find $v(t)$ for $t > 0$.

□ **8** For a critically damped parallel RLC circuit in which $v_C(0) = 0$, show that v_C reaches a maximum at $t_m = 1/\alpha$ and that the total energy stored in the circuit then is $5e^{-2}$ times the energy stored in the inductor at $t = 0$.

□ **9** The natural response of a source-free parallel RLC circuit for $t > 0$ is $v(t) = 100e^{-600t} \cos 400t$ V. If the initial energy stored in the capacitor is $^1/_{30}$ J, find R, L, C, and the initial inductor current.

☐ **10** Let the capacitance in Fig. 7-15 be changed to 13 μF. The switch is now left open for a long time and is then closed at $t = 0$. Find $i(t)$ for $t > 0$.

☐ **11** The switch in the circuit of Fig. 7-16 has been closed during a laboratory experiment. Seeing a maximum source voltage of only 10 V in the circuit, student *D* carelessly lets his hands roam all over it; in particular, across the inductor at $t = 0$ just as his newest enemy opens the switch. What is the maximum voltage to which he will be subjected? Sketch $v_L(t)$ vs. t.

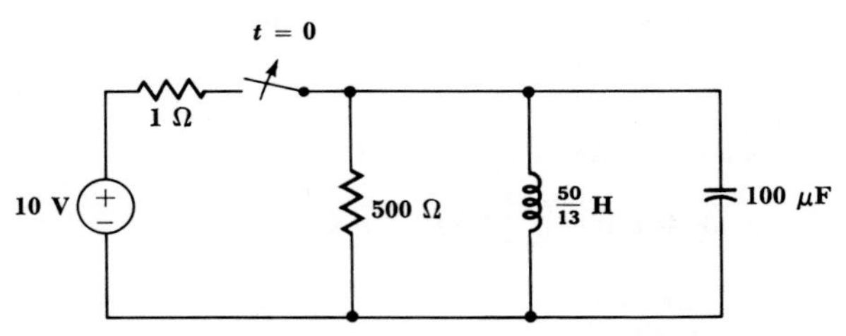

Fig. 7-16 See Prob. 11.

☐ **12** A parallel *RLC* circuit consists of a 100-pF capacitor, a 5-kΩ resistor, and a 2-mH inductor. If the capacitor voltage and inductor current (as sensed in Fig. 7-2) are 2 V and 1 mA, respectively, at $t = 0$, find $i_C(t)$ for $t > 0$.

☐ **13** A parallel *RLC* circuit contains a 750-Ω resistor, a $10/3$-μF capacitor, and an inductor L. Let the initial stored energies be $w_C(0) = 0$ and $w_L(0) = 0.05$ J. Find the maximum magnitude of the capacitor voltage, the time t_m at which it occurs, and the settling time t_s if $L =$: (*a*) 10 H; (*b*) 7.5 H; (*c*) 6 H.

☐ **14** A series *RLC* circuit contains a 2-μF capacitor, a $\frac{1}{2}$-H inductor, and a 2600-Ω resistor. If the inductor current and voltage ($v_L = L\, di_L/dt$) at $t = 0$ are $v_L = 30$ V, $i_L = 50$ mA, find $i_L(t)$ for $t > 0$.

☐ **15** The resistance of the underdamped circuit shown in Fig. 7-5 is changed from 14.85 Ω to 10.5 Ω. After finding the response $v(t)$, determine: (*a*) the time t_{m1} and voltage v_{m1} for the first maximum; (*b*) the time t_{01} for the first zero crossing; (*c*) the time t_{m2} and voltage v_{m2} for the first minimum; and (*d*) the settling time t_s.

☐ **16** The current circulating in a series *RLC* circuit is $i = 20e^{-1000t} - 10e^{-4000t}$ mA. Let the initial energy storage in the inductor be 0.2 mJ. Determine R, L, C, $v_C(0)$, and $i_C(0)$.

☐ **17** Find the current $i(t)$ for $t > 0$ in the circuit shown in Fig. 7-17.

☐ **18** It is desired to modify the circuit of Fig. 7-17 to produce critical damping by placing a resistance R_x in parallel with the resistor already present. (*a*) Find R_x. (*b*) Find $i(t)$ for the critically damped circuit.

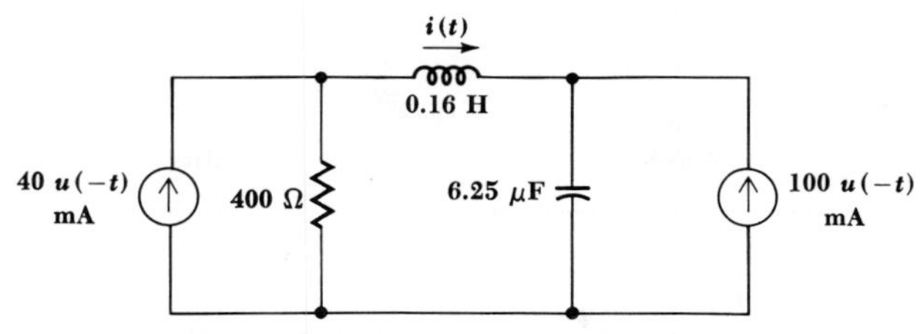

Fig. 7-17 See Probs. 17 and 18.

☐ **19** A source-free series RLC circuit has the current, $i(t) = 20e^{-900t} \cos 600t$ mA, established in it for $t > 0$. If the series resistance is 90 Ω, find L, C, and the initial energy stored in the capacitor and inductor.

☐ **20** After being closed for 1 s, the switch in the circuit of Fig. 7-18 is opened at $t = 0$. Find $i_1(t)$ for all t.

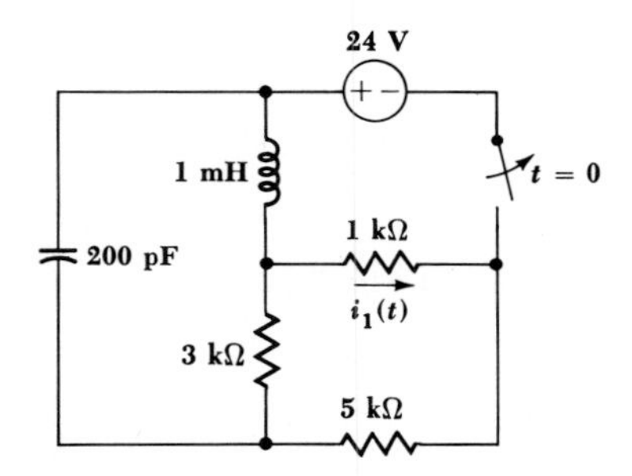

Fig. 7-18 See Prob. 20.

☐ **21** The circuit of Fig. 7-19 has been in the configuration shown for a long time. At $t = 0$, the 100-Ω resistor is connected into the circuit as the switch opens. Find $i(t)$ for all t.

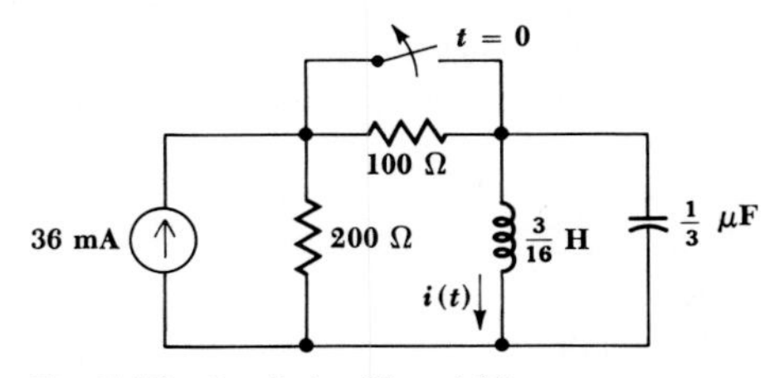

Fig. 7-19 See Probs. 21 and 22.

☐ **22** Refer to Fig. 7-19 and change the time of the switch opening to $t = -\infty$ and let the source become $36 + 36u(t)$ mA. Find $i(t)$ for all t.

☐ **23** Find $i_L(t)$ for all t for the circuit shown in Fig. 7-20.

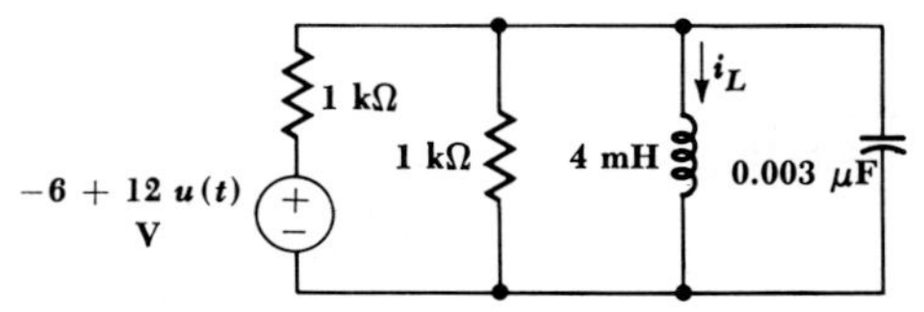

Fig. 7-20 See Prob. 23.

☐ **24** Find $v_x(t)$ for $t < 0$ and $t > 0$ in the circuit shown in Fig. 7-21. To what value must the voltage source be changed in order that $v_x(0^-) = -v_x(0^+)$?

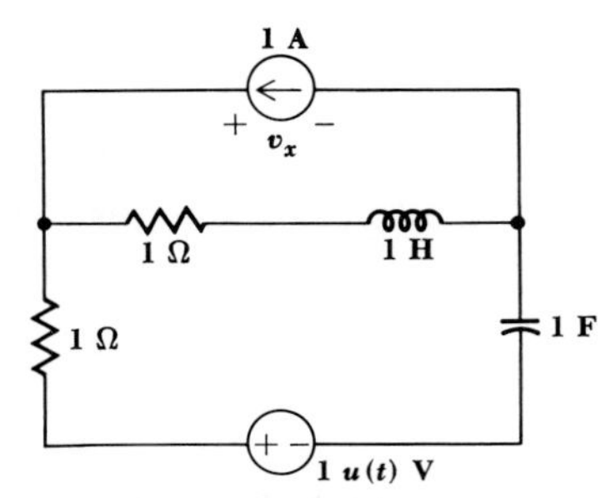

Fig. 7-21 See Prob. 24.

☐ **25** Find and sketch $v_R(t)$ for the circuit of Fig. 7-22.

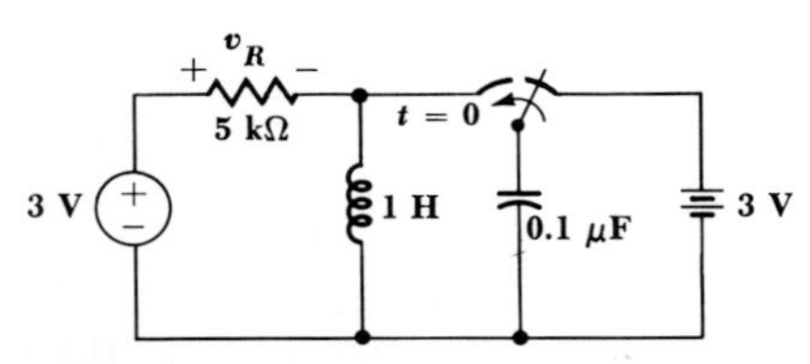

Fig. 7-22 See Prob. 25.

☐ **26** Write the single equation required to describe the circuit of Fig. 7-23, using i_C as the variable. By comparing this equation with the basic equations for the series and parallel *RLC* circuit, determine equivalent values for *R*, *L*, and *C*. Find $i_C(t)$ for $t > 0$.

Fig. 7-23 See Prob. 26.

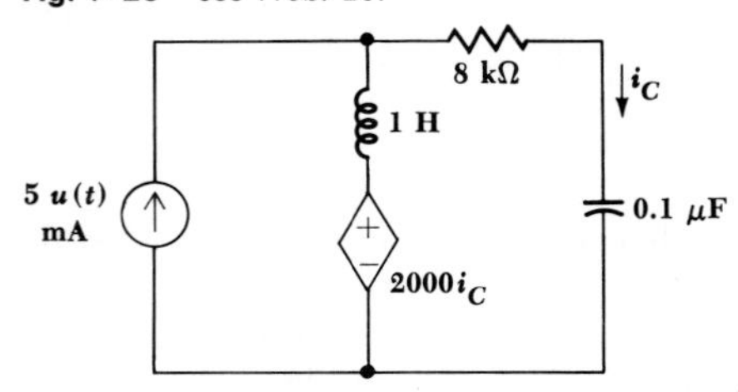

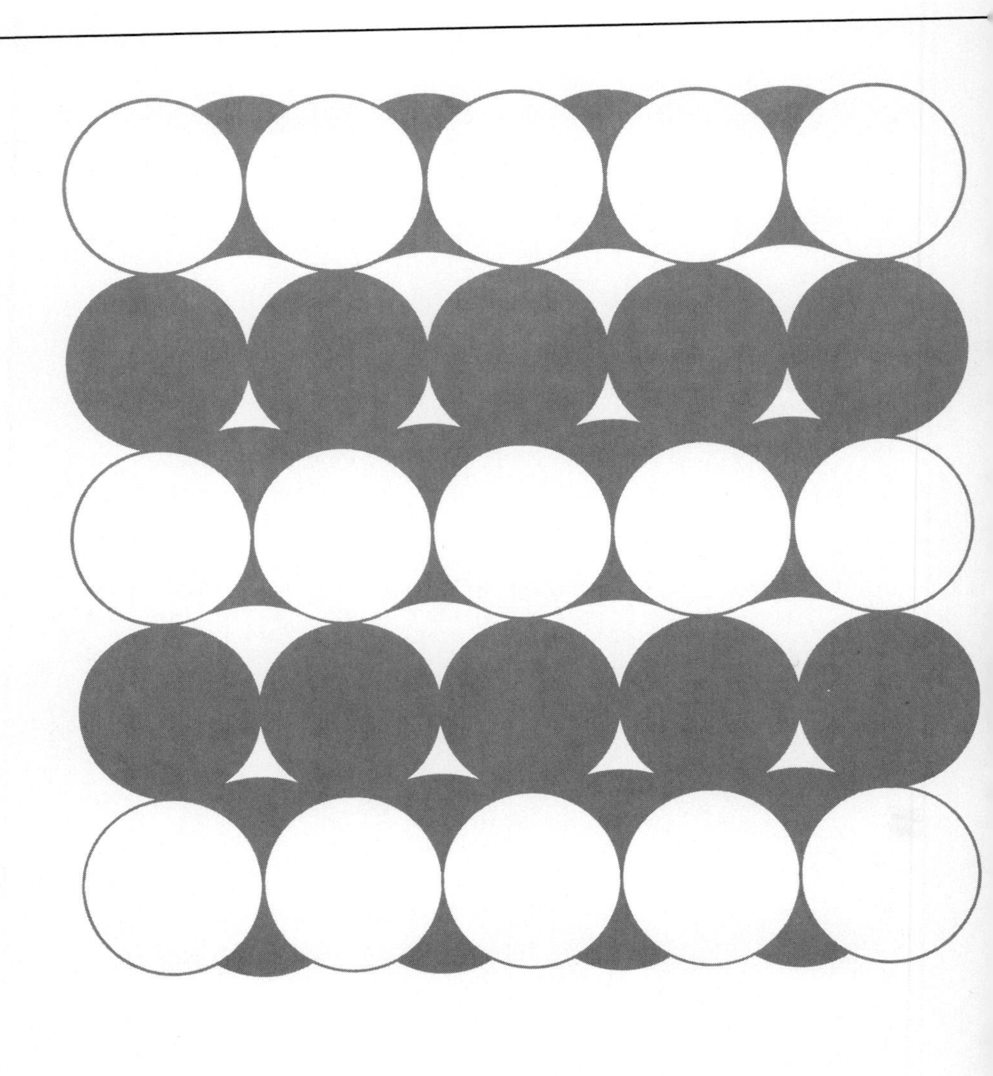

Part Three
SINUSOIDAL ANALYSIS

Chapter Eight
The Sinusoidal Forcing Function

8-1 INTRODUCTION

The complete response of a linear electric circuit is composed of two parts, the natural response and the forced response. The first part of our study was devoted to the resistive circuit, in which only the forced response is required or present. For simplicity, we usually restricted our forcing functions to dc sources, and we therefore became exceedingly familiar with the various techniques useful in finding the dc forced response. We then passed on to the next part and considered the natural response of a number of different circuits containing one or two energy-storage elements. Without undue strain we were then able to determine the complete response of these circuits by adding the natural response, which is characteristic of the circuit and not of the forcing function, to the forced response produced by dc forcing functions, the only forced response with which we are familiar. We are therefore now in a position where our mastery of the natural response is greater than our knowledge of the forced response.

In this third part of our study we shall extend our knowledge of the forced response by considering the sinusoidal forcing function.

Why should we select the sinusoidal forcing function as the second functional form to study? Why not the linear function, the exponential function, or a modified Bessel function of the second kind? There are many reasons for the choice of the sinusoid, and any one of them would probably be sufficient to lead us in this direction.

One of these reasons is apparent from the results of the preceding chapter; the natural response of an underdamped second-order system is a damped sinusoid, and if no losses are present it is a pure sinusoid. The sinusoid thus appears naturally (as does the negative exponential). Indeed, Nature in general seems to have a decidedly sinusoidal character; the motion of a pendulum, the bouncing of a ball, the vibration of a guitar string, the political atmosphere in any country, and the ripples on the surface of a stein of chocolate milk will always display a reasonably sinusoidal character.

Perhaps it was observations of these natural phenomena that led the great French mathematician Fourier to his discovery of the important analytical method embodied in the Fourier theorem. In Chap. 17 we shall see that this theorem enables us to represent most of the useful mathematical functions of time which repeat themselves f_0 times a second by the sum of an infinite number of sinusoidal time functions with frequencies that are integral multiples of f_0; the given periodic function $f(t)$ can also be approximated as closely as we wish by the sum of a finite number of such terms, even though the graph of $f(t)$ might look very nonsinusoidal. This decomposition of a periodic forcing function into a number of appropriately chosen sinusoidal forcing functions is a very powerful analytical method, for it enables us to superpose the partial responses produced in any linear circuit by each sinusoidal component in order to obtain the desired response caused by the given periodic forcing function. Thus, another reason for studying the response to a sinusoidal forcing function is found in the dependence of other forcing functions on sinusoidal analysis.

A third reason is found in an important mathematical property of the sinusoidal function. Its derivatives and integrals are also all sinusoids.[1] Since the forced response takes on the form of the forcing function, its integral, and its derivatives, the sinusoidal forcing function will produce

[1] We are using the term "sinusoid" collectively here to include cosinusoidal functions of time also. After all, a cosine function can be written as a sine function if the angle is increased by 90°.

a sinusoidal forced response throughout a linear circuit. The sinusoidal forcing function thus allows a much easier mathematical analysis than does almost every other forcing function.

Finally, the sinusoidal forcing function has important practical applications. It is an easy function to generate and is the waveform used predominantly throughout the electric power industry; and every electrical laboratory contains a number of sinusoidal generators which operate throughout a tremendous range of useful frequencies.

Drill Problem

8-1 The Fourier theorem, which we shall study in Chap. 17, shows that the periodic triangular waveform $v_1(t)$, shown in Fig. 8-1, and the infinite sum of cosine terms,

$$v_2(t) = \frac{12}{\pi^2}\left(\frac{\pi^2}{24} + \cos\frac{\pi t}{3} + \frac{1}{3^2}\cos\frac{3\pi t}{3} + \frac{1}{5^2}\cos\frac{5\pi t}{3} + \frac{1}{7^2}\cos\frac{7\pi t}{3} + \cdots\right)$$

are equal. Determine the ratio of v_2 to v_1 at $t = 6$ s if v_1 is approximated by only: (*a*) two terms of the infinite series: (*b*) four terms of the infinite series; (*c*) six terms of the infinite series.

Ans. 0.858; 0.950; 0.970

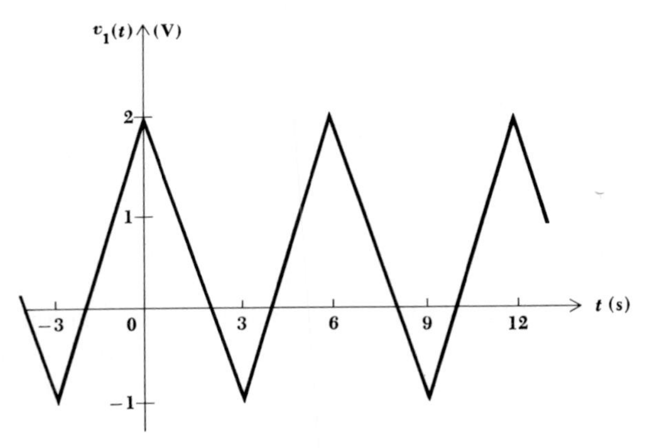

Fig. 8-1 See Drill Prob. 8-1.

8-2 CHARACTERISTICS OF SINUSOIDS

In this section we shall define the trigonometric nomenclature which is used to describe sinusoidal (or cosinusoidal) functions. The definitions should be familiar to most of us, and if we remember a little trigonometry, the section can be read over very rapidly.

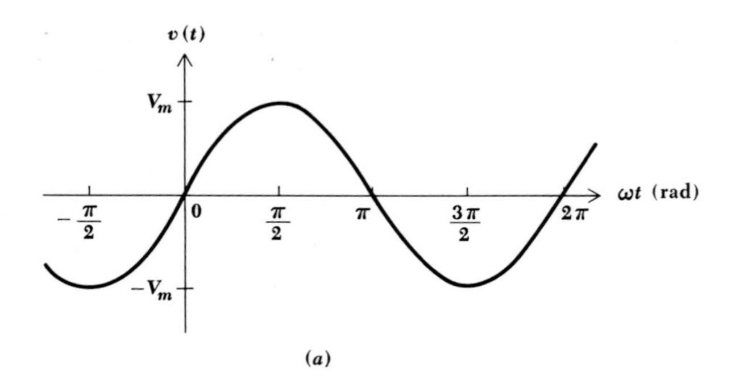

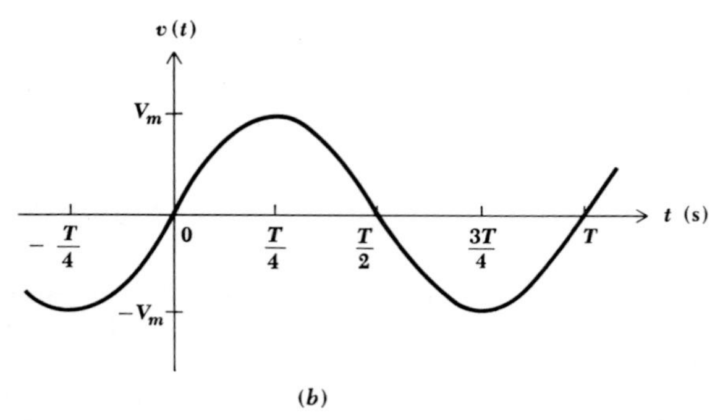

Fig. 8-2 The sinusoidal function $v(t) = V_m \sin \omega t$ is plotted versus ωt in (*a*) and versus t in (*b*).

Let us consider a sinusoidally varying voltage

$$v(t) = V_m \sin \omega t$$

shown graphically in Fig. 8-2*a* and *b*. The *amplitude* of the sine wave is V_m, and the *argument* is ωt. The *radian frequency* or *angular frequency* is ω. In Fig. 8-2*a*, $\sin \omega t$ is plotted as a function of the argument ωt, and the periodic nature of the sine wave is evident. The function repeats itself every 2π radians, and its *period* is therefore 2π radians. In Fig. 8-2*b*, $\sin \omega t$ is plotted as a function of t and the *period* is now T. The period may also be expressed in degrees, or occasionally in other units such as centimeters or inches. A sine wave having a period T must execute $1/T$ periods each second; its *frequency* f is $1/T$ hertz, abbreviated Hz. Thus, one hertz is identical to one cycle per second, a term whose use is now discouraged because so many people incorrectly used "cycle" for "cycle per second." Thus,

$$f = \frac{1}{T}$$

and since

$$\omega T = 2\pi$$

we obtain the common relationship between frequency and radian frequency,

$$\omega = 2\pi f$$

A more general form of the sinusoid

$$v(t) = V_m \sin(\omega t + \theta) \tag{1}$$

includes a *phase angle* θ in its argument $(\omega t + \theta)$. Equation (1) is plotted in Fig. 8-3 as a function of ωt, and the phase angle appears as the number of radians by which the original sine wave, shown as a broken line in the sketch, is shifted to the left, or earlier in time. Since corresponding points on the sinusoid $V_m \sin(\omega t + \theta)$ occur θ rad, or θ/ω s, earlier, we say that $V_m \sin(\omega t + \theta)$ *leads* $V_m \sin \omega t$ by θ rad. Conversely, it is correct to describe $\sin \omega t$ as *lagging* $\sin(\omega t + \theta)$ by θ rad, as leading $\sin(\omega t + \theta)$ by $-\theta$ rad, or as leading $\sin(\omega t - \theta)$ by θ rad.

In electrical engineering, the phase angle is commonly given in degrees, rather than radians, and no confusion will arise if the degree symbol is always used. Thus, instead of writing

$$v = 100 \sin\left(2\pi 1000t - \frac{\pi}{6}\right)$$

we customarily use

$$v = 100 \sin(2\pi 1000t - 30^\circ)$$

Two sinusoidal waves that are to be compared in phase must both be written as sine waves, or both as cosine waves; both waves must be written with

Fig. 8-3 The sine wave $V_m \sin(\omega t + \theta)$ leads $V_m \sin \omega t$ by θ rad.

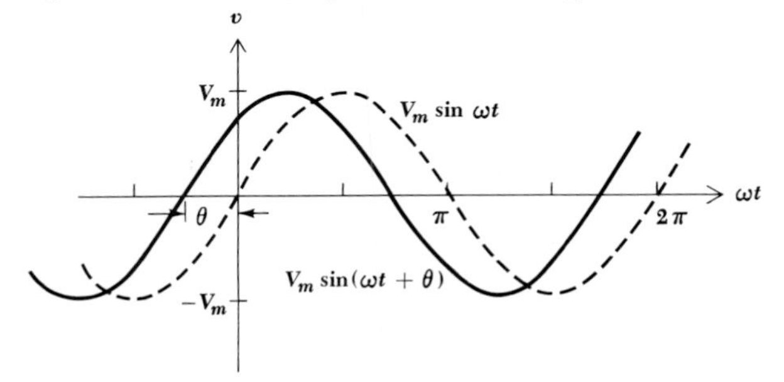

positive amplitudes; and each must be of the same frequency. It is also evident that multiples of 360° may be added to or subtracted from the argument of any sinusoidal function without changing the value of the function. Hence, we may say that

$$v_1 = V_{m1} \sin (5t - 30°)$$

lags

$$\begin{aligned} v_2 &= V_{m2} \cos (5t + 10°) \\ &= V_{m2} \sin (5t + 90° + 10°) \\ &= V_{m2} \sin (5t + 100°) \end{aligned}$$

by 130°, or it is also correct to say that v_1 leads v_2 by 230°, since v_2 may be written as

$$v_2 = V_{m2} \sin (5t - 260°)$$

V_{m1} and V_{m2} are each assumed to be positive quantities. Normally, the difference in phase between two sinusoids is expressed by that angle which is less than or equal to 180° in magnitude.

The concept of a leading or lagging relationship between two sinusoids will be used extensively, and the relationship should be recognizable both mathematically and graphically.

Drill Problems

8-2 Find the angle by which v_2 leads v_1 if $v_1 = 4 \cos (1000t - 40°)$ and $v_2 =$: (*a*) $3 \sin (1000t + 40°)$; (*b*) $-2 \cos (1000t - 120°)$; (*c*) $5 \sin (1000t - 180°)$.

Ans. $-10°$; $100°$; $130°$

8-3 Express each of the following in the form of $A \cos (\omega t + \theta)$ by specifying A and θ: (*a*) $6 \cos \omega t - 4.5 \sin \omega t$; (*b*) $4 \cos (\omega t - 45°) + 3 \sin (\omega t - 45°)$; (*c*) $5 \cos (\omega t - 36.87°) + 2 \cos \omega t - 6 \sin \omega t$.

Ans. 5, $-81.9°$; 6.71, 26.6°; 7.5, 36.9°

8-3 FORCED RESPONSE TO SINUSOIDAL FORCING FUNCTIONS

Now that we are familiar with the mathematical characteristics of sinusoids and can describe and compare them intelligently, we are ready to apply a sinusoidal forcing function to a simple circuit and to obtain the forced response. We shall first write the differential equation which applies to the given circuit. The complete solution of this equation is composed of two parts, the complementary solution (which we refer to as the natural response) and the particular integral (or forced response). The natural response is independent of the mathematical form of the forcing function and depends only upon the type of circuit, the element values, and the initial conditions. We find it by setting all the forcing functions equal to zero, thus reducing the equation to the simpler linear *homogeneous* differential equation. We have already determined the natural response of many RL, RC, and RLC circuits.

The forced response has the mathematical form of the forcing function, plus all its derivatives and its first integral. From this knowledge, it is apparent that one of the methods by which the forced response may be found is by assuming a solution composed of such a sum of functions, where each function has an unknown amplitude to be determined by direct substitution into the differential equation. This is a lengthy method, but it is the one which we shall use in this chapter to introduce sinusoidal analysis because it involves a minimum of new concepts. If the simpler method to be described in the following chapters were not available, circuit analysis would be an impractical, useless art.

The term *steady-state* response is used synonymously with forced response, and the circuits we are about to analyze are commonly said to be in the "sinusoidal steady state." Unfortunately, steady state carries the connotation of "not changing with time" in the minds of many students. This is true for dc forcing functions, but the sinusoidal steady-state response is definitely changing with time. The steady state simply refers to the condition which is reached after the transient or natural response has died out.

Now let us consider the series RL circuit shown in Fig. 8-4. The sinusoidal source voltage $v_s = V_m \cos \omega t$ has been switched into the circuit at some remote time in the past, and the natural response has died out completely. We seek the forced response, or steady-state response, and it must satisfy the differential equation

$$L\frac{di}{dt} + Ri = V_m \cos \omega t$$

Fig. 8-4 A series RL circuit for which the forced response is desired.

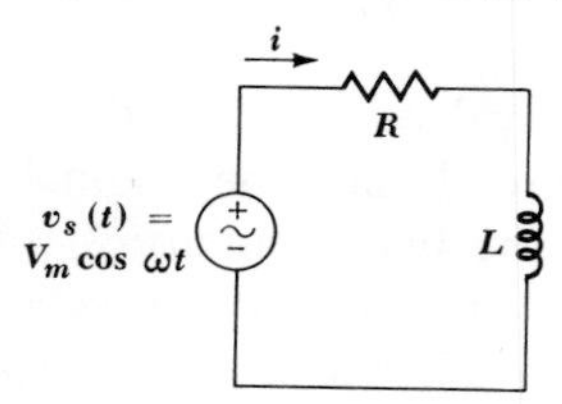

The functional form of the forced response is next obtained by integration and repeated differentiation of the forcing function. Only two different forms are obtained, $\sin \omega t$ and $\cos \omega t$. The forced response must therefore have the general form

$$i(t) = I_1 \cos \omega t + I_2 \sin \omega t$$

where I_1 and I_2 are real constants whose values depend upon V_m, R, L, and ω. No constant or exponential function can be present. Substituting the assumed form for the solution into the differential equation yields

$$L(-I_1\omega \sin \omega t + I_2\omega \cos \omega t) + R(I_1 \cos \omega t + I_2 \sin \omega t) = V_m \cos \omega t$$

If we collect the cosine and sine terms, we obtain

$$(-LI_1\omega + RI_2) \sin \omega t + (LI_2\omega + RI_1 - V_m) \cos \omega t = 0$$

This equation must be true for all values of t, and this can be achieved only if the factors multiplying $\cos \omega t$ and $\sin \omega t$ are each zero. Thus,

$$-\omega LI_1 + RI_2 = 0 \qquad \omega LI_2 + RI_1 - V_m = 0$$

and simultaneous solution for I_1 and I_2 leads to

$$I_1 = \frac{RV_m}{R^2 + \omega^2 L^2} \qquad I_2 = \frac{\omega LV_m}{R^2 + \omega^2 L^2}$$

Thus, the forced response is obtained,

$$i(t) = \frac{RV_m}{R^2 + \omega^2 L^2} \cos \omega t + \frac{\omega LV_m}{R^2 + \omega^2 L^2} \sin \omega t \tag{2}$$

This expression is slightly cumbersome, however, and a clearer picture of the response can be obtained by expressing the response as a single sinusoid or cosinusoid with a phase angle. Let us select the cosinusoid in anticipation of the method in the following chapter,

$$i(t) = A \cos (\omega t - \theta) \tag{3}$$

At least two methods of obtaining the values of A and θ should suggest themselves. We might substitute (3) directly into the original differential equation, or we could simply equate the two solutions (2) and (3). Let us select the latter method, since the former makes an excellent problem for the end of the chapter, and equate (2) and (3) after expanding the function $\cos(\omega t - \theta)$,

$$A \cos\theta \cos\omega t + A \sin\theta \sin\omega t = \frac{RV_m}{R^2 + \omega^2 L^2} \cos\omega t + \frac{\omega L V_m}{R^2 + \omega^2 L^2} \sin\omega t$$

Thus, again collecting and setting the coefficients of $\cos\omega t$ and $\sin\omega t$ equal to zero, we find

$$A \cos\theta = \frac{RV_m}{R^2 + \omega^2 L^2} \qquad \text{and} \qquad A \sin\theta = \frac{\omega L V_m}{R^2 + \omega^2 L^2}$$

To find A and θ, we divide one equation by the other,

$$\frac{A \sin\theta}{A \cos\theta} = \tan\theta = \frac{\omega L}{R}$$

and by drawing a small triangle, as shown in Fig. 8-5, we have

$$\cos\theta = \frac{R}{\sqrt{R^2 + \omega^2 L^2}}$$

Thus,

$$A = \frac{1}{\cos\theta} \frac{RV_m}{R^2 + \omega^2 L^2} \qquad \text{or} \qquad A = \frac{V_m}{\sqrt{R^2 + \omega^2 L^2}}$$

The alternative form of the forced response therefore becomes

$$i(t) = \frac{V_m}{\sqrt{R^2 + \omega^2 L^2}} \cos\left(\omega t - \tan^{-1}\frac{\omega L}{R}\right) \tag{4}$$

The electrical characteristics of the response $i(t)$ should now be considered. The amplitude of the response is proportional to the amplitude of the forcing function; if it were not, the linearity concept would have to be discarded. The amplitude of the response also decreases as R, L, or ω is increased, but not proportionately. This is confirmed by the differ-

Fig. 8-5 A right triangle displays the relationship, $\tan\theta = \omega L/R$, and enables us to write $\cos\theta = R/\sqrt{R^2 + \omega^2 L^2}$.

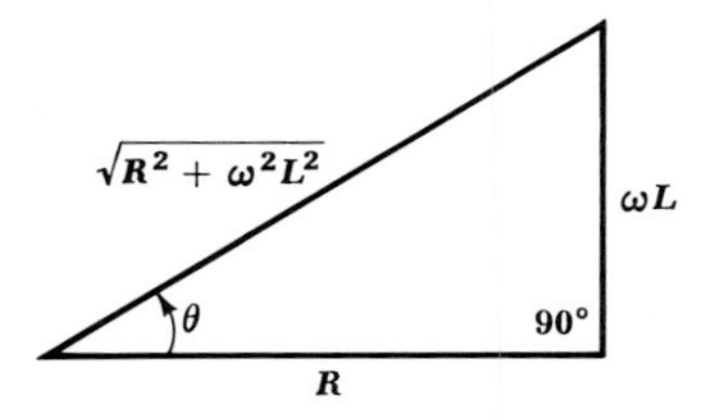

ential equation, for an increase in R, L, or di/dt requires a decrease in current amplitude if the source-voltage amplitude is not changed. The current is seen to lag the applied voltage by $\tan^{-1}(\omega L/R)$, an angle between 0 and 90°. When $\omega = 0$ or $L = 0$, the current must be in phase with the voltage; since the former situation is direct current and the latter provides a resistive circuit, the result is expected. If $R = 0$, the current lags the voltage by 90°; then $v_s = L(di/dt)$, and the derivative-integral relationship between the sine and cosine indicates the validity of the 90° phase difference.

The applied voltage and the resultant current are both plotted on the same ωt axis in Fig. 8-6, but arbitrary current and voltage ordinates are assumed. The fact that the current lags the voltage in this simple RL circuit is now visually apparent. We shall later be able to show easily that this result is typical for all inductive circuits, that is, circuits composed only of inductors and resistors. The phase difference between the current and voltage depends upon the ratio of the quantity ωL to R. We call ωL the *inductive reactance* of the inductor; it is measured in ohms, and it is a measure of the opposition which is offered by the inductor to the passage of a sinusoidal current. Much more will be said about reactance in the following chapter.

The method by which we have found the sinusoidal steady-state response for this simple series RL circuit has not been a trivial problem. We might think of the analytical complications as arising through the presence of the inductor; if both the passive elements had been resistors, the analysis would have been ridiculously easy, even with the sinusoidal forcing function present. The reason the analysis would be so easy results from the simple voltage-current relationship specified by Ohm's law. The voltage-current relationship for an inductor is not as simple, however; instead of solving an algebraic equation, we were faced with a nonhomogeneous differential equation. It would be quite impractical to analyze every circuit by the method described above, and in the following chapter we shall therefore take steps to simplify the analysis. Our result

Fig. 8-6 The applied sinusoidal forcing function (solid) and the resultant sinusoidal current response (broken) of the series RL circuit shown in Fig. 8-4.

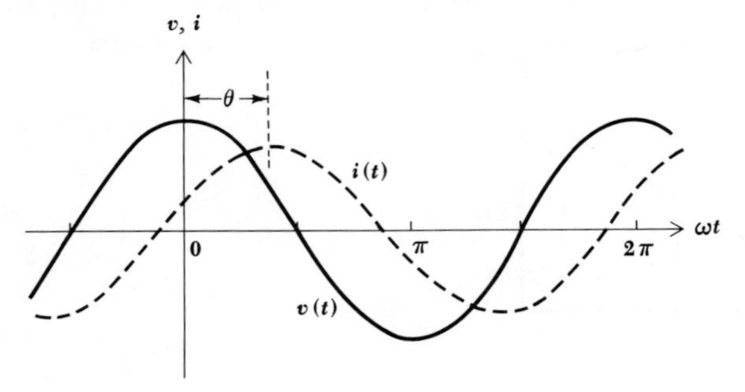

will be an algebraic relationship between sinusoidal current and sinusoidal voltage for inductors and capacitors as well as resistors, and we shall be able to produce a set of algebraic equations for a circuit of any complexity. The constants and the variables in the algebraic equations will be complex numbers rather than real numbers, but the analysis of any circuit in the sinusoidal steady state becomes almost as easy as the analysis of a similar resistive circuit.

Drill Problems

8-4 Find $i(t)$ for the circuit of Fig. 8-4 if $R = 100\ \Omega$, $V_m = 180$ V, and: (*a*) $\omega = 400$ rad/s, $L = 250$ mH; (*b*) $\omega = 200$ rad/s, $L = 500$ mH; (*c*) $\omega = 200$ rad/s, $L = 50$ mH.

Ans. $1.27 \cos(200t - 45°)$; $1.27 \cos(400t - 45°)$; $1.79 \cos(200t - 5.7°)$ A

8-5 A sinusoidal current source, $10 \cos 1000t$ A, is in parallel both with a 20-Ω resistor and the series combination of a 10-Ω resistor and a 10-mH inductor. Remembering the usefulness of Thévenin's theorem, find the voltage across the: (*a*) inductor; (*b*) 10-Ω resistor; (*c*) current source.

Ans. $63.2 \cos(1000t - 18.4°)$; $63.2 \cos(1000t + 71.6°)$; $89.4 \cos(1000t + 26.6°)$ V

PROBLEMS

☐ **1** A certain sinusoidal current waveform crosses the $i = 0$ axis at $t = -2.1$ ms with $i' > 0$; the adjacent zero crossing occurs at $t = 5.9$ ms with $i' < 0$. (*a*) Find T, f, and ω for this waveform. (*b*) If the current amplitude is 12 mA, write $i(t)$ as a simple function of time. (*c*) By how many degrees does this current waveform lag the voltage $12 \cos \omega t$, where voltage and current are at the same frequency?

☐ **2** Carry out the exercise threatened in the text by substituting the assumed current response (3), $i(t) = A \cos(\omega t - \theta)$, directly into the differential equation, $L(di/dt) + Ri = V_m \cos \omega t$, to show that values for A and θ are obtained which agree with (4).

☐ **3** The current, $i = 4 \cos(200t - 45°)$ A, is the response produced in a series RL circuit by the source, $v_s = 150 \cos(200t + 15°)$ V. Find R and L.

☐ **4** A voltage source, $V_m \cos 100t$, a 20-Ω resistor, and a physical coil which we may model with a 5-Ω resistor in series with a 0.5-H inductor are all in series. What is the ratio of the sinusoidal amplitude of the voltage across the physical coil to that across the 20-Ω resistor?

☐ **5** The voltage source in a series RL circuit is $v_s(t) = 50 \cos 200t + 100 \sin 300t$ V. If $R = 4\ \Omega$ and $L = 10$ mH, find $i(t)$.

☐ **6** The source voltage, $v_s(t) = 100 \cos 20t$ V, is applied to a series RL circuit containing a 5-Ω resistor and an inductor L. Find L if: (*a*) the amplitude of the current is 10 A; (*b*) the current waveform lags the source voltage by 30°.

☐ **7** A series circuit consists of a voltage source, $v_s = 20 \sin 50t$ V, a normally open switch, a 6-Ω resistor, and a 0.16-H inductor. The switch closes at $t = 0$. The desired response is the current. (*a*) Find the functional form of the natural response. (*b*) Find the forced response. (*c*) Use the initial conditions to evaluate the single unknown constant and determine the complete response for $t > 0$. (*d*) Find i at $t = 40$ ms.

☐ **8** In Chap. 6, the circuit of Fig. 6-12 was used as an example of the technique of finding the complete response to dc excitation. (*a*) Find $i(t)$ if the $50u(t)$-V source is changed to $50 \cos (2t/3)$ V. (*b*) Find $i(t)$ for $t > 0$ if the $50u(t)$-V source is changed to $50 \cos (2t/3)\ u(t)$ V.

☐ **9** A current source $I_m \cos \omega t$, a resistor R, and an inductor L are all in parallel. (*a*) Write the single integrodifferential equation in terms of the voltage between nodes. (*b*) Differentiate the above equation with respect to t and obtain the first-order linear differential equation. (*c*) Use the method of Sec. 8-3 to determine the forced response v.

☐ **10** A current source $I_m \cos \omega t$, a resistor R, and a capacitor C are all in parallel. (*a*) Write the single linear differential equation in terms of the voltage v between nodes. (*b*) Assume a suitable general form for the forced response v, substitute into the above differential equation, and determine the exact form of the forced response.

☐ **11** A voltage source $60 \cos 1000t$ V is in series with a 2-kΩ resistor and a 1-μF capacitor. Find i_{forced}.

☐ **12** A 1-μF capacitor, a 10-mH inductor, and a current source $0.2 \sin 8000t$ A are in parallel. The circuit is in a steady-state condition. Find the ratio of the amplitudes of the inductor and source currents.

Chapter Nine
The Phasor Concept

9-1 INTRODUCTION

Throughout the earlier portions of our study of circuit analysis, we devoted our entire attention to the resistive circuit. However, we might remember that we were often promised that those methods which we were applying to resistive circuits would later prove applicable to circuits containing inductors and capacitors as well. In this chapter we shall lay the descriptive groundwork which will make this prediction come true. We shall develop a method for representing a sinusoidal forcing function or a sinusoidal response by a complex-number symbolism called a phasor transform, or simply a phasor. This is nothing more than a number which, by specifying both the magnitude and phase angle of a sinusoid, characterizes that sinusoid just as completely as if it were expressed as an analytical function of time. By working with phasors, rather than with derivatives and integrals of sinusoids as we did in the preceding chapter, we shall effect a truly remarkable simplification in the steady-state sinusoidal analysis of general *RLC* circuits. This simplification should become apparent toward the end of this chapter.

The use of a mathematical transformation to simplify a problem should not be a new idea to us. For example, we have all used logarithms to simplify arithmetic multiplication and division. In order to multiply several numbers together, we first determined the logarithm of each of the numbers, or "transformed" the numbers into an alternative mathematical description. We might now describe that operation as obtaining the "logarithmic transform." We then added all the logarithms to obtain the logarithm of the desired product. Then, finally, we found the antilogarithm, a process which might be termed an inverse transformation; the antilogarithm was our desired answer. Our solution carried us from the domain of everyday numbers to the logarithmic domain, and back again.

Other familiar examples of transform operations may be found in the alternative representations of a circle as a mathematical equation, as a geometric figure on a rectangular-coordinate plane, or merely as a set of three numbers, where it is understood that the first is the x-coordinate value of the center, the second is the y-coordinate value, and the third is the magnitude of the radius. Each of the three representations contains exactly the same information, and once the rules of the transformations are laid down in analytic geometry, we find no difficulty in passing from the algebraic domain to the geometric domain or to the "domain of the ordered triplet."

Few other transforms with which we are familiar provide the simplification that can be achieved with the phasor concept.

9-2 THE COMPLEX FORCING FUNCTION

We are now ready to think about applying a complex forcing function (that is, one that has both a real and an imaginary part) to an electrical network.[1] It may seem strange, but we shall find that the use of complex quantities in sinusoidal steady-state analysis leads to methods which are much simpler than those involving purely real quantities. We should expect a complex forcing function to produce a complex response; we might even suspect, and suspect correctly, that the real part of the forcing function will produce

[1]Appendix 3 defines the complex number and related terms, describes the arithmetic operations on complex numbers, and develops Euler's identity and the exponential and polar forms.

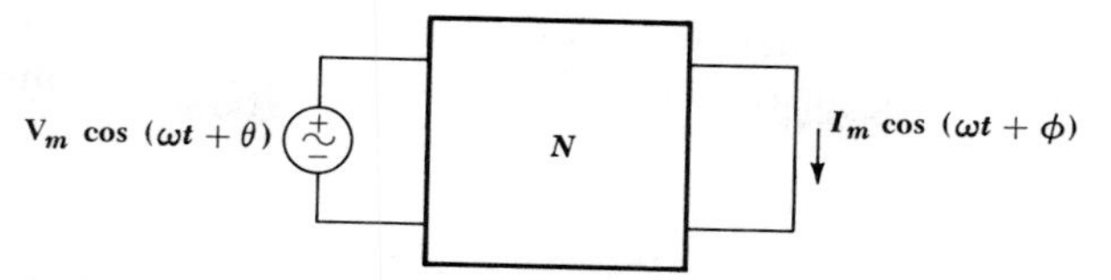

Fig. 9-1 The sinusoidal forcing function $V_m \cos(\omega t + \theta)$ produces the steady-state sinusoidal response $I_m \cos(\omega t + \phi)$.

the real part of the response, while the imaginary portion of the forcing function will result in the imaginary portion of the response. Our goal in this section is to prove, or at least demonstrate, that these suspicions are correct.

Let us first discuss the problem in rather general terms, thus indicating the method by which we might prove our assertions if we were to construct a general network and analyze it by means of a system of simultaneous equations. In Fig. 9-1, a sinusoidal source

$$V_m \cos(\omega t + \theta) \tag{1}$$

is connected to a general network, which we shall arbitrarily assume to be passive in order to avoid complicating our use of the superposition principle later. A current response in some other branch of the network is to be determined. The parameters appearing in (1) are all real quantities.

The discussion in Chap. 8 of the method whereby the response to a sinusoidal forcing function may be determined, through the assumption of a sinusoidal form with arbitrary amplitude and arbitrary phase angle, shows that the response may be represented by

$$I_m \cos(\omega t + \phi) \tag{2}$$

A sinusoidal forcing function always produces a sinusoidal forced response.

Now let us change our time reference by shifting the phase of the forcing function by 90° (or temporarily replacing t by $t' - \pi/2\omega$). Thus, the forcing function

$$V_m \cos(\omega t + \theta - 90°) = V_m \sin(\omega t + \theta) \tag{3}$$

when applied to the same network will produce a corresponding response

$$I_m \cos(\omega t + \phi - 90°) = I_m \sin(\omega t + \phi) \tag{4}$$

We must next depart from physical reality by applying an imaginary forcing function, one which cannot be applied in the laboratory but can be applied mathematically.

We construct an imaginary source very simply; it is only necessary to multiply the source, expressed by (3), by j, the imaginary operator. We thus apply

$$jV_m \sin(\omega t + \theta) \tag{5}$$

What is the response? If we had doubled the source, then the principle of linearity would require that we double the response; multiplication of the forcing function by a constant k would result in the multiplication of the response by the same constant k. The fact that this constant is the imaginary operator j does not destroy this relationship, even though our earlier definition and discussion of linearity did not specifically include complex constants. It is now more realistic to conclude that it did not specifically *exclude* them, for the entire discussion is equally applicable if all the constants in the equations are complex. The response to the imaginary source of (5) is thus

$$jI_m \sin(\omega t + \phi) \tag{6}$$

The imaginary source and response are indicated in Fig. 9-2.

We have applied a real source and obtained a real response; we have also applied an imaginary source and obtained an imaginary response. Now we may use the superposition theorem to find the response to that complex forcing function which is the sum of the real and imaginary forcing functions. The applicability of superposition, of course, is guaranteed by the linearity of the circuit and does not depend on the form of the forcing functions. Thus, the sum of the forcing functions of (1) and (5),

$$V_m \cos(\omega t + \theta) + jV_m \sin(\omega t + \theta) \tag{7}$$

must therefore produce a response which is the sum of (2) and (6),

$$I_m \cos(\omega t + \phi) + jI_m \sin(\omega t + \phi) \tag{8}$$

The complex source and response may be represented more simply by applying Euler's identity. The source of (7) thus becomes

Fig. 9-2 The imaginary sinusoidal forcing function $jV_m \sin(\omega t + \theta)$ produces the imaginary sinusoidal response $jI_m \sin(\omega t + \phi)$.

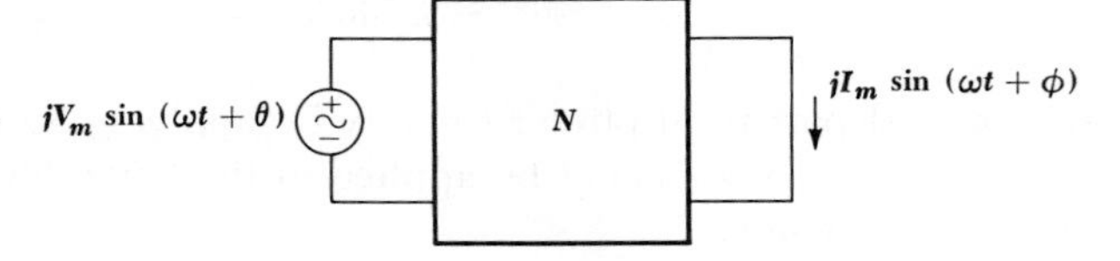

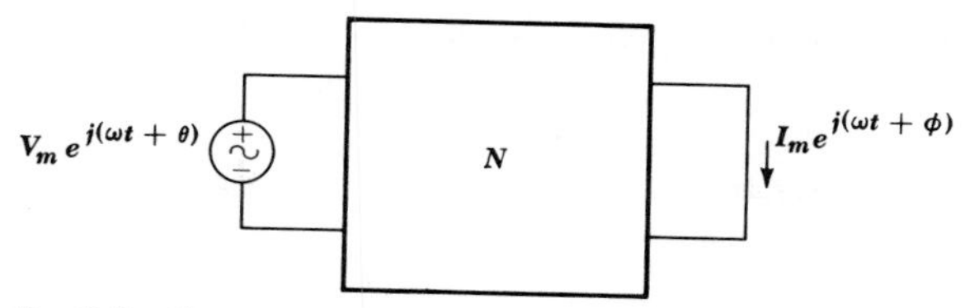

Fig. 9-3 The complex forcing function $V_m e^{j(\omega t+\theta)}$ produces the complex response $I_m e^{j(\omega t+\phi)}$ in the network of Fig. 9-1.

$$V_m e^{j(\omega t+\theta)} \tag{9}$$

and the response of (8) is

$$I_m e^{j(\omega t+\phi)} \tag{10}$$

The complex source and response are illustrated in Fig. 9-3.

There are several important conclusions to be drawn from this general example. A real, an imaginary, or a complex forcing function will produce a real, an imaginary, or a complex response, respectively. Moreover, a complex forcing function may be considered, by the use of the superposition theorem, as the sum of a real and an imaginary forcing function; thus the real part of the complex response is produced by the real part of the complex forcing function, while the imaginary part of the response is caused by the imaginary part of the complex forcing function.

Instead of applying a real forcing function to obtain the desired real response, we apply a complex forcing function whose real part is the given real forcing function; we obtain a complex response whose real part is the desired real response. Through this procedure, the integrodifferential equations describing the steady-state response of a circuit will become simple algebraic equations.

Let us try out this idea on the simple RL series circuit shown in Fig. 9-4. The real source $V_m \cos \omega t$ is applied; the real response $i(t)$ is desired.

We first construct the complex forcing function which, upon the appli-

Fig. 9-4 A simple circuit in the sinusoidal steady state is to be analyzed by the application of a complex forcing function.

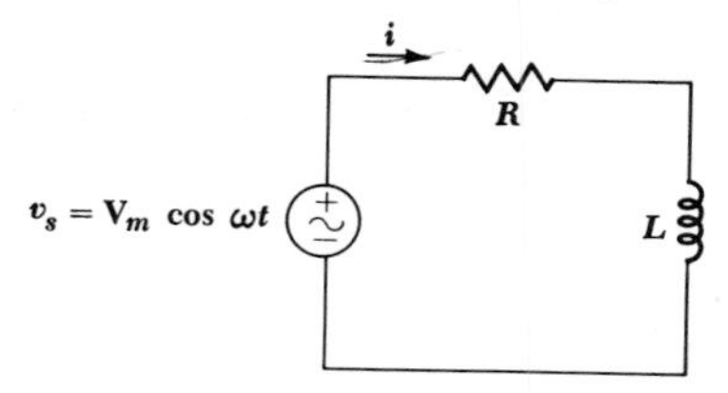

cation of Euler's identity, yields the given real forcing function. Since

$$\cos \omega t = \text{Re } e^{j\omega t}$$

then the necessary complex source is

$$V_m e^{j\omega t}$$

The complex response which results is expressed in terms of an unknown amplitude I_m and an unknown phase angle ϕ,

$$I_m e^{j(\omega t + \phi)}$$

Writing the differential equation for this particular circuit,

$$Ri + L\frac{di}{dt} = v_s$$

we insert our complex expressions for v_s and i,

$$RI_m e^{j(\omega t+\phi)} + L\frac{d}{dt}(I_m e^{j(\omega t+\phi)}) = V_m e^{j\omega t}$$

take the indicated derivative,

$$RI_m e^{j(\omega t+\phi)} + j\omega L I_m e^{j(\omega t+\phi)} = V_m e^{j\omega t}$$

and obtain a complex *algebraic* equation. In order to determine the value of I_m and ϕ, we divide throughout by the common factor $e^{j\omega t}$,

$$RI_m e^{j\phi} + j\omega L I_m e^{j\phi} = V_m \tag{11}$$

factor the left side,

$$I_m e^{j\phi}(R + j\omega L) = V_m$$

rearrange,

$$I_m e^{j\phi} = \frac{V_m}{R + j\omega L}$$

and identify I_m and ϕ by expressing the right side of the equation in exponential or polar form,

$$I_m e^{j\phi} = \frac{V_m}{\sqrt{R^2 + \omega^2 L^2}} e^{j[-\tan^{-1}(\omega L/R)]} \tag{12}$$

Thus,

$$I_m = \frac{V_m}{\sqrt{R^2 + \omega^2 L^2}}$$

and

$$\phi = -\tan^{-1}\frac{\omega L}{R}$$

The complex response is given by (12). The real response $i(t)$ may be obtained by reinserting the $e^{j\omega t}$ factor on both sides of (12) and taking the real part, easily obtained by applying Euler's omnipotent formula. Thus,

$$i(t) = I_m \cos(\omega t + \phi)$$

$$= \frac{V_m}{\sqrt{R^2 + \omega^2 L^2}} \cos\left(\omega t - \tan^{-1}\frac{\omega L}{R}\right)$$

which agrees with the response obtained for this same circuit in the previous chapter, Eq. (8-4).

Although we have successfully worked a sinusoidal steady-state problem by applying a complex forcing function and obtaining a complex response, we have not taken advantage of the full power of the complex representation. In order to do so, we must carry the concept of the complex source or response one additional step and define the quantity called a "phasor."

Drill Problem

9-1 Find the complex voltage which results when the complex current $0.12e^{j(4000t-30^\circ)}$ A is applied to: (*a*) a 50-Ω resistor; (*b*) a 0.01-H inductor; (*c*) a 5-μF capacitor.

Ans. $4.8e^{j(4000t+60^\circ)}$; $6e^{j(4000t-30^\circ)}$; $6e^{j(4000t-120^\circ)}$ V

9-3 THE PHASOR

A sinusoidal current or voltage *at a given frequency* is characterized by only two parameters, an amplitude and a phase angle. The complex representation of the voltage or current is also characterized by these same two parameters. For example, the assumed sinusoidal form of the current response in the example above was

$$I_m \cos(\omega t + \phi)$$

and the corresponding representation of this current in complex form is

$$I_m e^{j(\omega t+\phi)}$$

Once I_m and ϕ are specified, the current is exactly defined. Throughout any linear circuit operating in the sinusoidal steady state at a single frequency ω, every current and voltage may be characterized completely by a knowledge of its amplitude and phase angle. Moreover, the complex representation of every voltage and current will contain the same factor $e^{j\omega t}$. The factor is superfluous; since it is the same for every quantity, it contains no useful information. Of course, the value of the frequency may be recognized by inspecting one of these factors, but it is a lot simpler to write down the value of the frequency near the circuit diagram once and for all and avoid carrying redundant information throughout the solution. Thus, we could simplify the voltage source and the current response of the example by representing them concisely as

$$V_m \qquad \text{or} \qquad V_m e^{j0^\circ} \qquad \text{and} \qquad I_m e^{j\phi}$$

These complex quantities are usually written in polar form rather than exponential form in order to achieve a slight additional saving of time and effort. Thus, the source voltage

$$v(t) = V_m \cos \omega t$$

we now express in complex form as

$$V_m \underline{/0^\circ}$$

and the current response

$$i(t) = I_m \cos (\omega t + \phi)$$

becomes

$$I_m \underline{/\phi}$$

This abbreviated complex representation is called a *phasor*. Let us review the steps by which a real sinusoidal voltage or current is transformed into a phasor, and then we shall be able to define a phasor more meaningfully and to assign a symbol to represent it.

A real sinusoidal current

$$i(t) = I_m \cos (\omega t + \phi)$$

is expressed as the real part of a complex quantity by Euler's identity

$$i(t) = \text{Re}\,(I_m e^{j(\omega t + \phi)})$$

We then represent the current as a complex quantity by dropping the instruction Re, thus adding an imaginary component to the current without affecting the real component; further simplification is achieved by suppressing the factor $e^{j\omega t}$,

$$\mathbf{I} = I_m e^{j\phi}$$

and writing the result in polar form,

$$\mathbf{I} = I_m \underline{/\phi}$$

This abbreviated complex representation is the *phasor representation;* phasors are complex quantities and hence are printed in boldface type. Capital letters are used for the phasor representation of an electrical quantity because it is no longer an instantaneous function of time; the phasor contains only amplitude and phase information. We recognize this difference in viewpoint by referring to $i(t)$ as a *time-domain representation* and terming the phasor **I** a *frequency-domain representation.* It should be noted that the frequency-domain expression of a current or voltage does not explicitly include the frequency; however, we might think of the frequency as being so fundamental in the frequency domain that it is emphasized by its omission.[2]

The process by which we change $i(t)$ into **I** is called a phasor transformation from the time domain to the frequency domain. The steps in the time-domain to frequency-domain transformation are as follows:

1. Given the sinusoidal function $i(t)$ in the time domain, write $i(t)$ as a cosine wave with a phase angle. For example, $\sin \omega t$ should be written as $\cos(\omega t - 90°)$.
2. Express the cosine wave as the real part of a complex quantity by using Euler's identity.
3. Drop Re.
4. Suppress $e^{j\omega t}$.

As an example, let us transform the time-domain voltage

$$v(t) = 100 \cos(400t - 30°)$$

[2]Very little local mail in this country includes "U.S.A." in its address.

into the frequency domain. The time-domain expression is already in the form of a cosine wave with a phase angle, and thus we need take only the real part of the complex representation,

$$v(t) = \text{Re}\,(100e^{j(400t-30°)})$$

and drop Re and suppress $e^{j\omega t}$,

$$\mathbf{V} = 100\underline{/-30°}$$

In a similar fashion, the time-domain current

$$i(t) = 5 \sin (377t + 150°)$$

transforms into the phasor

$$\mathbf{I} = 5\underline{/60°}$$

Before we consider the analysis of circuits in the sinusoidal steady state through the use of phasors, it is necessary to learn how to shift our transformation smoothly into reverse to return to the time domain from the frequency domain. The process is exactly the reverse of the sequence given above. Thus, the steps in the frequency-domain to time-domain transformation are as follows:

1. Given the phasor current **I** in the frequency domain, write the complex expression in exponential form.
2. Reinsert (multiply by) the factor $e^{j\omega t}$.
3. Replace the real-part operator Re.
4. Obtain the time-domain representation by applying Euler's identity. The resultant cosine-wave expression may be changed to a sine wave, if desired, by increasing the argument by 90°.

Thus, given the phasor voltage

$$\mathbf{V} = 115\underline{/-45°}$$

we change from polar to exponential form,

$$\mathbf{V} = 115e^{-j45°}$$

pass from the frequency domain to the time domain by reinserting $e^{j\omega t}$ and the real-part operator,

$$v(t) = \text{Re}\,(115e^{j(\omega t-45°)})$$

and finally invoke Euler's identity,

$$v(t) = 115 \cos(\omega t - 45°)$$

As a sinusoid, the answer could be written

$$v(t) = 115 \sin(\omega t + 45°)$$

We have been tracing the transformation process by a fairly detailed sequence of steps, but with a little practice it will soon be possible to make the transformation in either direction in a single step. Even after we have achieved this facility, however, it will be well to know the exact steps and the mathematical justification for each one.

Before considering the methods of applying phasors in the analysis of the sinusoidal steady-state circuit, we may treat ourselves to a quick preview by returning to the example of the *RL* series circuit. A number of steps after writing the applicable differential equation, we arrived at (11), rewritten below,

$$RI_m e^{j\phi} + j\omega L I_m e^{j\phi} = V_m$$

If we substitute phasors for the current,

$$\mathbf{I} = I_m \underline{/\phi}$$

and the voltage

$$\mathbf{V} = V_m \underline{/0°}$$

we obtain

$$R\mathbf{I} + j\omega L\mathbf{I} = \mathbf{V}$$

or

$$(R + j\omega L)\mathbf{I} = \mathbf{V} \tag{13}$$

a complex algebraic equation in which the current and voltage are expressed in phasor form. This equation is only slightly more complicated than Ohm's law for a single resistor. The next time we analyze this circuit, we shall begin with (13).

Drill Problems

9-2 Give each voltage in phasor form: (*a*) $20 \cos(\omega t + 40°)$; (*b*) $12 \cos(\omega t + 360°) + 10 \cos(\omega t + 70°)$; (*c*) $-20 \sin(\omega t - 10°) + 20 \cos(\omega t - 50°)$ V.

Ans. $15.32 + j12.86$; $15.42 + j9.40$; $16.33 + j4.38$ V

9-3 Transform each of the following currents to the time domain: (*a*) $6 - j8$ A; (*b*) $-8 + j6$ A; (*c*) $-j10$ A.

Ans. $10 \cos(\omega t - 90°)$; $10 \cos(\omega t - 53.1°)$; $10 \cos(\omega t + 143.1°)$ A

9-4 PHASOR RELATIONSHIPS FOR *R*, *L*, AND *C*

Now that we are able to transform into and out of the frequency domain, we can proceed to our simplification of sinusoidal steady-state analysis by establishing the relationship between the phasor voltage and phasor current for each of the three passive elements. We shall begin with the defining equation for each of the elements, a time-domain relationship, and then let both the current and the voltage become complex quantities. After suppressing $e^{j\omega t}$ throughout the equation, the desired relationship between the phasor voltage and phasor current will then become apparent.

The resistor provides the simplest case. In the time domain, as indicated by Fig. 9-5*a*, the defining equation is

$$v(t) = Ri(t) \tag{14}$$

Now let us apply the complex voltage

$$V_m e^{j(\omega t+\theta)} = V_m \cos(\omega t + \theta) + jV_m \sin(\omega t + \theta) \tag{15}$$

and assume the complex current

$$I_m e^{j(\omega t+\phi)} = I_m \cos(\omega t + \phi) + jI_m \sin(\omega t + \phi) \tag{16}$$

and obtain

$$V_m e^{j(\omega t+\theta)} = RI_m e^{j(\omega t+\phi)}$$

By dividing throughout by $e^{j\omega t}$ (or suppressing $e^{j\omega t}$ on both sides of the equation), we find

$$V_m e^{j\theta} = RI_m e^{j\phi}$$

or, in polar form,

$$V_m \angle \theta = RI_m \angle \phi$$

But $V_m \angle \theta$ and $I_m \angle \phi$ merely represent the general voltage and current phasors **V** and **I**. Thus,

$$\mathbf{V} = R\mathbf{I} \tag{17}$$

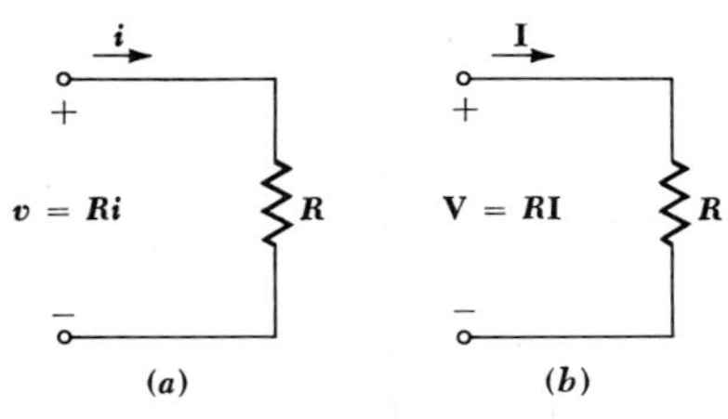

Fig. 9-5 A resistance R at which are present a voltage and current in: (a) the time domain, $v = Ri$; (b) the frequency domain, $\mathbf{V} = R\mathbf{I}$.

The voltage-current relationship in phasor form for a resistance has the same form as the relationship between the time-domain voltage and current. The defining equation in phasor form is illustrated in Fig. 9-5b. The equality of the angles θ and ϕ is apparent, and the current and voltage are thus in phase.

As an example of the use of both the time-domain and frequency-domain relationships, let us assume a voltage of $8 \cos (100t - 50°)$ V across a 4-Ω resistor. Working in the time domain, we find that the current must be

$$i(t) = \frac{v(t)}{R} = 2 \cos (100t - 50°)$$

The phasor form of the same voltage is $8\underline{/-50°}$ V, and therefore

$$\mathbf{I} = \frac{\mathbf{V}}{R} = 2\underline{/-50°} \quad \text{A}$$

If we transform this answer back to the time domain, it is evident that the same expression for the current is obtained.

It is apparent that there is no saving in time or effort when a resistive circuit is analyzed in the frequency domain. As a matter of fact, if it is necessary to transform a given time-domain source to the frequency domain and then translate the desired response back to the time domain, we should be much better off working completely in the time domain. This definitely does not apply to any circuit containing resistance and either inductance or capacitance.

Let us now turn to the inductor. The time-domain network is shown in Fig. 9-6a, and the defining equation, a time-domain expression, is

$$v(t) = L\frac{di(t)}{dt} \tag{18}$$

After substituting the complex voltage equation (15) and complex current

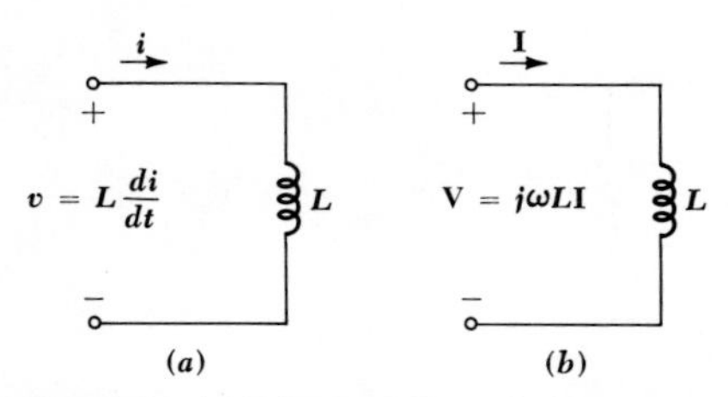

Fig. 9-6 An inductance L at which are present a voltage and current in: (a) the time domain, $v = L\,di/dt$; (b) the frequency domain, $\mathbf{V} = j\omega L\mathbf{I}$.

equation (16) into (18), we have

$$V_m e^{j(\omega t+\theta)} = L\frac{d}{dt}\left(I_m e^{j(\omega t+\phi)}\right)$$

Taking the indicated derivative

$$V_m e^{j(\omega t+\theta)} = j\omega L I_m e^{j(\omega t+\phi)}$$

and suppressing $e^{j\omega t}$,

$$V_m e^{j\theta} = j\omega L I_m e^{j\phi}$$

we obtain the desired phasor relationship

$$\mathbf{V} = j\omega L\mathbf{I} \tag{19}$$

The time-domain differential equation (18) has become an algebraic equation in the frequency domain. The phasor relationship is indicated in Fig. 9-6*b*.

As an illustration of the phasor relationship, let us apply the voltage $8\underline{/-50°}$ V at a frequency $\omega = 100$ rad/s to a 4-H inductor. From (19), the phasor current is

$$\mathbf{I} = \frac{\mathbf{V}}{j\omega L} = \frac{8\underline{/-50°}}{j100(4)} = -j0.02\underline{/-50°}$$

or

$$\mathbf{I} = 0.02\underline{/-140°} \quad \text{A}$$

If we express this current in the time domain, it becomes

$$i(t) = 0.02\cos(100t - 140°) \quad \text{A}$$

This response is also easily obtained by working entirely in the time domain; it is not so easily obtained if resistance or capacitance is combined with the inductance.

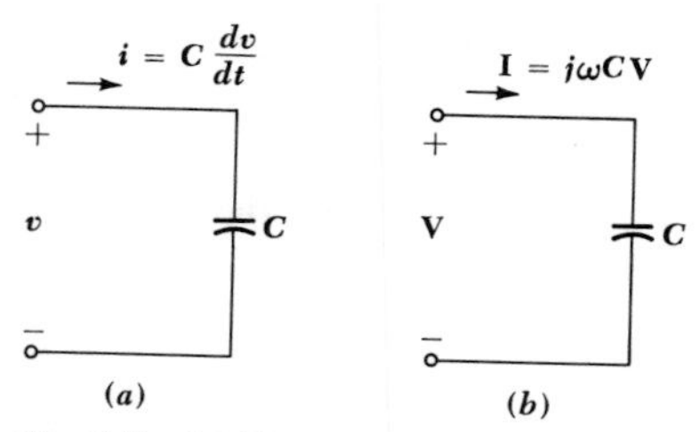

Fig. 9-7 (*a*) The time-domain and (*b*) the frequency-domain relationship between capacitor current and voltage.

The final element we must consider is the capacitor. The definition of capacitance, a familiar time-domain expression, is

$$i(t) = C\frac{dv(t)}{dt} \tag{20}$$

The equivalent expression in the frequency domain is obtained once more by letting $v(t)$ and $i(t)$ be the complex quantities of (15) and (16), taking the indicated derivative, suppressing $e^{j\omega t}$, and recognizing the phasors **V** and **I.** It is

$$\mathbf{I} = j\omega C\mathbf{V} \tag{21}$$

If the phasor voltage $8\underline{/-50^\circ}$ V is applied to a 4-F capacitor at $\omega = 100$ rad/s, the phasor current is

$$\mathbf{I} = j100(4)(8\underline{/-50^\circ}) = 3200\underline{/40^\circ} \quad \text{A}$$

The magnitude of the current is tremendous, but the assumed size of the capacitor is also unrealistic. If a 4-F capacitor were constructed of two flat plates separated 1 mm in air, each plate would have the area of about 85,000 football fields.[3] The time-domain and frequency-domain representations are compared in Fig. 9-7*a* and *b*.

We have now obtained the **V-I** relationships for the three passive elements. Each expression is algebraic in nature. Each is also a linear equation, and the equations relating to inductance and capacitance bear a great similarity to Ohm's law. We shall use them as we use Ohm's law.

Before we do so, we must show that phasors satisfy Kirchhoff's two laws. Kirchhoff's voltage law in the time domain is

$$v_1(t) + v_2(t) + \cdots + v_N(t) = 0$$

We now use Euler's identity to replace each real voltage by the complex

[3]Including both end zones.

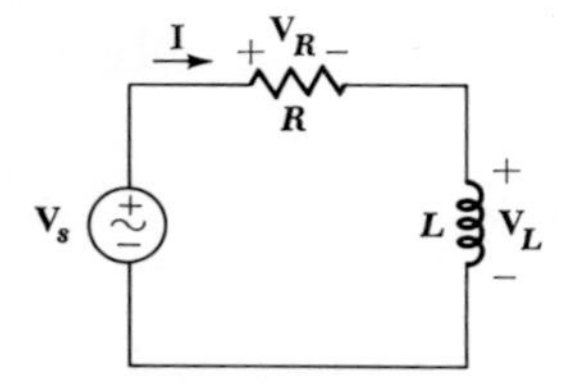

Fig. 9-8 The series RL circuit with a phasor voltage applied.

voltage having the same real part, suppress $e^{j\omega t}$ throughout, and obtain

$$\mathbf{V}_1 + \mathbf{V}_2 + \cdots + \mathbf{V}_N = 0$$

Kirchhoff's current law is shown to hold for phasor currents by a similar argument.

Now let us look briefly at the series RL circuit that we have considered several times before. The circuit is shown in Fig. 9-8, and a phasor current and several phasor voltages are indicated. We may obtain the desired response, a time-domain current, by first finding the phasor current. The method is similar to that used in analyzing our first single-loop resistive circuit. From Kirchhoff's voltage law,

$$\mathbf{V}_R + \mathbf{V}_L = \mathbf{V}_s$$

and the recently obtained **V-I** relationships for the elements

$$R\mathbf{I} + j\omega L\mathbf{I} = \mathbf{V}_s$$

the phasor current is found in terms of the source voltage $\mathbf{V}_s$,

$$\mathbf{I} = \frac{\mathbf{V}_s}{R + j\omega L}$$

Let us select a source-voltage magnitude of V_m and phase angle of 0°; the latter merely represents the simplest possible choice of a reference. Thus,

$$\mathbf{I} = \frac{V_m\underline{/0°}}{R + j\omega L}$$

The current may be transformed to the time domain by first writing it in polar form,

$$\mathbf{I} = \frac{V_m}{\sqrt{R^2 + \omega^2 L^2}} \underline{\Big/-\tan^{-1}\frac{\omega L}{R}}$$

and then following the familiar sequence of steps.

Drill Problems

9-4 Given a phasor current, $30 - j10$ mA at $\omega = 1000$ rad/s, assume a passive sign convention and find the phasor voltage present across a: (*a*) 40-Ω resistor; (*b*) 30-mH inductor; (*c*) 40-μF capacitor.

Ans. $-0.25 - j0.75$; $0.3 + j0.9$; $1.2 - j0.4$ V

9-5 For each part of Drill Prob. 9-4, find the voltage across the element at $t = 1$ ms.

Ans. -0.595; 0.496; 0.985 V

9-5 IMPEDANCE

The current-voltage relationships for the three passive elements in the frequency domain are

$$\mathbf{V} = R\mathbf{I} \qquad \mathbf{V} = j\omega L\mathbf{I} \qquad \mathbf{V} = \frac{\mathbf{I}}{j\omega C}$$

If these equations are written as phasor-voltage phasor-current ratios

$$\frac{\mathbf{V}}{\mathbf{I}} = R \qquad \frac{\mathbf{V}}{\mathbf{I}} = j\omega L \qquad \frac{\mathbf{V}}{\mathbf{I}} = \frac{1}{j\omega C}$$

then we find that these ratios are simple functions of the element values, and frequency also, in the case of inductance and capacitance. We treat these ratios in the same manner we treat resistances, with the exception that they are complex quantities and all algebraic manipulations must be those appropriate for complex numbers.

Let us define the ratio of the phasor voltage to the phasor current as *impedance*, symbolized by the letter **Z**. The impedance is a complex quantity having the dimensions of ohms. Impedance is not a phasor and cannot be transformed to the time domain by multiplying by $e^{j\omega t}$ and taking the real part. Instead, we think of an inductor L as being represented in the time domain by its inductance L and in the frequency domain by its impedance $j\omega L$. A capacitor in the time domain is a capacitance C and an impedance $1/j\omega C$ in the frequency domain. Impedance is a part of the frequency domain and not a concept which is a part of the time domain.

The validity of Kirchhoff's two laws in the frequency domain enables it to be easily demonstrated that impedances may be combined in series and parallel by the same rules we have already established for resistances. For example, at $\omega = 10^4$ rad/s, a 5-mH inductor in series with a 100-μF capacitor may be replaced by the single impedance which is the sum of the individual impedances. The impedance of the inductor is

$$\mathbf{Z}_L = j\omega L = j50 \quad \Omega$$

the impedance of the capacitor is

$$\mathbf{Z}_C = \frac{1}{j\omega C} = -j1 \quad \Omega$$

and the impedance of the series combination is therefore

$$\mathbf{Z}_{eq} = j50 - j1 = j49 \quad \Omega$$

The impedance of inductors and capacitors is a function of frequency, and this equivalent impedance is thus applicable only at the single frequency at which it was calculated, $\omega = 10{,}000$. At $\omega = 5000$, $\mathbf{Z}_{eq} = j23\ \Omega$.

The parallel combination of these same two elements at $\omega = 10^4$ yields an impedance which is the product over the sum,

$$\mathbf{Z}_{eq} = \frac{(j50)(-j1)}{j50 - j1} = \frac{50}{j49} = -j1.02 \quad \Omega$$

At $\omega = 5000$, the parallel equivalent is $-j2.17\ \Omega$.

The complex number or quantity representing impedance may be expressed in either polar or rectangular form. In polar form, an impedance, such as $100\underline{/-60°}$, is described as having an impedance magnitude of 100 Ω and a phase angle of $-60°$. The same impedance in rectangular form, $50 - j86.6$, is said to have a *resistive component*, or *resistance*, of 50 Ω and a *reactive component*, or *reactance*, of -86.6 Ω. The resistive component is the real part of the impedance, and the reactive component is the imaginary component of the impedance, including sign, but of course excluding the imaginary operator. It is important to note that the resistive component of the impedance is not necessarily equal to the resistance of the resistor which is present in the network. For example, a 20-Ω resistance in series with a 5-H coil at $\omega = 4$ produces an equivalent impedance $\mathbf{Z} = 20 + j20\ \Omega$, or in polar form, $28.3\underline{/45°}\ \Omega$. In this case, the resistive component of the impedance is equal to the resistance of the series resistance because the network is a simple series network. However, if these same two elements are placed in parallel, the equivalent impedance is $20(j20)/(20 + j20)$, or $10 + j10\ \Omega$. The resistive component of the impedance is now 10 Ω.

No special symbol is assigned for impedance magnitude or phase angle. A general form for an impedance in polar form might be

$$\mathbf{Z} = |\mathbf{Z}|\underline{/\theta}$$

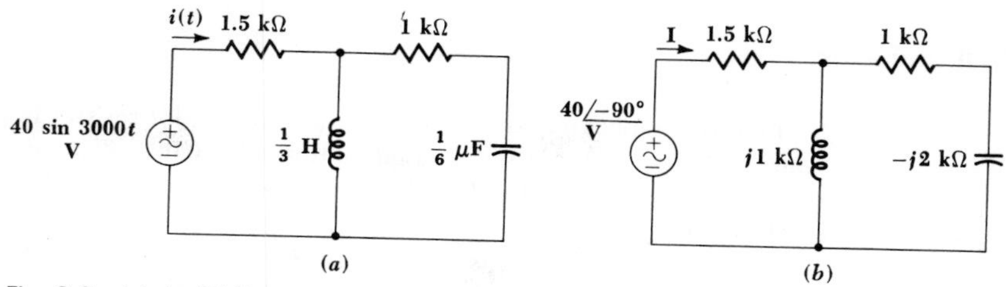

Fig. 9-9 (*a*) An *RLC* circuit for which the sinusoidal forced response $i(t)$ is desired. (*b*) The frequency-domain equivalent of the given circuit at $\omega = 3000$ rad/s.

In rectangular form, the resistive component is represented by R and the reactive component by X. Thus,

$$\mathbf{Z} = R + jX$$

Let us now use the impedance concept to analyze an *RLC* circuit, shown in Fig. 9-9*a*. The circuit is shown in the time domain, and the time-domain response is required. However, the analysis should be carried out in the frequency domain. We therefore begin by drawing a frequency-domain circuit; the source is transformed to the frequency domain, becoming $40\underline{/-90^\circ}$ V; the response is transformed to the frequency domain, being represented as $\mathbf{I}$; and the impedances of the inductor and capacitor, determined at $\omega = 3000$, are $j1$ and $-j2$ kΩ, respectively. The frequency-domain circuit is shown in Fig. 9-9*b*.

The equivalent impedance offered to the source is now calculated:

$$\begin{aligned}
\mathbf{Z}_{eq} &= 1.5 + \frac{(j1)(1-j2)}{j1+1-j2} = 1.5 + \frac{2+j1}{1-j1} \\
&= 1.5 + \frac{2+j1}{1-j1}\,\frac{1+j1}{1+j1} = 1.5 + \frac{1+j3}{2} \\
&= 2 + j1.5 = 2.5\underline{/36.9^\circ} \quad \text{k}\Omega
\end{aligned}$$

The phasor current is thus

$$\mathbf{I} = \frac{\mathbf{V}_s}{\mathbf{Z}_{eq}} = \frac{40\underline{/-90^\circ}}{2.5\underline{/36.9^\circ}} = 16\underline{/-126.9^\circ} \quad \text{mA}$$

Upon transforming the current to the time domain, the desired response is obtained,

$$i(t) = 16\cos(3000t - 126.9^\circ) \quad \text{mA}$$

If the capacitor current is desired, current division should be applied in the frequency domain.

Drill Problems

9-6 Find the impedance at $\omega = 1$ Mrad/s of: (*a*) a 2-kΩ resistor in parallel with a 1-mH inductor; (*b*) the series combination of a 0.001-μF capacitor and the network of (*a*) above; (*c*) the parallel combination of a 0.002-μF capacitor and the network of (*b*) above.

Ans. $0.277\underline{/-56.3^\circ}$; $0.447\underline{/-26.6^\circ}$; $0.894\underline{/63.4^\circ}$ kΩ

9-7 In the network described in Drill Prob. 9-6, part (*c*), let a current source, $20 \cos 10^6 t$ mA, be applied and find the amplitude of the sinusoidal voltage across the: (*a*) current source; (*b*) 0.001-μF capacitor; (*c*) 1-mH inductor.

Ans. 5.54; 11.1; 12.4 V

9-6 ADMITTANCE

Just as conductance, the reciprocal of resistance, proved to be a useful quantity in the analysis of resistive circuits, so does the reciprocal of impedance offer some convenience in the sinusoidal steady-state analysis of a general *RLC* circuit. We define *admittance* $\mathbf{Y}$ as the ratio of phasor current to phasor voltage:

$$\mathbf{Y} = \frac{\mathbf{I}}{\mathbf{V}}$$

and thus

$$\mathbf{Y} = \frac{1}{\mathbf{Z}}$$

The real part of the admittance is the *conductance* G, and the imaginary part of the admittance is the *susceptance* B. Thus,

$$\mathbf{Y} = G + jB = \frac{1}{\mathbf{Z}} = \frac{1}{R + jX} \tag{22}$$

Equation (22) should be scrutinized carefully; it does *not* state that the real part of the admittance is equal to the reciprocal of the real part of the impedance or that the imaginary part of the admittance is equal to the reciprocal of the imaginary part of the impedance. Admittance, conductance, and susceptance are all measured in mhos.

An impedance

$$\mathbf{Z} = 1 - j2 \quad \Omega$$

which might be represented, for example, by a 1-ohm resistance in series with a suitable capacitance, say 0.1 μF if $\omega = 5$ Mrad/s, possesses an admittance

$$\mathbf{Y} = \frac{1}{\mathbf{Z}} = \frac{1}{1 - j2} = \frac{1}{1 - j2}\frac{1 + j2}{1 + j2}$$
$$= 0.2 + j0.4 \quad \mho$$

Without stopping to inspect a formal proof, it should be apparent that the equivalent admittance of a network consisting of a number of parallel branches is the sum of the admittances of the individual branches. Thus, the numerical value of the admittance above might be obtained from a conductance of 0.2 ℧ in parallel with a positive susceptance of 0.4 ℧. The former could be represented by a 5-Ω resistor and the latter by a 0.08-μF capacitor at $\omega = 5$ Mrad/s, since the admittance of a capacitor is evidently $j\omega C$.

As a check on our analysis, let us compute the impedance of this latest network, a 5-Ω resistor in parallel with a 0.08-μF capacitor at $\omega =$ 5 Mrad/s. The equivalent impedance is

$$\mathbf{Z} = \frac{5(1/j\omega C)}{5 + 1/j\omega C} = \frac{5(-j2.5)}{5 - j2.5} = 1 - j2 \quad \Omega$$

as before. These two networks represent only two of an infinite number of different networks which possess this same impedance and admittance at this frequency. They do, however, represent the only two two-element networks, and thus might be considered to be the two simplest networks having an impedance of $1 - j2$ Ω and an admittance of $0.2 + j0.4$ ℧ at $\omega = 5 \times 10^6$ rad/s.

The term *immittance*, a combination of the words "impedance" and "admittance," is often used as a general term for both impedance and admittance. For example, it is evident that a knowledge of the phasor voltage across a known immittance enables the current through the immittance to be calculated.

Drill Problems

9-8 At 2000 rad/s, find the admittance of: (*a*) a 5-Ω resistor in series with a 100-μF capacitor; (*b*) the parallel combination of a 2.5-mH inductor and the network of (*a*) above; (*c*) the series combination of a 100-μF capacitor and the network of (*b*) above.

Ans. $0.1 - j0.1$; $0.1 + j0.1$; 0.2 ℧

9-9 Find the input admittance of the network shown in Fig. 9-10 at $\omega =$: (*a*) 5000 rad/s; (*b*) 10,000 rad/s; (*c*) 20,000 rad/s.

Ans. $0.271\underline{/-22^\circ}$; $0.271\underline{/22^\circ}$; 0.182 ℧

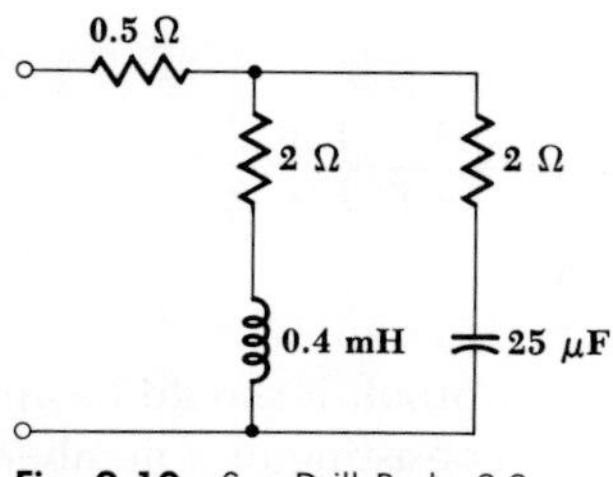

Fig. 9-10 See Drill Prob. 9-9.

PROBLEMS

☐ **1** Application of the complex forcing function $30e^{j120\pi t}$ V to a certain network produces the source current $5e^{j(120\pi t - 36.9°)}$ A. If the current arrow is directed out of the plus-marked source terminal: (*a*) What power is entering the network at $t = 0$? (*b*) What is the maximum instantaneous power entering the network?

☐ **2** A black box containing only passive elements possesses four terminals, labeled *A*, *B*, *C*, and *D*. If a voltage source $\mathbf{v}_{AB} = 100e^{j(4t+30°)}$ V is applied, the response, $\mathbf{v}_{CD} = 40e^{j(4t-120°)}$ V, is obtained. Find $\mathbf{v}_{CD}$ if $\mathbf{v}_{AB} =$: (*a*) $50e^{j(4t+30°)}$; (*b*) $50e^{j(4t-30°)}$; (*c*) $20 \cos 4t$; (*d*) $30 \sin (4t - 45°)$; (*e*) $100e^{j(7t+30°)}$ V.

☐ **3** A yellow box containing only passive elements possesses four terminals, labeled 1, 2, 3, and 4. If a current source, $i_{in,1} = 10 \cos (1000t - 20°)$ A, is applied between terminals 1 and 2, then the response, $i_{out,3} = 2 \sin (1000t + 50°)$ A, is obtained at terminals 3 and 4. Find $\mathbf{i}_{out,3}$ if $\mathbf{i}_{in,1} =$: (*a*) $5 \sin (1000t - 30°)$; (*b*) $6e^{j1000t}$; (*c*) $j5e^{j(1000t+20°)}$; (*d*) $(8 + j6)e^{j(1000t+30°)}$ A.

☐ **4** Given four nodes 1, 2, 3, and 4, and the phasor voltages $\mathbf{V}_{12} = 20 + j50$, $\mathbf{V}_{32} = -40 + j30$, and $\mathbf{V}_{34} = 30\underline{/45°}$ V, find v_{14} at $\omega t = 30°$.

☐ **5** Find $v_1(t)$ and $v_2(t)$ if $(2 - j1)\mathbf{V}_1 - j3\mathbf{V}_2 = 10$, $-j3\mathbf{V}_1 + 4\mathbf{V}_2 = -6$, and $\omega = 1000$ rad/s.

☐ **6** If three currents entering a node are $10 \cos (\omega t - 40°)$, $8 \cos (\omega t - 100°)$, and $15 \sin (\omega t + 30°)$ A, find the current $i_4(t)$ leaving the node in the fourth conductor.

☐ **7** The phasor current, $\mathbf{I} = 10 - j5$ A at $\omega = 500$ rad/s, is present in a certain circuit element. Find the power being delivered to the element at $t = \pi$ ms if the element is: (*a*) a 2-Ω resistor; (*b*) a 4-mH inductor; (*c*) a 1000-μF capacitor; (*d*) a voltage source, $\mathbf{V} = 10 + j20$ V.

☐ **8** The voltage $20 \cos (2000t + 60°)$ V is applied to a circuit element, and the current $I_m \cos (\omega t + \phi)$ results. (*a*) Find ω, ϕ, and L if the element is an inductor and $I_m = 0.1$ A. (*b*) Find ω, ϕ, and C if the element is a capacitor and $I_m = 0.1$ A.

☐ **9** A 200-Ω resistor, a 0.04-H inductor, and a 0.25-μF capacitor are connected in series. Find the phasor voltage across the combination if the phasor current, $30\angle 45°$ mA, is applied at $\omega =$: (*a*) 8000 rad/s; (*b*) 10,000 rad/s; (*c*) 12,500 rad/s.

☐ **10** A 0.5-μF capacitor, an 80-mH inductor, and a 500-Ω resistor are in parallel with the voltage source, $12 - j6$ V. Find the source current as a phasor if $\omega =$: (*a*) 500 rad/s; (*b*) 5000 rad/s; (*c*) 50,000 rad/s.

☐ **11** If the remaining two terminals are left open-circuited and $\omega = 1$ krad/s, find the input impedance of the network shown in Fig. 9-11 at terminals: (*a*) *a-b*; (*b*) *c-d*; (*c*) *a-c*. (*d*) Find $\mathbf{Z}_{in}$ at *c-d* if *a* and *b* are short-circuited.

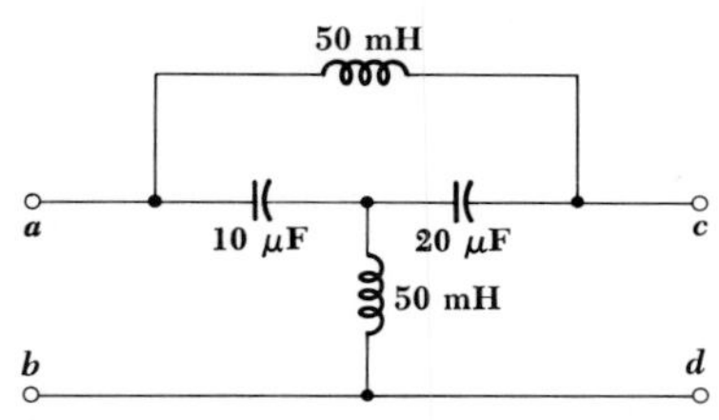

Fig. 9-11 See Prob. 11.

☐ **12** Find the impedance seen at terminals 1-2 in Fig. 9-12 at $\omega = 0.1$ Mrad/s if the network is cut at point: (*a*) *x*; (*b*) *y*; (*c*) *z*. (*d*) What is the impedance of the network as shown?

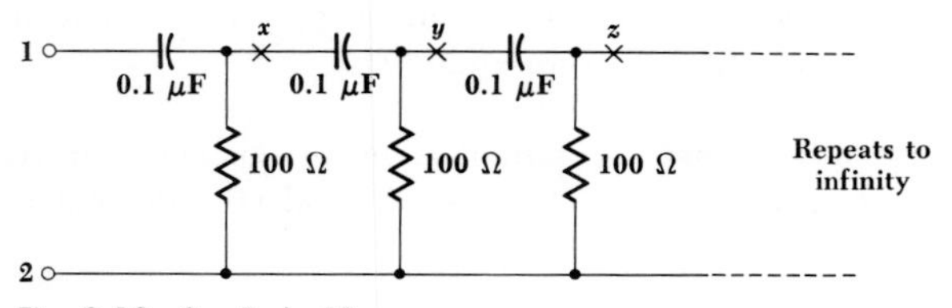

Fig. 9-12 See Prob. 12.

☐ **13** A 60-Ω resistor is in series with a 5-mH inductor. What size capacitor should be placed in parallel with this series combination so that the impedance of the parallel combination has zero reactance at $\omega = 24$ krad/s?

☐ **14** A 20-Ω resistor is in parallel with an inductor *L*; $\omega = 10^3$ rad/s. Find *L* so that the: (*a*) resistive part of the impedance of the parallel combination is 8 Ω; (*b*) reactive part of the impedance of the parallel combination is 10 Ω; (*c*) magnitude of the impedance of the parallel combination is 10 Ω.

☐ **15** Let each inductor in the circuit shown in Fig. 9-13 be 2 H, and let each capacitor be ⅓ μF. Find $\mathbf{Z}_{in}$ at $\omega = 1000$ rad/s if $R =$: (*a*) 0; (*b*) ∞; (*c*) 1 kΩ.

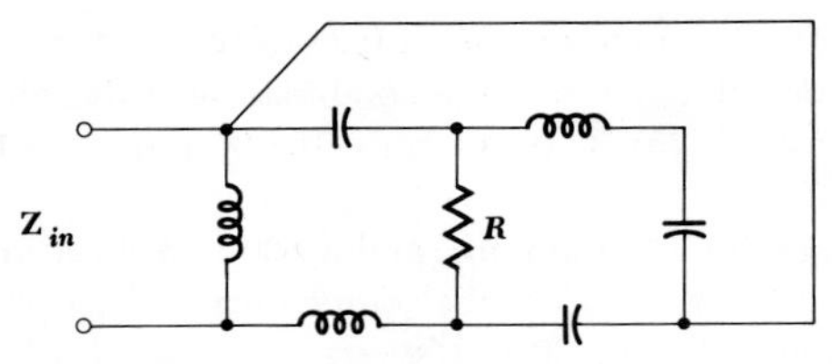

Fig. 9-13 See Prob. 15.

☐ **16** Find $\mathbf{Y}_{in}$ for the network shown in Fig. 9-14 at $\omega = 10^4$ rad/s if $G =$: (*a*) ∞; (*b*) 0.

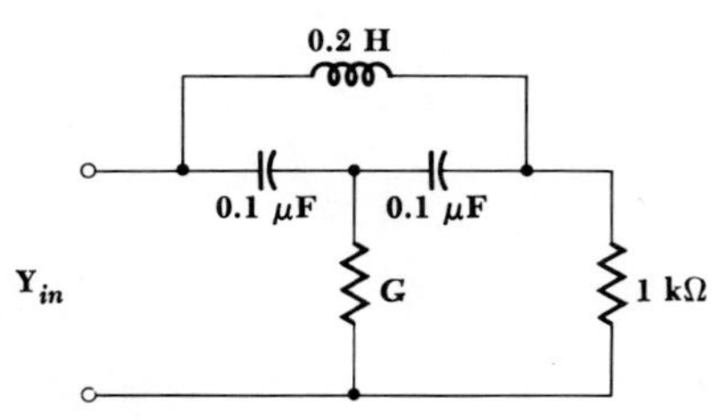

Fig. 9-14 See Prob. 16.

☐ **17** A 5-kΩ resistor and a capacitor C are in series. Let $\omega = 10^7$ rad/s. Find C so that: (*a*) the conductive part of the input admittance is 100 μ℧; (*b*) the susceptive part of $\mathbf{Y}_{in}$ is 50 μ℧; (*c*) the angle of $\mathbf{Y}_{in}$ is 60°.

☐ **18** Between terminals *a* and *b* is a capacitance C in series with the parallel combination of a 2000-Ω resistor and a 2-mH coil. Find C so that the susceptive part of the input admittance is zero at 20 kHz.

☐ **19** If $\omega = 10^4$ rad/s, what two circuit elements connected in parallel would provide the same admittance as a 500-Ω resistor and a 0.1-μF capacitor connected in series?

☐ **20** Find L and C so that the input impedance of the circuit shown in Fig. 9-15 is $2R$ at $\omega = \omega_0$.

☐ **21** The admittance and the impedance of the network shown in Fig. 9-16 are equal at every frequency. Find R and C.

Fig. 9-15 See Prob. 20.

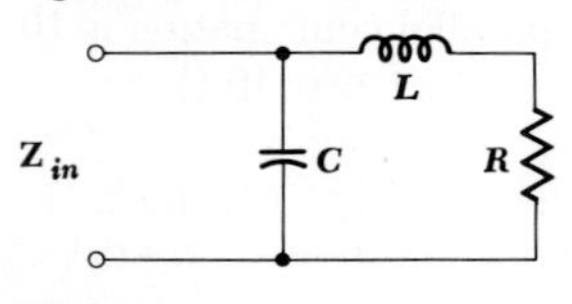

Fig. 9-16 See Prob. 21.

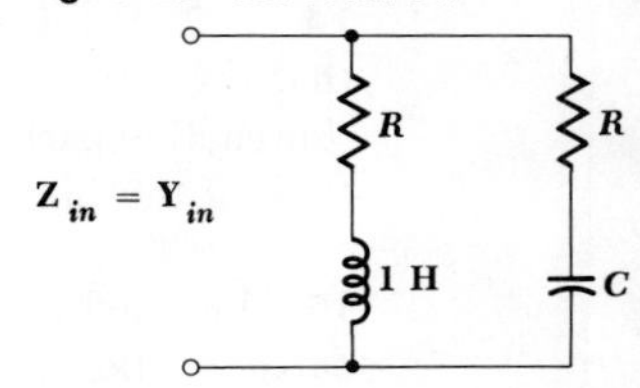

Chapter Ten The Sinusoidal Steady-state Response

10-1 INTRODUCTION

In Chap. 2 and, particularly, Chap. 3 we learned a number of methods which are useful in analyzing resistive circuits. No matter what the complexity of the resistive circuit is, we are able to determine any desired response by using nodal, mesh, or loop analysis, superposition, source transformations, or Thévenin's or Norton's theorems. Sometimes one method is sufficient, but more often we find it convenient to combine several methods to obtain the response in the most direct manner. We now wish to extend these techniques to the analysis of circuits in the sinusoidal steady state, and we have already seen that impedances combine in the same manner as do resistances. The extension of the techniques of resistive circuit analysis has been promised several times, and we must now find out why the extension is justified and practice its use.

10-2 NODAL, MESH, AND LOOP ANALYSIS

Let us first review the arguments by which we accepted nodal analysis for a purely resistive circuit. After designating a reference node and assigning voltage variables between each of the $N - 1$ remaining nodes and the reference, we applied Kirchhoff's current law to each of these $N - 1$ nodes. The application of Ohm's law to all the resistors then led to $N - 1$ equations in $N - 1$ unknowns if no voltage sources or dependent sources were present; if they were, additional equations were written in accordance with the definitions of the types of sources involved.

Now, is a similar procedure valid in terms of phasors and impedances for the sinusoidal steady state? We already know that both of Kirchhoff's laws are valid for phasors; also, we have an Ohm-like law for the passive elements, $\mathbf{E} = \mathbf{ZI}$. In other words, the laws upon which nodal analysis rests are true for phasors, and we may proceed, therefore, to analyze circuits by nodal techniques in the sinusoidal steady state. It is also evident that mesh- and loop-analysis methods are valid as well.

As an example of nodal analysis, consider the frequency-domain circuit shown in Fig. 10-1. Each passive element is specified by its impedance, although the analysis might be simplified slightly by using admittance values. Two current sources are given as phasors, and phasor node voltages $\mathbf{V}_1$ and $\mathbf{V}_2$ are indicated. At the left node we apply Kirchhoff's current law and $\mathbf{I} = \mathbf{E}/\mathbf{Z}$,

$$\frac{\mathbf{V}_1}{5} + \frac{\mathbf{V}_1}{-j10} + \frac{\mathbf{V}_1 - \mathbf{V}_2}{-j5} + \frac{\mathbf{V}_1 - \mathbf{V}_2}{j10} = 1 + j0$$

At the right node,

$$\frac{\mathbf{V}_2 - \mathbf{V}_1}{-j5} + \frac{\mathbf{V}_2 - \mathbf{V}_1}{j10} + \frac{\mathbf{V}_2}{j5} + \frac{\mathbf{V}_2}{10} = -(-j0.5)$$

Fig. 10-1 A frequency-domain circuit for which node voltages $\mathbf{V}_1$ and $\mathbf{V}_2$ are identified.

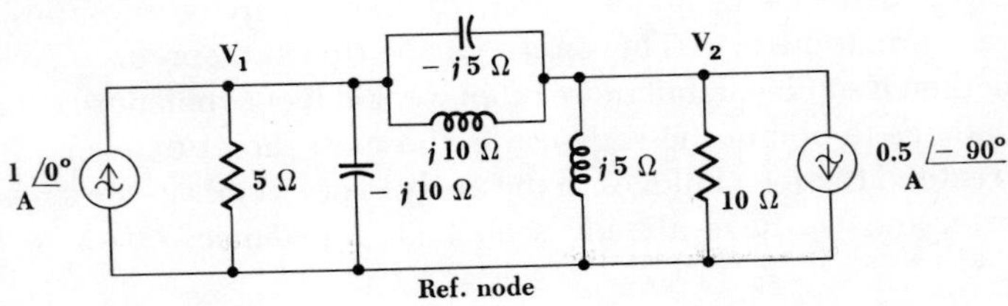

Combining terms, we have

$$(0.2 + j0.2)\mathbf{V}_1 - j0.1\mathbf{V}_2 = 1 \tag{1}$$

and

$$-j0.1\mathbf{V}_1 + (0.1 - j0.1)\mathbf{V}_2 = j0.5 \tag{2}$$

Using determinants to solve (1) and (2), we obtain

$$\mathbf{V}_1 = \frac{\begin{vmatrix} 1 & -j0.1 \\ j0.5 & (0.1 - j0.1) \end{vmatrix}}{\begin{vmatrix} (0.2 + j0.2) & -j0.1 \\ -j0.1 & (0.1 - j0.1) \end{vmatrix}} = \frac{0.1 - j0.1 - 0.05}{0.02 - j0.02 + j0.02 + 0.02 + 0.01}$$

$$= \frac{0.05 - j0.1}{0.05} = 1 - j2 \quad \text{V}$$

$$\mathbf{V}_2 = \frac{\begin{vmatrix} (0.2 + j0.2) & 1 \\ -j0.1 & j0.5 \end{vmatrix}}{0.05} = \frac{-0.1 + j0.1 + j0.1}{0.05} = -2 + j4 \quad \text{V}$$

The time-domain solutions are therefore obtained by expressing $\mathbf{V}_1$ and $\mathbf{V}_2$ in polar form,

$$\mathbf{V}_1 = 2.24\underline{/-63.4^\circ} \qquad \mathbf{V}_2 = 4.47\underline{/116.6^\circ}$$

and passing to the time domain:

$$v_1(t) = 2.24 \cos(\omega t - 63.4^\circ) \qquad v_2(t) = 4.47 \cos(\omega t + 116.6^\circ) \quad \text{V}$$

Note that the value of ω would have to be known in order to compute the impedance values given on the circuit diagram. Also, both sources are assumed to operate at the same frequency.

As an example of loop or mesh analysis, we consider the circuit given in Fig. 10-2*a*. Noting from the left source that $\omega = 10^3$ rad/s, we draw the frequency-domain circuit of Fig. 10-2*b* and assign mesh currents $\mathbf{I}_1$ and $\mathbf{I}_2$. Around mesh 1,

$$3\mathbf{I}_1 + j4(\mathbf{I}_1 - \mathbf{I}_2) = 10\underline{/0^\circ}$$

or

$$(3 + j4)\mathbf{I}_1 - j4\mathbf{I}_2 = 10$$

Fig. 10-2 (*a*) A time-domain circuit containing a dependent source. (*b*) The corresponding frequency-domain circuit on which mesh currents $\mathbf{I}_1$ and $\mathbf{I}_2$ are indicated.

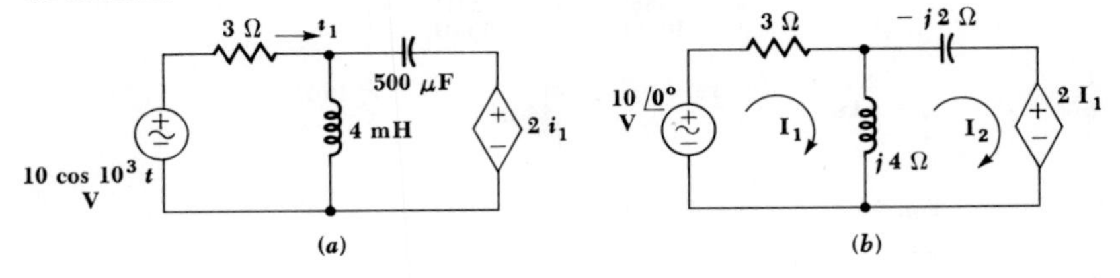

while mesh 2 leads to

$$j4(\mathbf{I}_2 - \mathbf{I}_1) - j2\mathbf{I}_2 + 2\mathbf{I}_1 = 0$$

or

$$(2 - j4)\mathbf{I}_1 + j2\mathbf{I}_2 = 0$$

Solving,

$$\mathbf{I}_1 = \frac{14 + j8}{13} = 1.24\underline{/29.8°} \quad \text{A}$$

$$\mathbf{I}_2 = \frac{20 + j30}{13} = 2.77\underline{/56.3°} \quad \text{A}$$

or

$$i_1(t) = 1.24 \cos (10^3 t + 29.8°) \quad \text{A}$$

$$i_2(t) = 2.77 \cos (10^3 t + 56.3°) \quad \text{A}$$

The solutions for either of the problems above could be checked by working entirely in the time domain, but it would be quite an undertaking and one which is typical of those we may safely ignore because the phasor method is providentially available.

Drill Problems

10-1 Use nodal analysis on the circuit shown in Fig. 10-3 to evaluate the phasor voltage: (*a*) $\mathbf{V}_1$; (*b*) $\mathbf{V}_2$; (*c*) $\mathbf{V}_3$.

Ans. $-105 - j15$; $15 - j105$; $90 + j120$ V

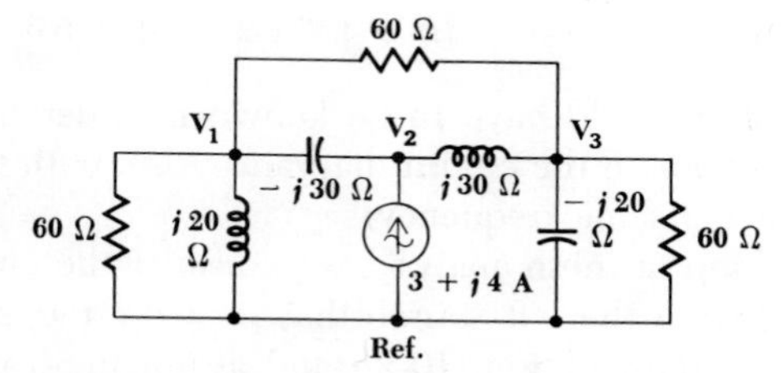

Fig. 10-3 See Drill Prob. 10-1.

10-2 In the circuit shown in Fig. 10-4, find: (*a*) $\mathbf{I}_1$; (*b*) $\mathbf{I}_2$; (*c*) $\mathbf{I}_3$.

Ans. $-3 - j2$; $-1 + j2$; $6 + j0$ A

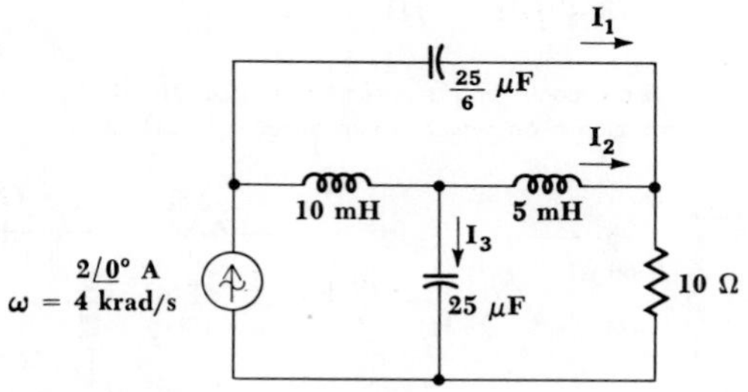

Fig. 10-4 See Drill Prob. 10-2.

10-3 SUPERPOSITION, SOURCE TRANSFORMATIONS, AND THÉVENIN'S THEOREM

After inductance and capacitance were introduced in Chap. 4, we found that circuits containing these elements were still linear, and that the benefits of linearity were again available. Included among these were the superposition principle and Thévenin's and Norton's theorems; source transformations we now recognize as a simple case of the latter two theorems. Thus, we know that these methods may be used on the circuits we are now considering; the fact that we happen to be applying sinusoidal sources and are seeking only the forced response is immaterial. The fact that we are analyzing the circuits in terms of phasors is also immaterial; they are still linear circuits. We might also remember that linearity and superposition were invoked when we combined real and imaginary sources to obtain a complex source.

We therefore shall consider several examples in which answers are obtained more readily through applying superposition, source transformations, or Thévenin's or Norton's theorems.

First we look again at the circuit of Fig. 10-1, redrawn as Fig. 10-5 with each pair of parallel impedances replaced by a single equivalent impedance. That is, 5 and $-j10$ in parallel yield $4 - j2\ \Omega$, $j10$ in parallel with $-j5$ gives $-j10\ \Omega$, and 10 in parallel with $j5$ provides $2 + j4\ \Omega$. To find $\mathbf{V}_1$, we first activate only the left source and find the partial response

$$\mathbf{V}_{1L} = 1\underline{/0°}\,\frac{(4 - j2)(-j10 + 2 + j4)}{4 - j2 - j10 + 2 + j4} = \frac{-4 - j28}{6 - j8} = 2 - j2$$

With only the right source active, current division helps us to obtain

$$\mathbf{V}_{1R} = (-0.5\underline{/-90°})\left(\frac{2 + j4}{4 - j2 - j10 + 2 + j4}\right)(4 - j2) = \frac{-6 + j8}{6 - j8} = -1$$

Fig. 10-5 $\mathbf{V}_1$ and $\mathbf{V}_2$ may be found by using superposition of the separate phasor responses.

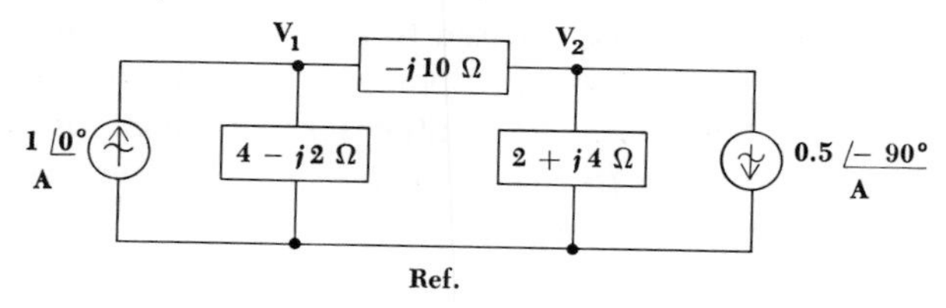

Summing, then

$$\mathbf{V}_1 = 2 - j2 - 1 = 1 - j2 \quad \text{V}$$

which agrees with our previous result.

We might also see whether or not Thévenin's theorem can help the analysis of this circuit (Fig. 10-5). Suppose we determine the Thévenin equivalent faced by the $-j10$-Ω impedance; the open-circuit voltage (+ reference to left) is

$$\mathbf{V}_{oc} = (1\underline{/0^\circ})(4 - j2) + (0.5\underline{/-90^\circ})(2 + j4)$$
$$= 4 - j2 + 2 - j1 = 6 - j3$$

The impedance of the inactive circuit, as viewed from the load terminals, is simply the sum of the two remaining impedances. Hence,

$$\mathbf{Z}_{th} = 6 + j2$$

Thus, when we reconnect the circuit, the current directed from node 1 toward node 2 through the $-j10$-Ω load is

$$\mathbf{I}_{12} = \frac{6 - j3}{6 + j2 - j10} = 0.6 + j0.3$$

Subtracting this from the left source current, the downward current through the $4 - j2$-Ω branch is found:

$$\mathbf{I}_1 = 1 - 0.6 - j0.3 = 0.4 - j0.3$$

and, thus,

$$\mathbf{V}_1 = (0.4 - j0.3)(4 - j2) = 1 - j2 \quad \text{V}$$

We might have been cleverer and used Norton's theorem on the three right elements, assuming that our chief interest is in $\mathbf{V}_1$. Source transformations can also be used repeatedly to simplify the circuit. Thus, all the shortcuts and tricks that arose in Chaps. 2 and 3 are available for circuit analysis in the frequency domain. The slight additional complexity that is apparent now arises from the necessity of using complex numbers and not from any more involved theoretical considerations.

Finally, we should also be pleased to hear that these same techniques will be applicable to the forced response of circuits driven by damped sinusoidal forcing functions, exponential forcing functions, and forcing functions having a *complex frequency* in general. Thus, we shall meet these same techniques again in Chap. 13.

Drill Problems

10-3 Using superposition to find **I** in the circuit shown in Fig. 10-6, what is the partial (phasor) response produced by the: (*a*) left source? (*b*) right source? (*c*) upper source?

Ans. $1.414;\ 1 - j1;\ 1 + j1$ A

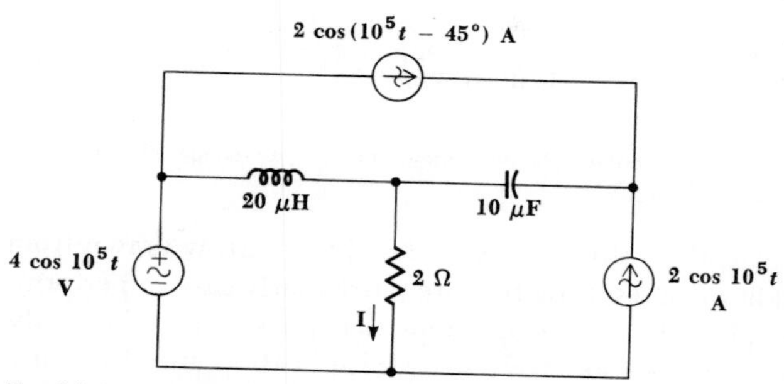

Fig. 10-6 See Drill Probs. 10-3 and 10-4.

10-4 Assume that the 2-Ω resistor is the load in Fig. 10-6 and determine the Thévenin equivalent of the circuit to which it is connected. Then calculate **I** if the load attached to the equivalent circuit is an impedance, $\mathbf{Z}_L =$: (*a*) $2 + j0\ \Omega$; (*b*) $2 - j2\ \Omega$; (*c*) $2 + j2\ \Omega$.

Ans. $2.16\underline{/-18.4^\circ}$; $3.41\underline{/0^\circ}$; $4.83\underline{/45^\circ}$ A

10-4 PHASOR DIAGRAMS

The phasor diagram is a name given to a sketch in the complex plane of the phasor voltages and phasor currents throughout a specific circuit. It provides a graphical method for solving certain problems in which the complex algebraic calculations are tedious; it serves as a check on more exact analytical methods; and it proves to be of considerable help in simplifying the analytical work in certain symmetrical problems by enabling the symmetry to be recognized and helpfully applied. In the following chapter we shall encounter similar diagrams which display the complex power relationships in the sinusoidal steady state. The use of other complex planes will also appear in connection with complex frequency in Chap. 13.

We are already familiar with the use of the complex plane in the graphical identification of a complex number and in their addition and subtraction. Since phasor voltages and currents are complex numbers, they may also be identified as points in a complex plane. For example, the phasor voltage $\mathbf{V}_1 = 6 + j8 = 10\underline{/53.1^\circ}$ is identified on the complex voltage plane shown in Fig. 10-7. The axes are the real voltage axis and the imaginary

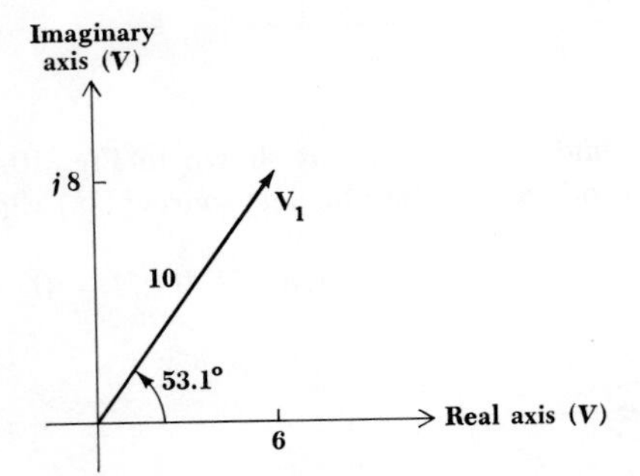

Fig. 10-7 A simple phasor diagram shows the single voltage phasor $\mathbf{V}_1 = 6 + j8 = 10\underline{/53.1^\circ}$ V.

voltage axis; the voltage $\mathbf{V}_1$ is located by an arrow drawn from the origin. Since addition and subtraction are particularly easy to perform and display on a complex plane, it is apparent that phasors may be easily added and subtracted in a phasor diagram. Multiplication and division result in the addition and subtraction of *angles* and a change of amplitude; the latter is less clearly shown, since the amplitude change depends on the amplitude of each phasor and on the scale of the diagram. Figure 10-8*a* shows the sum of $\mathbf{V}_1$ and a second phasor voltage $\mathbf{V}_2 = 3 - j4 = 5\underline{/-53.1^\circ}$ and Fig. 10-8*b* shows the current $\mathbf{I}_1$, which is the product of $\mathbf{V}_1$ and the admittance $\mathbf{Y} = 1 + j1$.

This last phasor diagram shows both current and voltage phasors on the same complex plane; it is understood that each will have its own amplitude scale, but a common angle scale. For example, a phasor voltage 1 cm long might represent 100 V, while a phasor current 1 cm long could indicate 3 mA.

Fig. 10-8 (*a*) A phasor diagram showing the sum of $\mathbf{V}_1 = 6 + j8$ and $\mathbf{V}_2 = 3 - j4$, $\mathbf{V}_1 + \mathbf{V}_2 = 9 + j4 = 9.85\underline{/24.0^\circ}$. (*b*) The phasor diagram shows $\mathbf{V}_1$ and $\mathbf{I}_1$, where $\mathbf{I}_1 = \mathbf{Y}\mathbf{V}_1$ and $\mathbf{Y} = 1 + j1$ ℧.

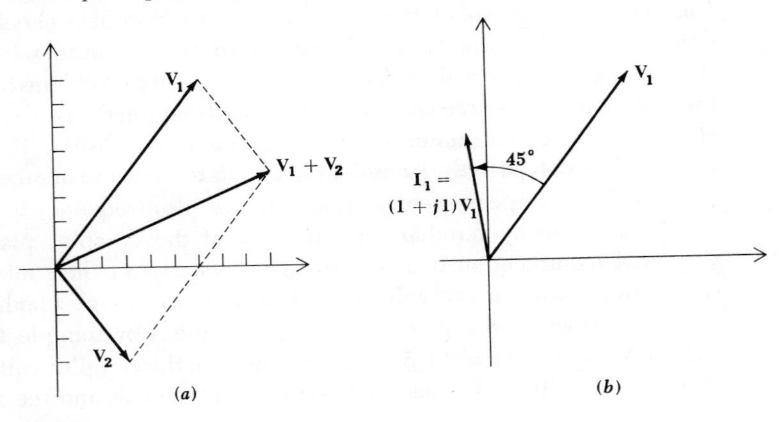

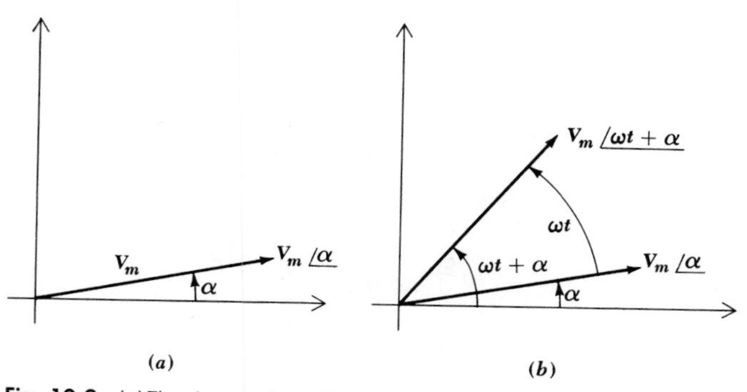

Fig. 10-9 (*a*) The phasor voltage $V_m\angle\alpha$. (*b*) The complex voltage $V_m\angle\omega t + \alpha$ is shown as a phasor at a particular instant of time. This phasor leads $V_m\angle\alpha$ by ωt rad.

The phasor diagram also offers an interesting interpretation of the time-domain to frequency-domain transformation, since the diagram may be interpreted from either the time- or frequency-domain viewpoint. Up to this time, it is obvious that we have been using the frequency-domain interpretation, because we have been showing phasors directly on the phasor diagram. However, let us proceed to a time-domain viewpoint by first showing the phasor $\mathbf{V} = V_m\angle\alpha$, as sketched in Fig. 10-9*a*. In order to transform $\mathbf{V}$ to the time domain, the next necessary step is the multiplication of the phasor by $e^{j\omega t}$; thus we have the complex voltage $V_m e^{j\alpha} e^{j\omega t} = V_m\angle\omega t + \alpha$. This voltage may also be interpreted as a phasor, one which possesses a phase angle which increases linearly with time. On a phasor diagram it therefore represents a rotating line segment, the instantaneous position being ωt rad ahead (counterclockwise) of $V_m\angle\alpha$. Both $V_m\angle\alpha$ and $V_m\angle\omega t + \alpha$ are shown on the phasor diagram of Fig. 10-9*b*.

The passage to the time domain is now completed by taking the real part of $V_m\angle\omega t + \alpha$. The real part of this complex quantity, however, is merely the projection of $V_m\angle\omega t + \alpha$ on the real axis. In summary, then, the frequency-domain phasor appears on the phasor diagram, and the transformation to the time domain is accomplished by allowing the phasor to rotate in a counterclockwise direction at an angular velocity of ω rad/s and then visualizing the projection on the real axis. It is helpful to think of the arrow representing the phasor $\mathbf{V}$ on the phasor diagram as the snapshot, taken at $\omega t = 0$, of a rotating arrow whose projection on the real axis is the instantaneous voltage $v(t)$.

Let us now construct the phasor diagrams for several simple circuits. The series *RLC* circuit shown in Fig. 10-10*a* has several different voltages associated with it, but only a single current. The phasor diagram is constructed most easily by employing the single current as the reference phasor. Let us arbitrarily select $\mathbf{I} = I_m\angle 0°$ and place it along the real axis of the phasor diagram, Fig. 10-10*b*. The resistor, capacitor, and inductor

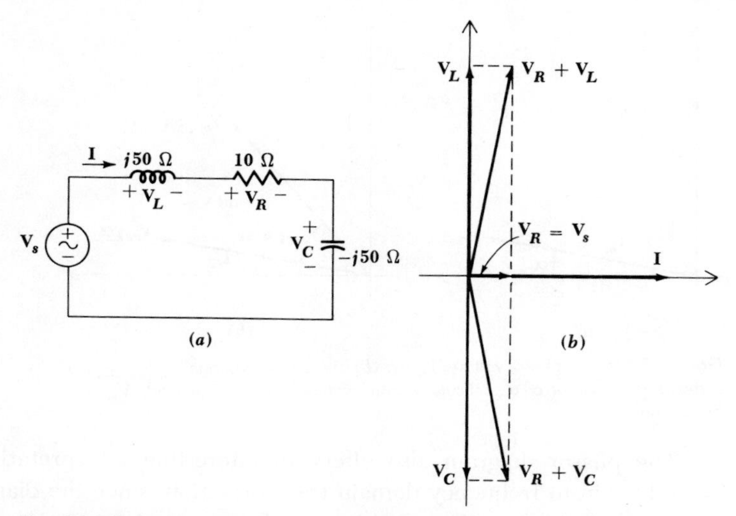

Fig. 10-10 (*a*) A series *RLC* circuit shown in the frequency domain. (*b*) The associated phasor diagram, drawn with the single mesh current as the reference phasor.

voltages may next be calculated and placed on the diagram, where the 90° phase relationships stand out clearly. The sum of these three voltages is the source voltage, and for this circuit, which is in the resonant condition[1] where $\mathbf{Z}_C = -\mathbf{Z}_L$, the source voltage and resistor voltage are equal. The total voltage across the resistance and inductance or resistance and capacitance is easily obtained from the phasor diagram.

The phasor diagram may be interpreted in the time domain by allowing all the phasors to rotate synchronously with a counterclockwise angular velocity of ω rad/s, and then considering the projections on the real axis.

Figure 10-11*a* shows a simple parallel circuit in which it is logical to use the single voltage between the two nodes as a reference phasor. Suppose that $\mathbf{V} = 1\underline{/0°}$ V. The resistor current is in phase with this voltage, $\mathbf{I}_R = 0.2\underline{/0°}$ A, and the capacitor current leads the reference voltage by 90°, $\mathbf{I}_C = j0.1$ A. After these two currents are added to the phasor diagram, shown as Fig. 10-11*b*, they may be summed to obtain the source current. The result is $\mathbf{I}_s = 0.2 + j0.1$ A.

If the source current were specified initially as, for example, $1\underline{/0°}$ A, and the node voltage is not initially known, it is still convenient to begin construction of the phasor diagram by assuming a node voltage, say $\mathbf{V} = 1\underline{/0°}$ V once again, and using it as the reference phasor. The diagram is then completed as before, and the source current which flows as a result of the assumed node voltage is again found to be $0.2 + j0.1$ A. The true source current is $1\underline{/0°}$ A, however, and thus the true node voltage is greater by the factor $1\underline{/0°}/(0.2 + j0.1)$; the true node voltage is therefore $4 - j2$ V.

[1]Resonance will be defined in Chap. 14.

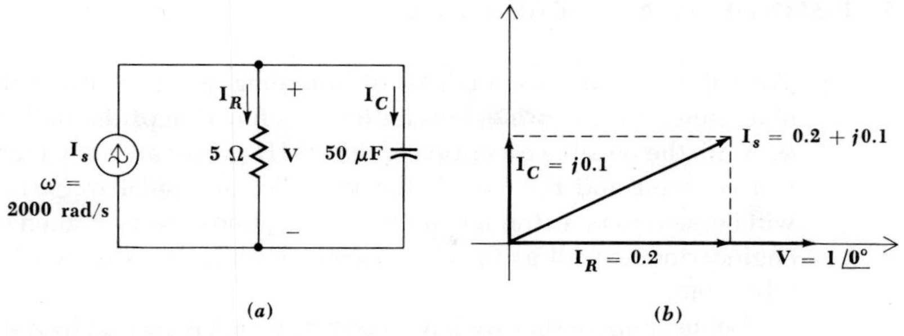

Fig. 10-11 (*a*) A parallel *RC* circuit. (*b*) The phasor diagram for this circuit; the node voltage $\mathbf{V}$ is used as a convenient reference phasor.

The assumed voltage leads to a phasor diagram which differs from the true phasor diagram by a change of scale (the assumed diagram is smaller by a factor of $1/\sqrt{20}$) and an angular rotation (the assumed diagram is rotated counterclockwise through 26.6°).

Phasor diagrams are usually very simple to construct, and most sinusoidal steady-state analyses will be more meaningful if such a diagram is included. Additional examples of the use of phasor diagrams will appear frequently throughout the remainder of our study.

Drill Problems

10-5 Analyze the circuit shown in Fig. 10-12*a* by constructing a reasonably accurate phasor diagram with the voltage $\mathbf{V}_1$ as the reference phasor. Let the arrow representing $\mathbf{V}_1$ be 2 in. long at an angle of 0°, and let the $\mathbf{I}_1$ arrow be 3 in. long. What is the length of the arrow for: (*a*) $\mathbf{I}_3$; (*b*) $\mathbf{I}_2$; (*c*) $\mathbf{V}_s$?

Ans. 1.50; 2.24; 3.35 in.

10-6 A phasor diagram is constructed for the circuit shown in Fig. 10-12*b*. The scale used is 20 V/in. and 0.4 A/in. If $\mathbf{V}_1$ is represented by an arrow 1.6 in. long at an angle of −90°, $\mathbf{V}_2$ by a 2-in. long arrow at an angle ϕ, and $\mathbf{I}_s$ by a 1.8-in. long arrow at an angle of −90°, find the impedance of the: (*a*) capacitor; (*b*) resistor; (*c*) inductor.

Ans. 25; −j33.3; j33.3 Ω

Fig. 10-12 (*a*) See Drill Prob. 10-5. (*b*) See Drill Prob. 10-6.

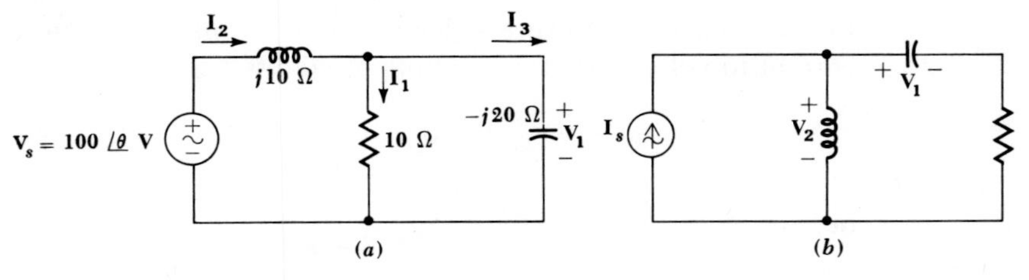

10-5 RESPONSE AS A FUNCTION OF ω

We will now consider methods of obtaining and presenting the response of a circuit with sinusoidal excitation as a function of the radian frequency ω. With the possible exception of the 60-Hz power area in which frequency is a constant and the load is the variable, sinusoidal frequency response will be seen to be extremely important in almost every branch of electrical engineering as well as in related areas, such as the theory of mechanical vibrations.

Let us suppose that we have a circuit which is excited by a single source $\mathbf{V}_s = V_s\underline{/\theta}$. This phasor voltage may also be transformed into the time-domain source voltage $V_s \cos(\omega t + \theta)$. Somewhere in the circuit exists the desired response, the current $\mathbf{I}$. As we know, this phasor response is a complex number, and its value cannot be specified in general without the use of two quantities: either a real part and an imaginary part, or a magnitude and a phase angle. The latter pair of quantities is more useful and more easily determined experimentally, and it is the information which we shall obtain analytically as a function of frequency. The data may be presented as two curves, the magnitude of the response as a function of ω and the phase angle of the response as a function of ω. We often normalize the curves by plotting the magnitude of the current-voltage ratio and the phase angle of the current-voltage ratio versus ω. It is evident that an alternative description of the resultant curves is the magnitude and phase angle of an admittance as a function of frequency. A normalized voltage response may be similarly presented as the magnitude and phase angle of an impedance versus ω. Other possibilities are voltage-voltage ratios (voltage gains) or current-current ratios (current gains). Let us consider the details of this process by thoroughly discussing several examples.

For the first example, let us select the series RL circuit. The phasor voltage $\mathbf{V}_s$ is therefore applied to this simple circuit, and the phasor current $\mathbf{I}$ is selected as the desired response. We are dealing with the forced response only, and the familiar phasor methods enable the current to be obtained:

$$\mathbf{I} = \frac{\mathbf{V}_s}{R + j\omega L}$$

Let us immediately express this result in normalized form as a ratio of current to voltage, that is, as an input admittance:

$$\mathbf{Y} = \frac{\mathbf{I}}{\mathbf{V}_s}$$

or

$$\mathbf{Y} = \frac{1}{R + j\omega L} \tag{3}$$

If we like, we may consider the admittance as the current produced by a source voltage $1\underline{/0^\circ}$ V. The magnitude of the response is

$$|\mathbf{Y}| = \frac{1}{\sqrt{R^2 + \omega^2 L^2}} \tag{4}$$

while the angle of the response is found to be

$$\text{ang } \mathbf{Y} = -\tan^{-1}\frac{\omega L}{R} \tag{5}$$

Equations (4) and (5) are the analytical expressions for the magnitude and phase angle of the response as a function of ω; we now desire to present this same information graphically.

We shall first consider the magnitude curve. The first important factor to note is that we are plotting the absolute magnitude of some quantity versus ω, and the entire curve must therefore lie *above* the ω axis. The response curve is constructed by noting that the value of the response at zero frequency is $1/R$, that the initial slope is zero, and that the response approaches zero as frequency approaches infinity; the graph of the magnitude of the response as a function of ω is shown in Fig. 10-13*a*.

For the sake of generality and completeness, the response is shown for both positive and negative values of frequency; the symmetry results from the fact that (4) indicates that $|\mathbf{Y}|$ has the same value when ω is replaced by $(-\omega)$. The physical interpretation of a negative radian frequency, such as $\omega = -100$ rad/s, depends on the time-domain function, and it may always be obtained by inspection of the time-domain expression. Suppose, for example, that we consider the voltage $v(t) = 50\cos(\omega t + 30^\circ)$. At $\omega = 100$, the voltage is $v(t) = 50\cos(100t + 30^\circ)$, while at $\omega = -100$, $v(t) = 50\cos(-100t + 30^\circ)$ or $50\cos(100t - 30^\circ)$. Any sinusoidal response may be treated in a similar manner.

The second part of the response, the phase angle of $\mathbf{Y}$ versus ω, is an inverse tangent function. The tangent function itself is quite familiar, and we should have no difficulty in turning that curve on its side; asymptotes of plus and minus 90° are helpful. The response curve is shown in Fig. 10-13*b*. The points at which $\omega = \pm R/L$ are marked on both the magnitude and phase curves. At these frequencies the magnitude is 0.707 of the maximum magnitude at zero frequency and the phase angle has a magnitude of 45°. At the frequency at which the admittance magnitude is 0.707 times its maximum value, the current magnitude is 0.707 times its maximum value, and the average power supplied by the source is 0.707^2 or 0.5 times its maximum value. It is not very strange that $\omega = R/L$ is identified as a *half-power frequency*.

As a second example, let us select a parallel LC circuit driven by a

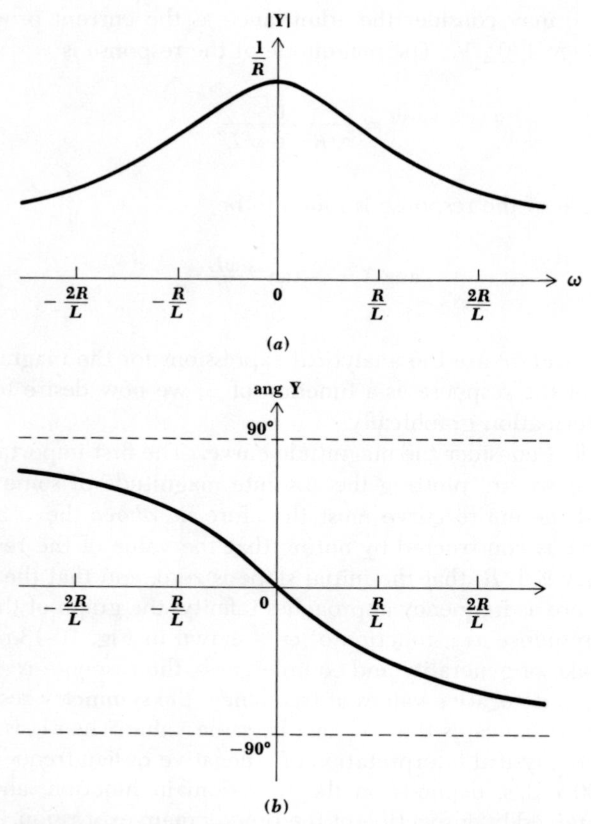

Fig. 10-13 (*a*) The magnitude of $\mathbf{Y} = \mathbf{I}/\mathbf{V}_s$ and (*b*) the angle of $\mathbf{Y}$ are plotted as functions of ω for a series RL circuit with sinusoidal excitation.

sinusoidal current source, as illustrated in Fig. 10-14*a*. The voltage response $\mathbf{V}$ is easily obtained:

$$\mathbf{V} = \mathbf{I}_s \frac{(j\omega L)(1/j\omega C)}{j\omega L - j(1/\omega C)}$$

and it may be expressed as an input impedance

$$\mathbf{Z} = \frac{\mathbf{V}}{\mathbf{I}_s} = \frac{L/C}{j(\omega L - 1/\omega C)}$$

or

$$\mathbf{Z} = -j\frac{1}{C}\frac{\omega}{\omega^2 - 1/LC} \tag{6}$$

By letting

$$\omega_0 = \frac{1}{\sqrt{LC}}$$

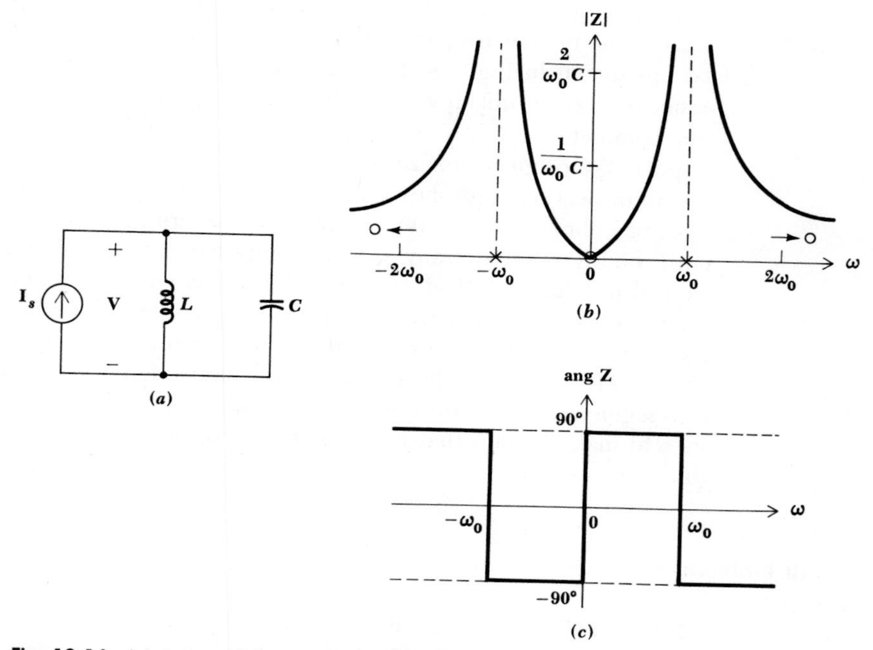

Fig. 10-14 (a) A sinusoidally excited parallel LC circuit. (b) The magnitude of the input impedance, $\mathbf{Z} = \mathbf{V}/\mathbf{I}_s$, and (c) the angle of the input impedance are plotted as functions of ω.

and factoring the expression for the input impedance,

$$|\mathbf{Z}| = \frac{1}{C} \frac{|\omega|}{|(\omega - \omega_0)(\omega + \omega_0)|} \tag{7}$$

the magnitude of the impedance may be written in a form which enables those frequencies to be identified at which the response is zero or infinite. Such frequencies are termed *critical frequencies,* and their early identification simplifies the construction of the response curve showing amplitude versus frequency. We note first that a zero-amplitude response occurs at $\omega = 0$; such a response as well as the frequency at which it occurs is called a *zero.* Response of infinite amplitude is noted at $\omega = \omega_0$ and $\omega = -\omega_0$; these frequencies are called *poles,* as is the infinite-amplitude response itself. Finally, we note that the response approaches zero as $\omega \to \infty$, and thus $\omega = \pm\infty$ is also a zero.[2]

The locations of the critical frequencies should be marked on the ω

[2]It is customary to consider plus infinity and minus infinity as being the same point. The phase angle of the response at very large positive and negative values of ω need not be the same, however.

axis, by using small circles for the zeros and crosses for the poles. Poles or zeros at infinite frequency should be indicated by an arrow near the axis, as shown in Fig. 10-14*b*. The actual drawing of the graph is made easier by adding broken vertical lines as asymptotes at each pole location. The completed graph of magnitude versus ω is shown in Fig. 10-14*b*; the slope at the origin is *not* zero.

An inspection of (6) shows that the phase angle of the input impedance must be either $+90$ or $-90°$; no other values are possible, as must apparently be the case for any circuit composed entirely of inductors and capacitors. An analytical expression for ang **Z** would therefore consist of a series of statements that the angle is $+90$ or $-90°$ in certain frequency ranges. It is simpler to present the information graphically, as shown in Fig. 10-14*c*. Although this curve is only a collection of horizontal straight line segments, errors are often made in its construction, and it is a good idea to make certain that it can be drawn directly from an inspection of (6).

Drill Problems

10-7 A 4-kΩ resistor, a 10-μF capacitor, and a sinusoidal current source $\mathbf{I}_s$ are in parallel. Sketch as a function of ω: (*a*) the magnitude of the voltage across the source; (*b*) the magnitude of the ratio of the voltage across the source to the source current; (*c*) the phase angle of the ratio of the voltage across the source to the source current. As a check on the results, the answers given below apply to $\omega = 25$ rad/s.

Ans. 2.83 kΩ; 2830 $|\mathbf{I}_s|$ V; $-45°$

10-8 A 3-μF capacitor, a sinusoidal source $\mathbf{V}_s$, and the parallel combination of 1 μF and ¼ H are all in series. Let the desired response be the ratio of the 3-μF capacitor voltage to the source voltage. Sketch the magnitude and phase angle of the response as a function of ω and determine all critical frequencies of the response.

Ans. ± 1000; ± 2000 rad/s

PROBLEMS

□ **1** (*a*) Find $\mathbf{V}_2$ in the circuit shown in Fig. 10-15. (*b*) What is $v_2(t)$ if $\omega = 500$ rad/s? (*c*) Find $\mathbf{V}_2$ if the 2-Ω resistor is replaced by a dependent current source, $0.5\mathbf{V}_2$, reference arrow directed upward.

□ **2** For the ladder network shown in Fig. 10-16, find $\mathbf{V}_{out}$ if $\mathbf{V}_{in} = 100\underline{/0°}$ mV at $\omega = 100$ rad/s.

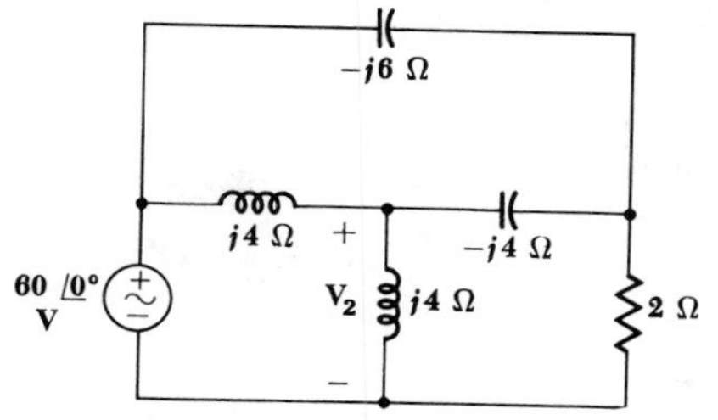

Fig. 10-15 See Prob. 1.

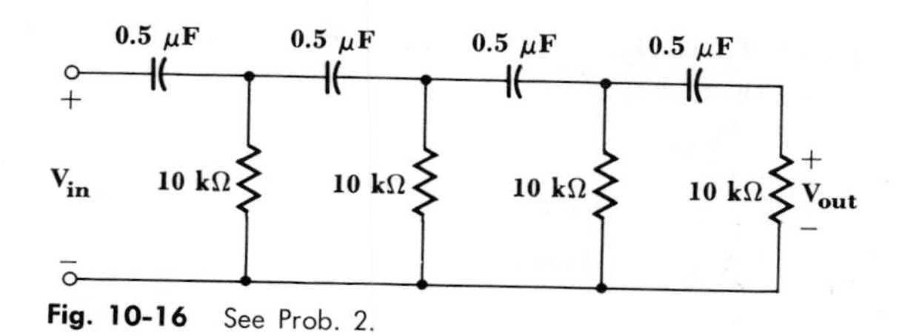

Fig. 10-16 See Prob. 2.

☐ **3** Find $v_x(t)$ in the circuit shown in Fig. 10-17.

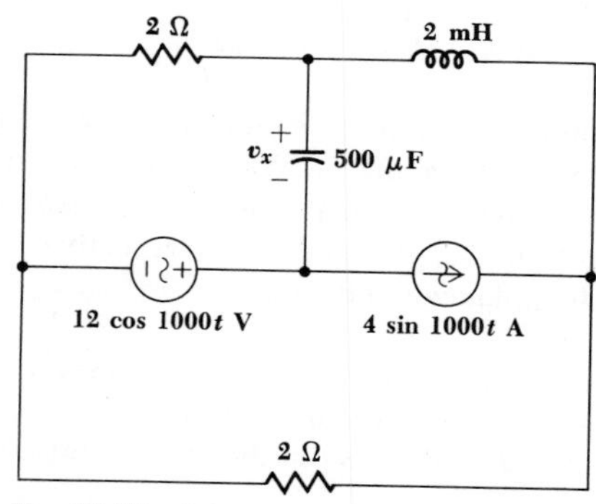

Fig. 10-17 See Prob. 3.

☐ **4** If $\omega = 2 \times 10^5$ rad/s, find $v_x(t)$ for the circuit of Fig. 10-18.

Fig. 10-18 See Probs. 4 and 7.

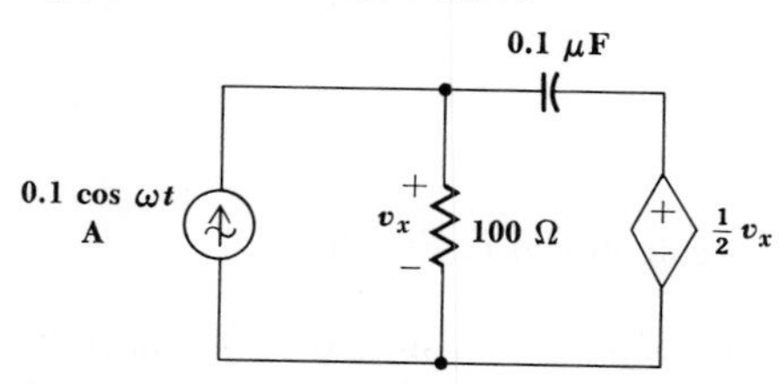

☐ **5** Find i for the circuit shown in Fig. 10-19.

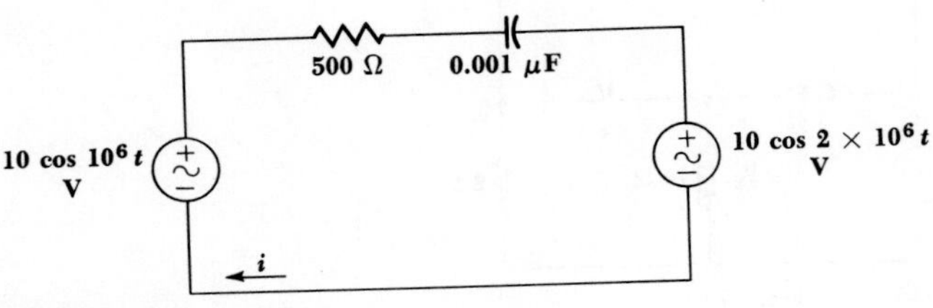

Fig. 10-19 See Prob. 5.

☐ **6** Find the ratio of $|\mathbf{I}_2|$ to $|\mathbf{I}_s|$ in the circuit shown in Fig. 10-20 if $\omega =$: (*a*) 200; (*b*) 2000; (*c*) 20,000 rad/s.

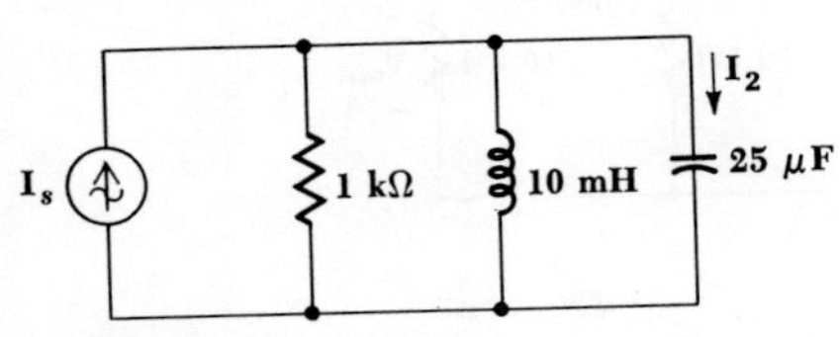

Fig. 10-20 See Prob. 6.

☐ **7** Find the Thévenin equivalent of that portion of the circuit of Fig. 10-18 that is to the right of the current source. Let $\omega = 10^5$ rad/s.

☐ **8** A linear network containing one or more sinusoidal sources operating at 2 krad/s has two accessible terminals, a and b. When a 50-Ω resistor, a 2.5-μF capacitor, and a 50-mH inductor are independently placed between the terminals, the magnitude of $\mathbf{V}_{ab}$ is found to be 25, 100, and 50 V, respectively. Determine the Thévenin equivalent of the unknown network.

☐ **9** With reference to terminals a-b and load $\mathbf{Z}_L$ in the circuit shown in Fig. 10-21, find the open-circuit voltage, the short circuit current, and draw the Norton equivalent circuit. What value of $\mathbf{Z}_L$ will result in the greatest magnitude of load voltage?

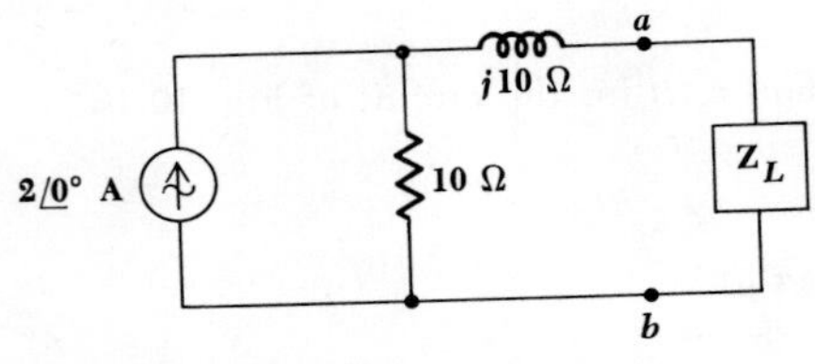

Fig. 10-21 See Prob. 9.

☐ **10** Find i_1 and v_2 in the circuit shown in Fig. 10-22.

☐ **11** (*a*) What is the maximum voltage present across the capacitor and resistor

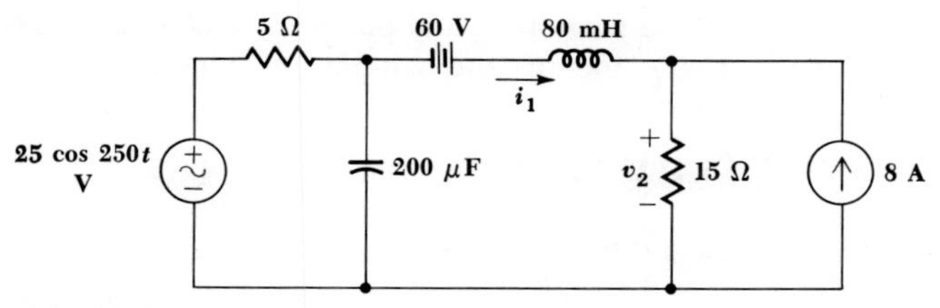

Fig. 10-22 See Prob. 10.

in the circuit of Fig. 10-23? (*b*) Replace the current source and capacitor by an equivalent series combination of a voltage source and a capacitor. Again determine the two maximum voltages.

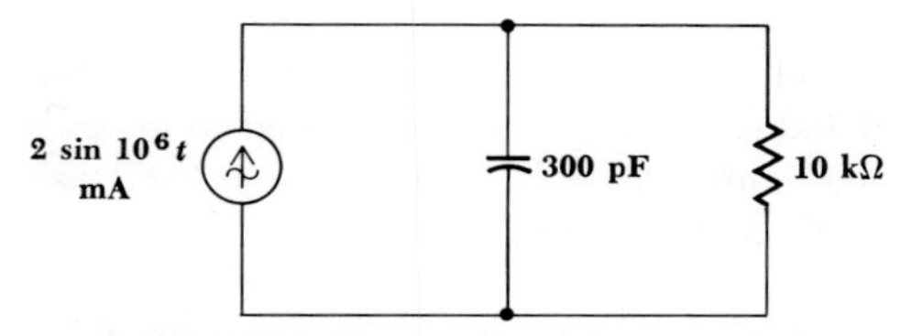

Fig. 10-23 See Prob. 11.

☐ **12** In the circuit shown in Fig. 10-24, the values of R_1, R_2, and $\mathbf{V}_s$ are unknown. Determine the value of C so that $i_x = 0$ at $\omega = \omega_1$.

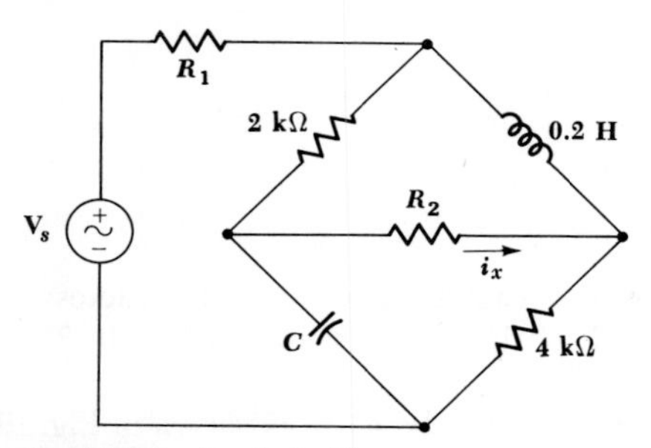

Fig. 10-24 See Prob. 12.

☐ **13** A voltage 100 cos 500*t* V is applied to $\mathbf{Z}_1$ and $\mathbf{Z}_2$ in series. The amplitudes of the voltages across $\mathbf{Z}_1$ and $\mathbf{Z}_2$ are found to be 100 V and 125 V, respectively. If it is known that $\mathbf{Z}_1$ is composed of R_1 and L_1 in series, where $R_1 = 1\ \Omega$: (*a*) construct a phasor diagram and use it to determine L_1; (*b*) find L_1 without using a phasor diagram.

☐ **14** A parallel *RLC* circuit with $R = 200\ \Omega$, $L = 10$ mH, and $C = 1\ \mu$F is excited by a parallel current source, $\mathbf{I}_s = 2\underline{/0°}$ mA. Show the element currents and voltages on a phasor diagram for: (*a*) $\omega = 10^4$; (*b*) $\omega = 5000$ rad/s.

☐ **15** Sketch curves of the magnitude and phase angle of $\mathbf{V}_1$ as functions of ω for the circuit of Fig. 10-25. Identify the half-power frequency.

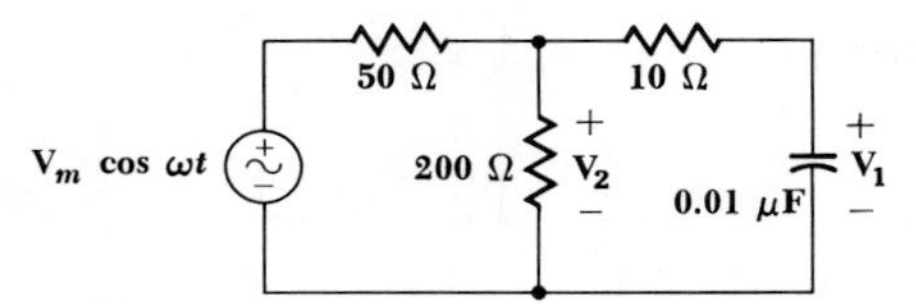

Fig. 10-25 See Probs. 15 and 16.

☐ **16** (*a*) Sketch curves of the magnitude and phase angle of $\mathbf{V}_2$ as functions of ω for the circuit of Fig. 10-25. (*b*) What is the greatest difference in phase angle between $\mathbf{V}_2$ and the source voltage?

☐ **17** A certain physical coil is modeled by the series combination of a 200-Ω resistance and a 50-mH inductance. The coil is placed in series with a 0.2-μF capacitance and a voltage source, $10 \cos \omega t$ mV. What is the maximum voltage that appears across the: (*a*) inductor; (*b*) coil?

☐ **18** For the network shown in Fig. 10-26, sketch $|\mathbf{Z}_{in}|$ and ang $\mathbf{Z}_{in}$ versus ω if the 4-μF capacitor is: (*a*) short-circuited; (*b*) left as is.

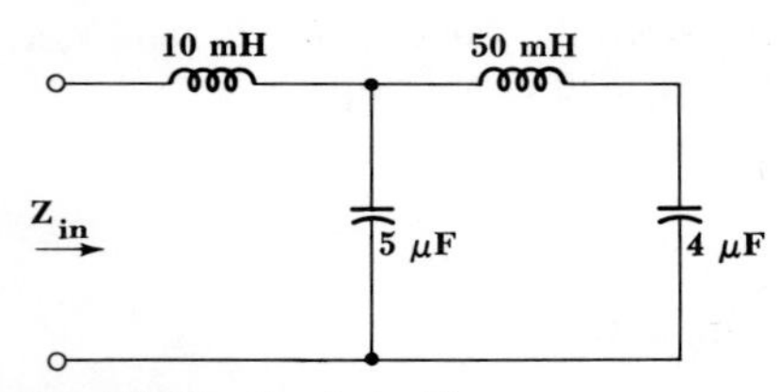

Fig. 10-26 See Prob. 18.

☐ **19** Sketch a curve showing the magnitude of the voltage across the source as a function of ω for $0 \leq \omega \leq 5$ for the circuit shown in Fig. 10-27.

☐ **20** Sketch a curve of $|\mathbf{Y}_{in}|$ versus ω for the network shown in Fig. 10-28 if $k =$: (*a*) 0; (*b*) 100.

Fig. 10-27 See Prob. 19.

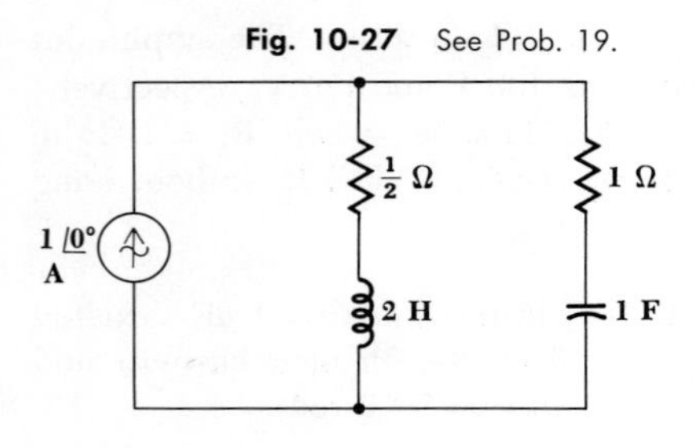

Fig. 10-28 See Prob. 20.

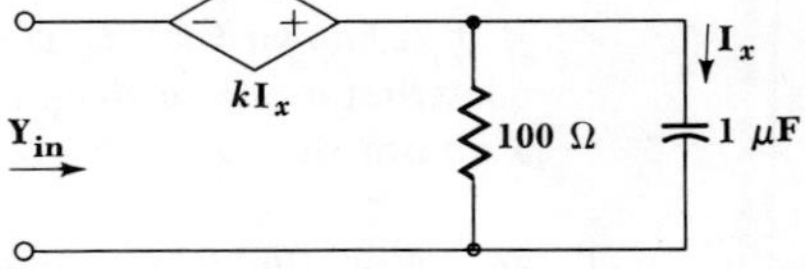

Chapter Eleven
Average Power and RMS Values

11-1 INTRODUCTION

Nearly all problems in circuit analysis are concerned with applying one or more sources of electrical energy to a circuit and then quantitatively determining the consequent response throughout the circuit. The response may be a current or a voltage, but we are also interested in the amount of energy supplied from the sources, in the amount of energy dissipated or stored within the circuit, and in the manner in which energy is delivered to the points at which the responses are determined. Primarily, however, we are concerned with the *rate* at which energy is being generated and absorbed; our attention must now be directed to *power*.

We shall begin by considering instantaneous power, the product of the time-domain voltage and time-domain current associated with the element in which we are interested. The instantaneous power is sometimes quite useful in its own right, because its maximum value might have to be limited in order to avoid exceeding the safe or useful operating range of a physical device. For example, transistor and vacuum-tube power amplifiers both produce a distorted output when the peak power exceeds a certain limiting value. However, we are mainly interested in instantaneous power for the simple reason that it provides us with the means to calculate a more important quantity, the average power. In a similar way, the progress of a cross-country automobile trip is best described by the average velocity; our interest in the instantaneous velocity is limited to the avoidance of maximum velocities which will endanger our safety or arouse the highway patrol.

In practical problems we shall deal with values of average power which range from the small fraction of a picowatt available in a telemetry signal from outer space, the few watts of audio power supplied to the speakers in a high-fidelity stereo system, the several hundred watts required to invigorate the morning coffeepot, to the millions of watts needed to supply all the electrical needs of a large city.

Our discussion will not be concerned entirely with the average power delivered by a sinusoidal current or voltage; we shall establish a mathematical measure of the effectiveness of other waveforms in delivering power, which we shall call the effective value. Our study of power will be completed by considering the descriptive quantities, power factor and complex power, two concepts which introduce the practical and economic aspects associated with the distribution of electric power.

11-2 INSTANTANEOUS POWER

The power delivered to any device as a function of time is given by the product of the instantaneous voltage across the device and the instantaneous current through it, as we well know; the passive sign convention is assumed. Thus,

$$p = vi \tag{1}$$

A knowledge of both the current and the voltage is presumed. If the device in question is a resistor R, then the power may be expressed solely in terms of either the current or the voltage,

$$p = vi = i^2 R = \frac{v^2}{R} \tag{2}$$

If the voltage and current are associated with a device which is entirely inductive, then

$$p = vi = Li\frac{di}{dt} = \frac{1}{L}v\int_{-\infty}^{t} v\,dt \qquad (3)$$

where we have arbitrarily assumed that the voltage is zero at $t = -\infty$. In the case of a capacitor,

$$p = vi = Cv\frac{dv}{dt} = \frac{1}{C}i\int_{-\infty}^{t} i\,dt \qquad (4)$$

where a like assumption about the current is made. This listing of equations for power in terms of only a current or voltage soon becomes unwieldy, however, as we begin to consider more general networks. The listing is also quite unnecessary, for we need only find *both* the current and voltage at the network terminals. As an example, we may consider the series RL circuit, as shown in Fig. 11-1, excited by a step-voltage source. The familiar current response is

$$i(t) = \frac{V_0}{R}(1 - e^{-Rt/L})u(t)$$

and thus the total power delivered by the source or absorbed by the passive network is

$$p = vi = \frac{V_0^2}{R}(1 - e^{-Rt/L})u(t)$$

since the square of the unit-step function is obviously the unit-step function itself.

The power delivered to the resistor is

$$p_R = i^2R = \frac{V_0^2}{R}(1 - e^{-Rt/L})^2u(t)$$

Fig. 11-1 The power that is delivered to R is $p_R = i^2R = (V_0^2/R)(1 - e^{-Rt/L})^2u(t)$.

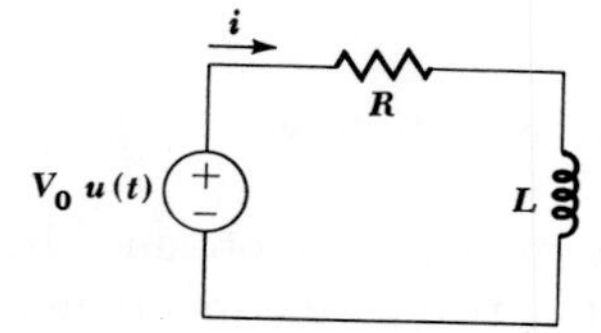

In order to determine the power absorbed by the inductor, we first obtain the inductor voltage

$$\begin{aligned} v_L &= L\frac{di}{dt} \\ &= V_0 e^{-Rt/L} u(t) + \frac{LV_0}{R}(1 - e^{-Rt/L})\frac{du(t)}{dt} \\ &= V_0 e^{-Rt/L} u(t) \end{aligned}$$

since $du(t)/dt$ is zero for $t > 0$ and $(1 - e^{-Rt/L})$ is zero at $t = 0$. The power absorbed by the inductor is thus

$$p_L = v_L i = \frac{V_0^2}{R} e^{-Rt/L}(1 - e^{-Rt/L})u(t)$$

Only a few algebraic manipulations are required to show that

$$p = p_R + p_L$$

which serves to check our work.

The majority of the problems which involve power calculations are perhaps those which deal with circuits excited by sinusoidal forcing functions in the steady state; as we have been told previously, even when periodic forcing functions which are not sinusoidal are employed, it is possible to resolve the problem into a number of subproblems in which the forcing functions *are* sinusoidal. The special case of the sinusoid therefore deserves special attention.

Let us change the voltage source in the circuit of Fig. 11-1 to the sinusoidal source $V_m \cos \omega t$. The familiar time-domain response is

$$i(t) = I_m \cos(\omega t + \theta)$$

where

$$I_m = \frac{V_m}{\sqrt{R^2 + \omega^2 L^2}} \qquad \text{and} \qquad \theta = -\tan^{-1}\frac{\omega L}{R}$$

The instantaneous power delivered to the entire circuit in the sinusoidal steady state is, therefore,

$$p = vi = V_m I_m \cos(\omega t + \theta) \cos \omega t$$

which we shall find convenient to rewrite in a form obtained by using the trigonometric identity for the product of two cosine functions. Thus,

$$p = \frac{V_m I_m}{2}[\cos(2\omega t + \theta) + \cos\theta]$$

$$= \frac{V_m I_m}{2}\cos\theta + \frac{V_m I_m}{2}\cos(2\omega t + \theta)$$

The last equation possesses several characteristics which are true in general for circuits in the sinusoidal steady state. One term, the first, is not a function of time; and a second term is included which has a cyclic variation at *twice* the applied frequency. Since this term is a cosine wave, and since sine waves and cosine waves have average values which are zero (when averaged over an integral number of periods), this introductory example may serve to indicate that the *average* power is $\frac{1}{2}V_m I_m \cos\theta$. This is true, and we shall now establish this relationship in more general terms.

Drill Problems

11-1 A current source, 8 cos 500t A, a 5-Ω resistor, and a 20-mH inductor are in parallel. At $t = 1$ ms find the power absorbed by the: (*a*) resistor; (*b*) inductor; (*c*) source.

Ans. −*143; 60; 83* W

11-2 A current source, $8u(t)$ A, a 5-Ω resistor, and a 20-mH inductor are in parallel. At $t = 3$ ms, find the power absorbed by the: (*a*) resistor; (*b*) inductor; (*c*) source.

Ans. −*151; 71; 80* W

11-3 AVERAGE POWER

When we speak of an average value for the instantaneous power, the time interval over which the averaging process takes place must be clearly defined. Let us first select a general interval of time from t_1 to t_2. We may then obtain the average value by integrating $p(t)$ from t_1 to t_2 and dividing the result by the time interval $t_2 - t_1$. Thus,

$$P = \frac{1}{t_2 - t_1}\int_{t_1}^{t_2} p(t)\,dt \tag{5}$$

The average value is denoted by the capital letter P since it is not a function of time, and it usually appears without subscripts. Although P is not a function of time, it is a function of t_1 and t_2, the two instants of time which define the interval of integration. This dependence of P on a specific time interval may be expressed in a simpler manner if $p(t)$ is a periodic function. We shall consider this important case first.

Let us assume that our forcing function and the circuit responses are all periodic; a steady-state condition has been reached, although not necessarily the sinusoidal steady state. We may define a *periodic* function mathematically by requiring that

$$f(t) = f(t + T) \tag{6}$$

where T is the period. We now show that the average value of the instantaneous power as expressed by (5) may be computed over an interval of one period having an arbitrary beginning.

A general periodic waveform is shown in Fig. 11-2 and identified as $p(t)$. We first compute the average power by integrating from t_1 to a time t_2 which is one period later, $t_2 = t_1 + T$,

$$P_1 = \frac{1}{T}\int_{t_1}^{t_1+T} p(t)\,dt$$

and then by integrating from some other time t_x to $t_x + T$,

$$P_x = \frac{1}{T}\int_{t_x}^{t_x+T} p(t)\,dt$$

The equality of P_1 and P_x should be evident from the graphical interpretation of the integrals; the area which represents the integral to be evaluated in determining P_x is smaller by the area from t_1 to t_x, but greater by the area from $t_1 + T$ to $t_x + T$, and the periodic nature of the curve requires these two areas to be equal. Thus, the average power may be computed by integrating the instantaneous power over any interval which is one period in length and then dividing by the period

$$P = \frac{1}{T}\int_{t_x}^{t_x+T} p\,dt \tag{7}$$

It is important to note that we might also integrate over any integral number of periods, provided that we divide by this same integral number of periods. Thus,

$$P = \frac{1}{nT}\int_{t_x}^{t_x+nT} p\,dt \qquad n = 1,2,3, \ldots \tag{8}$$

If we carry this concept to the extreme by integrating over all time, another

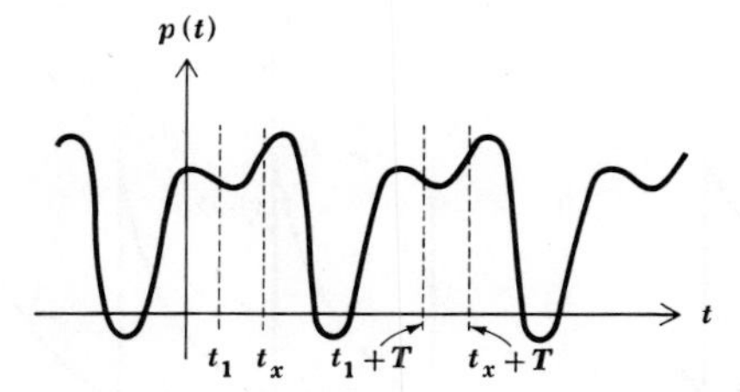

Fig. 11-2 The average value P of a periodic power function $p(t)$ is the same over any period T.

useful result is obtained. We first provide ourselves with symmetrical limits on the integral

$$P = \frac{1}{nT}\int_{-nT/2}^{nT/2} p\,dt$$

and then take the limit as n becomes infinite,

$$P = \lim_{n\to\infty} \frac{1}{nT}\int_{-nT/2}^{nT/2} p\,dt$$

If $p(t)$ is a mathematically well-behaved function, as all *physical* forcing functions and responses are, it is apparent that if a large integer n is replaced by a slightly larger number which is not an integer, then the value of the integral and of P is changed by a negligible amount; moreover, the error decreases as n increases. Without justifying this step rigorously, we therefore replace the discrete variable nT by the continuous variable τ,

$$P = \lim_{\tau\to\infty} \frac{1}{\tau}\int_{-\tau/2}^{\tau/2} p\,dt \tag{9}$$

We shall find it convenient on several occasions to integrate periodic functions over this "infinite period." Examples of the use of (7), (8), and (9) are given below.

Let us illustrate the calculation of the average power of a periodic wave by finding the average power delivered by the (periodic) sawtooth current waveform shown in Fig. 11-3a to a resistor R. We have

$$i(t) = \frac{I_m}{T}t \qquad 0 < t \le T$$

$$i(t) = \frac{I_m}{T}(t - T) \qquad T < t \le 2T$$

etc.

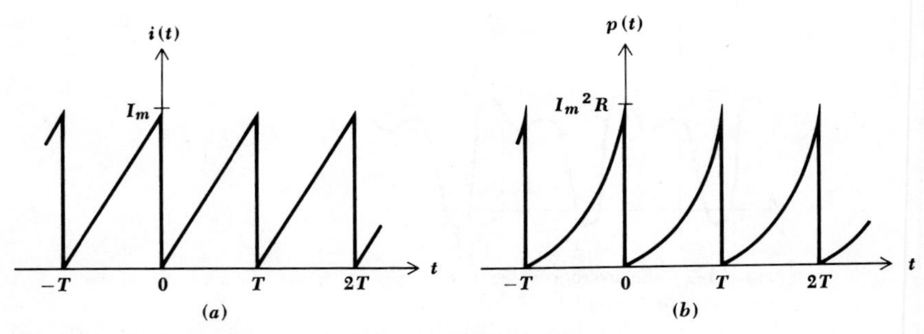

Fig. 11-3 (*a*) A sawtooth current waveform and (*b*) the instantaneous power waveform it produces in a resistor R.

and
$$p(t) = \frac{1}{T^2} I_m{}^2 R t^2 \qquad 0 < t \leq T$$

$$p(t) = \frac{1}{T^2} I_m{}^2 R (t - T)^2 \qquad T < t \leq 2T$$

$$\text{etc.}$$

as sketched in Fig. 11-3*b*. Integrating over the simplest range of one period, from $t = 0$ to $t = T$, we have

$$P = \frac{1}{T} \int_0^T \frac{I_m{}^2 R}{T^2} t^2 \, dt = \tfrac{1}{3} I_m{}^2 R$$

The selection of other ranges of one period, such as from $t = 0.1T$ to $t = 1.1T$, would produce the same answer. Integration from 0 to $2T$ and division by $2T$, that is, the application of (8) with $n = 2$ and $t_x = 0$, would also provide the same answer.

Now let us obtain the general result for the sinusoidal steady state. We shall assume the general sinusoidal voltage

$$v(t) = V_m \cos (\omega t + \alpha)$$

and current

$$i(t) = I_m \cos (\omega t + \alpha - \theta)$$

associated with the device in question. The instantaneous power is

$$p(t) = V_m I_m \cos (\omega t + \alpha) \cos (\omega t + \alpha - \theta)$$

Again expressing the product of two cosine functions as one-half the sum of the cosine of the difference angle and the cosine of the sum angle,

$$p(t) = \tfrac{1}{2}V_m I_m \cos\theta + \tfrac{1}{2}V_m I_m \cos(2\omega t + 2\alpha - \theta) \qquad (10)$$

we may save ourselves some integration by an inspection of the result. The first term is a constant, independent of t. The remaining term is a cosine function; $p(t)$ is therefore periodic, and its period is $\frac{1}{2}T$. Note that the period T is associated with the given current and voltage, and not with the power; the power function has a period $\frac{1}{2}T$. However, we may integrate over an interval of T to determine the average value if we wish; it is only necessary to divide also by T. Our familiarity with cosine and sine waves, however, shows that the average value of either over a period is zero. There is thus no need to integrate (10) formally; by inspection, the average value of the second term is zero over a period T (or $\frac{1}{2}T$), and the average value of the first term, a constant, must be that constant itself. Thus,

$$P = \tfrac{1}{2}V_m I_m \cos\theta \qquad (11)$$

This important result, introduced in the previous section for a specific circuit, is therefore quite general for the sinusoidal steady state. The average power is one-half the product of the crest amplitude of the voltage, the crest amplitude of the current, and the cosine of the phase-angle difference between the current and the voltage; the sense of the difference is immaterial.

As a numerical illustration, let us assume that a voltage

$$v(t) = 4\cos\frac{\pi t}{6}$$

or

$$\mathbf{V} = 4\underline{/0^\circ}$$

is applied across an impedance $\mathbf{Z} = 2\underline{/60^\circ}\ \Omega$. The phasor current is therefore $2\underline{/-60^\circ}$ A, and the average power is

$$P = \tfrac{1}{2}(4)(2)\cos 60^\circ = 2 \quad \text{W}$$

The time-domain voltage

$$v(t) = 4\cos\frac{\pi t}{6}$$

time-domain current

$$i(t) = 2\cos\left(\frac{\pi t}{6} - 60^\circ\right)$$

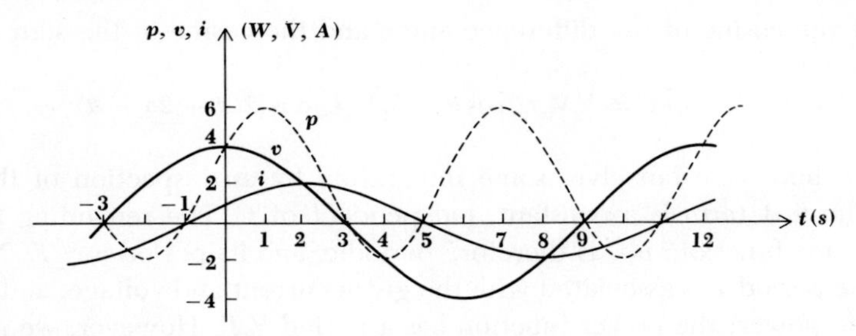

Fig. 11-4 Curves of $v(t)$, $i(t)$, and $p(t)$ are plotted as a function of time for a simple circuit in which the phasor voltage $\mathbf{V} = 4\underline{/0°}$ V is applied to the impedance $\mathbf{Z} = 2\underline{/60°}\ \Omega$ at $\omega = \pi/6$.

and instantaneous power

$$p(t) = 8 \cos \frac{\pi t}{6} \cos \left(\frac{\pi t}{6} - 60° \right)$$

$$= 2 + 4 \cos \left(\frac{\pi t}{3} - 60° \right)$$

are all sketched on the same time axis in Fig. 11-4. Both the 2-W average value of the power and its period of 6 s, one-half the period of either the current or the voltage, are evident. The zero value of the instantaneous power at each instant when either the voltage or current is zero is also apparent.

Two special cases are worth isolating for consideration, the average power delivered to an ideal resistor and that to an ideal reactor (any combination of only capacitors and inductors). The phase-angle difference between the current through and the voltage across a pure resistor is zero, and therefore

$$P_R = \tfrac{1}{2} V_m I_m$$

or

$$P_R = \tfrac{1}{2} I_m^{\ 2} R \tag{12}$$

or

$$P_R = \frac{V_m^{\ 2}}{2R} \tag{13}$$

The last two formulas, enabling us to determine the average power delivered to a pure resistance from a knowledge of either the sinusoidal current or voltage, are simple and important. They are often misused. The most common error is made in trying to apply them in cases where, say, the voltage included in (13) is *not the voltage across the resistor.* If care is taken to use the current through the resistor in (12) and the voltage across the resistor in (13), satisfactory operation is guaranteed.

The average power delivered to any device which is purely reactive

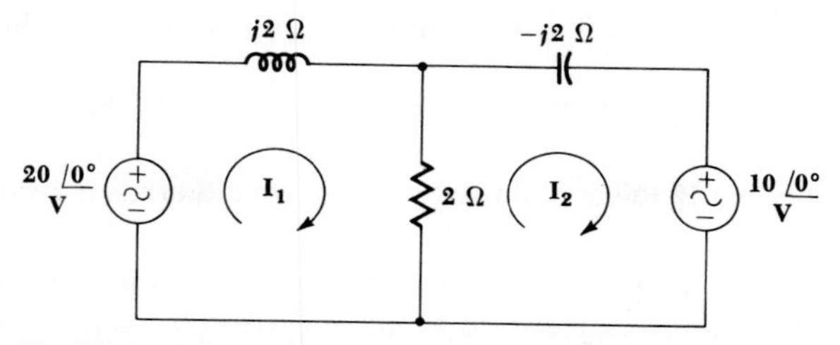

Fig. 11-5 The average power delivered to each reactive element is zero in the sinusoidal steady state.

must be zero. This is evident from the 90° phase difference which must exist between current and voltage; hence, $\cos\theta = 0$ and

$$P_x = 0$$

The *average* power delivered to any network composed entirely of ideal inductors and capacitors is zero; the instantaneous power is zero only at specific instants. Thus, power flows into the network for a part of the cycle and out of the network during another portion of the cycle.

As an example illustrating these relationships, let us consider the circuit shown in Fig. 11-5. The values of $\mathbf{I}_1$ and $\mathbf{I}_2$ are found by any of several methods, such as mesh analysis, nodal analysis, or superposition:

$$\mathbf{I}_1 = 5 - j10 = 11.18\underline{/-63.45^\circ}$$

$$\mathbf{I}_2 = 5 - j5 = 7.07\underline{/-45^\circ}$$

The current through the 2-Ω resistor is

$$\mathbf{I}_1 - \mathbf{I}_2 = -j5 = 5\underline{/-90^\circ}$$

and the resistor voltage is, therefore,

$$\mathbf{V}_R = 10\underline{/-90^\circ}$$

This current and voltage satisfy the passive sign convention, and the average power absorbed by the resistor is, therefore,

$$P_R = \tfrac{1}{2}(10)(5)\cos(-90^\circ + 90^\circ) = 25 \quad \text{W}$$

a result which may be checked by using (12) or (13). Turning to the left source, the voltage $20\underline{/0^\circ}$ and current $11.18\underline{/-63.45^\circ}$ satisfy the *active* sign convention, and thus the power *delivered* by this source is

$$P_{\text{left}} = \tfrac{1}{2}(20)(11.18)\cos(0^\circ + 63.45^\circ) = 50 \quad \text{W}$$

In a similar manner, we find the power *absorbed* by the right source,

$$P_{\text{right}} = \tfrac{1}{2}(10)(7.07)\cos(0° + 45°) = 25 \quad \text{W}$$

The power delivered to each of the two reactive elements is zero; the power relations check.

In Sec. 4 of Chap. 3 we considered the maximum-power-transfer theorem as it applies to resistive loads and resistive source impedances. For a Thévenin source $\mathbf{V}_s$ and impedance $\mathbf{Z}_{th} = R_{th} + jX_{th}$ connected to a load $\mathbf{Z}_L = R_L + jX_L$, it may be shown readily (Prob. 6) that the average power delivered to the load is a maximum when $R_L = R_{th}$ and $X_L = -X_{th}$. This result is often dignified by calling it the *maximum-power-transfer theorem for the sinusoidal steady state.* It is apparent that the resistive condition considered earlier is merely a special case.

We must now pay some attention to *nonperiodic* functions. One practical example of a nonperiodic power function for which an average power value is desired is the power output of a radio telescope which is directed toward a "radio star." Another is the sum of a number of periodic functions, each function having a different period, such that no greater common period can be found for the combination. For example, the current

$$i(t) = \sin t + \sin \pi t \tag{14}$$

is nonperiodic because the ratio of the periods of the two sine waves is an irrational number. At $t = 0$, both terms are zero and increasing. But the first term is zero and increasing only when $t = 2\pi n$, where n is an integer, and thus periodicity demands that πt or $\pi(2\pi n)$ must equal $2\pi m$, where m is also an integer. No solution (integral values for both m and n) for this equation is possible. It may be illuminating to compare the nonperiodic expression (14) with the *periodic* function

$$i(t) = \sin t + \sin 3.14t \tag{15}$$

where 3.14 is an exact decimal expression and is not intended to be interpreted as 3.141592 . . . . With a little effort, it can be shown that the period of this current wave is 100π s.

The average value of the power delivered to a 1-Ω resistor by either the periodic current (15) or the nonperiodic current (14) may be found by integrating over an infinite interval; much of the actual integration can be avoided because of our thorough knowledge of the average value of simple functions. We therefore obtain the instantaneous power delivered by the current in (14) by applying (9),

$$P = \lim_{T \to \infty} \frac{1}{T} \int_{-T/2}^{T/2} (\sin^2 t + \sin^2 \pi t + 2 \sin t \sin \pi t)\, dt$$

We now consider P as the sum of three average values. The average value of $\sin^2 t$ over an infinite interval is found by replacing $\sin^2 t$ by $(\frac{1}{2} - \frac{1}{2}\cos 2t)$; it is obviously $\frac{1}{2}$. Similarly, the average value of $\sin^2 \pi t$ is also $\frac{1}{2}$. And the last term can be expressed as the sum of two cosine functions, each of which must certainly have an average value of zero. Thus,

$$P = \tfrac{1}{2} + \tfrac{1}{2} = 1 \quad \text{W}$$

An identical result is obtained for the periodic current (15).

Applying this same method to a current function which is the sum of several sinusoids *of different periods* and arbitrary amplitudes,

$$i(t) = I_{m1}\cos\omega_1 t + I_{m2}\cos\omega_2 t + \cdots + I_{mN}\cos\omega_N t \tag{16}$$

we find the average power delivered to a resistance R,

$$P = \tfrac{1}{2}(I_{m1}{}^2 + I_{m2}{}^2 + \cdots + I_{mN}{}^2)R \tag{17}$$

The result is unchanged if an arbitrary phase angle is assigned to each component of the current. This important result is surprisingly simple when we think of the steps required for its derivation: squaring the current function, integrating, and taking the limit. The result is also just plain surprising, because it shows that, *in this special case of a current such as* (16), superposition is applicable to power. Superposition is *not* applicable for a current which is the sum of two direct currents, nor is it applicable for a current which is the sum of two sinusoids of the same period.

Drill Problems

11-3 A sinusoidal current source $\mathbf{I}_s = 10\underline{/0^\circ}$ A is connected to each of the networks described below. Find the average power delivered by the source in each case: (*a*) 2 Ω in parallel with *j*4 Ω; (*b*) 4 Ω in series with a capacitive reactance of 2 Ω; (*c*) a current source $5\underline{/90^\circ}$ A (sensed in the same direction as $\mathbf{I}_s$) in parallel with *j*4 Ω.

Ans. −*100; 80; 200* W

11-4 Find the average power delivered to a 12-Ω resistor by each of the three periodic current waveforms displayed in Fig. 11-6.

Ans. 96; 144; 216 W

11-5 In the circuit of Fig. 11-7, find the average power received by the: (*a*) resistor; (*b*) dependent source; (*c*) independent source.

Ans. −*188; 35; 153* mW

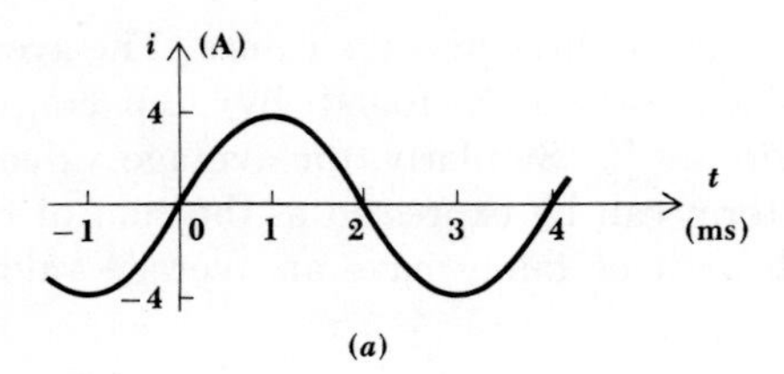

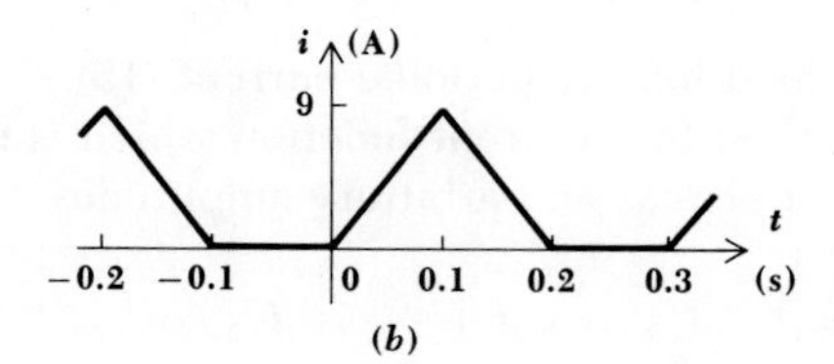

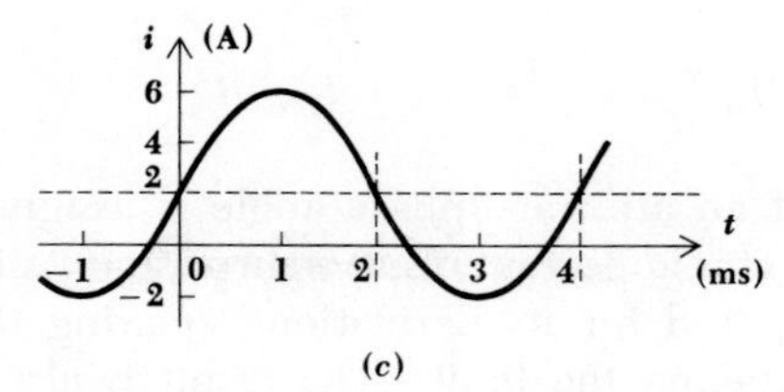

Fig. 11-6 See Drill Probs. 11-4 and 11-7.

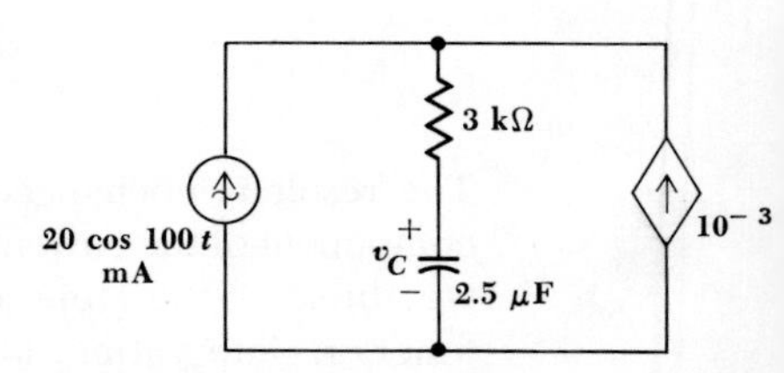

Fig. 11-7 See Drill Prob. 11-5.

11-6 Determine the (average) power delivered to the series combination of a 5-Ω resistor and a 10-mH inductor by the voltage: (*a*) $20 \cos 1000t - 15 \sin 1000t$ V; (*b*) $20 \cos 1000t - 15 \sin 500t$ V; (*c*) $20 \cos 1000t - 15 \sin (1000t - 36.9°)$ V.

Ans. 12.5; 19.25; 19.7 W

11-4 EFFECTIVE VALUES OF CURRENT AND VOLTAGE

Most of us are aware that the voltage available at the power outlets in our homes is a sinusoidal voltage having a frequency of 60 Hz and a "voltage" of 115 V. But what is meant by "115 volts"? This is certainly not the instantaneous value of the voltage, for the voltage is not a constant. The value of 115 V is also not the maximum value which we have been symbolizing as V_m; if we displayed the voltage waveform on a calibrated cathode-ray oscilloscope, we should find that the maximum value of this voltage at one of our ac outlets is $115\sqrt{2}$ or 162.6 V. We also cannot fit the concept of an average value to the 115 V because the average value of the sine wave is zero. We might come a little closer by trying the magnitude of the average over a positive or negative half cycle; by using a rectifier-type voltmeter at the outlet, we should measure 103.5 V. As it turns out, however, the 115 V is the *effective value* of this sinusoidal voltage. This value is a measure of the effectiveness of a voltage source in delivering power to a resistive load.

Let us now proceed to define the effective value of any periodic waveform representing either a current or voltage. We shall consider the sinusoidal waveform as only a special, albeit practically important, case. Let us arbitrarily define effective value in terms of a current waveform, although a voltage could equally well be selected. The *effective value* of any periodic current is equal to the value of the direct current which, flowing through a resistance R, delivers the same power to R as the periodic current does. In other words, we allow the given periodic current to flow through an arbitrary resistance R, determine the instantaneous power i^2R, and then find the average value of i^2R over a period; this is the average power. We then cause a direct current to flow through this same resistance and adjust the value of the direct current until the same value of average power is obtained. The magnitude of the direct current is equal to the effective value of the given periodic current.

The general mathematical expression for the effective value of $i(t)$ is now easily obtained. The average power delivered to the resistor by the periodic current $i(t)$ is

$$P = \frac{1}{T}\int_0^T i^2R\,dt = \frac{R}{T}\int_0^T i^2\,dt$$

where the period of $i(t)$ is T. The power delivered by the direct current is

$$P = I^2R = I_{eff}{}^2R$$

Equating the power expressions and solving for I_{eff},

$$I_{eff} = \sqrt{\frac{1}{T}\int_0^T i^2\,dt} \tag{18}$$

The result is independent of the resistance R, as it must be to provide us with a worthwhile concept. A similar expression is obtained for the effective value of a periodic voltage by replacing i and I_{eff} by v and V_{eff}, respectively.

Notice that the effective value is obtained by first squaring the time function, then taking the average value of the squared function over a period, and finally taking the square root of the average of the squared function. In abbreviated language, the operation involved in finding an effective value is the (square) *root* of the *mean* of the *square;* for this reason, the effective value is often called the *root-mean-square* value, or simply the rms value.

The most important special case is that of the sinusoidal waveform. Let us select the sinusoidal current

$$i(t) = I_m \cos(\omega t - \theta)$$

which has a period

$$T = \frac{2\pi}{\omega}$$

and substitute into (18) to obtain the effective value

$$\begin{aligned} I_{eff} &= \sqrt{\frac{1}{T}\int_0^T I_m^2 \cos^2(\omega t - \theta)\,dt} \\ &= I_m\sqrt{\frac{\omega}{2\pi}\int_0^{2\pi/\omega} [\tfrac{1}{2} + \tfrac{1}{2}\cos(2\omega t - 2\theta)]\,dt} \\ &= I_m\sqrt{\frac{\omega}{4\pi}[t]_0^{2\pi/\omega}} = \frac{I_m}{\sqrt{2}} \end{aligned}$$

Thus the effective value of a sinusoidal current is a real quantity which is independent of the phase angle and numerically equal to 0.707 times its maximum value. A current $\sqrt{2}\cos(\omega t - \theta)$, therefore, has an effective value of 1 A and will deliver the same power to any resistor as will a direct current of 1 A.

The use of the effective value also simplifies slightly the expression for the average power delivered by a sinusoidal current or voltage. For example, the average power delivered to a resistance R by a sinusoidal current is

$$P = \tfrac{1}{2}I_m^2 R$$

If we replace I_m by $\sqrt{2}\, I_{eff}$, the average power may be written

$$P = I_{eff}^2 R \tag{19}$$

The other familiar power expressions may also be written in terms of effective values:

$$P = V_{eff} I_{eff} \cos\theta \tag{20}$$

and

$$P = \frac{V_{eff}^2}{R} \tag{21}$$

The fact that the effective value is defined in terms of an equivalent dc quantity provides us with average power formulas for resistive circuits which are identical with those used in dc analysis.

Although we have succeeded in eliminating the factor of one-half from our average-power relationships, we must now take care to determine whether a sinusoidal quantity is expressed in terms of its maximum value or its effective value. In practice, the effective value is usually used in the fields of power transmission or distribution and of rotating machinery; in the areas of electronics and communications, the maximum value is more often used. We shall assume that the maximum value is specified unless the term rms is explicitly used.

In the sinusoidal steady state, phasor voltages and currents may be given as either effective values or maximum values; the two expressions differ only by a factor $\sqrt{2}$. The voltage $50\underline{/30^\circ}$ V is expressed in terms of a maximum value; as an rms voltage, we should write $35.35\underline{/30^\circ}$ V rms.

In order to determine the effective value of a periodic or nonperiodic waveform which is composed of the sum of a number of sinusoids of different frequencies, we may use the appropriate average power relationship (17) developed in the previous section, rewritten in terms of the effective values of the several components,

$$P = (I_{1\,eff}^2 + I_{2\,eff}^2 + \cdots + I_{N\,eff}^2)R \tag{22}$$

These results indicate that if a sinusoidal current of 5 A rms at 60 Hz flows through a 2-Ω resistor, an average power of 50 W is absorbed by the resistor; if a second current, say 3 A rms at 120 Hz, is also present, the absorbed power is 68 W; however, if the second current is also at 60 Hz, then the absorbed power may have any value between 8 and 128 W, depending on the relative phase of the two current components.

We therefore have found the effective value of a current which is composed of any number of sinusoidal currents of *different* frequencies,

$$I_{eff} = \sqrt{I_{1\,eff}^2 + I_{2\,eff}^2 + \cdots + I_{N\,eff}^2} \tag{23}$$

The total current may or may not be periodic; the result is the same. The effective value of the sum of the 60- and 120-Hz currents in the example above is 5.83 A; the effective value of the sum of the two 60-Hz currents may have any value between 2 and 8 A.

Drill Problems

11-7 Find the effective values of the three periodic current waveforms shown in Fig. 11-6.

Ans. 2.83; 3.46; 4.24 A

11-8 Find the effective value of: (*a*) $2 \cos (1000t + 0.1\pi) - \sin (1000t - 0.15\pi)$ V; (*b*) $2 \cos 1000t + \sin (500t + 0.2\pi)$ V; (*c*) $v = 1.2$ V for $0 \le t < 2$ ms, 2.4 V for $2 \le t < 4$ ms, 3.0 V for $4 \le t < 5$ ms, where $T = 5$ ms.

Ans. *1.98; 2.16; 1.58* V

11-5 APPARENT POWER AND POWER FACTOR

Historically, the introduction of the concepts of apparent power and power factor can be traced to the electric-power industry, where large amounts of electrical energy must be transferred from one point to another; the efficiency with which this transfer is effected is related directly to the cost of the electrical energy, which is eventually paid by the consumer. A customer who provides a load which results in a relatively poor transmission efficiency must pay a greater price for each kilowatthour (kWh) of electrical energy he actually receives and uses. In a similar way, a customer who requires a costlier investment in transmission and distribution equipment by the power company will also pay more for each kilowatthour.

Let us first define apparent power and power factor and then show briefly how these terms are related to the economic situations mentioned above. We shall assume that the sinusoidal voltage

$$v = V_m \cos (\omega t + \alpha)$$

is applied to a network, and the resultant sinusoidal current is

$$i = I_m \cos (\omega t + \beta)$$

The phase angle by which the voltage leads the current is therefore

$$\theta = \alpha - \beta$$

The average power delivered to the network, assuming a passive sign convention at its input terminals, may be expressed either in terms of the maximum values,

$$P = \tfrac{1}{2} V_m I_m \cos \theta$$

or in terms of the effective values,

$$P = V_{eff} I_{eff} \cos \theta$$

If our applied voltage and current responses had been dc quantities, the average power delivered to the network would have been given simply by the product of the voltage and the current. Applying this dc technique to the sinusoidal problem, we should obtain a value for the absorbed power which is "apparently" given by the product $V_{eff}I_{eff}$. This product of the effective values of the voltage and current is not the average power; we define it as the *apparent power*. Dimensionally, apparent power must be measured in the same units as real power, since $\cos\theta$ is dimensionless, but in order to avoid confusion the term voltamperes, or VA, or kVA is applied to apparent power. Since $\cos\theta$ cannot have a magnitude greater than unity, it is evident that the magnitude of the real power can never be greater than the magnitude of the apparent power.

Apparent power is not a concept which is limited to sinusoidal forcing functions and responses. It may be determined for any current and voltage waveshape by simply taking the product of the effective values of the current and voltage. This extension of the definition of apparent power need not concern us now, but it will receive further amplification in connection with general periodic functions when Fourier analysis is discussed in Chap. 17.

The ratio of the real or average power to the apparent power is called the *power factor*, symbolized by PF. Hence,

$$\text{PF} = \frac{\text{average power}}{\text{apparent power}} = \frac{P}{V_{eff}I_{eff}}$$

In the sinusoidal case, the power factor is simply $\cos\theta$, where θ is the angle by which the voltage leads the current. This relationship is the reason why the angle θ is often referred to as the *PF angle*.

For a purely resistive load, the voltage and current are in phase, θ is zero, and the PF is unity. The apparent power and the average power are equal. A unity PF, however, may also be achieved for loads which contain both inductance and capacitance if the element values and the operating frequency are selected to provide an input impedance having a zero phase angle.

A purely reactive load, that is, one containing no resistance, will cause a phase difference between the voltage and current of either plus or minus 90°, and the PF is therefore zero.

Between these two extreme cases there are the general networks for which the PF can range from zero to unity. A PF of 0.5, for example, indicates a load having an input impedance with a phase angle of either 60° or −60°; the former describes an inductive load, since the voltage leads the current by 60°, while the latter refers to a capacitive load. The ambiguity in the exact nature of the load is resolved by referring to a leading PF or a lagging PF, the term leading or lagging referring to the *phase of the current with respect to the voltage.* Thus, an inductive load will have a lagging PF and a capacitive load a leading PF.

The practical importance of these new terms is shown by the several examples which follow. Let us first assume that we have a sinusoidal ac generator, which is a rotating machine driven by some other device whose output is a mechanical torque such as a steam turbine, an electric motor, or an internal-combustion engine. We shall let our generator produce an output voltage of 200 V rms at 60 Hz. Suppose now that an additional rating of the generator is stated as a maximum power output of 1 kW. The generator would therefore be capable of delivering an rms current of 5 A to a resistive load. If, however, a load requiring 1 kW at a lagging power factor of 0.5 is connected to the generator, then an rms current of 10 A is necessary. As the PF decreases, greater and greater currents must be delivered to the load if operation at 200 V and 1 kW is maintained. If our generator were correctly and economically designed to furnish safely a maximum current of 5 A, then these greater currents would cause unsatisfactory operation. The rating of the generator is more informatively given in terms of apparent power in volt-amperes. Thus a 1000-VA rating at 200 V indicates that the generator can deliver a maximum current of 5 A at rated voltage; the power it delivers depends on the load, and in an extreme case might be zero. An apparent power rating is equivalent to a current rating when operation is at a constant voltage.

When electric power is being supplied to large industrial consumers by a power company, the company will frequently include a PF clause in its rate schedules. Under this clause, an additional charge is made to the consumer whenever his PF drops below a certain specified value, usually about 0.85 lagging. Very little industrial power is consumed at leading PFs because of the nature of the typical industrial loads. There are several reasons that force the power company to make this additional charge for low PFs. In the first place, it is apparent that larger current-carrying capacity must be built into its generators in order to provide the larger currents that go with lower PF operation at constant power and constant voltage. Another reason is found in the increased losses in its transmission and distribution system.

As an example, let us suppose that a certain consumer is using an average power of 11 kW at unity PF and 220 V rms. We may also assume a total resistance of 0.2 Ω in the transmission lines through which the power is delivered to the consumer. An rms current of 50 A therefore flows in the load and in the lines, producing a line loss of 500 W. In order to supply 11 kW to the consumer, the power company must generate 11.5 kW (at the higher voltage, 230 V). Since the energy is necessarily metered at the location of each consumer, this consumer would be billed for 95.6 per cent of the energy which the power company actually produced.

Now let us hypothesize another consumer, also requiring 11 kW, but at a PF angle of 60° lagging. This consumer forces the power company to push 100 A through his load and (of particular interest to the company) through the line resistance. The line losses are now found to be 2 kW, and the customer's meter indicates only 84.6 per cent of the actual energy generated. This figure departs from 100 per cent by more than the power company will tolerate; this costs it money. Of course, the transmission losses might be reduced by using heavier transmission lines which have lower resistance, but this costs more money too. The power company's solution to this problem is to encourage operation at PFs which exceed 0.9 lagging by offering slightly reduced rates and to discourage operation at PFs which are less than 0.85 lagging by invoking increased rates.

The power drawn by most homes is used at reasonably high PFs (and reasonably small power levels); no charge is customarily made for low PF operation.

Besides paying for the actual energy consumed and for operation at excessively low PFs, industrial consumers are also billed for inordinate *demand*. An energy of 100 kWh is delivered much more economically as 5 kW for 20 h than it is as 20 kW for 5 h, particularly if everyone else demands large amounts of power at the same time.

Drill Problem

11-9 Determine the PF associated with a load: (*a*) consisting of a 6.25-μF capacitor in parallel with the series combination of 400-Ω resistance and 1 H at $\omega = 400$ rad/s; (*b*) that is inductive and draws 33 A rms and 6.9 kW at 230 V rms; (*c*) composed of parallel loads, one of which draws 10 kVA at 0.8 PF lagging and the other 8 kVA at 0.9 PF leading.

Ans. 0.707 lead; 0.909 lag; 0.987 lag

11-6 COMPLEX POWER

Some simplification in power calculations is achieved if power is considered to be a complex quantity. The magnitude of the complex power will be found to be the apparent power, and the real part of the complex power will be shown to be the (real) average power. The new quantity, the imaginary part of the complex power, we shall call *reactive power*.

We define complex power with reference to a general sinusoidal voltage $\mathbf{V}_{eff} = V_{eff}\underline{/\theta_v}$ across a pair of terminals and a general sinusoidal current $\mathbf{I}_{eff} = I_{eff}\underline{/\theta_i}$ flowing into one of the terminals in such a way as to satisfy the passive sign convention. The average power P absorbed by the two-terminal network is thus

$$P = V_{eff} I_{eff} \cos(\theta_v - \theta_i)$$

Complex nomenclature is next introduced by making use of Euler's formula in the same way as we did in introducing phasors. We express P as

$$P = V_{eff} I_{eff} \, \text{Re} \, [e^{j(\theta_v - \theta_i)}]$$

or

$$P = \text{Re} \, [V_{eff} e^{j\theta_v} I_{eff} e^{-j\theta_i}]$$

The phasor voltage may now be recognized as the first two factors within the brackets in the equation above, but the second two factors do not quite correspond to the phasor current because the angle includes a minus sign which is not present in the expression for the phasor current. In other words, the phasor current is

$$\mathbf{I}_{eff} = I_{eff} e^{j\theta_i}$$

and we therefore must make use of conjugate notation,

$$\mathbf{I}^*_{eff} = I_{eff} e^{-j\theta_i}$$

Hence

$$P = \text{Re} \, (\mathbf{V}_{eff} \mathbf{I}^*_{eff})$$

and we may now let power become complex by defining the complex power $\mathbf{P}$ as

$$\mathbf{P} = \mathbf{V}_{eff} \mathbf{I}^*_{eff} \tag{24}$$

If we first inspect the polar or exponential form of the complex power

$$\mathbf{P} = V_{eff} I_{eff} e^{j(\theta_v - \theta_i)}$$

it is evident that the magnitude of $\mathbf{P}$ is the apparent power, and the angle of $\mathbf{P}$ is the PF angle, that is, the angle by which the voltage leads the current. In rectangular form,

$$\mathbf{P} = P + jQ \tag{25}$$

where P is the real average power, as before.[1] The imaginary part of the complex power is symbolized as Q and is termed the *reactive power*. The dimensions of Q are obviously the same as those of the real power P, the

[1]It should be noted that P is not the magnitude of $\mathbf{P}$, although such nomenclature has been used previously. For example, V_{eff} is the magnitude of $\mathbf{V}_{eff}$, Z_L is the magnitude of $\mathbf{Z}_L$, and so forth.

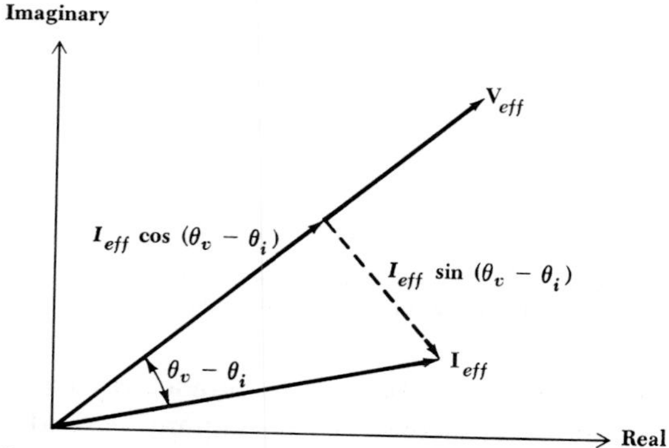

Fig. 11-8 The current phasor $\mathbf{I}_{eff}$ is resolved into two components, one in phase with the voltage phasor $\mathbf{V}_{eff}$ and the other 90° out of phase with the voltage phasor. This latter component is called a quadrature component.

complex power **P**, and the apparent power $|\mathbf{P}|$. In order to avoid confusion with these other quantities, the units of Q are defined as *vars*, standing for voltamperes reactive. From (24), it is seen that

$$Q = V_{eff}I_{eff}\sin(\theta_v - \theta_i)$$

Another interpretation of the reactive power may be seen by constructing a phasor diagram containing $\mathbf{V}_{eff}$ and $\mathbf{I}_{eff}$, as shown in Fig. 11-8. If the phasor current is resolved into two components, one in phase with the voltage, having a magnitude $I_{eff}\cos(\theta_v - \theta_i)$, and one 90° out of phase with the voltage, with magnitude $I_{eff}\sin(\theta_v - \theta_i)$, then it is clear that the real power is given by the product of the magnitude of the voltage phasor and the component of the phasor current which is in phase with the voltage. Moreover, the product of the magnitude of the voltage phasor and the component of the phasor current which is 90° out of phase with the voltage is the reactive power Q. It is common to speak of the component of a phasor which is 90° out of phase with some other phasor as a *quadrature component*. Thus Q is simply V_{eff} times the quadrature component of $\mathbf{I}_{eff}$; Q is also known as the quadrature power.

The sign of the reactive power characterizes the nature of a passive load at which $\mathbf{V}_{eff}$ and $\mathbf{I}_{eff}$ are specified. If the load is inductive, then $\theta_v - \theta_i$ is an angle between 0 and 90°, the sine of this angle is positive, and the reactive power is positive. A capacitive load results in a negative reactive power.

Just as a wattmeter[2] reads the average real power drawn by a load, a varmeter will read the average reactive power Q drawn by the load. Both quantities may be metered simultaneously. In addition, watthour-

[2]The wattmeter is discussed in Sec. 12-5.

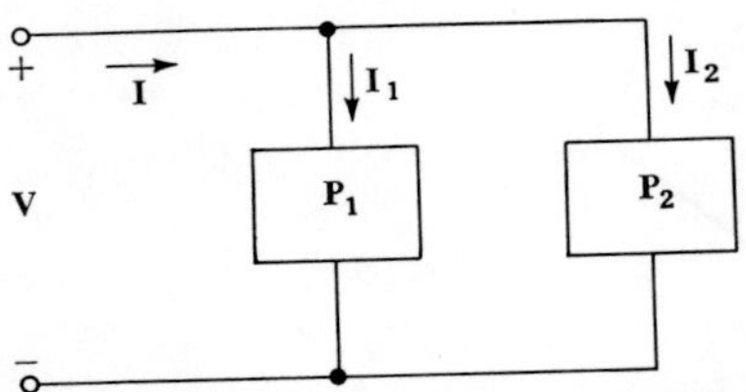

Fig. 11-9 A circuit used to show that the complex power drawn by two parallel loads is the sum of the complex powers drawn by the individual loads.

meters and varhour-meters may be used simultaneously to record real and reactive energy used by any consumer during any desired time interval. From these records the average PF may be determined and the consumer's bill may be adjusted accordingly.

It is easy to show that the complex power delivered to several interconnected loads is the sum of the complex powers delivered to each of the individual loads, no matter how the loads are interconnected. For example, consider the two loads shown connected in parallel in Fig. 11-9. If rms values are assumed, the complex power drawn by the combined loads is

$$\mathbf{P} = \mathbf{V}\mathbf{I}^* = \mathbf{V}(\mathbf{I}_1 + \mathbf{I}_2)^* = \mathbf{V}(\mathbf{I}_1^* + \mathbf{I}_2^*)$$

and thus

$$\mathbf{P} = \mathbf{V}\mathbf{I}_1^* + \mathbf{V}\mathbf{I}_2^*$$

as stated.

These new ideas can be clarified by a practical numerical example. Let us suppose that an industrial consumer is operating a 1-kW induction motor at a lagging PF of 0.8. In order to obtain lower electrical rates, he wishes to raise his PF to 0.95 lagging. Although the PF might be raised by increasing his real power and maintaining the reactive power constant, this would not result in a lower bill; this cure therefore does not interest the consumer. A purely reactive load must be added to the system, and it is clear that it must be added in parallel, since the supply voltage to the induction motor must not change. The circuit of Fig. 11-9 is thus applicable if we interpret $\mathbf{P}_1$ as the induction motor power and $\mathbf{P}_2$ as the complex power drawn by the corrective device. Let us assume a voltage of $200\underline{/0^\circ}$ V rms.

The complex power supplied to the induction motor must have a real part of 1000 W and an angle of $\cos^{-1}(0.8)$. Hence,

$$\mathbf{P}_1 = \frac{1000\underline{/\cos^{-1}(0.8)}}{0.8} = 1000 + j750$$

In order to achieve a PF of 0.95, the total complex power must become

$$\mathbf{P} = \frac{1000}{0.95} \underline{/\cos^{-1}(0.95)} = 1000 + j329$$

Thus, the complex power drawn by the corrective load is

$$\mathbf{P}_2 = -j421$$

The necessary load impedance $\mathbf{Z}_2$ may be found in several simple steps. The current drawn by $\mathbf{Z}_2$ is

$$\mathbf{I}_2^* = \frac{\mathbf{P}_2}{\mathbf{V}} = \frac{-j421}{200} = -j2.105$$

or

$$\mathbf{I}_2 = j2.105$$

and, therefore,

$$\mathbf{Z}_2 = \frac{\mathbf{V}}{\mathbf{I}_2} = \frac{200}{j2.105} = -j95.0 \quad \Omega$$

If the operating frequency is 60 Hz, this load can be provided by a 27.9-μF capacitor. The load may also be simulated by a synchronous capacitor, a type of rotating machine, although this is usually economical only for much smaller capacitive reactances. Whatever device is selected, its initial costs, maintenance, and depreciation must be covered by the reduction in the electric bill.

Drill Problem

11-10 Find the value of an impedance which: (*a*) absorbs a complex power of $4600\underline{/30^\circ}$ VA at 230 V rms; (*b*) absorbs a complex power of $5000\underline{/45^\circ}$ VA when the rms current through it is 12.5 A; (*c*) requires -1500 vars at 230 V rms and 10 A rms.

Ans. $9.95 + j5.75$; $17.4 - j15.0$; $22.6 + j22.6$ Ω

PROBLEMS

☐ **1** For the circuit shown in Fig. 11-10, determine the instantaneous powers delivered to the resistor if $v_s =$: (*a*) $30 \cos 2500t$ V; (*b*) $30u(t)$ V; (*c*) $30 \cos 2500t + 30u(t)$ V.

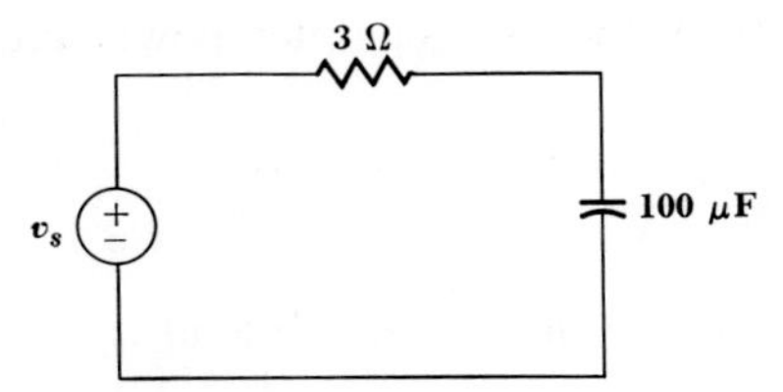

Fig. 11-10 See Prob. 1.

☐ **2** The current through a 50-Ω resistor is given as $i = 3e^{-2t}$ A for $t > 0$ and $i = 0$ for $t \leq 0$. Find the average power delivered to the resistor during the time interval: (*a*) $0 \leq t \leq 0.5$ s; (*b*) $-0.5 \leq t \leq 0.5$ s; (*c*) $0 \leq t \leq 5$ s.

☐ **3** The voltage v is across the series combination of 6 Ω and 20 mH. Find the average power delivered to this impedance if $v =$: (*a*) $10 \cos 400t$ V; (*b*) $10 \sin 400t$ V; (*c*) $10 \sin (400t + 20°)$ V; (*d*) $6 + 10 \cos 400t$ V.

☐ **4** Find R_1 and $|\mathbf{I}_s|$ in the circuit shown in Fig. 11-11 if the average power delivered to each resistor is 1.5 W.

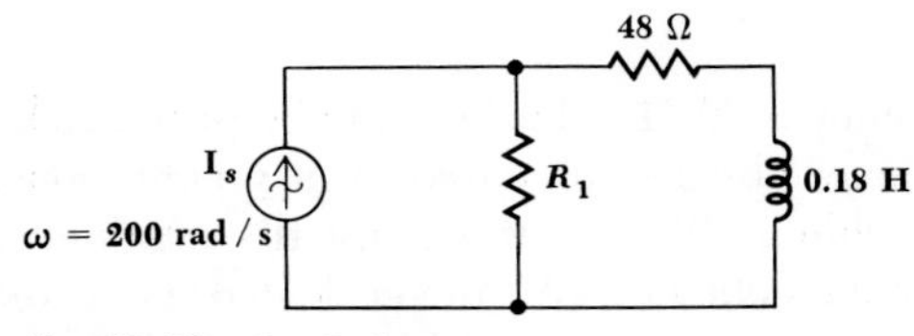

Fig. 11-11 See Prob. 4.

☐ **5** Find the average power delivered to the 3-Ω resistor in the circuit shown in Fig. 11-12.

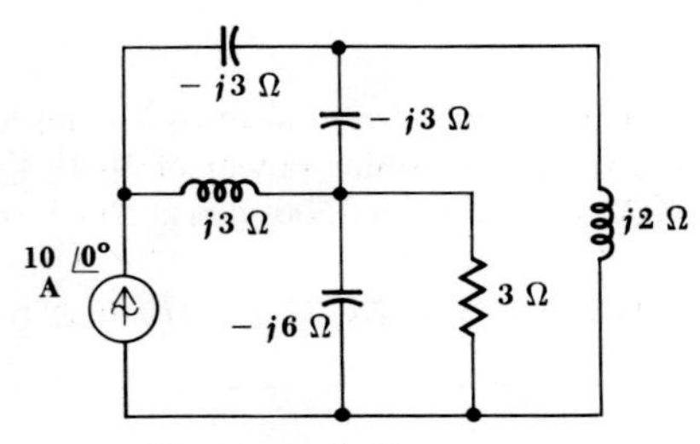

Fig. 11-12 See Prob. 5.

☐ **6** Assume a Thévenin equivalent source composed of $\mathbf{V}_s$ in series with $\mathbf{Z}_{eq} = R_{eq} + jX_{eq}$. (*a*) Prove that maximum average power is delivered to a load $\mathbf{Z}_L = R_L + jX_L$ when $R_L = R_{eq}$ and $X_L = -X_{eq}$. (*b*) Prove that maximum average power is delivered to a load R_L when $R_L = |\mathbf{Z}_{eq}|$.

☐ **7** The Thévenin equivalent of a sinusoidal voltage source is the ideal voltage source $v_s = V_m \cos \omega t$ in series with a resistance R_{eq}. (*a*) What is the maximum power that can be delivered to an external load resistance? (*b*) If the source has an internal impedance $\mathbf{Z}_{eq} = R_{eq} + jX_{eq}$, what is the maximum power that can be delivered to an external load impedance?

☐ **8** An ideal voltage source, $100\underline{/0°}$ V, is in series with a 200-Ω resistor and a capacitive reactance of -100 Ω. (*a*) What impedance $\mathbf{Z}_L$ in parallel with the capacitance will absorb maximum average power, and what is this maximum power? (*b*) What resistive load R_L in parallel with the capacitance will absorb maximum average power, and what is this maximum power?

☐ **9** Find the effective value of each of the following voltages: (*a*) $20 \cos 800t$ V; (*b*) $20 \cos (800t + 20°)$ V; (*c*) $12 \cos (800t + 20°) + 16 \cos (800t - 70°)$ V; (*d*) $4 \cos (800t + 20°) + 12\sqrt{2} \cos (800t + 65°)$; (*e*) $15 \cos 800t + 10 \sin 801t - 5\sqrt{3} \sin 802t$ V.

☐ **10** Find the effective value of the periodic voltage v if: $v = 2,\ 0 < t < 4$; $v = t - 2,\ 4 \le t < 10$; $T = 10$.

☐ **11** A current waveform is I_0 for a time t_1 and then kI_0 for a like time. If the period is $2t_1$, determine k such that the effective value is $2I_0$.

☐ **12** Let $i = 4 - 8[u(t-2) - u(t-3)] + tu(t-2)$ for a period interval $0 < t \le 5$. For this current, determine: (*a*) the positive and negative peak values; (*b*) the average value; (*c*) the effective value.

☐ **13** A resistance R and reactance X in series are connected to a 115-V 60-Hz voltage supply. Good laboratory instruments are used to show that the reactor voltage is 75 V and the total power supplied to the circuit is 190 W. Find R, X, and the rms current.

☐ **14** With reference to the circuit of Fig. 11-13, what should be the value of R_L to absorb a maximum power? Find the value of this maximum power.

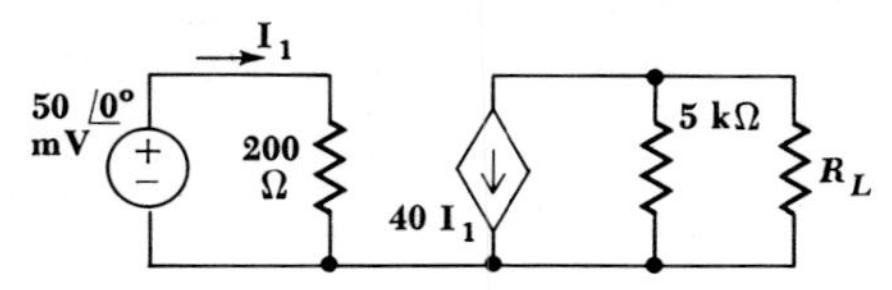

Fig. 11-13 See Prob. 14.

☐ **15** In the circuit shown in Fig. 11-14, $I_L = 12$ A rms and $I_s = 15$ A rms. Find the average power delivered to the resistor without knowing the value of ω.

☐ **16** The PF of the total load supplied by the current source of Fig. 11-14 is 0.5 lagging. What is the operating frequency in radians per second?

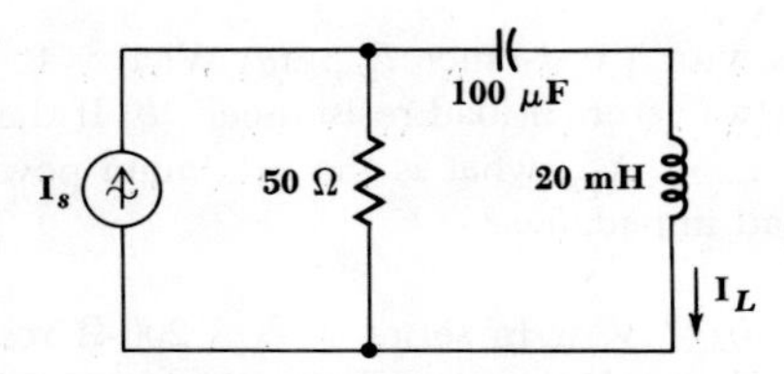

Fig. 11-14 See Probs. 15, 16, and 17.

☐ **17** The operating frequency of the circuit shown in Fig. 11-14 is 1000 rad/s. Find the real, reactive, and apparent power supplied by the source if $\mathbf{I}_s = 2.6\underline{/12^\circ}$ A rms.

☐ **18** A source, 120 V rms at 60 Hz, supplies 2400 VA to a load operating at a 0.707 lagging PF. Find the value of capacitance and its necessary kilovoltampere rating such that the PF will be raised to 0.95 lagging when it is placed in parallel with the load.

☐ **19** Three parallel loads drawing 3 kW, 5kVA at 0.866 PF lagging and 4 kVA at 0.707 PF lagging are connected to a 230-V 60 Hz source. How many kilovolt-amperes are required by the total load?

☐ **20** A voltage source, $40\underline{/20^\circ}$ V, is in series with an impedance $\mathbf{Z}_1$ and the parallel combination of impedances $\mathbf{Z}_2$ and $\mathbf{Z}_3$. Find the complex power delivered to every impedance if: (*a*) $\mathbf{Z}_1 = j10\ \Omega$, $\mathbf{Z}_2 = 6\ \Omega$, $\mathbf{Z}_3 = -j8\ \Omega$; (*b*) $\mathbf{Z}_1 = 3 + j4\ \Omega$, $\mathbf{Z}_2 = 2 - j4\ \Omega$, $\mathbf{Z}_3 = 1 + j8\ \Omega$.

☐ **21** The voltage across a device is labeled **V** and the current entering the plus-marked terminal is **I**. The complex power delivered to the device is **P**, and $\omega = 10^3$ rad/s. (*a*) If $\mathbf{V} = 2 - j6$ V and $\mathbf{P} = 20 - j10$ VA, determine two passive series elements that are equivalent to the device. (*b*) If $\mathbf{I} = 3 + j3$ A and $\mathbf{P} = 20 + j40$ VA, determine two passive parallel elements that are equivalent to the device.

☐ **22** Three passive loads, $\mathbf{Z}_1$, $\mathbf{Z}_2$, and $\mathbf{Z}_3$, are receiving the complex power values, $2 + j3$, $3 - j1$, and $1 + j2$ VA, respectively. What total complex power is received if the three loads are: (*a*) in series with a voltage source $100\underline{/30^\circ}$ V; (*b*) in parallel with a current source $1200\underline{/-17^\circ}$ A?

Chapter Twelve Polyphase Circuits

12-1 INTRODUCTION

One of the reasons for studying the sinusoidal steady state is that most household and industrial electric power is utilized as alternating current. The sinusoidal waveform may characterize a special mathematical function, but it represents a very common and very useful forcing function. A *polyphase* source is even more specialized, but we again consider it because almost the entire output of the electric power industry in this country is generated and distributed as polyphase power at a 60-Hz frequency. Before defining our terms carefully, let us look briefly at the most common polyphase system, a balanced three-phase system. The source has perhaps three terminals, and voltmeter measurements will show that sinusoidal voltages of equal magnitude are present between any two terminals. However, these voltages are not in phase; it will be easily shown later that each of the three voltages is 120° out of phase with each of the other two, the sign of the phase angle depending on the sense of the voltages. A balanced load draws power equally from the three phases, but when one of the voltages is instantaneously zero, the phase relationship shows that the other two must each be at half amplitude. At no instant does the instantaneous power drawn by the total load reach zero; as a matter of fact, this total instantaneous power is constant. This is an advantage in rotating machinery, for it keeps the torque much more constant than it would be if a single-phase source were used. There is less vibration.

There are also advantages in using rotating machinery to generate three-phase power rather than single-phase power, and there are economical advantages in favor of the transmission of power in a three-phase system.

The use of a higher number of phases, such as 6- and 12-phase systems, is limited almost entirely to the supply of power to large rectifiers. Here, the rectifiers convert the alternating current to direct current, which is required for certain processes such as electrolysis. The rectifier output is a direct current plus a smaller pulsating component, or ripple, which decreases as the number of phases increases.

Almost without exception, polyphase systems in practice contain sources which may be closely approximated by ideal voltage sources or by ideal voltage sources in series with small internal impedances. Three-phase current sources are extremely rare.

It is convenient to describe polyphase voltages and currents using a double-subscript notation. With this notation, a voltage or current, such as $\mathbf{V}_{ab}$ or $\mathbf{I}_{aA}$, has more meaning than if it were indicated simply as $\mathbf{V}_3$ or $\mathbf{I}_x$. By definition, let the voltage of point a with respect to point b be $\mathbf{V}_{ab}$. Thus, the plus sign is located at a, as indicated in Fig. 12-1a. With reference to Fig. 12-1b, it is now obvious that $\mathbf{V}_{ad} = \mathbf{V}_{ab} + \mathbf{V}_{cd}$. The power of the double-subscript notation lies in the fact that Kirchhoff's voltage

Fig. 12-1 (a) The definition of the voltage $\mathbf{V}_{ab}$. (b) $\mathbf{V}_{ad} = \mathbf{V}_{ab} + \mathbf{V}_{bc} + \mathbf{V}_{cd} = \mathbf{V}_{ab} + \mathbf{V}_{cd}$.

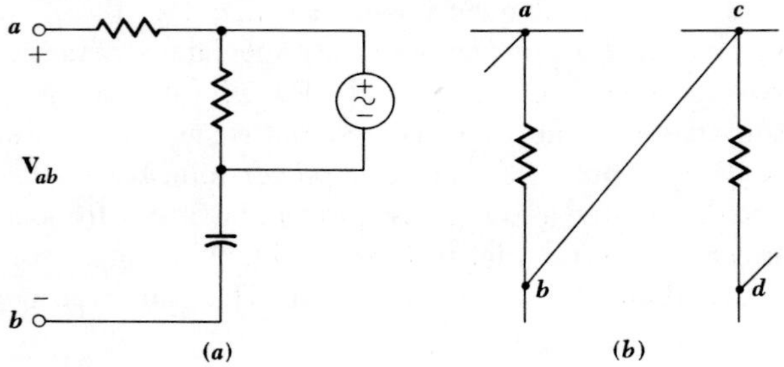

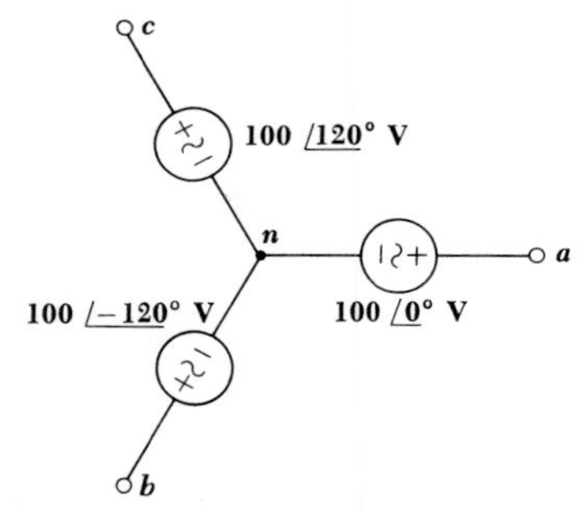

Fig. 12-2 A network used as a numerical example of double-subscript voltage notation.

law requires the voltage between two points to be the same, regardless of the path chosen between the points, and thus $\mathbf{V}_{ad} = \mathbf{V}_{ab} + \mathbf{V}_{bd} = \mathbf{V}_{ac} + \mathbf{V}_{cd} = \mathbf{V}_{ab} + \mathbf{V}_{bc} + \mathbf{V}_{cd}$, etc. It is apparent that Kirchhoff's voltage law may be satisfied without reference to the circuit diagram; correct equations may be written even though a point, or subscript letter, is included which is not marked on the diagram. For example, we might also have written, above, $\mathbf{V}_{ad} = \mathbf{V}_{ax} + \mathbf{V}_{xd}$.

One possible representation of a three-phase system of voltages[1] will be found to be that of Fig. 12-2. Let us assume that the voltages $\mathbf{V}_{an}$, $\mathbf{V}_{bn}$, and $\mathbf{V}_{cn}$ are known,

$$\begin{aligned} \mathbf{V}_{an} &= 100\underline{/0^\circ} \quad \text{V rms} \\ \mathbf{V}_{bn} &= 100\underline{/-120^\circ} \\ \mathbf{V}_{cn} &= 100\underline{/-240^\circ} \end{aligned}$$

and thus the voltage $\mathbf{V}_{ab}$ may be found, with an eye on the subscripts,

$$\begin{aligned} \mathbf{V}_{ab} = \mathbf{V}_{an} + \mathbf{V}_{nb} &= \mathbf{V}_{an} - \mathbf{V}_{bn} \\ &= 100\underline{/0^\circ} - 100\underline{/-120^\circ} \\ &= 100 - (-50 - j86.6) \\ &= 173.2\underline{/30^\circ} \end{aligned}$$

The three given voltages and the construction of the phasor $\mathbf{V}_{ab}$ are shown on the phasor diagram of Fig. 12-3.

A double-subscript notation may also be applied to currents. We define the current $\mathbf{I}_{ab}$ as the current flowing from a to b *by the direct path.* In every complete circuit we consider, there must of course be at least two possible paths between the points a and b, and we agree that we shall not use double-subscript notation unless it is obvious that one path is much

[1]Rms values of currents and voltages will be used throughout this chapter.

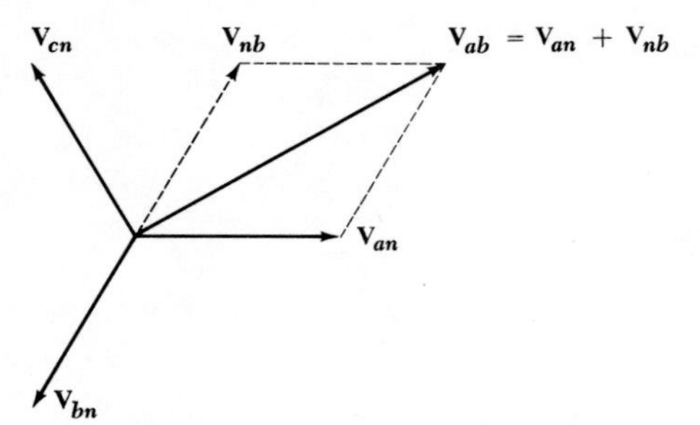

Fig. 12-3 This phasor diagram illustrates the graphical use of the double-subscript voltage convention to obtain $\mathbf{V}_{ab}$ for the network of Fig. 12-2.

shorter, or much more direct. Usually this path is through a single element. Thus, the current $\mathbf{I}_{ab}$ is correctly indicated in Fig. 12-4, but the mere identification of a current as $\mathbf{I}_{cd}$ would cause confusion.

Before considering polyphase systems, we shall make use of double-subscript notation to help with the analysis of a special single-phase system.

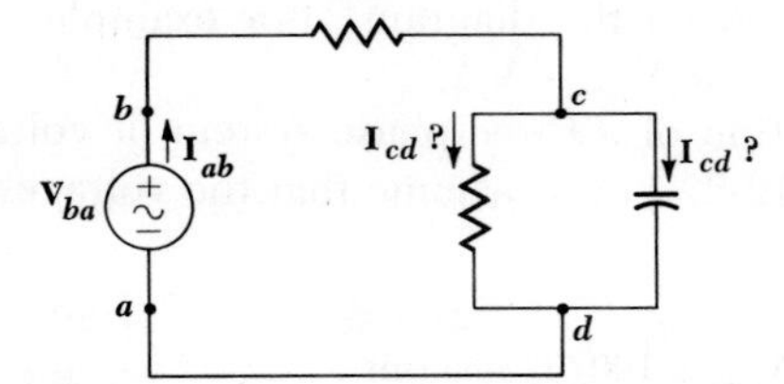

Fig. 12-4 An illustration of the use and misuse of the double-subscript convention for current notation.

Drill Problems

12-1 Given that $\mathbf{V}_{12} = 6\underline{/30^\circ}$, $\mathbf{V}_{23} = 2\underline{/0^\circ}$, $\mathbf{V}_{24} = 6\underline{/-60^\circ}$, and $\mathbf{V}_{15} = 4\underline{/90^\circ}$, find: (*a*) $\mathbf{V}_{14}$; (*b*) $\mathbf{V}_{43}$; (*c*) $\mathbf{V}_{35}$.

Ans. $5.29\underline{/100.9^\circ}$; $7.26\underline{/172.1^\circ}$; $8.48\underline{/-15^\circ}$

Fig. 12-5 See Drill Prob. 12-2.

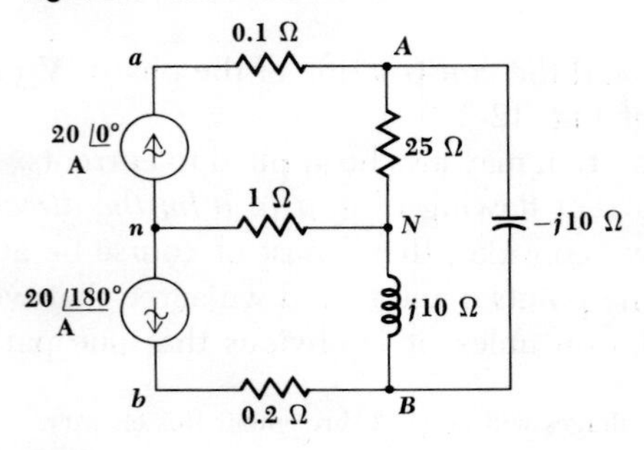

12-2 In the circuit shown in Fig. 12-5, determine: (*a*) $\mathbf{I}_{nN}$; (*b*) $\mathbf{I}_{AN}$; (*c*) $\mathbf{I}_{AB}$.
Ans. 0; $8\underline{/-90°}$; $21.5\underline{/21.8°}$ A

Fig. 12-6 (*a*) A single-phase three-wire source. (*b*) The representation of a single-phase three-wire source by two identical voltage sources.

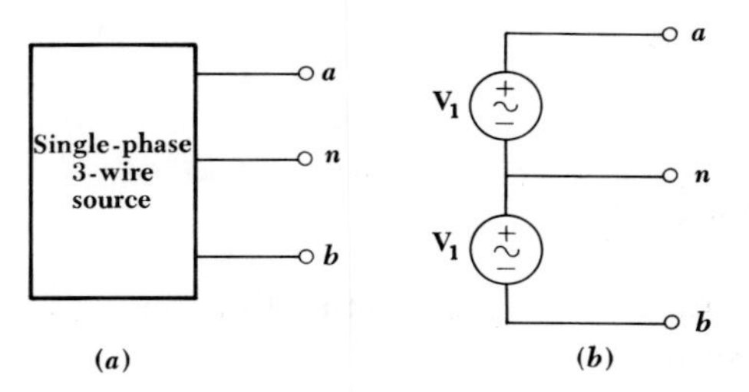

12-2 SINGLE-PHASE THREE-WIRE SYSTEMS

A single-phase three-wire source is defined as a source having three output terminals, such as *a*, *n*, and *b* in Fig. 12-6*a*, at which the phasor voltages $\mathbf{V}_{an}$ and $\mathbf{V}_{nb}$ are equal. The source may therefore be represented by the combination of two identical voltage sources; in Fig. 12-6*b*, $\mathbf{V}_{an} = \mathbf{V}_{nb} = \mathbf{V}_1$. It is apparent that $\mathbf{V}_{ab} = 2\mathbf{V}_{an} = 2\mathbf{V}_{nb}$, and we therefore have a source to which loads operating at either of two voltages may be connected. The normal household system is single-phase three-wire, permitting the operation of both 115-V and 230-V appliances. The higher-voltage appliances are normally those drawing a larger power, and thus they cause a current in the lines which is only half that which operation at the same power and half the voltage would produce. Smaller-diameter wire may consequently be used safely in the appliance, the household distribution system, and the distribution system of the utility company.

The name single phase arises because the voltages $\mathbf{V}_{an}$ and $\mathbf{V}_{nb}$, being equal, must have the same phase angle. From another viewpoint, however, the voltages between the outer wires and the central wire, which is usually referred to as the *neutral*, are exactly 180° out of phase. That is, $\mathbf{V}_{an} = -\mathbf{V}_{bn}$ and $\mathbf{V}_{an} + \mathbf{V}_{bn} = 0$. In a following section we shall see that balanced polyphase systems are characterized by possessing a set of voltages of equal *magnitude* whose (phasor) sum is zero. From this viewpoint, then, the single-phase three-wire system is really a balanced two-phase system. "Two phase," however, is a term that is traditionally reserved for a relatively unimportant system utilizing two voltage sources 90° out of phase; we shall not discuss it further.

Let us now consider a single-phase three-wire system which contains identical loads $\mathbf{Z}_p$ between each outer wire and the neutral (Fig. 12-7).

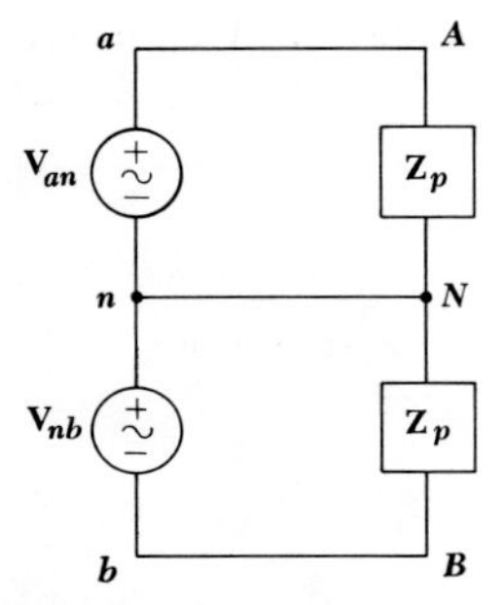

Fig. 12-7 A simple single-phase three-wire system. The two loads are identical and the neutral current is zero.

We shall first assume that the wires connecting the source to the load are perfect conductors. Since

$$\mathbf{V}_{an} = \mathbf{V}_{nb}$$

then,

$$\mathbf{I}_{aA} = \frac{\mathbf{V}_{an}}{\mathbf{Z}_p} = \mathbf{I}_{Bb} = \frac{\mathbf{V}_{nb}}{\mathbf{Z}_p}$$

and, therefore,

$$\mathbf{I}_{nN} = \mathbf{I}_{Bb} + \mathbf{I}_{Aa} = \mathbf{I}_{Bb} - \mathbf{I}_{aA} = 0$$

Thus there is no current in the neutral wire, and it could be removed without changing any current or voltage in the system. This result is achieved through the equality of the two loads and of the two sources.

We next consider the effect of a finite impedance in each of the wires. If lines aA and bB each have the same impedance, this impedance may be added to $\mathbf{Z}_p$, resulting in two equal loads once more and zero neutral current. Now let us allow the neutral wire to possess some impedance $\mathbf{Z}_n$. Without carrying out any detailed analysis, superposition should show us that the symmetry of the circuit will still cause zero neutral current. Moreover, the addition of any impedance connected directly from one of the outer lines to the other outer line also yields a symmetrical circuit and zero neutral current. Thus, zero neutral current is a consequence of a balanced, or symmetrical, load; any impedance in the neutral wire does not destroy the symmetry.

The most general single-phase three-wire system will contain unequal loads between each outside line and the neutral and another load directly between the two outer lines; the impedances of the two outer lines may be expected to be approximately equal, but the neutral impedance may be slightly larger. An example of such a system is shown in Fig. 12-8. The analysis of the circuit may be achieved by assigning mesh currents

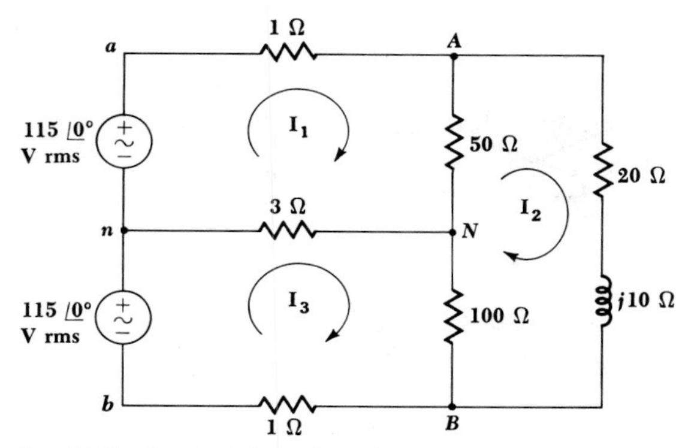

Fig. 12-8 A typical single-phase three-wire system.

and writing the appropriate equations. The results of this labor are

$$\mathbf{I}_1 = 11.2\underline{/-19.8^\circ} \quad \text{A rms}$$
$$\mathbf{I}_2 = 9.38\underline{/-24.5^\circ}$$
$$\mathbf{I}_3 = 10.4\underline{/-21.8^\circ}$$

The currents in the outer lines are thus

$$\mathbf{I}_{aA} = \mathbf{I}_1 = 11.2\underline{/-19.8^\circ} \quad \text{A rms}$$
$$\mathbf{I}_{bB} = -\mathbf{I}_3 = 10.4\underline{/158.2^\circ}$$

and the smaller neutral current is

$$\mathbf{I}_{nN} = \mathbf{I}_3 - \mathbf{I}_1 = 0.95\underline{/-177.9^\circ} \quad \text{A rms}$$

The power drawn by each load may be determined,

$$P_{50} = |\mathbf{I}_1 - \mathbf{I}_2|^2(50) = 207 \quad \text{W}$$

which could represent two 100-W lamps in parallel,

$$P_{100} = |\mathbf{I}_3 - \mathbf{I}_2|^2(100) = 117 \quad \text{W}$$

which might represent one 100-W lamp,

$$P_{20+j10} = |\mathbf{I}_2|^2(20) = 1763 \quad \text{W}$$

which we may think of as a 2-hp induction motor. The total load power is 2087 W. The loss in each of the wires is next found:

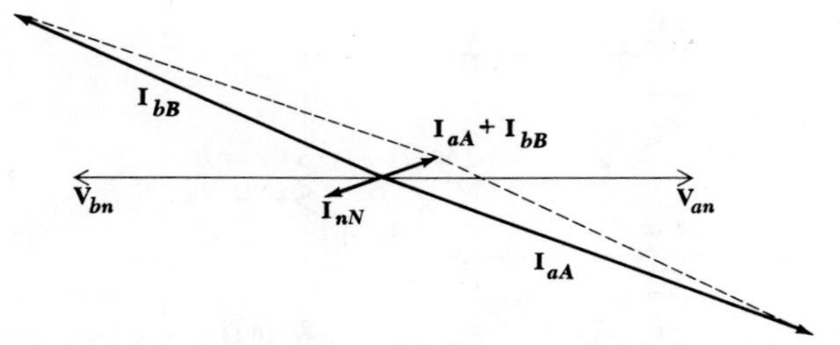

Fig. 12-9 The source voltages and three of the currents in the circuit of Fig. 12-8 are shown on a phasor diagram. Note that $\mathbf{I}_{aA} + \mathbf{I}_{bB} + \mathbf{I}_{nN} = 0$.

$$P_{aA} = |\mathbf{I}_1|^2(1) = 126 \quad \text{W}$$
$$P_{bB} = |\mathbf{I}_3|^2(1) = 108 \quad \text{W}$$
$$P_{nN} = |\mathbf{I}_{nN}|^2(3) = 2.7 \quad \text{W}$$

or a total line loss of 237 W. The wires are evidently quite long; otherwise, the relatively high power loss in the two outer lines would cause a dangerous temperature rise. The total generated power must therefore be 2324 W, and this may be checked by finding the power delivered by each voltage source:

$$P_{an} = 115(11.2) \cos 19.8° = 1218 \quad \text{W}$$
$$P_{bn} = 115(10.4) \cos 21.8° = 1108 \quad \text{W}$$

or a total of 2326 W. The transmission efficiency for this system is

$$\text{Eff.} = \frac{2087}{2087 + 237} = 89.8\%$$

This value would be unbelievable for a steam engine or an internal-combustion engine, but it is too low for a well-designed distribution system. Larger-diameter wires should be used if the source and the load cannot be placed closer to each other.

A phasor diagram showing the two source voltages, the currents in the outer lines, and the current in the neutral is constructed in Fig. 12-9. The fact that $\mathbf{I}_{aA} + \mathbf{I}_{bB} + \mathbf{I}_{nN} = 0$ is indicated on the diagram.

Drill Problem

12-3 Replace the 50-Ω load by a 10-Ω load in the circuit shown in Fig. 12-8, and then find the average power delivered to the: (*a*) 10-Ω load; (*b*) 100-Ω load; (*c*) 20 + *j*10-Ω load.

Ans. *157; 623; 1670* W

12-3 THREE-PHASE Y-Y CONNECTION

Three-phase sources have three terminals, called the *line* terminals, and they may or may not have a fourth terminal, the *neutral* connection. We shall begin by discussing a three-phase source which does have a neutral

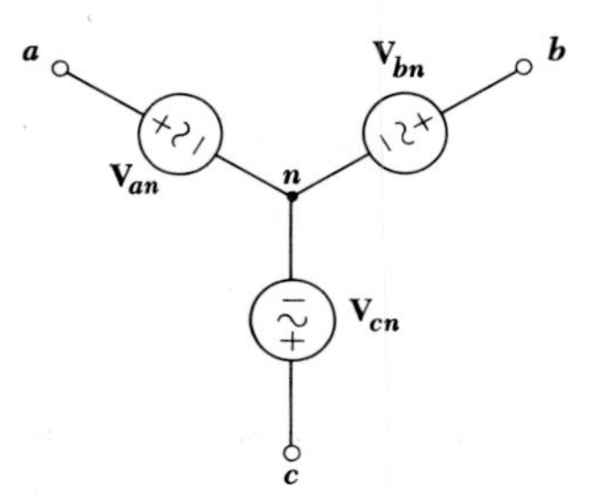

Fig. 12-10 A Y-connected three-phase source.

connection. It may be represented by three ideal voltage sources connected in a **Y**, as shown in Fig. 12-10; terminals *a*, *b*, *c*, and *n* are available. We shall consider only balanced three-phase sources which may be defined as having

$$|\mathbf{V}_{an}| = |\mathbf{V}_{bn}| = |\mathbf{V}_{cn}|$$

and

$$\mathbf{V}_{an} + \mathbf{V}_{bn} + \mathbf{V}_{cn} = 0$$

These three voltages, each existing between one line and the neutral, are called *phase voltages*. If we arbitrarily choose $\mathbf{V}_{an}$ as the reference,

$$\mathbf{V}_{an} = V_p\underline{/0^\circ}$$

where we shall consistently use V_p to represent the rms *magnitude* of any of the phase voltages, then the definition of the three-phase source indicates that either

$$\mathbf{V}_{bn} = V_p\underline{/-120^\circ} \qquad \mathbf{V}_{cn} = V_p\underline{/-240^\circ}$$

or

$$\mathbf{V}_{bn} = V_p\underline{/120^\circ} \qquad \mathbf{V}_{cn} = V_p\underline{/240^\circ}$$

The former is called *positive phase sequence,* or *abc* phase sequence, and is shown in Fig. 12-11*a*; the latter is termed *negative phase sequence,* or *cba* phase sequence, and is indicated by the phasor diagram of Fig. 12-11*b*. It is apparent that the phase sequence of a physical three-phase source depends on the arbitrary choice of the three terminals to be lettered *a*, *b*, and *c*. They may always be chosen to provide positive phase sequence,

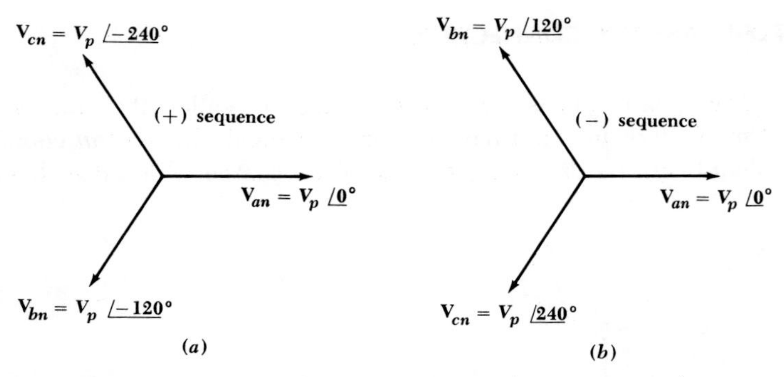

Fig. 12-11 (a) Positive, or abc, phase sequence. (b) Negative, or cba, phase sequence.

and we shall assume that this has been done in most of the systems we consider.

Let us next find the line-to-line voltages (or simply "line" voltages) which are present when the phase voltages are those of Fig. 12-11a. It is easiest to do this with the help of a phasor diagram, since the angles are all multiples of 30°. The necessary construction is shown in Fig. 12-12; the results are

$$\mathbf{V}_{ab} = \sqrt{3}\ V_p \underline{/30°} \qquad \mathbf{V}_{bc} = \sqrt{3}\ V_p \underline{/-90°} \qquad \mathbf{V}_{ca} = \sqrt{3}\ V_p \underline{/-210°}$$

Kirchhoff's voltage law requires the sum of these three voltages to be zero, and it is zero.

Denoting the magnitude of any of the line voltages by V_L, then one of the important characteristics of the **Y**-connected three-phase source may be expressed as

$$V_L = \sqrt{3}\ V_p$$

Fig. 12-12 A phasor diagram which is used to determine the line voltages from the given phase voltages.

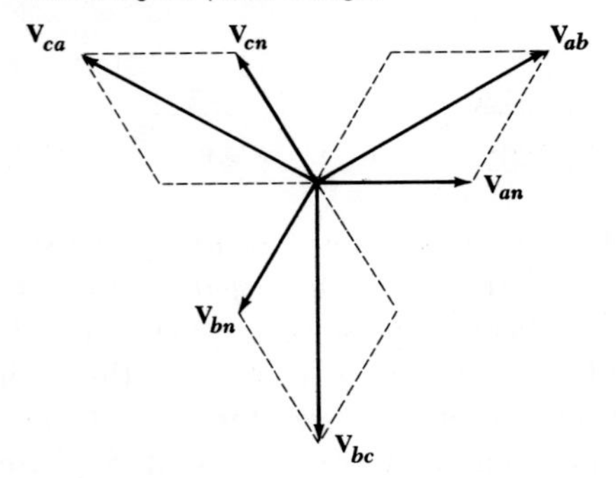

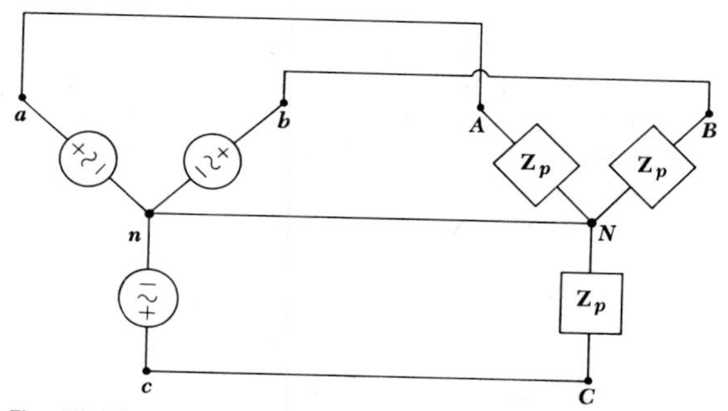

Fig. 12-13 A balanced three-phase system, connected Y-Y, and including a neutral.

Note that, with positive phase sequence, $\mathbf{V}_{an}$ leads $\mathbf{V}_{bn}$ and $\mathbf{V}_{bn}$ leads $\mathbf{V}_{cn}$, in each case by 120°, and also that $\mathbf{V}_{ab}$ leads $\mathbf{V}_{bc}$ and $\mathbf{V}_{bc}$ leads $\mathbf{V}_{ca}$, again by 120°. The statement is true for negative sequence if "lags" is substituted for "leads."

Now let us connect a balanced **Y**-connected three-phase load to our source, using three lines and a neutral, as drawn in Fig. 12-13. The load is represented by an impedance $\mathbf{Z}_p$ between each line and the neutral. The three line currents are found very easily, since we really have three single-phase circuits which possess one common lead:

$$\mathbf{I}_{aA} = \frac{\mathbf{V}_{an}}{\mathbf{Z}_p}$$

$$\mathbf{I}_{bB} = \frac{\mathbf{V}_{bn}}{\mathbf{Z}_p} = \frac{\mathbf{V}_{an}\underline{/-120^\circ}}{\mathbf{Z}_p} = \mathbf{I}_{aA}\underline{/-120^\circ}$$

$$\mathbf{I}_{cC} = \mathbf{I}_{aA}\underline{/-240^\circ}$$

and thus

$$\mathbf{I}_{Nn} = \mathbf{I}_{aA} + \mathbf{I}_{bB} + \mathbf{I}_{cC} = 0$$

Thus, the neutral carries no current if the source and load are both balanced and if the four wires have zero impedance. How will this change if an impedance $\mathbf{Z}_L$ is inserted in series with each of the three lines and an impedance $\mathbf{Z}_n$ is inserted in the neutral? Evidently, the line impedances may be combined with the three load impedances; this effective load is still balanced, and a perfectly conducting neutral wire could be removed. Thus, if no change is produced in the system with a short circuit or an open circuit between n and N, any impedance may be inserted in the neutral and the neutral current will remain zero.

It follows that, if we have balanced sources, balanced loads, and balanced line impedances, a neutral wire of any impedance may be replaced by any other impedance, including a short circuit and an open circuit.

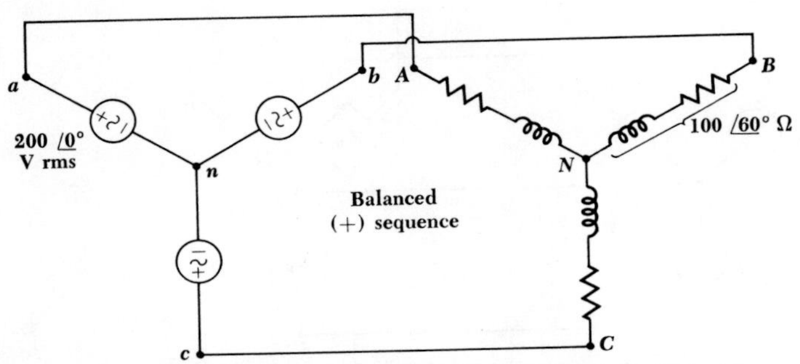

Fig. 12-14 A balanced three-phase three-wire Y-Y connected system.

It is often helpful to visualize a short circuit between the two neutral points; the problem is then reduced to three single-phase problems, all identical. We say that we thus work the problem on a "per-phase" basis.

Let us work several problems involving a balanced three-phase system having a Y-Y connection. A straightforward problem is suggested by the circuit of Fig. 12-14; we are asked to find the several currents and voltages throughout the circuit and to find the total power.

Since one of the source phase voltages is given, and since positive phase sequence is assumed, the three phase voltages are

$$\mathbf{V}_{an} = 200\underline{/0^\circ} \qquad \mathbf{V}_{bn} = 200\underline{/-120^\circ} \qquad \mathbf{V}_{cn} = 200\underline{/-240^\circ}$$

The line voltage is $200\sqrt{3}$, or 346 V rms; the phase angle of each line voltage can be determined by constructing a phasor diagram, as before. As a matter of fact, the phasor diagram of Fig. 12-12 is applicable, and $\mathbf{V}_{ab}$ is $346\underline{/30^\circ}$ V.

Let us work with phase A. The line current is

$$\mathbf{I}_{aA} = \frac{\mathbf{V}_{an}}{\mathbf{Z}_p} = \frac{200\underline{/0^\circ}}{100\underline{/60^\circ}} = 2\underline{/-60^\circ} \quad \text{A rms}$$

and the power absorbed by this phase is, therefore,

$$P_{AN} = 200(2)\cos(0^\circ + 60^\circ) = 200 \quad \text{W}$$

Thus, the total power drawn by the three-phase load is 600 W. The problem is completed by drawing a phasor diagram and reading from it the appropriate phase angles which apply to the other line voltages and currents. The completed diagram is shown in Fig. 12-15.

We may also use per-phase methods to work problems in what might be called the backward direction. Suppose that we have a balanced

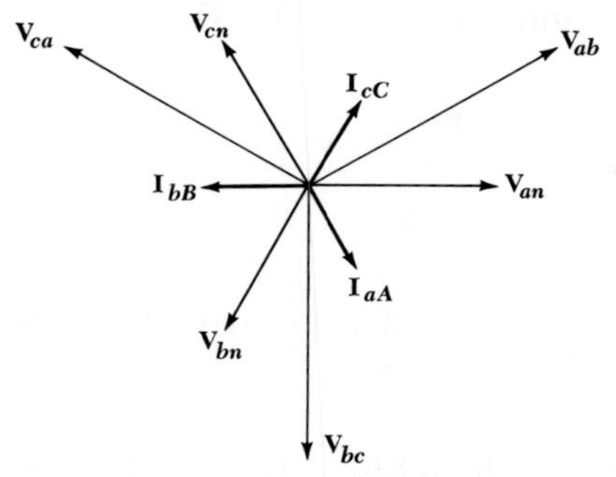

Fig. 12-15 The phasor diagram that is applicable to the circuit of Fig. 12-14.

three-phase system with a line voltage of 300 V rms, and we know that it is supplying a balanced Y-connected load with 1200 W at a leading PF of 0.8. What is the line current and the per-phase load impedance? It is evident that the phase voltage is $300/\sqrt{3}$ V rms and the per-phase power is 400 W. Thus the line current may be found from the power relationship

$$400 = \frac{300}{\sqrt{3}}(I_L)(0.8)$$

and the line current is therefore 2.89 A rms. The phase impedance is given by

$$|\mathbf{Z}_p| = \frac{V_p}{I_L} = \frac{300/\sqrt{3}}{2.89} = 60 \quad \Omega$$

Since the PF is 0.8, leading, the impedance phase angle is $-36.8°$, and $\mathbf{Z}_p = 60\underline{/-36.8°}\ \Omega$.

More complicated loads can be easily handled, since the problems reduce to simpler single-phase problems. Suppose that a balanced 600-W lighting load is added (in parallel) to the system above. A suitable per-phase circuit is first sketched, as shown in Fig. 12-16.

Fig. 12-16 The per-phase circuit that is used to analyze a balanced three-phase example.

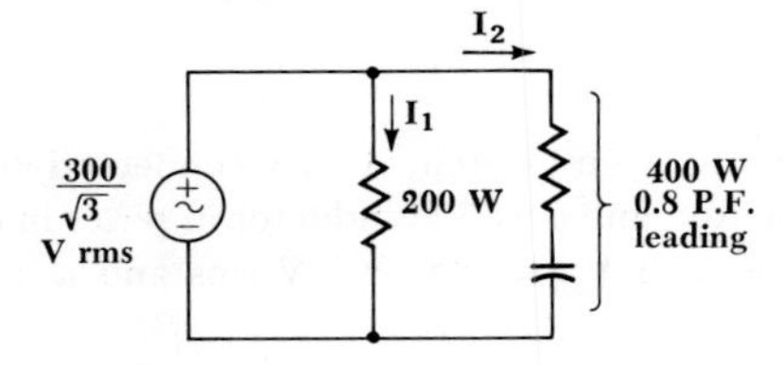

The magnitude of the lighting current is determined by

$$200 = \frac{300}{\sqrt{3}} |\mathbf{I}_1| \cos 0°$$

and

$$|\mathbf{I}_1| = 1.155$$

In a similar way, the magnitude of the capacitive load current is found to be unchanged from its previous value,

$$|\mathbf{I}_2| = 2.89$$

If we assume that the phase with which we are working has a phase voltage with an angle of 0°, then

$$\mathbf{I}_1 = 1.155\underline{/0°} \qquad \mathbf{I}_2 = 2.89\underline{/+36.8°}$$

and the line current is

$$\mathbf{I}_L = \mathbf{I}_1 + \mathbf{I}_2 = 3.87\underline{/+26.6°} \quad \text{A rms}$$

The power generated by this phase of the source is, therefore,

$$P_p = \frac{300}{\sqrt{3}} 3.87 \cos(+26.6°) = 600 \quad \text{W}$$

which checks with the original hypothesis.

If an *unbalanced* Y-connected load is present in an otherwise balanced three-phase system, the circuit may still be analyzed on a per-phase basis *if* the neutral wire is present and *if* it has zero impedance. If either of these conditions is not met, other methods must be used. An engineer who spends most of his time with unbalanced three-phase systems will find the use of *symmetrical components* a great timesaver. We shall not discuss this method here.

Drill Problems

12-4 A balanced three-phase three-wire system has a Y-connected load containing a 50-Ω resistor, a 5-μF capacitor, and a 0.56-H inductor in series in each phase. Using positive phase sequence with $\mathbf{V}_{an} = 390\underline{/30°}$ V rms and $\omega = 500$ rad/s, find: (*a*) $\mathbf{V}_{cn}$; (*b*) $\mathbf{V}_{bc}$; (*c*) $\mathbf{V}_{ac}$.

Ans. $390\underline{/150°}$; $675\underline{/0°}$; $675\underline{/-60°}$ *V rms*

12-5 For the circuit described in Drill Prob. 12-4 above, find: (*a*) $\mathbf{I}_{aA}$; (*b*) $\mathbf{I}_{bB}$; (*c*) $\mathbf{I}_{Cc}$.

Ans. $3\underline{/-22.6°}$; $3\underline{/37.4°}$; $3\underline{/97.4°}$ A *rms*

12-6 A 440-V rms (line voltage) three-phase three-wire system feeds two balanced Y-connected loads. One load is an induction motor which may be represented by an impedance of $10 + j5\ \Omega$ per phase. The other is a lighting load equivalent to $15\ \Omega$ per phase. Find the average power: (*a*) delivered to the lighting load; (*b*) delivered to the induction motor; (*c*) provided by one phase of the source.

Ans. 9.46; 12.9; 15.5 kW

12-7 A balanced three-phase three-wire system supplies two balanced Y-connected loads. The first draws 6 kW at 0.8 PF lagging while the other requires 12 kW at 0.833 PF leading. If the current in each line is 8 A rms, find the current in the: (*a*) first load; (*b*) second load; (*c*) source phase.

Ans. 3.28; 6.29; 8 A *rms*

12-4 THE DELTA (Δ) CONNECTION

A three-phase load is more apt to be found Δ-connected than Y-connected. One reason for this, at least for the case of an unbalanced load, is the flexibility with which loads may be added or removed on a single phase. This is difficult (or impossible) to do with a Y-connected three-wire load.

Let us consider a balanced Δ-connected load which consists of an impedance $\mathbf{Z}_p$ inserted between each pair of lines. We shall assume a three-wire system for obvious reasons. With reference to Fig. 12-17, let us assume known line voltages

$$V_L = |\mathbf{V}_{ab}| = |\mathbf{V}_{bc}| = |\mathbf{V}_{ca}|$$

or known phase voltages

$$V_p = |\mathbf{V}_{an}| = |\mathbf{V}_{bn}| = |\mathbf{V}_{cn}|$$

Fig. 12-17 A balanced Δ-connected load is present on a three-wire three-phase system. The source happens to be Y-connected.

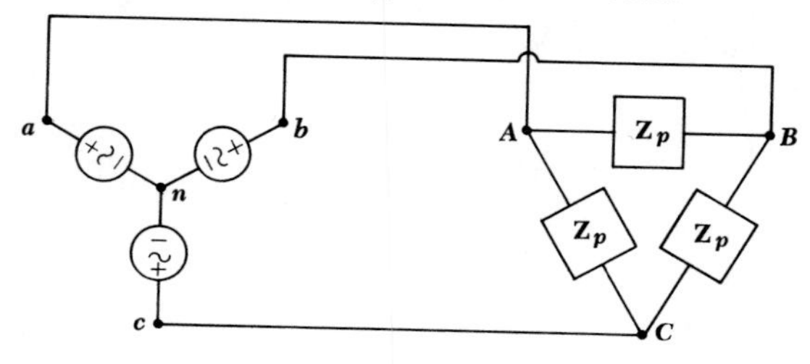

where

$$V_L = \sqrt{3}\ V_p \qquad \text{and} \qquad \mathbf{V}_{ab} = \sqrt{3}\ \mathbf{V}_{an}\underline{/30^\circ}$$

and so forth, as before. Since the voltage across each branch of the Δ is known, the *phase currents* are found,

$$\mathbf{I}_{AB} = \frac{\mathbf{V}_{ab}}{\mathbf{Z}_p} \qquad \mathbf{I}_{BC} = \frac{\mathbf{V}_{bc}}{\mathbf{Z}_p} \qquad \mathbf{I}_{CA} = \frac{\mathbf{V}_{ca}}{\mathbf{Z}_p}$$

and their differences provide us with the line currents, such as

$$\mathbf{I}_{aA} = \mathbf{I}_{AB} - \mathbf{I}_{CA}$$

The three phase currents are of equal magnitude,

$$I_p = |\mathbf{I}_{AB}| = |\mathbf{I}_{BC}| = |\mathbf{I}_{CA}|$$

The line currents are also equal in magnitude. This is due to having phase currents which are equal in magnitude and necessarily 120° out of phase. The symmetry is apparent from the phasor diagram of Fig. 12-18. We thus have

$$I_L = |\mathbf{I}_{aA}| = |\mathbf{I}_{bB}| = |\mathbf{I}_{cC}|$$

and

$$I_L = \sqrt{3}\ I_p$$

Let us disregard the source for the moment and consider only the balanced load. If the load is Δ-connected, then the phase voltage and the line voltage are indistinguishable, but the line current is larger than the phase current by a factor of $\sqrt{3}$; with a Y-connected load, however, the phase

Fig. 12-18 A phasor diagram which could apply to the circuit of Fig. 12-17 if $\mathbf{Z}_p$ is an inductive impedance.

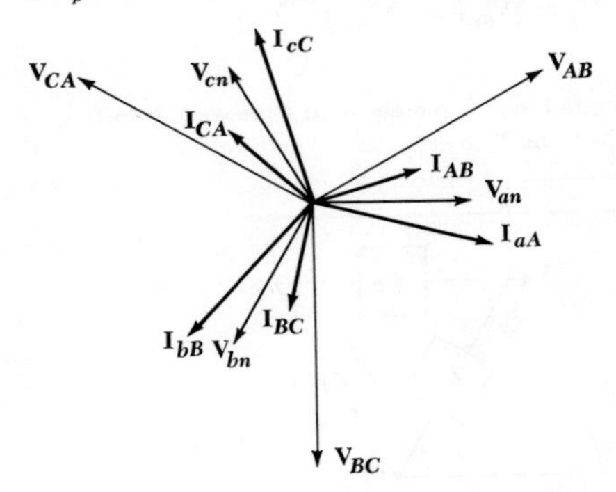

current and the line current refer to the same current, and the line voltage is greater than the phase voltage by a factor of $\sqrt{3}$.

The solution of three-phase problems will be speedily accomplished if the $\sqrt{3}$'s are used properly. Let us consider a typical numerical example. We are to determine the magnitude of the line current in a 300-V rms three-phase system which supplies 1200 W to a Δ-connected load at a lagging PF of 0.8. Let us again consider a single phase. It draws 400 W, 0.8 lagging PF, at a 300-V rms line voltage. Thus,

$$400 = 300(I_p)(0.8)$$

and

$$I_p = 1.667 \quad \text{A rms}$$

and the relationship between phase currents and line currents yields

$$I_L = \sqrt{3}\,(1.667) = 2.89 \quad \text{A rms}$$

Furthermore, the impedance in each phase must be

$$\mathbf{Z}_p = \frac{300}{1.667}\underline{/\cos^{-1} 0.8} = 180\underline{/36.8^\circ} \quad \Omega$$

Now let us change the statement of the problem: the load is Y-connected instead of Δ-connected. On a per-phase basis, we now have a phase voltage of $300/\sqrt{3}$ V rms, a power of 400 W, and a lagging PF of 0.8. Thus,

$$400 = \frac{300}{\sqrt{3}}(I_p)(0.8)$$

and

$$I_p = 2.89 \qquad \text{or} \qquad I_L = 2.89 \quad \text{A rms}$$

The impedance in each phase of the Y is

$$\mathbf{Z}_p = \frac{300/\sqrt{3}}{2.89}\underline{/\cos^{-1} 0.8} = 60\underline{/36.8^\circ} \quad \Omega$$

The $\sqrt{3}$ factor not only relates phase and line quantities but also appears in a useful expression for the total power drawn by any balanced three-phase load. If we assume a Y-connected load with a power-factor angle θ, then the power taken by any phase is

$$P_p = V_p I_p \cos\theta$$
$$= V_p I_L \cos\theta$$
$$= \frac{V_L}{\sqrt{3}} I_L \cos\theta$$

and the total power is

$$P = 3P_p$$
$$= \sqrt{3}\, V_L I_L \cos\theta$$

In a similar way, the power delivered to each phase of a Δ-connected load is

$$P_p = V_p I_p \cos\theta$$
$$= V_L I_p \cos\theta$$
$$= V_L \frac{I_L}{\sqrt{3}} \cos\theta$$

or a total power

$$P = 3P_p$$

or

$$P = \sqrt{3}\, V_L I_L \cos\theta \tag{1}$$

Thus (1) enables us to calculate the total power delivered to a balanced load from a knowledge of the magnitude of the line voltage, of the line current, and of the phase angle of the load impedance (or admittance). The numerical example above can be worked in one line:

$$1200 = \sqrt{3}(300)(I_L)(0.8)$$

Therefore

$$I_L = \frac{5}{\sqrt{3}} = 2.89 \quad \text{A rms}$$

The source may also be connected in Δ. This is not typical, however, for a slight unbalance in the source phases can lead to large currents circulating around the Δ loop. As an example, let us call the three single-phase sources $\mathbf{V}_{ab}$, $\mathbf{V}_{bc}$, and $\mathbf{V}_{cd}$. Before closing the Δ by connecting d to a, let us determine the unbalance by measuring the sum $\mathbf{V}_{ab} + \mathbf{V}_{bc} + \mathbf{V}_{cd}$. Suppose that the magnitude of the resultant is only 1 per cent of the line voltage. The circulating current is thus approximately ⅓ per cent of the line voltage divided by the internal impedance of any source. How large is this impedance apt to be? It must depend on the current that the source is expected to deliver with a negligible drop in terminal voltage. If we assume that this maximum current causes a 1 per cent drop in the terminal voltage, then it is seen that the circulating current is one-third of the maximum current. This reduces the useful current capacity of the source and also increases the losses in the system.

Drill Problems

12-8 A balanced three-phase three-wire system has a Δ-connected load with a 50-Ω resistor, a 5-μF capacitor, and a 0.56-H inductor in series in each phase. Using positive phase sequence with $\mathbf{V}_{an} = 390\underline{/30°}$ V rms and $\omega = 500$ rad/s, find: (*a*) $\mathbf{I}_{BC}$; (*b*) $\mathbf{I}_{aA}$; (*c*) $\mathbf{I}_{cC}$.

Ans. $5.20\underline{/7.4°}$; $9\underline{/-142.6°}$; $9\underline{/97.4°}$ *A rms*

12-9 A balanced Δ-connected load contains $8 + j4$ Ω per phase, while a balanced Y-connected load consists of $2 - j1$ Ω per phase. Both loads are connected in parallel to a three-phase three-wire system in which $V_L = 120$ V rms at the loads. The three lines extending from the loads to the source each have a resistance of 0.2 Ω. Find the total power: (*a*) delivered to the Δ-connected load; (*b*) delivered to the Y-connected load; (*c*) lost in the wires.

Ans. 1.43; 4.32; 5.76 *kW*

12-5 USE OF THE WATTMETER

Before discussing the specialized techniques used to measure power in three-phase systems (Sec. 12-6), it will be to our advantage to consider first how a wattmeter is used in a single-phase circuit.

Power measurement is most often accomplished at frequencies below a few hundred Hz through the use of a wattmeter which contains two separate coils. One of these coils is made of heavy wire, having a very low resistance, and is called the *current coil;* the second coil is composed of a much greater number of turns of fine wire, with relatively high resistance, and is termed the *potential coil,* or *voltage coil.* Additional resistance may also be inserted internally or externally in series with the potential coil. The torque applied to the moving system and the pointer is proportional to the instantaneous product of the currents flowing in the two coils. The mechanical inertia of the moving system, however, causes a deflection which is proportional to the *average* value of this torque.

The wattmeter is used by connecting it into a network in such a way that the current flowing in the current coil is the current flowing into the network and the voltage across the potential coil is the voltage across the two terminals of the network. The current in the potential coil is thus the input voltage divided by the resistance of the potential coil. The wattmeter deflection is therefore proportional to the average power delivered to the network.

It is apparent that the wattmeter has four available terminals, and correct connections must be made to these terminals in order to obtain an upscale reading on the meter. To be specific, let us assume that we are measuring the power absorbed by a passive network. The current coil

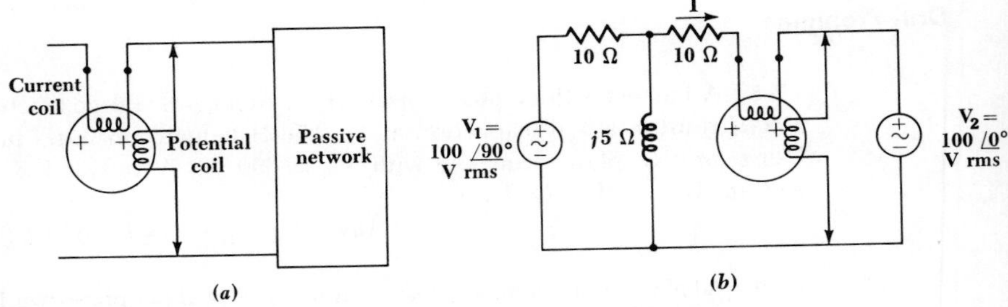

Fig. 12-19 (*a*) A wattmeter connection that will ensure an upscale reading for the power absorbed by the passive network. (*b*) An example in which the wattmeter is installed to give an upscale indication of the power absorbed by the right source.

is inserted in series with one of the two conductors connected to the load, and the potential coil is installed between the two conductors, usually on the "load side" of the current coil. The potential coil terminals are often indicated by arrows, as shown in Fig. 12-19*a*. Each coil has two terminals, and the proper relationship between the sense of the current and voltage must be observed. One end of each coil is usually marked (+), and an upscale reading is obtained if a positive current is flowing into the (+) end of the current coil while the (+) terminal of the potential coil is positive with respect to the unmarked end. The wattmeter shown in the network of Fig. 12-19*a* therefore gives an upscale deflection when the network to the right is absorbing power.

A reversal of either coil, but not both, will cause the meter to *try* to deflect downscale; a reversal of both coils will never affect the reading.

As an example of the use of such a wattmeter in measuring average power, let us consider the circuit shown in Fig. 12-19*b*. The connection of the wattmeter is such that an upscale reading corresponds to a positive absorbed power for the network to the right of the meter, that is, the right source. The power absorbed by this source is given by

$$P = |\mathbf{V}_2||\mathbf{I}| \cos (\text{ang}\, \mathbf{V}_2 - \text{ang}\, \mathbf{I})$$

Using superposition or mesh analysis, the current is found,

$$\mathbf{I} = 11.18\underline{/153.4^\circ} \quad \text{A rms}$$

and thus the absorbed power is

$$P = (100)(11.18) \cos (0^\circ - 153.4^\circ) = -1000 \quad \text{W}$$

The pointer therefore rests against the downscale stop. In practice, the potential coil may be reversed more quickly than may the current coil, and this reversal provides an upscale reading of 1000 W.

Drill Problem

12-10 In the circuit of Fig. 12-20, determine whether or not the potential coil must be reversed to obtain an upscale reading and find that reading for the ideal wattmeter: (*a*) *A*; (*b*) *B*; (*c*) *C*.

Ans. reversed, 15 W; as is, 35 W; as is, 70 W

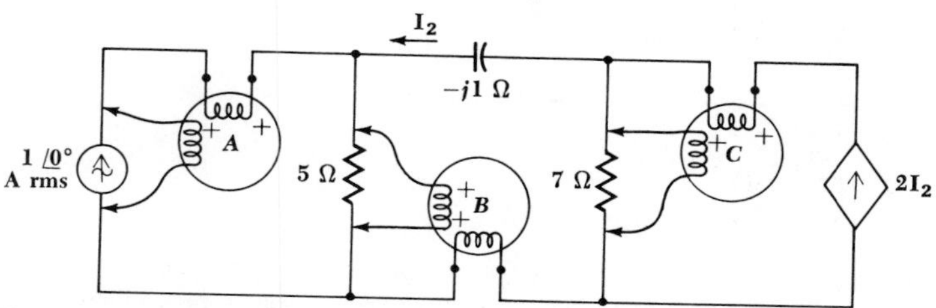

Fig. 12-20 See Drill Prob. 12-10.

12-6 POWER MEASUREMENT IN THREE-PHASE SYSTEMS

At first glance, the measurement of the power drawn by a three-phase load seems to be a simple problem. We need place only one wattmeter in each of the three phases and add the results. For example, the proper connections for a Y-connected load are shown in Fig. 12-21*a*. Each wattmeter has its current coil inserted in one phase of the load and its potential coil connected between the line side of that load and the neutral. In a similar

Fig. 12-21 Three wattmeters are connected in such a way that each reads the power taken by one phase of a three-phase load, and the sum of the readings is the total power. (*a*) A Y-connected load. (*b*) A Δ-connected load. Neither the loads nor the source need be balanced.

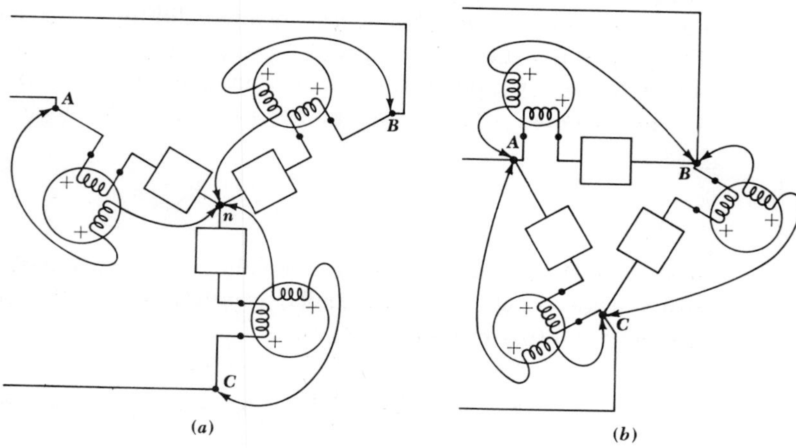

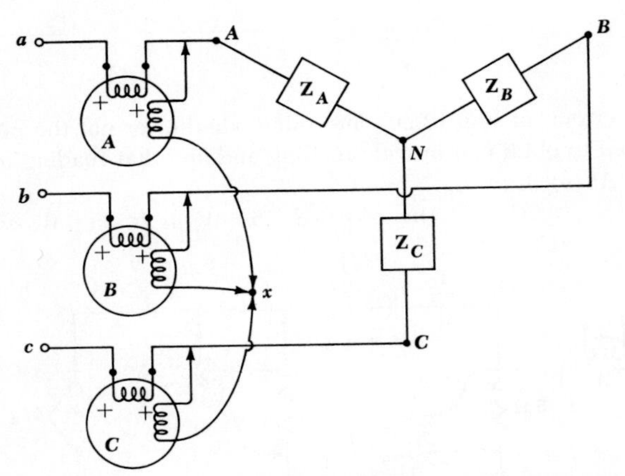

Fig. 12-22 A method of connecting three wattmeters to measure the total power taken by a three-phase load. Only the three terminals of the load are accessible.

way, three wattmeters may be connected as shown in Fig. 12-21*b* to measure the total power taken by a Δ-connected load. The methods are theoretically correct, but they may be useless in practice because the neutral of the Y is usually not accessible and the phases of the Δ are not available. A three-phase rotating machine, for example, has only three accessible terminals, those we have been calling *A*, *B*, and *C*.

It is obvious that we need a method for measuring the total power drawn by a three-phase load having only three accessible terminals; measurements may be made on the "line" side of these terminals, but not on the "load" side. Such a method is available, and it is capable of measuring the power taken by an *unbalanced* load from an *unbalanced* source. Let us connect three wattmeters in such a way that each has its current coil in one line and its voltage coil between that line and some common point x, as shown in Fig. 12-22. Although a system with a Y-connected load is illustrated, the arguments presented below are equally valid for a Δ-connected load. The point x may be some unspecified point in the three-phase system, or it may be merely a point in space at which the three potential coils have a common node. The average power indicated by wattmeter A must be

$$P_A = \frac{1}{T}\int_0^T v_{Ax} i_{aA}\, dt$$

where T is the period of all the source voltages. The readings of the other two wattmeters are given by similar expressions, and the total average power drawn by the load is therefore

$$P = P_A + P_B + P_C = \frac{1}{T}\int_0^T (v_{Ax} i_{aA} + v_{Bx} i_{bB} + v_{Cx} i_{cC})\, dt$$

Each of the three voltages in the above expression may be written in terms of a phase voltage and the voltage between point x and the neutral,

$$v_{Ax} = v_{AN} + v_{Nx}$$
$$v_{Bx} = v_{BN} + v_{Nx}$$
$$v_{Cx} = v_{CN} + v_{Nx}$$

and, therefore,

$$P = \frac{1}{T}\int_0^T (v_{AN}i_{aA} + v_{BN}i_{bB} + v_{CN}i_{cC})\,dt + \frac{1}{T}\int_0^T v_{Nx}(i_{aA} + i_{bB} + i_{cC})\,dt$$

However, the entire three-phase load may be considered to be a supernode, and Kirchhoff's current law requires

$$i_{aA} + i_{bB} + i_{cC} = 0$$

Thus

$$P = \frac{1}{T}\int_0^T (v_{AN}i_{aA} + v_{BN}i_{bB} + v_{CN}i_{cC})\,dt$$

Reference to the circuit diagram shows that this sum is indeed the sum of the average powers taken by each phase of the load, and the sum of the readings of the three wattmeters therefore represents the total average power drawn by the entire load.

Let us illustrate this procedure by a numerical example before we discover that one of these three wattmeters is really superfluous. We shall assume a balanced source,

$$\mathbf{V}_{ab} = 100\underline{/0^\circ} \quad \text{V rms}$$
$$\mathbf{V}_{bc} = 100\underline{/-120^\circ}$$
$$\mathbf{V}_{ca} = 100\underline{/-240^\circ}$$

or

$$\mathbf{V}_{an} = \frac{100}{\sqrt{3}}\underline{/-30^\circ}$$
$$\mathbf{V}_{bn} = \frac{100}{\sqrt{3}}\underline{/-150^\circ}$$
$$\mathbf{V}_{cn} = \frac{100}{\sqrt{3}}\underline{/-270^\circ}$$

and an unbalanced load,

$$\mathbf{Z}_A = -j10 \quad \Omega$$
$$\mathbf{Z}_B = j10$$
$$\mathbf{Z}_C = 10$$

Let us assume ideal wattmeters, connected as illustrated in Fig. 12-22, with point x located on the neutral of the source n. The three line currents may be obtained by mesh analysis,

$$\mathbf{I}_{aA} = \frac{\begin{vmatrix} 100\angle 0^\circ & -j10 \\ 100\angle -120^\circ & 10 + j10 \end{vmatrix}}{\begin{vmatrix} 0 & -j10 \\ -j10 & 10 + j10 \end{vmatrix}} = 19.3\angle 15^\circ \quad \text{A rms}$$

and by similar methods,

$$\mathbf{I}_{bB} = 19.3\angle 165^\circ \qquad \mathbf{I}_{cC} = 10\angle -90^\circ$$

The voltage between the neutrals is

$$\mathbf{V}_{nN} = \mathbf{V}_{nb} + \mathbf{V}_{BN} = \mathbf{V}_{nb} + \mathbf{I}_{bB}(j10) = 157.8\angle -90^\circ$$

Thus, the average power indicated by each wattmeter may be calculated,

$$P_A = V_p I_{aA} \cos (\text{ang } \mathbf{V}_{an} - \text{ang } \mathbf{I}_{aA})$$

$$= \frac{100}{\sqrt{3}} 19.3 \cos (15^\circ + 30^\circ) = 788 \quad \text{W}$$

$$P_B = \frac{100}{\sqrt{3}} 19.3 \cos (165^\circ + 150^\circ) = 788 \quad \text{W}$$

$$P_C = \frac{100}{\sqrt{3}} 10 \cos (-90^\circ + 270^\circ) = -577 \quad \text{W}$$

or a total power of 999 W. Since an rms current of 10 A flows through the resistive load, the total power drawn by the load is

$$P = 10^2(10) = 1000 \quad \text{W}$$

and the three-wattmeter method checks within slide-rule accuracy.

It is interesting to note that the reading of one of the wattmeters is negative. The discussion of the use of the wattmeter in the previous section indicates that an upscale reading on that meter can be obtained only after either the potential coil or the current coil is reversed.

We have proved that point x, the common connection of the three potential coils, may be located any place we wish without affecting the algebraic sum of the three wattmeter readings. Let us now consider the effect of placing point x, this common connection of the three wattmeters, directly on one of the lines. If, for example, one end of each potential coil is returned to B, then there is no voltage across the potential coil of wattmeter B and this meter must read zero. It may therefore be removed, and the algebraic sum of the remaining two wattmeter readings is still the total power drawn by the load. When the location of x is selected in this way, we describe the method of power measurement as the *two-wattmeter* method. The sum of the readings indicates the total power, regardless of (1) load unbalance, (2) source unbalance, (3) differences in the two wattmeters, and (4) the waveform of the periodic source. The only assumption we have made is that wattmeter corrections are sufficiently small so that we can ignore them. In Fig. 12-22, for example, the current coil of each meter has passing through it the line current drawn by the load plus the current taken by the potential coil. Since the latter current is usually quite small, its effect may be estimated from a knowledge of the resistance of the potential coil and the voltage across it. These two quantities enable a close estimate to be made of the power dissipated in the potential coil.

In the numerical example described above, let us now assume that two wattmeters are used, one with current coil in line A and potential coil between lines A and B, the other with current coil in line C and potential coil between C and B. The first meter reads

$$\begin{aligned} P_1 &= V_{AB}I_{aA} \cos (\text{ang } \mathbf{V}_{AB} - \text{ang } \mathbf{I}_{aA}) \\ &= 1000(19.3) \cos (15° - 0°) \\ &= 1866 \quad \text{W} \end{aligned}$$

and the second

$$\begin{aligned} P_2 &= V_{CB}I_{cC} \cos (\text{ang } \mathbf{V}_{CB} - \text{ang } \mathbf{I}_{cC}) \\ &= 100(10) \cos (-90° - 60°) \\ &= -866 \quad \text{W} \end{aligned}$$

and, therefore,

$$P = P_1 + P_2 = 1866 - 866 = 1000 \quad \text{W}$$

This we know is the correct answer.

In the case of a balanced load, the two-wattmeter method enables the PF angle to be determined, as well as the total power drawn by the load. Let us assume a load impedance with a phase angle θ; either a Y or Δ connection may be used and we shall assume the Δ connection shown in Fig. 12-23. The construction of a standard phasor diagram, such as that of Fig. 12-18, enables us to determine the proper phase angle between the several line voltages and line currents. We therefore determine the readings

$$\begin{aligned} P_1 &= |V_{AB}|\,|I_{aA}| \cos (\text{ang } \mathbf{V}_{AB} - \text{ang } \mathbf{I}_{aA}) \\ &= V_L I_L \cos (30° + \theta) \end{aligned}$$

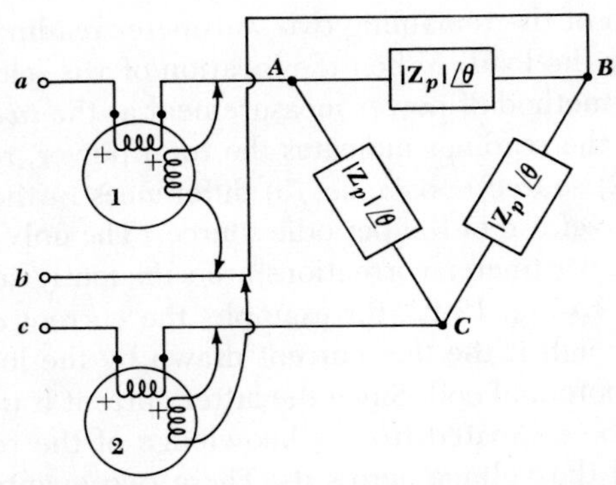

Fig. 12-23 Two wattmeters connected to read the total power drawn by a balanced three-phase load.

and
$$P_2 = |V_{CB}|\,|I_{cC}| \cos(\text{ang } \mathbf{V}_{CB} - \text{ang } \mathbf{I}_{cC})$$
$$= V_L I_L \cos(30° - \theta)$$

The ratio of the two readings is

$$\frac{P_1}{P_2} = \frac{\cos(30° + \theta)}{\cos(30° - \theta)} \tag{2}$$

If we expand the cosine terms, this equation is solved easily for $\tan\theta$,

$$\tan\theta = \sqrt{3}\frac{P_2 - P_1}{P_2 + P_1} \tag{3}$$

Thus, equal wattmeter readings indicate a unity PF load, equal and opposite readings indicate a purely reactive load, a reading of P_2 which is (algebraically) greater than P_1 indicates an inductive impedance, and a reading of P_2 which is less than P_1 signifies a capacitive load. How can we tell which wattmeter reads P_1 and which reads P_2? It is true that P_1 is in line A, and P_2 is in line C, and our positive phase-sequence system forces $\mathbf{V}_{an}$ to lag $\mathbf{V}_{cn}$. This is enough information to differentiate between the two wattmeters, but it is confusing to apply in practice. Even if we were unable to distinguish the two, we know the magnitude of the phase angle, but not its sign. This is often sufficient information; if the load is an induction motor, the angle must be positive and we do not need to make any tests to determine which reading is which. If no previous knowledge of the load is assumed, then there are several methods of resolving the ambiguity. Perhaps the simplest method is that which involves adding a high-impedance reactive load, say a three-phase capacitor, across the unknown load. The load must become more capacitive. Thus, if the magnitude of $\tan\theta$ (or the magnitude of θ) decreases, then the load was inductive, whereas an increase in the magnitude of $\tan\theta$ signifies an original capacitive impedance.

Let us suppose that one meter indicates 30 W while the other reads 100 W. Since we are assuming a passive load, the sign of the larger reading must be positive, but we must determine the sign of the smaller reading by a careful inspection of the location of the coil terminals marked (+). Let us assume that we know that it actually represents a negative power. Thus, P_1 is either -30 or 100, and P_2 is either 100 or -30. The value of $\tan\theta$ is therefore either 3.22 or -3.22, and the power-factor angle is either 72.7° or $-72.7°$. We then place three capacitors across the load in a balanced arrangement and find that the reading which was -30 W is now -27 W, while the 100-W reading has increased to 104 W. Since neither reading has changed appreciably, we decide that the capacitors have an impedance which is sufficiently high to produce only an incremental change, and we may calculate the new phase angle; it is either $+71.3°$ or $-71.3°$. The load must now be more capacitive than it was previously, and the only conclusion that can be drawn is that the phase angle was originally 72.7°, whereas now it is 71.3°.

Drill Problems

12-11 Three ideal wattmeters are arranged as shown in Fig. 12-22. Let $\mathbf{V}_{ab} = 125\underline{/0°}$ V rms, $\mathbf{Z}_A = \mathbf{Z}_B = \mathbf{Z}_C = 24 - j7\ \Omega$, point x be on point C, and the phase sequence be positive. Find: (a) P_A; (b) P_B; (c) P_C.

Ans. 0; 249; 351 W

12-12 Identify the source and load sides of a three-phase three-wire transmission system by a, b, c and A, B, C, respectively, and assume positive phase sequence with $\mathbf{V}_{ab} = 125\underline{/0°}$ V rms. One phase current in the balanced Δ-connected load is $\mathbf{I}_{AB} = 8\underline{/-45°}$ A rms. The current coil of an ideal wattmeter is inserted in line aA, (+) terminal toward a. Find the wattmeter reading if the (+) and (−) terminals of the potential coil are connected to: (a) A and B; (b) A and C; (c) B and C, all respectively.

Ans. 448; 1225; 1673 W

12-13 The two-wattmeter method is used to measure the total power taken by a balanced three-phase load, known to be passive. What is the PF of the load if: (a) one wattmeter reads 0; (b) the magnitude of one wattmeter reading is twice the magnitude of the other; (c) the magnitudes of the two wattmeter readings are equal?

Ans. 0 or 1.000; 0.189 or 0.866; 0.500

PROBLEMS

☐ **1** Given a balanced six-phase system in which $\mathbf{V}_{an} = 20\underline{/0^\circ}$, $\mathbf{V}_{bn} = 20\underline{/-60^\circ}$, . . . , $\mathbf{V}_{fn} = 20\underline{/-300^\circ}$, find: (*a*) $\mathbf{V}_{ab}$; (*b*) $\mathbf{V}_{de}$; (*c*) $\mathbf{V}_{bf}$; (*d*) $\mathbf{V}_{ad}$.

☐ **2** A single-phase three-wire system operates with $\mathbf{V}_{an} = \mathbf{V}_{nb} = 120\underline{/0^\circ}$ V rms. Lines *aA* and *bB* each have 0.1 Ω of resistance while the neutral wire has 1.0 Ω. (*a*) If $\mathbf{Z}_{AN} = \mathbf{Z}_{BN} = 6\underline{/0^\circ}$ Ω and $\mathbf{Z}_{AB} = 8\underline{/0^\circ}$ Ω, find $\mathbf{I}_{AN}$. (*b*) Find $\mathbf{I}_{AN}$ if $\mathbf{Z}_{BN}$ is open-circuited. (*c*) Find $\mathbf{I}_{AN}$ if both $\mathbf{Z}_{BN}$ and $\mathbf{Z}_{AB}$ are open-circuited.

☐ **3** A single-phase three-wire system is fed by $\mathbf{V}_{an} = \mathbf{V}_{nb} = 100\underline{/0^\circ}$ V rms. The two outside line currents are $\mathbf{I}_{aA} = 10 + j0$ and $\mathbf{I}_{bB} = -9 + j1$ A rms. If load $\mathbf{Z}_{AN}$ is $25 + j0$ Ω, find $\mathbf{Z}_{BN}$ and $\mathbf{Z}_{AB}$. Assume zero line resistance.

☐ **4** A common 120/240-V rms single-phase three-wire household circuit has 2.4-kVA loads across each 120-V line, one at unity PF and the other at 0.8 PF lagging. The 240-V load is 4.5 kW at 0.6 PF lagging. Find both line currents and the neutral current.

☐ **5** A balanced three-phase voltage source, $\mathbf{V}_{an} = 200\underline{/0^\circ}$ V rms, positive phase sequence, is connected by four lossless conductors to an unbalanced load: $\mathbf{Z}_{AN} = 20 + j0$, $\mathbf{Z}_{BN} = 10 + j10$, $\mathbf{Z}_{CN} = 10 - j10$ Ω. Find the three phasor line currents and the neutral current.

☐ **6** With reference to the circuit shown in Fig. 12-24, let $\mathbf{V}_{ab} = 150\underline{/0^\circ}$ V rms and assume positive phase sequence. If $\mathbf{Z}_p = 3 - j4$ Ω, find the rms line current and the total power delivered to the load if: (*a*) $R_W = 0$; (*b*) $R_W = 0.2$ Ω.

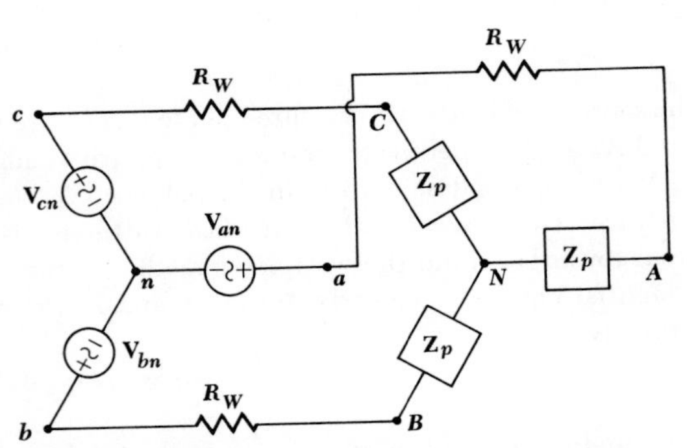

Fig. 12-24 See Probs. 6 and 8.

☐ **7** A three-phase induction motor requires 6 kW at 0.8 PF lagging. Determine the values for a Y of capacitors that will produce net unity PF operation when placed in parallel with the motor on a balanced 250-V rms (line voltage) system for which $f = 60$ Hz.

☐ **8** In the balanced three-phase circuit shown in Fig. 12-24, $\mathbf{V}_{an} = 120\underline{/0^\circ}$ V rms and the source provides 3600 VA at PF = 0.75 lagging. Assuming positive phase sequence with $R_W = 0.5\ \Omega$, find: (*a*) $\mathbf{Z}_p$; (*b*) the transmission efficiency.

☐ **9** A balanced three-phase voltage source, $\mathbf{V}_{an} = 200\underline{/0^\circ}$ V rms, positive phase sequence, is connected to an unbalanced Δ load: $\mathbf{Z}_{AB} = 60 + j0$, $\mathbf{Z}_{BC} = 30 + j30$, $\mathbf{Z}_{CA} = 30 - j30\ \Omega$. Find: (*a*) the three phasor line currents; (*b*) the power delivered to each phase of the load; (*c*) the power provided by each phase of the source.

☐ **10** In the circuit shown in Fig. 12-17, $\mathbf{V}_{an} = 150\underline{/0^\circ}$ V rms with positive phase sequence. If $\mathbf{Z}_p = 3 - j4\ \Omega$, find the rms line current and the total power delivered to the load.

☐ **11** Find the line current, the total power delivered to the load, and the total power loss in the lines for the circuit of Fig. 12-25 if $\mathbf{V}_{ab} = 180\underline{/0^\circ}$ V rms, $\mathbf{Z}_p = 7.5 + j3\ \Omega$, $R_W = 0.5\ \Omega$, and positive phase sequence is assumed.

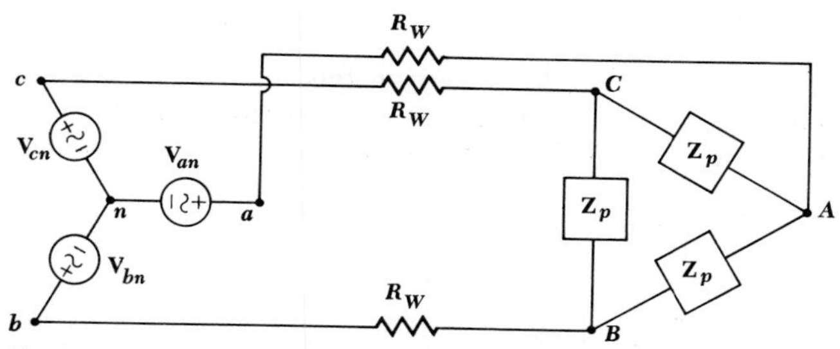

Fig. 12-25 See Prob. 11.

☐ **12** A three-phase motor draws 9 kVA at a lagging PF of 0.8 from a balanced system with line voltages of 300 V rms at 60 Hz. Three capacitors of what size should be arranged as a parallel Δ-connected load to produce unity PF operation?

☐ **13** Given a balanced three-phase system for which $V_L = 450$ V rms, two loads are present; one is Y-connected with an impedance of $20 - j10\ \Omega$ per phase, and the other is Δ-connected with $15 + j30\ \Omega$ per phase. Find the line current, the power delivered to each load, and the overall PF.

☐ **14** A Δ-connected source is slightly unbalanced. Each source phase is a practical voltage source with a resistance of 0.1 Ω, but the three voltages are $100\underline{/0^\circ}$, $101\underline{/-120^\circ}$, and $99\underline{/120^\circ}$ V rms. Find the source phase currents under no-load conditions and the total power loss in the source.

☐ **15** Determine the impedance **Z** of the capacitor in the circuit shown in Fig. 12-26 so that the wattmeter reading will be: (*a*) zero; (*b*) 18 W; (*c*) 3.6 W after potential coil reversal.

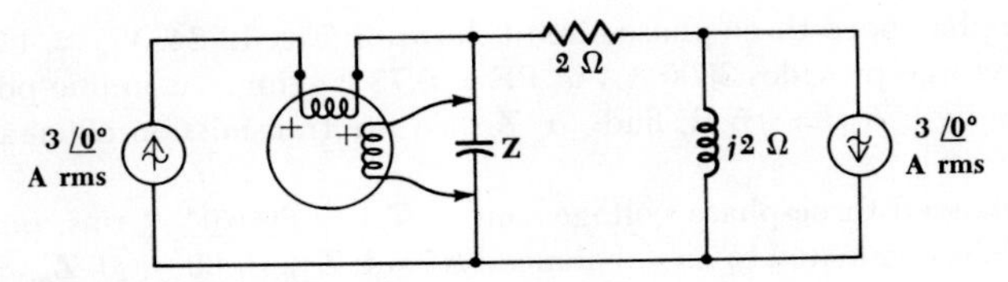

Fig. 12-26 See Prob. 15.

☐ **16** Determine the wattmeter reading in the circuit shown in Fig. 12-27. What is the physical interpretation of this reading?

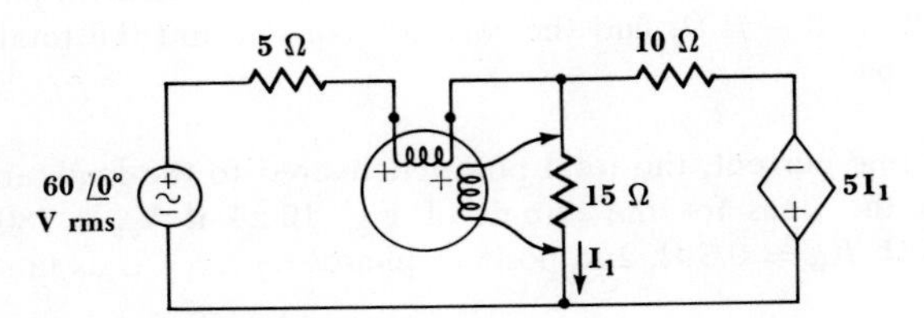

Fig. 12-27 See Prob. 16.

☐ **17** Show whether or not the two wattmeters in the circuit of Fig. 12-28 measure the total power taken by the general load consisting of $\mathbf{Z}_1$, $\mathbf{Z}_2$, and $\mathbf{Z}_3$: (*a*) if $\mathbf{V}_1 = \mathbf{V}_2$; (*b*) if $\mathbf{V}_1$ and $\mathbf{V}_2$ have unequal magnitudes and phase angles.

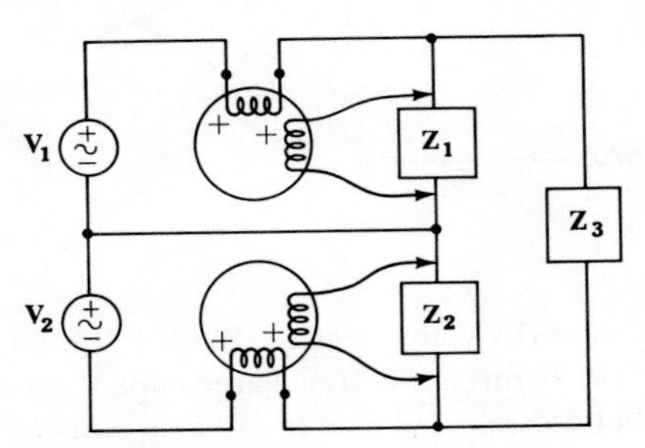

Fig. 12-28 See Probs. 17 and 18.

☐ **18** Refer to Fig. 12-28 and find the power indicated by each wattmeter if $\mathbf{Z}_1 = 4 + j2$, $\mathbf{Z}_2 = 4 - j2$, $\mathbf{Z}_3 = 4\ \Omega$, $\mathbf{V}_1 = 120\underline{/0°}$, and $\mathbf{V}_2 = 120\underline{/120°}$ V rms. What power is absorbed by each load?

☐ **19** A balanced Δ-connected load is present on a 240-V rms (line voltage) three-phase system. The line current is 10 A rms. A wattmeter with its current coil in one line and its potential coil between the two remaining lines reads 1500 W. Describe the load impedance.

☐ **20** Under what conditions will a single wattmeter read the total power taken by a balanced three-phase load in a three-wire system where the current coil is in one line and the potential coil is between the other two lines?

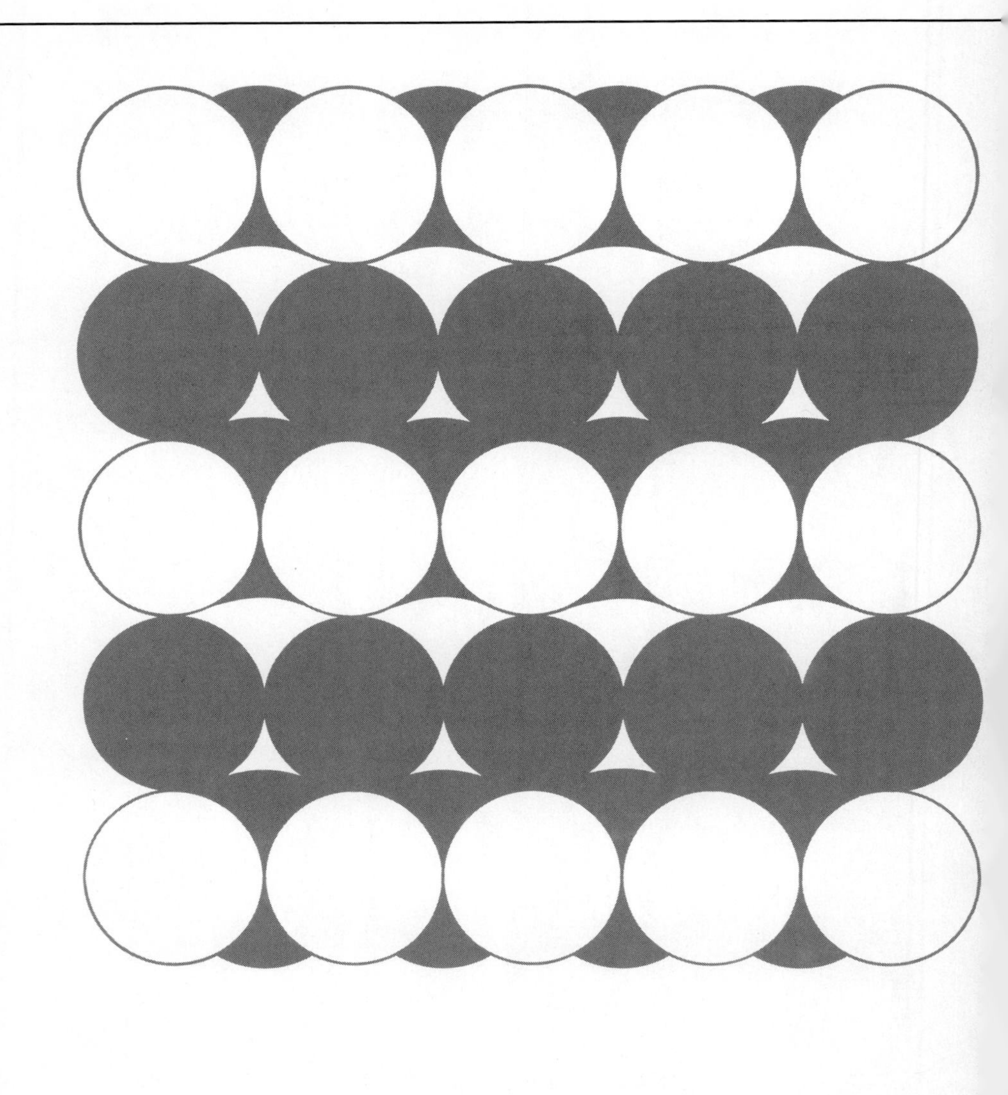

Part Four
COMPLEX FREQUENCY

Chapter Thirteen
Complex Frequency

13-1 INTRODUCTION

We are now about to begin the fourth major portion of our study of circuit analysis, a discussion of the concept of complex frequency. This, we shall see, is a remarkably unifying concept which will enable us to tie together all our previously developed analytical techniques into one neat package. Resistive circuit analysis, steady-state sinusoidal analysis, transient analysis, the forced response, the complete response, and the analysis of circuits excited by exponential forcing functions and exponentially damped sinusoidal forcing functions will all become special cases of the general techniques which are associated with the complex-frequency concept.

We shall introduce the concept of a complex frequency by considering an exponentially damped sinusoidal function, such as the voltage

$$v(t) = V_m e^{\sigma t} \cos(\omega t + \theta) \tag{1}$$

where σ (sigma) is a real quantity and is usually negative. Although we refer to this function as being "damped," it is possible that the sinusoidal amplitude may increase if σ is positive; the more practical case is that of the damped function. Our work with the natural response of the *RLC* circuit also indicates that σ is the negative of the exponential damping coefficient.

We may first construct a constant voltage from (1) by letting both σ and ω be zero,

$$v(t) = V_m \cos\theta = V_0 \tag{2}$$

If we set only σ equal to zero, then we obtain a general sinusoidal voltage

$$v(t) = V_m \cos(\omega t + \theta) \tag{3}$$

And if $\omega = 0$, we have the exponential voltage

$$v(t) = V_m \cos\theta\, e^{\sigma t} = V_0 e^{\sigma t} \tag{4}$$

Thus, the damped sinusoid (1) includes as special cases the dc (2), sinusoidal (3), and exponential (4) forcing functions.

Some additional insight into the significance of σ can be obtained by comparing the exponential function (4) with the complex representation of a sinusoidal forcing function with a zero-degree phase angle,

$$v(t) = V_0 e^{j\omega t} \tag{5}$$

It is apparent that the two functions, (4) and (5), have much in common. Their only difference is in the presence of a real and an imaginary exponent. The similarity between the two functions is emphasized by describing σ as a "frequency." This choice of terminology will be discussed in detail in the following sections, but for now we need merely note that σ is specifically termed the real part of the complex frequency. It should not be called the "real frequency," however, for this is a term which is more suitable for f (or, loosely, for ω). We shall also refer to σ as the *neper frequency,* the name arising from the dimensionless unit of the exponent of e. Thus, given e^{2t}, the dimensions of $2t$ are nepers (Np), and 2 is the neper frequency in nepers per second. The neper itself was named after

Napier and his napierian logarithm system; the spelling of his name is historically uncertain.

A forcing function of damped sinusoidal or exponential form raises some questions with regard to its amplitude for large negative values of t when $\sigma < 0$. That is, we have considered the forced response as that produced by a forcing function applied since $t = -\infty$; its application at some finite time gives rise to a transient response in addition to the forced response. Since the infinite amplitude of the forcing function at $t = -\infty$ may make us somewhat uncomfortable, we might note that initial conditions can be established in any circuit so that the application of a specified forcing function at a specified instant of time produces no transient response. Examples of this will appear later. From a practical point of view, we know that, although it may be impossible to generate such forcing functions in the lab which are accurate for all time, we can produce approximations that are satisfactory for circuits whose transient response does not last very long.

The response of a network to this general forcing function (1) is found very simply by using a method almost identical with that used for the sinusoidal forcing function; we shall discuss the method in Sec. 13-4. When we are able to find the response to this damped sinusoid, we should realize that we shall also have found the response to a dc voltage, an exponential voltage, and a sinusoidal voltage. Now let us see how we may consider σ and ω as the real and imaginary parts of a complex frequency.

13-2 COMPLEX FREQUENCY

Let us first provide ourselves with a purely mathematical definition of complex frequency and then gradually develop a physical interpretation in the next few sections. We shall say that any function which may be written in the form

$$\mathbf{f}(t) = \mathbf{K}e^{\mathbf{s}t} \tag{6}$$

where $\mathbf{K}$ and $\mathbf{s}$ are complex constants (independent of time), is characterized by the *complex frequency* $\mathbf{s}$. The complex frequency $\mathbf{s}$ is therefore simply the factor which multiplies t in this complex exponential representation. Until we are able to determine the complex frequency of a given function by inspection, it is necessary to write the function in the form of (6).

We may apply this definition first to the more familiar forcing functions. For example, a constant voltage

$$v(t) = V_0$$

may be written in the form

$$v(t) = V_0 e^{(0)t}$$

The complex frequency of a dc voltage or current is thus zero; $\mathbf{s} = 0$. The next simple case is the exponential function

$$v(t) = V_0 e^{\sigma t}$$

which is already in the required form; the complex frequency of this voltage is therefore σ; $\mathbf{s} = \sigma + j0$. Finally, let us consider a sinusoidal voltage, one which may provide a slight surprise. Given

$$v(t) = V_m \cos(\omega t + \theta)$$

it is necessary to find an equivalent expression in terms of the complex exponential. From our past experience, we therefore use the identity we derived from Euler's formula,

$$\cos(\omega t + \theta) = \tfrac{1}{2}(e^{j(\omega t+\theta)} + e^{-j(\omega t+\theta)})$$

and obtain

$$v(t) = \tfrac{1}{2}V_m(e^{j(\omega t+\theta)} + e^{-j(\omega t+\theta)})$$

or

$$v(t) = (\tfrac{1}{2}V_m e^{j\theta})e^{j\omega t} + (\tfrac{1}{2}V_m e^{-j\theta})e^{-j\omega t}$$

We have the *sum* of two complex exponentials, and *two* complex frequencies are therefore present, one for each term. The complex frequency of the first term is $\mathbf{s} = \mathbf{s}_1 = j\omega$ and that of the second term is $\mathbf{s} = \mathbf{s}_2 = -j\omega$. These two values of $\mathbf{s}$ are conjugates, or $\mathbf{s}_2 = \mathbf{s}_1^*$, and the two values of $\mathbf{K}$ are also conjugates. The entire first term and the entire second term are therefore conjugates, which we should expect inasmuch as their sum must be a real quantity.

Now let us determine the complex frequency or frequencies associated with the exponentially damped sinusoidal function (1). We again use Euler's formula to obtain a complex exponential representation:

$$\begin{aligned} v(t) &= V_m e^{\sigma t}\cos(\omega t + \theta) \\ &= \tfrac{1}{2}V_m e^{\sigma t}(e^{j(\omega t+\theta)} + e^{-j(\omega t+\theta)}) \end{aligned}$$

and thus

$$v(t) = \tfrac{1}{2}V_m e^{j\theta}e^{(\sigma+j\omega)t} + \tfrac{1}{2}V_m e^{-j\theta}e^{(\sigma-j\omega)t}$$

We therefore find once again that a conjugate complex pair of frequencies is required to describe the exponentially damped sinusoid, $\mathbf{s}_1 = \sigma + j\omega$, and $\mathbf{s}_2 = \mathbf{s}_1^* = \sigma - j\omega$. In general, neither σ nor ω is zero, and we see that the exponentially varying sinusoidal waveform is the general case; the constant, sinusoidal, and exponential waveforms are special cases.

As numerical illustrations, we should now recognize at sight the complex frequencies associated with these voltages:

$$v(t) = 100 \qquad \mathbf{s} = 0$$

$$v(t) = 5e^{-2t} \qquad \mathbf{s} = -2 + j0$$

$$v(t) = 2 \sin 500t \qquad \begin{cases} \mathbf{s}_1 = j500 \\ \mathbf{s}_2 = \mathbf{s}_1^* = -j500 \end{cases}$$

$$v(t) = 4e^{-3t} \sin (6t + 10°) \qquad \begin{cases} \mathbf{s}_1 = -3 + j6 \\ \mathbf{s}_2 = \mathbf{s}_1^* = -3 - j6 \end{cases}$$

The reverse type of example is also worth consideration. Given a complex frequency or a pair of conjugate complex frequencies, we must be able to identify the nature of the function with which they are associated. The most special case, $\mathbf{s} = 0$, defines a constant or dc function. With reference to the defining functional form (6), it is apparent that the constant $\mathbf{K}$ must be real if the function is to be real.

Let us next consider real values of $\mathbf{s}$. A positive real value, such as $\mathbf{s} = 5 + j0$, identifies an exponentially increasing function $\mathbf{K}e^{5t}$, where again $\mathbf{K}$ must be real if the function is to be a physical one. A negative real value for $\mathbf{s}$, such as $\mathbf{s} = -5 + j0$, refers to an exponentially decreasing function $\mathbf{K}e^{-5t}$.

A purely imaginary value of $\mathbf{s}$, for example, $j10$, can never be associated with a real quantity; the functional form is $\mathbf{K}e^{j10t}$, which can also be written as $\mathbf{K}(\cos 10t + j \sin 10t)$, and obviously possesses both real and imaginary parts. Each part is sinusoidal in nature. In order to construct a real function, it is necessary to consider conjugate values of $\mathbf{s}$, such as $\mathbf{s}_{1,2} = \pm j10$, with which must be associated conjugate values of $\mathbf{K}$. Loosely speaking, however, we may identify either of the complex frequencies $\mathbf{s}_1 = j10$ or $\mathbf{s}_2 = -j10$ with a sinusoidal voltage at the radian frequency of 10 rad/s. The presence of the conjugate complex frequency is understood. The amplitude and phase angle of the sinusoidal voltage will depend on the choice of $\mathbf{K}$ for each of the two frequencies. Thus, selecting $\mathbf{s}_1 = j10$ and $\mathbf{K}_1 = 6 - j8$, where $\mathbf{s}_2 = \mathbf{s}_1^*$ and $\mathbf{K}_2 = \mathbf{K}_1^*$, we obtain the real sinusoid $20 \cos (10t - 53.1°)$.

In a similar manner, a general value for $\mathbf{s}$, such as $3 - j5$, can be associated with a real quantity only if it is accompanied by its conjugate $3 + j5$. Speaking loosely again, we may think of either of these two conjugate frequencies as describing an exponentially increasing sinusoidal function $e^{3t} \cos 5t$; the specific amplitude and phase angle will again depend on the specific values of the conjugate complex $\mathbf{K}$'s.

By now we should have achieved some appreciation of the physical nature of the complex frequency **s**; in general, it describes an exponentially varying sinusoid. The real part of **s** is associated with the exponential variation; if it is negative, the function decays; if positive, the function increases; and if it is zero, the sinusoidal amplitude is constant. The imaginary part of **s** describes the sinusoidal variation; it is specifically the radian frequency. It is customary to call the real part of **s**, σ, and the imaginary part ω (*not* $j\omega$):

$$\mathbf{s} = \sigma + j\omega \tag{7}$$

The radian frequency is sometimes referred to as the "real frequency," but this terminology can be very confusing when we find that we must then say that "the real frequency is the imaginary part of the complex frequency"! When we need to be specific, we shall call **s** the complex frequency, σ the neper frequency, ω the radian frequency, and f the cyclic frequency; when no confusion seems likely, it is permissible to use "frequency" to refer to any of these four quantities. The neper frequency is measured in nepers per second, radian frequency is measured in radians per second, and complex frequency **s** is measured in units which are variously termed complex nepers per second or complex radians per second.

Drill Problems

13-1 Determine all the complex frequencies associated with the voltage waveform: (*a*) $v(t) = -0.1(1 - 0.2e^{-0.6t} \cos 100t)$; (*b*) $v(t) = 2 + 4e^{-0.6t} - 6e^{-0.4t}$; (*c*) $v(t) = 10e^{-0.6t} + 8 \sin (100t + 15°)$.

Ans. $-0.6,\ -j100,\ j100;\ 0,\ -0.6 - j100,\ -0.6 + j100;\ 0,\ -0.6,\ -0.4\quad s^{-1}$

13-2 Find the complex frequencies associated with the natural response of a source-free parallel *RLC* circuit in which $L = 10/9$ H, $C = 10\ \mu$F, and $R =$: (*a*) ∞; (*b*) 100 Ω; (*c*) $625/3\ \Omega$.

Ans. $-900,\ -100;\ -240 - j180,\ -240 + j180;\ -j300,\ j300\quad s^{-1}$

13-3 A FURTHER INTERPRETATION OF COMPLEX FREQUENCY

The preceding discussion has shown us that the real and imaginary parts of the complex frequency describe, respectively, the exponential and sinusoidal variation of an exponentially varying sinusoid. It is apparent that the significance of a complex frequency is more general than is that of the everyday variety of frequency which describes the number of times some phenomenon repeats itself per unit time. Our ordinary concept of frequency, however, actually carries with it another connotation in addition to "repetitions per second." It also tells us something about the *rate of change* of the function being considered. That is, high frequency means "rapidly varying." Let us see if we can relate this connotation of frequency to the complex frequency of a complex exponential function of time. We again take

$$\mathbf{f}(t) = \mathbf{K}e^{\mathbf{s}t} \tag{8}$$

After obtaining the time rate of change of $\mathbf{f}(t)$,

$$\frac{d\mathbf{f}}{dt} = \mathbf{s}\mathbf{K}e^{\mathbf{s}t}$$

we normalize by dividing by $\mathbf{f}(t)$,

$$\frac{d\mathbf{f}/dt}{\mathbf{f}} = \mathbf{s} \tag{9}$$

This normalized rate of change is a constant, independent of time; it is identically equal to the complex frequency **s**. Thus, we may also interpret complex frequency as the *normalized time rate of change* of the complex exponential function (8). The use of this alternative definition of **s** on familiar functions which are *not* in the form of the complex exponential function (8) may lead to some curious results. Thus, although the complex frequency associated with the function e^{j5t} is $\mathbf{s} = j5$, the normalized rate of change associated with $\cos 5t$ must be $(-5)\tan 5t$. If we try to treat this result, which is a function of time, as a complex frequency, then we are led to a complex frequency which is a function of time. The correct answer to our problem is obtained only when we recognize that a conjugate *pair* of complex frequencies is required to characterize $\cos 5t$; one complex frequency is insufficient.

Drill Problem

13-3 A network is composed of three different resistors and one capacitor in some configuration which is immaterial to this problem. No sources are present, but initial energy is present in the capacitor. The voltage waveform across the capacitor is displayed on an oscilloscope. At $t = 30$ ms, $v_C = 50$ V, and it is decreasing at the rate of 1000 V/s. After finding the complex frequency which characterizes v_C, write a suitable functional form for the voltage and determine the instant of time at which: (*a*) $v_C = 25$ V; (*b*) the rate of change of v_C is -100 V/s; (*c*) the rate of change of v_C is -10 kV/s.

Ans. -85; 64.7; 145 ms

13-4 THE DAMPED SINUSOIDAL FORCING FUNCTION

We have devoted enough time to the definition and introductory interpretation of complex frequency; it is now time to put this concept to work and become familiar with it by seeing what it will do and how it is used.

The general exponentially varying sinusoid, which we may represent as a voltage for the moment,

$$v(t) = V_m e^{\sigma t} \cos (\omega t + \theta) \tag{10}$$

is expressible in terms of the complex frequency **s** by making use of Euler's identity as before:

$$v(t) = \text{Re} \, (V_m e^{\sigma t} e^{j(\omega t + \theta)}) \tag{11}$$

or

$$v(t) = \text{Re} \, (V_m e^{\sigma t} e^{j(-\omega t - \theta)}) \tag{12}$$

Either representation is suitable, and the two expressions should remind us that a pair of conjugate complex frequencies is associated with a sinusoid or an exponentially damped sinusoid. Equation (11) is more directly related to the given damped sinusoid, and we shall concern ourselves principally with it. Collecting factors,

$$v(t) = \text{Re} \, (V_m e^{j\theta} e^{(\sigma + j\omega)t})$$

we now substitute $\mathbf{s} = \sigma + j\omega$, and obtain

$$v(t) = \text{Re} \, (V_m e^{j\theta} e^{\mathbf{s}t}) \tag{13}$$

Before we apply a forcing function of this form to any circuit, we should note the resemblance of this last representation of the damped sinusoid to the corresponding representation of the *undamped* sinusoid,

$$\text{Re} \, (V_m e^{j\theta} e^{j\omega t})$$

The only difference is that we now have **s** where we previously had $j\omega$. Instead of restricting ourselves to sinusoidal forcing functions and their angular frequencies, we have now extended our notation to include the damped sinusoidal forcing function at a complex frequency. It should be no surprise at all to see later in this section and the following one that we shall develop a *frequency-domain* description of the exponentially damped sinusoid in exactly the same way that we did for the sinusoid; we shall simply omit the Re notation and suppress $e^{\mathbf{s}t}$.

We are now ready to apply the exponentially damped sinusoid, as given by (10), (11), (12), or (13), to an electrical network. The forced response, say a current in some branch of the network, is the desired response. Since the forced response has the form of the forcing function, its integral, and its derivatives, the response may be assumed to be

$$i(t) = I_m e^{\sigma t} \cos(\omega t + \phi)$$

or

$$i(t) = \text{Re}\,(I_m e^{j\phi} e^{st})$$

where the complex frequency of the source and the response must be identical.

If we now recall that the real part of a complex forcing function produces the real part of the response, while the imaginary part of the complex forcing function causes the imaginary part of the response, then we are again led to the application of a *complex* forcing function to our network. We shall obtain a complex response whose real part is the desired real response. Actually, we shall work with the Re notation omitted, but we should realize that it may be reinserted at any time and that it *must* be reinserted whenever we desire the time-domain response. Thus, given the real forcing function

$$v(t) = \text{Re}\,(V_m e^{j\theta} e^{st})$$

we apply the complex forcing function $V_m e^{j\theta} e^{st}$; the resultant complex response $I_m e^{j\phi} e^{st}$ must have as its real part the desired time-domain response

$$i(t) = \text{Re}\,(I_m e^{j\phi} e^{st})$$

The solution of our circuit-analysis problem must consist of the determination of the unknown response amplitude I_m and phase angle ϕ.

Before we actually carry out the details of an analysis problem and see how exactly the procedure follows that which was used in the sinusoidal analysis, it is worthwhile outlining the steps of the basic method. We must first characterize the circuit with a set of loop or nodal integrodifferential equations. The given forcing functions, in complex form, and the assumed responses, also in complex form, are then substituted into the equations and the indicated integrations and differentiations performed. Each term in every equation will then contain the same factor e^{st}. We shall therefore divide throughout by this factor, or "suppress e^{st}," understanding that it must be reinserted if a time-domain description of any response function is desired. Now that the Re symbol and the e^{st} factor have disappeared, we have converted all the voltages and currents from the time domain to the frequency domain. The integrodifferential equations have become algebraic equations, and their solution is obtained just as easily as they were

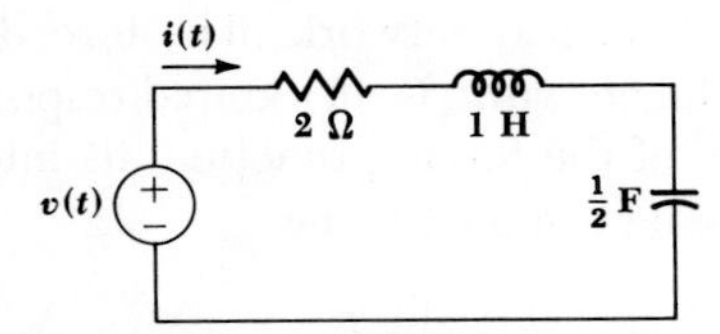

Fig. 13-1 A series RLC circuit to which a damped sinusoidal forcing function is applied. A frequency-domain solution for $i(t)$ is desired.

in the sinusoidal steady state. Let us illustrate the basic method by a numerical example.

We shall apply the forcing function

$$v(t) = 60e^{-t} \cos (2t + 10°)$$

to the series RLC circuit shown in Fig. 13-1, and we desire the forced response

$$i(t) = I_m e^{-t} \cos (2t + \phi)$$

We first express the forcing function in Re notation,

$$\begin{aligned} v(t) &= 60e^{-t} \cos (2t + 10°) \\ &= \text{Re}\,(60e^{-t}e^{j(2t+10°)}) \\ &= \text{Re}\,(60e^{j10°}e^{(-1+j2)t}) \end{aligned}$$

or

$$v(t) = \text{Re}\,(\mathbf{V}e^{\mathbf{s}t})$$

where

$$\mathbf{V} = 60\underline{/10°} \qquad \text{and} \qquad \mathbf{s} = -1 + j2$$

After dropping Re, we are left with the complex forcing function

$$60\underline{/10°}e^{\mathbf{s}t}$$

In a similar manner, we represent the unknown response by the complex quantity

$$\mathbf{I}e^{\mathbf{s}t}$$

where

$$\mathbf{I} = I_m\underline{/\phi}$$

Since we have not as yet extended the impedance concept to the exponentially damped sinusoid and complex frequency, our next step must be the integrodifferential equation for our circuit. From Kirchhoff's voltage law we obtain

$$v(t) = Ri + L\frac{di}{dt} + \frac{1}{C}\int i\,dt$$

$$= 2i + \frac{di}{dt} + 2\int i\,dt$$

and we substitute the given complex forcing function and the assumed complex response into this equation,

$$60\underline{/10°}\,e^{\mathbf{s}t} = 2\mathbf{I}e^{\mathbf{s}t} + \mathbf{s}\mathbf{I}e^{\mathbf{s}t} + \frac{2}{\mathbf{s}}\mathbf{I}e^{\mathbf{s}t}$$

The common factor $e^{\mathbf{s}t}$ is next suppressed:

$$60\underline{/10°} = 2\mathbf{I} + \mathbf{s}\mathbf{I} + \frac{2}{\mathbf{s}}\mathbf{I}$$

and

$$\mathbf{I} = \frac{60\underline{/10°}}{2 + \mathbf{s} + 2/\mathbf{s}}$$

Before we evaluate this complex current, it is worth examining the form of this equation. The left side of the equation is a current, the numerator of the right side is a voltage, and the denominator must therefore have the dimensions of ohms; it is apparent that we shall soon interpret it as an impedance. Moreover, we might make some pretty astute guesses as to the impedance of each of the three passive elements at a complex frequency **s**. The impedance of the 2-Ω resistor is simply 2 Ω; the impedance of the 1-H inductor is 1**s** or **s***L*; and the impedance of the 0.5-F capacitor is 2/**s** or 1/**s***C*. These statements will be proved in the following section.

Going back to our numerical example, we now let $\mathbf{s} = -1 + j2$ in the last equation and solve for the complex current **I**:

$$\mathbf{I} = \frac{60\underline{/10°}}{2 + (-1 + j2) + 2/(-1 + j2)}$$

After manipulating the complex numbers for a few minutes, we find

$$\mathbf{I} = 44.8\underline{/-53.4°}$$

Thus, I_m is 44.8 A and ϕ is $-53.4°$. The desired forced response is thus

$$i(t) = 44.8e^{-t}\cos(2t - 53.4°)$$

Now we must show that the impedance concept may be extended to complex frequencies. Except for the presence of complex numbers, the actual solution of the circuit equations will then be no different than the procedure used for purely resistive circuits.

Drill Problems

13-4 Express each of the following currents in the frequency domain: (*a*) $4e^{-20t} \cos (1000t + 60°)$ mA; (*b*) $4 \sin (800t + 60°)$ mA; (*c*) $-4e^{-5t} \sin (1000t - 60°)$ mA.

Ans. $4\underline{/-30°}$; $4\underline{/30°}$; $4\underline{/60°}$ mA

13-5 If $\mathbf{V} = 64\underline{/80°}$ V, find $v(0.001)$ if $\mathbf{s} =$: (*a*) $-800 + j600$; (*b*) $-j600$; (*c*) $-800 - j600$.

Ans. -11.9; 20.1; 44.8 V

13-5 Z(s) AND Y(s)

In order to apply Kirchhoff's laws directly to the complex forcing functions and complex responses, it is necessary to know the constant of proportionality between the complex voltage across an element and the complex current through it. This proportionality constant is the impedance or admittance of the element; it is easily determined for the resistor, inductor, and capacitor.

Let us consider the inductor carefully and then merely present the results for the other elements. Suppose that a voltage source

$$v(t) = V_m e^{\sigma t} \cos (\omega t + \theta)$$

is applied to an inductor L; the current response must have the form

$$i(t) = I_m e^{\sigma t} \cos (\omega t + \phi)$$

The passive sign convention is employed, as indicated in Fig. 13-2*a*, the time-domain circuit. If we represent the voltage as

$$v(t) = \text{Re}\,(V_m e^{j\theta} e^{\mathbf{s}t}) = \text{Re}\,(\mathbf{V}e^{\mathbf{s}t})$$

and the current as

$$i(t) = \text{Re}\,(I_m e^{j\phi} e^{\mathbf{s}t}) = \text{Re}\,(\mathbf{I}e^{\mathbf{s}t})$$

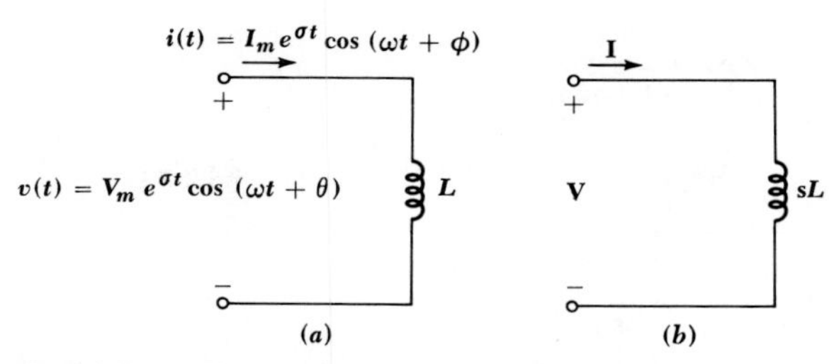

Fig. 13-2 (*a*) The time-domain inductor voltage and current are related by $v = L\, di/dt$. (*b*) The (complex) frequency-domain inductor voltage and current are related by $\mathbf{V} = \mathbf{s}L\mathbf{I}$.

then the substitution of these expressions into the defining equation of an inductor,

$$v(t) = L\frac{di(t)}{dt}$$

leads to

$$\text{Re}\,(\mathbf{V}e^{\mathbf{s}t}) = \text{Re}\,(\mathbf{s}L\mathbf{I}e^{\mathbf{s}t})$$

We now drop Re, thus considering the complex response to a complex forcing function, and suppress the superfluous factor $e^{\mathbf{s}t}$:

$$\mathbf{V} = \mathbf{s}L\mathbf{I}$$

The ratio of the complex voltage to the complex current is once again the impedance. Since it depends in general upon the complex frequency **s**, this functional dependence is sometimes indicated by writing

$$\mathbf{Z}(\mathbf{s}) = \frac{\mathbf{V}}{\mathbf{I}} = \mathbf{s}L$$

In a similar manner, the admittance of an inductor L is

$$\mathbf{Y}(\mathbf{s}) = \frac{1}{\mathbf{s}L}$$

We shall still call **V** and **I** *phasors*. These complex quantities have an amplitude and phase angle which, along with a specific complex-frequency value, enable us to characterize the exponentially varying sinusoidal waveform completely. The phasor is still a frequency-domain description, but its application is not limited to the realm of radian frequencies. The frequency-domain equivalent of Fig. 13-2*a* is shown in Fig. 13-2*b*; phasor currents, phasor voltages, and impedances or admittances are now used.

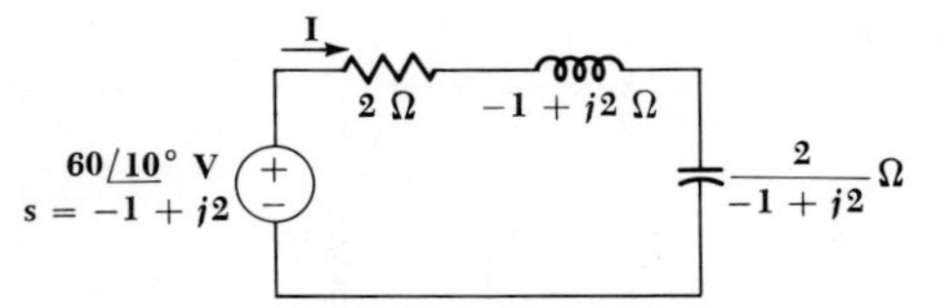

Fig. 13-3 The frequency-domain equivalent of the series *RLC* circuit shown in Fig. 13-1.

The impedances of a resistor R and a capacitor C at a complex frequency $\mathbf{s}$ are obtained by a similar sequence of steps. Without going through the details, the results for a resistor are

$$\mathbf{Z}(\mathbf{s}) = R \qquad \mathbf{Y}(\mathbf{s}) = G = \frac{1}{R}$$

and for a capacitor

$$\mathbf{Z}(\mathbf{s}) = \frac{1}{\mathbf{s}C} \qquad \mathbf{Y}(\mathbf{s}) = \mathbf{s}C$$

If we now reconsider the series *RLC* example of Fig. 13-1 in the frequency domain, the source voltage

$$v(t) = 60e^{-t}\cos(2t + 10°)$$

is transformed to the phasor voltage

$$\mathbf{V} = 60\underline{/10°}$$

a phasor current $\mathbf{I}$ is assumed, and the impedance of each element at the complex frequency $\mathbf{s} = -1 + j2$ is determined and placed on a frequency-domain diagram, Fig. 13-3. The unknown current is now easily obtained by dividing the phasor voltage by the sum of the three impedances:

$$\mathbf{I} = \frac{60\underline{/10°}}{2 + (-1 + j2) + 2/(-1 + j2)}$$

Thus, the previous result is obtained, but much more easily and rapidly.

It is hardly necessary to say that all the techniques which we have used in the past to simplify frequency-domain analysis, such as superposition, source transformations, duality, Thévenin's theorem, and Norton's theorem, are still valid and useful. For example, the Thévenin equivalent of the network shown in Fig. 13-4*a* is obtained as follows:

The complex frequency is

$$\mathbf{s} = -5 + j10$$

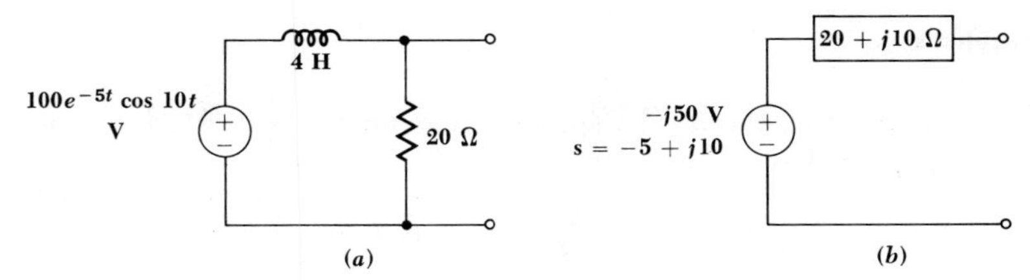

Fig. 13-4 (*a*) A given two-terminal network. (*b*) The frequency-domain Thévenin equivalent.

The frequency-domain source voltage is

$$\mathbf{V}_s = 100\underline{/0^\circ}$$

The inductor impedance is

$$\mathbf{Z}_L(\mathbf{s}) = 4(-5 + j10) = -20 + j40$$

The resistor impedance is

$$\mathbf{Z}_R(\mathbf{s}) = 20$$

The Thévenin impedance is the parallel equivalent of $\mathbf{Z}_L$ and $\mathbf{Z}_R$:

$$\mathbf{Z}_{th} = \frac{20(-20 + j40)}{20 - 20 + j40} = 20 + j10$$

The open-circuit voltage is

$$\mathbf{V}_{oc} = 100\underline{/0^\circ}\frac{20}{20 - 20 + j40} = -j50$$

The frequency-domain Thévenin equivalent network is thus as shown in Fig. 13-4*b*. The return to the time domain is taken after some desired response has been determined in the frequency domain. If another 4-H inductor is placed across the open circuit, for example, the frequency-domain current is

$$\mathbf{I} = \frac{-j50}{20 + j10 - 20 + j40} = -1$$

corresponding to the time-domain current

$$i(t) = -e^{-5t}\cos 10t$$

Drill Problems

13-6 Find the input impedance $\mathbf{Z}(\mathbf{s})$ for the network shown in Fig. 13-5 as presented at terminal pair: (*a*) *a-b*; (*b*) *c-d*; (*c*) *a-c*.

Ans. $\dfrac{3(\mathbf{s}+4)}{2(\mathbf{s}^2+4\mathbf{s}+3)}$; $\dfrac{2(\mathbf{s}^2+3)}{\mathbf{s}^2+4\mathbf{s}+3}$; $\dfrac{\mathbf{s}(4\mathbf{s}+3)}{2(\mathbf{s}^2+4\mathbf{s}+3)}$ Ω

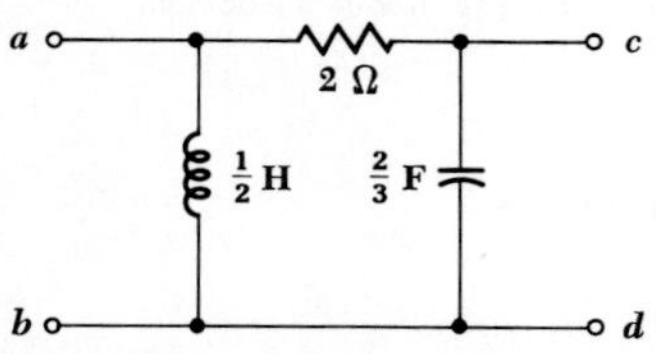

Fig. 13-5 See Drill Probs. 13-6 to 13-8.

13-7 For part (*c*) of Drill Prob. 6 above, find $\mathbf{Z}(\mathbf{s})$ at $\mathbf{s} =$: (*a*) 0; (*b*) ∞; (*c*) $-2 + j1$.

Ans. $-6 + j4$; 2; 2 Ω

13-8 For the network shown in Fig. 13-5, find $v_{cd}(t)$ if a current source connected at terminals *a-b* delivers at node *a* a current of: (*a*) $8e^{-2t}$ A; (*b*) 8 cos $3t$ A; (*c*) $8e^{-2t}$ cos $3t$ A.

Ans. $24e^{-2t}$; $2.68 \cos(3t - 26.6°)$; $4.33e^{-2t} \cos(3t - 56.3°)$ V

13-6 FREQUENCY RESPONSE AS A FUNCTION OF σ

We have already considered circuit response as a function of the radian frequency ω, representing such quantities as impedance, admittance, specific voltages or currents, voltage and current gain, and transfer impedances and admittances as functions of ω, determining their poles and zeros, and sketching the response. Before we discuss the more general problem of frequency response as a function of the complex frequency $\mathbf{s}$ in the next section, let us devote a little time to the simpler problem of frequency response as a function of σ.

As a simple example we may select the series RL circuit excited by the frequency-domain voltage source $V_m\underline{/0°}$. The current is obtained as a function of $\mathbf{s}$ by dividing the source voltage by the input impedance,

$$\mathbf{I} = \frac{V_m\underline{/0°}}{R + \mathbf{s}L}$$

We now set $\omega = 0$, $\mathbf{s} = \sigma + j0$, thus restricting ourselves to time-domain sources of the form

$$v_s = V_m e^{\sigma t}$$

and thus

$$I = \frac{V_m}{R + \sigma L}$$

or

$$I = \frac{V_m}{L} \frac{1}{\sigma + R/L} \tag{14}$$

Transforming to the time domain,

$$i(t) = \frac{V_m}{R + \sigma L} e^{\sigma t} \tag{15}$$

The necessary information about the response, however, is all contained in the frequency-domain description (14). As the neper frequency σ varies, a qualitative description of the response is easily provided. When σ is a large negative number, corresponding to a rapidly decreasing exponential function, the current response (14) is negative and relatively small in amplitude; it is, of course, also a rapidly decreasing (in magnitude) exponential function. As σ increases, becoming a smaller negative number, the magnitude of the negative response increases. When σ is exactly equal to $-R/L$, a pole of the response, the current amplitude is infinite.

As σ continues to increase, the next noteworthy point occurs when $\sigma = 0$. Since $v_s = V_m$, we are now faced with the dc case, and the response is obviously V_m/R, which agrees with the response indicated by (14). Positive values of σ must all provide positive amplitude responses, the larger amplitudes arising when σ is smaller. Finally, an infinite value of σ provides a zero-amplitude response, and thus establishes a zero. The only critical frequencies are the pole at $\sigma = -R/L$ and the zero at $\sigma = \pm\infty$.

This information may be presented quite easily by plotting $|I|$, the current magnitude, as a function of σ, as shown in Fig. 13-6. In preparation for the more general response information with changing **s** in the next section, we show the magnitude of I versus σ; the phase (not shown) is, of course, either 0 or 180°.

Let us now turn our attention to the two critical frequencies of this response. The only finite critical frequency is the pole at $\sigma = -R/L$. It is worthwhile to determine why an infinite response is obtained when the circuit is excited at this frequency by the voltage

$$v = V_m e^{-Rt/L} \tag{16}$$

The forcing function (16) has a familiar form; it has all the characteristics

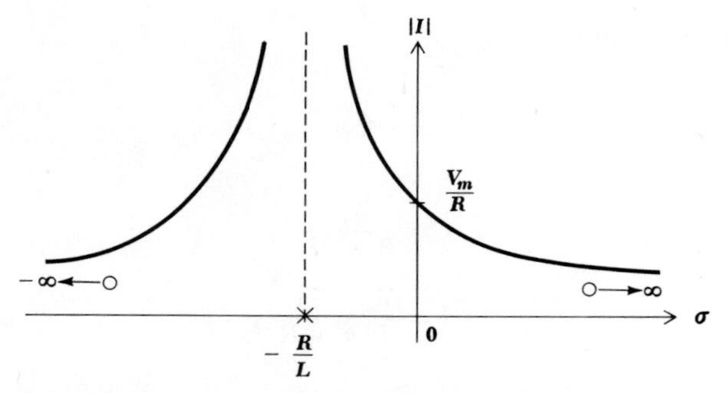

Fig. 13-6 A plot of the magnitude of I versus neper frequency σ for a series RL circuit excited by an exponential voltage source, $V_m e^{\sigma t}$. The only pole is located at $\sigma = -R/L$; the zero occurs at $\sigma = \pm\infty$.

of the natural response. As a matter of fact, if we reduce the amplitude of this forcing function to zero, thus "exciting" the circuit by a short circuit, then the current

$$i = I_0 e^{-Rt/L} \tag{17}$$

would flow if an initial current I_0 were assumed. Although this is a natural response, it is informative to interpret the result as a forced response; we find that a zero-amplitude forcing function then produces a non-zero-amplitude response. Since the circuit is linear, a non-zero-amplitude forcing function must produce an infinite response.

We shall find that this result is quite general; when any circuit is excited at a frequency which is a pole of the response, then an infinite response must result; the frequencies of the poles and zeros are directly related to the natural response of the circuit. This relationship will be seen to arise in several examples discussed in this section, but we shall not disucuss it thoroughly until the following section.

As a second example, let us consider the circuit drawn in Fig. 13-7a. The exponential current source

$$i = I_m e^{\sigma t}$$

is applied to a series RC circuit. The voltage across the source is given by

$$V = IZ(\sigma) = I_m\left(R + \frac{1}{\sigma C}\right)$$

or

$$V = I_m R \frac{\sigma + 1/RC}{\sigma} \tag{18}$$

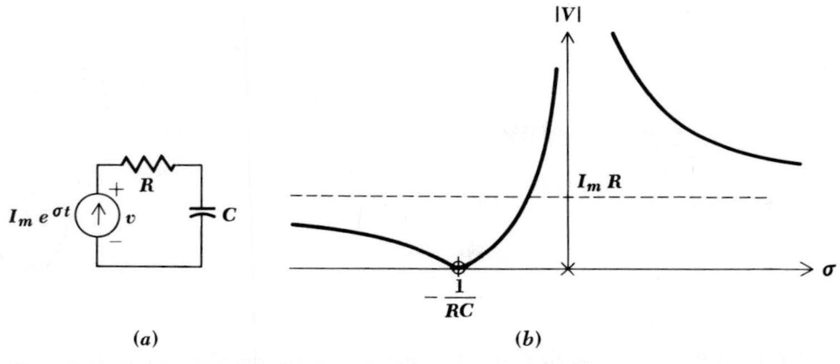

Fig. 13-7 (*a*) A series RC circuit excited by an exponential current source. (*b*) The magnitude of the voltage response possesses a pole at $\sigma = 0$ and a zero at $\sigma = -1/RC$.

This last form of the voltage response, written as a constant times the ratio of factors having the form $(\sigma + \sigma_1)$, is obviously well suited for the quick determination of the poles and zeros of the response. The voltage response indicates a pole at $\sigma = 0$ and a zero at $\sigma = -1/RC$; infinite frequency is not a critical frequency. The response magnitude $|V|$ is plotted as a function of frequency in Fig. 13-7*b*. The reason for the pole at zero frequency may again be explained on physical grounds; if $I_m = 0$, the current source is effectively an open circuit and the response is the constant initial capacitor voltage; a nonzero I_m must therefore produce an infinite voltage response. In other words, if a constant current has been applied to the network forever, then the capacitor must have charged to an infinite voltage. All poles are of course a consequence of the ideal models we are assuming to represent the physical devices; a real capacitor would only permit the voltage to increase to some large value before its dielectric broke down.

A more complicated response curve is obtained for the circuit shown in Fig. 13-8*a*. The current is easily found:

$$I = \frac{100}{6 + \sigma + 5/\sigma}$$

or

$$I = 100\frac{\sigma}{(\sigma + 1)(\sigma + 5)} \tag{19}$$

The response curve is most easily obtained by first indicating the locations of all poles and zeros on the σ axis, and placing vertical asymptotes at the poles. When this is done, we find that a relative minimum of the response must exist between the two poles, and a relative maximum must be present at some frequency greater than zero. By differentiation, the locations of this minimum and maximum may be determined as $\sigma = -\sqrt{5}$ and $\sigma = \sqrt{5}$, respectively. The value of the response at the minimum turns

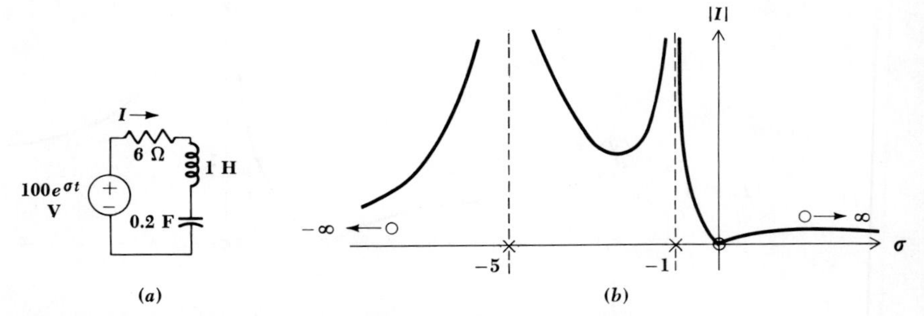

Fig. 13-8 (*a*) A series *RLC* circuit is driven by an exponential forcing function. (*b*) The resultant current-magnitude response curve shows zeros at $\sigma = 0$ and $\sigma = \pm\infty$, and poles at $\sigma = -5$ and -1 Np/s.

out to be 65.5 and that of the maximum, 9.55. To emphasize the relationship between the time domain and frequency domain once again, the response at $\sigma = -3$ is found from (19) to be 75 A; hence, excitation of the network by the forcing function $v(t) = 100e^{-3t}$ produces the forced current response $i(t) = 75e^{-3t}$.

The two poles may again be identified as the natural resonant frequencies of the circuit. That is, the transient response of an *RLC* circuit has the form

$$i_n = A_1 e^{\mathbf{s}_1 t} + A_2 e^{\mathbf{s}_2 t}$$

where $\mathbf{s}_1$ and $\mathbf{s}_2$ are given by

$$\mathbf{s}_{1,2} = -\frac{R}{2L} \pm \sqrt{\left(\frac{R}{2L}\right)^2 - \frac{1}{LC}}$$
$$= -1 \text{ and } -5$$

Thus, a zero-amplitude forcing function at either of these two frequencies, $\sigma = -1$ or $\sigma = -5$, is associated with a finite amplitude response, the natural response of the circuit; a non-zero-amplitude forcing function provides an infinite response or a pole.

Drill Problems

13-9 If v_s represents an exponential forcing function in the circuit shown in Fig. 13-9*a*, find all critical frequencies in the frequency-domain response I_1 and sketch $|I_1|$ versus σ.

Ans. -0.4; -1.23; -0.271; 0 Np/s

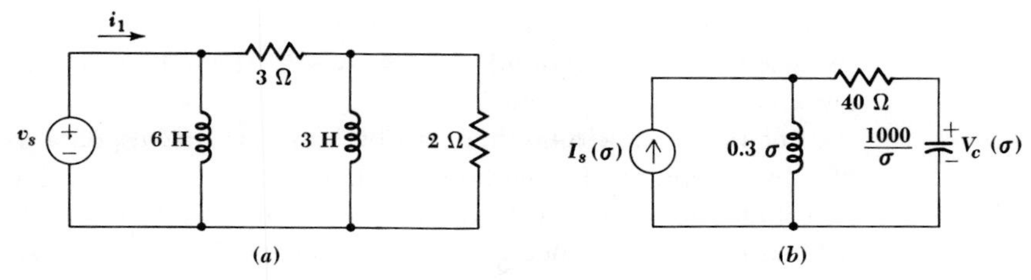

Fig. 13-9 (*a*) See Drill Prob. 13-9. (*b*) See Drill Prob. 13-10.

13-10 For the circuit shown in Fig. 13-9*b*, find all critical frequencies of $V_C(\sigma)$ and sketch $|V_C|$ versus σ.

Ans. -100; -33.3; 0; $\pm\infty$ Np/s

13-11 A two-terminal network is driven by the source $0.4e^{\sigma t}$ A. The voltage across the source is the response, and it has a pole at $\sigma = -2$ and a zero at $\sigma = -1$ Np/s. When $\sigma = 0$, $V = 4$ V. Write an expression for $V(\sigma)$ and evaluate it at $\sigma =$: (*a*) -1.6; (*b*) -6; (*c*) $-\infty$ Np/s.

Ans. -12; 8; 10 V

13-12 The impedance $Z(\sigma)$ may be viewed as a response by considering it as the voltage produced by a current of unit amplitude. Assume that $Z(\sigma)$ has a zero at $\sigma = -5$ and a pole at $\sigma = -1$; the impedance is 12 Ω at infinite frequency. Find: (*a*) $Z(0)$; (*b*) $Z(-3)$; (*c*) $Z(-6)$.

Ans. -12; 2.4; 60 Ω

13-7 THE COMPLEX-FREQUENCY PLANE

Now that we have considered the response of a circuit as ω varies (with $\sigma = 0$) and as σ varies (with $\omega = 0$), we are prepared to develop a more general graphical presentation by graphing quantities as functions of **s**; that is, we wish to show the response simultaneously as functions of both σ and ω.

Such a graphical portrayal of the response as a function of the complex frequency **s** is a useful, enlightening technique in the analysis of circuits, as well as in the design or synthesis of circuits. After we have developed the concept of the complex-frequency plane, or **s** plane, we shall see how quickly the behavior of a circuit can be approximated from a graphical representation of its critical frequencies in this **s** plane. The converse procedure is also very useful; if we are given a desired response curve (the frequency response of a filter, for example), it will be possible to decide upon the necessary location of its poles and zeros in the **s** plane and then to synthesize the filter. This synthesis problem is a subject for detailed study in subsequent courses and is not one which we shall do more than briefly examine here. The **s** plane is also the basic tool with which the possible presence of undesired oscillations is investigated in feedback amplifiers and automatic control systems.

Let us develop a method of obtaining circuit response as a function of **s** by extending the methods we have been using to find the response as a function of either σ or ω. To review these methods, let us obtain the driving-point impedance of a network composed of a 3-Ω resistor in series with a 4-H inductor. As a function of **s**, we have

$$\mathbf{Z}(\mathbf{s}) = 3 + 4\mathbf{s}$$

If we wish to obtain a graphical interpretation of the impedance variation with σ, we let $\mathbf{s} = \sigma + j0$,

$$Z(\sigma) = 3 + 4\sigma$$

and recognize a zero at $\sigma = -\frac{3}{4}$ and a pole at infinity. These critical frequencies are marked on a σ axis, and after identifying the value of $Z(\sigma)$ at some convenient noncritical frequency [perhaps $Z(0) = 3$], it is easy to sketch $|Z(\sigma)|$ versus σ.

In order to plot the response as a function of the radian frequency ω, we let $\mathbf{s} = 0 + j\omega$,

$$\mathbf{Z}(j\omega) = 3 + j4\omega$$

and then obtain both the magnitude and phase angle of $\mathbf{Z}(j\omega)$ as functions of ω:

$$|\mathbf{Z}(j\omega)| = \sqrt{9 + 16\omega^2}$$

$$\text{ang } \mathbf{Z}(j\omega) = \tan^{-1}\frac{4\omega}{3}$$

The magnitude function shows a single pole at infinity and a minimum at $\omega = 0$; it can be readily sketched as a curve of $|\mathbf{Z}(j\omega)|$ versus ω. The phase angle is an inverse tangent function, zero at $\omega = 0$ and $\pm 90°$ at $\omega = \pm\infty$; it is also easily presented as a plot of ang $\mathbf{Z}(j\omega)$ versus ω.

In graphing the response $\mathbf{Z}(j\omega)$ as a function of ω, two two-dimensional plots are required, magnitude and phase angle as functions of ω. When exponential excitation is assumed, we could present all the information on a single two-dimensional graph by permitting both positive and negative values of $Z(\sigma)$ versus σ. However, we chose to plot the magnitude of $Z(\sigma)$ in order that our sketches would compare more closely with those depicting the magnitude of $\mathbf{Z}(j\omega)$. The phase angle ($\pm 180°$ only) of $Z(\sigma)$ was largely ignored. The important point to note is that there is only one independent variable, σ in the case of exponential excitation and ω in the sinusoidal case. Now let us consider what alternatives are available to us if we wish to plot a response as a function of **s**.

The complex frequency **s** requires two parameters, σ and ω, for its complete specification. The response is also a complex function, and we must therefore consider sketching both the magnitude and phase angle as functions of **s**. Either of these quantities, for example, the magnitude, is a function of the two parameters σ and ω, and we can only plot it in two dimensions as a family of curves, such as magnitude versus ω, with σ as parameter. Conversely, we could also show the magnitude versus σ, with ω as the parameter. Such a family of curves represents a tremendous amount of work, however, and this is what we are trying to avoid; it is also questionable whether we could ever draw any useful conclusions from the family of curves even after they were obtained.

A better method of representing the magnitude of some complex response graphically involves using a *three*-dimensional model. Although such a model is difficult to draw on a two-dimensional sheet of paper, we shall find that the model is not difficult to visualize; most of the drawing will be done mentally, where few supplies are needed and construction, correction, and erasures are quickly accomplished. Let us think of a σ axis and a $j\omega$ axis, perpendicular to each other, laid out on a horizontal surface such as the floor. The floor now represents a *complex-frequency plane*, or **s** *plane*, as sketched in Fig. 13-10. To each point in this plane there corresponds exactly one value of **s**, and to each value of **s** we may associate a single point in this complex plane.

Since we are already quite familiar with the type of time-domain function associated with a particular value of the complex frequency **s**, it is now possible to associate the functional form of a forcing function or response with the various regions in the **s** plane. The origin, for example, must represent a dc quantity. Points lying on the σ axis must represent exponential functions, decaying for $\sigma < 0$, increasing for $\sigma > 0$. Pure sinusoids are associated with points on the positive or negative $j\omega$ axis. The right half of the **s** plane, usually referred to simply as the RHP, contains points describing frequencies with positive real parts and thus corresponds

Fig. 13-10 The complex-frequency plane, or **s** plane.

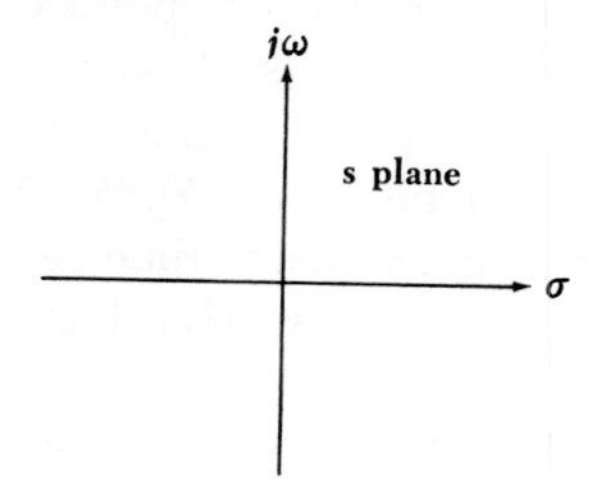

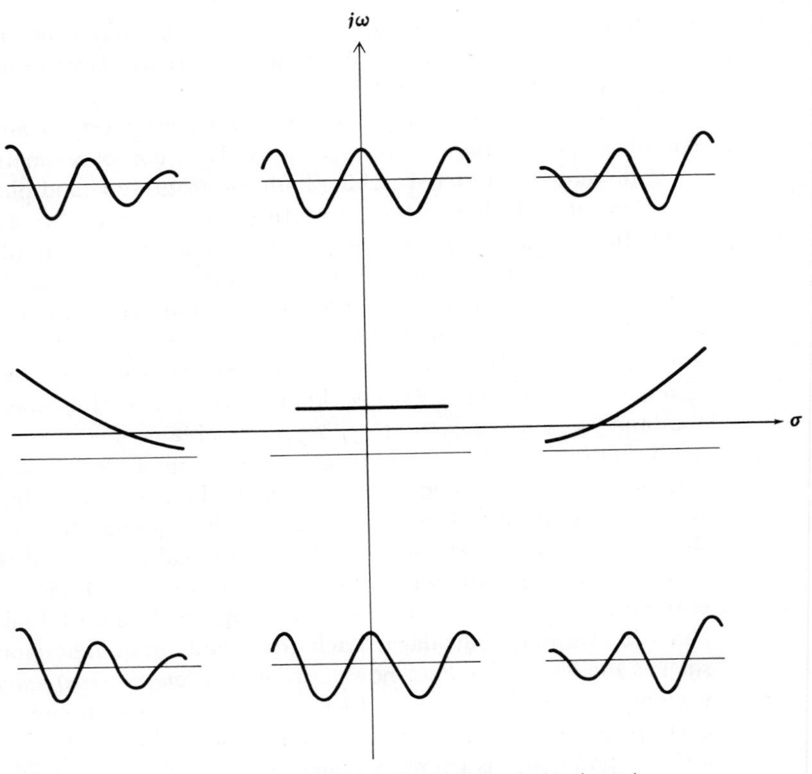

Fig. 13-11 The nature of the time-domain function is sketched in the region of the complex-frequency plane to which it corresponds.

to time-domain quantities which are exponentially increasing sinusoids, except on the σ axis. Correspondingly, points in the left half of the **s** plane (LHP) describe the frequencies of exponentially decreasing sinusoids, again with the exception of the negative σ axis. Figure 13-11 summarizes the relationship between the time domain and the various regions of the **s** plane.

Let us now return to our search for an appropriate method of representing a response graphically as a function of the complex frequency **s**. The magnitude of the response may be represented by constructing, say, a plaster model whose height above the floor at every point corresponds to the magnitude of the response at that value of **s**. In other words, we have added a third axis, perpendicular to both the σ axis and the $j\omega$ axis and passing through the origin; this axis is labeled $|\mathbf{Z}|$, $|\mathbf{Y}|$, $|\mathbf{V}_2/\mathbf{V}_1|$, or with whatever symbol is appropriate. The response magnitude is determined for every value of **s**, and the resultant plot is a *surface* lying above (or just touching) the **s** plane.

Let us try out these preliminary ideas by seeing what such a plaster

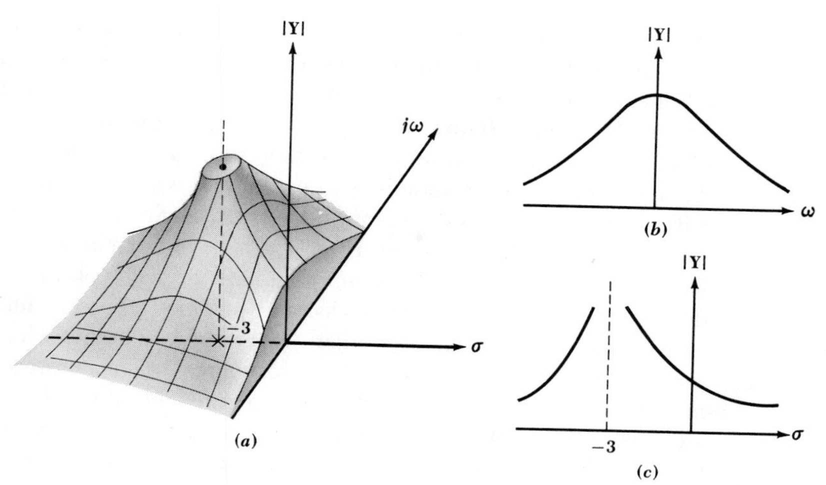

Fig. 13-12 (*a*) A cutaway view of a plaster model whose top surface represents $|\mathbf{Y}(\mathbf{s})|$ for the series combination of a 1-H inductor and a 3-Ω resistor. (*b*) $|\mathbf{Y}(\mathbf{s})|$ as a function of ω. (*c*) $|\mathbf{Y}(\mathbf{s})|$ as a function of σ.

model might look like. As an example, we may consider the admittance of the series combination of a 1-H inductor and a 3-Ω resistor,

$$\mathbf{Y}(\mathbf{s}) = \frac{1}{\mathbf{s} + 3}$$

In terms of both σ and ω we have, as the magnitude,

$$|\mathbf{Y}(\mathbf{s})| = \frac{1}{\sqrt{(\sigma + 3)^2 + \omega^2}}$$

When $\mathbf{s} = -3 + j0$, the response magnitude is infinite; and when $\mathbf{s}$ is infinite, the magnitude of $\mathbf{Y}(\mathbf{s})$ is zero. Thus our model must be infinitely high over the point $(-3 + j0)$, and it must be zero height at all points infinitely far away from the origin. A cutaway view of such a model is shown in Fig. 13-12*a*.

Once the model is constructed, it is simple to visualize the variation of $|\mathbf{Y}|$ as a function of ω (with $\sigma = 0$) by cutting the model with a perpendicular plane containing the $j\omega$ axis. The model shown in Fig. 13-12*a* happens to be cut along this plane, and the desired plot of $|\mathbf{Y}|$ versus ω can be seen; the curve is also drawn in Fig. 13-12*b*. In a similar manner, a vertical plane containing the σ axis enables us to obtain $|\mathbf{Y}|$ versus σ (with $\omega = 0$), shown in Fig. 13-12*c*.

How might we obtain some qualitative response information without doing all this work? After all, most of us have neither the time nor the inclination to be good plasterers, and some more practical method is needed. Let us visualize the **s** plane once again as the floor and then imagine a larger rubber sheet laid on it. We now fix our attention on all the poles and zeros of the response. At each zero, the response is zero, the height of the sheet must be zero, and we therefore tack the sheet to the floor. At the value of **s** corresponding to each pole, we may prop up the sheet with a thin vertical rod. Zeros and poles at infinity must be treated by using a large radius clamping ring or a high circular fence, respectively. If we have used an infinitely large, weightless, perfectly elastic sheet, tacked down with vanishingly small tacks, and propped up with infinitely long, zero-diameter rods, then the rubber sheet assumes a height which is exactly proportional to the magnitude of the response. Less accurate rubber-sheet models may actually be constructed in the laboratory, but their main advantage lies in the ease by which their construction may be visualized from a knowledge of the pole-zero locations of the response.

These comments may be illustrated by considering the configuration of the poles and zeros, sometimes called a *pole-zero constellation,* which locates all the critical frequencies of some frequency-domain quantity, say the impedance $\mathbf{Z}(\mathbf{s})$. Such a pole-zero constellation is shown in Fig. 13-13*a*. If we visualize a rubber-sheet model, tacked down at $\mathbf{s} = -2 + j0$ and propped up at $\mathbf{s} = -1 + j5$ and $-1 - j5$, we should see a terrain whose distinguishing features are two mountains and one conical crater or depression. The portion of the model for the upper left half of the **s** plane is shown in Fig. 13-13*b*.

Let us now build up the expression for $\mathbf{Z}(\mathbf{s})$ which leads to this pole-zero configuration. The zero requires a factor of $(\mathbf{s} + 2)$ in the numerator, and the two poles require the factors $(\mathbf{s} + 1 - j5)$ and $(\mathbf{s} + 1 + j5)$ in the denominator. Except for a multiplying constant k, we now know the form of $\mathbf{Z}(\mathbf{s})$:

Fig. 13-13 (*a*) The pole-zero constellation of some impedance $\mathbf{Z}(\mathbf{s})$. (*b*) A portion of the rubber-sheet model of the magnitude of $\mathbf{Z}(\mathbf{s})$.

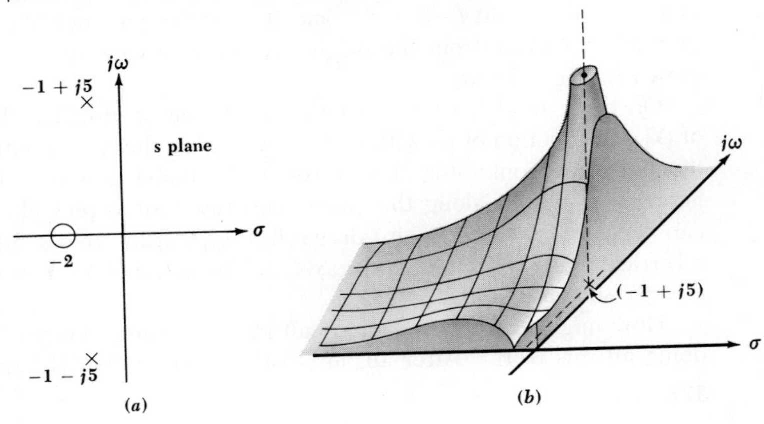

$$\mathbf{Z}(\mathbf{s}) = k\frac{\mathbf{s}+2}{(\mathbf{s}+1-j5)(\mathbf{s}+1+j5)}$$

or

$$\mathbf{Z}(\mathbf{s}) = k\frac{\mathbf{s}+2}{\mathbf{s}^2+2\mathbf{s}+26} \tag{20}$$

Let us select k by assuming a single additional fact about $\mathbf{Z}(\mathbf{s})$; let $\mathbf{Z}(0) = 1$. By direct substitution in (20), we find that k is 13, and therefore

$$\mathbf{Z}(\mathbf{s}) = 13\frac{\mathbf{s}+2}{\mathbf{s}^2+2\mathbf{s}+26} \tag{21}$$

The plots $|Z(\sigma)|$ versus σ and $|\mathbf{Z}(j\omega)|$ versus ω may be obtained exactly from (21), but the general form of the function is apparent from the pole-zero configuration and the rubber-sheet analogy. Portions of these two curves appear at the sides of the model shown in Fig. 13-13*b*.

Thus far, we have been using the **s** plane and the rubber-sheet model to obtain *qualitative* information about the variation of the *magnitude* of the frequency-domain function with frequency. It is possible, however, to get *quantitative* information concerning the variation of both the *magnitude* and *phase angle*. The method provides us with a powerful new tool.

Consider the representation of a complex frequency in polar form, as suggested by an arrow drawn from the origin of the **s** plane to the complex frequency under consideration. The length of the arrow is the magnitude of the frequency, and the angle that the arrow makes with the positive direction of the σ axis is the angle of the complex frequency. The frequency $\mathbf{s}_1 = -3 + j4 = 5\underline{/126.9^\circ}$ is indicated in Fig. 13-14*a*.

Fig. 13-14 (*a*) The complex frequency $\mathbf{s}_1 = -3 + j4$ is indicated by drawing an arrow from the origin to $\mathbf{s}_1$. (*b*) The frequency $\mathbf{s} = j7$ is also represented vectorially. (*c*) The difference $\mathbf{s} - \mathbf{s}_1$ is represented by the vector drawn from $\mathbf{s}_1$ to **s**.

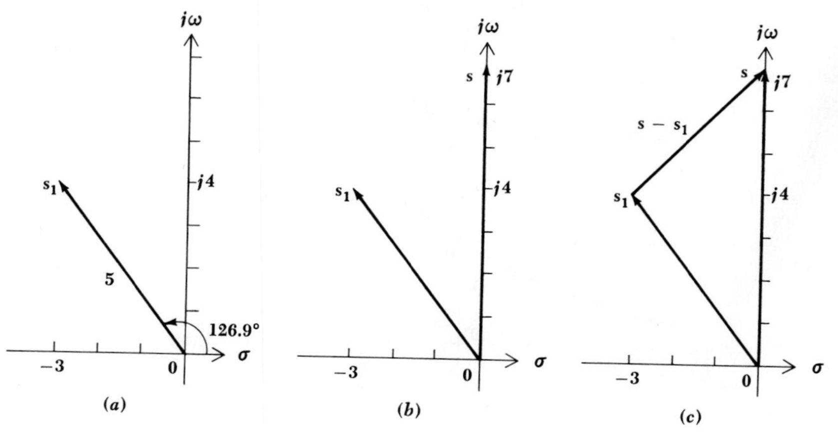

It is also necessary to represent the difference between two values of s as an arrow or vector on the complex plane. Let us select a value of s that corresponds to a sinusoid $\mathbf{s} = j7$ and indicate it also as a vector, as shown in Fig. 13-14*b*. The *difference* $\mathbf{s} - \mathbf{s}_1$ is seen to be the vector drawn from the last-named point $\mathbf{s}_1$ to the first-named point s; the vector $\mathbf{s} - \mathbf{s}_1$ is drawn in Fig. 13-14*c*. Note that $\mathbf{s}_1 + (\mathbf{s} - \mathbf{s}_1) = \mathbf{s}$. Numerically, $\mathbf{s} - \mathbf{s}_1 = j7 - (-3 + j4) = 3 + j3 = 4.24\underline{/45°}$, and this value agrees with the graphical difference.

Let us see how this graphical interpretation of the difference $(\mathbf{s} - \mathbf{s}_1)$ enables us to determine frequency response. Consider the admittance

$$\mathbf{Y}(\mathbf{s}) = \mathbf{s} + 2$$

Fig. 13-15 (*a*) The vector representing the admittance $\mathbf{Y}(\mathbf{s}) = \mathbf{s} + 2$ is shown for $\mathbf{s} = j\omega$. (*b*) Sketches of $|\mathbf{Y}(j\omega)|$ and ang $\mathbf{Y}(j\omega)$ as they might be obtained from the performance of the vector as s moves up or down the $j\omega$ axis from the origin.

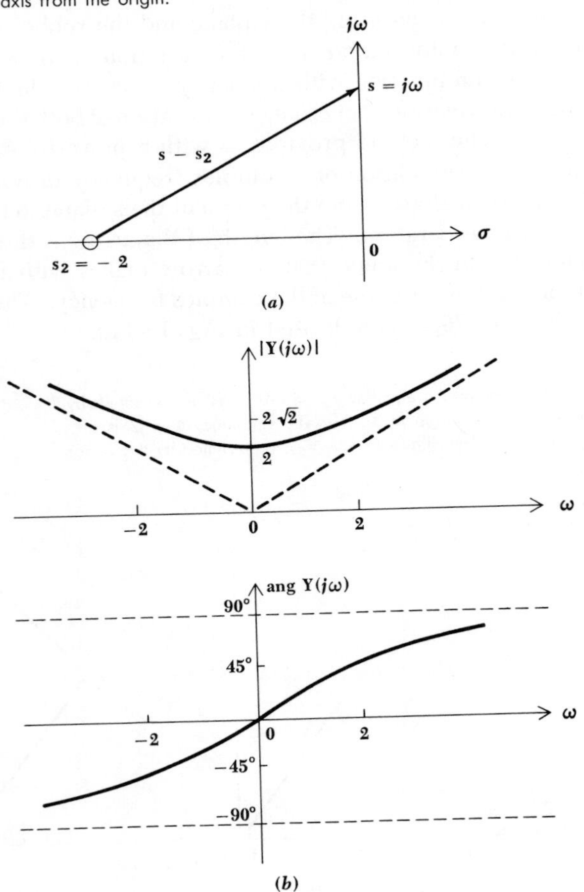

This expression may be interpreted as the difference between some frequency of interest $\mathbf{s}$ and a zero location. Thus, the zero is present at $\mathbf{s}_2 = -2 + j0$, and the factor $\mathbf{s} + 2$, which may be written as $\mathbf{s} - \mathbf{s}_2$, is represented by the vector drawn from the zero location $\mathbf{s}_2$ to the frequency $\mathbf{s}$ at which the response is desired. If the sinusoidal response is desired, $\mathbf{s}$ must lie on the $j\omega$ axis, as illustrated in Fig. 13-15*a*. The magnitude of $\mathbf{s} + 2$ may now be visualized as ω varies from zero to infinity. When $\mathbf{s}$ is zero, the vector has a magnitude of 2 and an angle of 0°. Thus $\mathbf{Y}(0) = 2$. As ω increases, the magnitude increases, slowly at first, and then almost linearly with ω; the phase angle increases almost linearly at first, and then gradually approaches 90° as ω becomes infinite. At $\omega = 7$, $\mathbf{Y}(j7)$ has a magnitude of $\sqrt{2^2 + 7^2}$ and has a phase angle of $\tan^{-1}(3.5)$. The magnitude and phase of $\mathbf{Y}(\mathbf{s})$ are sketched as functions of ω in Fig. 13-15*b*.

Let us now construct a more realistic example by considering a frequency-domain function given by the quotient of two factors,

$$\mathbf{V}(\mathbf{s}) = \frac{\mathbf{s} + 2}{\mathbf{s} + 3}$$

We again select a value of $\mathbf{s}$ which corresponds to sinusoidal excitation and draw the vectors $\mathbf{s} + 2$ and $\mathbf{s} + 3$, the first from the zero to the chosen point on the $j\omega$ axis and the second from the pole to the chosen point. The two vectors are sketched in Fig. 13-16*a*. The quotient of these two vectors has a magnitude equal to the quotient of the magnitudes and a phase angle equal to the difference of the numerator and denominator phase angles. An investigation of the variation of the magnitude of $\mathbf{V}(\mathbf{s})$ versus ω is made by allowing $\mathbf{s}$ to move from the origin up the $j\omega$ axis and considering the ratio of the distance from the zero to $\mathbf{s} = j\omega$ and the distance from the pole to the same point on the $j\omega$ axis. The ratio evidently is $2/3$ at $\omega = 0$ and approaches unity as ω becomes infinite. A consideration of the difference of the two phase angles shows that ang $\mathbf{V}(j\omega)$ is 0° at $\omega = 0$, increases at first as ω increases since the angle of the vector $\mathbf{s} + 2$ is greater than that of $\mathbf{s} + 3$, and then decreases with a further increase in ω, finally approaching 0° at infinite frequency, where both vectors possess 90° angles. These results are sketched in Fig. 13-16*b*. Although no quantitative markings are present on these sketches, it is important to note that they may be obtained easily. For example, the complex response at $\mathbf{s} = j4$ must be given by the ratio

$$\mathbf{V}(j4) = \frac{\sqrt{4 + 16}\,\underline{/\tan^{-1}(4/2)}}{\sqrt{9 + 16}\,\underline{/\tan^{-1}(4/3)}}$$

$$\mathbf{V}(j4) = \sqrt{20/25}\;\underline{/\tan^{-1} 2 - \tan^{-1}(4/3)}$$

$$= 0.894\underline{/10.3^\circ}$$

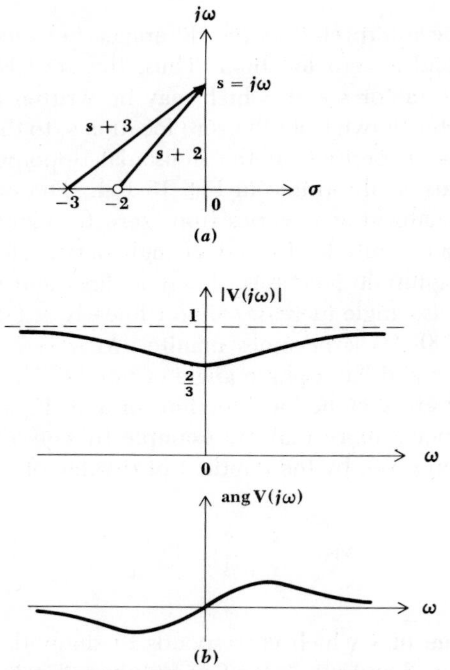

Fig. 13-16 (*a*) Vectors are drawn from the two critical frequencies of the voltage response $\mathbf{V}(s) = (s + 2)/(s + 3)$. (*b*) Sketches of the magnitude and phase angle of $\mathbf{V}(j\omega)$ as obtained from the quotient of the two vectors shown in (*a*).

In designing networks to produce some desired response, the behavior of the vectors drawn from each critical frequency to a general point on the $j\omega$ axis is an important aid. For example, if it were necessary to increase the hump in the phase response of Fig. 13-16*b*, we can see that we must provide a greater difference in the angles of the two vectors. This may be achieved in Fig. 13-16*a* either by moving the zero closer to the origin or by locating the pole farther from the origin, or both.

The ideas we have been discussing to help in the graphical determination of the magnitude and angular variation of some frequency-domain function with frequency will be needed in the following chapter when we investigate the frequency performance of highly selective filters, or resonant circuits. These concepts are fundamental in obtaining a quick, clear understanding of the behavior of electrical networks and other engineering systems. The procedure is briefly summarized as follows:

1 Draw the pole-zero constellation of the frequency-domain function under consideration in the s plane, and locate a test point corresponding to the frequency at which the function is to be evaluated.

2 Draw an arrow from each pole and zero to the test point.
3 Determine the length of each pole arrow and zero arrow and the value of each pole-arrow angle and zero-arrow angle.
4 Divide the product of the zero-arrow lengths by the product of the pole-arrow lengths. This quotient is the magnitude of the frequency-domain function for the assumed frequency of the test point [within a multiplying constant, since $\mathbf{F}(\mathbf{s})$ and $k\mathbf{F}(\mathbf{s})$ have the same pole-zero constellations].
5 Subtract the sum of the pole-arrow angles from the sum of the zero-arrow angles. The resultant difference is the angle of the frequency-domain function, evaluated at the frequency of the test point. The angle does not depend upon the value of the real multiplying constant k.

Drill Problems

13-13 An impedance consists of a 2.5-mH inductor in series with the parallel combination of a 5-Ω resistor and a 50-μF capacitor. Find all the critical frequencies of $\mathbf{Z}(\mathbf{s})$ and draw the pole-zero configuration.

Ans. $-2 - j2;\ -2 + j2;\ -4;\ \infty$ *krad/s*

13-14 For the pole-zero constellations shown in Fig. 13-17, assume that they represent a voltage ratio $\mathbf{V}_2/\mathbf{V}_1$, and that the value of the ratio at infinite frequency is 20. Express the ratios as ratios of polynomials in $\mathbf{s}$.

Ans. $\dfrac{20\mathbf{s}^2 - 80}{\mathbf{s}^2 + 4};\ \dfrac{20\mathbf{s} + 200}{\mathbf{s} + 20};\ \dfrac{20\mathbf{s}^3 + 1600\mathbf{s}^2 + 40{,}000\mathbf{s}}{\mathbf{s}^3 + 20\mathbf{s}^2 + 2500\mathbf{s} + 50{,}000}$

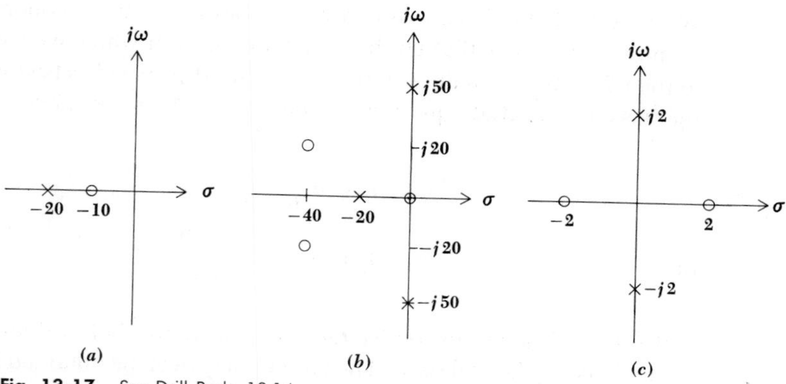

Fig. 13-17 See Drill Prob. 13-14.

13-15 In a certain network, the output current $\mathbf{I}_2$ produced by an input voltage $\mathbf{V}_1$ is found to have poles at $-3 - j4$ and $-3 + j4$, and zeros at $-1 + j0$ and

$-4 + j0$. A plaster model of the response function $\mathbf{I}_2/\mathbf{V}_1$ has a height of 8 cm at the origin. What is its height at $\mathbf{s} =$: (*a*) -2; (*b*) $-2 - j2$; (*c*) $2 + j2$?

Ans. *5.88; 23.2; 27.1 cm*

13-16 A pole-zero configuration shows zeros at $-1 \pm j4$ and poles at $-1 \pm j2$. By sketching this constellation and visualizing the pertinent vectors, estimate the positive radian frequency at which the response: (*a*) magnitude is a minimum; (*b*) magnitude is a maximum; (*c*) phase angle is a negative maximum ($\doteq -90°$).

Ans. *1.5, 3, 4 (approx.); 1.43, 3.02, 4.47 (exact) rad/s*

13-8 NATURAL RESPONSE AND THE s PLANE

There is a tremendous amount of information contained in the pole-zero plot of some response in the **s** plane. We shall find out how the complete current response, natural plus forced, produced by an arbitrary voltage forcing function can be quickly written from the pole-zero configuration of the impedance offered to the voltage source and from the initial conditions; the method is similarly effective for the dual problem, the complete voltage response produced by a current source.

Let us introduce the method by considering the simplest example, a series RL circuit as shown in Fig. 13-18. A general voltage source $v(t)$ causes the current $i(t)$ to flow after closure of the switch at $t = 0$. The complete response $i(t)$ is composed of a natural response and a forced response:

$$i(t) = i_n(t) + i_f(t) \tag{22}$$

We may find the forced response by working in the frequency domain, assuming, of course, that $v(t)$ has a functional form which we can transform to the frequency domain; if $v(t) = t^2$, we must proceed as best we can from the basic differential equation for the circuit. Here we have

$$\mathbf{I}_f(\mathbf{s}) = \frac{\mathbf{V}(\mathbf{s})}{R + \mathbf{s}L}$$

or

$$\mathbf{I}_f(\mathbf{s}) = \frac{1}{L}\frac{\mathbf{V}(\mathbf{s})}{\mathbf{s} + R/L} \tag{23}$$

and $i_f(t)$ is obtained by replacing **s**, L, and R by their values, reinserting $e^{\mathbf{s}t}$, and taking the real part. The answer may even be obtained as a function of a general ω, σ, R, and L if desired.

Now let us consider the natural response. Of course, we know that the form will be a decaying exponential with the time constant L/R, but we may pretend that we are finding it for the first time. The natural response or *source-free* response is, by definition, of a form independent

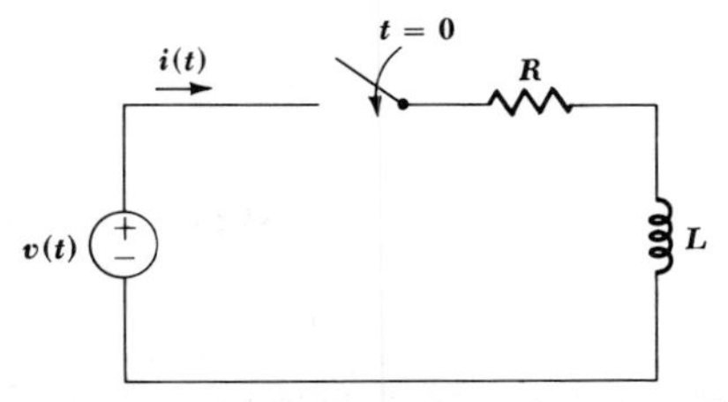

Fig. 13-18 An example which illustrates the determination of the complete response through a knowledge of the critical frequencies of the impedance faced by the source.

of the forcing function; the forcing function contributes, along with the other initial conditions, only to the magnitude of the natural response. To find the proper form, replace all independent sources by their internal impedances; here, $v(t)$ is replaced by a short circuit. Now let us try to obtain this natural response as a limiting case of the forced response; we return to the frequency-domain expression (23) and obediently set $\mathbf{V}(\mathbf{s}) = 0$. On the surface, it appears that $\mathbf{I}(\mathbf{s})$ must also be zero, but this is not necessarily true; the denominator may be zero. In other words, if we apply no voltage at the frequency $\mathbf{s} = -R/L$, some current at this frequency may flow. It is of course necessary that there be some energy stored initially in the circuit for this to occur.

Let us inspect this new idea from a slightly different vantage point. The forced response to the given voltage source is

$$\mathbf{I}_f(\mathbf{s}) = \frac{\mathbf{V}(\mathbf{s})}{\mathbf{Z}(\mathbf{s})}$$

If we happen to apply a voltage at the exact frequency of one of the zeros of $\mathbf{Z}(\mathbf{s})$, then an infinite current will flow. This is true even though only 1 μV is applied. We then conclude that a finite current at this frequency may flow even though no voltage is applied. Since the circuit is now source-free, that current is of the form of the source-free or natural response.

Returning to our series RL circuit, we see by (23) that infinite current results when the operating frequency is $\mathbf{s} = -R/L + j0$. A finite current at this frequency thus represents the natural response

$$\mathbf{I}(\mathbf{s}) = A \qquad \text{at } \mathbf{s} = -\frac{R}{L} + j0$$

Transforming this natural response to the time domain,

$$i_n(t) = \text{Re}\,(Ae^{-Rt/L})$$

or

$$i_n(t) = Ae^{-Rt/L}$$

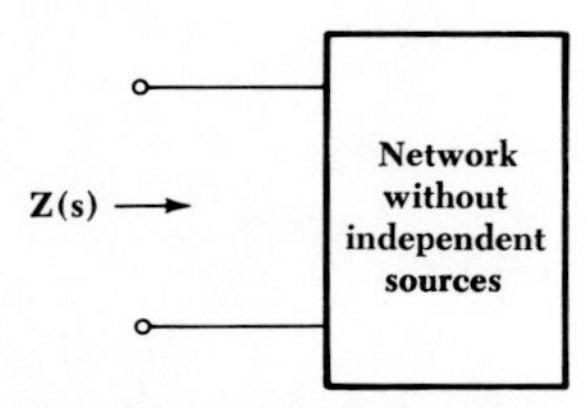

Fig. 13-19 The zeros of $\mathbf{Z}(\mathbf{s})$ determine the form of the natural response of the current which would be present in a short circuit across the input terminals.

To complete this example, the total response is then

$$i(t) = Ae^{-Rt/L} + i_f(t)$$

and A may be determined once the initial conditions are specified for this circuit.

Now let us generalize these results. Figure 13-19 shows a general two-terminal network that contains no independent sources. The input impedance may be written in a form which displays all the critical frequencies:

$$\mathbf{Z}(\mathbf{s}) = k\frac{(\mathbf{s} - \mathbf{s}_1)(\mathbf{s} - \mathbf{s}_3)\cdots}{(\mathbf{s} - \mathbf{s}_2)(\mathbf{s} - \mathbf{s}_4)\cdots} \tag{24}$$

It is customary to designate the zeros of an impedance or the poles of an admittance by odd-numbered subscripts; impedance poles and admittance zeros receive even-numbered subscripts. Hence, the zeros of $\mathbf{Z}(\mathbf{s})$ occur at $\mathbf{s} = \mathbf{s}_1$, $\mathbf{s} = \mathbf{s}_3$, etc., and if a voltage is applied to the input terminals at any of these frequencies, infinite current will flow. Thus, a finite current at each of these frequencies must be a possible functional form for the natural response. Therefore, we think of a zero-volt source (which is just a short circuit) applied to the input terminals; the natural response which occurs when the input terminals are short-circuited must thus have the form

$$i_n(t) = \mathbf{A}_1 e^{\mathbf{s}_1 t} + \mathbf{A}_3 e^{\mathbf{s}_3 t} + \cdots$$

where the **A**'s must be evaluated in terms of the initial conditions (including the initial value of any voltage source applied at the input terminals).

Let us now consider the dual of this problem. Again we may refer to the network shown in Fig. 13-19, but let us suppose for the moment that a current source is applied at the input terminals. The input voltage response is then

$$\mathbf{V}(\mathbf{s}) = \mathbf{I}(\mathbf{s})\mathbf{Z}(\mathbf{s})$$

If we happen to apply a current at the frequency of one of the poles of **Z**(s), an infinite-amplitude input voltage results; it follows then that if we apply a zero-amplitude current (an open circuit) at this frequency, a finite response may be present at the same frequency. This is a source-free response, and it is identifiable as the natural response. Hence, if we assume the input impedance is expressed in the form of (24), then the functional form of the natural voltage response across the open-circuited input terminals must be

$$v_n(t) = \mathbf{A}_2 e^{\mathbf{s}_2 t} + \mathbf{A}_4 e^{\mathbf{s}_4 t} + \cdots$$

The open-circuit natural voltage response is composed of a sum of terms at the frequencies of the poles of the input impedance or, stated slightly differently, at the frequencies of the zeros of the input admittance. The short-circuit natural current response is composed of a sum of terms at the frequencies of the zeros of the input impedance or the poles of the input admittance.

Let us now try out these techniques on two examples. The first circuit is shown in Fig. 13-20, and it is identical with that of Fig. 5-13, a network whose natural response was found earlier with a great deal of difficulty. Let us again find the form of i_1 and i_2. Considering i_1 first, we must find a suitable location for the short circuit through which this current flows; let us use the points marked x and x'. The input impedance viewed from these terminals when they are open-circuited is

$$\mathbf{Z}(\mathbf{s}) = 2\mathbf{s} + 1 + \frac{6\mathbf{s}}{3\mathbf{s} + 2}$$

or

$$\mathbf{Z}(\mathbf{s}) = 2\frac{(\mathbf{s} + 2)(\mathbf{s} + \frac{1}{6})}{\mathbf{s} + \frac{2}{3}}$$

Thus, i_1 must be of the form

$$i_1(t) = A_1 e^{-2t} + A_3 e^{-t/6}$$

Fig. 13-20 The circuit shown in Fig. 5-13 is redrawn here. The natural responses i_1 and i_2 are desired.

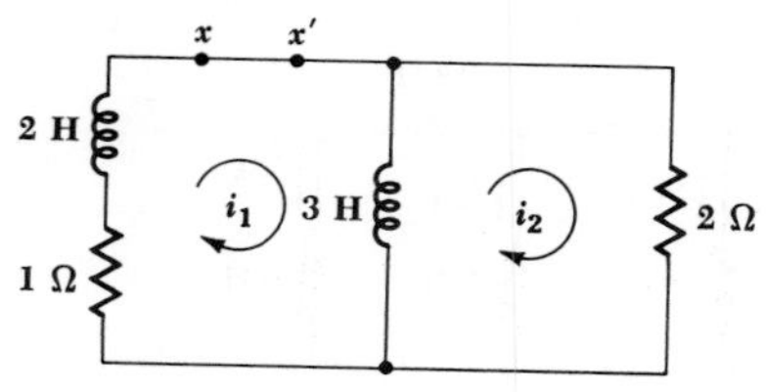

The solution is completed by using the given initial conditions (i_1 and i_2 are both 11 A at $t = 0$) to establish the values of A_1 and A_3. Although we have been able to obtain the *form* of the solution much more rapidly than we did before, we cannot shorten the procedure for evaluating the unknown constants; any saving in effort here must come later with a study of the Laplace transform. For practice, let us complete the solution for i_1. Since $i_1(0)$ is given as 11, then

$$11 = A_1 + A_3$$

The necessary additional equation is obtained by differentiating i_1:

$$\left.\frac{di_1}{dt}\right|_{t=0} = -2A_1 - \tfrac{1}{6}A_3$$

But $2di_1/dt$ is the voltage across the 2-H inductor, and Kirchhoff's voltage law shows that this must be the sum of the initial voltages across the two resistors. Thus

$$\left.\frac{di_1}{dt}\right|_{t=0} = \frac{11 + 22}{2} = -2A_1 - \tfrac{1}{6}A_3$$

Thus, $A_1 = 3$ and $A_3 = 8$, and the desired solution is

$$i_1(t) = 3e^{-2t} + 8e^{-t/6}$$

The solution for i_2 obtained in Chap. 5 may be checked by inspecting the impedance obtained when the right-hand mesh is broken.

As our last example we shall find the complete response $v(t)$ of the circuit shown in Fig. 13-21. The switch is closed prior to $t = 0$, and thus all currents and voltages to the right of the switch are initially zero. At $t = 0$ the switch is opened, and the voltage across the switch is to be found. This response is composed of both a forced and a natural response,

$$v(t) = v_n(t) + v_f(t)$$

Fig. 13-21 A circuit whose complete response is to be found through an investigation of its critical frequencies.

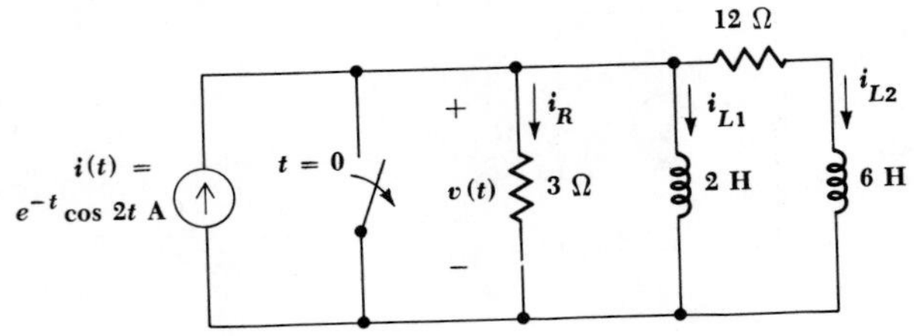

Each may be found through a knowledge of the pole-zero configuration of the input impedance or admittance of the portion of the network to the right of the switch. Remembering that an inactive current source is an open circuit, and that the switch is also an open circuit for $t > 0$, the natural response $v_n(t)$ is thus composed of a sum of terms, each term corresponding to a pole of the input impedance or a zero of the input admittance. We have for $\mathbf{Y}(\mathbf{s})$

$$\mathbf{Y}(\mathbf{s}) = \frac{1}{3} + \frac{1}{2\mathbf{s}} + \frac{1}{6\mathbf{s} + 12}$$

or, after combining and factoring,

$$\mathbf{Y}(\mathbf{s}) = \frac{1}{3}\frac{(\mathbf{s} + 1)(\mathbf{s} + 3)}{\mathbf{s}(\mathbf{s} + 2)} \tag{25}$$

The form of the natural response may now be written

$$v_n(t) = A_2e^{-t} + A_4e^{-3t}$$

In order to find the forced response, the frequency-domain current source $\mathbf{I}(\mathbf{s}) = 1$ at $\mathbf{s} = -1 + j2$ may be divided by the input admittance, evaluated at $\mathbf{s} = -1 + j2$,

$$\mathbf{V}(\mathbf{s}) = \frac{\mathbf{I}(\mathbf{s})}{\mathbf{Y}(\mathbf{s})} = \frac{1}{\mathbf{Y}(-1 + j2)}$$

$$= 3\frac{(-1 + j2)(1 + j2)}{j2(2 + j2)}$$

and thus

$$\mathbf{V}(\mathbf{s}) = 1.875\sqrt{2}\,\underline{/45^\circ}$$

Transforming to the time domain, we have

$$v_f(t) = 1.875\sqrt{2}e^{-t}\cos(2t + 45^\circ)$$

The complete response is therefore

$$v(t) = A_2e^{-t} + A_4e^{-3t} + 1.875\sqrt{2}e^{-t}\cos(2t + 45^\circ)$$

Since the current through both inductors is initially zero, the initial source current of 1 A must flow through the 3-Ω resistor. Thus

$$v(0) = 3 = A_2 + A_4 + \frac{1.875\sqrt{2}}{\sqrt{2}} \tag{26}$$

Again it is necessary to differentiate and then to obtain an initial condition for dv/dt. We first find

$$\left.\frac{dv}{dt}\right|_{t=0} = 1.875\sqrt{2}\left(-\frac{2}{\sqrt{2}} - \frac{1}{\sqrt{2}}\right) - A_2 - 3A_4$$

or

$$\left.\frac{dv}{dt}\right|_{t=0} = -5.625 - A_2 - 3A_4 \tag{27}$$

The initial value of this rate of change is obtained by analyzing the circuit. However, those rates of change which are most easily found are the derivatives of the inductor currents, for $v = L\,di/dt$, and the initial values of the inductor voltages should not be difficult to find. We therefore express the response $v(t)$ in terms of the resistor current,

$$v(t) = 3i_R$$

and then apply Kirchhoff's current law,

$$v(t) = 3i - 3i_{L1} - 3i_{L2}$$

Now we may take the derivative,

$$\frac{dv}{dt} = 3\frac{di}{dt} - 3\frac{di_{L1}}{dt} - 3\frac{di_{L2}}{dt}$$

Differentiation of the source function and evaluation at $t = 0$ provide a value of -3 V/s for the first term; the second term is numerically $3/2$ of the initial voltage across the 2-H inductor, or -4.5 V/s; and the last term is -1.5 V/s. Thus,

$$\left.\frac{dv}{dt}\right|_{t=0} = -9$$

and we may now use (26) and (27) to determine the unknown amplitudes

$$A_2 = 0 \qquad A_4 = 1.125$$

The complete response is therefore

$$v(t) = 1.125e^{-3t} + 1.875\sqrt{2}e^{-t}\cos(2t + 45°)$$

In spite of the detailed process which we must pursue to evaluate the amplitude coefficients of the natural response, except in those cases where the initial values of the desired response and its derivatives are obvious, we should not lose sight of the ease and rapidity with which the form of the natural response can be obtained.

Drill Problems

13-17 Find the frequency components present in the natural response $i(t)$ of Fig. 13-22 when a charged 3-μF capacitor is suddenly connected to terminals: (*a*) *a*-*b*; (*b*) *b*-*c*; (*c*) *a*-*c*.

Ans. -698, -79.6; -500, -111; -50 kNp/s

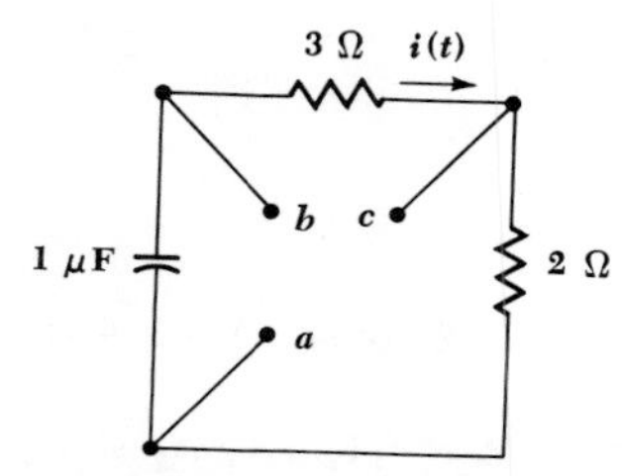

Fig. 13-22 See Drill Prob. 13-17.

13-18 The circuit shown in Fig. 13-23 contains stored energy at $t = 0$. Find the complex frequencies present in the response $v(t)$ for $t > 0$.

Ans. -1; -2; -3 Np/s

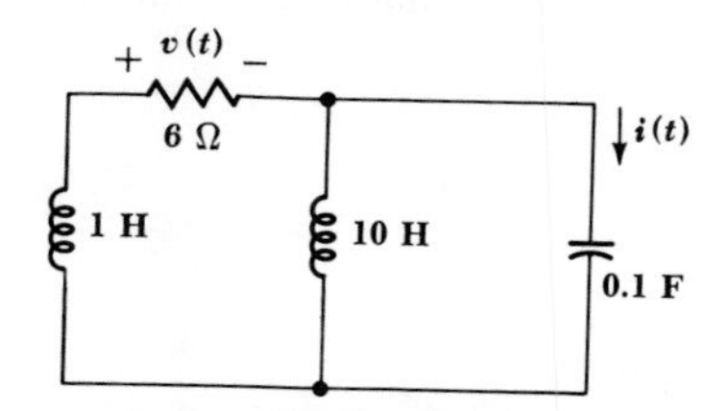

Fig. 13-23 See Drill Prob. 13-18 and Prob. 13-22.

13-19 Find all the complex frequencies present in the complete voltage response $v(t)$ in the circuit of Fig. 13-24 for $t > 0$ if $v_s(t) = 20(\cos 1000\pi t)u(t)$ V.

Ans. $\pm j1000\pi$; $\pm j1000$; $\pm j2000$ s^{-1}

Fig. 13-24 See Drill Prob. 13-19.

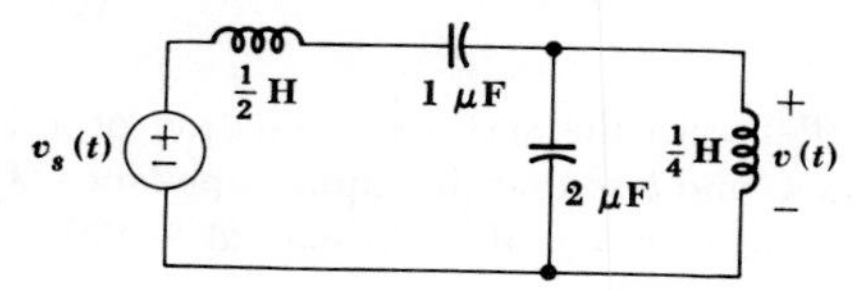

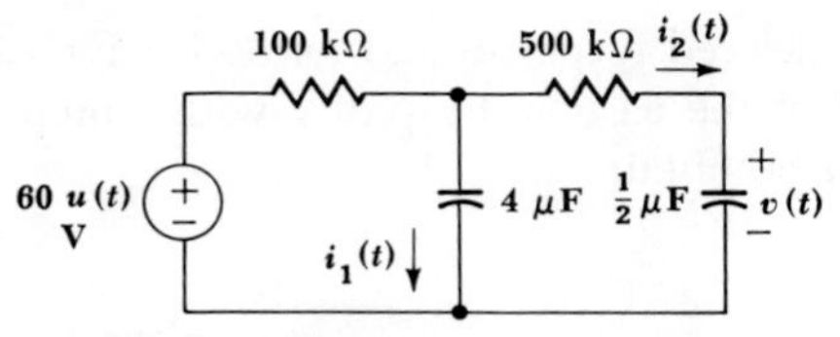

Fig. 13-25 See Drill Prob. 13-20.

13-20 For the circuit shown in Fig. 13-25, find the complete response: (*a*) $v(t)$; (*b*) $i_2(t)$; (*c*) $i_1(t)$.

Ans. $0.1(e^{-2t} - e^{-5t})u(t)$ mA; $(0.4e^{-2t} + 0.2e^{-5t})u(t)$ mA; $(60 - 100e^{-2t} + 40e^{-5t})u(t)$ V

PROBLEMS

☐ **1** If $\mathbf{V}(\mathbf{s}) = 2\mathbf{s} + 10/\mathbf{s}$, find $v(t)$ for $\mathbf{s} =$: (*a*) -2; (*b*) $j4$; (*c*) $-2 + j4$; (*d*) $-2 - j4$.

☐ **2** If $i(t)$ is represented by the sum $\mathbf{A}_1 e^{\mathbf{s}_1 t} + \mathbf{A}_2 e^{\mathbf{s}_2 t}$, find $\mathbf{A}_1$, $\mathbf{A}_2$, $\mathbf{s}_1$, and $\mathbf{s}_2$ if $i(t) =$: (*a*) $6e^{-2t} \cos 3t$; (*b*) $6e^{-2t} \sin (3t + 30°)$; (*c*) $6 \cos 3t$; (*d*) $6e^{-2t}$.

☐ **3** The current, $i_s = 0.3e^{\sigma t} \cos \omega t$ A, is present in the admittance $\mathbf{Y}(\mathbf{s}) = 0.01 + 0.6/\mathbf{s}$. Find the resultant voltage $v(t)$ if: (*a*) $\sigma = -10$, $\omega = 0$; (*b*) $\sigma = 0$, $\omega = 30$; (*c*) $\sigma = -10$, $\omega = 30$.

☐ **4** Let $i_s = 2e^{-3000t} \cos (4000t + 21°)$ mA in the circuit shown in Fig. 13-26. Find: (*a*) $v_1(t)$; (*b*) $v_2(t)$.

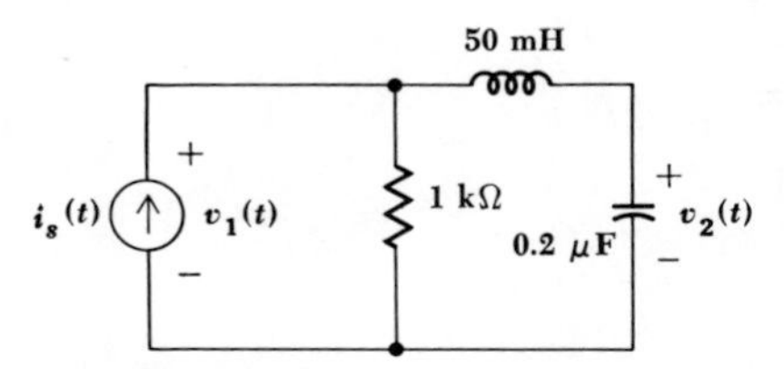

Fig. 13-26 See Prob. 4.

☐ **5** The voltage $80e^{-3t} \sin 4t$ V is present across the series combination of a 20-Ω resistor and a 4-H inductor. Assume steady-state conditions and determine the total energy delivered to the combination after $t = 0$.

☐ **6** A capacitor C is in parallel with the series combination of a 10-Ω resistor and an inductor L. Determine C and L so that the input impedance of the parallel combination possesses a: (*a*) zero at $\mathbf{s} = -100$; (*b*) pole at $-100 + j20$ s^{-1}.

☐ **7** Given an *RLC* series circuit, determine the element values so that the input admittance has poles at $-20 \pm j140\ \mathrm{s}^{-1}$ and $|\mathbf{Y}(j100)| = 0.1\ \mho$.

☐ **8** Sketch $\mathbf{Z}(\sigma)$ and $|\mathbf{Z}(\sigma)|$ versus σ for the network shown in Fig. 13-27 if the unknown element is: (*a*) a short circuit; (*b*) an open circuit; (*c*) a 6-Ω resistor; (*d*) a 2-H inductor; (*e*) a 10^4-μF capacitor.

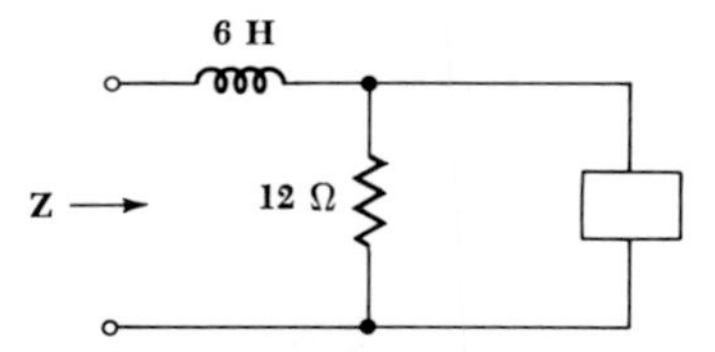

Fig. 13-27 See Probs. 8 and 9.

☐ **9** Make the sketches requested in Prob. 8 if the unknown element in the circuit of Fig. 13-27 is a dependent voltage source, positive reference on top, labeled $2i_{in}$, where i_{in} is the current to the right in the 6-H inductor.

☐ **10** For each of the three networks shown in Fig. 13-28, determine all the critical frequencies of the input impedance and indicate their locations on the **s** plane.

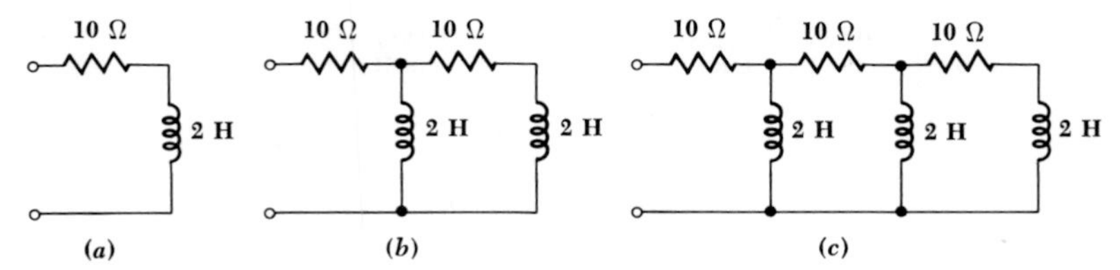

Fig. 13-28 See Prob. 10.

☐ **11** The series branches R_1-L_1 and R_2-C_2 are connected in parallel. Find all critical frequencies of the admittance of the parallel combination if $R_1 = R_2 = 50\ \Omega$, $C_2 = 1\ \mu\mathrm{F}$, and $L_1 =$: (*a*) 1.6 mH; (*b*) 2.5 mH; (*c*) 5 mH.

☐ **12** A 10-Ω resistor is in series with the parallel combination of a 20-Ω resistor and a 5-H inductor. Make a pole-zero plot of $\mathbf{Z}(\mathbf{s})$ on the **s** plane and prepare a qualitative sketch of $|\mathbf{Z}|$ versus σ and $|\mathbf{Z}|$ versus ω.

☐ **13** Make an **s**-plane plot of the poles and zeros of the transfer admittance, $\mathbf{Y} = \mathbf{I}_1/\mathbf{V}_s$, for the circuit shown in Fig. 13-29. Using this plot, sketch $|\mathbf{Y}|$ versus ω; indicate values at $\omega = 0$ and $\omega = \infty$.

☐ **14** Determine the voltage gain $\mathbf{V}_0(\mathbf{s})/\mathbf{V}_1(\mathbf{s})$ for the transistor-amplifier equivalent circuit shown in Fig. 13-30. Locate all critical frequencies of the ratio on the **s** plane and sketch the gain magnitude as a function of ω.

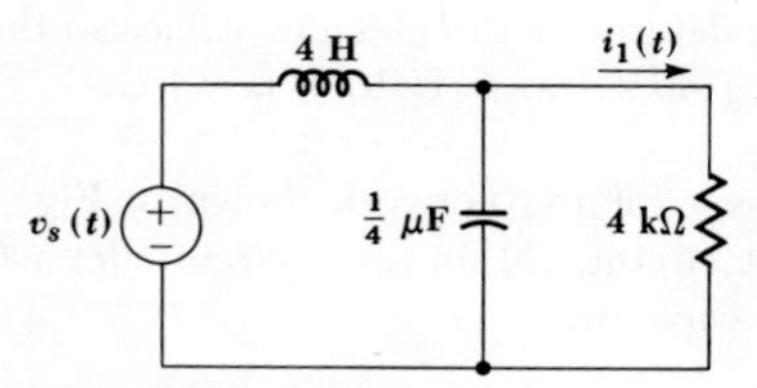

Fig. 13-29 See Prob. 13.

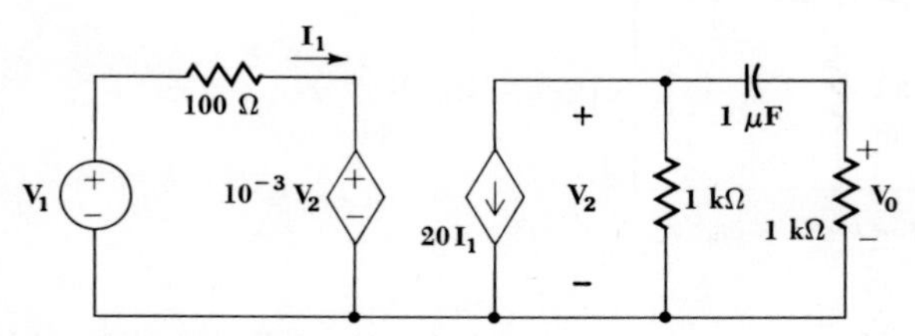

Fig. 13-30 See Prob. 14.

☐ **15** The circuit of Fig. 13-31 shows an equivalent circuit for a transformer (Chap. 15). Determine the poles and zeros and sketch the magnitude versus ω curve for the ratio: (*a*) $\mathbf{I}_2/\mathbf{I}_1$; (*b*) $\mathbf{V}_0/\mathbf{V}_1$.

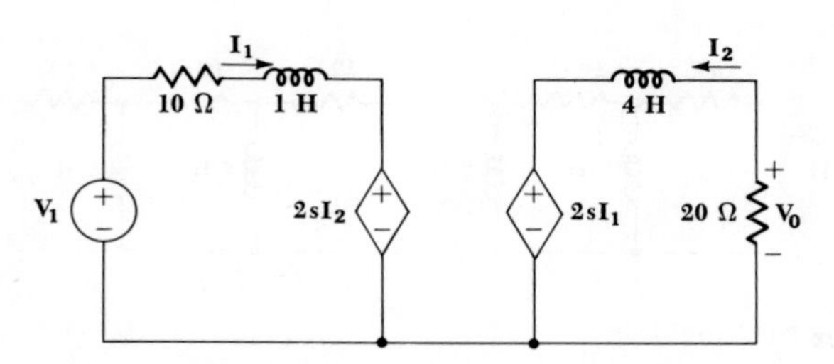

Fig. 13-31 See Prob. 15.

☐ **16** An admittance $\mathbf{Y}(s)$ has only one pole and one zero. When $s = -1, -3$, and $\pm\infty$, $\mathbf{Y}(s) = 3, 1$, and 5 ℧, respectively. Find an expression for $\mathbf{Y}(s)$ and locate the critical frequencies in the s plane.

☐ **17** The voltage ratio $\mathbf{V}_2/\mathbf{V}_1$ for the circuit shown in Fig. 13-32 has a pole at $-100 + j700$. If $R = 500\ \Omega$, find L and C.

Fig. 13-32 See Prob. 17.

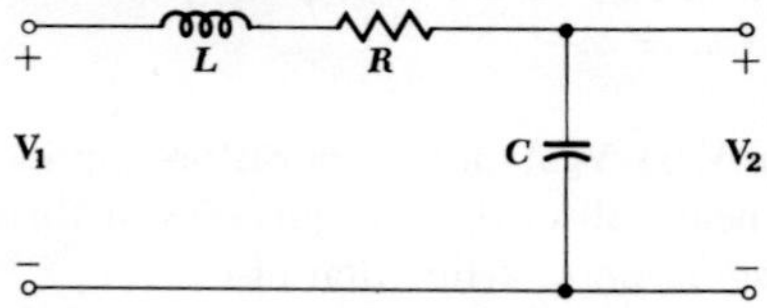

☐ **18** The critical frequencies of a certain impedance are zeros at $s = -1 \pm j10$ and poles at $s = 0$ and ∞. Let $|\mathbf{Z}|$ be 2 Ω at $s = -1 + j0$. (*a*) Determine the expression for $\mathbf{Z}(s)$. (*b*) Sketch $|\mathbf{Z}(j\omega)|$ versus ω. (*c*) By considering the form of $\mathbf{Z}(s)$, find the simplest circuit that will provide this impedance.

☐ **19** A battery V_0, an open switch that closes at $t = 0$, and an impedance $\mathbf{Z}(s) = 2(s + 2)/(s + 4)$ are in series. (*a*) Determine the appropriate form for the current in the circuit after $t = 0$. (*b*) Evaluate the unknown amplitudes by making use of the information that $i(0^+) = 6$ A and $di/dt = 12$ A/s at $t = 0^+$. (*c*) After being closed for a long time, now assume that the switch is opened at $t = 0$. Determine $v(t)$, the voltage across the impedance, if $v(0^+) = -12$ V.

☐ **20** With reference to the circuit shown in Fig. 13-33, let $\mathbf{Z}(s) = 2(s + 6)/(s + 2)$. (*a*) Determine the form for $v(t)$, $t > 0$, except for unknown amplitudes. (*b*) If $v(0^+) = 1$ V and $v'(0^+) = 4$ V/s, determine $v(t)$ for $t > 0$. (*c*) Assume the switch has been open for a very long time and let it close at $t = 0$. Assume $i(0^+) = 1$ A and find $i(t)$, $t > 0$.

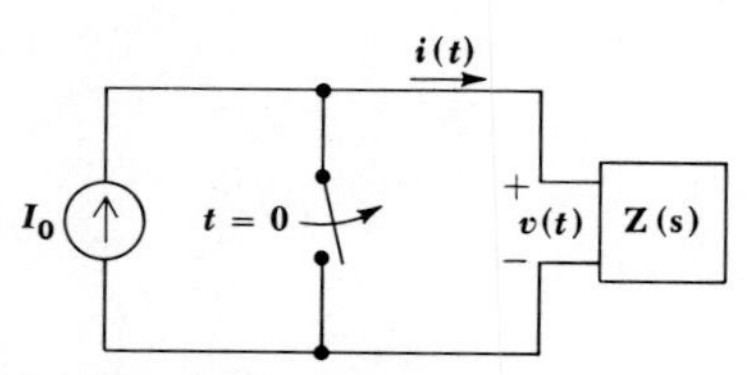

Fig. 13-33 See Prob. 20.

☐ **21** An admittance $\mathbf{Y}(s)$ has a zero at $s = -6$ and a pole at $s = -2$. Furthermore, $\mathbf{Y}(\infty) = 0.2$ ℧. (*a*) Find $\mathbf{Y}(s)$. (*b*) A 12-V battery and a switch are placed in series with $\mathbf{Y}(s)$. The switch is closed at $t = 0$. Find $i(t)$ if $i(0^+) = 2.4$ A.

☐ **22** Find $v(t)$ for $t > 0$ in the circuit of Fig. 13-23 if the given initial conditions are: (*a*) $v(0) = 1$, $v'(0) = -2$, $v''(0) = 6$; (*b*) $v(0) = 2$, $i(0) = 3/20$, $i'(0) = 1/60$.

Chapter Fourteen
Frequency Response

14-1 INTRODUCTION

Frequency response is a subject that has come up for consideration several times before. In Chap. 10 we discussed plots of admittance, impedance, current, and voltage as functions of ω, and the pole-zero concept was introduced as an aid in constructing and interpreting response curves. Response as a function of the neper frequency σ was discussed from the same standpoint in the last chapter. At that time, we also broadened our concept of frequency and introduced the complex frequency $\mathbf{s}$ and the $\mathbf{s}$ plane. We found that a plot of the critical frequencies of a response on the complex-frequency plane enabled us to tie together the forced response and the natural response; the critical frequencies themselves presented us almost directly with the form of the natural response, and the visualization of a three-dimensional rubber-sheet model or the performance of vectors drawn from each critical frequency to some test frequency gave us valuable information concerning the variation of the forced response with frequency.

In this chapter, we shall concentrate again on the forced response, and we shall particularly consider its variation with the radian frequency ω.

Why should we be so interested in the response to sinusoidal forcing functions when we so seldom encounter them in practice as such? The electric power industry is an exception, for the sinusoidal waveform appears throughout, although it is sometimes necessary to consider other frequencies introduced by the nonlinearity of some devices. But in most other electrical systems, the forcing functions and responses are not sinusoidal. In any system in which information is to be transmitted, the sinusoid by itself is almost valueless; it contains no information because its future values are exactly predictable from its past values. Moreover, once one period has been completed, any periodic nonsinusoidal waveform also contains no additional information.

Sinusoidal analysis, however, provides us with the response of a network as a function of ω, and the later work in Chaps. 18 and 19 will develop methods of determining network response to aperiodic signals (which *can* have a high information content) from the known sinusoidal frequency response.

The frequency response of a network provides useful information in its own right, however. Let us suppose that a certain forcing function is found to contain sinusoidal components having frequencies within the range of 10 to 100 Hz. Now let us imagine this forcing function being applied to a network which has the property that all sinusoidal voltages with frequencies from zero to 200 Hz applied at the input terminals appear doubled in magnitude at the output terminals, with no change in phase angle. The output function is therefore an undistorted facsimile of the input function, but with twice the amplitude. If, however, the network has a frequency response such that the magnitudes of input sinusoids between 10 and 50 Hz are multiplied by a different factor than are those between 50 and 100 Hz, then the output would in general be distorted; it would no longer be a magnified version of the input. This distorted output might be desirable in some cases and undesirable in others. That is, the network frequency response might be chosen deliberately to reject some frequency components of the forcing function, or to emphasize others.

The importance of frequency response is not limited to electrical systems, however; an understanding of electrical frequency response is certainly helpful in analyzing the frequency response of, say, a mechanical system. Suppose that we consider the launching of a space vehicle. The booster will subject the entire vehicle to extreme nonsinusoidal vibration, the mechanical forcing function, with frequency components from a few Hz up to perhaps 50 Hz. The structural members of every stage will be deflected or distorted, the mechanical response, by an amount which depends on the sinusoidal components of the vibrational forces and on the frequency response of the structure. From this information, a prediction as to whether or not the space vehicle can survive the rigors of the launch operation can be made. If the mechanical response at some vibration frequency is found to be excessive, then that component of the booster vibration must be reduced, a redesign of the supporting structure must be undertaken, or some vibration dampers (filters) must be provided.

This example should not make us think that it is undesirable to have the network response at some particular frequency be much larger than the response at all other frequencies. Such behavior is characteristic of tuned circuits or resonant circuits, as we shall see in this chapter. In discussing resonance we shall be able to apply all the methods we have discussed in presenting frequency response.

14-2 PARALLEL RESONANCE

In this section we shall begin the study of a very important phenomenon which may occur in circuits containing both inductors and capacitors. The phenomenon is called resonance, and it may be loosely described as the condition existing in any physical system when a fixed-amplitude sinusoidal forcing function produces a response of maximum amplitude. However, we often speak of resonance as occurring even when the forcing function is not sinusoidal. The resonant system may be electrical, mechanical, hydraulic, acoustic, or some other kind, but we shall restrict our attention to electrical systems. We shall define resonance more exactly below.

Resonance is a familiar phenomenon. By jumping up and down on the bumper of an automobile, for example, the vehicle can be put into rather large oscillatory motion if the jumping is done *at the proper frequency* (about one jump per second), and if the shock absorbers are somewhat decrepit. However, if the jumping frequency is increased or decreased, the vibrational response of the automobile will be considerably less than it was before. A further illustration is furnished in the case of an opera singer who is able to shatter crystal goblets by means of a well-formed note *at the proper frequency*. In each of these examples, we are thinking of frequency as being adjusted until resonance occurs; it is also possible to adjust the size, shape, and material of the mechanical object being vibrated, but this may not be so easily accomplished physically.

The condition of resonance may or may not be desirable, depending upon the purpose which the physical system is to serve. In the automotive

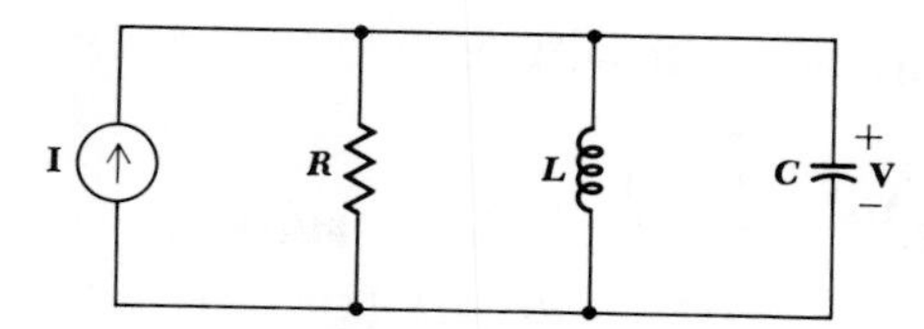

Fig. 14-1 The parallel combination of a resistor, an inductor, and a capacitor, often referred to as a parallel resonant circuit.

example above, a large amplitude of vibration may help to separate locked bumpers, but it would be somewhat disagreeable at 60 mi/h.

Let us now define resonance more carefully. In a two-terminal electrical network containing at least one inductor and one capacitor, *resonance* is the condition which exists when the input impedance of the network is purely resistive. Thus, a network is said to be in resonance (or *resonant*) when the voltage and current at the network input terminals are in phase. We shall also find that a maximum-amplitude response is produced in the network when it is in the resonant condition or *almost in the resonant condition.*

We shall apply the definition of resonance to the parallel *RLC* network shown in Fig. 14-1. In many practical situations, this circuit is a very good approximation to the circuit we might build in the laboratory by connecting a physical inductor in parallel with a physical capacitor, where this parallel combination is driven by an energy source having a very high output impedance. The admittance offered to the ideal current source is

$$\mathbf{Y} = \frac{1}{R} + j\left(\omega C - \frac{1}{\omega L}\right) \tag{1}$$

and thus resonance occurs when

$$\omega C - \frac{1}{\omega L} = 0$$

The resonant condition may be achieved by adjusting L, C, or ω; we shall devote our attention to the case for which ω is the variable. Hence, the resonant frequency ω_0 is

$$\omega_0 = \frac{1}{\sqrt{LC}} \tag{2}$$

or

$$f_0 = \frac{1}{2\pi\sqrt{LC}} \tag{3}$$

The pole-zero configuration of the admittance function can also be used to considerable advantage here. Given $\mathbf{Y}(\mathbf{s})$,

$$\mathbf{Y}(\mathbf{s}) = \frac{1}{R} + \frac{1}{L\mathbf{s}} + C\mathbf{s}$$

or

$$\mathbf{Y}(\mathbf{s}) = C\frac{\mathbf{s}^2 + \mathbf{s}/RC + 1/LC}{\mathbf{s}} \tag{4}$$

we may display the zeros of $\mathbf{Y}(\mathbf{s})$ by factoring the numerator,

$$\mathbf{Y}(\mathbf{s}) = C\frac{(\mathbf{s} + \alpha - j\omega_d)(\mathbf{s} + \alpha + j\omega_d)}{\mathbf{s}}$$

where α and ω_d represent the same quantities that they did when we discussed the natural response of the parallel RLC circuit in Sec. 7-2. That is, α is the exponential damping coefficient,

$$\alpha = \frac{1}{2RC}$$

and ω_d is the natural resonant frequency (not the resonant frequency ω_0),

$$\omega_d = \sqrt{\omega_0^2 - \alpha^2}$$

The pole-zero constellation shown in Fig. 14-2*a* follows directly from the factored form.

In view of the relationship among α, ω_d, and ω_0, it is apparent that

Fig. 14-2 (*a*) The pole-zero constellation of the input admittance of a parallel resonant circuit is shown on the $\mathbf{s}$ plane; $\omega_0^2 = \alpha^2 + \omega_d^2$. (*b*) The pole-zero constellation of the input impedance.

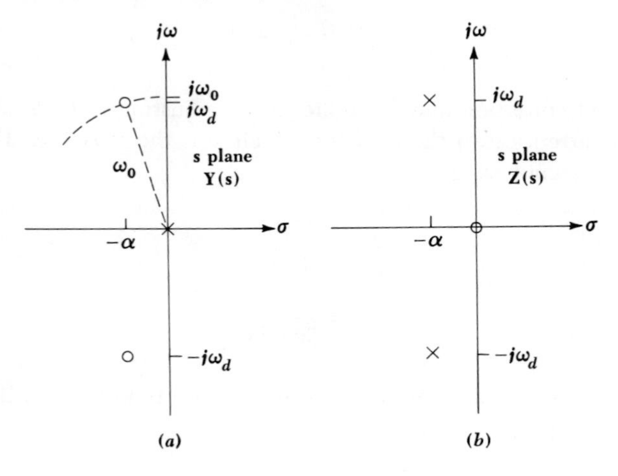

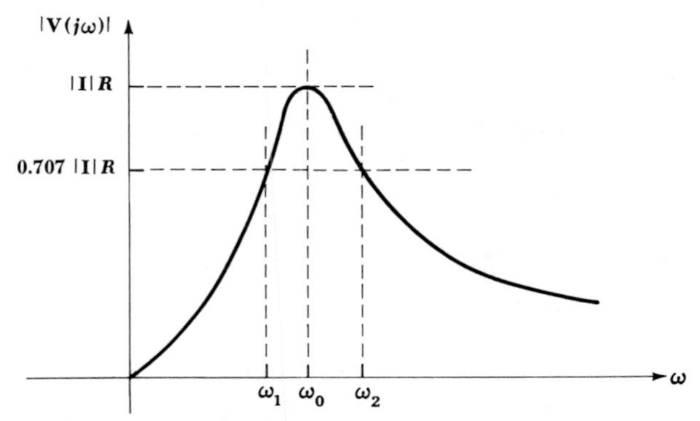

Fig. 14-3 The magnitude of the voltage response of a parallel resonant circuit is shown as a function of frequency.

the distance from the origin of the **s** plane to one of the admittance zeros is numerically equal to ω_0. Given the pole-zero configuration, the resonant frequency may therefore be obtained by purely graphical methods. We merely swing an arc, using the origin of the **s** plane as a center, through one of the zeros. The intersection of this arc and the positive $j\omega$ axis locates the point $\mathbf{s} = j\omega_0$. It is evident that ω_0 is slightly greater than the natural resonant frequency ω_d, but their ratio approaches unity as the ratio of ω_d to α increases.

Next let us examine the magnitude of the response, the voltage $\mathbf{V}(\mathbf{s})$ indicated in Fig. 14-1, as the frequency of the forcing function is varied. If we assume a constant-amplitude sinusoidal current source, the voltage response is proportional to the input impedance. This response can therefore be obtained from the pole-zero plot of the impedance $\mathbf{Z}(\mathbf{s})$, shown in Fig. 14-2*b*. The response obviously starts at zero, reaches a maximum value in the vicinity of the natural resonant frequency, and then drops again to zero as ω becomes infinite. The frequency response is sketched in Fig. 14-3. The maximum value of the response is indicated as R times the amplitude of the source current, implying that the maximum value of the circuit impedance is simply R; moreover, the response maximum is shown to occur *exactly* at the resonant frequency ω_0. The two frequencies ω_1 and ω_2, which we shall see later will give us a measure of the width of the response curve, are also identified. Let us first show that the maximum impedance is R and that this maximum impedance occurs at resonance.

The admittance, as specified by (1), possesses a constant conductance and a susceptance which has a minimum magnitude (zero) at resonance. The minimum admittance magnitude therefore occurs at resonance, and it is $1/R$. Hence, the maximum impedance magnitude is R, and it occurs at resonance.

The maximum value of the response magnitude and the frequency at which it occurs are not always found so easily. In less standard resonant circuits, we may find it necessary to express the magnitude of the response in analytical form, usually as the square root of the sum of the real part squared and the imaginary part squared; then we should differentiate this expression with respect to frequency, equate the derivative to zero, solve for the frequency of maximum response, and finally substitute this frequency into the magnitude expression to obtain the maximum-amplitude response. This procedure may be carried out for this simple case merely as a corroborative exercise, but as we have seen, it is not necessary.

It should be emphasized that, although the height of the response curve of Fig. 14-3 depends only upon the value of R for constant-amplitude excitation, the width of the curve or the steepness of the sides depends upon the other two element values also. We shall shortly relate the "width of the response curve" to a more carefully defined quantity, the bandwidth, but it will be helpful to express this relationship in terms of a very important parameter, the *quality factor* Q.

We shall find that the sharpness of the response curve of any resonant circuit is determined by the maximum amount of energy that can be stored in the circuit, compared with the energy that is lost during one complete period of the response. We define Q as

$$Q = 2\pi \frac{\text{maximum energy stored}}{\text{total energy lost per period}} \tag{5}$$

The proportionality constant 2π is included in the definition in order to simplify the more useful expressions for Q which we shall now obtain. Since energy can be stored only in the inductor and the capacitor, and can be lost only in the resistor, we may express Q in terms of the instantaneous energy associated with each of the reactive elements and the average power dissipated in the resistor:

$$Q = 2\pi \frac{[w_L(t) + w_C(t)]_{\max}}{P_R T}$$

Now let us apply this definition to the parallel RLC circuit and determine the value of Q at the resonant frequency. This value of Q is denoted by Q_0. We select the current forcing function

$$i(t) = I_m \cos \omega_0 t$$

and obtain the corresponding voltage response at resonance,

$$v(t) = Ri(t) = RI_m \cos \omega_0 t$$

Then the energy stored in the capacitor is

$$w_C(t) = \tfrac{1}{2}Cv^2 = \frac{I_m{}^2R^2C}{2}\cos^2 \omega_0 t$$

The instantaneous energy stored in the inductor becomes

$$w_L(t) = \tfrac{1}{2}Li_L{}^2 = \tfrac{1}{2}L\left(\frac{1}{L}\int_0^t v\,dt\right)^2$$

Thus

$$w_L(t) = \frac{I_m{}^2R^2C}{2}\sin^2 \omega_0 t$$

The total *instantaneous* stored energy is therefore constant,

$$w(t) = w_L(t) + w_C(t) = \frac{I_m{}^2R^2C}{2}$$

and this constant value must also be the maximum value. In order to find the energy lost in the resistor in one period, we take the average power absorbed by the resistor,

$$P_R = \tfrac{1}{2}I_m{}^2R$$

and multiply by one period, obtaining

$$P_RT = \frac{1}{2f_0}I_m{}^2R$$

We thus find the quality factor at resonance:

$$Q_0 = 2\pi\frac{I_m{}^2R^2C/2}{I_m{}^2R/2f_0}$$

or

$$Q_0 = 2\pi f_0 RC = \omega_0 RC \tag{6}$$

Equivalent expressions for Q_0 which are often quite useful may be obtained by simple substitution:

$$Q_0 = R\sqrt{\frac{C}{L}} = \frac{R}{X_{C0}} = \frac{R}{X_{L0}} \tag{7}$$

It is apparent that Q_0 is a dimensionless constant which is a function of all three circuit elements in the parallel resonant circuit. The concept of Q, however, is not limited to electric circuits or even to electrical systems; it is useful in describing any resonant phenomenon. For example, let us consider a bouncing golf ball. If we assume a weight W and release the golf ball from a height h_1 above a very hard (lossless) horizontal surface, then the ball rebounds to some lesser height h_2. The energy stored initially is Wh_1, and the energy lost in one period is $W(h_1 - h_2)$. The Q_0 is therefore

$$Q_0 = 2\pi \frac{h_1 W}{(h_1 - h_2)W} = \frac{2\pi h_1}{h_1 - h_2}$$

A perfect golf ball would rebound to its original height and have an infinite Q_0; a more typical value is 35. It should be noted that the Q in this mechanical example has been calculated from the natural response and not from the forced response. The Q of an electric circuit may also be determined from a knowledge of the natural response, as illustrated by Prob. 12.

Let us now relate the various parameters which we have associated with a parallel resonant circuit. The three parameters α, ω_d, and ω_0 were introduced much earlier in connection with the natural response. Resonance, by definition, is fundamentally associated with the forced response since it is defined in terms of a purely resistive input impedance, a sinusoidal steady-state concept. The two most important parameters of a resonant circuit are perhaps the resonant frequency ω_0 and the quality factor Q_0. Both the exponential damping coefficient and the natural resonant frequency may be expressed in terms of ω_0 and Q_0:

$$\alpha = \frac{1}{2RC} = \frac{1}{2(Q_0/\omega_0 C)C}$$

or

$$\alpha = \frac{\omega_0}{2Q_0} \tag{8}$$

and

$$\omega_d = \sqrt{\omega_0^2 - \alpha^2}$$

or

$$\omega_d = \omega_0 \sqrt{1 - \left(\frac{1}{2Q_0}\right)^2} \tag{9}$$

In terms of these two resonance parameters, the natural response of the parallel resonant circuit in the underdamped case

$$v(t) = Ae^{-\alpha t} \cos \omega_d t$$

becomes

$$v(t) = Ae^{-(\omega_0/2Q_0)t} \cos \omega_0 \sqrt{1 - \left(\frac{1}{2Q_0}\right)^2}\, t$$

The three critical frequencies (of the input admittance, for example) may also be expressed in terms of ω_0 and Q_0. We have the pole at the origin,

$$\mathbf{s}_1 = 0$$

and the pair of conjugate complex zeros,

$$\mathbf{s}_{2,4} = -\alpha \pm j\omega_d$$

or

$$\mathbf{s}_{2,4} = -\frac{\omega_0}{2Q_0} \pm j\omega_0\sqrt{1 - \left(\frac{1}{2Q_0}\right)^2}$$

The resonance parameters ω_0 and Q_0 may therefore be used to describe the natural response in either the time domain or the frequency domain. The knowledge provided by these same two parameters about the pole-zero constellation also leads to a further interpretation of the settling time and the oscillatory behavior of the parallel RLC circuit. For example, we know that the settling time, which is the time taken for the natural response to "die out," is approximately inversely proportional to α; to the same degree of approximation, then, the settling time is directly proportional to Q_0. If both ω_0 and Q_0 are increased by the same factor, however, no appreciable change in settling time can occur. It is also evident that large values of Q_0 cause the natural resonant frequency and the resonant frequency to be nearly equal.

Let us now interpret Q_0 in terms of the pole-zero locations of the admittance $\mathbf{Y}(\mathbf{s})$ of the parallel RLC circuit. We shall keep ω_0 constant; this may be done, for example, by changing R while holding L and C constant. As Q_0 is increased, the relationship between α, Q_0, and ω_0 indicates that the two zeros must move closer to the $j\omega$ axis. The relationship between ω_d, ω_0, and Q_0 shows that the zeros must simultaneously move away from the σ axis. The exact nature of the movement becomes clearer when we remember that the point at which $\mathbf{s} = j\omega_0$ could be located on the $j\omega$ axis by swinging an arc, centered at the origin, through one of the zeros and over to the positive $j\omega$ axis; since ω_0 is to be held constant, the radius must be constant, and the zeros must therefore move along this arc toward the positive $j\omega$ axis as Q_0 increases. It is evident that ω_d and ω_0 are becoming more nearly equal; if Q_0 is 5, these two frequencies differ by about one-half of 1 per cent.

The "width" of the resonant response curve, such as the one shown in Fig. 14-3, may now be defined more carefully and related to Q_0. Let us first define the two *half-power frequencies* ω_1 and ω_2 as those frequencies at which the magnitude of the input admittance of a parallel resonant circuit is greater than the magnitude at resonance by a factor of $\sqrt{2}$. Since the response curve of Fig. 14-3 displays the voltage produced across the parallel circuit by a sinusoidal current source as a function of frequency, the half-power frequencies also locate those points at which the voltage response is $1/\sqrt{2}$, or 0.707, times its maximum value. A similar relationship holds for the impedance magnitude. We shall select ω_1 as the *lower half-power frequency* and ω_2 as the *upper half-power frequency*. These names arise from the fact that a voltage which is 0.707 times the resonant voltage is equivalent to a squared voltage which is *one-half* the squared voltage at resonance.

The (half-power) bandwidth of a resonant circuit is defined as the difference of these two half-power frequencies,

$$\mathcal{B} = \omega_2 - \omega_1 \tag{10}$$

We think of this bandwidth as the "width" of the response curve, even though the curve actually extends from $\omega = 0$ to $\omega = \infty$. More exactly, the half-power bandwidth is measured by that portion of the response curve which is equal to or greater than 70.7 per cent of the maximum value.

Now let us express the bandwidth $\mathcal{B}$ in terms of Q_0 and the resonant frequency. In order to do so, we first express the admittance of the parallel *RLC* circuit,

$$\mathbf{Y} = \frac{1}{R} + j\left(\omega C - \frac{1}{\omega L}\right)$$

in terms of Q_0,

$$\mathbf{Y} = \frac{1}{R} + j\frac{1}{R}\left(\frac{\omega\omega_0 CR}{\omega_0} - \frac{\omega_0 R}{\omega\omega_0 L}\right)$$

or

$$\mathbf{Y} = \frac{1}{R}\left[1 + jQ_0\left(\frac{\omega}{\omega_0} - \frac{\omega_0}{\omega}\right)\right] \tag{11}$$

We note again that the magnitude of the admittance at resonance is $1/R$, and then realize that an admittance magnitude of $\sqrt{2}/R$ can occur only when a frequency is selected such that the imaginary part of the bracketed quantity has a magnitude of unity. Thus

$$Q_0\left(\frac{\omega_2}{\omega_0} - \frac{\omega_0}{\omega_2}\right) = 1 \qquad \text{and} \qquad Q_0\left(\frac{\omega_1}{\omega_0} - \frac{\omega_0}{\omega_1}\right) = -1$$

Solving, we have

$$\omega_1 = \omega_0\left[\sqrt{1 + \left(\frac{1}{2Q_0}\right)^2} - \frac{1}{2Q_0}\right] \tag{12}$$

$$\omega_2 = \omega_0\left[\sqrt{1 + \left(\frac{1}{2Q_0}\right)^2} + \frac{1}{2Q_0}\right] \tag{13}$$

Although these expressions are somewhat unwieldy, their difference provides a very simple formula for the bandwidth:

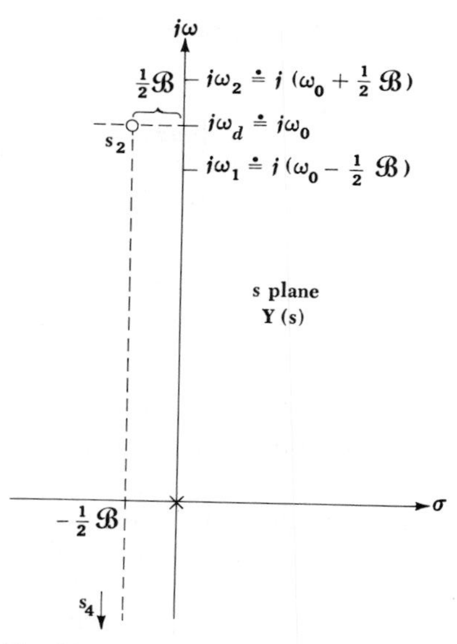

Fig. 14-4 The pole-zero constellation of $\mathbf{Y}(\mathbf{s})$ for a parallel *RLC* circuit. The two zeros are exactly $\frac{1}{2}\mathcal{B}$ Np/s (or rad/s) to the left of the $j\omega$ axis and approximately $j\omega_0$ rad/s (or Np/s) from the σ axis. The upper and lower half-power frequencies are separated exactly $\mathcal{B}$ rad/s, and each is approximately $\frac{1}{2}\mathcal{B}$ rad/s away from the resonant frequency and the natural resonant frequency.

$$\mathcal{B} = \omega_2 - \omega_1 = \frac{\omega_0}{Q_0} \tag{14}$$

Circuits possessing a higher Q_0 have a narrower bandwidth, or a sharper response curve; they have greater *frequency selectivity* or higher quality.

Many resonant circuits are deliberately designed to have a large Q_0 in order to take advantage of the narrow bandwidth and high-frequency selectivity associated with such circuits. When Q_0 is larger than about 5, it is possible to make some very useful approximations in the expressions for the upper and lower half-power frequencies and in the general expressions for the response in the neighborhood of resonance. Let us arbitrarily refer to a *high-Q* circuit as one in which Q_0 is equal to or greater than 5. The pole-zero configuration of $\mathbf{Y}(\mathbf{s})$ for a parallel *RLC* circuit having a Q_0 of about 5 is shown in Fig. 14-4. Since

$$\alpha = \frac{\omega_0}{2Q_0}$$

then

$$\alpha = \tfrac{1}{2}\mathcal{B}$$

and the locations of the two zeros may be approximated:

$$\mathbf{s}_{2,4} \doteq -\tfrac{1}{2}\mathcal{B} \pm j\omega_0$$

Moreover, the locations of the two half-power frequencies (on the positive $j\omega$ axis) may also be determined in a concise approximate form:

$$\omega_{1,2} = \omega_0 \left[\sqrt{1 + \left(\frac{1}{2Q_0}\right)^2} \mp \frac{1}{2Q_0}\right] \doteq \omega_0 \left(1 \mp \frac{1}{2Q_0}\right)$$

or

$$\omega_{1,2} \doteq \omega_0 \mp \tfrac{1}{2}\mathcal{B} \tag{15}$$

In a high-Q circuit, therefore, each half-power frequency is located approximately one-half bandwidth from the resonant frequency; this is indicated in Fig. 14-4.

Now let us visualize a test point slightly above $j\omega_0$ on the $j\omega$ axis. In order to determine the admittance offered by the parallel RLC network at this frequency, we construct the three vectors from the critical frequencies to the test point. If the test point is close to $j\omega_0$, then the vector from the pole is approximately $j\omega_0$ and that from the lower zero is nearly $j2\omega_0$. The admittance is therefore given approximately by

$$\mathbf{Y}(\mathbf{s}) \doteq C\frac{(j2\omega_0)(\mathbf{s} - \mathbf{s}_2)}{j\omega_0} \doteq 2C(\mathbf{s} - \mathbf{s}_2) \tag{16}$$

where the multiplicative constant C is determined from (4). In order to determine a useful approximation for the vector $(\mathbf{s} - \mathbf{s}_2)$, let us consider an enlarged view of that portion of the $\mathbf{s}$ plane in the neighborhood of the zero $\mathbf{s}_2$ (Fig. 14-5). We define the angle of the admittance as θ_Y,

$$\theta_Y = \text{ang } \mathbf{Y}$$

and note from (16) that θ_Y may also be identified approximately as the angle of $\mathbf{s} - \mathbf{s}_2$. Thus

$$\mathbf{s} - \mathbf{s}_2 \doteq \tfrac{1}{2}\mathcal{B} + j\tfrac{1}{2}\mathcal{B}\tan\theta_Y \doteq \tfrac{1}{2}\mathcal{B}(1 + j\tan\theta_Y)$$

or

$$\mathbf{s} - \mathbf{s}_2 \doteq \tfrac{1}{2}\mathcal{B}\sqrt{1 + \tan^2\theta_Y}\ \underline{/\theta_Y}$$

Finally

$$\mathbf{Y}(\mathbf{s}) \doteq 2C(\tfrac{1}{2}\mathcal{B})\sqrt{1 + \tan^2\theta_Y}\ \underline{/\theta_Y}$$

or

$$\mathbf{Y}(\mathbf{s}) \doteq \frac{1}{R}\sqrt{1 + \tan^2\theta_Y}\ \underline{/\theta_Y} \tag{17}$$

where

$$\theta_Y \doteq \tan^{-1}\frac{\omega - \omega_0}{\tfrac{1}{2}\mathcal{B}} \tag{18}$$

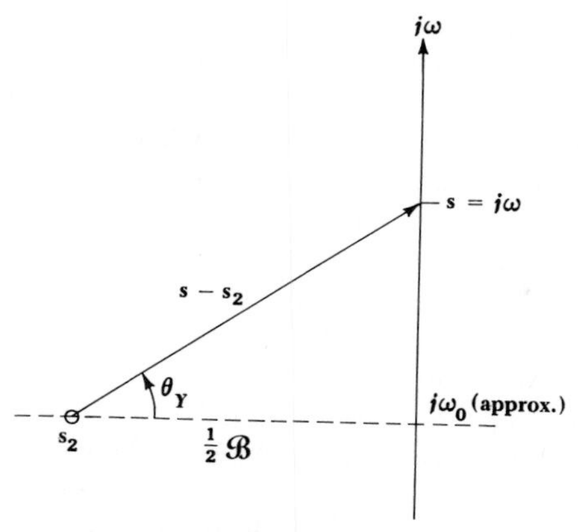

Fig. 14-5 An enlarged portion of the pole-zero constellation for $\mathbf{Y}(\mathbf{s})$ of a high-Q_0 parallel RLC circuit.

Thus, the angle of the admittance at a given frequency near resonance is obtained by calculating the number of half bandwidths by which the frequency is off resonance and then taking the inverse tangent of this number, being careful to preserve the algebraic sign. The magnitude of the admittance at this frequency is given approximately by the product of the conductance $1/R$ and the square root of one plus the square of the number of half bandwidths off resonance.

As an example of the use of these approximations, let us determine the approximate value of the admittance of a parallel RLC network for which $R = 25$ kΩ, $L = 1$ H, and $C = \frac{1}{25}$ μF. We find that $Q_0 = 5$, $\omega_0 = 5$ krad/s, $\mathcal{B} = 1$ krad/s, and $\mathcal{B}/2 = 0.5$ krad/s. Let us evaluate the admittance at $\omega = 5.25$ krad/s, a frequency which is one-half of a half bandwidth above resonance. Thus,

$$\theta_Y \doteq \tan^{-1}(0.5) = 26.6°$$

and

$$\mathbf{Y}(j5250) \doteq 0.04\sqrt{1 + (0.5)^2}\,\underline{/26.6°} \doteq 0.0447\underline{/26.6°} \quad \text{m℧}$$

An exact calculation of the admittance shows that

$$\mathbf{Y}(j5250) = 0.0445\underline{/26.0°} \quad \text{m℧}$$

The approximate method therefore leads to values of admittance magnitude and angle that are too high by about 0.5 and 2 per cent, respectively.

Our intention is to use these approximations for high-Q circuits near resonance. We have already agreed that we shall let "high-Q" infer $Q_0 \geq 5$, but how near is "near"? It can be shown that the error in magnitude or phase is less than 5 per cent if $Q_0 \geq 5$ and $0.9\omega_0 \leq \omega \leq 1.1\omega_0$.

Although this narrow band of frequencies may seem to be prohibitively small, it is usually more than sufficient to contain the range of frequencies in which we are most interested. For example, a home radio may contain a circuit tuned to a resonant frequency of 455 kHz with a half-power bandwidth of 10 kHz. This circuit must then have a value of 45.5 for Q_0, and the half-power frequencies are about 450 and 460 kHz. Our approximations, however, are valid from 409.5 to 500.5 kHz (with errors less than 5 per cent), a range which covers essentially all the peaked portion of the response curve; only in the remote "tails" of the response curve do the approximations lead to reasonably large errors.[1]

Let us conclude our coverage of the parallel resonant circuit by reviewing the various conclusions we have reached. The resonant frequency ω_0 is the frequency at which the imaginary part of the input admittance becomes zero or the admittance angle θ_Y becomes zero. Then, $\omega_0 = 1/\sqrt{LC}$. The circuit's figure of merit Q_0 is defined as 2π times the ratio of the maximum energy stored in the circuit to the energy lost each period in the circuit. From this definition, we find that $Q_0 = \omega_0 RC$. The two half-power frequencies ω_1 and ω_2 are defined as the frequencies at which the admittance magnitude is $\sqrt{2}$ times the minimum admittance magnitude. These are also the frequencies at which the voltage response is 70.7 per cent of the maximum response. The exact and approximate (for high Q_0) expressions for these two frequencies are

$$\omega_{1,2} = \omega_0 \left[\sqrt{1 + \left(\frac{1}{2Q_0} \right)^2} \mp \frac{1}{2Q_0} \right] \doteq \omega_0 \mp \tfrac{1}{2}\mathcal{B}$$

where $\mathcal{B}$ is the difference between the upper and the lower half-power frequencies. This half-power bandwidth is given by

$$\mathcal{B} = \omega_2 - \omega_1 = \frac{\omega_0}{Q_0}$$

The input admittance may also be expressed in an exact or approximate (for high Q_0) form

$$\mathbf{Y} = \frac{1}{R}\sqrt{1 + \tan^2 \theta_Y} \,\underline{/\theta_Y} \doteq \frac{1}{R}\sqrt{1 + \left(\frac{\omega - \omega_0}{\tfrac{1}{2}\mathcal{B}} \right)^2} \,\underline{/\tan^{-1} [(\omega - \omega_0)/\tfrac{1}{2}\mathcal{B}]}$$

The approximations are valid for frequencies which do not differ from the resonant frequency by more than one-tenth of the resonant frequency.

[1] At frequencies remote from resonance, we are often satisfied with very rough results; greater accuracy is not always necessary.

Drill Problems

14-1 Determine ω_0 for the parallel resonant circuit described by: (*a*) $L = \frac{1}{12}$ H, $C = 3\ \mu$F, $R = 1$ kΩ; (*b*) $\alpha = 40{,}000$ s^{-1}, $R = 100$ kΩ, $L = 80\ \mu$H; (*c*) $\omega_d = 100\sqrt{21}$ s^{-1}, $C = 20\ \mu$F, $\alpha = 200$ s^{-1}.

Ans. 500; 2000; 10,000,000 rad/s

14-2 Find Q_0 for each of the networks described in Drill Prob. 14-1.

Ans. 1.25; 6; 125

14-3 Specify the locations of the zeros of $\mathbf{Y}(s)$ for a parallel *RLC* circuit for which: (*a*) $Q_0 = 5$, $L = 0.4$ H, $C = 10\ \mu$F; (*b*) $Q_0 = 4$, $\omega_0 = 480$ rad/s, $R = 172\ \Omega$; (*c*) $Q_0 = 3$, $R = 500\ \Omega$, $C = 20\ \mu$F.

Ans. $-50 \pm j296$; $-50 \pm j498$; $-60 \pm j476$ s^{-1}

14-4 Find $\mathcal{B}$, ω_1, and ω_2 for the parallel *RLC* circuit of Drill Prob. 14-3: (*a*) 1*a*; (*b*) 1*b*; (*c*) 1*c*.

Ans. 400, 339, 739; 333, 1840, 2174; 8×10^4, 9.96×10^6, 10.04×10^6 rad/s

14-5 Given a parallel circuit, $R = 100\ \Omega$, $L = 5$ mH, $C = 50\ \mu$F, and a current source, $i_s = 0.01 \cos \omega t$ A, find the amplitude of the capacitor voltage if $\omega =$: (*a*) 2000; (*b*) 1850; (*c*) 2120 rad/s.

Ans. 0.55; 0.64; 1.00 V

14-3 SERIES RESONANCE

Although we probably find less use for the series *RLC* circuit than we do for the parallel *RLC* circuit, it is still worthy of our attention. We shall consider the circuit shown in Fig. 14-6. It should be noted that the various circuit elements are given *s* (for series) subscripts for the time being in order to avoid confusing them with the parallel elements when the circuits are to be compared.

Our discussion of parallel resonance required a section of considerable length. We could now give the series *RLC* circuit the same kind of treat-

Fig. 14-6 A series resonant circuit.

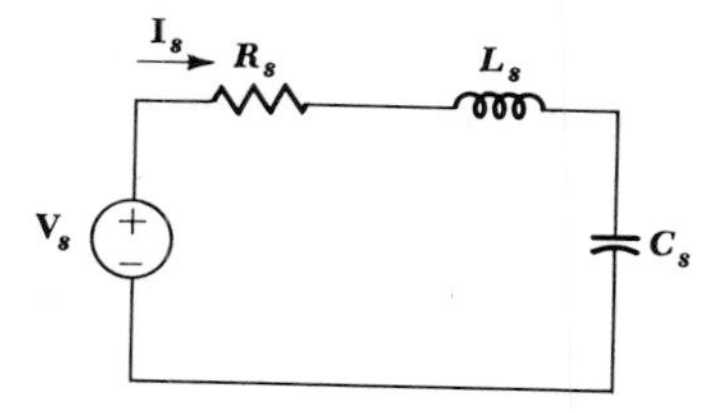

ment, but it is much cleverer to avoid such needless repetition and use the duality principle. For simplicity, let us concentrate on the conclusions presented in the last paragraph of the preceding section on parallel resonance. The important results are contained there, and the use of dual language enables us to transcribe this paragraph to present the important results for the series *RLC* circuit.

Let us conclude our coverage of the series resonant circuit by summarizing the more important conclusions. The resonant frequency ω_{0s} is the frequency at which the imaginary part of the input impedance becomes zero or the impedance phase angle θ_Z becomes zero. Then, $\omega_{0s} = 1/\sqrt{L_s C_s}$. The circuit's figure of merit Q_{0s} is defined as 2π times the ratio of the maximum energy stored in the circuit to the energy lost each period in the circuit. From this definition, we find that $Q_{0s} = \omega_{0s} L_s / R_s$. The two half-power frequencies ω_{1s} and ω_{2s} are defined as the frequencies at which the impedance magnitude is $\sqrt{2}$ times the minimum impedance magnitude. These are also the frequencies at which the current response is 70.7 per cent of the maximum response. The exact and approximate (for high Q_{0s}) expressions for these two frequencies are

$$\omega_{1s,2s} = \omega_{0s}\left[\sqrt{1 + \left(\frac{1}{2Q_{0s}}\right)^2} \mp \frac{1}{2Q_{0s}}\right] \doteq \omega_{0s} \mp \tfrac{1}{2}\mathcal{B}_s$$

where $\mathcal{B}_s$ is the difference between the upper and the lower half-power frequencies. This half-power bandwidth is given by

$$\mathcal{B}_s = \omega_{2s} - \omega_{1s} = \frac{\omega_{0s}}{Q_{0s}}$$

The input impedance may also be expressed in an exact or approximate (for high Q_{0s}) form

$$\mathbf{Z}_s = R_s\sqrt{1 + \tan^2\theta_Z}\,/\theta_Z \doteq R_s\sqrt{1 + \left(\frac{\omega - \omega_{0s}}{\frac{1}{2}\mathcal{B}_s}\right)^2}\;/\tan^{-1}[(\omega - \omega_{0s})/\tfrac{1}{2}\mathcal{B}_s]$$

The approximations are valid for frequencies which do not differ from the resonant frequency by more than one-tenth of the resonant frequency.

The series resonant circuit is characterized by a low impedance at resonance, while the parallel resonant circuit produces a high resonant impedance. The latter circuit provides inductor currents and capacitor currents at resonance which have amplitudes Q_0 times as great as the source current; the series resonant circuit provides inductor voltages and capacitor voltages which are greater than the source voltage by the factor Q_{0s}.

From this point on, we shall no longer identify series resonant circuits by use of an *s* subscript.

Drill Problems

14-6 Find the half-power bandwidth and the Q_0 of a series *RLC* circuit containing: (*a*) $R = 2\ \Omega$, $L = 10$ mH, $C = 100\ \mu$F; (*b*) $R = 25\ \Omega$, $L = 50$ mH, $C = 20\ \mu$F; (*c*) $R = 10\ \Omega$, $L = 0.2$ H, $C = 5\ \mu$F.

Ans. 50 rad/s, 20; 200 rad/s, 5; 500 rad/s, 2

14-7 Let each network described in Drill Prob. 6 be excited by the forcing function $2 \cos \omega t$ V. Find the magnitude of the capacitor voltage at $\omega = 960$ rad/s.

Ans. 4.1; 9.6; 21.8 V

14-4 OTHER RESONANT FORMS

The parallel and series *RLC* circuits of the previous two sections represent *idealized* resonant circuits; they are no more than useful approximate representations of a physical circuit which might be constructed by combining a coil of wire, a carbon resistor, and a tantalum capacitor in parallel or series. The degree of accuracy with which the idealized model fits the actual circuit depends on the operating frequency range, the Q of the circuit, the materials present in the physical elements, the element sizes, and many other factors. We are not studying the techniques of determining the best model for a given physical circuit, for this requires some knowledge of electromagnetic field theory and the properties of materials; we are, however, concerned with the problem of reducing a more complicated model to one of the two simpler models with which we are more familiar.

The network shown in Fig. 14-7*a* is a reasonably accurate model for the parallel combination of a physical inductor, capacitor, and resistor. The resistor R_1 represents the ohmic losses, core losses, and radiation losses of the physical coil. The losses in the dielectric within the physical capacitor are accounted for by the resistor R_2, as well as for the resistance of

Fig. 14-7 (*a*) A useful model of a physical network which consists of a physical inductor, capacitor, and resistor in parallel. (*b*) A network which can be equivalent to (*a*) over a narrow frequency band.

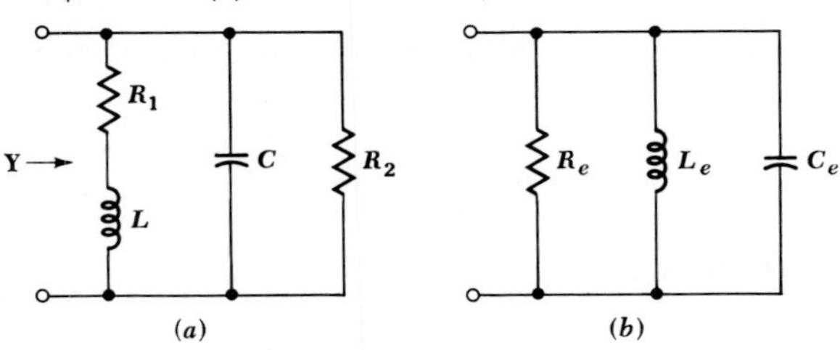

the physical resistor in the given *RLC* circuit. In this model, there is no way to combine elements and produce a simpler model which is equivalent to the original model *for all frequencies.* We shall show, however, that a simpler equivalent may be constructed which is valid over a frequency band which is usually large enough to include all frequencies of interest. The equivalent will take the form of the network shown in Fig. 14-7*b*.

Before we learn how to develop such an equivalent circuit, let us first consider the given circuit, Fig. 14-7*a*. The resonant radian frequency for this network is *not* $1/\sqrt{LC}$, although if R_1 is sufficiently small it may be very close to this value. The definition of resonance is unchanged, and we may determine the resonant frequency by setting the imaginary part of the input admittance equal to zero:

$$\text{Im}\,[\mathbf{Y}(j\omega)] = \text{Im}\left(\frac{1}{R_2} + j\omega C + \frac{1}{R_1 + j\omega L}\right) = 0$$

Thus,

$$C = \frac{L}{R_1^{\,2} + \omega^2 L^2}$$

and

$$\omega_0 = \sqrt{\frac{1}{LC} - \left(\frac{R_1}{L}\right)^2}$$

We note that ω_0 is less than $1/\sqrt{LC}$, but sufficiently small values of the ratio R_1/L may result in a negligible difference between ω_0 and $1/\sqrt{LC}$.

The maximum magnitude of the input impedance also deserves consideration. It is *not* R_2, and it does *not* occur at ω_0 (or at $\omega = 1/\sqrt{LC}$). The proof of these statements will not be shown because the expressions soon become algebraically cumbersome; the theory, however, is straightforward. Let us be content with a numerical example. We select the simple values $R_1 = 2\ \Omega$, $L = 1$ H, $C = \frac{1}{8}$ F, and $R_2 = 3\ \Omega$, and find the resonant frequency

$$\omega_0 = 2 \quad \text{rad/s}$$

and the input impedance at resonance

$$\mathbf{Z}(j2) = 1.714 \quad \Omega$$

At the frequency which would be the resonant frequency *if* R_1 were zero,

$$\frac{1}{\sqrt{LC}} = 2.83 \quad \text{rad/s}$$

the input impedance is

$$\mathbf{Z}(j2.83) = 1.947\underline{/-13.26^\circ} \quad \Omega$$

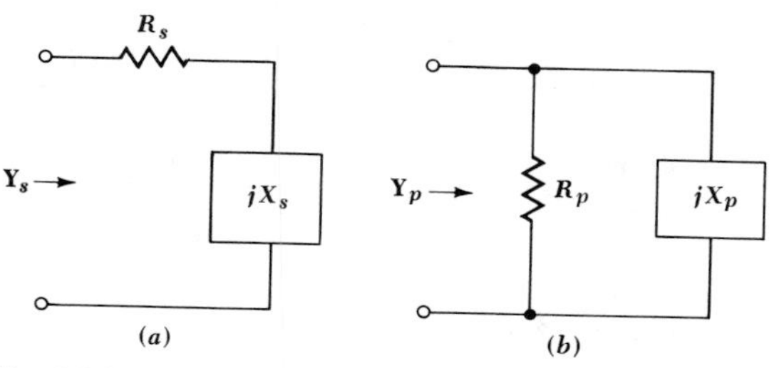

Fig. 14-8 (*a*) A series network which consists of a resistance R_s and an inductive or capacitive reactance X_s may be transformed into (*b*) a parallel network such that $\mathbf{Y}_s = \mathbf{Y}_p$ at one specific frequency. The reverse transformation is equally possible.

However, the frequency at which the maximum impedance magnitude occurs, indicated by ω_m, is found to be

$$\omega_m = 3.26 \quad \text{rad/s}$$

and the impedance having the maximum magnitude is

$$\mathbf{Z}(j3.26) = 1.98\underline{/-21.4^\circ} \quad \Omega$$

The impedance magnitude at resonance and the maximum magnitude differ by about 13 per cent. Although it is true that such an error may be neglected occasionally in practice, it is too large to neglect on a test. The later work in this section will show that the Q of the inductor-resistor combination at 2 rad/s is unity; this low value accounts for the 13 per cent discrepancy.

In order to transform the given circuit of Fig. 14-7*a* into an equivalent of the form of that shown in Fig. 14-7*b*, we must discuss the Q of a simple series or parallel combination of a resistor and a reactor (inductor or capacitor). We first consider the series circuit shown in Fig. 14-8*a*. The Q of this network is again defined as 2π times the ratio of the maximum stored energy to the energy lost each period, but the Q may be evaluated at any frequency we choose. In other words, Q is a function of ω. It is true that we shall choose to evaluate it at a frequency which is, or apparently is, the resonant frequency of some network of which the series arm is a part. This frequency, however, is not known until a more complete circuit is available. The eager reader is encouraged to show that the Q of this series arm is $|X_s|/R_s$, whereas the Q of the parallel network of Fig. 14-8*b* is $R_p/|X_p|$.

Let us now carry out the details necessary to find values for R_p and X_p so that the parallel network of Fig. 14-8*b* is equivalent to the series network of Fig. 14-8*a* at some single specific frequency. We equate $\mathbf{Y}_s$ and $\mathbf{Y}_p$,

$$\mathbf{Y}_s = \frac{1}{R_s + jX_s} = \frac{R_s - jX_s}{R_s^2 + X_s^2} = \mathbf{Y}_p = \frac{1}{R_p} - j\frac{1}{X_p}$$

and obtain

$$R_p = \frac{R_s^2 + X_s^2}{R_s} \qquad X_p = \frac{R_s^2 + X_s^2}{X_s}$$

Dividing these two expressions, we find

$$\frac{R_p}{X_p} = \frac{X_s}{R_s}$$

It follows that the Q's of the series and parallel networks must be equal:

$$Q_p = Q_s = Q$$

The transformation equations may therefore be simplified:

$$R_p = R_s(1 + Q^2) \tag{19}$$

$$X_p = X_s\left(1 + \frac{1}{Q^2}\right) \tag{20}$$

It is apparent that R_s and X_s may also be found if R_p and X_p are the given values; the transformation in either direction may be performed.

If $Q \geq 5$, little error is introduced by using the approximate relationships

$$R_p \doteq Q^2 R_s \tag{21}$$

$$X_p \doteq X_s \qquad (C_p \doteq C_s \text{ or } L_p \doteq L_s) \tag{22}$$

As an example, consider the series combination of a 100-mH inductor and a 5-Ω resistor. We shall perform the transformation at a frequency of 1000 rad/s, a value selected because it is approximately the resonant frequency of the network (not shown) of which this series arm is a part. We find that X_s is 100 Ω and Q is 20. Since the Q is sufficiently high, we use (21) and (22) to obtain

$$R_p \doteq Q^2 R_s = 2000 \quad \Omega \qquad L_p \doteq L_s = 100 \quad \text{mH}$$

The conclusion is that a 100-mH inductor in series with a 5-Ω resistor provides essentially the same input impedance as does a 100-mH inductor in parallel with a 2000-Ω resistor *at the frequency* 1000 rad/s. In order to check the accuracy of the equivalence, let us evaluate the input impedance for each network at 1000 rad/s. We find

$$\mathbf{Z}_s(j1000) = 5 + j100 = 100.1\underline{/87.1^\circ}$$

$$\mathbf{Z}_p(j1000) = \frac{2000(j100)}{2000 + j100} = 99.9\underline{/87.1^\circ}$$

and conclude that the approximation is exceedingly accurate at the transformation frequency. The accuracy at 900 rad/s is also reasonably good, because

$$\mathbf{Z}_s(j900) = 90.1\underline{/86.8^\circ}$$

$$\mathbf{Z}_p(j900) = 89.9\underline{/87.4^\circ}$$

If this inductor and series resistor had been used as part of a series RLC circuit for which the resonant frequency was 1000 rad/s, then the half-power bandwidth would have been

$$\mathcal{B} = \frac{\omega_0}{Q_0} = \frac{1000}{20} = 50$$

and the frequency of 900 rad/s would have represented a frequency that was 4 half bandwidths off resonance. Thus the equivalent networks that we evaluated above would have been adequate for reproducing essentially all the peaked portion of the response curve.

As a further example of the replacement of a more complicated resonant circuit by an equivalent series or parallel RLC circuit, let us consider a problem in electronic instrumentation. The simple series RLC network in Fig. 14-9a is excited by a sinusoidal voltage source at the resonant frequency. The effective value of the source voltage is 0.5 V, and we wish to measure the effective value of the voltage across the capacitor with an electronic voltmeter (VM) having an internal resistance of 100,000 Ω. That is, an equivalent representation of the voltmeter is an ideal voltmeter in parallel with a 100,000-Ω resistor.

Before the voltmeter is connected, we find that the resonant frequency is 10^5 rad/s, $Q_0 = 50$, the current is 25 mA, and the rms capacitor voltage is 25 V. As indicated at the end of Sec. 14-3, this voltage is Q_0 times the applied voltage. Thus, if the voltmeter were ideal, it would read 25 V when connected across the capacitor.

However, when the actual voltmeter is connected, the circuit shown in Fig. 14-9b results. In order to obtain a series RLC circuit, it is now necessary to replace the parallel RC network by a series RC network. Let us assume that the Q of this RC network is sufficiently high so that the equivalent series capacitor will be the same as the given parallel capacitor. We do this in order to approximate the resonant frequency of the final series RLC circuit. Thus, if the series RLC circuit also contains a 0.01-μF capacitor, the resonant frequency remains 10^5 rad/s. We need to know

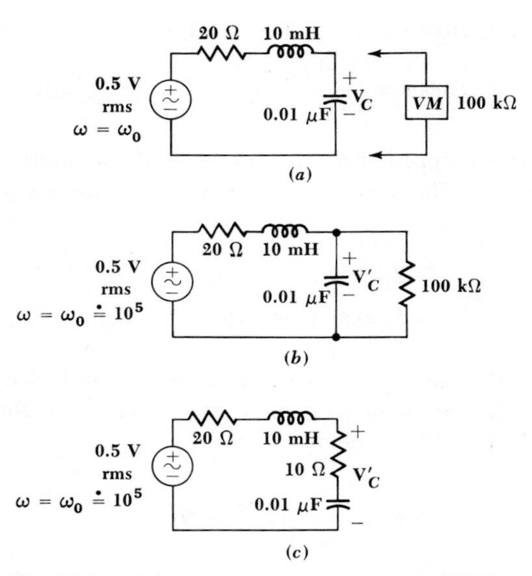

Fig. 14-9 *(a)* A given series-resonant circuit in which the capacitor voltage is to be measured by a nonideal electronic voltmeter. *(b)* The effect of the voltmeter is included in the circuit; it reads V_c' volts. *(c)* A series resonant circuit is obtained when the parallel RC network in *(b)* is replaced by the series RC network which is equivalent at 10^5 rad/s.

this estimated resonant frequency in order to calculate the Q of the parallel RC network; it is

$$Q = \frac{R_p}{|X_p|} = \omega R_p C_p = 10^5(10^5)(10^{-8}) = 100$$

Since this value is greater than 5, our vicious circle of assumptions is justified, and the equivalent series RC network consists of the capacitor

$$C_s = 0.01 \quad \mu\text{F}$$

and the resistor

$$R_s \doteq \frac{R_p}{Q^2} = 10 \quad \Omega$$

Hence, the equivalent circuit of Fig. 14-9*c* is obtained. The resonant Q of this circuit is now only 33.3, and thus the voltage across the capacitor in the circuit of Fig. 14-9*c* is $16\frac{2}{3}$ V. But we need to find $|\mathbf{V}_c'|$, the voltage across the series RC combination; we obtain

$$|\mathbf{V}_c'| = \frac{0.5}{30}|10 - j1000| = 16.7 \quad \text{V}$$

The capacitor voltage and $|\mathbf{V}_c'|$ are essentially equal since the voltage across the 10-Ω resistor is quite small.

The final conclusion must be that an apparently good voltmeter may produce a severe effect on the response of a high-Q resonant circuit. A similar effect may occur when a nonideal ammeter is inserted in the circuit.

Drill Problems

14-8 Given a 10-Ω resistor in series with a 10-μF capacitor, determine the two-element parallel equivalent if $\omega =$: (*a*) 200; (*b*) 1000; (*c*) 5000 rad/s.

Ans. 50 Ω, 8 μF; 1 kΩ, 10 μF; 25 kΩ, 10 μF

14-9 At $\omega = 10^3$ rad/s, find the effective Q of the two-terminal RC networks shown in Fig. 14-10*a*, *b*, and *c*.

Ans. 2; 10; 20

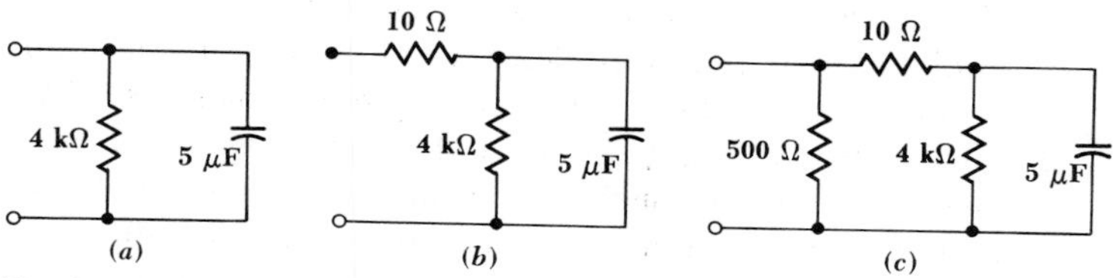

Fig. 14-10 See Drill Prob. 14-9.

14-10 A certain physical inductor is modeled by an ideal 10-mH inductor in series with a 10-Ω resistor, while a capacitor is approximated by an ideal ⅑-μF capacitor in series with 5 Ω. If the two physical elements are connected in parallel, find the approximate: (*a*) parallel resonant frequency; (*b*) lower half-power frequency; (*c*) upper half-power frequency.

Ans. 29,250; 30,000; 30,750 rad/s

14-5 SCALING

Some of the examples and problems which we have been solving have involved circuits containing element values ranging around a few ohms, a few henrys, and a few farads. The applied frequencies were a few radians per second. These particular numerical values were used not because they are those commonly met in practice, but because arithmetic manipulations are so much easier than they would be if it were necessary to carry along various powers of 10 throughout the calculations. The scaling procedures that will be discussed in this section enable us to analyze networks composed of practical-sized elements by scaling the element values to

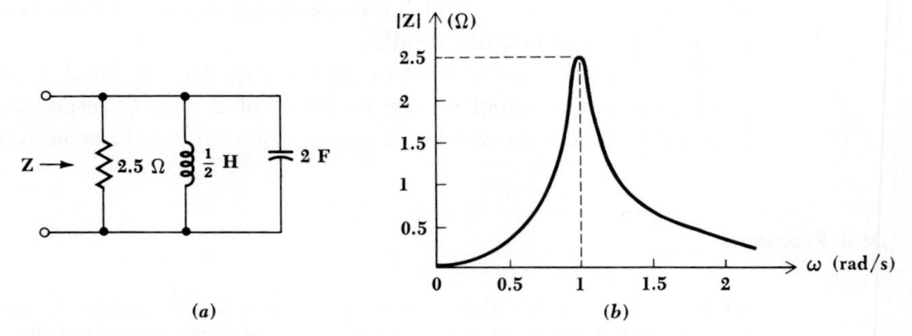

Fig. 14-11 (*a*) A parallel resonant circuit used as an example to illustrate magnitude and frequency scaling. (*b*) The magnitude of the input impedance is shown as a function of frequency.

permit more convenient numerical calculations. We shall consider both magnitude scaling and frequency scaling.

Let us select the parallel resonant circuit shown in Fig. 14-11*a* as our example. The impractical element values lead to the unlikely response curve drawn as Fig. 14-11*b*; the maximum impedance is 2.5 Ω, the resonant frequency is 1 rad/s, Q_0 is 5, and the bandwidth is 0.2 rad/s. These numerical values are much more characteristic of the electrical analog of some mechanical system than they are of any basically electrical device. We have convenient numbers with which to calculate but an impractical circuit to construct.

Let us assume that our goal is to scale this network in such a way as to provide an impedance maximum of 5000 Ω at a resonant frequency of 5×10^6 rad/s, or 796 kHz. In other words, we may use the same response curve shown in Fig. 14-11*b* if every number on the ordinate scale is increased by a factor of 2000 and every number on the abscissa scale is increased by a factor of 5×10^6. We shall treat this as two problems: (1) scaling in magnitude by a factor of 2000 and (2) scaling in frequency by a factor of 5×10^6.

Magnitude scaling is defined as the process by which the *impedance* of a two-terminal network is increased by a factor of K_m, the frequency remaining constant. The factor K_m is real and positive; it may be greater or smaller than unity. We shall understand that the shorter statement, "the network is scaled in magnitude by a factor of 2," infers that the impedance of the new network is to be *twice* that of the old network at any frequency. Let us now determine how we must scale each type of passive element. To increase the input impedance of a network by a factor of K_m, it is sufficient to increase the impedance of each element in the network by this same factor. Thus, a resistor R must be replaced by a resistor K_mR. Each inductor must also exhibit an impedance which is K_m times as great

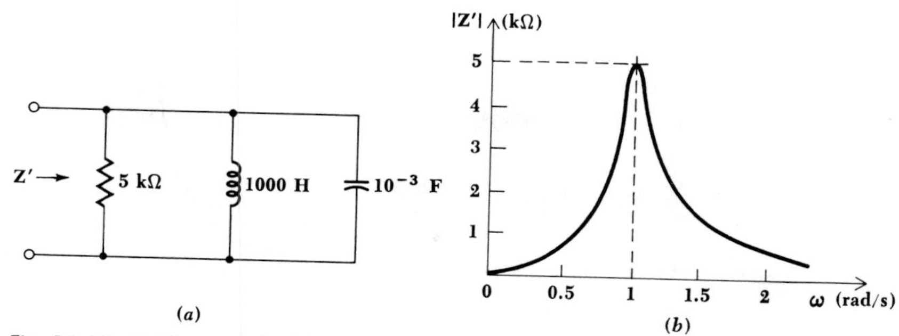

Fig. 14-12 (*a*) The network of Fig. 14-11*a* after being scaled in magnitude by a factor $K_m = 2000$. (*b*) The corresponding response curve.

at any frequency. In order to increase an impedance sL by a factor of K_m when s remains constant, the inductor L must be replaced by an inductor K_mL. In a similar manner, each capacitor C must be replaced by a capacitor C/K_m. In summary, these changes will produce a network which is scaled in magnitude by a factor of K_m:

$$\left.\begin{aligned} R &\rightarrow K_m R \\ L &\rightarrow K_m L \\ C &\rightarrow \frac{C}{K_m} \end{aligned}\right\} \text{magnitude scaling}$$

When each element in the network of Fig. 14-11*a* is scaled in magnitude by a factor of 2000, the network shown in Fig. 14-12*a* results. The response curve shown in Fig. 14-12*b* indicates that no change in the previously drawn response curve need be made other than a change in the scale of the ordinate.

Let us now take this new network and scale it in frequency. We define *frequency scaling* as the process by which the frequency at which any impedance occurs is increased by a factor of K_f. Again, we shall make use of the shorter expression, "the network is scaled in frequency by a factor of 2," to infer that the same impedance is now obtained at a frequency twice as great. Again, frequency scaling is accomplished by scaling each passive element in frequency. It is apparent that no resistor is affected. The impedance of any inductor is sL, and if this same impedance is to be obtained at a frequency K_f times as great, then the inductor L must be replaced by one having an inductance of L/K_f. Similarly, a capacitor C is to be replaced by one having a capacitance C/K_f. Thus, if a network is to be scaled in frequency by a factor of K_f, then the changes necessary in each passive element are

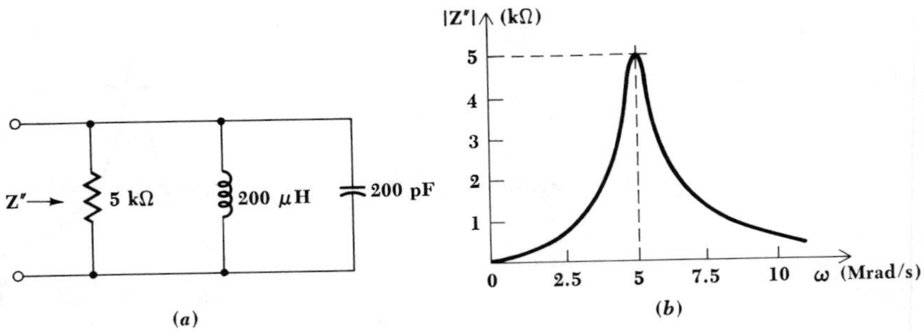

Fig. 14-13 (*a*) The network of Fig. 14-12*a* after being scaled in frequency by a factor $K_f = 5 \times 10^6$. (*b*) The corresponding response curve.

$$\left.\begin{aligned} R &\rightarrow R \\ L &\rightarrow \frac{L}{K_f} \\ C &\rightarrow \frac{C}{K_f} \end{aligned}\right\} \text{frequency scaling}$$

When each element of the magnitude-scaled network of Fig. 14-12*a* is scaled in frequency by a factor of 5×10^6, the network of Fig. 14-13*a* is obtained. The corresponding response curve is shown in Fig. 14-13*b*.

The circuit elements in this last network have values which are easily achieved in physical circuits; the network can actually be built and tested. It follows that, if the original network of Fig. 14-11*a* were actually an analog of some mechanical resonant system, we could have scaled this analog in both magnitude and frequency in order to achieve a network which we might construct in the laboratory; tests that are expensive or inconvenient to run on the mechanical system could then be made on the scaled electrical system, and the results should then be "unscaled" and converted into mechanical units to complete the analysis.

The effect of magnitude scaling or frequency scaling on the pole-zero constellation of an impedance is not very difficult to ascertain, but its determination offers such an excellent opportunity to review the meaning and significance of a pole-zero plot in the **s** plane that it forms the basis for one of the drill problems below.

An impedance which is given as a function of **s** may also be scaled in magnitude or frequency, and this may be done without any knowledge of the specific elements out of which the two-terminal network is composed. In order to scale **Z**(**s**) in magnitude, the definition of magnitude scaling shows that it is only necessary to multiply **Z**(**s**) by K_m in order to obtain the magnitude-scaled impedance. Thus, the impedance of the parallel resonant circuit shown in Fig. 14-11*a* is

$$\mathbf{Z}(s) = \frac{s}{2s^2 + 0.4s + 2}$$

or

$$\mathbf{Z}(s) = 0.5\frac{s}{(s + 0.1 + j0.995)(s + 0.1 - j0.995)}$$

The impedance $\mathbf{Z}'(s)$ of the magnitude-scaled network is

$$\mathbf{Z}'(s) = K_m\mathbf{Z}(s)$$

and, therefore,

$$\mathbf{Z}'(s) = 1000\frac{s}{(s + 0.1 + j0.995)(s + 0.1 - j0.995)}$$

If $\mathbf{Z}'(s)$ is now to be scaled in frequency by a factor of 5×10^6, then $\mathbf{Z}''(s)$ and $\mathbf{Z}'(s)$ are to provide identical values of impedance if $\mathbf{Z}''(s)$ is evaluated at a frequency K_f times that at which $\mathbf{Z}'(s)$ is evaluated. After some careful cerebral activity, this conclusion may be stated concisely in functional notation:

$$\mathbf{Z}''\left(\frac{s}{K_f}\right) = \mathbf{Z}'(s)$$

Remember that the correspondence results when a larger value of s is used in $\mathbf{Z}''$ (assuming that $K_f > 1$). In other words, we obtain $\mathbf{Z}''(s)$ by replacing every s in $\mathbf{Z}'(s)$ by s/K_f. The analytic expression for the impedance of the network shown in Fig. 14-13*a* must therefore be

$$\mathbf{Z}''(s) = 1000\frac{s/(5 \times 10^6)}{[s/(5 \times 10^6) + 0.1 + j0.995][s/(5 \times 10^6) + 0.1 - j0.995]}$$

or

$$\mathbf{Z}''(s) = 5 \times 10^9\frac{s}{(s + 0.5 \times 10^6 + j4.975 \times 10^6)(s + 0.5 \times 10^6 - j4.975 \times 10^6)}$$

Drill Problems

14-11 A parallel resonant circuit has a resonant frequency of 2500 rad/s, a bandwidth of 100 rad/s, and an inductance of 200 mH. Find the new bandwidth and capacitance if the circuit is scaled: (*a*) in magnitude by a factor of 5; (*b*) in frequency by a factor of 5; (*c*) in magnitude and frequency by factors of 5.
Ans. *100 rad/s, 0.16 μF; 500 rad/s, 0.032 μF; 500 rad/s, 0.16 μF*

14-12 A scaled series resonant circuit consists of a 2-Ω resistor, a 5-H inductor, and a 0.8-F capacitor. Find K_m and K_f if the original circuit contained a: (*a*)

50-Ω resistor and a 16-μF capacitor; (*b*) 50-Ω resistor and a 1-H inductor; (*c*) 1-H inductor and a 16-μF capacitor.

Ans. 10^{-2}, 2×10^{-3}; 4×10^{-2}, 5×10^{-4}; 4×10^{-2}, 8×10^{-3}

14-13 The magnitude of an admittance has been modeled on the s plane by an elastic sheet, as described in Sec. 13-7. The height of the sheet above the origin is 28 cm, and the location of a certain zero is 40 cm distant from the origin in a direction 110° counterclockwise from the $+\sigma$ axis. Find the distance, direction, and height if the admittance is scaled by: (*a*) $K_m = 2$, $K_f = 1$; (*b*) $K_m = 1$, $K_f = 2$; (*c*) $K_m = 2$, $K_f = 2$.

Ans. 40 cm, 110°, 56 cm; 80 cm, 110°, 28 cm; 80 cm, 110°, 56 cm

14-14 A voltage $\mathbf{V}_1(\mathbf{s})$ applied to a given network produces an output response $\mathbf{I}_2(\mathbf{s}) = (2\mathbf{s} + 5)/(3\mathbf{s}^2 + 4\mathbf{s} + 6)$ A. Find $\mathbf{I}_2(\mathbf{s})$ if the network is scaled in: (*a*) frequency by a factor of 2; (*b*) magnitude by a factor of 2; (*c*) frequency and magnitude by factors of 2.

Ans. $\dfrac{\mathbf{s} + 2.5}{3\mathbf{s}^2 + 4\mathbf{s} + 6}$; $\dfrac{4\mathbf{s} + 20}{3\mathbf{s}^2 + 8\mathbf{s} + 24}$; $\dfrac{2\mathbf{s} + 10}{3\mathbf{s}^2 + 8\mathbf{s} + 24}$ A

PROBLEMS

☐ **1** A parallel network contains $R = 1$ kΩ, $L = 1/12$ H, and $C = 1/3$ μF. Find ω_0, f_0, Q_0, α, ω_d, and the locations of the impedance poles.

☐ **2** If the independent current source, $6 \cos \omega_0 t$ mA, is connected to the network described in Prob. 1, find $\mathbf{V}$, $\mathbf{I}_R$, $\mathbf{I}_L$, $\mathbf{I}_C$, P_R, $w_L(t)$, $w_C(t)$, and the ratios of $|\mathbf{I}_L|$ and $|\mathbf{I}_C|$ to the source current amplitude.

☐ **3** Find the values of the three elements comprising a parallel resonant circuit if: (*a*) $Q_0 = 10$, $L = 2$ H, $R = 2$ kΩ; (*b*) $W_0 = 0.2$ J, $C = 0.1$ μF, $\omega_0 = 2000$ rad/s, $P_{R0} = 400$ W.

☐ **4** A parallel resonant circuit exhibits a natural response in which successive positive maxima have a ratio of 0.99 and the zero crossings are 0.4 ms apart. Find α, ω_d, ω_0, and Q_0.

☐ **5** If $\omega_0 = 1$ krad/s in a parallel resonant circuit, find Q_0 if the lower half-power frequency is: (*a*) 0.99 krad/s; (*b*) 0.8 krad/s; (*c*) 0.1 krad/s.

☐ **6** A parallel resonant circuit driven by a current source, $0.2 \cos \omega t$ A, shows a maximum voltage response amplitude of 80 V at $\omega = 2500$ rad/s and 40 V at 2200 rad/s. Find R, L, C, Q_0, and the half-power frequencies.

☐ **7** Measurements on a pendulum making one complete swing every 1.6 s indicate that the oscillation amplitude drops to 50 per cent of its initial value in 2 min. What is the Q_0?

☐ **8** Sketch a curve of $|\mathbf{V}|$ versus ω for the circuit shown in Fig. 14-14 if $\omega_0 = 10^4$ rad/s and $C =$: (*a*) 1 μF; (*b*) 0.1 μF.

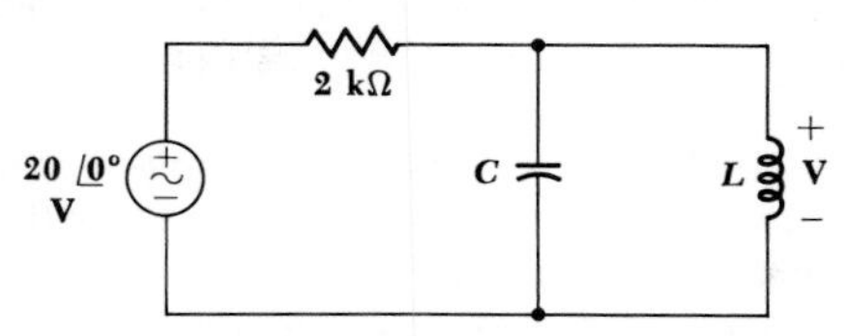

Fig. 14-14 See Probs. 8 and 9.

☐ **9** Find ω_0, Q_0, $\mathcal{B}$, and $\mathbf{V}$ at $\omega = \omega_0$ in the circuit shown in Fig. 14-14 if $C = 2\ \mu$F, $L = 5$ mH, and there is a 500-Ω resistor placed in parallel with the capacitor.

☐ **10** A voltage source $\mathbf{V}_s$ is in series with a series *RLC* network. (*a*) Show that maximum amplitude of the capacitor voltage is $Q_0|\mathbf{V}_s|/\sqrt{1 - 1/(4Q_0^2)}$ and that it occurs at $\omega_0\sqrt{1 - 1/(2Q_0^2)}$. (*b*) Show that the maximum inductor voltage equals the maximum capacitor voltage, but that it occurs at $\omega_0/\sqrt{1 - 1/(2Q_0^2)}$.

☐ **11** Find Q_0, ω_0, ω_1, ω_2, and $\mathcal{B}$ for a series *RLC* network in which: (*a*) $R = 8$ Ω, $L = 0.02$ H, $C = 0.5\ \mu$F; (*b*) $R = 400$ Ω, $L = 0.2$ H, $C = 5\ \mu$F.

☐ **12** The Q of a series or parallel *RLC* network may be obtained by applying the definition of Q to the natural response. For example, assume a high-Q series circuit and a natural response of the form $i(t) = I_0 e^{-\alpha t} \cos \omega_d t$. Determine the stored energy at $t = 0$ and $t \doteq 2\pi/\omega_d \doteq 2\pi/\omega_0$, the time at which i has its first maximum after $t = 0$. From these facts, find Q_0.

☐ **13** A two-terminal network has an impedance whose pole-zero constellation is shown in Fig. 14-15. If the minimum impedance magnitude at any radian frequency is 72 Ω, find the magnitude of the impedance at: (*a*) $\omega = 4000$; (*b*) $\omega = 3600$; (*c*) $\omega = 8000$; (*d*) $\mathbf{s} = -200 + j0$; (*e*) $\mathbf{s} = -200 + j200\ \text{s}^{-1}$.

Fig. 14-15 See Prob. 13.

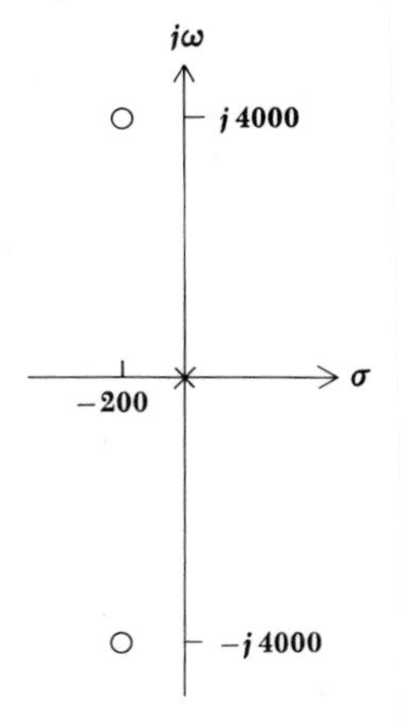

☐ **14** A voltage source, 2 cos ωt V, is available and a current response as a function of frequency is desired that has a maximum of 2 mA at $\omega_0 = 10^5$ rad/s and falls off to $\sqrt{2}$ mA at 2000 rad/s above or below ω_0. Design a series resonant circuit to do this.

☐ **15** The circuit shown in Fig. 14-16 is a simplified example of a Q meter. A physical inductor, which we represent as a resistor R_L in series with an inductor L, is mounted between terminals a-b. The capacitor is adjusted to maximize $|\mathbf{V}_C|$. What restrictions must be placed on R_L and L in order that $|\mathbf{V}_C|_{\text{in } mV} = Q_0 = \omega_0 L/R_L$ with less than 1 per cent error?

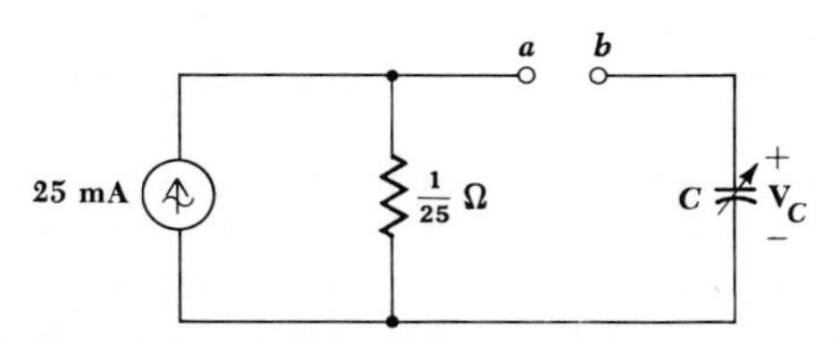

Fig. 14-16 See Prob. 15.

☐ **16** A source $V_0 \cos \omega t$ is applied to a high-Q RLC series network. Assume that an ideal voltmeter can be connected across L and C in series (but not R), and sketch $|\mathbf{V}_{LC}|$ versus ω. Indicate ω_0, ω_1, ω_2, and V_0 on the sketch and specify the shape of the curve near $\omega = \omega_0$.

☐ **17** For the circuit shown in Fig. 14-17, determine the resonant frequency, the Q_0, the bandwidth, and sketch $|\mathbf{I}|$ versus ω. What would the Q_0 of the circuit be if the 20-Ω resistor were short-circuited? Or if the 200-kΩ resistor were open-circuited?

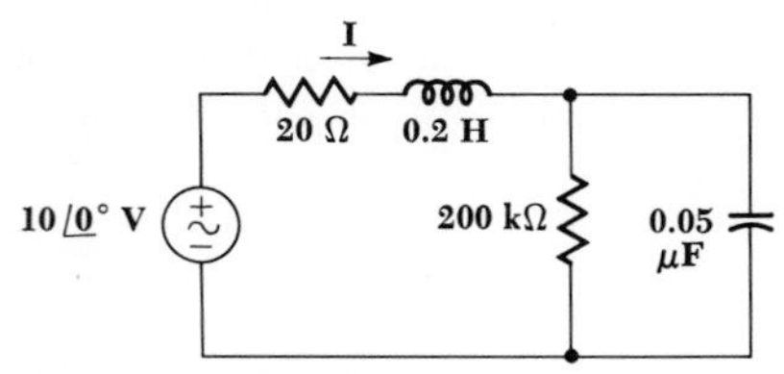

Fig. 14-17 See Prob. 17.

☐ **18** Determine the maximum amplitude of the capacitor voltage in the circuit shown in Fig. 14-18, the frequency at which it occurs, and indicate by a simple sketch its variation with frequency near resonance, near $\omega = 0$, and as ω approaches infinity.

Fig. 14-18 See Prob. 18.

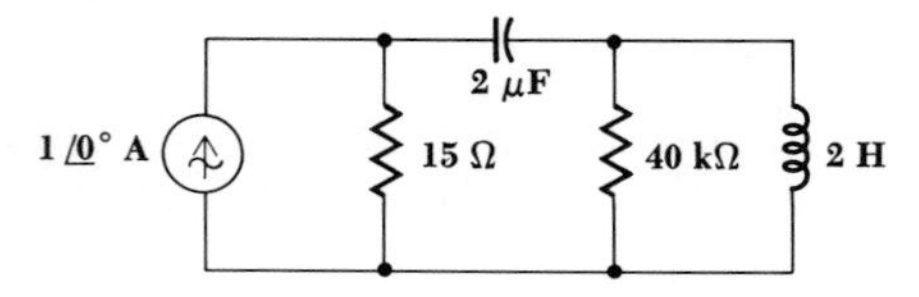

☐ **19** A 0.01-μF capacitor having an effective series resistance of 20 Ω is placed in parallel with a 10-mH inductor having an effective series resistance of 10 Ω. Two other parallel elements, a 5-mA sinusoidal current source and a 300-kΩ resistor, complete the circuit. Describe the variation of the voltage across the source as a function of frequency.

☐ **20** As an indication of the fact that active elements may be used to increase the Q of resonant circuits, find Q_0 for the circuit shown in Fig. 14-19 by an inspection of the input admittance presented to the independent source if $k =$: (*a*) 0; (*b*) 4×10^{-5}; (*c*) -2×10^{-4}.

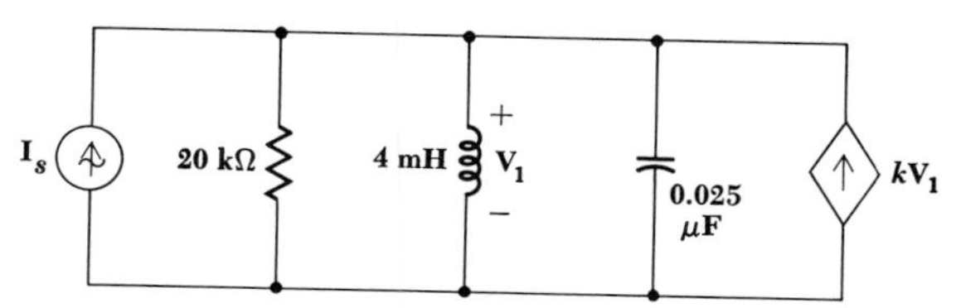

Fig. 14-19 See Prob. 20.

☐ **21** (*a*) Express the resonant frequency and Q_0 of the network shown in Fig. 14-20 as functions of R, L, C, and k. (*b*) Repeat if the positions of R and C are interchanged.

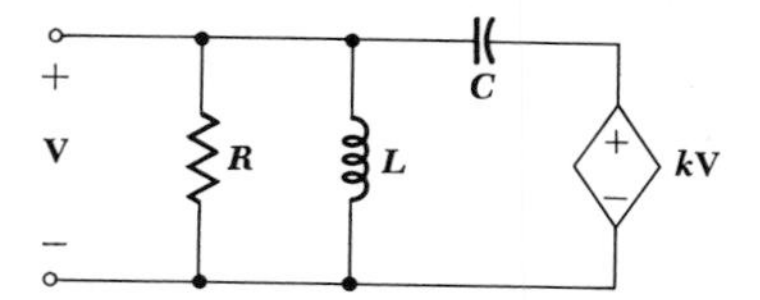

Fig. 14-20 See Prob. 21.

☐ **22** (*a*) If the network shown in Fig. 14-21 is excited by a forcing function $V_m \cos \omega t$, determine the source current at $\omega = 0$, 1, 11, 12, 13, and 30 rad/s and sketch the current amplitude as a function of ω, $0 < \omega < 30$. Check your results by approximate methods. (*b*) If the forcing function is $V_0 u(t)$, find the source current as a function of time and sketch the current as a function of time, $0 < t < 3$ s.

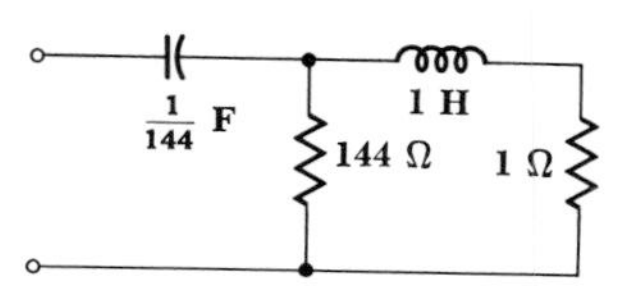

Fig. 14-21 See Probs. 22 and 23.

☐ **23** Refer to the circuit shown in Fig. 14-21 and scale it: (*a*) in magnitude so that $L = 2$ mH; (*b*) in frequency so that $L = 2$ mH; (*c*) in magnitude and frequency so that $L = 2$ mH and $C = 0.2\ \mu$F.

☐ **24** A parallel resonant circuit in which $R = 60\ \text{k}\Omega$ has $\omega_0 = 2 \times 10^5$ rad/s and $Q_0 = 80$. Find the new element values after the circuit is scaled so that $R' = 1\ \Omega$, and $\omega_0' = 1$ rad/s.

☐ **25** A series resonant circuit in which $\alpha = 2$, $\omega_d = 30\ \text{s}^{-1}$, and $L = 20$ H is scaled in such a way that α becomes $10^3\ \text{s}^{-1}$ while the new capacitance is $0.01\ \mu\text{F}$. Determine the new element values.

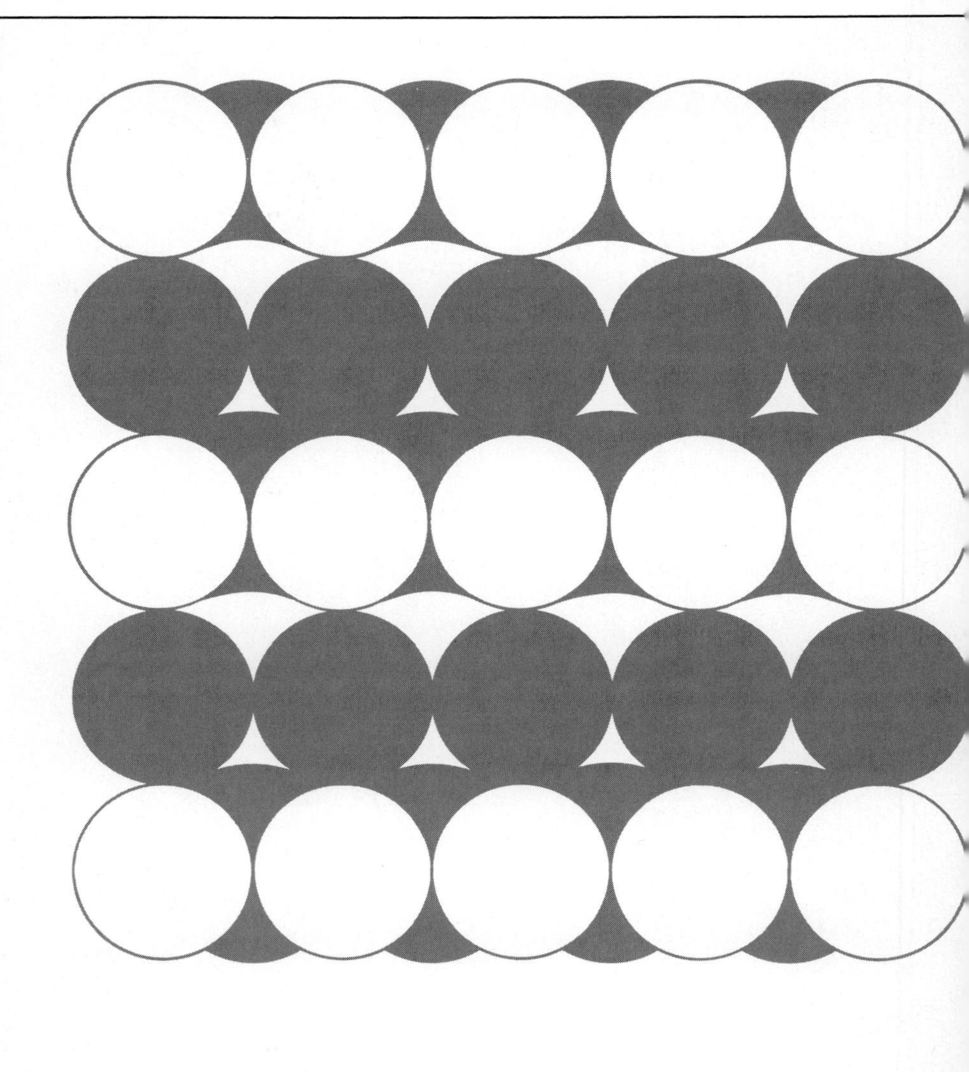

Part Five
TWO-PORT NETWORKS

Chapter Fifteen Magnetically Coupled Circuits

15-1 INTRODUCTION

Several hundred pages ago the inductor was introduced as a circuit element and defined in terms of the voltage across it and the time rate of change of the current through it. Strictly speaking, our definition was of "self-inductance," but, loosely speaking, "inductance" is the commonly used term. Now we need to consider mutual inductance, a property which is associated mutually with two or more coils which are physically close together. A circuit element called the "mutual inductor" does not exist; furthermore, mutual inductance is not a property which is associated with a single pair of terminals, but instead is defined with reference to two pairs of terminals.

Mutual inductance results through the presence of a common magnetic flux which links two coils. It may be defined in terms of this common magnetic flux, just as we might have defined self-inductance in terms of the magnetic flux about the single coil. However, we have agreed to confine our attention to circuit concepts, and such quantities as magnetic flux and flux linkages are mentioned only in passing; we cannot define these quantities easily or accurately at this time, and we must accept them as nebulous concepts which are useful in establishing some background only.

The physical device whose operation is based inherently on mutual inductance is the transformer. The 60-Hz power systems use many transformers, ranging in size from the dimensions of a living room to those of the living-room wastebasket. They are used to change the magnitude of the voltage, increasing it for more economical transmission, and then decreasing it for safer operation of home or industrial electrical equipment. Most radios contain one or more transformers, as do television receivers, hi-fi systems, some telephones, automobiles, and electrified railroads (Penn Central and Lionel both).

We must next define mutual inductance and study the methods whereby its effects are included in the circuit equations. We shall conclude with a study of the important characteristics of a linear transformer and an important approximation to a good iron-core transformer which is known as an ideal transformer.

15-2 MUTUAL INDUCTANCE

When we defined inductance, we did so by specifying the relationship between the terminal voltage and current,

$$v(t) = L\frac{di(t)}{dt}$$

We did learn, however, that the physical basis for such a current-voltage characteristic rests upon (1) the production of a magnetic flux by a current, the flux being proportional to the current in linear inductors; and (2) the production of a voltage by the time-varying magnetic field, the voltage being proportional to the time rate of change of the magnetic field or the magnetic flux. The proportionality between voltage and time rate of change of current thus becomes evident.

Mutual inductance results from a slight extension of this same argument. A current flowing in one coil establishes a magnetic flux about that coil and also about a second coil which is in its vicinity; the time-varying flux surrounding the second coil produces a voltage across the terminals of this second coil; this voltage is proportional to the time rate of change of the current flowing through the first coil. Figure 15-1*a* shows a simple model of two coils L_1 and L_2 sufficiently close together that the flux produced by a current $i_1(t)$ flowing through L_1 establishes an open-circuit voltage $v_2(t)$ across the terminals of L_2. Without considering the proper algebraic sign for the relationship, we define the *coefficient of mutual inductance,* or simply *mutual inductance* M_{21},

$$v_2(t) = M_{21}\frac{di_1(t)}{dt} \tag{1}$$

The order of the subscripts on M_{21} indicates that a voltage response is produced at L_2 by a current source at L_1. If the system is reversed, as indicated in Fig. 15-1*b*, and a voltage response is produced at L_1 by a current source at L_2, then we have

$$v_1(t) = M_{12}\frac{di_2(t)}{dt} \tag{2}$$

Two coefficients of mutual inductance are not necessary, however; we shall use energy relationships a little later to prove that M_{12} and M_{21} are equal. Thus, $M_{12} = M_{21} = M$. The existence of mutual coupling between two coils is indicated by a double-headed arrow, as shown in Fig. 15-1*a* and *b*.

Mutual inductance is measured in henrys and, like resistance, inductance, and capacitance, is always positive.[1] The voltage $M\,di/dt$, however, may appear as either a positive or a negative quantity in the same way that $v = -Ri$ is useful.

The inductor is a two-terminal element, and we are able to use the passive sign convention in order to select the correct sign for the voltage $L\,di/dt$, $j\omega L\mathbf{I}$, or $sL\mathbf{I}$. If the current enters the terminal at which the positive voltage reference is located, then the positive sign is used. Mutual induc-

Fig. 15-1 (*a*) A current i_1 at L_1 produces an open-circuit voltage v_2 at L_2. (*b*) A current i_2 at L_2 produces an open-circuit voltage v_1 at L_1.

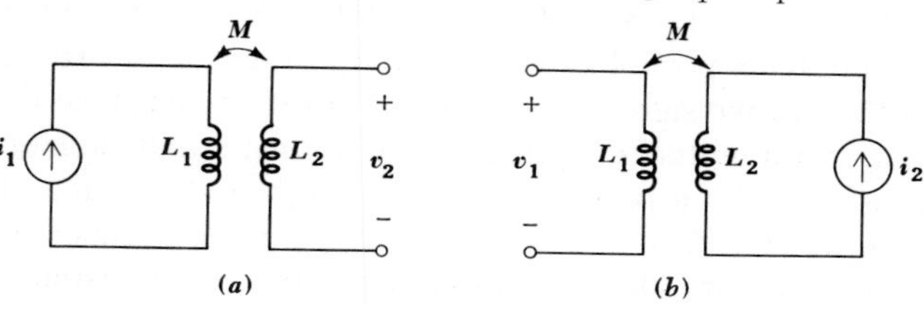

[1]Mutual inductance is not universally assumed to be positive. It is particularly convenient to allow it to "carry its own sign" when three or more coils are involved and each coil interacts with each other coil. We shall restrict our attention to the more important simple case of two coils.

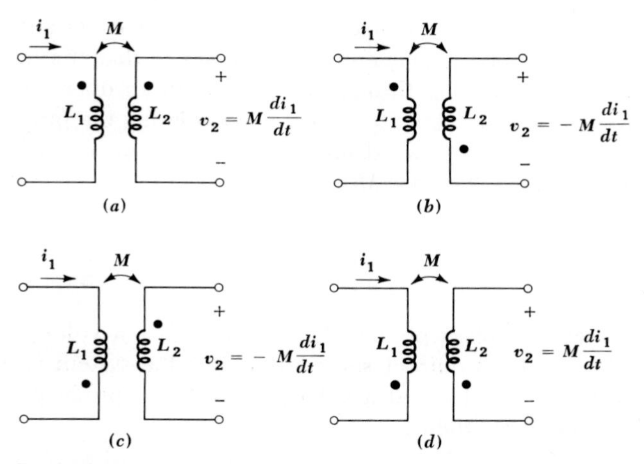

Fig. 15-2 Current entering the dotted terminal of one coil produces a voltage which is sensed positively at the dotted terminal of the second coil. Current entering the undotted terminal of one coil produces a voltage which is sensed positively at the undotted terminal of the second coil.

tance, however, cannot be treated in exactly the same way because four terminals are involved. The choice of a correct sign is established by use of one of several possibilities which include the "dot convention," or an extension of the dot convention which involves the use of a larger variety of special symbols, or by an examination of the particular way in which each coil is wound. We shall use the dot convention and merely look briefly at the physical construction of the coils; the use of other special symbols is not necessary when only two coils are coupled.

The dot convention makes use of a large dot placed at one end of each of the two coils which are mutually coupled. A current entering the dotted terminal of one coil produces an open-circuit voltage between the terminals of the second coil which is sensed in the direction indicated by a positive voltage reference at the dotted terminal of this second coil. Thus, in Fig. 15-2*a*, i_1 enters the dotted terminal of L_1, v_2 is sensed positively at the dotted terminal of L_2, and thus $v_2 = M\,di_1/dt$. We have found previously that it is often not possible to select voltages or currents throughout a circuit so that the passive sign convention is everywhere satisfied; the same situation arises with mutual coupling. For example, it may be more convenient to represent v_2 by a positive voltage reference at the undotted terminal, as shown in Fig. 15-2*b*: then $v_2 = -M\,di_1/dt$. Currents which enter the dotted terminal are also not always available, as indicated by Fig. 15-2*c* and *d*. Such a current provides a voltage which is positive at the undotted terminal of the second coil.

We have as yet considered only a mutual voltage present across an *open-circuited* coil. In general, a nonzero current will be flowing in each

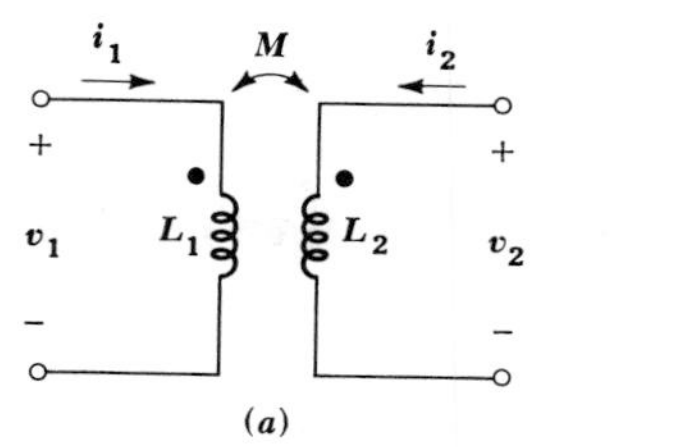

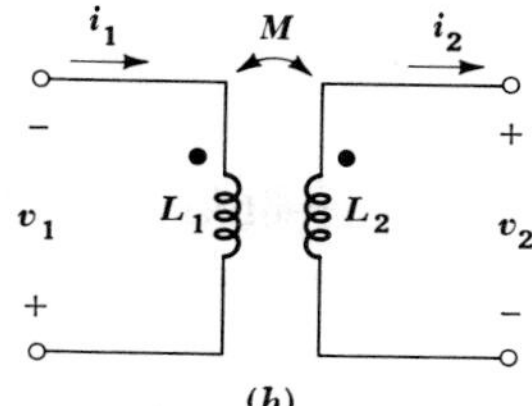

Fig. 15-3 (*a*) Since the pairs v_1, i_1 and v_2, i_2 each satisfy the passive sign convention, the voltages of self-induction are both positive; since i_1 and i_2 each enter dotted terminals, and since v_1 and v_2 are both positively sensed at the dotted terminals, the voltages of mutual induction are also both positive. (*b*) Since the pairs v_1, i_1 and v_2, i_2 are not sensed according to the passive sign convention, the voltages of self-induction are both negative; since i_1 enters the dotted terminals and v_2 is positively sensed at the dotted terminal, the mutual term of v_2 is positive; and since i_2 enters the undotted terminal and v_1 is positively sensed at the undotted terminal, the mutual term of v_1 is also positive.

of the two coils, and a mutual voltage will be produced in each coil because of the current flowing in the other coil. *This mutual voltage is present independently of any voltage of self-induction.* In other words, the voltage across the terminals of L_1 will be composed of two terms $L_1\,di_1/dt$ and $M\,di_2/dt$, each carrying a sign depending on the current directions, the assumed voltage sense, and the placement of the two dots. In the portion of a circuit drawn in Fig. 15-3*a*, currents i_1 and i_2 are shown, each arbitrarily assumed entering the dotted terminal. The voltage across L_1 is thus composed of two parts,

$$v_1 = L_1\frac{di_1}{dt} + M\frac{di_2}{dt}$$

as is the voltage across L_2,

$$v_2 = L_2\frac{di_2}{dt} + M\frac{di_1}{dt}$$

In Fig. 15-3*b* the currents and voltages are not selected with the object of obtaining all positive terms for v_1 and v_2. By inspecting only the reference symbols for i_1 and v_1, it is apparent that the passive sign convention is not satisfied and the sign of $L_1\,di_1/dt$ must therefore be negative. An identical conclusion is reached for the term $L_2\,di_2/dt$. The mutual term of v_2 is signed by inspecting the direction of i_1 and v_2; since i_1 enters the dotted terminal and v_2 is positive at the dotted terminal, the sign of $M\,di_1/dt$ must be positive. Finally, i_2 enters the undotted terminal of L_2 and v_1 is positive at the undotted terminal of L_1; hence, the mutual portion of v_1, $M\,di_2/dt$, must also be positive. Thus, we have

$$v_1 = -L_1\frac{di_1}{dt} + M\frac{di_2}{dt} \qquad v_2 = -L_2\frac{di_2}{dt} + M\frac{di_1}{dt}$$

The same considerations lead to identical choices of signs for excitation at a complex frequency **s**,

$$\mathbf{V}_1 = -\mathbf{s}L_1\mathbf{I}_1 + \mathbf{s}M\mathbf{I}_2 \qquad \mathbf{V}_2 = -\mathbf{s}L_2\mathbf{I}_2 + \mathbf{s}M\mathbf{I}_1$$

or at a real frequency $\mathbf{s} = j\omega$,

$$\mathbf{V}_1 = -j\omega L_1\mathbf{I}_1 + j\omega M\mathbf{I}_2 \qquad \mathbf{V}_2 = -j\omega L_2\mathbf{I}_2 + j\omega M\mathbf{I}_1$$

Before we apply the dot convention to the analysis of a numerical example, we can gain a more complete understanding of the dot symbolism by looking at the physical basis for the convention. The meaning of the dots is now interpreted in terms of magnetic flux. Two coils are shown wound on a cylindrical form in Fig. 15-4, and the direction of each winding is evident. Let us assume that the current i_1 is positive and increasing with time. The magnetic flux that i_1 produces within the form has a direction which may be found by the right-hand rule: when the right hand is wrapped around the coil with the fingers pointing in the direction of current flow, the thumb indicates the direction of the flux within the coil. Thus i_1 produces a flux which is directed downward; since i_1 is increasing with time, the flux, which is proportional to i_1, is also increasing with time. Turning now to the second coil, let us also think of i_2 as positive and increasing; the application of the right-hand rule shows that i_2 also produces a magnetic flux which is directed downward and is increasing. In other words, the assumed currents i_1 and i_2 produce *additive* fluxes.

Fig. 15-4 The physical construction of two mutually coupled coils. From a consideration of the direction of magnetic flux produced by each coil, it is shown that dots may be placed either on the upper terminal of each coil or on the lower terminal of each coil.

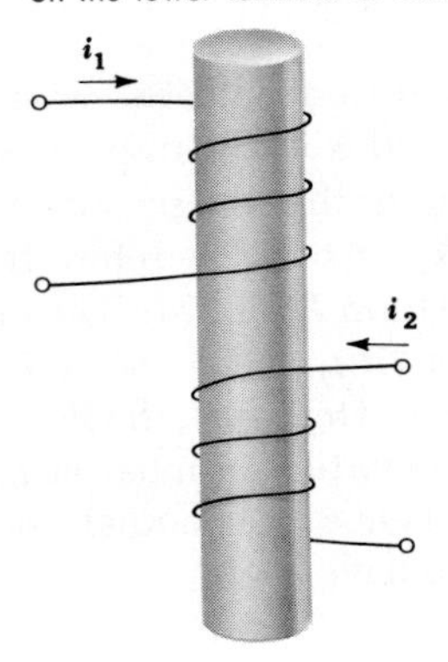

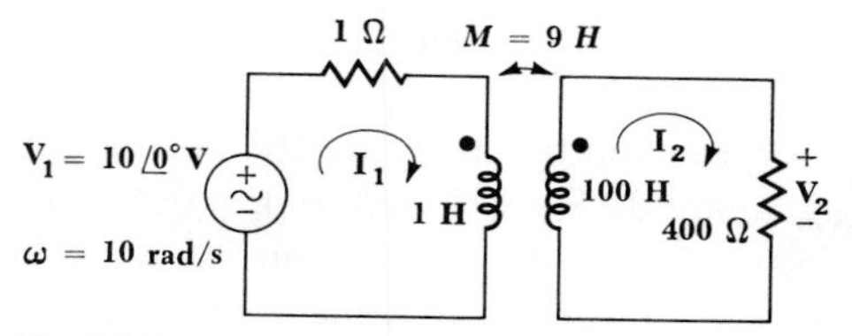

Fig. 15-5 A circuit containing mutual inductance in which the voltage ratio $\mathbf{V}_2/\mathbf{V}_1$ is desired.

The voltage across the terminals of any coil results from the time rate of change of the flux linking that coil. The voltage across the terminals of the first coil is therefore greater with i_2 flowing than it would be if i_2 were zero. Thus i_2 induces a voltage in the first coil which has the same sense as the self-induced voltage in that coil. The sign of the self-induced voltage is known from the passive sign convention, and the sign of the mutual voltage is thus obtained.

The dot convention merely enables us to suppress the physical construction of the coils by placing a dot at one terminal of each coil such that currents entering dot-marked terminals produce additive fluxes. It is apparent that there are always two possible locations for the dots, because both dots may always be moved to the other ends of the coils and additive fluxes will still result.

The alternative method of selecting the correct signs for the mutual terms therefore consists of first noting whether or not both currents enter dot-marked terminals. If they do, then the sign of the mutual voltage at each coil is the same as the sign of the self-induced voltage at that same coil. The same result occurs if both currents leave dot-marked terminals. However, if neither of these conditions occurs, then subtractive fluxes result and the sign of the mutual voltage at each coil is the opposite of the sign of the self-induced voltage at that same coil.

In the example of Fig. 15-3*b*, the currents do not both enter either the dotted or undotted terminals, and the signs of the mutual terms must be the opposite of the signs of the self-induced terms. The passive sign convention indicates that both of the self-induced voltages carry negative signs, and thus both of the mutual voltages are positive. This agrees with the previous set of equations obtained for this circuit; either method may be used to find the correct signs for the $M\,di/dt$ terms.

Let us now apply these methods to the analysis of the circuit outlined in Fig. 15-5. We desire the ratio of the output voltage across the 400-Ω resistor to the source voltage. Two conventional mesh currents are established, and Kirchhoff's voltage law must next be applied to each mesh. In the left mesh, the sign of the mutual term may be determined, for example, by applying the dot convention directly. Since $\mathbf{I}_2$ leaves the dot-marked terminal of L_2, the mutual voltage across L_1 must have the positive reference at the undotted terminal. Thus,

$$\mathbf{I}_1(1 + j10) - j90\mathbf{I}_2 = 10$$

For variety, the sign of the mutual term in the second mesh may be determined by another method. Since $\mathbf{I}_1$ enters the dot-marked terminal while $\mathbf{I}_2$ leaves the dot-marked terminal, the mutual term in each mesh must have the opposite sign from the self-inductance term. Thus, we may write

$$\mathbf{I}_2(400 + j1000) - j90\mathbf{I}_1 = 0$$

The two equations may be solved by determinants (or a simple elimination of $\mathbf{I}_1$),

$$\mathbf{I}_2 = \frac{\begin{vmatrix} 1 + j10 & 10 \\ -j90 & 0 \end{vmatrix}}{\begin{vmatrix} 1 + j10 & -j90 \\ -j90 & 400 + j1000 \end{vmatrix}}$$

and

$$\mathbf{I}_2 = 0.1725\underline{/-16.7^\circ}$$

and thus

$$\frac{\mathbf{V}_2}{\mathbf{V}_1} = \frac{400(0.1725\underline{/-16.7^\circ})}{10}$$

or

$$\frac{\mathbf{V}_2}{\mathbf{V}_1} = 6.90\underline{/-16.7^\circ}$$

The output voltage is greater in magnitude than the input voltage, and thus a voltage gain is possible with mutual coupling just as it is in a resonant circuit. The voltage gain available in this circuit, however, is present over a relatively wide range of frequency; in a moderately high Q resonant circuit, the voltage step-up is proportional to the Q and occurs only over a range of frequency which is inversely proportional to Q. Let us see if this point can be made clear by reference to the complex plane.

We may find $\mathbf{I}_2(\mathbf{s})$ for this particular circuit,

$$\mathbf{I}_2(\mathbf{s}) = \frac{\begin{vmatrix} 1 + \mathbf{s} & \mathbf{V}_1 \\ -9\mathbf{s} & 0 \end{vmatrix}}{\begin{vmatrix} 1 + \mathbf{s} & -9\mathbf{s} \\ -9\mathbf{s} & 400 + 100\mathbf{s} \end{vmatrix}} = \frac{9\mathbf{s}\mathbf{V}_1}{19\mathbf{s}^2 + 500\mathbf{s} + 400}$$

and thus obtain the ratio of output to input voltage as a function of $\mathbf{s}$,

$$\frac{\mathbf{V}_2}{\mathbf{V}_1} = \frac{3600\mathbf{s}}{19\mathbf{s}^2 + 500\mathbf{s} + 400} = 189.5\frac{\mathbf{s}}{(\mathbf{s} + 0.826)(\mathbf{s} + 25.5)}$$

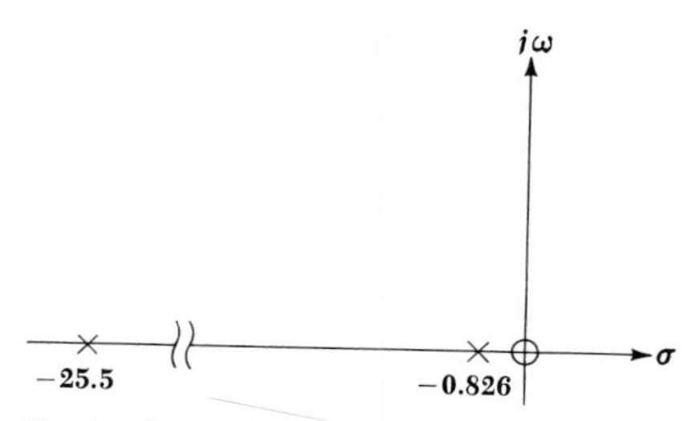

Fig. 15-6 A pole-zero plot of the transfer function $\mathbf{V}_2/\mathbf{V}_1$ for the circuit shown in Fig. 15-5. The plot is useful in showing that the magnitude of the transfer function is relatively large from $\omega = 1$ or 2 to $\omega = 15$ or 20.

The pole-zero plot of this transfer function is shown in Fig. 15-6. The location of the pole at $\mathbf{s} = -25.5$ is distorted in order to show both poles clearly; the ratio of the distances of the two poles from the origin is actually about 30:1. An inspection of this plot shows that the transfer function is zero at zero frequency but that as soon as ω is greater than, say, 1 or 2 the ratio of the distances from the zero and the pole nearer the origin is essentially unity. Thus the voltage gain is affected only by the distant pole on the negative σ axis, and this distance does not increase appreciably until ω approaches 15 or 20.

These tentative conclusions are verified by the response curve of Fig. 15-7, which shows that the magnitude of the voltage gain is greater than 0.707 of its maximum value from $\omega = 0.78$ to $\omega = 26$.

The circuit is still passive, except for the voltage source, and the voltage gain must not be mistakenly interpreted as a power gain. At $\omega = 10$, the

Fig. 15-7 The voltage gain $|\mathbf{V}_2/\mathbf{V}_1|$ of the circuit shown in Fig. 15-5 is plotted as a function of ω. The voltage gain is greater than 5 from about $\omega = 0.75$ to approximately $\omega = 28$.

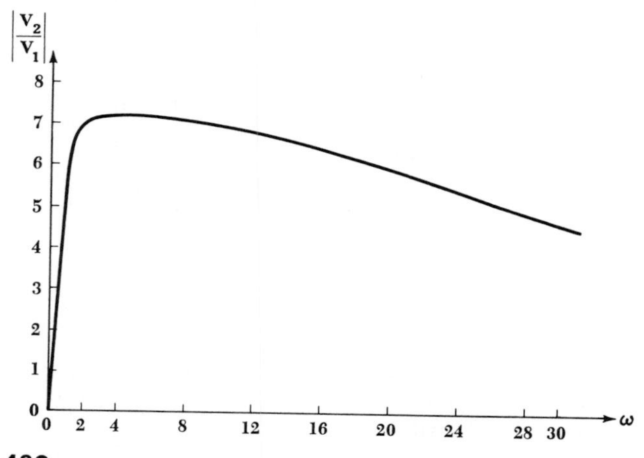

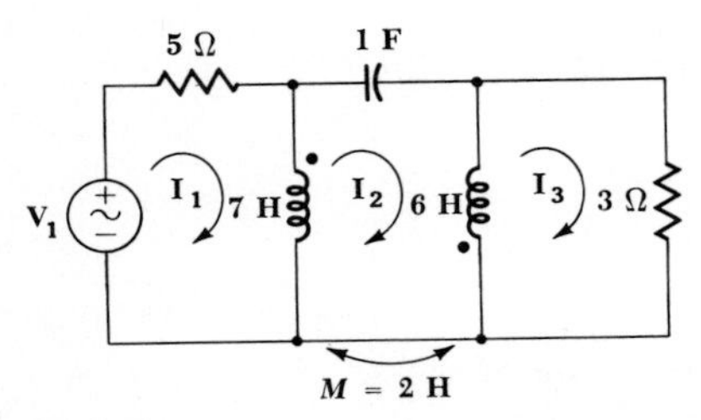

Fig. 15-8 A three-mesh circuit with mutual coupling may be analyzed most easily by loop or mesh methods.

voltage gain is 6.90, but the ideal voltage source, possessing a voltage of 10 V, delivers a total power of 8.08 W of which only 5.95 W reaches the 400-Ω resistor. The ratio of the output power to the source power, which we may define as the power gain, is thus 0.737.

Let us consider briefly one additional example, illustrated in Fig. 15-8. The circuit contains three meshes, and three mesh currents are assigned. Applying Kirchhoff's voltage law to the first mesh, a positive sign for the mutual term is assured by selecting $(\mathbf{I}_3 - \mathbf{I}_2)$ as the current through the second coil. Thus,

$$5\mathbf{I}_1 + 7\mathbf{s}(\mathbf{I}_1 - \mathbf{I}_2) + 2\mathbf{s}(\mathbf{I}_3 - \mathbf{I}_2) = \mathbf{V}_1$$

or

$$(5 + 7\mathbf{s})\mathbf{I}_1 - 9\mathbf{s}\mathbf{I}_2 + 2\mathbf{s}\mathbf{I}_3 = \mathbf{V}_1 \tag{3}$$

The second mesh requires two self-inductance terms and two mutual-inductance terms; the equation cannot be written carelessly. We obtain

$$7\mathbf{s}(\mathbf{I}_2 - \mathbf{I}_1) + 2\mathbf{s}(\mathbf{I}_2 - \mathbf{I}_3) + \frac{1}{\mathbf{s}}\mathbf{I}_2 + 6\mathbf{s}(\mathbf{I}_2 - \mathbf{I}_3) + 2\mathbf{s}(\mathbf{I}_2 - \mathbf{I}_1) = 0$$

or

$$-9\mathbf{s}\mathbf{I}_1 + \left(17\mathbf{s} + \frac{1}{\mathbf{s}}\right)\mathbf{I}_2 - 8\mathbf{s}\mathbf{I}_3 = 0 \tag{4}$$

Finally, for the third mesh

$$6\mathbf{s}(\mathbf{I}_3 - \mathbf{I}_2) + 2\mathbf{s}(\mathbf{I}_1 - \mathbf{I}_2) + 3\mathbf{I}_3 = 0$$

or

$$2\mathbf{s}\mathbf{I}_1 - 8\mathbf{s}\mathbf{I}_2 + (3 + 6\mathbf{s})\mathbf{I}_3 = 0 \tag{5}$$

Equations (3) to (5) may be solved by any of the conventional methods.

Drill Problems

15-1 Refer to the coupled coils shown in Fig. 15-3*a* and let $L_1 = 4$ H, $L_2 = 3$ H, and $M = 2$ H. Find v_2 if: (*a*) $i_1 = 5 \cos 6t$ A, $i_2 = 0$; (*b*) $i_1 = 0$, $i_2 = 3 \cos 6t$ A; (*c*) $i_1 = 5 \cos 6t$, $i_2 = 3 \cos 6t$ A.

Ans. $-114 \sin 6t$; $-60 \sin 6t$; $-54 \sin 6t$ V

15-2 In the circuit shown in Fig. 15-9, find: (*a*) $\mathbf{I}_2$; (*b*) $\mathbf{I}_1$; (*c*) $\mathbf{I}_R$.
Ans. $17.9\underline{/-153.4°}$; $1.79\underline{/26.6°}$; $2.53\underline{/-18.4°}$ mA

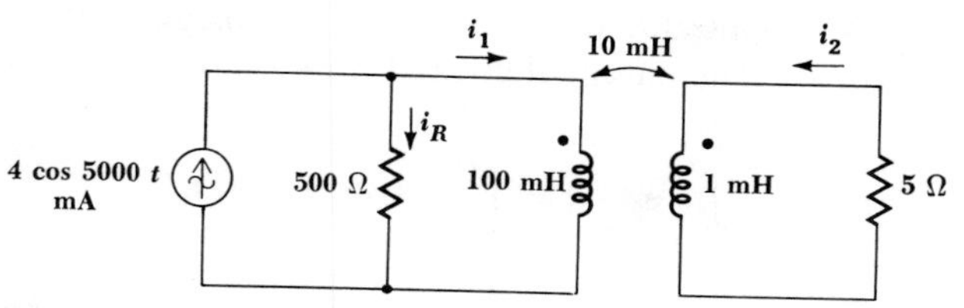

Fig. 15-9 See Drill Probs. 15-2 and 15-3.

15-3 For the circuit of Fig. 15-9, let the current source be $\mathbf{I}_s$ and the voltage across the 5-Ω resistor (positive reference up) be $\mathbf{V}_2$. Find $\mathbf{V}_2/\mathbf{I}_s$ as a function of **s** and evaluate for **s** = : (*a*) -2000; (*b*) $j5000$; (*c*) $-2500 + j5000$ s^{-1}.
Ans. $22.4\underline{/26.6°}$; $27.9\underline{/26.6°}$; -100 Ω

15-3 ENERGY CONSIDERATIONS

Let us now consider the energy stored in a pair of mutually coupled inductors. The results will be useful in several different ways. We shall first justify our assumption that $M_{12} = M_{21}$, and we may then determine the maximum possible value of the mutual inductance between two given inductors. Finally, we shall provide ourselves with a basis from which it is possible to establish the initial conditions in magnetically coupled circuits, although we shall consider only the more elementary cases.

A pair of coupled coils is shown in Fig. 15-3*a* with currents, voltages, and polarity dots indicated. Let us first show that reciprocity is valid, or that $M_{12} = M_{21}$. In order to do so, we shall begin by letting all currents and voltages be zero, thus establishing zero initial energy storage in the network. We first open-circuit the right terminal pair and increase i_1 from zero to some constant value I_1. The power entering the network from the left at any instant is

$$v_1 i_1 = L_1 \frac{di_1}{dt} i_1$$

and that entering from the right is

$$v_2 i_2 = 0$$

since $i_2 = 0$.

The energy stored within the network when $i_1 = I_1$ is thus

$$\int_0^{t_1} v_1 i_1\,dt = \int_0^{I_1} L_1 i_1\,di_1 = \tfrac{1}{2}L_1 I_1^2$$

We now hold i_1 constant, $i_1 = I_1$, and let i_2 change from zero to some constant value I_2. The energy delivered from the right source is thus

$$\int_{t_1}^{t_2} v_2 i_2\,dt = \int_0^{I_2} L_2 i_2\,di_2 = \tfrac{1}{2}L_2 I_2^2$$

However, even though the value of i_1 remains constant, the left source also delivers energy to the network during this time interval,

$$\begin{aligned}\int_{t_1}^{t_2} v_1 i_1\,dt &= \int_{t_1}^{t_2} M_{12}\frac{di_2}{dt} i_1\,dt \\ &= M_{12} I_1 \int_0^{I_2} di_2 \\ &= M_{12} I_1 I_2\end{aligned}$$

The total energy stored in the network when both i_1 and i_2 have reached a constant value is

$$W_{\text{total}} = \tfrac{1}{2}L_1 I_1^2 + \tfrac{1}{2}L_2 I_2^2 + M_{12} I_1 I_2$$

Now we may establish the same final currents in this network by allowing the currents to reach their final values in the reverse order, that is, first increasing i_2 from zero to I_2 and then holding i_2 constant while i_1 increases from zero to I_1. If the total energy stored is calculated for this experiment, the result is found to be

$$W_{\text{total}} = \tfrac{1}{2}L_1 I_1^2 + \tfrac{1}{2}L_2 I_2^2 + M_{21} I_1 I_2$$

The only difference is the interchange of the mutual inductances M_{21} and M_{12}. The initial and final conditions in the network are the same, however, and the two values of the stored energy must be identical. Thus,

$$M_{12} = M_{21} = M$$

and

$$W = \tfrac{1}{2}L_1 I_1^2 + \tfrac{1}{2}L_2 I_2^2 + M I_1 I_2 \tag{6}$$

If one current enters a dot-marked terminal while the other leaves a dot-marked terminal, the sign of the mutual energy term is reversed:

$$W = \tfrac{1}{2}L_1 I_1^2 + \tfrac{1}{2}L_2 I_2^2 - M I_1 I_2 \tag{7}$$

Although (6) and (7) were derived by treating the final values of the two currents as constants, it is apparent that these "constants" may have any value, and the energy expressions correctly represent the energy stored when the *instantaneous* values of i_1 and i_2 are I_1 and I_2, respectively. In other words, lowercase symbols might just as well be used,

$$w(t) = \tfrac{1}{2}L_1[i_1(t)]^2 + \tfrac{1}{2}L_2[i_2(t)]^2 \pm M[i_1(t)][i_2(t)] \tag{8}$$

The only assumption upon which (8) is based is the logical establishment of a zero-energy reference level when both currents are zero.

Equation (8) may now be used to establish an upper limit for the value of M. Since $w(t)$ represents the energy stored within a *passive* network, it cannot be negative for any values of i_1, i_2, L_1, L_2, or M. Let us assume first that i_1 and i_2 are either both positive or both negative; their product is therefore positive. From (8), the only case in which the energy could possibly be negative is

$$w = \tfrac{1}{2}L_1 i_1^2 + \tfrac{1}{2}L_2 i_2^2 - M i_1 i_2$$

which we may write, by completing the square,

$$w = \tfrac{1}{2}(\sqrt{L_1}i_1 - \sqrt{L_2}i_2)^2 + \sqrt{L_1L_2}i_1i_2 - Mi_1i_2$$

Now the energy cannot be negative; the right side of this equation therefore cannot be negative. The first term, however, may be as small as zero, and thus the sum of the last two terms cannot be negative. Hence,

$$\sqrt{L_1L_2} \geq M$$

or

$$M \leq \sqrt{L_1L_2} \tag{9}$$

There is, therefore, an upper limit to the possible magnitude of the mutual inductance; it can be no larger than the geometric mean of the inductances of the two coils between which the mutual inductance exists. Although we have derived this inequality on the assumption that i_1 and i_2 carried the same algebraic sign, a similar development is possible if the signs are opposite; it is only necessary to select the positive sign in (8).

We might also have demonstrated the truth of (9) from a physical consideration of the magnetic coupling; if we think of i_2 as being zero and the current i_1 as establishing the magnetic flux linking both L_1 and L_2, it is apparent that the flux linking L_2 cannot be greater than the flux linking L_1, which represents the total flux. Qualitatively, then, there is an upper limit to the magnitude of the mutual inductance possible between two given inductors. For example, if $L_1 = 1$ H and $L_2 = 10$ H, then $M \leq 3.16$ H.

The degree to which M approaches its maximum value is exactly described by the coefficient of coupling. We define the *coefficient of coupling,* symbolized as k,

$$k = \frac{M}{\sqrt{L_1 L_2}} \tag{10}$$

It is evident that

$$0 \le k \le 1$$

The larger values of the coefficient of coupling are obtained with coils which are physically closer, which are wound or oriented to provide a larger common magnetic flux, or which are provided with a common path through a material which serves to concentrate and localize the magnetic flux (a high-permeability material). Coils having a large coefficient of coupling are said to be tightly coupled.

The complete response of a circuit possessing mutual coupling must of course be made up of the sum of a forced and a natural response. We are now equipped to find the forced response for the more familiar forcing functions, and we shall consider this portion of the response in detail for several interesting circuits in the remainder of this chapter. If we can find the forced response, then we certainly can locate the poles or zeros of any immittance function associated with the circuit. This knowledge of the pole-zero locations then enables us to establish the functional form of the natural response. The remaining problem, the determination of the arbitrary constants through the values of the initial conditions in the network, is not as easily resolved with the analytical methods we possess. The use of the initial conditions turns out to be much more straightforward when the operational method of the Laplace transform is employed, and this topic is one which we shall investigate in Chap. 19. Rather than make a simple problem appear difficult, we shall therefore largely neglect the transient analysis of mutually coupled circuits at this time. Let us consider only the single straightforward problem of finding the complete response when a constant voltage is suddenly applied to one coil while the second coil is connected to a resistive load.

The circuit to be analyzed is shown in Fig. 15-10; the currents i_1 and i_2 are desired. The forced response may be determined readily, for it will have the form of the forcing function, that is, a constant times $u(t)$. Since neither current is changing, there can be no mutual voltage. Thus, i_2 is flowing in a source-free circuit and this *forced* response must be zero. The 10-V source present in the first mesh, however, produces a forced response of 5 A for i_1. Thus,

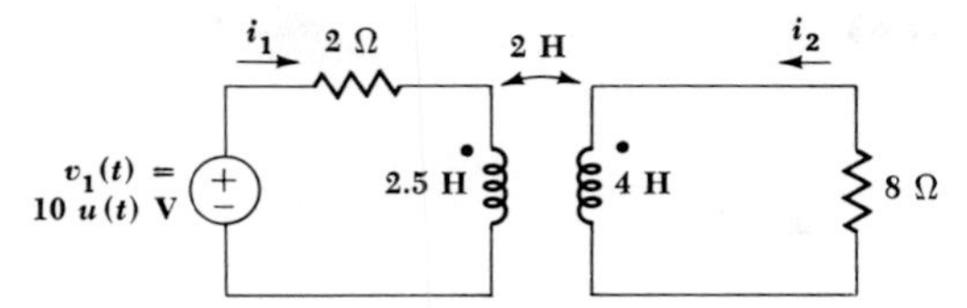

Fig. 15-10 A circuit containing mutual inductance for which the complete response is desired.

$$i_{1f} = 5u(t) \qquad i_{2f} = 0$$

The form of the natural response is now obtained by determining the zeros of the impedance offered to the source. After writing the Kirchhoff voltage equation about each mesh,

$$(2 + 2.5\mathbf{s})\mathbf{I}_1 + 2\mathbf{s}\mathbf{I}_2 = \mathbf{V}_1$$

$$2\mathbf{s}\mathbf{I}_1 + (8 + 4s)\mathbf{I}_2 = 0$$

we determine $\mathbf{Z}_{\text{in}}$ as the ratio of $\mathbf{V}_1$ to $\mathbf{I}_1$,

$$\mathbf{Z}_{\text{in}} = 2 + 2.5\mathbf{s} - \frac{4\mathbf{s}^2}{8 + 4\mathbf{s}}$$

or

$$\mathbf{Z}_{\text{in}} = 1.5\frac{(\mathbf{s} + \frac{2}{3})(\mathbf{s} + 4)}{\mathbf{s} + 2}$$

and thus the form of the natural response is established for both i_1 and i_2,

$$i_{1n} = A_1e^{-2t/3} + B_1e^{-4t}$$

$$i_{2n} = A_2e^{-2t/3} + B_2e^{-4t}$$

The complete responses are therefore

$$i_1 = 5u(t) + A_1e^{-2t/3} + B_1e^{-4t} \qquad (11)$$

$$i_2 = A_2e^{-2t/3} + B_2e^{-4t} \qquad (12)$$

We must next establish the initial values of the two currents and their first derivatives from physical considerations. There is no energy stored initially in the circuit, and there can be no discontinuous change in the stored energy without the presence of an impulse source. Thus, the energy stored remains zero at $t = 0^+$, and this requires that both i_1 and i_2 be zero at $t = 0^+$. The initial values of the two derivatives are most easily found by using the time-domain statements of Kirchhoff's voltage law for each mesh:

$$2i_1 + 2.5\frac{di_1}{dt} + 2\frac{di_2}{dt} = 10u(t)$$

$$2\frac{di_1}{dt} + 8i_2 + 4\frac{di_2}{dt} = 0$$

At $t = 0^+$, we have

$$0 + 2.5\frac{di_1}{dt}\bigg|_{0^+} + 2\frac{di_2}{dt}\bigg|_{0^+} = 10$$

$$2\frac{di_1}{dt}\bigg|_{0^+} + 0 + 4\frac{di_2}{dt}\bigg|_{0^+} = 0$$

Solving simultaneously,

$$\frac{di_1}{dt}\bigg|_{0^+} = \frac{20}{3} \qquad \text{and} \qquad \frac{di_2}{dt}\bigg|_{0^+} = -\frac{10}{3}$$

Taking the derivative of (11) and (12) and evaluating at $t = 0^+$, we therefore find

$$\frac{di_1}{dt}\bigg|_{0^+} = \frac{20}{3} = -\frac{2A_1}{3} - 4B_1$$

$$\frac{di_2}{dt}\bigg|_{0^+} = -\frac{10}{3} = -\frac{2A_2}{3} - 4B_2$$

Since each current is also initially zero,

$$A_1 + B_1 + 5 = 0 \qquad A_2 + B_2 = 0$$

These last four equations may be solved easily for the four arbitrary constants. When the results are inserted in (11) and (12), we obtain

$$i_1 = 5 - 4e^{-2t/3} - e^{-4t}$$

$$i_2 = -e^{-2t/3} + e^{-4t}$$

for $t > 0$.

The source current is quite similar to the response of the simple *RL* series circuit having a time constant of approximately 4 s. The value of 4 s, however, is determined by both resistors, both self-inductances, and the mutual inductance. The load current is a negative pulse having a maximum magnitude of about 0.6 A, occurring about 0.5 s after $t = 0$.

The principles on which this analysis is based are not difficult to comprehend, but one does suffer from fatigue in carrying out the analytical details. The use of operational methods serves to reduce greatly the number of trivial steps required, as well as to organize the solution in a more systematic fashion.

Drill Problems

15-4 If currents i_1 and i_2 enter the dotted terminals of a pair of coupled coils, $L_1 = 0.2$ H, $L_2 = 1.25$ H, $k = 0.8$, find the total energy stored in the system at $t = 0$ if: (*a*) $i_1 = i_2 = 4 \cos \omega t$ A; (*b*) $i_1 = -i_2 = 4 \cos \omega t$ A; (*c*) $i_1 = 2e^{-\sigma t}$, $i_2 = 2 \cos \omega t$ A.

Ans. 4.5; 5.2; 18 J

15-5 In the circuit shown in Fig. 15-10, change M to 0.8 H and find the initial values ($t = 0^+$) of: (*a*) i_1; (*b*) di_1/dt; (*c*) d^2i_1/dt^2.

Ans. -4.24 A/s^2; 0; 4.27 A/s

15-6 Two inductors, $L_1 = 25L_2 = 1$ H, are mutually coupled with $k = 1$. An impedance $\mathbf{Z}_L$ is connected across L_2. Find the input impedance at the terminals of L_1 at $\omega = 1000$ rad/s if $\mathbf{Z}_L =$: (*a*) 4; (*b*) $j4$; (*c*) $-j4$ Ω.

Ans. $-j111.1$; $j90.9$; $99.0 + j9.90$ Ω

15-4 THE LINEAR TRANSFORMER

We are now ready to apply our knowledge of magnetic coupling to an analytical description of the performance of two specific practical devices, each of which may be represented by a model containing mutual inductance. Both of the devices are *transformers*, a term which we may define as a network containing two or more coils which are deliberately coupled magnetically. In this section we shall consider the linear transformer, which is, even in practice, a linear device finding its greatest application at radio frequencies, or higher frequencies. In the following section we shall consider the ideal transformer, which is an idealized unity-coupled model of a physical transformer that has a core made of some magnetic material, usually iron alloy.

In Fig. 15-11 there is shown a transformer with two mesh currents identified. The first mesh, usually containing the source, is called the *primary*, while the second mesh, usually containing the load, is known as the *secondary*. The inductors L_1 and L_2 are also referred to as the primary and secondary, respectively, of the transformer. We shall assume that the transformer is linear. This merely infers that no magnetic material is employed to cause most of the magnetic flux produced by the primary winding to link the secondary winding. Without such material, it is difficult

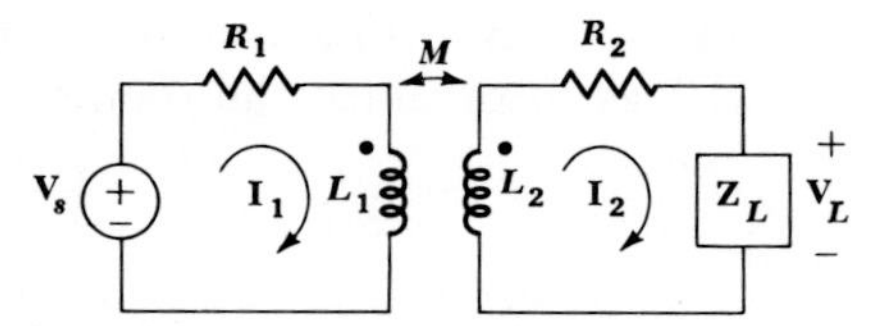

Fig. 15-11 A linear transformer containing a source in the primary circuit and a load in the secondary circuit. Resistance is also included in both the primary and secondary.

to achieve a coefficient of coupling greater than a few tenths. The two resistors serve to account for the resistance of the wire out of which the primary and secondary coils are wound, and any other losses.

In many applications, the linear transformer is used with a tuned, or resonant, secondary; the primary winding is also operated often in a resonant condition by replacing the ideal voltage source by a current source in parallel with a large resistance and a capacitance. The analysis of such a single-tuned or double-tuned circuit is a fairly lengthy process, and we shall not undertake it at this time. It may be pointed out, however, that the secondary response is characterized by the familiar resonance curve for relatively small coefficients of coupling but that a greater control of the shape of the response curve as a function of frequency becomes possible for larger coefficients of coupling. Response curves which possess flatter tops and sharper drops on each side may be achieved in the double-tuned circuit.

Let us consider only the input impedance offered at the terminals of the primary circuit. The two mesh equations are

$$\mathbf{V}_s = \mathbf{I}_1(R_1 + sL_1) - \mathbf{I}_2 sM \tag{13}$$

$$0 = -\mathbf{I}_1 sM + \mathbf{I}_2(R_2 + sL_2 + \mathbf{Z}_L) \tag{14}$$

We may simplify by defining

$$\mathbf{Z}_{11} = R_1 + sL_1 \qquad \mathbf{Z}_{22} = R_2 + sL_2 + \mathbf{Z}_L$$

and thus

$$\mathbf{V}_s = \mathbf{I}_1\mathbf{Z}_{11} - \mathbf{I}_2 sM \tag{15}$$

$$0 = -\mathbf{I}_1 sM + \mathbf{I}_2\mathbf{Z}_{22} \tag{16}$$

Solving the second equation for $\mathbf{I}_2$ and inserting this in the first equation enables us to find the input impedance,

$$\mathbf{Z}_{\text{in}} = \frac{\mathbf{V}_s}{\mathbf{I}_1} = \mathbf{Z}_{11} - \frac{s^2M^2}{\mathbf{Z}_{22}} \tag{17}$$

Before manipulating this expression any further, we can draw several exciting conclusions. In the first place, this result is independent of the location of the dots on either winding, for if either dot is moved to the other end of the coil, the result is a change in sign of each term involving M in (13) to (16). This same effect could be obtained by replacing M by $(-M)$, and such a change cannot affect the input impedance, as (17) demonstrates. We also may note in (17) that the input impedance is simply $\mathbf{Z}_{11}$ if the coupling is reduced to zero. As the coupling is increased from zero, the input impedance differs from $\mathbf{Z}_{11}$ by an amount $-\mathbf{s}^2M^2/\mathbf{Z}_{22}$, termed the *reflected impedance*. This change can be inspected more closely by letting $\mathbf{s} = j\omega$,

$$\mathbf{Z}_{\text{in}}(j\omega) = \mathbf{Z}_{11}(j\omega) + \frac{\omega^2 M^2}{R_{22} + jX_{22}}$$

and rationalizing the reflected impedance,

$$\mathbf{Z}_{\text{in}} = \mathbf{Z}_{11} + \frac{\omega^2 M^2 R_{22}}{R_{22}^2 + X_{22}^2} + \frac{-j\omega^2 M^2 X_{22}}{R_{22}^2 + X_{22}^2}$$

Since $\omega^2M^2R_{22}/(R_{22}^2 + X_{22}^2)$ must be positive, it is evident that the presence of the secondary increases the losses in the primary circuit. In other words, the presence of the secondary might be accounted for in the primary circuit by increasing the value of R_1. Moreover, the reactance which the secondary *reflects* into the primary circuit has a sign which is opposite to that of X_{22}, the net reactance around the secondary loop. This reactance X_{22} is the sum of ωL_2 and X_L; it is necessarily positive for inductive loads and either positive or negative for capacitive loads, depending on the magnitude of the load reactance.

Let us consider the effects of this reflected reactance and resistance by considering the special case in which both the primary and secondary are identical series-resonant circuits. Thus, $R_1 = R_2 = R$, $L_1 = L_2 = L$, and the load impedance $\mathbf{Z}_L$ is produced by a capacitance C, identical to a capacitance inserted in series in the primary circuit. The series-resonant frequency of either the primary or secondary alone is thus $\omega_0 = 1/\sqrt{LC}$. At this resonant frequency, the net secondary reactance is zero, the net primary reactance is zero, and no reactance is reflected into the primary by the secondary. The input impedance is therefore a pure resistance; a resonant condition is present. At a slightly higher frequency, the net primary and secondary reactances are both inductive, and the reflected reactance is therefore capacitive. If the magnetic coupling is sufficiently large, the input impedance may once again be a pure resistance; another resonant condition is achieved, but at a frequency slightly higher than ω_0. A similar condition will also occur at a frequency slightly below ω_0. Each circuit alone is capacitive, the reflected reactance is inductive, and cancellation may occur.

Although we are leaping at conclusions, it seems possible that these three adjacent resonances, each similar to a series resonance, may permit a relatively large primary current to flow from the voltage source. The large primary current in turn provides a large induced voltage in the secondary circuit, and a large secondary current. The secondary current is present over a band of frequencies extending from slightly below to slightly above ω_0, and thus a maximum response is achieved over a wider range of frequencies than is possible in a simple resonant circuit. Such a response curve is obviously desirable if the primary source is some intelligence signal containing energy distributed throughout a band of frequencies, rather than at a single frequency. Such signals are present in AM and FM radio, television, telemetry, radar, and all other communication systems.

It is often convenient to replace a transformer by an equivalent network in the form of a T or π. If we separate the primary and secondary resistances from the transformer, only the pair of mutually coupled inductors remains, as shown in Fig. 15-12. The differential equations describing this circuit are, once again,

$$v_1 = L_1\frac{di_1}{dt} + M\frac{di_2}{dt} \tag{18}$$

and

$$v_2 = M\frac{di_1}{dt} + L_2\frac{di_2}{dt} \tag{19}$$

The form of these two equations is familiar and may be easily interpreted in terms of mesh analysis. Let us select a clockwise i_1 and a counterclockwise i_2 so that i_1 and i_2 are exactly identifiable with the currents in Fig. 15-12. The terms $M\,di_2/dt$ in (18) and $M\,di_1/dt$ in (19) indicate that the two meshes must then have a common *self*-inductance M. Since the total inductance around the left mesh is L_1, a self-inductance of $L_1 - M$ must be inserted in the first mesh, but not in the second mesh. Similarly, a self-inductance of $L_2 - M$ is required in the second mesh, but not in the first mesh. The resultant equivalent network is shown in Fig. 15-13. The equivalence is guaranteed by the identical pairs of equations relating v_1, i_1, v_2, and i_2 for the two networks.

Fig. 15-12 A given transformer which may be replaced by the equivalent networks shown either in Fig. 15-13 or 15-14.

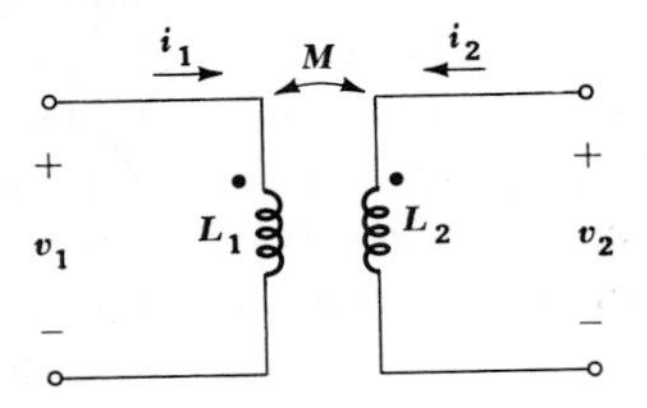

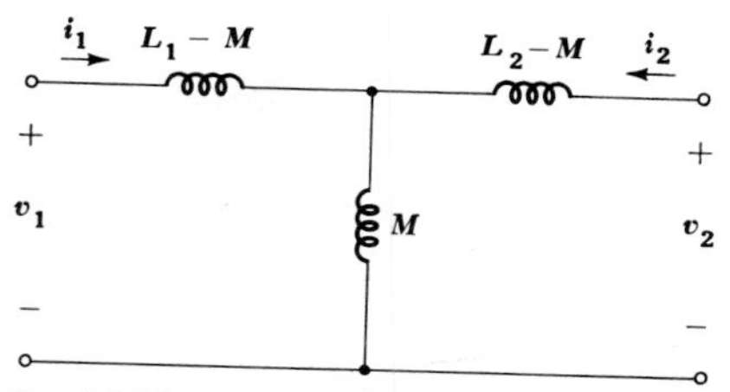

Fig. 15-13 The T equivalent of the transformer shown in Fig. 15-12.

If either of the dots on the windings of the given transformer is placed on the opposite end of its coil, the sign of the mutual terms in (18) and (19) will be negative. This is analogous to replacing M by $-M$, and such a replacement in the network of Fig. 15-13 leads to the correct equivalent for this case.

The inductances in the T equivalent are all self-inductances; no mutual inductance is present. It is possible that negative values of inductance may be obtained for the equivalent circuit, but this is immaterial if our only desire is a mathematical analysis; the actual construction of the equivalent network is of course impossible in any form involving a negative inductance.

The equivalent π network is not obtained as easily; it is more complicated, and it is not used as much. We develop it by solving (19) for di_2/dt and substituting the result into (18),

$$v_1 = L_1 \frac{di_1}{dt} + \frac{M}{L_2} v_2 - \frac{M^2}{L_2} \frac{di_1}{dt}$$

or

$$\frac{di_1}{dt} = \frac{L_2}{L_1 L_2 - M^2} v_1 - \frac{M}{L_1 L_2 - M^2} v_2$$

If we now integrate from 0 to t, we obtain

$$i_1 - i_1(0)u(t) = \frac{L_2}{L_1 L_2 - M^2} \int_0^t v_1 \, dt - \frac{M}{L_1 L_2 - M^2} \int_0^t v_2 \, dt \tag{20}$$

In a similar fashion, we also have

$$i_2 - i_2(0)u(t) = \frac{-M}{L_1 L_2 - M^2} \int_0^t v_1 \, dt + \frac{L_1}{L_1 L_2 - M^2} \int_0^t v_2 \, dt \tag{21}$$

Equations (20) and (21) may be interpreted as a pair of nodal equations. A step-current source must be installed at each node in order to provide the proper initial conditions. The factors multiplying each integral are evidently the inverses of certain equivalent inductances. Thus, the second coefficient in (20), $M/(L_1 L_2 - M^2)$, is $1/L_B$, or the reciprocal of the inductance extending between nodes 1 and 2, as shown on the equivalent

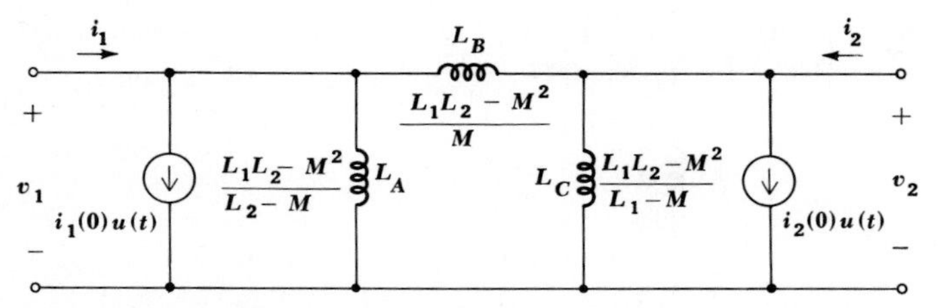

Fig. 15-14 The π network which is equivalent to the transformer shown in Fig. 15-12.

π network, Fig. 15-14. The first coefficient in (20), $L_2/(L_1L_2 - M^2)$, is $1/L_A + 1/L_B$. Thus,

$$\frac{1}{L_A} = \frac{L_2}{L_1L_2 - M^2} - \frac{M}{L_1L_2 - M^2}$$

or

$$L_A = \frac{L_1L_2 - M^2}{L_2 - M}$$

Finally,

$$L_C = \frac{L_1L_2 - M^2}{L_1 - M}$$

No magnetic coupling is present among the inductances in the equivalent circuit, and the initial currents *in the three self-inductances* are zero.

We may compensate for a reversal of either dot in the given transformer by merely changing the sign of M in the equivalent network. Also, just as we found in the equivalent T, negative self-inductances may appear in the equivalent network.

Drill Problems

15-7 Find the three inductances appearing in the T equivalent of the transformer shown in Fig. 15-15 if $L_2 = 4$ H, and the input inductance at A-B is 6 H with C-D open-circuited and 2 H with C-D short-circuited.

Ans. -4; 8; 10 H

Fig. 15-15 See Drill Probs. 15-7 to 15-9.

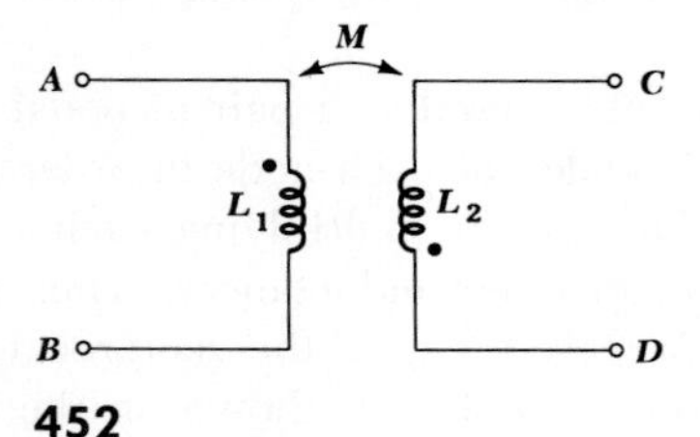

15-8 Find the three inductances appearing in the π equivalent of the transformer shown in Fig. 15-15 if $L_1 = 2$ mH, $L_2 = 8$ mH, and $k = 0.1$.

Ans. -39.6; 1.89; 6.60 mH

15-9 Let $L_1 = 0.1$ H, $L_2 = 0.4$ H, and $M = 0.12$ H in the transformer of Fig. 15-15. Find the input inductance at terminals *A-B* if: (*a*) *C* is connected to *D*; (*b*) *C* is connected to *A* and *D* to *B*; (*c*) *C* is connected to *B* and *D* to *A*.

Ans. 34.6; 64.0; 98.5 mH

15-5 THE IDEAL TRANSFORMER

An ideal transformer is a useful approximation of a very tightly coupled transformer in which the coefficient of coupling is almost unity and both the primary and secondary inductive reactances are extremely large in comparison with the terminating impedances. These characteristics are closely approached by most well-designed iron-core transformers over a reasonable range of frequencies for a reasonable range of terminal impedances. The approximate analysis of a circuit containing an iron-core transformer may be achieved very simply by replacing that transformer by an ideal transformer; the ideal transformer may be thought of as a first-order model of an iron-core transformer.

One new concept arises with the ideal transformer, the *turns ratio a*. The self-inductance of either the primary or secondary coil is proportional to the square of the number of turns of wire forming the coil. This relationship is valid only if all the flux established by the current flowing in the coil links all the turns. In order to develop this result logically, it is necessary to utilize magnetic-field concepts, a subject which is not included in our discussion of circuit analysis. However, a qualitative argument may suffice. If a current I flows through a coil of N turns, then N times the magnetic flux of a single-turn coil will be produced. If we think of the N turns as being coincident, then all the flux certainly links all the turns. As the current and flux change with time, a voltage is then induced *in each turn* which is N times larger than that caused by a single-turn coil. Finally, the voltage induced *in the N-turn coil* must be N^2 times the single-turn voltage. Thus, the proportionality between inductance and the square of the number of turns arises. It follows that

$$\frac{L_2}{L_1} = \frac{N_2^2}{N_1^2} = a^2$$

where

$$a = \frac{N_2}{N_1}$$

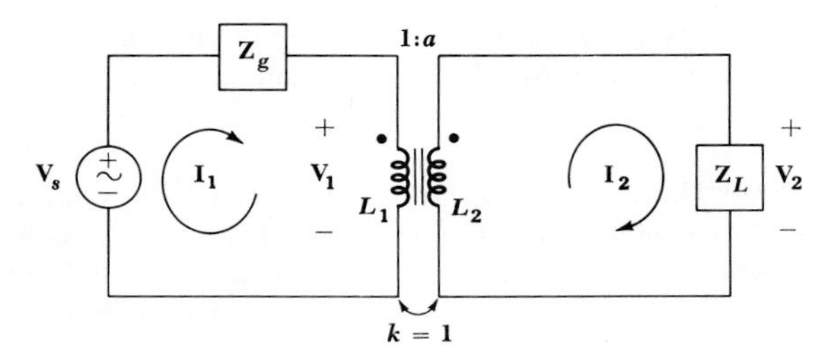

Fig. 15-16 An ideal transformer driven by a practical voltage source is connected to a general load impedance.

Figure 15-16 shows an ideal transformer to which a secondary load and a primary source, including a source impedance, are connected. The ideal nature of the transformer is established by several conventions, the use of the vertical lines between the two coils to indicate the iron laminations present in many iron-core transformers, the unity value of the coupling coefficient, and the presence of the symbol $1{:}a$, suggesting a turns ratio of N_1 to N_2.

Let us analyze this transformer in the sinusoidal steady state in order that we may interpret our assumptions in the simplest context. The two mesh equations are

$$\mathbf{V}_s = \mathbf{I}_1(\mathbf{Z}_g + j\omega L_1) - \mathbf{I}_2 j\omega M \tag{22}$$

$$0 = -\mathbf{I}_1 j\omega M + \mathbf{I}_2(\mathbf{Z}_L + j\omega L_2) \tag{23}$$

We first determine the input impedance of an ideal transformer. Although we shall let the self-inductance of each winding become infinite, the input impedance will remain finite. By solving (23) for $\mathbf{I}_2$ and substituting into (22), we obtain

$$\mathbf{V}_s = \mathbf{I}_1(\mathbf{Z}_g + j\omega L_1) + \mathbf{I}_1 \frac{\omega^2 M^2}{\mathbf{Z}_L + j\omega L_2}$$

and

$$\mathbf{Z}_{\text{in}} = \mathbf{Z}_g + j\omega L_1 + \frac{\omega^2 M^2}{\mathbf{Z}_L + j\omega L_2}$$

Since $k = 1$, $M^2 = L_1 L_2$, and

$$\mathbf{Z}_{\text{in}} = \mathbf{Z}_g + j\omega L_1 + \frac{\omega^2 L_1 L_2}{\mathbf{Z}_L + j\omega L_2}$$

We now must let both L_1 and L_2 tend to infinity. Their ratio, however, remains finite, as specified by the turns ratio. Thus,

$$L_2 = a^2 L_1$$

and
$$\mathbf{Z}_{\text{in}} = \mathbf{Z}_g + j\omega L_1 + \frac{\omega^2 a^2 L_1^2}{\mathbf{Z}_L + j\omega a^2 L_1}$$

Now if we let L_1 become infinite, both of the last two terms in the above expression become infinite, and the result is indeterminate. It is necessary to first combine these two terms,

$$\mathbf{Z}_{\text{in}} = \mathbf{Z}_g + \frac{j\omega L_1 \mathbf{Z}_L - \omega^2 a^2 L_1^2 + \omega^2 a^2 L_1^2}{\mathbf{Z}_L + j\omega a^2 L_1} \tag{24}$$

or
$$\mathbf{Z}_{\text{in}} = \mathbf{Z}_g + \frac{j\omega L_1 \mathbf{Z}_L}{\mathbf{Z}_L + j\omega a^2 L_1} \tag{25}$$

Now as L_1 becomes infinite, it is apparent that $\mathbf{Z}_{\text{in}}$ becomes

$$\mathbf{Z}_{\text{in}} = \mathbf{Z}_g + \frac{\mathbf{Z}_L}{a^2} \tag{26}$$

for finite $\mathbf{Z}_L$.

This result has some interesting implications, and at least one of them appears to contradict one of the characteristics of the linear transformer. It should not, of course, since the linear transformer represents the more general case. The input impedance of the ideal transformer is the series combination of the generator impedance and an impedance which is proportional to the load impedance, the proportionality constant being the reciprocal of the square of the turns ratio. In other words, if the load impedance is a capacitive impedance, then the input impedance is the generator impedance plus a capacitive impedance. In the linear transformer, however, the reflected impedance suffered a sign change in its reactive part; a capacitive load led to an inductive contribution to the input impedance. The explanation of this occurrence is achieved by first realizing that $\mathbf{Z}_L/a^2$ is *not* the reflected impedance, although it is often loosely called by that name. The true reflected impedance is infinite in the ideal transformer; otherwise it could not "cancel" the infinite impedance of the primary inductance. This cancellation occurs in the numerator of the fraction in (24). The impedance $\mathbf{Z}_L/a^2$ represents a small term which is the amount by which an exact cancellation does not occur. The true reflected impedance in the ideal transformer *does* change sign in its reactive part; as the primary and secondary inductances become infinite, however, the effect of the infinite primary-coil reactance and the infinite, but negative, reflected reactance of the secondary coil is one of cancellation.

The first important characteristic of the ideal transformer is therefore its capability to change the magnitude of an impedance, or to change impedance level. An ideal transformer having 100 primary turns and 10,000 secondary turns has a turns ratio of 10,000/100, or 100. Any impedance placed across the secondary then appears at the primary terminals reduced in magnitude by a factor of 100^2, or 10,000. A 20,000-Ω resistor looks like 2 Ω, a 200-mH inductor looks like 20 μH, and a 100-pF capacitor looks like 1 μF. If the primary and secondary windings are interchanged, then $a = 0.01$ and the load impedance is apparently increased in magnitude. In practice, this exact change in magnitude does not always occur, for we must remember that as we took the last step in our derivation and allowed L_1 to become infinite in (25), it was necessary to neglect $\mathbf{Z}_L$ in comparison with $j\omega L_2$. Since L_2 can never be infinite, it is evident that the ideal transformer model will begin to fail for large load impedances.

A practical example of the use of an iron-core transformer as an impedance-level changing device is in the output of an audio power amplifier which must be connected to a loudspeaker. In order to achieve maximum power transfer, we know that the resistance of the load should be equal to the internal resistance of the source; the speaker usually has an impedance magnitude (often assumed to be a resistance) of only a few ohms, while the power amplifier possesses an internal resistance of several thousand ohms. An ideal transformer in which $N_2 < N_1$ is called for. For example, if the amplifier internal impedance is 4000 Ω and the loudspeaker impedance is 8 Ω, then we desire that

$$\mathbf{Z}_g = 4000 = \frac{\mathbf{Z}_L}{a^2} = \frac{8}{a^2}$$

or

$$a = \frac{1}{22.4}$$

and thus

$$\frac{N_1}{N_2} = 22.4$$

There is also a simple relationship between the primary and secondary currents $\mathbf{I}_1$ and $\mathbf{I}_2$ in an ideal transformer. From (23)

$$\frac{\mathbf{I}_2}{\mathbf{I}_1} = \frac{j\omega M}{\mathbf{Z}_L + j\omega L_2}$$

We allow L_2 to become infinite,

$$\frac{\mathbf{I}_2}{\mathbf{I}_1} = \frac{j\omega M}{j\omega L_2} = \sqrt{\frac{L_1}{L_2}}$$

or

$$\frac{\mathbf{I}_2}{\mathbf{I}_1} = \frac{1}{a} \tag{27}$$

The ratio of the primary and secondary currents is the turns ratio. If we have $N_2 > N_1$, then $a > 1$, and it is apparent that the larger current flows in the winding with the fewer number of turns. In other words,

$$N_1\mathbf{I}_1 = N_2\mathbf{I}_2$$

It should also be noted that the current ratio is the negative of the turns ratio if either current is reversed or if either dot location is changed. In the example above in which an ideal transformer was used to change the impedance level to match a loudspeaker efficiently to a power amplifier, an rms current of 50 mA at 1000 Hz in the primary causes an rms current of 1.12 A at 1000 Hz in the secondary. The power delivered to the loudspeaker is $1.12^2(8)$, or 10 W, and the power delivered to the transformer by the power amplifier is $(0.05)^2 4000$, or 10 W. The result is comforting, since the ideal transformer does not contain either an active device which can deliver power or any resistance to absorb power.

Since the power delivered to the ideal transformer is identical with that delivered to the load, whereas the primary and secondary currents are related by the turns ratio, it is obvious that the primary and secondary voltages must also be related to the turns ratio. If we define the secondary voltage, or load voltage,

$$\mathbf{V}_2 = \mathbf{I}_2\mathbf{Z}_L$$

and the primary voltage as the voltage across L_1, then

$$\mathbf{V}_1 = \mathbf{I}_1(\mathbf{Z}_{\text{in}} - \mathbf{Z}_g) = \mathbf{I}_1\frac{\mathbf{Z}_L}{a^2}$$

The ratio of the two voltages is

$$\frac{\mathbf{V}_2}{\mathbf{V}_1} = a^2\frac{\mathbf{I}_2}{\mathbf{I}_1}$$

or

$$\frac{\mathbf{V}_2}{\mathbf{V}_1} = a = \frac{N_2}{N_1} \tag{28}$$

The ratio of the secondary to primary voltage is equal to the turns ratio. This ratio may also be negative if either voltage is reversed or either dot location is changed.

Combining the voltage and current ratios, (27) and (28),

$$\mathbf{V}_2\mathbf{I}_2 = \mathbf{V}_1\mathbf{I}_1$$

and we see that the primary and secondary complex voltamperes are equal. The magnitude of this product is usually specified as a maximum allowable value on power transformers. If the load has a phase angle θ,

$$\mathbf{Z}_L = |\mathbf{Z}_L|\underline{/\theta}$$

then $\mathbf{V}_2$ leads $\mathbf{I}_2$ by an angle θ. Moreover, the input impedance *at the terminals* of L_1 is $\mathbf{Z}_L/a^2$, and thus $\mathbf{V}_1$ also leads $\mathbf{I}_1$ by the same angle θ. If we let the voltage and current represent rms values, then we see that $|\mathbf{V}_2||\mathbf{I}_2| \cos \theta$ must equal $|\mathbf{V}_1||\mathbf{I}_1| \cos \theta$, and all the power delivered to the primary terminals reaches the load; none is absorbed by or delivered to the ideal transformer.

The characteristics of the ideal transformer which we have obtained have all been determined by frequency-domain analysis. They are certainly true in the sinusoidal steady state, but we have no reason to believe that they are correct for the complete response. Actually, they are applicable in general, and the demonstration that this statement is true is much simpler than the frequency-domain analysis we have just completed. Our analysis, however, has served to point out the exact nature of approximations which must be made on a more exact model of an actual transformer in order to obtain an ideal transformer. For example, we have seen that the reactance of the secondary winding must be much greater in magnitude than the impedance of any load which is connected to the secondary. Some feeling for those operating conditions under which a transformer ceases to behave as an ideal transformer is thus achieved.

Returning to the circuit shown in Fig. 15-12 and the two equations, (18) and (19), describing it, we may solve the second equation for di_2/dt and substitute into the first equation,

$$v_1 = L_1 \frac{di_1}{dt} + \frac{M}{L_2} v_2 - \frac{M^2}{L_2} \frac{di_1}{dt}$$

However, for unity coupling, $M^2 = L_1 L_2$, and thus

$$v_1 = \frac{M}{L_2} v_2 = \sqrt{\frac{L_1}{L_2}} v_2 = \frac{1}{a} v_2$$

The relationship between primary and secondary voltage is thus found to apply to the complete time-domain response.

An expression relating primary and secondary current is most quickly obtained by dividing (18) throughout by L_1,

$$\frac{v_1}{L_1} = \frac{di_1}{dt} + \frac{M}{L_1} \frac{di_2}{dt} = \frac{di_1}{dt} + a \frac{di_2}{dt}$$

and then invoking one of the hypotheses underlying the ideal transformer: L_1 must be infinite. If we assume that v_1 is not infinite, then

$$\frac{di_1}{dt} = -a\frac{di_2}{dt}$$

Integrating,

$$i_1 = -ai_2 + A$$

where A is a constant of integration which does not vary with time. Thus, if we neglect any direct currents in the two windings and fix our attention only on the time-varying portion of the response,

$$i_1 = -ai_2$$

The minus sign arises, of course, from the placement of the dots and selection of the current directions in Fig. 15-12.

The same current and voltage relationships are obtained in the time domain as were obtained previously in the frequency domain, provided that dc components are ignored. The time-domain results are more general, but they have been obtained by a less informative process.

The characteristics of the ideal transformer which we have established may be utilized to simplify circuits in which ideal transformers appear. Let us assume, for purposes of illustration, that everything to the left of the primary terminals has been replaced by its Thévenin equivalent, as has the network to the right of the secondary terminals. We thus consider the circuit shown in Fig. 15-17. Excitation at any complex frequency **s** is assumed.

Thévenin's or Norton's theorems may now be used to achieve an equivalent circuit which does not contain a transformer. For example, let us determine the Thévenin equivalent of the network to the left of the secondary terminals. Open-circuiting the secondary, $\mathbf{I}_2 = 0$ and therefore $\mathbf{I}_1 = 0$ (remember L_1 is infinite). No voltage appears across $\mathbf{Z}_{g1}$, and thus $\mathbf{V}_1 = \mathbf{V}_{s1}$ and $\mathbf{V}_{2oc} = a\mathbf{V}_{s1}$. The Thévenin impedance is obtained by killing $\mathbf{V}_{s1}$ and utilizing the square of the turns ratio, being careful to use the reciprocal turns ratio since we are looking in at the secondary terminals. Thus, $\mathbf{Z}_{th2} = \mathbf{Z}_{g1}a^2$. As a check on our equivalent, let us also determine the short-circuit secondary current $\mathbf{I}_{2sc}$. With the secondary short-circuited,

Fig. 15-17 The networks connected to the primary and secondary terminals of an ideal transformer are represented by their Thévenin equivalents.

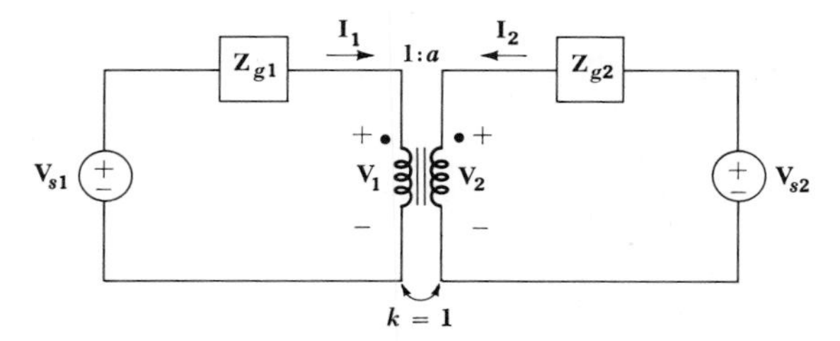

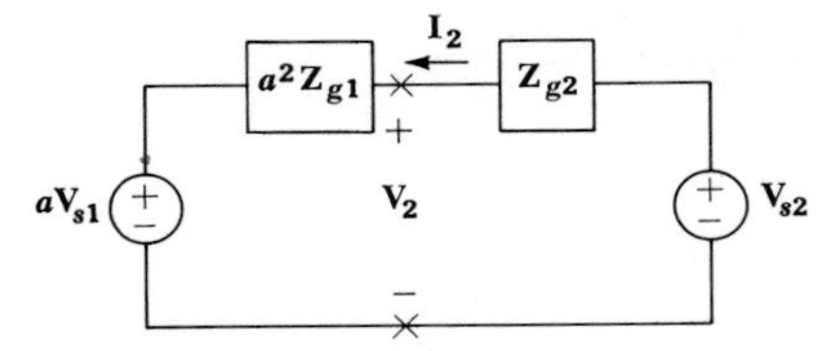

Fig. 15-18 The Thévenin equivalent of the network to the left of the secondary terminals in Fig. 15-17 is used to simplify that circuit.

the primary generator faces an impedance of $\mathbf{Z}_{g1}$ and, therefore, $\mathbf{I}_1 = \mathbf{V}_{s1}/\mathbf{Z}_{g1}$. Thus, $\mathbf{I}_{2sc} = -\mathbf{V}_{s1}/a\mathbf{Z}_{g1}$. The ratio of the open-circuit voltage to the short-circuit current is $-a^2\mathbf{Z}_{g1}$, as it should be. The Thévenin equivalent of the transformer and the primary circuit is shown in the circuit of Fig. 15-18.

Each primary voltage may therefore be multiplied by the turns ratio, each primary current divided by the turns ratio, and each primary impedance multiplied by the square of the turns ratio, these modified voltages, currents, and impedances replacing the given voltages, currents, and impedances plus the transformer. If either dot is interchanged, the equivalent may be obtained by using the negative of the turns ratio.

A similar analysis of the transformer and the secondary network shows that everything to the right of the primary terminals may be replaced by an identical network without the transformer, each voltage being divided by a, each current being multiplied by a, and each impedance being divided by a^2. A reversal of either winding corresponds to the use of a turns ratio of $-a$.

As a simple example of this application of equivalent circuits, consider the circuit given in Fig. 15-19. Let $a = 10$. The input impedance is 10,000/100, or 100 Ω. Thus $\mathbf{I}_1 = 0.25$ A, $\mathbf{V}_1 = 25$ V, and the source delivers 12.5 W, of which 6.25 W is dissipated in the internal resistance of the source and 6.25 W is delivered to the load. This is the condition for maximum power transfer to the load. If the secondary circuit and the ideal transformer are removed by the use of the Thévenin equivalent, the simplified circuit of Fig. 15-20a is obtained. The primary current and voltage are now immediately evident. If, instead, the network to the left

Fig. 15-19 A simple circuit in which a resistive load is matched to the source impedance by means of an ideal transformer.

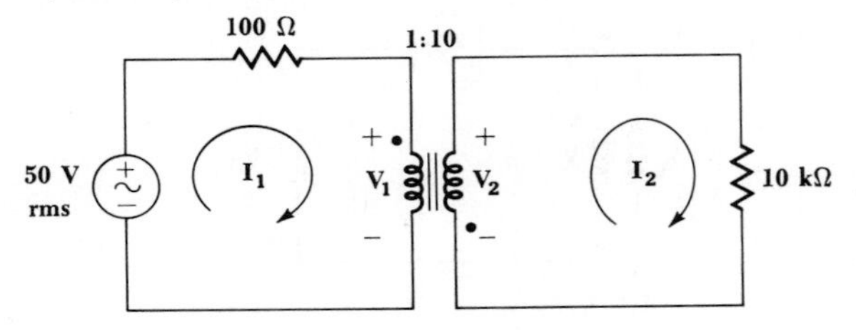

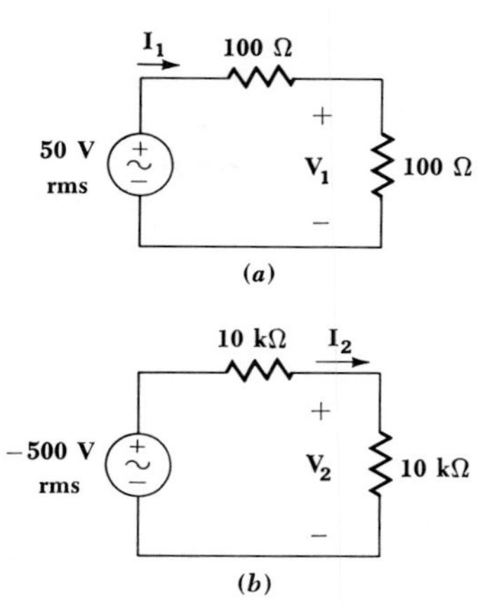

Fig. 15-20 The circuit of Fig. 15-19 is simplified by replacing (*a*) the transformer and secondary circuit by its Thévenin equivalent and (*b*) the transformer and primary circuit by its Thévenin equivalent.

of the secondary terminals is replaced by its Thévenin equivalent, the simpler circuit of Fig. 15-20*b* is obtained. The presence of the minus sign on the equivalent source should be verified. The corresponding Norton equivalents may also be obtained easily.

Drill Problems

15-10 The primary voltage of an ideal power distribution transformer is 4600 V rms, the turns ratio is 0.05, and parallel loads are connected to the secondary, one drawing 11.5 kW at unity PF and the other 9.2 kW at 0.6 PF lagging. Find the rms current amplitude in: (*a*) the lagging load; (*b*) the unity PF load; (*c*) the primary.

Ans. 5.23; 50; 66.7 A

15-11 Find the turns ratio a for the transformer in the circuit of Fig. 15-21 so that: (*a*) the power delivered to the 10-Ω resistor is 25 per cent of that delivered to the 2-Ω resistor for any independent sinusoidal source connected to x-y; (*b*) $\mathbf{V}_2 = \mathbf{V}_s$ when an independent source $\mathbf{V}_s$ is connected to x-y; (*c*) the input resistance at x-y is 8 Ω.

Ans. 1.38 or 3.62; 4.47; 40.0

Fig. 15-21 See Drill Prob. 15-11.

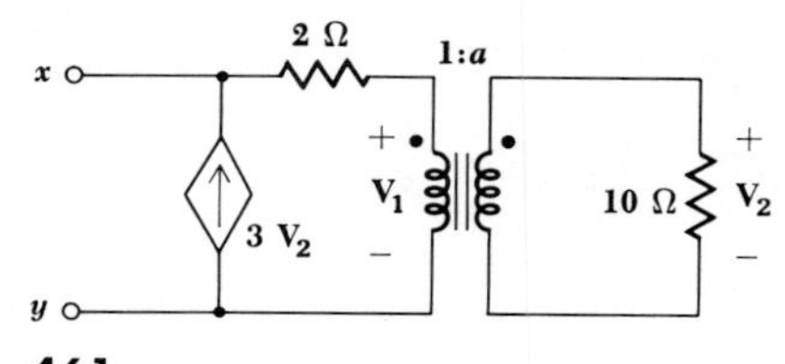

15-12 A Thévenin source, $40\underline{/0^\circ}$ V rms in series with 50 Ω, is connected to the primary of an ideal transformer, turns ratio of 1:2. The primary of a second ideal transformer, turns ratio 1:5, is connected to the secondary of the first transformer. The load at the output is 3000 Ω. Find the load power if: (*a*) the circuit is as described; (*b*) a 100-Ω resistor is inserted in series with the primary of the second transformer; (*c*) a 100-Ω resistor is connected across the primary of the second transformer.

Ans. *2.45; 4.35; 7.50* W

PROBLEMS

☐ **1** The physical construction of three pairs of coupled coils is shown in Fig. 15-22. Show two different possible locations for the two dots in each pair of coils.

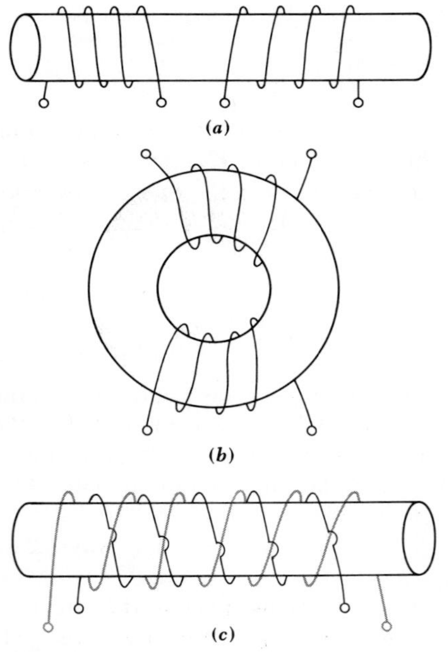

Fig. 15-22 See Prob. 1.

☐ **2** Find $v_2(t)$ in the circuit shown in Fig. 15-23 and sketch it as a function of time.

Fig. 15-23 See Probs. 2 and 12.

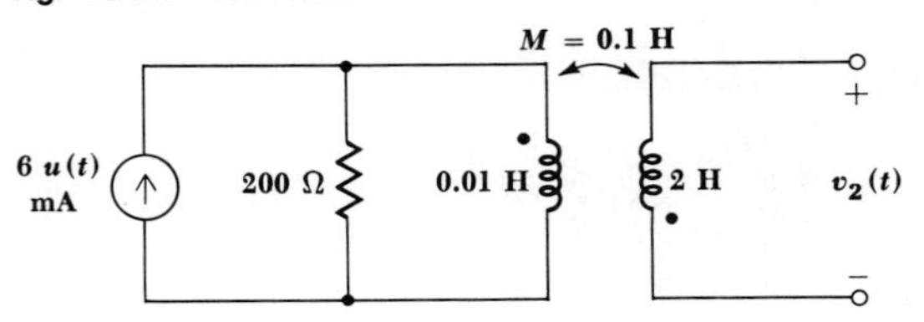

☐ **3** In the circuit shown in Fig. 15-24, find: (*a*) $v_{ac}(t)$; (*b*) $v_{ab}(t)$; (*c*) $v_{bc}(t)$.

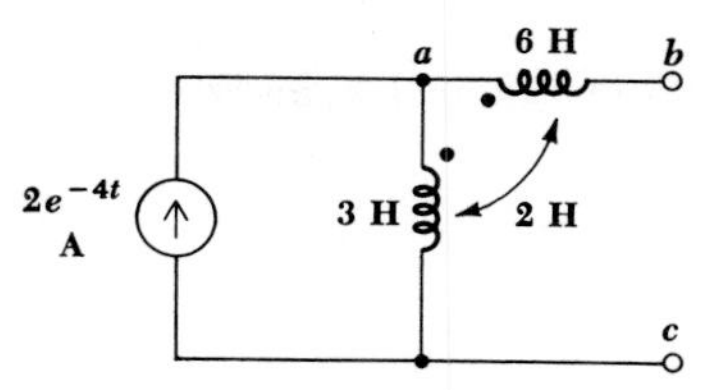

Fig. 15-24 See Prob. 3.

☐ **4** Find i_1, i_2, and v in the circuit shown in Fig. 15-25. Assume that the average value of $i_1(t)$ is zero.

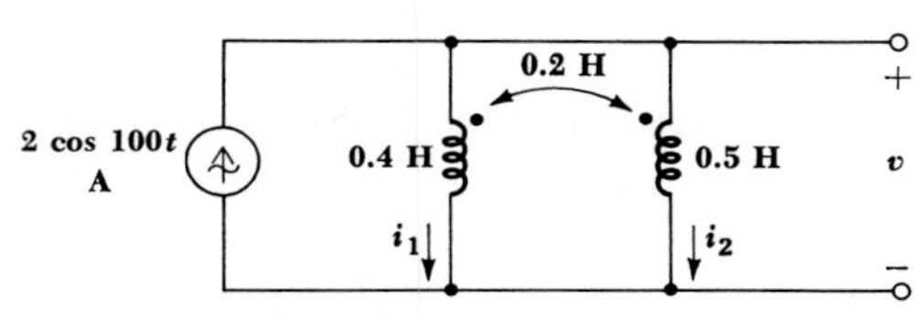

Fig. 15-25 See Prob. 4.

☐ **5** It is possible to arrange three coils physically in such a way that there is mutual coupling between coils *A* and *B* and between *B* and *C*, but not between *A* and *C*. Such an arrangement is illustrated in Fig. 15-26. Find $v_a(t)$ if *x-y* is: (*a*) left open-circuited; (*b*) short-circuited.

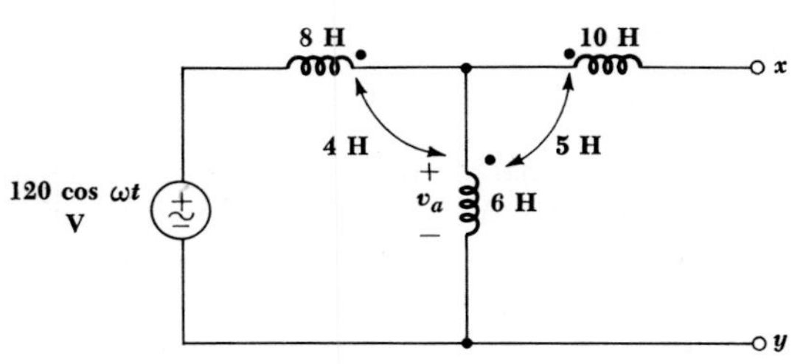

Fig. 15-26 See Prob. 5.

☐ **6** Determine values for *C* and *M* in the circuit shown in Fig. 15-27 so that the input impedance $\mathbf{Z}_{ab}$ has a pole at $\omega = 1$ rad/s and a zero at $\omega = 2$ rad/s.

Fig. 15-27 See Prob. 6.

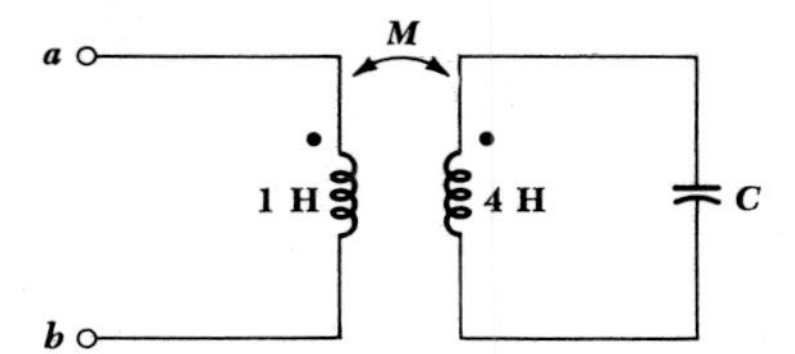

☐ **7** In the circuit shown in Fig. 15-3*a*, $L_1 = 2$ H, $L_2 = 5$ H, $M = 3$ H, and $v_1 = v_2 = 15 \sin 6t$ V. Find $i_1(t)$ and $i_2(t)$.

☐ **8** For the circuit shown in Fig. 15-28, determine $\mathbf{I}_2(s)$ and sketch: (*a*) $|\mathbf{I}_2(\sigma)|$ versus σ; (*b*) $|\mathbf{I}_2(j\omega)|$ versus ω.

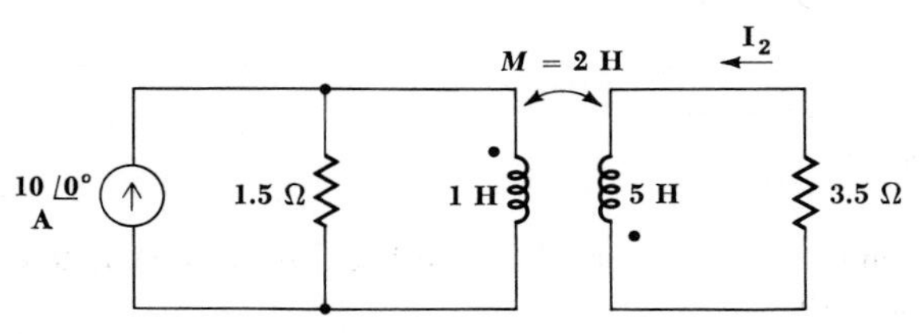

Fig. 15-28 See Probs. 8 and 17.

☐ **9** Find $\mathbf{V}_0(s)$ for the circuit shown in Fig. 15-29. Identify all critical frequencies on the $j\omega$ axis.

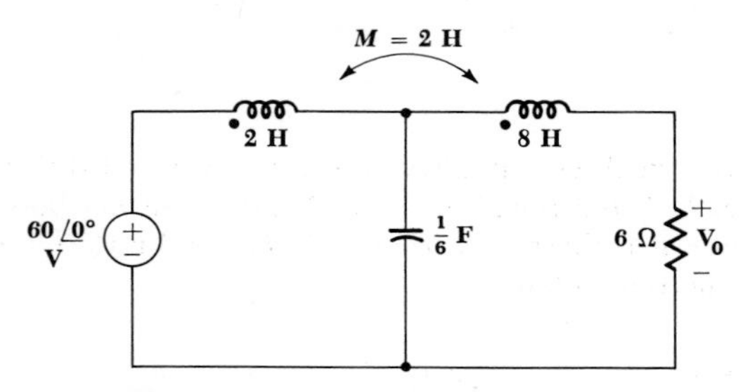

Fig. 15-29 See Prob. 9.

☐ **10** If i_1 and i_2 both enter dot-marked terminals of a pair of coupled coils for which $L_1 = 20$ mH, $L_2 = 60$ mH, and $M = 30$ mH, find the maximum instantaneous energy storage if: (*a*) $i_1 = i_2 = 0.6 \cos \omega t$ A; (*b*) $i_1 = -i_2 = 0.6 \cos \omega t$ A; (*c*) $i_1 = 2i_2 = \cos \omega t$ A; (*d*) $i_1 = \cos \omega t$, $i_2 = 0.5 \sin \omega t$ A.

☐ **11** The only energy stored in the circuit of Fig. 15-30 at $t = 0$ is 5 mJ in the capacitor. At what time is the energy stored in the magnetic fields of the coils a maximum, and what is its magnitude?

Fig. 15-30 See Prob. 11.

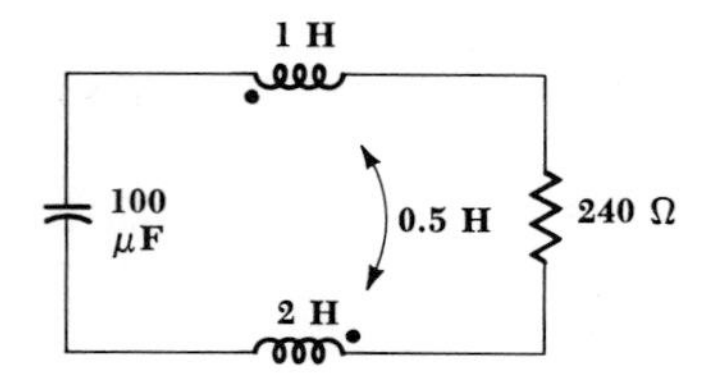

☐ **12** The source shown in Fig. 15-23 is replaced with a current source, $i_s = 4 \cos 10^4 t$ A. (*a*) Find $v_2(t)$. (*b*) How much energy is dissipated in any one period?

☐ **13** Two coupled coils are inside a sealed box with the ends brought out to terminals *A*, *B*, *C*, and *D*. Assume the coils are lossless. A voltage source, $v_s = 10 \cos 400t$ V, is applied as v_{AB} and the resultant current is $(1/240) \sin 400t$ A, while $v_{CD} = 20 \cos 400t$ V. If *C* and *D* are then short-circuited, the short-circuit current into *D* is $(1/80) \sin 400t$ A. Find: (*a*) L_{AB}; (*b*) *M*, including identification of the dot-marked terminals; (*c*) L_{CD}.

☐ **14** With reference to the circuit illustrated in Fig. 15-10, find $i_1(t)$ if the mutual inductance is: (*a*) 0; (*b*) 0.8 H; (*c*) 2 H, but with the dot located at the lower end of the 4-H coil.

☐ **15** The value of *i* at $t = 0$ in the circuit illustrated in Fig. 15-31 is 50 mA. Find $i(t)$ for $t > 0$.

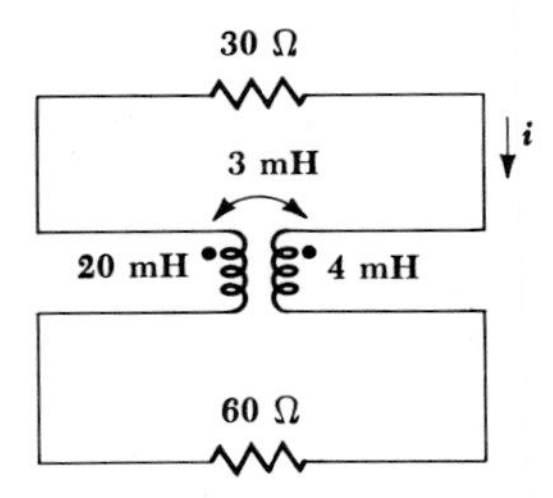

Fig. 15-31 See Prob. 15.

☐ **16** A sinusoidal source, $10\underline{/0°}$ V rms at $\omega = 1$ krad/s, is connected to the primary of a linear transformer and a 3-Ω resistor is across the secondary. Let $L_1 = 1$ mH, $L_2 = 4$ mH, and the coefficient of coupling be *k*. Plot a graph of the average power delivered to the 3-Ω resistor as a function of *k*.

☐ **17** Refer to the circuit shown in Fig. 15-28 and determine the Thévenin equivalent of the network to the left of the 3.5-Ω resistor at a complex frequency **s**. Use the equivalent circuit to find $\mathbf{I}_2(j2)$.

☐ **18** In the circuit shown in Fig. 15-32, the switch closes at $t = 0$. Find $v_2(t)$ for $t > 0$.

Fig. 15-32 See Prob. 18.

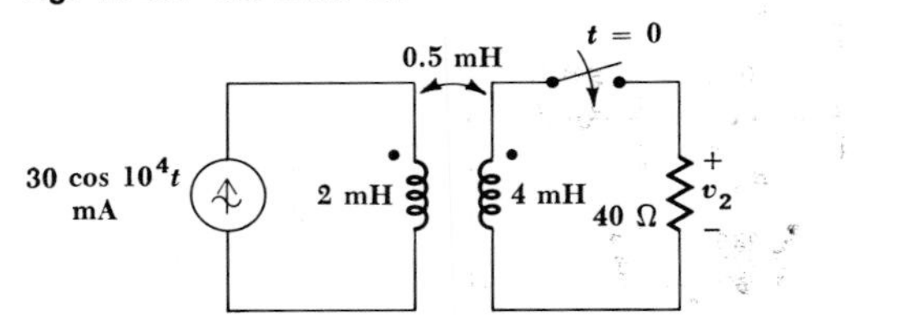

☐ **19** Let $C_1 = C_2 = 0.1\ \mu\text{F}$, $L_1 = L_2 = 1$ mH, and $R_1 = R_2 = 2\ \Omega$, in the transformer shown in Fig. 15-33. Determine $|\mathbf{I}_0|$ at $\omega = 10^5$ rad/s as a function of k, find the value of k that leads to maximum current amplitude, and sketch $|\mathbf{I}_0|$ versus k.

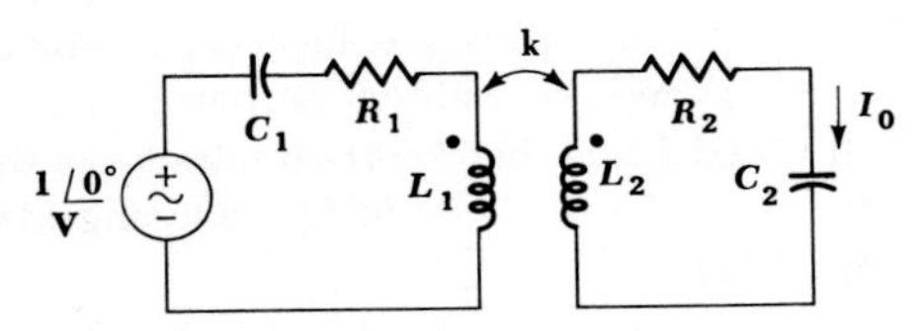

Fig. 15-33 See Prob. 19.

☐ **20** Show how two transformers may be used to match a generator having a Thévenin equivalent impedance of $2800 + j0\ \Omega$ to a load consisting of three 8-Ω speakers, two of which must be supplied with equal power while the third receives 50 per cent more than either of the other two alone. Draw a suitable circuit and specify the required turns ratios.

☐ **21** Find the average power dissipated in each of the three resistors in the circuit of Fig. 15-34.

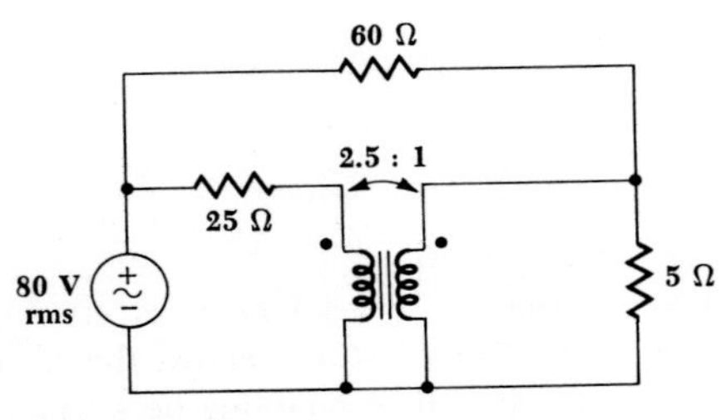

Fig. 15-34 See Prob. 21.

☐ **22** The circuit shown in Fig. 15-35 contains two ideal transformers. Find $\mathbf{I}_x$.

Fig. 15-35 See Prob. 22.

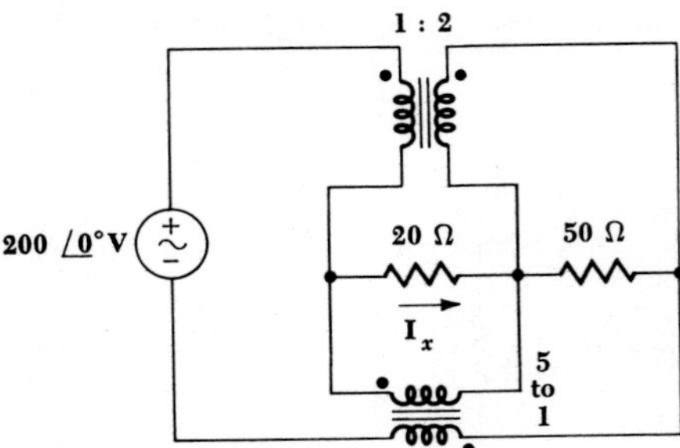

☐ **23** A transformer having primary and secondary inductances L_1 and L_2 and a coefficient of coupling k has a 1-μF capacitor as load. Assume that there are no resistances associated with any element. To show the effect of noninfinite L_1 and L_2 and nonunity k on the ideal transformer approximations, find the equivalent input capacitance at $\omega = 10^4$ rad/s for the following cases: (*a*) $L_1 = \infty$, $L_2 = \infty$, $k = 1$, $N_2/N_1 = 2$; (*b*) $L_1 = 0.1$ H, $L_2 = 0.4$ H, $k = 1$; (*c*) $L_1 = 0.1$ H, $L_2 = 0.4$ H, $k = 0.99$.

☐ **24** A transformer whose nameplate data is 2300/230 V, 25 kVA, operates with primary and secondary voltages of 2300 V and 230 V rms, respectively, and can supply 25 kVA from its secondary winding. If this transformer is supplied with 2300 V rms and is connected to secondary loads requiring 5 kW at 0.9 PF lagging and 12 kW at 0.8 PF lagging: (*a*) How many kilowatts of lighting load can it still supply? (*b*) What is the primary current?

Chapter Sixteen
Two-port Networks

16-1 INTRODUCTION

In defining and using mutual inductance in the previous chapter, we were forced to consider a phenomenon associated with two pairs of terminals. A general network having two pairs of terminals, one perhaps labeled the "input terminals" and the other the "output terminals," is a very important building block in electronic systems, communication systems, automatic control systems, transmission and distribution systems, or other systems in which an electrical signal or electrical energy enters the input terminals, is acted upon by the network, and leaves via the output terminals. The output terminal pair may very well connect with the input terminal pair

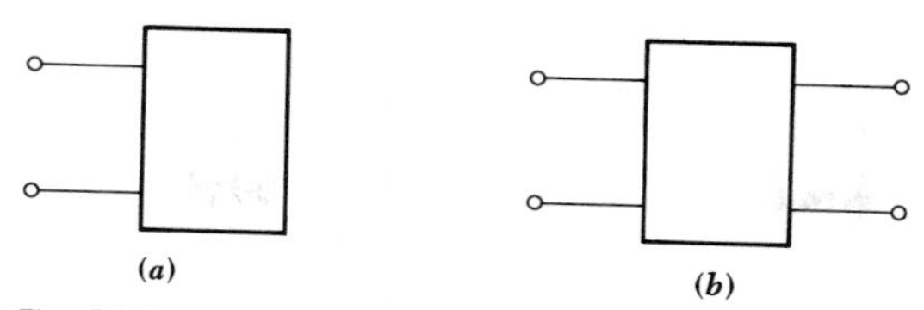

Fig. 16-1 (*a*) A one-port network. (*b*) A two-port network.

of another network. A pair of terminals at which a signal may enter or leave a network is called a *port*, and a network having only one such pair of terminals is called a *one-port network*, or simply a *one port*. When more than one pair of terminals is present, the network is known as a *multiport network*. A one-port network is shown in Fig. 16-1*a*, and the two-port network to which this chapter is principally devoted is shown in Fig. 16-1*b*. We shall assume that sources and loads are connected directly across the two terminals of a port; for example, we should not expect any device to be connected between terminals *a* and *c* of the two-port network in Fig. 16-1*b*. If such a circuit must be analyzed, general loop or nodal equations should usually be written.

The special methods of analysis which are developed for two-port networks, or simply two ports, emphasize the current and voltage relationships at the terminals of the networks and suppress the specific nature of the currents and voltages within the networks. Our introductory study should serve to acquaint us with a number of important parameters and their use in simplifying and systematizing linear two-port network analysis.

16-2 ONE-PORT NETWORKS

Some of the introductory study of one- and two-port networks is accomplished best by using a generalized network notation and the abbreviated nomenclature for determinants introduced in Appendix 1. Thus, if we write a set of loop equations,

$$\begin{aligned}
\mathbf{Z}_{11}\mathbf{I}_1 + \mathbf{Z}_{12}\mathbf{I}_2 + \mathbf{Z}_{13}\mathbf{I}_3 + \cdots + \mathbf{Z}_{1N}\mathbf{I}_N &= \mathbf{V}_1 \\
\mathbf{Z}_{21}\mathbf{I}_1 + \mathbf{Z}_{22}\mathbf{I}_2 + \mathbf{Z}_{23}\mathbf{I}_3 + \cdots + \mathbf{Z}_{2N}\mathbf{I}_N &= \mathbf{V}_2 \\
\mathbf{Z}_{31}\mathbf{I}_1 + \mathbf{Z}_{32}\mathbf{I}_2 + \mathbf{Z}_{33}\mathbf{I}_3 + \cdots + \mathbf{Z}_{3N}\mathbf{I}_N &= \mathbf{V}_3 \\
\cdots\cdots\cdots\cdots\cdots\cdots\cdots\cdots \\
\mathbf{Z}_{N1}\mathbf{I}_1 + \mathbf{Z}_{N2}\mathbf{I}_2 + \mathbf{Z}_{N3}\mathbf{I}_3 + \cdots + \mathbf{Z}_{NN}\mathbf{I}_N &= \mathbf{V}_N
\end{aligned} \tag{1}$$

then the coefficient of each current will be an impedance $\mathbf{Z}_{ij}(\mathbf{s})$, and the circuit determinant, or determinant of the coefficients, is

$$\Delta_Z = \begin{vmatrix} \mathbf{Z}_{11} & \mathbf{Z}_{12} & \mathbf{Z}_{13} & \cdots & \mathbf{Z}_{1N} \\ \mathbf{Z}_{21} & \mathbf{Z}_{22} & \mathbf{Z}_{23} & \cdots & \mathbf{Z}_{2N} \\ \mathbf{Z}_{31} & \mathbf{Z}_{32} & \mathbf{Z}_{33} & \cdots & \mathbf{Z}_{3N} \\ \cdots & \cdots & \cdots & \cdots & \cdots \\ \mathbf{Z}_{N1} & \mathbf{Z}_{N2} & \mathbf{Z}_{N3} & \cdots & \mathbf{Z}_{NN} \end{vmatrix} \tag{2}$$

where N loops have been assumed, the currents appear in subscript order in each equation, and the order of the equations is the same as that of the currents. We also assume that Kirchhoff's voltage law is applied so that the sign of each $\mathbf{Z}_{ii}$ term ($\mathbf{Z}_{11}, \mathbf{Z}_{22}, \ldots, \mathbf{Z}_{NN}$) is positive; the sign of any $\mathbf{Z}_{ij}$ ($i \neq j$) or mutual term may be either positive or negative, depending on the reference directions assigned to $\mathbf{I}_i$ and $\mathbf{I}_j$.

The use of minor notation (Appendix 1) enables the input or driving-point impedance at the terminals of a *one-port* network to be expressed very concisely. The result is also applicable to a *two-port* network if one of the two ports is terminated in a passive impedance.

Let us suppose that the one-port network shown in Fig. 16-2*a* is composed entirely of passive elements and dependent sources; linearity is also assumed. An ideal voltage source $\mathbf{V}_1$ is connected to the port, and the source current is identified as the current in loop 1. By the familiar procedure, then,

$$\mathbf{I}_1 = \frac{\begin{vmatrix} \mathbf{V}_1 & \mathbf{Z}_{12} & \mathbf{Z}_{13} & \cdots & \mathbf{Z}_{1N} \\ 0 & \mathbf{Z}_{22} & \mathbf{Z}_{23} & \cdots & \mathbf{Z}_{2N} \\ 0 & \mathbf{Z}_{32} & \mathbf{Z}_{33} & \cdots & \mathbf{Z}_{3N} \\ \cdots & \cdots & \cdots & \cdots & \cdots \\ 0 & \mathbf{Z}_{N2} & \mathbf{Z}_{N3} & \cdots & \mathbf{Z}_{NN} \end{vmatrix}}{\begin{vmatrix} \mathbf{Z}_{11} & \mathbf{Z}_{12} & \mathbf{Z}_{13} & \cdots & \mathbf{Z}_{1N} \\ \mathbf{Z}_{21} & \mathbf{Z}_{22} & \mathbf{Z}_{23} & \cdots & \mathbf{Z}_{2N} \\ \mathbf{Z}_{31} & \mathbf{Z}_{32} & \mathbf{Z}_{33} & \cdots & \mathbf{Z}_{3N} \\ \cdots & \cdots & \cdots & \cdots & \cdots \\ \mathbf{Z}_{N1} & \mathbf{Z}_{N2} & \mathbf{Z}_{N3} & \cdots & \mathbf{Z}_{NN} \end{vmatrix}}$$

or, more concisely,

$$\mathbf{I}_1 = \frac{\mathbf{V}_1 \Delta_{11}}{\Delta_Z}$$

Thus,

$$\mathbf{Z}_{in} = \frac{\mathbf{V}_1}{\mathbf{I}_1}$$

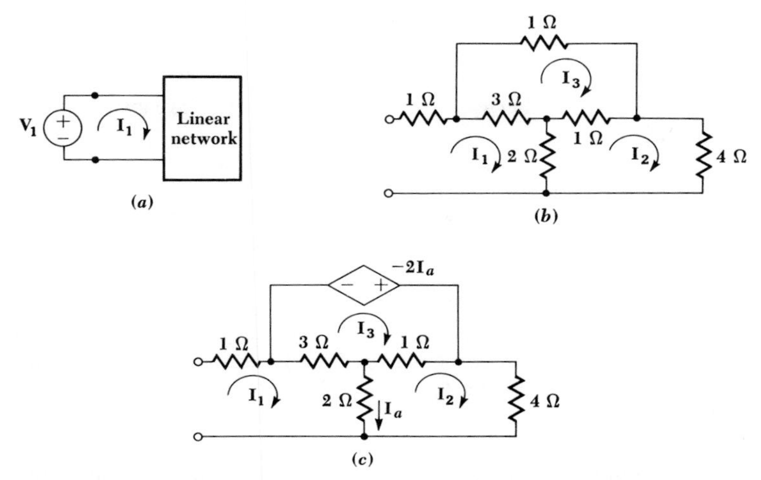

Fig. 16-2 (*a*) An ideal voltage source $\mathbf{V}_1$ is connected to the single port of a linear one-port network containing no independent sources; $\mathbf{Z}_{in} = \mathbf{\Delta_Z}/\mathbf{\Delta}_{11}$. (*b*) A resistive one port used as an example. (*c*) A one port containing a dependent source used as an example.

and

$$\mathbf{Z}_{in} = \frac{\mathbf{\Delta_Z}}{\mathbf{\Delta}_{11}} \tag{3}$$

For the one-port resistive network shown in Fig. 16-2*b*,

$$\mathbf{\Delta_Z} = \begin{vmatrix} 6 & -2 & -3 \\ -2 & 7 & -1 \\ -3 & -1 & 5 \end{vmatrix} = 109$$

and

$$\mathbf{\Delta}_{11} = \begin{vmatrix} 7 & -1 \\ -1 & 5 \end{vmatrix} = 34$$

Thus,

$$\mathbf{Z}_{in} = \frac{109}{34} = 3.21 \quad \Omega$$

Upon changing the circuit by including a dependent source, Fig. 16-2*c*, the three mesh equations are written,

$$\begin{aligned} 6\mathbf{I}_1 - 2\mathbf{I}_2 - 3\mathbf{I}_3 &= \mathbf{V}_1 \\ -2\mathbf{I}_1 + 7\mathbf{I}_2 - \mathbf{I}_3 &= 0 \\ -3\mathbf{I}_1 - \mathbf{I}_2 + 4\mathbf{I}_3 &= -2\mathbf{I}_a = -2(\mathbf{I}_1 - \mathbf{I}_2) \end{aligned}$$

and

$$\mathbf{\Delta_Z} = \begin{vmatrix} 6 & -2 & -3 \\ -2 & 7 & -1 \\ -1 & -3 & 4 \end{vmatrix} = 93$$

while

$$\Delta_{11} = \begin{vmatrix} 7 & -1 \\ -3 & 4 \end{vmatrix} = 25$$

giving

$$\mathbf{Z}_{in} = {}^{93}\!/_{25} = 3.72 \quad \Omega$$

Drill Problems

16-1 Connect an ideal source $\mathbf{V}_s$, with its positive reference at the upper terminal, to the circuit shown in Fig. 16-2*b*, and show that: (*a*) $\mathbf{\Delta}_{12} = -13$; (*b*) $\mathbf{I}_2 = -\mathbf{V}_s\mathbf{\Delta}_{12}/\mathbf{\Delta}_\mathbf{Z}$; (*c*) $\mathbf{I}_3 = \mathbf{V}_s\mathbf{\Delta}_{13}/\mathbf{\Delta}_\mathbf{Z}$.

Ans. Proof

16-2 Find the input impedance presented to an ideal source by the network of Fig. 16-3 if the source is inserted in series with the: (*a*) 1-Ω resistor; (*b*) 3-Ω resistor; (*c*) 4-Ω resistor.

Ans. 3.80; 5.00; 6.70 Ω

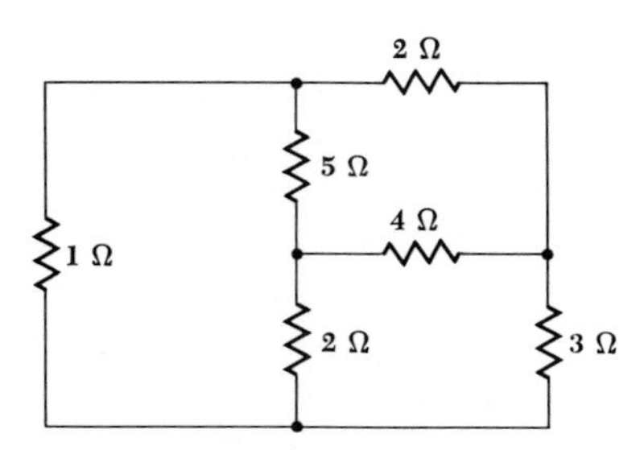

Fig. 16-3 See Drill Prob. 16-2.

16-3 Using the reference node and node-to-reference voltages indicated in Fig. 16-4, write the three necessary systematic equations and find the input admittance presented to an ideal source connected between node 1 and the reference by: (*a*) evaluating $\mathbf{\Delta}_\mathbf{Y}/\mathbf{\Delta}_{11(\mathbf{Y})}$; (*b*) determining the reciprocal of $\mathbf{Z}_{in} = \mathbf{\Delta}_\mathbf{Z}/\mathbf{\Delta}_{11(\mathbf{Z})}$; (*c*) combining elements.

Ans. 15/46; 15/46; 15/46 m℧

Fig. 16-4 Element values are given in m℧; see Drill Prob. 16-3.

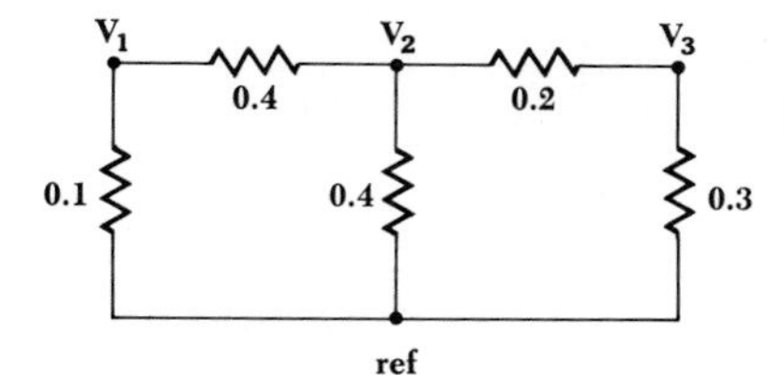

16-3 ADMITTANCE PARAMETERS

Let us now turn our attention to two-port networks. We shall assume in all that follows that the network is composed of linear elements and contains no independent sources; dependent sources are permissible. Further conditions will also be placed on the network in some special cases.

We shall consider the two port as it is shown in Fig. 16-5; the voltage and current at the input terminals are $\mathbf{V}_1$ and $\mathbf{I}_1$, and $\mathbf{V}_2$ and $\mathbf{I}_2$ are specified at the output port. The directions of $\mathbf{I}_1$ and $\mathbf{I}_2$ are both customarily selected as *into* the network at the upper conductors (and out at the lower conductors). Since the network is linear and contains no independent sources within it, $\mathbf{I}_1$ may be considered to be the superposition of two components, one caused by $\mathbf{V}_1$ and the other by $\mathbf{V}_2$. When the same argument is applied to $\mathbf{I}_2$, we may begin with the set of equations

$$\mathbf{I}_1 = \mathbf{y}_{11}\mathbf{V}_1 + \mathbf{y}_{12}\mathbf{V}_2 \tag{4}$$

$$\mathbf{I}_2 = \mathbf{y}_{21}\mathbf{V}_1 + \mathbf{y}_{22}\mathbf{V}_2 \tag{5}$$

where the y's are no more than proportionality constants, or unknown coefficients, for the present. They are called the y *parameters* and they are defined by (4) and (5).

The most useful and informative way to attach a physical meaning to the y parameters is through a direct inspection of (4) and (5). Consider (4), for example; if we let $\mathbf{V}_2$ be zero, then we see that $\mathbf{y}_{11}$ must be given by the ratio of $\mathbf{I}_1$ to $\mathbf{V}_1$. We therefore describe $\mathbf{y}_{11}$ as the admittance measured at the input terminals with the output terminals *short-circuited* ($\mathbf{V}_2 = 0$). Since there can be no question as to which terminals are short-circuited, $\mathbf{y}_{11}$ is best described as the *short-circuit input admittance*. Alternatively, we might describe $\mathbf{y}_{11}$ as the reciprocal of the input impedance measured with the output terminals short-circuited, but a description as an admittance is obviously more direct. It is not the *name* of the parameter that is important; rather it is the conditions which must be applied to (4) or (5), and hence to the network, that are most meaningful; when the conditions are determined, the parameter can be found directly from an analysis of the circuit (or by experiment on the physical circuit). Each of

Fig. 16-5 A general two port with terminal voltages and currents specified. The two port is composed of linear elements, including dependent sources, but not containing any independent sources.

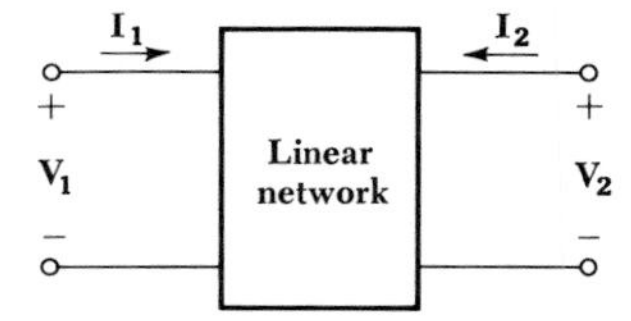

the y parameters may be described as a current-voltage ratio with either $\mathbf{V}_1 = 0$ (the input terminals short-circuited) or $\mathbf{V}_2 = 0$ (the output terminals short-circuited):

$$\mathbf{y}_{11} = \left.\frac{\mathbf{I}_1}{\mathbf{V}_1}\right|_{\mathbf{V}_2=0} \tag{6}$$

$$\mathbf{y}_{12} = \left.\frac{\mathbf{I}_1}{\mathbf{V}_2}\right|_{\mathbf{V}_1=0} \tag{7}$$

$$\mathbf{y}_{21} = \left.\frac{\mathbf{I}_2}{\mathbf{V}_1}\right|_{\mathbf{V}_2=0} \tag{8}$$

$$\mathbf{y}_{22} = \left.\frac{\mathbf{I}_2}{\mathbf{V}_2}\right|_{\mathbf{V}_1=0} \tag{9}$$

Because each parameter is an admittance which is obtained by short-circuiting either the output or input port, the y parameters are known as the *short-circuit admittance parameters*. The specific name of $\mathbf{y}_{11}$ is the *short-circuit input admittance,* $\mathbf{y}_{22}$ is the *short-circuit output admittance,* and $\mathbf{y}_{12}$ and $\mathbf{y}_{21}$ are the *short-circuit transfer admittances*.

Consider the resistive two port shown in Fig. 16-6*a*. The values of the parameters may be easily established by applying (6) to (9), which we obtained directly from the defining equations (4) and (5). To determine $\mathbf{y}_{11}$, we short-circuit the output and find the ratio of $\mathbf{I}_1$ to $\mathbf{V}_1$. This may be done by letting $\mathbf{V}_1 = 1$ V, for then $\mathbf{y}_{11} = \mathbf{I}_1$. By inspection of Fig. 16-6*a*, it is apparent that 1 V applied at the input with the output short-circuited will cause an input current of $(1/5 + 1/10)$, or 0.3 A. Hence,

$$\mathbf{y}_{11} = 0.3 \quad \mho$$

In order to find $\mathbf{y}_{12}$, we short-circuit the input terminals and apply 1 V at the output terminals. The input current flows through the short circuit and is $-1/10$ A. Thus

$$\mathbf{y}_{12} = -0.1 \quad \mho$$

By similar methods,

$$\mathbf{y}_{21} = -0.1 \quad \mho \qquad \mathbf{y}_{22} = 0.15 \quad \mho$$

The describing equations for this two port in terms of the admittance parameters are, therefore,

$$\mathbf{I}_1 = 0.3\mathbf{V}_1 - 0.1\mathbf{V}_2 \tag{10}$$

$$\mathbf{I}_2 = -0.1\mathbf{V}_1 + 0.15\mathbf{V}_2 \tag{11}$$

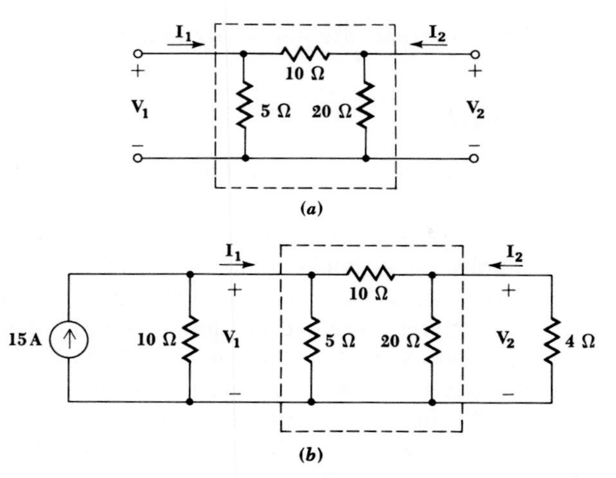

Fig. 16-6 (*a*) A resistive two port. (*b*) The resistive two port is terminated with specific one ports.

In order to see what use might be made of such a system of equations, let us now terminate each port with some specific network. As a simple example, shown in Fig. 16-6*b*, let us connect a general current source to the input port and a resistive load to the output port. A relationship must now exist between $\mathbf{V}_1$ and $\mathbf{I}_1$ at the input port because a specific network is present. This relationship may be determined solely from this external circuit. If we apply Kirchhoff's current law (or write a single nodal equation) at the input,

$$\mathbf{I}_1 = 15 - 0.1\mathbf{V}_1$$

For the output, Ohm's law yields

$$\mathbf{I}_2 = -0.25\mathbf{V}_2$$

Substituting these expressions for $\mathbf{I}_1$ and $\mathbf{I}_2$ into (10) and (11), we have

$$15 = 0.4\mathbf{V}_1 - 0.1\mathbf{V}_2$$
$$0 = -0.1\mathbf{V}_1 + 0.4\mathbf{V}_2$$

from which are obtained

$$\mathbf{V}_1 = 40 \quad \text{V} \qquad \mathbf{V}_2 = 10 \quad \text{V}$$

The input and output currents are also easily found,

$$\mathbf{I}_1 = 11 \quad \text{A} \qquad \mathbf{I}_2 = -2.5 \quad \text{A}$$

and the complete terminal characteristics of this resistive two port are then known.

The advantages of two-port analysis do not show up very strongly for such a simple example, but it should be apparent that once the y parameters are determined for a more complicated two port the performance of the two port for different terminal conditions is easily determined; it is only necessary to relate $\mathbf{V}_1$ to $\mathbf{I}_1$ at the input and $\mathbf{V}_2$ to $\mathbf{I}_2$ at the output.

In the example just concluded, $\mathbf{y}_{12}$ and $\mathbf{y}_{21}$ were both found to be -0.1 ℧. It is not difficult to show that this equality is also obtained if three general impedances $\mathbf{Z}_A$, $\mathbf{Z}_B$, and $\mathbf{Z}_C$ are contained in this π network. It is somewhat more difficult to determine the specific conditions which are necessary in order that $\mathbf{y}_{12} = \mathbf{y}_{21}$, but the use of determinant notation is of some help. Let us see if the relationships (6) to (9) can be expressed in terms of the impedance determinant and its minors.

Since our concern is with the two port and not with the specific networks with which it is terminated, we shall let $\mathbf{V}_1$ and $\mathbf{V}_2$ be represented by two ideal voltage sources. Equation (6) is applied by letting $\mathbf{V}_2 = 0$ (thus short-circuiting the output) and finding the input admittance. The network now, however, is simply a one port, and the input impedance of a one port was found in the previous section. We select mesh 1 to include the input terminals and let $\mathbf{I}_1$ be that mesh current; we identify $(-\mathbf{I}_2)$ as the mesh current in mesh 2 and assign the remaining loop or mesh currents in any convenient manner. Thus,

$$\mathbf{Z}_{in}|_{\mathbf{V}_2=0} = \frac{\Delta_\mathbf{Z}}{\Delta_{11}}$$

and, therefore,

$$\mathbf{y}_{11} = \frac{\Delta_{11}}{\Delta_\mathbf{Z}} \tag{12}$$

Similarly,

$$\mathbf{y}_{22} = \frac{\Delta_{22}}{\Delta_\mathbf{Z}} \tag{13}$$

In order to find $\mathbf{y}_{12}$, we let $\mathbf{V}_1 = 0$, and find $\mathbf{I}_1$ as a function of $\mathbf{V}_2$. We find that $\mathbf{I}_1$ is given by the ratio

$$\mathbf{I}_1 = \frac{\begin{vmatrix} 0 & \mathbf{Z}_{12} & \cdots & \mathbf{Z}_{1N} \\ -\mathbf{V}_2 & \mathbf{Z}_{22} & \cdots & \mathbf{Z}_{2N} \\ 0 & \mathbf{Z}_{32} & \cdots & \mathbf{Z}_{3N} \\ \cdots & \cdots & \cdots & \cdots \\ 0 & \mathbf{Z}_{N2} & \cdots & \mathbf{Z}_{NN} \end{vmatrix}}{\begin{vmatrix} \mathbf{Z}_{11} & \mathbf{Z}_{12} & \cdots & \mathbf{Z}_{1N} \\ \mathbf{Z}_{21} & \mathbf{Z}_{22} & \cdots & \mathbf{Z}_{2N} \\ \cdots & \cdots & \cdots & \cdots \\ \mathbf{Z}_{N1} & \mathbf{Z}_{N2} & \cdots & \mathbf{Z}_{NN} \end{vmatrix}}$$

Thus,

$$\mathbf{I}_1 = -\frac{(-\mathbf{V}_2)\mathbf{\Delta}_{21}}{\mathbf{\Delta_Z}}$$

and

$$\mathbf{y}_{12} = \frac{\mathbf{\Delta}_{21}}{\mathbf{\Delta_Z}} \tag{14}$$

In a similar manner, we may show that

$$\mathbf{y}_{21} = \frac{\mathbf{\Delta}_{12}}{\mathbf{\Delta_Z}} \tag{15}$$

The equality of $\mathbf{y}_{12}$ and $\mathbf{y}_{21}$ is thus contingent on the equality of the two minors of $\mathbf{\Delta_Z}$, $\mathbf{\Delta}_{12}$ and $\mathbf{\Delta}_{21}$. These two minors are

$$\mathbf{\Delta}_{21} = \begin{vmatrix} \mathbf{Z}_{12} & \mathbf{Z}_{13} & \mathbf{Z}_{14} & \cdots & \mathbf{Z}_{1N} \\ \mathbf{Z}_{32} & \mathbf{Z}_{33} & \mathbf{Z}_{34} & \cdots & \mathbf{Z}_{3N} \\ \mathbf{Z}_{42} & \mathbf{Z}_{43} & \mathbf{Z}_{44} & \cdots & \mathbf{Z}_{4N} \\ \cdots & \cdots & \cdots & \cdots & \cdots \\ \mathbf{Z}_{N2} & \mathbf{Z}_{N3} & \mathbf{Z}_{N4} & \cdots & \mathbf{Z}_{NN} \end{vmatrix} \qquad \mathbf{\Delta}_{12} = \begin{vmatrix} \mathbf{Z}_{21} & \mathbf{Z}_{23} & \mathbf{Z}_{24} & \cdots & \mathbf{Z}_{2N} \\ \mathbf{Z}_{31} & \mathbf{Z}_{33} & \mathbf{Z}_{34} & \cdots & \mathbf{Z}_{3N} \\ \mathbf{Z}_{41} & \mathbf{Z}_{43} & \mathbf{Z}_{44} & \cdots & \mathbf{Z}_{4N} \\ \cdots & \cdots & \cdots & \cdots & \cdots \\ \mathbf{Z}_{N1} & \mathbf{Z}_{N3} & \mathbf{Z}_{N4} & \cdots & \mathbf{Z}_{NN} \end{vmatrix}$$

Their equality is shown by first interchanging the rows and columns of one minor, say $\mathbf{\Delta}_{21}$, an operation which any college algebra book proves is valid, and then letting every mutual impedance $\mathbf{Z}_{ij}$ be replaced by $\mathbf{Z}_{ji}$. Thus, we set

$$\mathbf{Z}_{12} = \mathbf{Z}_{21} \qquad \mathbf{Z}_{23} = \mathbf{Z}_{32} \qquad \text{etc.}$$

This equality of $\mathbf{Z}_{ij}$ and $\mathbf{Z}_{ji}$ is certainly obvious for the three passive elements, the resistor, capacitor, and inductor, and it is also true for mutual inductance, as we proved in the preceding chapter. However, it is not true for every type of device which we may wish to include inside a two-port network. Specifically, it is not true in general for a dependent source, and it is not true for a device which is found in some microwave circuits, the gyrator. This device, over a narrow range of radian frequencies, provides an additional phase shift of 180° for a signal passing from the output to the input over that for a signal in the forward direction, and thus $\mathbf{y}_{12} = -\mathbf{y}_{21}$.

Any device for which $\mathbf{Z}_{ij} = \mathbf{Z}_{ji}$ is called a *bilateral element*, and a circuit which contains only bilateral elements is called a *bilateral circuit*. We have therefore shown that an important property of a bilateral two port is

$$\mathbf{y}_{12} = \mathbf{y}_{21}$$

and this property is glorified by stating it as the *reciprocity theorem:*

> In any passive linear bilateral network, if the single voltage source $\mathbf{V}_x$ in branch x produces the current response $\mathbf{I}_y$ in branch y, then the removal of the voltage source from branch x and its insertion in branch y will produce the current response $\mathbf{I}_y$ in branch x.

A simple way of stating the theorem is to say that the interchange of an ideal voltage source and an ideal ammeter in any passive linear bilateral circuit will not change the ammeter reading.

If we had been working with the admittance determinant of the circuit and had proved that the minors $\mathbf{\Delta}_{21}$ and $\mathbf{\Delta}_{12}$ of the admittance determinant $\mathbf{\Delta}_{\mathbf{Y}}$ were equal, then we should have obtained the reciprocity theorem in its dual form:

> In any passive linear bilateral network, if the single current source $\mathbf{I}_x$ between nodes x and x' produces the voltage response $\mathbf{V}_y$ between nodes y and y', then the removal of the current source from nodes x and x' and its insertion between nodes y and y' will produce the voltage response $\mathbf{V}_y$ between nodes x and x'.

In other words, the interchange of an ideal current source and an ideal voltmeter in any passive linear bilateral circuit will not change the voltmeter reading.

Two ports containing dependent sources receive emphasis in the following section.

Drill Problems

16-4 By applying the appropriate 1-V sources and short circuits to the resistive two port of Fig. 16-7, determine; (*a*) $\mathbf{y}_{11}$; (*b*) $\mathbf{y}_{12}$; (*c*) $\mathbf{y}_{22}$.

Ans. $-\frac{1}{14}$; $\frac{3}{14}$; $\frac{5}{14}$ ℧

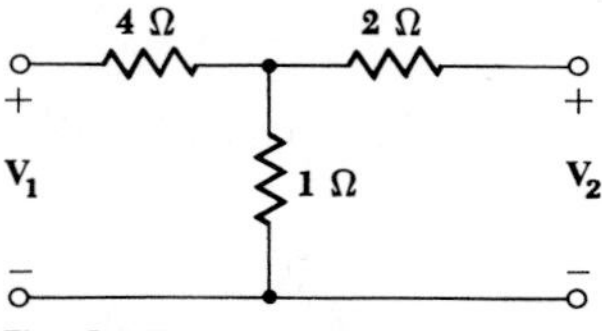

Fig. 16-7 See Drill Prob. 16-4.

16-5 With reference to the π network of Fig. 16-8, determine $\mathbf{Z}_A$, $\mathbf{Z}_B$, and $\mathbf{Z}_C$ so that $\mathbf{y}_{11} = 100\ \mu℧$, $\mathbf{y}_{22} = 500\ \mu℧$, and $\mathbf{y}_{12} = \mathbf{y}_{21} = -200\ \mu℧$.

Ans. −10; 3.33; 5 kΩ

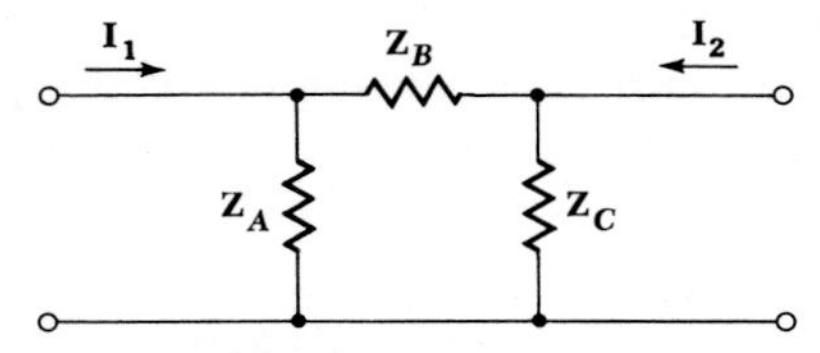

Fig. 16-8 See Drill Prob. 16-5.

16-6 Find $\mathbf{y}_{11}$, $\mathbf{y}_{12}$, and $\mathbf{y}_{21}$ for the network of Fig. 16-9.

Ans. $-\frac{1}{4}$; $-\frac{1}{12}$; $\frac{5}{12}$ ℧

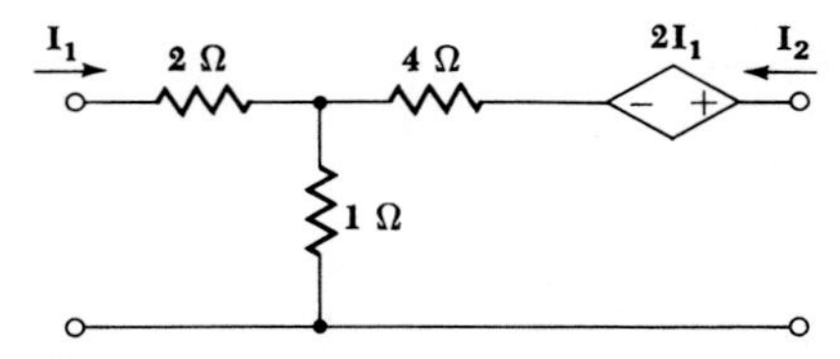

Fig. 16-9 See Drill Prob. 16-6.

16-4 SOME EQUIVALENT NETWORKS

The two basic equations which determine the short-circuit admittance parameters,

$$\mathbf{I}_1 = \mathbf{y}_{11}\mathbf{V}_1 + \mathbf{y}_{12}\mathbf{V}_2 \tag{16}$$

$$\mathbf{I}_2 = \mathbf{y}_{21}\mathbf{V}_1 + \mathbf{y}_{22}\mathbf{V}_2 \tag{17}$$

have the form of a pair of nodal equations written for a circuit containing two nonreference nodes. The determination of an equivalent circuit that leads to (16) and (17) is made more difficult by the inequality, in general, of $\mathbf{y}_{12}$ and $\mathbf{y}_{21}$; it helps to resort to a little trickery in order to obtain a pair of equations which will possess equal mutual coefficients. Let us add and subtract $\mathbf{y}_{12}\mathbf{V}_1$ (the term we would like to see present in the second equation above) on the right side of (17),

$$\mathbf{I}_2 = \mathbf{y}_{12}\mathbf{V}_1 + \mathbf{y}_{22}\mathbf{V}_2 + (\mathbf{y}_{21} - \mathbf{y}_{12})\mathbf{V}_1 \tag{18}$$

or

$$\mathbf{I}_2 - (\mathbf{y}_{21} - \mathbf{y}_{12})\mathbf{V}_1 = \mathbf{y}_{12}\mathbf{V}_1 + \mathbf{y}_{22}\mathbf{V}_2 \tag{19}$$

The right sides of (16) and (19) now show the proper symmetry for a bilateral circuit; the left side of (19) may be interpreted as the algebraic sum of two current sources, one an independent source $\mathbf{I}_2$ entering node 2, and the other a dependent source $(\mathbf{y}_{21} - \mathbf{y}_{12})\mathbf{V}_1$ leaving node 2.

Let us now "read" the equivalent network from (16) and (19). We first establish the reference node, and then a node labeled $\mathbf{V}_1$ and one labeled $\mathbf{V}_2$. From (16), we establish the current $\mathbf{I}_1$ flowing into node 1, we supply a mutual admittance $(-\mathbf{y}_{12})$ between nodes 1 and 2, and we supply an

Fig. 16-10 (*a*) and (*b*) Two ports which are equivalent to any general linear two port. The dependent source in (*a*) depends on $\mathbf{V}_1$, and that in (*b*) depends on $\mathbf{V}_2$. (*c*) An equivalent for a bilateral network.

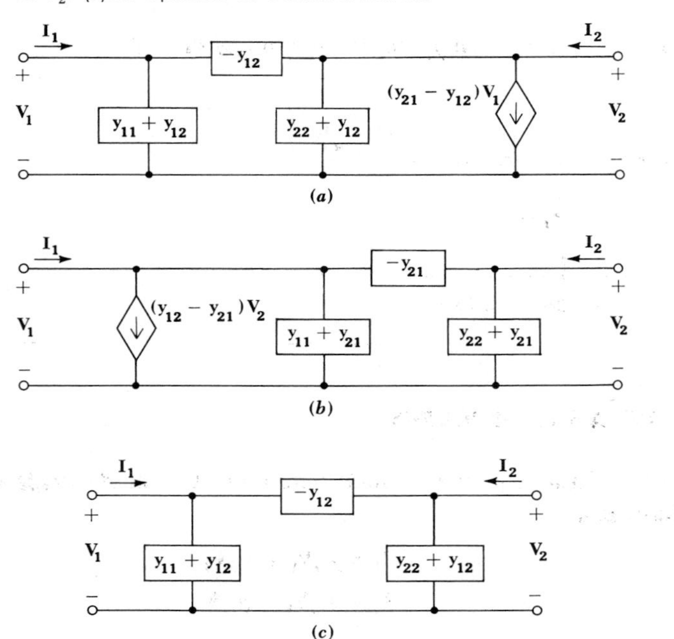

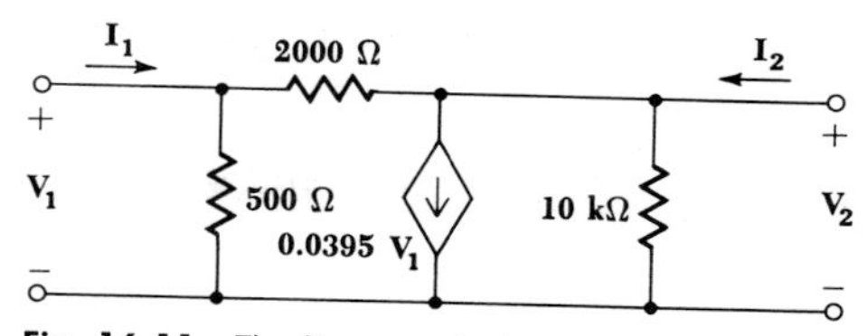

Fig. 16-11 The linear equivalent circuit of a transistor in common-emitter configuration with resistive feedback between collector and base. It is used as a two-port example.

admittance between node 1 and the reference node of $(\mathbf{y}_{11} + \mathbf{y}_{12})$. With $\mathbf{V}_2 = 0$, the ratio of $\mathbf{I}_1$ to $\mathbf{V}_1$ is then $\mathbf{y}_{11}$, as it should be. Now consider (19); we cause the current $\mathbf{I}_2$ to flow into the second node, we cause the current $(\mathbf{y}_{21} - \mathbf{y}_{12})\mathbf{V}_1$ to leave the node, we note that the proper admittance $(-\mathbf{y}_{12})$ exists between the nodes, and we complete the circuit by installing the admittance $(\mathbf{y}_{22} + \mathbf{y}_{12})$ from node 2 to the reference node. The completed circuit is shown in Fig. 16.10*a*.

Another form of equivalent network is obtained by subtracting and adding $\mathbf{y}_{21}\mathbf{V}_2$ in (16); this equivalent circuit is shown in Fig. 16-10*b*.

If the two port is bilateral, then $\mathbf{y}_{12} = \mathbf{y}_{21}$, and either of the equivalents reduces to a simple π network. The dependent source disappears. This equivalent of the bilateral two port is shown in Fig. 16-10*c*.

There are several uses to which these equivalent circuits may be put. In the first place, we have succeeded in showing that an equivalent of any complicated linear two port *exists*. It does not matter how many nodes or loops are contained within the network; the equivalent is no more complex than the circuits of Fig. 16-10. One of these may be much simpler to use than the given circuit if we are interested only in the terminal characteristics of the given network. We should also realize that linear transistor and vacuum-tube amplifiers must possess equivalent circuits in the form of Fig. 16-10*a* and *b*.

As an example of the use of an equivalent circuit with a dependent generator, let us consider the two port shown in Fig. 16-11. This circuit may be considered as an approximate linear equivalent of a transistor circuit in which the emitter terminal is the bottom node, the base terminal is the upper input node, and the collector terminal is the upper output node; a 2000-Ω resistor is connected between collector and base for some special application and makes the analysis of the circuit more difficult.

There are two ways we might think about this circuit. If we recognize it as being in the form of the equivalent circuit shown in Fig. 16-10*a*, then we may immediately determine the values of the $\mathbf{y}$ parameters. If recognition is not immediate, then the $\mathbf{y}$ parameters may be determined for the two port by applying the relationships (6) to (9). We also might avoid any use of two-port analysis methods and write equations directly for the circuit as it stands. Let us compare the network with the equivalent circuit. Thus, we first obtain

$$y_{12} = -\tfrac{1}{2000} = -0.5 \quad m\mho$$

and hence,

$$y_{11} = \tfrac{1}{500} - (-\tfrac{1}{2000}) = 2.5 \quad m\mho$$

Then

$$y_{22} = \frac{1}{10{,}000} - \left(-\frac{1}{2000}\right) = 0.6 \quad m\mho$$

and

$$y_{21} = 0.0395 + (-\tfrac{1}{2000}) = 39 \quad m\mho$$

The following equations must then apply:

$$\mathbf{I}_1 = 2.5\mathbf{V}_1 - 0.5\mathbf{V}_2 \quad \text{mA} \tag{20}$$

$$\mathbf{I}_2 = 39\mathbf{V}_1 + 0.6\mathbf{V}_2 \quad \text{mA} \tag{21}$$

where we are now using units of mA, m℧, kΩ, and V. We make use of these two equations by analyzing the performance of this two port when a current source of $1\underline{/0^\circ}$ mA is provided at the input and a 0.5-kΩ (2-m℧) load is connected to the output. The terminating networks thus give us the following specific information relating $\mathbf{I}_1$ to $\mathbf{V}_1$ and $\mathbf{I}_2$ to $\mathbf{V}_2$:

$$\mathbf{I}_1 = 1 \qquad \mathbf{I}_2 = -2\mathbf{V}_2$$

Substituting into (20) and (21), we may obtain two equations relating $\mathbf{V}_1$ and $\mathbf{V}_2$:

$$1 = 2.5\mathbf{V}_1 - 0.5\mathbf{V}_2 \qquad 0 = 39\mathbf{V}_1 + 2.6\mathbf{V}_2$$

Thus,

$$\mathbf{V}_1 = 0.1 \quad \text{V} \qquad \mathbf{V}_2 = -1.5 \quad \text{V}$$

$$\mathbf{I}_1 = 1 \quad \text{mA} \qquad \mathbf{I}_2 = 3 \quad \text{mA}$$

From these data, it is easy to calculate the voltage gain

$$\mathbf{G}_V = \frac{\mathbf{V}_2}{\mathbf{V}_1} = -15$$

the current gain

$$\mathbf{G}_I = \frac{\mathbf{I}_2}{\mathbf{I}_1} = 3$$

and, if sinusoidal excitation is assumed, the power gain

$$\mathbf{G}_P = \frac{P_{out}}{P_{in}} = \frac{-\frac{1}{2}\mathbf{V}_2\mathbf{I}_2^*}{\frac{1}{2}\mathbf{V}_1\mathbf{I}_1^*} = 45$$

The device might be termed either a voltage, a current, or a power amplifier, since all the gains are greater than unity. If the 2-kΩ resistor were removed, the power gain would rise to 354.

The input and output impedances of the amplifier are often desired in order that maximum power transfer may be achieved to or from an adjacent two port. The input impedance is the ratio of input voltage to current:

$$\mathbf{Z}_{in} = \frac{\mathbf{V}_1}{\mathbf{I}_1} = 0.1 \quad \text{k}\Omega$$

This is the impedance offered to the current source when the 500-Ω load is connected to the output. (With the output short-circuited, the input impedance is necessarily $1/\mathbf{y}_{11}$, or 400 Ω.) It should be noted that the input impedance *cannot* be determined by replacing every source by its internal impedance and then combining resistances or conductances. In the circuit given, this procedure would yield a value of 416 Ω. The error, of course, comes from treating the dependent source as an independent source. If we think of the input impedance as being numerically equal to the input voltage produced by an input current of 1 A, the application of the 1-A source produces some input voltage $\mathbf{V}_1$, and the strength of the dependent source ($0.0395\mathbf{V}_1$) cannot be zero. We should recall that, when we obtain the Thévenin equivalent impedance of a circuit containing a dependent source along with one or more independent sources, we must replace the independent sources by short circuits or open circuits, but a dependent source must not be killed. Of course, if the voltage or current on which the dependent source depends is zero, the dependent source will itself be inactive; occasionally a circuit may be simplified by recognizing such an occurrence.

The output impedance is just another term for the Thévenin impedance appearing in the Thévenin equivalent circuit of that portion of the network faced by the load. In our circuit, which we have assumed driven by a $1\underline{/0^\circ}$-mA current source, we therefore replace this source by an open circuit, leave the dependent source alone, and seek the *input* impedance seen looking to the left from the output terminals. This impedance may be found by several methods (none of which consists of combining all the elements in series or in parallel); let us apply $1\underline{/0^\circ}$ mA at the output terminals and determine $\mathbf{V}_2$. We place these requirements on (20) and (21), and obtain

$$0 = 2.5\mathbf{V}_1 - 0.5\mathbf{V}_2 \qquad 1 = 39\mathbf{V}_1 + 0.6\mathbf{V}_2$$

Solving,

$$\mathbf{V}_2 = 0.119 \quad \text{V}$$

and thus

$$\mathbf{Z}_{out} = 0.119 \quad \text{k}\Omega$$

An alternative procedure might be to find the open-circuit output voltage and the short-circuit output current; the ratio is the Thévenin impedance or $\mathbf{Z}_{out}$. With an input current $\mathbf{I}_1 = 1$ mA and with the load short-circuited ($\mathbf{V}_2 = 0$), we find that

$$\mathbf{I}_1 = 1 = 2.5\mathbf{V}_1 - 0 \qquad \mathbf{I}_2 = 39\mathbf{V}_1 + 0$$

and thus

$$\mathbf{I}_{2sc} = 15.6 \text{ mA}$$

Again, with $\mathbf{I}_1 = 1$ mA and with the output open-circuited ($\mathbf{I}_2 = 0$), we have

$$1 = 2.5\mathbf{V}_1 - 0.5\mathbf{V}_2 \qquad 0 = 39\mathbf{V}_1 + 0.6\mathbf{V}_2$$

and

$$\mathbf{V}_{2oc} = -1.857 \quad \text{V}$$

The assumed directions of $\mathbf{V}_2$ and $\mathbf{I}_2$ therefore result in a Thévenin or output impedance

$$\mathbf{Z}_{out} = -\frac{\mathbf{V}_{2oc}}{\mathbf{I}_{2sc}} = -\frac{-1.857}{15.6} = 0.119 \quad \text{k}\Omega$$

as before.

We now have enough information to enable us to draw the Thévenin or Norton equivalent of the two port of Fig. 16-11 (driven by a $1\underline{/0°}$-mA current source and terminated in a 500-Ω load, although this termination is of no consequence in the output equivalent circuits) as offered to the 500-Ω load. The Thévenin and Norton equivalents offered to the current

Fig. 16-12 (*a*) The Norton equivalent of the network (Fig. 16-11) to the left of the output terminals, with $\mathbf{I}_1 = 1\underline{/0°}$ mA. (*b*) The Thévenin equivalent of that portion of the network to the right of the input terminals, if $\mathbf{I}_2 = -2\mathbf{V}_2$ mA.

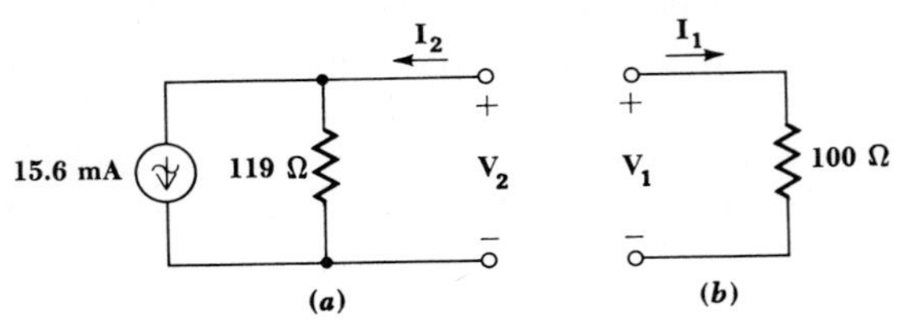

source are also readily drawn. Thus, the Norton equivalent presented to the load must contain a current source equal to the short-circuit current $\mathbf{I}_{2sc}$ in parallel with the output impedance; this equivalent is shown in Fig. 16-12*a*. The Thévenin equivalent offered to the $1\underline{/0°}$-mA input source must consist solely of the input impedance, as drawn in Fig. 16-12*b*.

Drill Problems

16-7 If the 2000-Ω resistor is removed from the circuit of Fig. 16-11, if $\mathbf{I}_1 = 1\underline{/0°}$ mA, and if a 500-Ω resistor is installed at the output, find: (*a*) $\mathbf{G}_V$; (*b*) $\mathbf{G}_I$; (*c*) $\mathbf{G}_P$.

Ans. -18.8; *18.8*; *354*

16-8 A two port is described by the following parameters: $\mathbf{y}_{11} = 5$ ℧, $\mathbf{y}_{12} = -1$ ℧, $\mathbf{y}_{21} = 30$ ℧, and $\mathbf{y}_{22} = 1.5$ ℧. Find the power gain $\mathbf{G}_P = (-\frac{1}{2}\mathbf{V}_2\mathbf{I}_2^*)/(\frac{1}{2}\mathbf{V}_1\mathbf{I}_1^*)$ if: (*a*) $\mathbf{I}_1 = 1\underline{/0°}$ A, and an output load, $R_L = \frac{2}{3}$ Ω, is used; (*b*) $\mathbf{I}_1 = 1\underline{/0°}$ A and $\mathbf{I}_2 = 0.5\underline{/0°}$ A; (*c*) a source $0.2\underline{/0°}$ V in series with a 0.1-Ω resistor is connected to the input and $R_L = 0.5$ Ω.

Ans. 6.88; 10.0; 10.8

16-9 For the circuit shown in Fig. 16-13, find: (*a*) $\mathbf{y}_{12}$; (*b*) $\mathbf{Y}_{in}$ if $\mathbf{V}_2 = -4\mathbf{I}_2$; (*c*) $\mathbf{Y}_{out}$ if $\mathbf{V}_1 = 0$.

Ans. $-\frac{1}{9}$; $\frac{1}{9}$; $\frac{17}{130}$ ℧

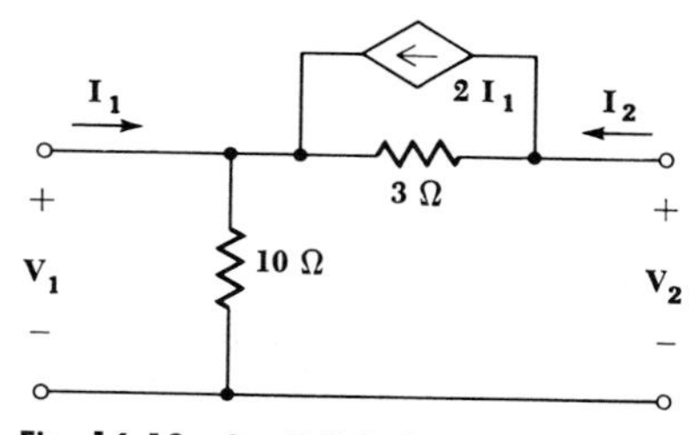

Fig. 16-13 See Drill Probs. 16-9, 16-10, and 16-16.

16-5 IMPEDANCE PARAMETERS

The concept of two-port parameters has been introduced in terms of the short-circuit admittance parameters. There are other sets of parameters, however, and each set is associated with a particular class of networks for which its use provides the simplest analysis. We shall consider only two other types of parameters, the open-circuit impedance parameters which are the subject of this section and the hybrid parameters which are discussed in the following section.

We begin again with the general linear two port which does not contain any independent sources; the currents and voltages are assigned as before (Fig. 16-5). Now let us consider the voltage $\mathbf{V}_1$ as the response produced by two current sources $\mathbf{I}_1$ and $\mathbf{I}_2$. We thus write for $\mathbf{V}_1$

$$\mathbf{V}_1 = \mathbf{z}_{11}\mathbf{I}_1 + \mathbf{z}_{12}\mathbf{I}_2 \tag{22}$$

and for $\mathbf{V}_2$

$$\mathbf{V}_2 = \mathbf{z}_{21}\mathbf{I}_1 + \mathbf{z}_{22}\mathbf{I}_2 \tag{23}$$

Of course, in using this pair of equations it is not necessary that $\mathbf{I}_1$ and $\mathbf{I}_2$ be current sources; nor is it necessary that $\mathbf{V}_1$ and $\mathbf{V}_2$ be voltage sources. In general, we may have any networks terminating the two port at either end. As the equations are written, we probably think of $\mathbf{V}_1$ and $\mathbf{V}_2$ as given quantities, or independent variables, and $\mathbf{I}_1$ and $\mathbf{I}_2$ as unknowns, or dependent variables. The various ways in which two equations may be written to relate these four quantities define the different systems of parameters.

The most informative description of the $\mathbf{z}$ parameters, defined in (22) and (23), is obtained by setting each of the currents equal to zero. Thus

$$\mathbf{z}_{11} = \left.\frac{\mathbf{V}_1}{\mathbf{I}_1}\right|_{\mathbf{I}_2=0} \tag{24}$$

$$\mathbf{z}_{12} = \left.\frac{\mathbf{V}_1}{\mathbf{I}_2}\right|_{\mathbf{I}_1=0} \tag{25}$$

$$\mathbf{z}_{21} = \left.\frac{\mathbf{V}_2}{\mathbf{I}_1}\right|_{\mathbf{I}_2=0} \tag{26}$$

$$\mathbf{z}_{22} = \left.\frac{\mathbf{V}_2}{\mathbf{I}_2}\right|_{\mathbf{I}_1=0} \tag{27}$$

Since zero current results from an open-circuit termination, the $\mathbf{z}$ parameters are known as the *open-circuit impedance parameters*. They are easily related to the short-circuit admittance parameters by solving (22) and (23) for $\mathbf{I}_1$ and $\mathbf{I}_2$:

$$\mathbf{I}_1 = \frac{\begin{vmatrix} \mathbf{V}_1 & \mathbf{z}_{12} \\ \mathbf{V}_2 & \mathbf{z}_{22} \end{vmatrix}}{\begin{vmatrix} \mathbf{z}_{11} & \mathbf{z}_{12} \\ \mathbf{z}_{21} & \mathbf{z}_{22} \end{vmatrix}}$$

or

$$\mathbf{I}_1 = \left(\frac{\mathbf{z}_{22}}{\mathbf{z}_{11}\mathbf{z}_{22} - \mathbf{z}_{12}\mathbf{z}_{21}}\right)\mathbf{V}_1 + \left(-\frac{\mathbf{z}_{12}}{\mathbf{z}_{11}\mathbf{z}_{22} - \mathbf{z}_{12}\mathbf{z}_{21}}\right)\mathbf{V}_2$$

Using determinant notation, and being careful that the subscript is a lowercase $\mathbf{z}$, we assume that $\mathbf{\Delta}_\mathbf{z} \neq 0$ and obtain

$$\mathbf{y}_{11} = \frac{\Delta_{11}}{\Delta_\mathbf{z}} \qquad \mathbf{y}_{12} = -\frac{\Delta_{21}}{\Delta_\mathbf{z}}$$

and from solving for $\mathbf{I}_2$,

$$\mathbf{y}_{21} = -\frac{\Delta_{12}}{\Delta_\mathbf{z}} \qquad \mathbf{y}_{22} = \frac{\Delta_{22}}{\Delta_\mathbf{z}}$$

In a similar manner, the **z** parameters may be expressed in terms of the admittance parameters. Transformations of this nature are possible between any of the various parameter systems, and quite a collection of occasionally useful formulas may be obtained. We shall not specifically consider any other transformations.[1]

If the two port is a bilateral network, reciprocity is present; it is easy to show that this results in the equality of $\mathbf{z}_{12}$ and $\mathbf{z}_{21}$.

Equivalent circuits may again be obtained from an inspection of (22) and (23); their construction is facilitated by adding and subtracting either $\mathbf{z}_{12}\mathbf{I}_1$ to (23) or $\mathbf{z}_{21}\mathbf{I}_2$ to (22). Each of these equivalent circuits contains a dependent voltage source.

Let us leave the derivation of such an equivalent to Prob. 15 and consider next an example of a rather general nature. Can we construct a general Thévenin equivalent of the two port, as viewed from the output terminals? It is necessary first to assume a specific input circuit configuration, and we shall select an independent voltage source $\mathbf{V}_s$ in series with a generator impedance $\mathbf{Z}_g$. Thus

$$\mathbf{V}_s = \mathbf{V}_1 + \mathbf{I}_1\mathbf{Z}_g$$

Combining this result with (22) and (23), we may eliminate $\mathbf{V}_1$ and $\mathbf{I}_1$, and obtain

$$\mathbf{V}_2 = \frac{\mathbf{z}_{21}}{\mathbf{z}_{11} + \mathbf{Z}_g}\mathbf{V}_s + \left(\mathbf{z}_{22} - \frac{\mathbf{z}_{12}\mathbf{z}_{21}}{\mathbf{z}_{11} + \mathbf{Z}_g}\right)\mathbf{I}_2$$

The Thévenin equivalent circuit may be drawn directly from this equation; it is shown in Fig. 16-14. The output impedance, expressed in terms of the **z** parameters, is

$$\mathbf{Z}_{out} = \mathbf{z}_{22} - \frac{\mathbf{z}_{12}\mathbf{z}_{21}}{\mathbf{z}_{11} + \mathbf{Z}_g}$$

[1]Except in the problems.

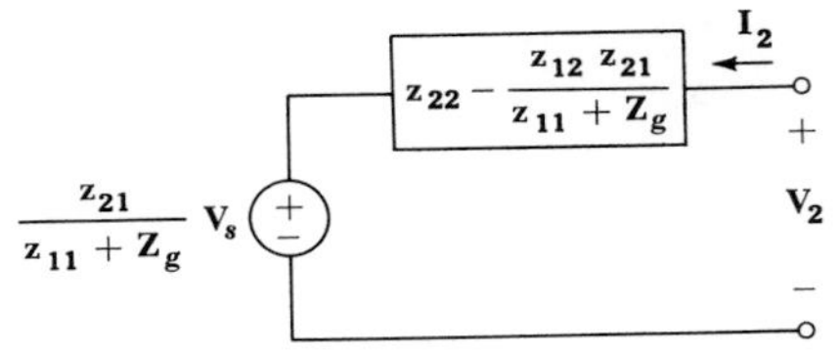

Fig. 16-14 The Thévenin equivalent of a general two port, as viewed from the output terminals, expressed in terms of the open-circuit impedance parameters.

If the generator impedance is zero, the simpler expression

$$\mathbf{Z}_{out} = \frac{\mathbf{z}_{11}\mathbf{z}_{22} - \mathbf{z}_{12}\mathbf{z}_{21}}{\mathbf{z}_{11}} = \frac{\Delta_z}{\Delta_{22}} = \frac{1}{\mathbf{y}_{22}} \qquad (\mathbf{Z}_g = 0)$$

is obtained. For this special case, the output *admittance* is identical to $\mathbf{y}_{22}$, as indicated by the basic relationship of (9).

As a numerical example, let us use a set of impedance parameters representative of a transistor operating in the grounded-emitter configuration,

$$\mathbf{z}_{11} = 10^3 \quad \Omega \qquad \mathbf{z}_{12} = 10 \quad \Omega$$
$$\mathbf{z}_{21} = -10^6 \quad \Omega \qquad \mathbf{z}_{22} = 10^4 \quad \Omega$$

and we shall consider the two port as driven by an ideal sinusoidal voltage source $\mathbf{V}_s$ in series with a 500-Ω resistor and terminated in a 10-kΩ load resistor. The two describing equations are

$$\mathbf{V}_1 = 10^3\mathbf{I}_1 + 10\mathbf{I}_2 \tag{28}$$
$$\mathbf{V}_2 = -10^6\mathbf{I}_1 + 10^4\mathbf{I}_2 \tag{29}$$

and the characterizing equations of the input and output networks are

$$\mathbf{V}_s = 500\mathbf{I}_1 + \mathbf{V}_1 \tag{30}$$
$$\mathbf{V}_2 = -10^4\mathbf{I}_2 \tag{31}$$

From these last four equations, we may easily obtain expressions for $\mathbf{V}_1$, $\mathbf{I}_1$, $\mathbf{V}_2$, and $\mathbf{I}_2$ in terms of $\mathbf{V}_s$,

$$\mathbf{V}_1 = 0.75\mathbf{V}_s \qquad \mathbf{I}_1 = \frac{\mathbf{V}_s}{2000}$$
$$\mathbf{V}_2 = -250\mathbf{V}_s \qquad \mathbf{I}_2 = \frac{\mathbf{V}_s}{40}$$

From this information, it is simple to determine the voltage gain,

$$\mathbf{G}_V = \frac{\mathbf{V}_2}{\mathbf{V}_1} = -333$$

the current gain,

$$\mathbf{G}_I = \frac{\mathbf{I}_2}{\mathbf{I}_1} = 50$$

the power gain,

$$\mathbf{G}_P = \frac{-\frac{1}{2}\mathbf{V}_2\mathbf{I}_2^*}{\frac{1}{2}\mathbf{V}_1\mathbf{I}_1^*} = 16{,}700$$

and the input impedance,

$$\mathbf{Z}_{in} = \frac{\mathbf{V}_1}{\mathbf{I}_1} = 1500 \quad \Omega$$

The output impedance may be obtained by referring to Fig. 16-14,

$$\mathbf{Z}_{out} = \mathbf{z}_{22} - \frac{\mathbf{z}_{12}\mathbf{z}_{21}}{\mathbf{z}_{11} + \mathbf{Z}_g} = 16.7 \quad \text{k}\Omega$$

Drill Problems

16-10 For the circuit shown in Fig. 16-13, find: (*a*) $\mathbf{z}_{12}$; (*b*) $\mathbf{z}_{21}$; (*c*) $\mathbf{z}_{22}$.
Ans. 4; 10; 13 Ω

16-11 The admittance parameters of a bilateral two port are: $\mathbf{y}_{11} = 0.25$ ℧, $\mathbf{y}_{12} = -0.05$ ℧, $\mathbf{y}_{22} = 0.1$ ℧. Find: (*a*) $\mathbf{z}_{11}$; (*b*) $\mathbf{z}_{21}$; (*c*) $\mathbf{z}_{22}$.
Ans. $^{20}/_{9}$; $^{40}/_{9}$; $^{100}/_{9}$ Ω

16-12 The T-equivalent circuit of a transistor in the common-base arrangement is shown in Fig. 16-15. Find: (*a*) $\mathbf{z}_{11}$; (*b*) $\mathbf{z}_{12}$; (*c*) $\mathbf{z}_{21}$.
Ans. r_b; $r_e + r_b$; $r_b + \alpha r_c$

Fig. 16-15 See Drill Probs. 16-12, 16-13, and 16-15.

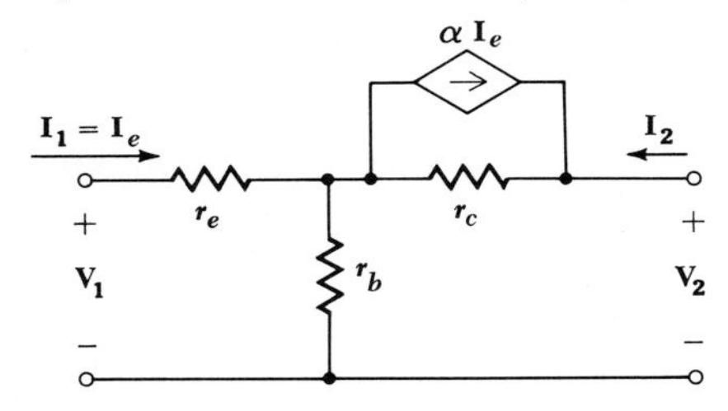

16-13 For the transistor considered in Drill Prob. 16-12 and Fig. 16-15, let r_e = 20 Ω, r_b = 800 Ω, r_c = 500 kΩ, and α = 0.98. A current source, $\mathbf{I}_1$ = 2 mA, is applied at the input and a load, R_L = 10 kΩ, is connected to the output. Find: (*a*) $\mathbf{G}_I = \mathbf{I}_2/\mathbf{I}_1$; (*b*) $\mathbf{G}_V = \mathbf{V}_2/\mathbf{V}_1$; (*c*) $\mathbf{G}_P = -\mathbf{G}_I\mathbf{G}_V$.

Ans. −*0.96; 180; 187*

16-6 HYBRID PARAMETERS

The use of the hybrid parameters is well suited to transistor circuits because these parameters are among the most convenient to measure experimentally for a transistor. The difficulty in measuring, say, the open-circuit impedance parameters arises when a parameter such as $\mathbf{z}_{21}$ must be measured. A known sinusoidal current is easily supplied at the input terminals, but because of the exceedingly high output impedance of the transistor circuit, it is difficult to open-circuit the output terminals and yet supply the necessary dc biasing voltages and measure the sinusoidal output voltage. A short-circuit current measurement at the output terminals is much simpler to instrument.

The hybrid parameters are defined by writing the pair of equations relating $\mathbf{V}_1$, $\mathbf{I}_1$, $\mathbf{V}_2$, and $\mathbf{I}_2$ as if $\mathbf{V}_1$ and $\mathbf{I}_2$ were the independent variables:

$$\mathbf{V}_1 = \mathbf{h}_{11}\mathbf{I}_1 + \mathbf{h}_{12}\mathbf{V}_2 \tag{32}$$

$$\mathbf{I}_2 = \mathbf{h}_{21}\mathbf{I}_1 + \mathbf{h}_{22}\mathbf{V}_2 \tag{33}$$

The nature of the parameters is made clear by first setting $\mathbf{V}_2 = 0$,

$$\mathbf{h}_{11} = \left.\frac{\mathbf{V}_1}{\mathbf{I}_1}\right|_{\mathbf{V}_2=0} = \text{short-circuit input impedance}$$

and

$$\mathbf{h}_{21} = \left.\frac{\mathbf{I}_2}{\mathbf{I}_1}\right|_{\mathbf{V}_2=0} = \text{short-circuit current gain}$$

and then letting $\mathbf{I}_1 = 0$,

$$\mathbf{h}_{12} = \left.\frac{\mathbf{V}_1}{\mathbf{V}_2}\right|_{\mathbf{I}_1=0} = \text{open-circuit reverse voltage gain}$$

and

$$\mathbf{h}_{22} = \left.\frac{\mathbf{I}_2}{\mathbf{V}_2}\right|_{\mathbf{I}_1=0} = \text{open-circuit output admittance}$$

Since the parameters represent an impedance, an admittance, a voltage gain, and a current gain, it is understandable that they are called the "hybrid" parameters.

In order to illustrate the ease with which these parameters may be evaluated, consider the bilateral resistive circuit drawn in Fig. 16-16. With

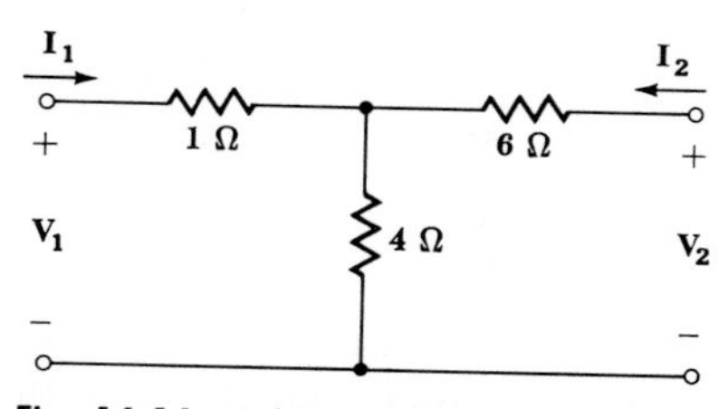

Fig. 16-16 A bilateral network for which the **h** parameters are found; $\mathbf{h}_{12} = -\mathbf{h}_{21}$.

the output short-circuited ($\mathbf{V}_2 = 0$), the application of a 1-A source at the input ($\mathbf{I}_1 = 1$ A) produces an input voltage of 3.4 V ($\mathbf{V}_1 = 3.4$ V); hence, $\mathbf{h}_{11} = 3.4\ \Omega$. Under these same conditions, the output current is easily obtained by current division, $\mathbf{I}_2 = -0.4$ A; thus, $\mathbf{h}_{21} = -0.4$. The remaining two parameters are obtained with the input open-circuited ($\mathbf{I}_1 = 0$). Let us apply a voltage of 1 V at the output terminals ($\mathbf{V}_2 = 1$ V). The response at the input terminals is 0.4 V ($\mathbf{V}_1 = 0.4$ V), and thus $\mathbf{h}_{12} = 0.4$. The current delivered by this source at the output terminals is 0.1 A ($\mathbf{I}_2 = 0.1$ A), and therefore $\mathbf{h}_{22} = 0.1$ ℧. It is a consequence of the reciprocity theorem that $\mathbf{h}_{12} = -\mathbf{h}_{21}$ for a bilateral network.

The circuit shown in Fig. 16-17 is a direct translation of the two defining equations (32) and (33). The first represents Kirchhoff's voltage law about the input loop, while the second is obtained from Kirchhoff's current law at the upper output node. This circuit is also a popular transistor equivalent circuit. Let us assume some reasonable values for the common-emitter configuration: $\mathbf{h}_{11} = 1200\ \Omega$, $\mathbf{h}_{12} = 2 \times 10^{-4}$, $\mathbf{h}_{21} = 50$, $\mathbf{h}_{22} = 50 \times 10^{-6}$ ℧, an input voltage source of $1\underline{/0°}$ mV in series with 800 Ω, and a 5-kΩ load. For the input,

$$10^{-3} = (1200 + 800)\mathbf{I}_1 + 2 \times 10^{-4}\mathbf{V}_2$$

and at the output,

$$\mathbf{I}_2 = -2 \times 10^{-4}\mathbf{V}_2 = 50\mathbf{I}_1 + 50 \times 10^{-6}\mathbf{V}_2$$

Fig. 16-17 The four **h** parameters are referred to a two port. The pertinent equations are: $\mathbf{V}_1 = \mathbf{h}_{11}\mathbf{I}_1 + \mathbf{h}_{12}\mathbf{V}_2$, $\mathbf{I}_2 = \mathbf{h}_{21}\mathbf{I}_1 + \mathbf{h}_{22}\mathbf{V}_2$.

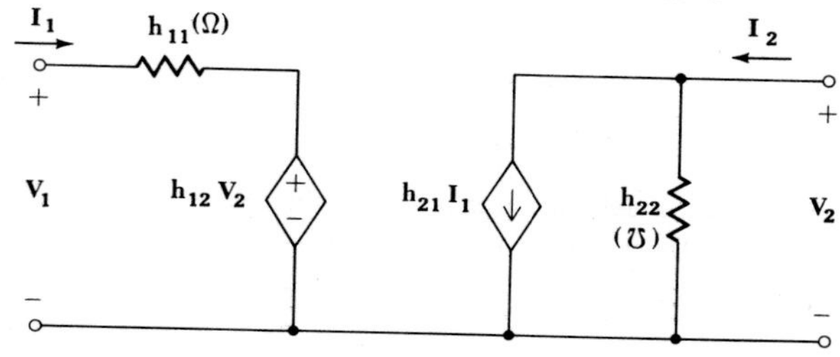

Solving,

$$\mathbf{I}_1 = 0.510 \quad \mu\text{A} \qquad \mathbf{V}_1 = 0.592 \quad \text{mV}$$

$$\mathbf{I}_2 = 20.4 \quad \mu\text{A} \qquad \mathbf{V}_2 = -102 \quad \text{mV}$$

Through the transistor we have a current gain of 40, a voltage gain of -172, and a power gain of 6880. The input impedance to the transistor is 1160 Ω, and a few more calculations show that the output impedance is 22.2 kΩ.

Drill Problems

16-14 For a bilateral two port, $\mathbf{y}_{11} = 0.5$ ℧, $\mathbf{y}_{12} = -0.4$ ℧, and $\mathbf{y}_{22} = 0.6$ ℧. Find: (*a*) $\mathbf{h}_{11}$; (*b*) $\mathbf{h}_{21}$; (*c*) $\mathbf{h}_{22}$.

Ans. -0.8; 0.28 ℧; 2 Ω

16-15 For the transistor T-equivalent circuit of Fig. 16-15, let $r_e = 20$ Ω, $r_b = 800$ Ω, $r_c = 500$ kΩ, and $\alpha = 0.98$. Find: (*a*) $\mathbf{h}_{12}$; (*b*) $\mathbf{h}_{21}$; (*c*) $\mathbf{h}_{22}$.

Ans. -0.98; 2×10^{-6} ℧; 1.6×10^{-3}

16-16 For the circuit shown in Fig. 16-13, find: (*a*) $\mathbf{h}_{11}$; (*b*) $\mathbf{h}_{12}$; (*c*) $\mathbf{h}_{21}$.

Ans. $^{90}/_{13}$; $-^{4}/_{13}$; $^{10}/_{13}$ Ω

16-17 For the two port terminated as shown in Fig. 16-18, find the average power: (*a*) supplied by the 1-V sinusoidal source; (*b*) delivered to the 10-Ω load resistor; (*c*) lost in the 5-Ω internal resistance of the source.

Ans. 4; 20; 200 mW

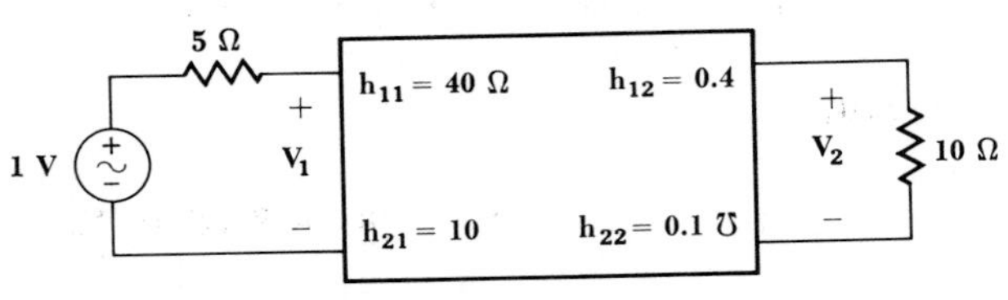

Fig. 16-18 See Drill Prob. 16-17.

PROBLEMS

□ **1** The three rows of a determinant are in order: 9, -2, -3; -2, 6, -1; and -3, -1, 4. (*a*) Find the input impedance offered to a source appearing in only the first mesh, assuming that the determinant is $\mathbf{\Delta_Z}$ (values in ohms). (*b*) Suppose that the reference direction for $\mathbf{I}_2$ had been selected in the opposite direction so that the three rows of the determinant $\mathbf{\Delta}_Z$ were 9, 2, −3; 2, 6, 1; −3, 1, 4; repeat part (*a*). (*c*) Find the input impedance offered to a source connected between the first node and the reference node, assuming that the original determinant is $\mathbf{\Delta}_Y$ (values in mhos).

☐ **2** (*a*) Find the input impedance of the one port illustrated in Fig. 16-19. (*b*) If a 10-V dc source is connected to the input terminals, find the power absorbed by each element.

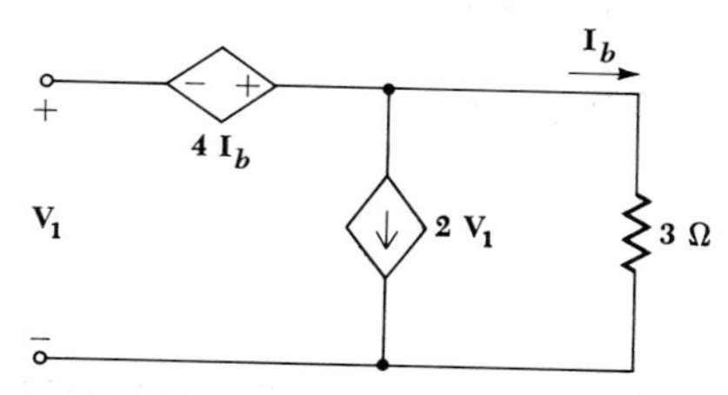

Fig. 16-19 See Prob. 2.

☐ **3** The admittance determinant of a circuit is given by

$$\mathbf{\Delta}_Y = \begin{vmatrix} 4 & -j2 & -1+j2 \\ -j2 & 1+j3 & j1 \\ -1+j2 & j1 & 2 \end{vmatrix}$$

Find the average power supplied by a source, $2\underline{/0°}$ A, connected between node 1 and the reference node.

☐ **4** The reciprocity theorem also holds for the complete response, forced plus natural, if there is no initial-energy storage. This restriction is necessary because the network must be passive, except for the single source specified. Demonstrate the truth of this extension of the reciprocity theorem by: (*a*) finding v_C in the circuit shown in Fig. 16-20 for $t < 0$ and $t > 0$; (*b*) interchanging the step-current source and the (instantaneous) voltmeter and finding the voltmeter response for $t < 0$ and $t > 0$.

Fig. 16-20 See Prob. 4.

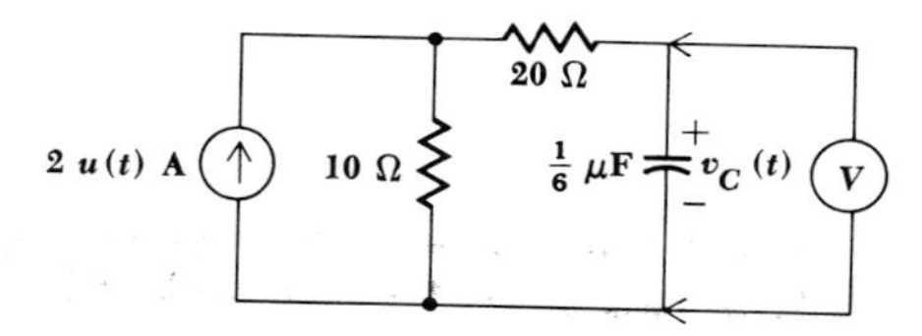

☐ **5** Measurements made on a two-port network yield the following data: for $\mathbf{V}_1 = -5$ V and $\mathbf{V}_2 = -8$ V, $\mathbf{I}_1 = -9$ A and $\mathbf{I}_2 = -17$ A; and for $\mathbf{V}_1 = -8$ V and $\mathbf{V}_2 = -5$ V, $\mathbf{I}_1 = -30$ A and $\mathbf{I}_2 = 4$ A. (*a*) Find the four short-circuit admittance parameters. (*b*) Find $\mathbf{I}_1$, $\mathbf{I}_2$, and $\mathbf{V}_2$ if $\mathbf{V}_1 = 10$ V and a 2-Ω resistor is connected to the output port.

☐ **6** Find the four short-circuit admittance parameters for the circuit shown in Fig. 16-21.

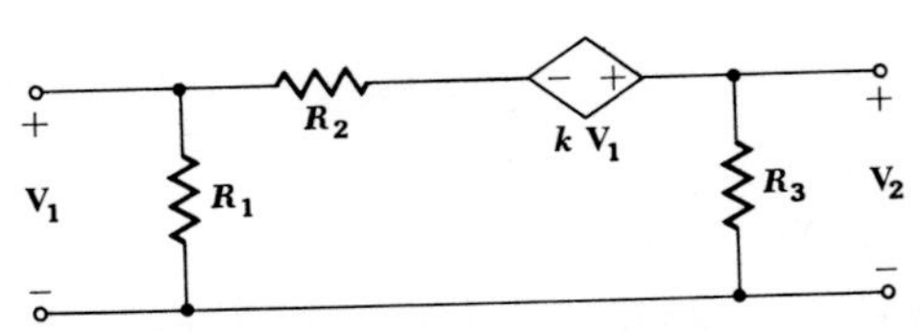

Fig. 16-21 See Probs. 6 and 21.

☐ **7** Two two ports, identified by *a* and *b* subscripts, have y parameters $\mathbf{y}_{11a}$, $\mathbf{y}_{11b}$, $\mathbf{y}_{12a}$, etc. Find the y parameters for the overall network if the output of network *a* is connected directly to the input of network *b*.

☐ **8** Find the y parameters for the: (*a*) π network shown in Fig. 16-22*a*; (*b*) T network shown in Fig. 16-22*b*. (*c*) Show that the corresponding y parameters for the two networks are equal if:

$$\mathbf{Z}_1 = \frac{\mathbf{Z}_a\mathbf{Z}_b}{\mathbf{Z}_a + \mathbf{Z}_b + \mathbf{Z}_c} \qquad \mathbf{Z}_a = \frac{\mathbf{Z}_1\mathbf{Z}_2 + \mathbf{Z}_2\mathbf{Z}_3 + \mathbf{Z}_3\mathbf{Z}_1}{\mathbf{Z}_2}$$

$$\mathbf{Z}_2 = \frac{\mathbf{Z}_b\mathbf{Z}_c}{\mathbf{Z}_a + \mathbf{Z}_b + \mathbf{Z}_c} \qquad \mathbf{Z}_b = \frac{\mathbf{Z}_1\mathbf{Z}_2 + \mathbf{Z}_2\mathbf{Z}_3 + \mathbf{Z}_3\mathbf{Z}_1}{\mathbf{Z}_3}$$

$$\mathbf{Z}_3 = \frac{\mathbf{Z}_c\mathbf{Z}_a}{\mathbf{Z}_a + \mathbf{Z}_b + \mathbf{Z}_c} \qquad \mathbf{Z}_c = \frac{\mathbf{Z}_1\mathbf{Z}_2 + \mathbf{Z}_2\mathbf{Z}_3 + \mathbf{Z}_3\mathbf{Z}_1}{\mathbf{Z}_1}$$

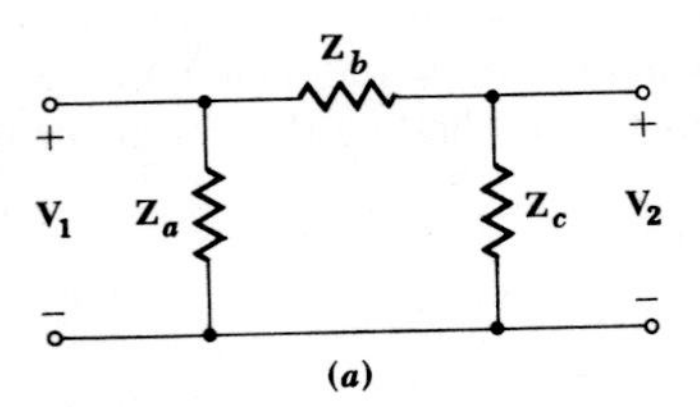

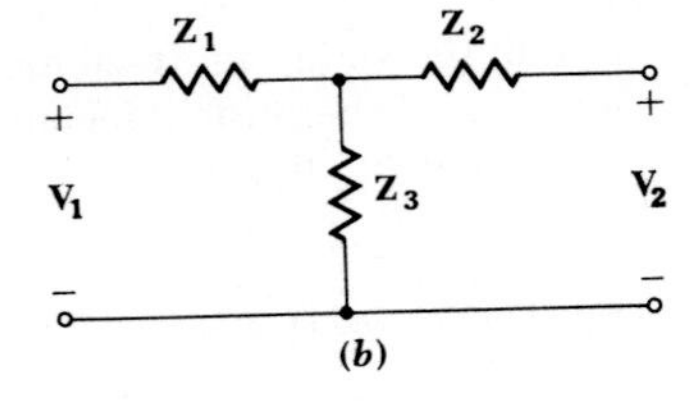

Fig. 16-22 See Prob. 8.

☐ **9** Given the equivalence between π and T networks expressed by the relationships in part (*c*) of Prob. 8, transform: (*a*) the π network of Fig. 16-23*a* to a T network; (*b*) the T network of Fig. 16-23*b* to a π network; (*c*) the one ports of Fig. 16-23*c* and *d* to single resistor equivalents.

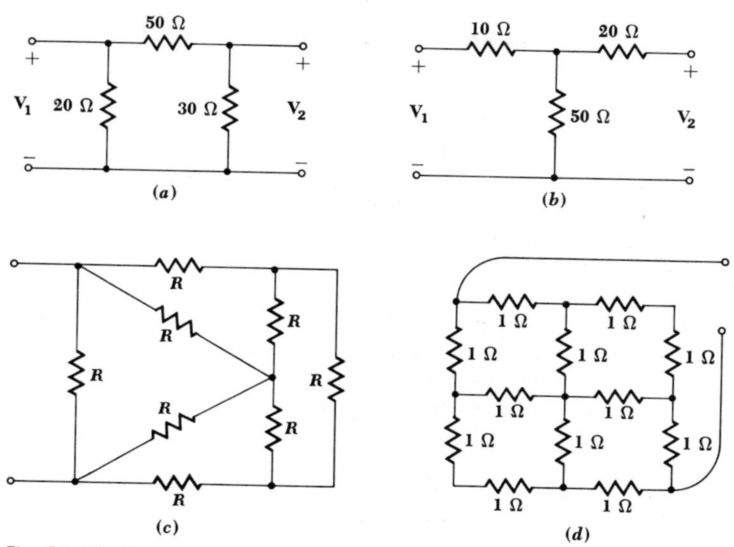

Fig. 16-23 See Prob. 9.

☐ **10** Figure 16-24*a* shows the equivalent circuit of a vacuum-tube circuit called a *cathode follower*, famous for its relatively low output impedance. Determine the output impedance for this cathode follower and compare it to that for a typical vacuum-tube amplifier, as illustrated by the equivalent circuit of Fig. 16-24*b*. Let $g_m = 2$ m℧, $r_p = 10$ kΩ.

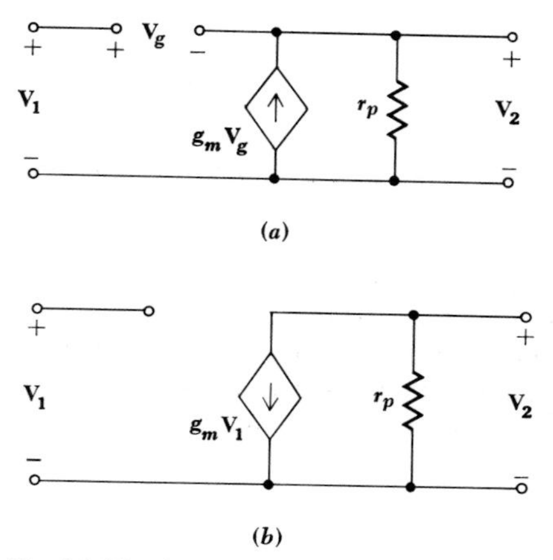

Fig. 16-24 See Prob. 10.

☐ **11** Determine the short-circuit admittance parameters for the circuit of: (*a*) Fig. 16-25*a*; (*b*) Fig. 16-25*b*.

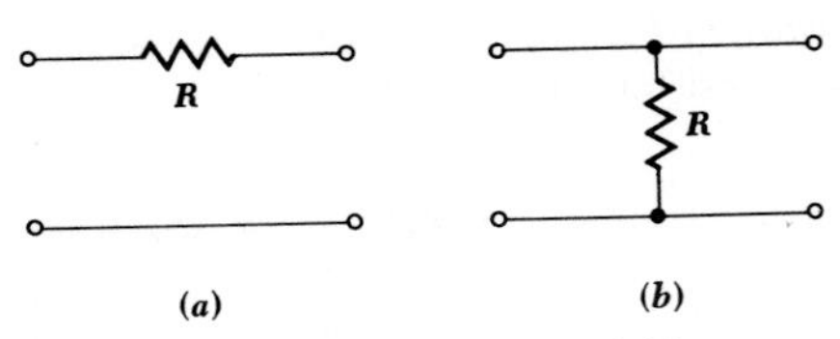

Fig. 16-25 See Probs. 11, 16, and 22.

☐ **12** Measurements on a linear two-port network show that $\mathbf{y}_{11} = 2\mathbf{y}_{22} = 0.1$ ℧, and $\mathbf{y}_{21} = -10\mathbf{y}_{12} = 0.4$ ℧. How do these values change if a 10-Ω resistor is installed: (*a*) across the input terminals? (*b*) across the output terminals? (*c*) in series with the input? (*d*) in series with the output?

☐ **13** A linear two port for which $\mathbf{y}_{11} = 2$ ℧, $\mathbf{y}_{12} = -0.5$ ℧, $\mathbf{y}_{21} = 20$ ℧, and $\mathbf{y}_{22} = 0.4$ ℧ is terminated in R_L. A 5-V sinusoidal source in series with 1 Ω is connected to the input. For each part below, determine the value of R_L required to maximize the magnitude of the specified quantity, and give the maximum value of that quantity: (*a*) $\mathbf{V}_2$; (*b*) $\mathbf{I}_2$; (*c*) P_{2av}; (*d*) $\mathbf{V}_1$.

☐ **14** Obtain expressions giving the open-circuit impedance parameters in terms of the short-circuit admittance parameters.

☐ **15** Determine values for the elements shown in Fig. 16-26 if the circuit is to be equivalent to any circuit described by: $\mathbf{V}_1 = \mathbf{z}_{11}\mathbf{I}_1 + \mathbf{z}_{12}\mathbf{I}_2$, $\mathbf{V}_2 = \mathbf{z}_{21}\mathbf{I}_1 + \mathbf{z}_{22}\mathbf{I}_2$.

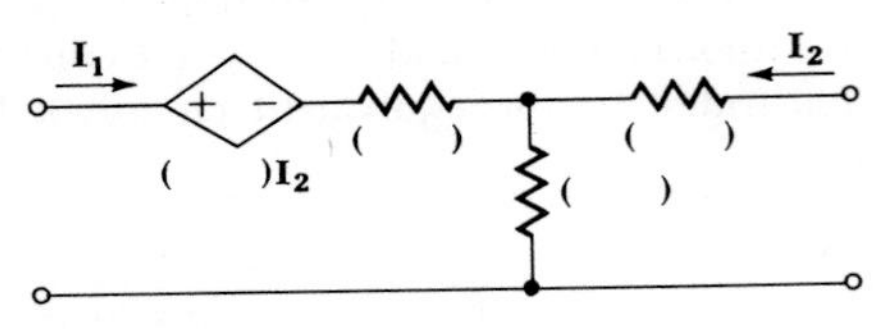

Fig. 16-26 See Prob. 15.

☐ **16** Determine the open-circuit impedance parameters for the circuit of: (*a*) Fig. 16-25*a*; (*b*) Fig. 16-25*b*.

☐ **17** A three-terminal network, with terminals labeled *a*, *b*, and *c*, is operated as a two-port network by connecting *a* to 1, *b* to 1′ and 2′, and *c* to 2, as shown in Fig. 16-27. For this arrangement $\mathbf{z}_{11} = 10$ Ω, $\mathbf{z}_{12} = 2$ Ω, $\mathbf{z}_{21} = -2$ Ω, and $\mathbf{z}_{22} = 6$ Ω. Find the **z** parameters that result from connecting: (*a*) *b* to 1, *c* to 1′ and 2′, and *a* to 2; (*b*) *c* to 1, *a* to 1′ and 2′, and *b* to 2.

Fig. 16-27 See Prob. 17.

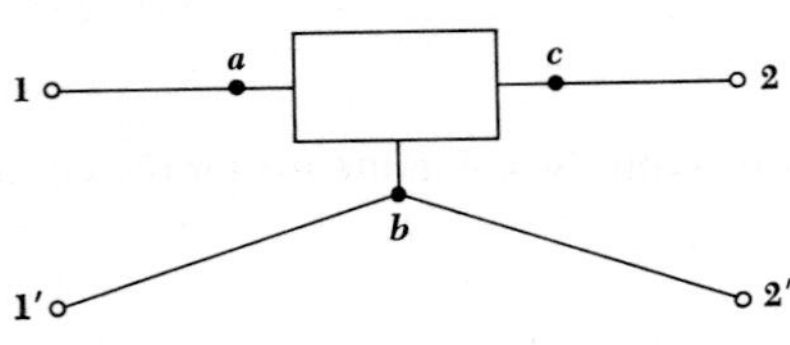

☐ **18** A two port is connected as shown in Fig. 16-28. Obtain an expression for: (*a*) the voltage gain, $\mathbf{G}_V = \mathbf{V}_2/\mathbf{V}_1$; (*b*) the current gain, $\mathbf{G}_I = \mathbf{I}_2/\mathbf{I}_1$; (*c*) the input impedance, $\mathbf{Z}_{in} = \mathbf{V}_1/\mathbf{I}_1$; (*d*) the power gain, $G_P = (\frac{1}{2}\text{ Re }\mathbf{V}_2\mathbf{I}_2^*)/(\frac{1}{2}\text{ Re }\mathbf{V}_1\mathbf{I}_1^*)$, for a pure-resistance network; (*e*) the output impedance $\mathbf{Z}_{th}$.

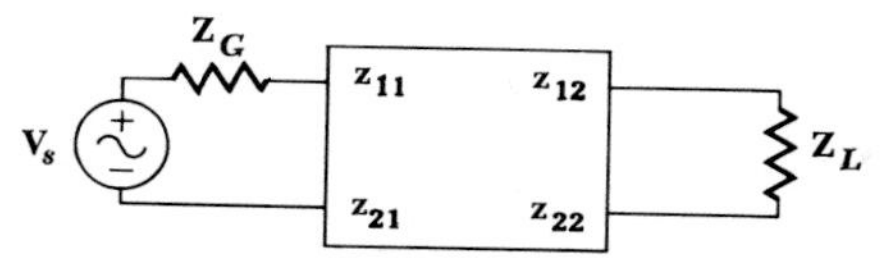

Fig. 16-28 See Prob. 18.

☐ **19** For the circuit shown in Fig. 16-29: (*a*) determine R_G so that the generator is matched ($R_G = R_{in}$) to the two port when $R_L = 4\ \Omega$; (*b*) determine R_L so that R_L is matched ($R_L = R_{th}$) to the two port when $R_G = 6\ \Omega$; (*c*) determine R_L and R_G so that matched conditions are achieved simultaneously at the input and the output.

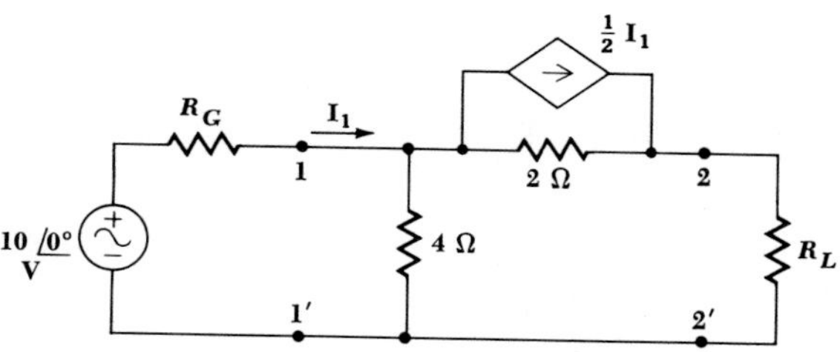

Fig. 16-29 See Prob. 19.

☐ **20** The most convenient measurements on a certain two-port device are the following: the open-circuit forward-voltage gain is −20; the short-circuit forward-current gain is 1000; the open-circuit input impedance is 10 kΩ; and the short-circuit input impedance is 8 kΩ. Find the four **z** parameters.

☐ **21** The **ABCD** parameters are defined by $\mathbf{V}_1 = \mathbf{AV}_2 - \mathbf{BI}_2$, $\mathbf{I}_1 = \mathbf{CV}_2 - \mathbf{DI}_2$. (*a*) Find the values of these parameters for the network shown in Fig. 16-21. (*b*) What condition is forced on these parameters when the network is reciprocal?

☐ **22** Determine the hybrid parameters for the circuit of: (*a*) Fig. 16-25*a*; (*b*) Fig. 16-25*b*.

☐ **23** Express each of the four hybrid parameters in terms of the admittance parameters.

☐ **24** Consider a transistor whose **h** parameters for common-base operation are $\mathbf{h}_{11} = 50\ \Omega$, $\mathbf{h}_{12} = 3 \times 10^{-3}$, $\mathbf{h}_{21} = -0.98$, and $\mathbf{h}_{22} = 4 \times 10^{-6}$ ℧. This transistor is to be used as a voltage amplifier between a 2-kΩ load and a 10-mV source having an internal resistance of 100 Ω. (*a*) Determine the load voltage. (*b*) The transistor is installed backwards; assuming that it does not burn out, find the load voltage.

☐ **25** A simplified linear equivalent circuit for a good transistor used at audio frequencies in the common-emitter arrangement is shown in Fig. 16-30. Two identical two ports are used in cascade (output of first connected directly to input of second). If the combination is terminated in $R_L = 10$ kΩ, and $\mathbf{h}_{11} = 10^3$ Ω, $\mathbf{h}_{21} = 10^2$, $\mathbf{h}_{22} = 10^{-5}$ ℧, find $\mathbf{V}_L/\mathbf{V}_1$.

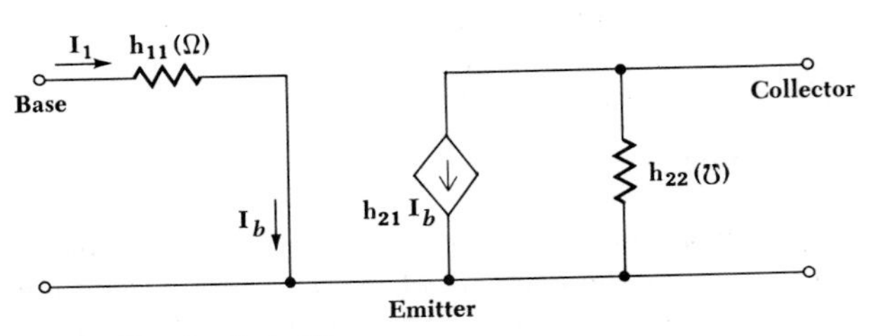

Fig. 16-30 See Prob. 25.

☐ **26** Find the four **h** parameters for the circuit shown in Fig. 16-31.

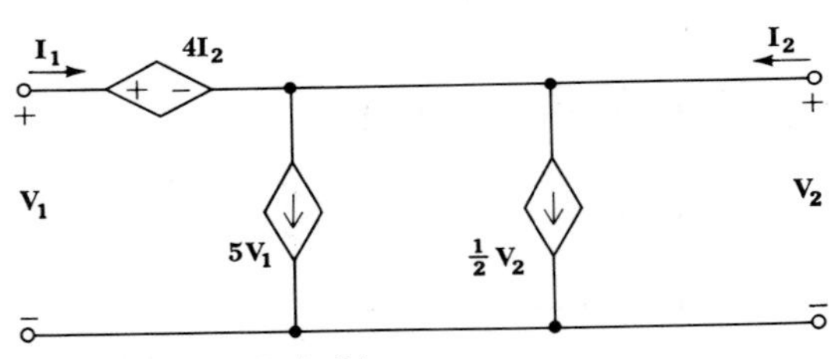

Fig. 16-31 See Prob. 26.

☐ **27** Find the **h** parameters for the parallel connection of two two ports, as illustrated in Fig. 16-32.

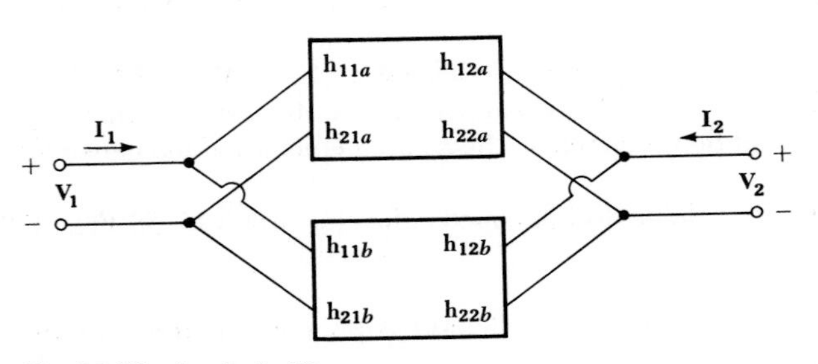

Fig. 16-32 See Prob. 27.

☐ **28** Scraps from an old lab notebook of Ben Franklin's apparently describe a network for which: "... transistor ... bias ... $\mathbf{z}_{21} = -160$ ... $\mathbf{h}_{12} = 0.2$ ... $\mathbf{y}_{22} = 1/200$ ... $\mathbf{z}_{12} = 2$. ..." Determine the open-circuit forward-voltage gain and the short-circuit reverse-current gain for this device.

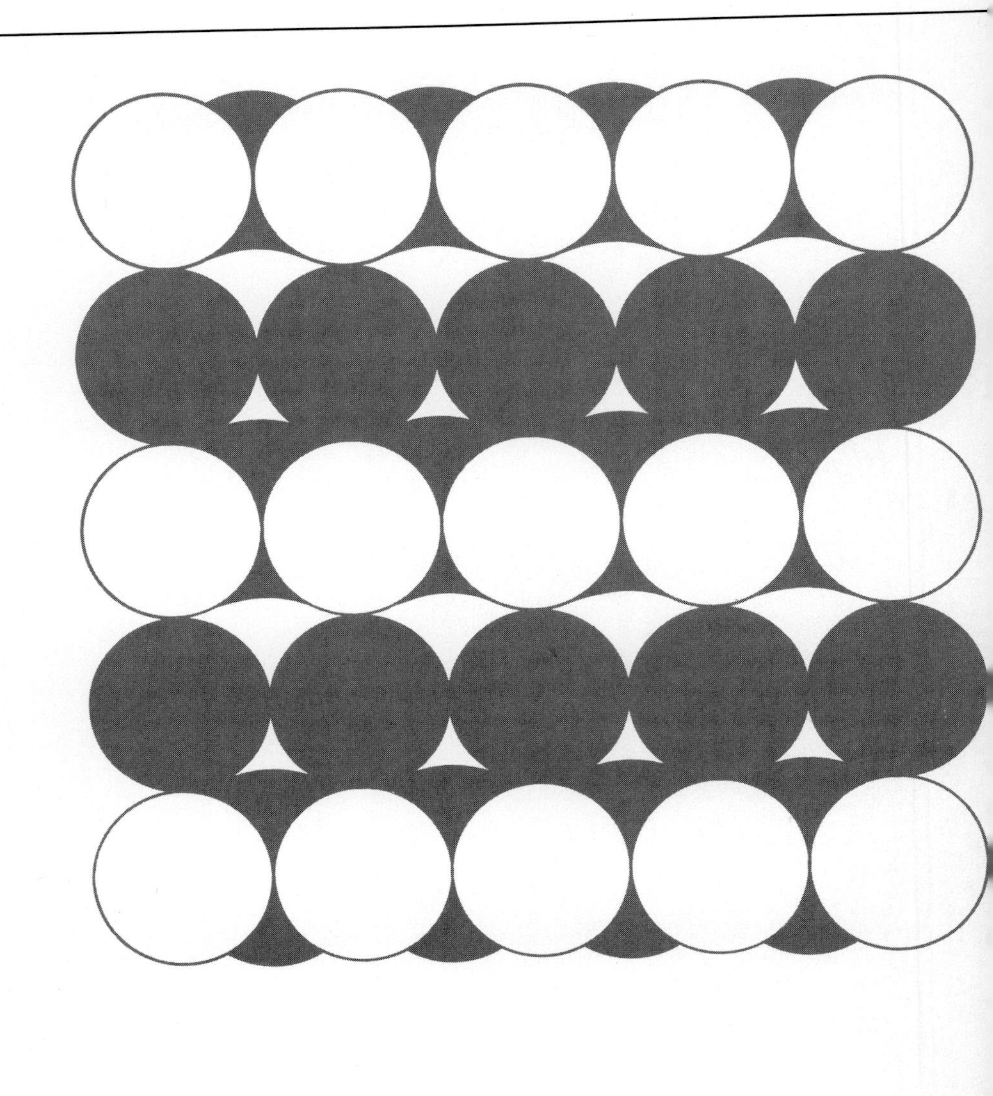

Part Six
NETWORK ANALYSIS

Chapter Seventeen
Fourier Analysis

17-1 INTRODUCTION

In this chapter we continue our introduction to circuit analysis by extending the frequency-domain concept to include forcing functions and responses which are not in general sinusoids, exponentials, or damped sinusoids. Our goal is a generalization of the process of determining the forced response.

The complete response of a linear circuit to an arbitrary forcing function is composed of the sum of a forced response and a natural response. The natural response was initially considered in Chaps. 5 to 7 but, with few exceptions, only simple series or parallel *RL*, *RC*, and *RLC* circuits were examined. However, the complex-frequency concept in Chap. 13 provided us with a general method of obtaining the natural response; we discovered that we could write the form of the natural response after inspecting the pole-zero constellation of an appropriate immittance or transfer function of the network. Thus, a powerful general method for determining the natural response became available.

Now let us consider our status with respect to the forced response. We are able to find the forced response in any purely resistive linear circuit, regardless of the nature of the forcing function, but this can hardly be classed as a scientific breakthrough. If the circuit includes energy-storage elements, then we can find the forced response only for those circuits and forcing functions to which we can apply the impedance concept; that is, the forcing function must be direct current, exponential, sinusoidal, or damped sinusoidal. This is the barrier which we seek to breech in this chapter.

We shall begin by considering forcing functions which are *periodic* and have functional natures which satisfy certain mathematical restrictions that are characteristic of any function which we can generate in the laboratory. Any such function may be represented as the sum of an infinite number of sine and cosine functions which are harmonically related. Therefore, since the forced response to each sinusoidal component may be determined easily by sinusoidal steady-state analysis, the response of the linear network to the general periodic forcing function may be obtained by superposing the partial responses.

Some feeling for the validity of representing a general periodic function by an infinite sum of sine and cosine functions may be gained by considering a simple example. Let us first assume a cosine function of radian frequency ω_0,

$$v_1(t) = 2 \cos \omega_0 t$$

where

$$\omega_0 = 2\pi f_0$$

and the period T is

$$T = \frac{1}{f_0} = \frac{2\pi}{\omega_0}$$

The *harmonics* of this sinusoid have frequencies $n\omega_0$, where ω_0 is the fundamental frequency and $n = 1, 2, 3, \ldots$. The frequency of the first harmonic is the fundamental frequency. Next let us select a third harmonic voltage

$$v_{3a}(t) = \cos 3\omega_0 t$$

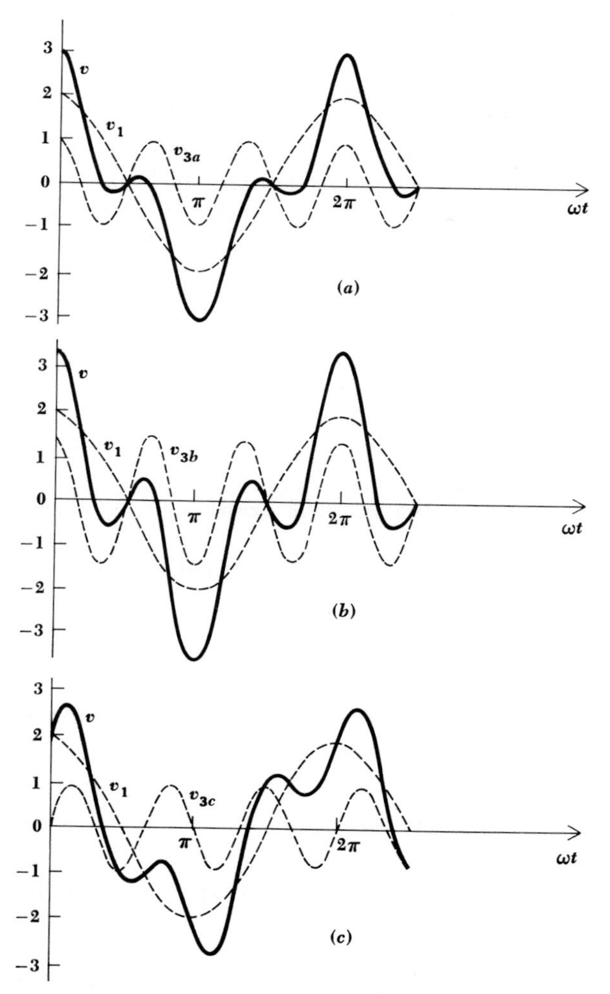

Fig. 17-1 Several of the infinite number of different waveforms which may be obtained by combining a fundamental and a third harmonic. The fundamental is $v_1 = 2 \cos \omega_0 t$ and the third harmonic is: (a) $v_{3a} = \cos 3\omega_0 t$; (b) $v_{3b} = 1.5 \cos 3\omega_0 t$; (c) $v_{3c} = \sin 3\omega_0 t$.

The fundamental $v_1(t)$, the third harmonic $v_{3a}(t)$, and the sum of these two waves are shown as functions of time in Fig. 17-1*a*. It should be noted that the sum is periodic with period $T = 2\pi/\omega_0$.

The form of the resultant periodic function changes as the phase and amplitude of the third harmonic component change. Thus, Fig. 17-1*b* shows the effect of combining $v_1(t)$ and a third harmonic of slightly larger amplitude,

$$v_{3b}(t) = 1.5 \cos 3\omega_0 t$$

By shifting the phase of the third harmonic,

$$v_{3c}(t) = \sin 3\omega_0 t$$

the sum, shown in Fig. 17-1*c*, takes on a still different character. In all cases, the period of the resultant waveform is the same as the period of the fundamental waveform. The nature of the waveform depends on the amplitude and phase of every possible harmonic component, and we shall find it possible to generate waveforms which have an extremely nonsinusoidal character by an appropriate combination of sinusoidal functions.

After we have become familiar with the use of an infinite sum of sine and cosine functions to represent a periodic waveform, we shall consider the frequency-domain representation of a general nonperiodic waveform in the next chapter.

17-2 TRIGONOMETRIC FORM OF THE FOURIER SERIES

We first consider a *periodic* function $f(t)$ defined in Sec. 11-3 by the functional relationship

$$f(t) = f(t + T)$$

where T is the period. We further assume that the function $f(t)$ satisfies the following properties:

1. $f(t)$ is single-valued everywhere.
2. The integral $\int_{t_0}^{t_0+T} |f(t)|\, dt$ exists (i.e., is not infinite) for any choice of t_0.
3. $f(t)$ has a finite number of discontinuities in any one period.
4. $f(t)$ has a finite number of maxima and minima in any one period.

We shall consider $f(t)$ to represent a voltage or current waveform, and any voltage or current waveform which we can actually produce must satisfy these conditions. Certain mathematical functions which we might hypothesize may not satisfy these conditions, but we shall assume that the four conditions listed above are always satisfied.

Given such a periodic function $f(t)$, the Fourier theorem[1] states that $f(t)$ may be represented by the infinite series

[1]Jean-Baptiste Joseph Fourier published this theorem in 1822. Some rather unbelievable pronunciations of this French name come from American students; it should rhyme with "poor today."

$$f(t) = a_0 + a_1 \cos \omega_0 t + a_2 \cos 2\omega_0 t + \cdots + b_1 \sin \omega_0 t + b_2 \sin 2\omega_0 t + \cdots$$

$$= a_0 + \sum_{n=1}^{\infty} (a_n \cos n\omega_0 t + b_n \sin n\omega_0 t) \qquad (1)$$

where the fundamental frequency ω_0 is related to the period T by

$$\omega_0 = \frac{2\pi}{T}$$

and where a_0 and the a_n and b_n are constants which depend upon n and $f(t)$. Equation (1) is the trigonometric form of the *Fourier series for* $f(t)$, and the process of determining the values of the constants a_0, a_n, and b_n is called *Fourier analysis*. Our object is not the proof of this theorem, but only a simple development of the procedures of Fourier analysis and a feeling that the theorem is plausible.

Before we discuss the evaluation of the constants appearing in the Fourier series, let us collect a set of useful trigonometric integrals. We shall let both n and k represent any of the set of integers 1, 2, 3, ... In the following integrals, we shall use 0 and T as the integration limits, but it is understood that any interval of one period is equally correct. Since the average value of a sinusoid over one period is zero,

$$\int_0^T \sin n\omega_0 t \, dt = 0 \qquad (2)$$

and

$$\int_0^T \cos n\omega_0 t \, dt = 0 \qquad (3)$$

It is also a simple matter to show that the following three definite integrals are zero:

$$\int_0^T \sin k\omega_0 t \cos n\omega_0 t \, dt = 0 \qquad (4)$$

$$\int_0^T \sin k\omega_0 t \sin n\omega_0 t \, dt = 0 \qquad k \neq n \qquad (5)$$

$$\int_0^T \cos k\omega_0 t \cos n\omega_0 t \, dt = 0 \qquad k \neq n \qquad (6)$$

Those cases which are excepted in (5) and (6) are also easily evaluated; we obtain

$$\int_0^T \sin^2 n\omega_0 t \, dt = \frac{T}{2} \qquad (7)$$

$$\int_0^T \cos^2 n\omega_0 t \, dt = \frac{T}{2} \qquad (8)$$

The evaluation of the unknown constants may now be accomplished readily. We first attack a_0. If we integrate each side of the Fourier series (1) over a full period, we obtain

$$\int_0^T f(t)\,dt = \int_0^T a_0\,dt + \int_0^T \sum_{n=1}^{\infty} (a_n \cos n\omega_0 t + b_n \sin n\omega_0 t)\,dt$$

But every term in the summation is of the form of (2) or (3), and thus

$$\int_0^T f(t)\,dt = a_0 T$$

or

$$a_0 = \frac{1}{T}\int_0^T f(t)\,dt \tag{9}$$

This constant a_0 is simply the average value of $f(t)$ over a period, and we therefore describe it as the dc component of $f(t)$.

To evaluate one of the cosine coefficients, say a_k, the coefficient of $\cos k\omega_0 t$, we first multiply each side of (1) by $\cos k\omega_0 t$ and then integrate both sides of the equation over a full period:

$$\int_0^T f(t) \cos k\omega_0 t\,dt = \int_0^T a_0 \cos k\omega_0 t\,dt + \int_0^T \sum_{n=1}^{\infty} a_n \cos k\omega_0 t \cos n\omega_0 t\,dt$$
$$+ \int_0^T \sum_{n=1}^{\infty} b_n \cos k\omega_0 t \sin n\omega_0 t\,dt$$

From (3), (4), and (6) we note that every term on the right side of the equation with one exception is zero. That term is evaluated by (8) and we obtain

$$a_k = \frac{2}{T}\int_0^T f(t) \cos k\omega_0 t\,dt \tag{10}$$

This result is twice the average value of the product $f(t) \cos k\omega_0 t$ over a period.

In a similar way, we obtain b_k by multiplying by $\sin k\omega_0 t$, integrating over a period, noting that all but one of the terms on the right side are zero, and performing that single integration by (7). The result is

$$b_k = \frac{2}{T}\int_0^T f(t) \sin k\omega_0 t\,dt \tag{11}$$

which is twice the average value of $f(t) \sin k\omega_0 t$ over a period.

Equations (9) to (11) now enable us to determine values for a_0 and all the a_n and b_n in the Fourier series (1).

Let us consider a numerical example. The "half-sinusoidal" waveform shown in Fig. 17-2*a* represents the voltage response obtained at the output of a half-wave rectifier circuit, a nonlinear circuit whose function is to convert a sinusoidal input voltage to a (pulsating) dc output voltage. In order to represent this voltage as a Fourier series, we must first determine the period and then express the graphical voltage as an analytical function of time. From the graph, the period is seen to be

$$T = 0.4 \text{ s}$$

and thus

$$f_0 = 2.5 \text{ Hz}$$

and

$$\omega_0 = 5\pi \text{ rad/s}$$

With these three quantities determined, we now seek an appropriate expression for $f(t)$ or $v(t)$ which is valid throughout the period. Obtaining this equation or set of equations proves to be the most difficult part of Fourier analysis for many students. The source of the difficulty is apparently either the inability to recognize the given curve, carelessness in determining multiplying constants within the functional expression, or

Fig. 17-2 (*a*) The output of a half-wave rectifier to which a sinusoidal input is applied. (*b*) The discrete line spectrum of the waveform in (*a*).

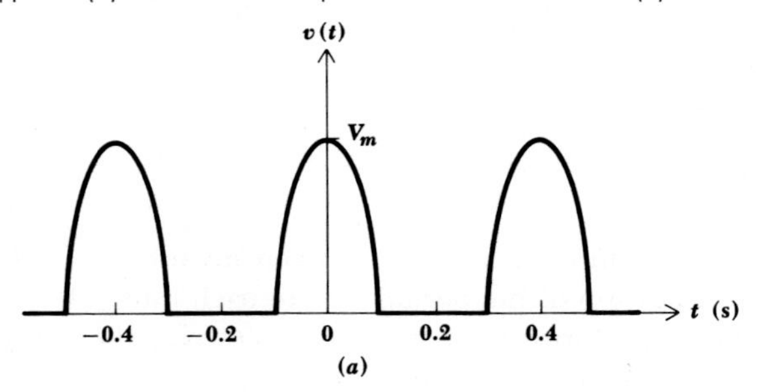

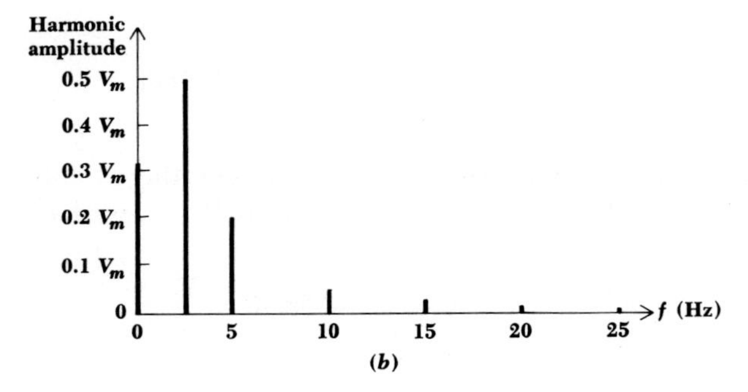

negligence in not writing the complete expression. In this example, the statement of the problem infers that the functional form is a sinusoid; the amplitude is V_m, the radian frequency has already been determined as 5π, and only the positive portion of the cosine wave is present. The functional expression for the period $t = 0$ to $t = 0.4$ is therefore

$$v(t) = \begin{cases} V_m \cos 5\pi t & 0 \le t \le 0.1 \\ 0 & 0.1 \le t \le 0.3 \\ V_m \cos 5\pi t & 0.3 \le t \le 0.4 \end{cases}$$

It is evident that the choice of the period extending from $t = -0.1$ to $t = 0.3$ will result in fewer equations and, hence, fewer integrals:

$$v(t) = \begin{cases} V_m \cos 5\pi t & -0.1 \le t \le 0.1 \\ 0 & 0.1 \le t \le 0.3 \end{cases} \tag{12}$$

This form is preferable, although either description will yield the correct results.

The zero-frequency component is easily obtained:

$$a_0 = \frac{1}{0.4} \int_{-0.1}^{0.3} v(t)\,dt$$

$$= \frac{1}{0.4} \left[\int_{-0.1}^{0.1} V_m \cos 5\pi t\,dt + \int_{0.1}^{0.3} (0)\,dt \right]$$

and

$$a_0 = \frac{V_m}{\pi} \tag{13}$$

Notice that the integral must be broken up into a number of integrals, each over a portion of the period, where each integral corresponds to one of the functional forms used to express $v(t)$ over the complete period.

The amplitude of a general cosine term is

$$a_k = \frac{2}{0.4} \int_{-0.1}^{0.1} V_m \cos 5\pi t \cos 5\pi k t\,dt$$

The form of the function we obtain upon integrating is different when k is unity than it is for any other choice of k. If $k = 1$, we have

$$a_1 = 5V_m \int_{-0.1}^{0.1} \cos^2 5\pi t\,dt = \frac{V_m}{2} \tag{14}$$

whereas if k is not equal to unity, we find

$$a_k = 5V_m \int_{-0.1}^{0.1} \cos 5\pi t \cos 5\pi kt \, dt$$

$$= 5V_m \int_{-0.1}^{0.1} \tfrac{1}{2}[\cos 5\pi(1 + k)t + \cos 5\pi(1 - k)t] \, dt$$

or

$$a_k = \frac{2V_m}{\pi} \frac{\cos(\pi k/2)}{1 - k^2} \qquad k \neq 1 \tag{15}$$

Some of the details of the integration have been left out for those who prefer to work out the small tedious steps for themselves. It should be pointed out, incidentally, that the expression for a_k when $k \neq 1$ will yield the correct result for $k = 1$ in the limit.

A similar integration shows that all the b_k are zero, and the Fourier series thus contains no sine terms. The Fourier series is therefore obtained from (13) to (15):

$$v(t) = \frac{V_m}{\pi} + \frac{V_m}{2} \cos 5\pi t + \frac{2V_m}{3\pi} \cos 10\pi t - \frac{2V_m}{15\pi} \cos 20\pi t + \frac{2V_m}{35\pi} \cos 30\pi t - \cdots \tag{16}$$

In Fig. 17-2*a*, $v(t)$ is shown as a function of time; in (12), $v(t)$ is expressed as an analytical function of time. Either of these representations is a time-domain representation. Equation (16), the Fourier series representation of $v(t)$, is also a time-domain expression, but it may be transformed easily into a frequency-domain representation. For example, we could locate the points in the **s** plane that represent the frequencies present in (16). The result would be a mark at the origin and symmetrical marks on the positive and negative $j\omega$ axis. A more customary method of presenting this information, and one which shows the amplitude of each frequency component, is by a *line spectrum*. A line spectrum for (16) is shown in Fig. 17-2*b*; the amplitude of each frequency component is indicated by the length of the vertical line located at the corresponding frequency. We also speak of this spectrum as a *discrete* spectrum because any finite frequency interval contains only a finite number of frequency components.

One note of caution must be injected. The example we have considered contains no sine terms, and the amplitude of the kth harmonic is therefore $|a_k|$. If b_k is not zero, then the amplitude of the component at a frequency $k\omega_0$ must be $\sqrt{a_k^2 + b_k^2}$. This is the general quantity which we must show in a line spectrum. When we discuss the complex form of the Fourier series, we shall see that this quantity is obtained more directly.

The Fourier series obtained for this example includes no sine terms and no odd harmonics (except the fundamental) among the cosine terms. It is possible to anticipate the absence of certain terms in a Fourier series before any integrations are performed, by an inspection of the symmetry of the given time function. We shall investigate the use of symmetry in the following section.

Drill Problems

17-1 In the Fourier series for $f(t)$, determine a_5, b_5, and $\sqrt{a_5{}^2 + b_5{}^2}$ if $f(t) =$: (*a*) 2, $0 < t < 0.1$; -1, $0.1 < t < 0.2$; $T = 0.2$; (*b*) -1, $0 < t < 0.1$; 2, $0.1 < t < 0.2$; $T = 0.2$; (*c*) 2, $-0.05 < t < 0.05$; -1, $0.05 < t < 0.15$; $T = 0.2$

Ans. 0, -0.382, 0.382; 0, 0.382, 0.382; 0.382, 0, 0.382

Fig. 17-3 See Drill Probs. 17-2 and 17-5.

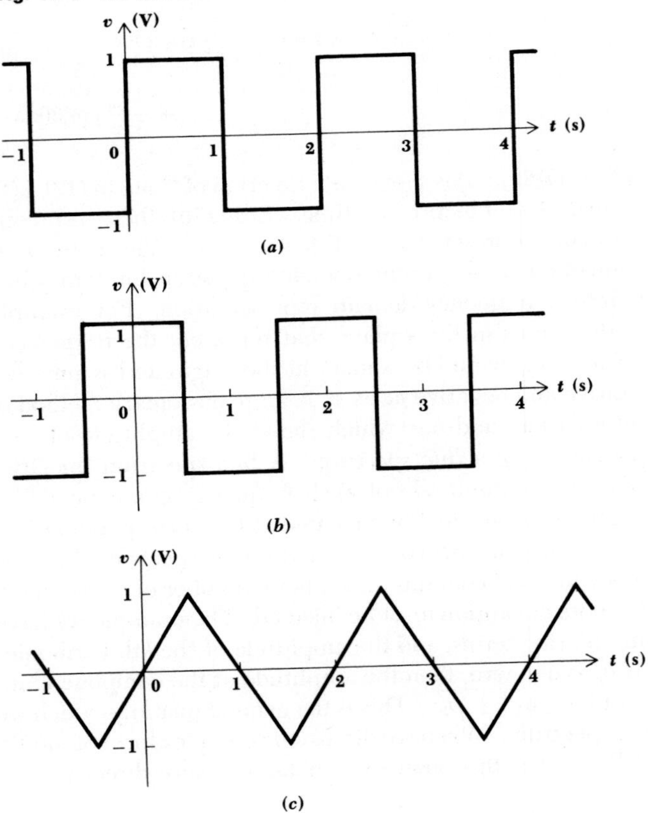

17-2 Write the Fourier series for the three voltage waveforms shown in Fig. 17-3.

Ans. $\frac{4}{\pi}(\sin \pi t + \frac{1}{3}\sin 3\pi t + \frac{1}{5}\sin 5\pi t + \cdots)$;
$\frac{4}{\pi}(\cos \pi t - \frac{1}{3}\cos 3\pi t + \frac{1}{5}\cos 5\pi t - \cdots)$;
$\frac{8}{\pi^2}(\sin \pi t - \frac{1}{9}\sin 3\pi t + \frac{1}{25}\sin 5\pi t - \cdots)$ V

17-3 THE USE OF SYMMETRY

The two types of symmetry which are most readily recognized are *even-function symmetry* and *odd-function symmetry,* or simply *even symmetry* and *odd symmetry.* We say that $f(t)$ possesses the property of even symmetry if

$$f(t) = f(-t) \tag{17}$$

Such functions as t^2, $\cos 3t$, $\ln(\cos t)$, $\sin^2 7t$, and a constant C all possess even symmetry; the replacement of t by $(-t)$ does not change the value of any of these functions. This type of symmetry may also be recognized graphically, for if $f(t) = f(-t)$ then mirror symmetry exists about the $f(t)$ axis. The function shown in Fig. 17-4*a* possesses even symmetry; if the figure were to be folded along the $f(t)$ axis, then the portions of the graph

Fig. 17-4 (*a*) A waveform showing even symmetry. (*b*) A waveform showing odd symmetry.

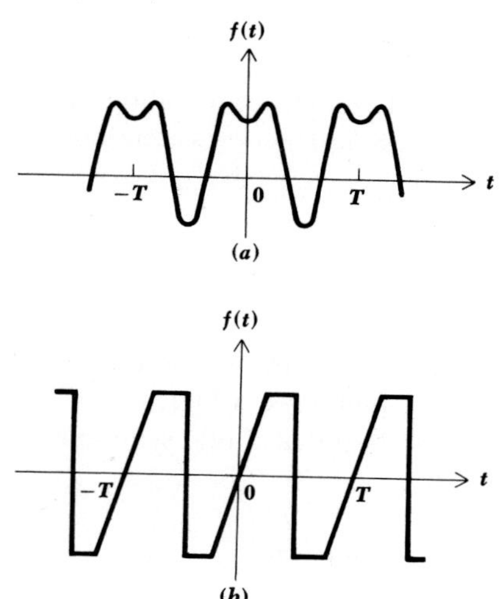

of the function for positive and negative time would fit exactly, one on top of the other.

Now let us investigate the effects that even symmetry produces in a Fourier series. If we think of the expression which equates an even function $f(t)$ and an infinite sum of sine and cosine functions, then it is apparent that the infinite sum must also be an even function. A sine wave, however, is an odd function, and no sum of sine waves can produce any even function other than zero (which is both even and odd). It is thus plausible that the Fourier series of any even function is composed of only a constant and cosine functions. Let us now show carefully that $b_k = 0$. We have

$$b_k = \frac{2}{T}\int_{-T/2}^{T/2} f(t) \sin k\omega_0 t \, dt$$

$$= \frac{2}{T}\left[\int_{-T/2}^{0} f(t) \sin k\omega_0 t \, dt + \int_{0}^{T/2} f(t) \sin k\omega_0 t \, dt\right]$$

Now let us replace the variable t in the first integral by $-\tau$, or $\tau = -t$, and make use of the fact that $f(t) = f(-t) = f(\tau)$:

$$b_k = \frac{2}{T}\left[-\int_{T/2}^{0} f(-\tau) \sin(-k\omega_0 \tau) \, d\tau + \int_{0}^{T/2} f(t) \sin k\omega_0 t \, dt\right]$$

$$= \frac{2}{T}\left[-\int_{0}^{T/2} f(\tau) \sin k\omega_0 \tau \, d\tau + \int_{0}^{T/2} f(t) \sin k\omega_0 t \, dt\right]$$

But the symbol we use to identify the variable of integration cannot affect the value of the integral. Hence,

$$b_k = 0 \tag{18}$$

Thus, no sine terms are present. A similar examination of the expression for a_k leads to an integral over the *half period* extending from $t = 0$ to $t = \frac{1}{2}T$:

$$a_k = \frac{4}{T}\int_{0}^{T/2} f(t) \cos k\omega_0 t \, dt \tag{19}$$

The fact that a_k may be obtained for an even function by taking "twice the integral over half the range" should seem logical.

We define odd symmetry by stating that if odd symmetry is a property of $f(t)$, then

$$f(t) = -f(-t) \tag{20}$$

In other words, if t is replaced by $(-t)$, then the negative of the given function is obtained; for example, t, $\sin t$, $t\cos 70t$, $t\sqrt{1+t^2}$, and the function sketched in Fig. 17-4*b* are all odd functions and possess odd symmetry. The graphical characteristics of odd symmetry are apparent if the portion of $f(t)$ for $t > 0$ is rotated about the positive t axis and the resultant figure is then rotated about the $f(t)$ axis; the two curves will fit exactly, one on top of the other. That is, we now have symmetry about the origin, rather than about the $f(t)$ axis as we did for even functions.

A function having odd symmetry can contain no constant term or cosine terms in its Fourier expansion. Let us prove the second part of this statement. We have

$$a_k = \frac{2}{T}\int_{-T/2}^{T/2} f(t)\cos k\omega_0 t\,dt$$

$$= \frac{2}{T}\left[\int_{-T/2}^{0} f(t)\cos k\omega_0 t\,dt + \int_{0}^{T/2} f(t)\cos k\omega_0 t\,dt\right]$$

and we now let $t = -\tau$ in the first integral,

$$a_k = \frac{2}{T}\left[-\int_{T/2}^{0} f(-\tau)\cos(-k\omega_0\tau)\,d\tau + \int_{0}^{T/2} f(t)\cos k\omega_0 t\,dt\right]$$

$$= \frac{2}{T}\left[\int_{0}^{T/2} f(-\tau)\cos k\omega_0\tau\,d\tau + \int_{0}^{T/2} f(t)\cos k\omega_0 t\,dt\right]$$

But $f(-\tau) = -f(\tau)$, and therefore

$$a_k = 0 \tag{21}$$

A similar, but simpler, proof shows that

$$a_0 = 0$$

Furthermore, the values for b_k may again be obtained by integrating over half the range:

$$b_k = \frac{4}{T}\int_{0}^{T/2} f(t)\sin k\omega_0 t\,dt \tag{22}$$

Examples of even and odd symmetry are afforded by Drill Prob. 17-2, preceding this section. In both parts (*a*) and (*b*), a square wave of the same amplitude and period is the given function. The time origin, however, is selected to provide odd symmetry in (*a*) and even symmetry in (*b*), and the resultant series contain, respectively, only sine terms and cosine terms.

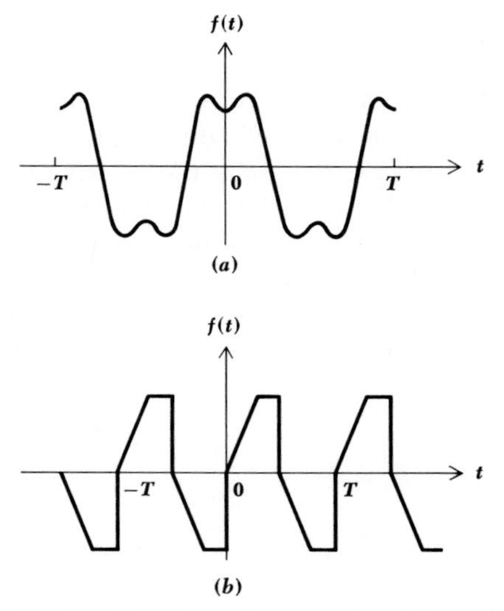

Fig. 17-5 (*a*) A waveform somewhat similar to the one shown in Fig. 17-4*a*, but possessing half-wave symmetry. (*b*) A waveform somewhat similar to the one shown in Fig. 17-4*b*, but possessing half-wave symmetry.

It is also worthwhile pointing out that the point at which $t = 0$ could be selected to provide neither even nor odd symmetry; the determination of the coefficients of the terms in the Fourier series then takes twice as long.

The Fourier series for either of these square waves also have one other interesting characteristic; neither contains any even *harmonics*.[2] That is, the only frequency components present in the series have frequencies which are odd multiples of the fundamental frequency; a_k and b_k are zero for k even. This result is caused by another type of symmetry, called half-wave symmetry. We shall say that $f(t)$ possesses *half-wave symmetry* if

$$f(t) = -f(t - \tfrac{1}{2}T)$$

Except for a change of sign, each half cycle is like the adjacent half cycles. Half-wave symmetry, unlike even and odd symmetry, is not a function of the choice of the point $t = 0$. Thus, we can state that the square wave (Fig. 17-3*a* or *b*) shows half-wave symmetry. Neither waveform shown in Fig. 17-4 has half-wave symmetry, but the two somewhat similar functions plotted in Fig. 17-5 do possess half-wave symmetry.

It may be shown that the Fourier series of any function which has

[2]Constant vigilance is required to avoid confusion between an even function and an even harmonic, or an odd function and an odd harmonic. For example, b_{10} is the coefficient of an even harmonic, and it is zero if $f(t)$ is an even function.

half-wave symmetry contains only odd harmonics. Let us consider the coefficients a_k. We have again

$$a_k = \frac{2}{T}\int_{-T/2}^{T/2} f(t) \cos k\omega_0 t \, dt$$
$$= \frac{2}{T}\left[\int_{-T/2}^{0} f(t) \cos k\omega_0 t \, dt + \int_{0}^{T/2} f(t) \cos k\omega_0 t \, dt\right]$$

which we may represent as

$$a_k = \frac{2}{T}(I_1 + I_2)$$

Now we substitute the new variable $\tau = t + \frac{1}{2}T$ into integral I_1:

$$I_1 = \int_0^{T/2} f(\tau - \tfrac{1}{2}T) \cos k\omega_0(\tau - \tfrac{1}{2}T) \, d\tau$$
$$= \int_0^{T/2} -f(\tau)\left(\cos k\omega_0\tau \cos \frac{k\omega_0 T}{2} + \sin k\omega_0\tau \sin \frac{k\omega_0 T}{2}\right) d\tau$$

But $\omega_0 T$ is 2π, and thus

$$\sin \frac{k\omega_0 T}{2} = \sin k\pi = 0$$

Hence

$$I_1 = -\cos k\pi \int_0^{T/2} f(\tau) \cos k\omega_0\tau \, d\tau$$

After noting the form of I_2, we therefore may write

$$a_k = \frac{2}{T}(1 - \cos k\pi) \int_0^{T/2} f(t) \cos k\omega_0 t \, dt$$

The factor $(1 - \cos k\pi)$ indicates that a_k is zero if k is even. When k is odd, we have

$$a_k = \frac{4}{T}\int_0^{T/2} f(t) \cos k\omega_0 t \, dt \qquad k \text{ odd} \tag{23}$$

A similar investigation shows that b_k is also zero for all even k, and when k is odd

$$b_k = \frac{4}{T}\int_0^{T/2} f(t) \sin k\omega_0 t \, dt \qquad k \text{ odd} \tag{24}$$

It should be noted that half-wave symmetry may be present in a waveform which also shows odd symmetry or even symmetry. The waveform sketched in Fig. 17-5a, for example, possesses both even symmetry and half-wave symmetry. Its Fourier series therefore contains only odd-harmonic cosine functions; the values of a_k may be determined from (23) [once the functional form for $f(t)$ is known]. It is always worthwhile spending a few moments investigating the symmetry of a function for which a Fourier series is to be determined.

Drill Problems

17-3 Sketch each of the functions described below and decide whether or not even symmetry, odd symmetry, and half-wave symmetry are present: (a) $v = 10$, $0 < t < 1; v = 20, 1 < t < 2; v = -10, 2 < t < 3; v = -20, 3 < t < 4$; repeats; ($b$) $v = 5t$, $-2 < t < 2$; repeats; (c) $v = 10 \cos 0.25\pi t$, $-2 < t < 2$; repeats.

Ans. Yes, no, no; no, yes, no; no, no, yes

17-4 Determine the Fourier series for each of the functions described in Drill Prob. 3.

$$\textit{Ans.}\ \frac{20}{\pi} - \frac{40}{\pi}\sum_{n=1}^{\infty}\frac{(-1)^n}{4n^2-1}\cos\frac{n\pi t}{2};\quad -\frac{20}{\pi}\sum_{n=1}^{\infty}\frac{1}{n}\cos n\pi\sin\frac{n\pi t}{2};$$

$$\sum_{\substack{n=1\\ \text{odd}}}^{\infty}\left[\frac{20}{n\pi}(-1)^{(n+1)/2}\cos\frac{n\pi t}{2} + \frac{60}{n\pi}\sin\frac{n\pi t}{2}\right]$$

17-4 COMPLETE RESPONSE TO PERIODIC FORCING FUNCTIONS

Through the use of the Fourier series, we may now express an arbitrary periodic forcing function as an infinite sum of sinusoidal forcing functions; the forced response to each of these functions may be determined by conventional steady-state analysis; the form of the natural response may be determined from the pole-zero constellation of the appropriate network function; the initial conditions existing throughout the network, including the initial value of the forced response, enable the amplitude of the natural response to be selected; and, finally, the complete response is obtained as the sum of the forced and natural responses. Let us illustrate this general procedure by a specific example.

We shall apply the square wave of Fig. 17-6a, which includes a dc component, to the series RL circuit, as shown in Fig. 17-6b. The forcing function is applied at $t = 0$, and the current is the desired response. Its

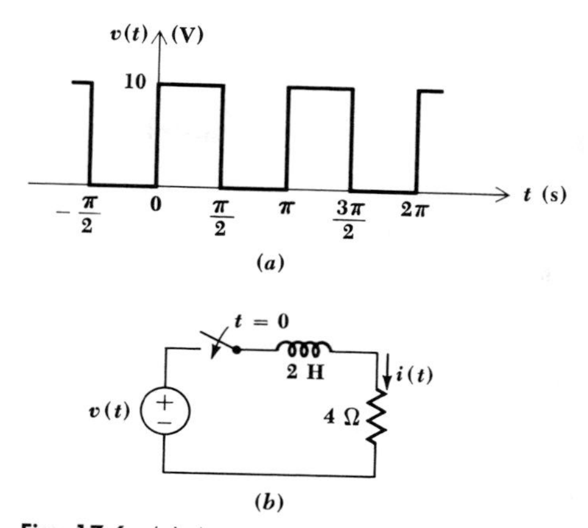

Fig. 17-6 (*a*) A square-wave voltage forcing function. (*b*) The forcing function of (*a*) is applied to this series RL circuit at $t = 0$; the complete response $i(t)$ is desired.

initial value is zero. The forcing function has a fundamental frequency $\omega_0 = 2$, and its Fourier series may be written down by comparison with the solution of Drill Prob. 17-2*a*:

$$v(t) = 5 + \frac{20}{\pi} \sum_{\substack{n=1 \\ \text{odd}}}^{\infty} \frac{\sin 2nt}{n}$$

We shall find the forced response of the kth harmonic by working in the frequency domain. Thus,

$$v_k(t) = \frac{20}{\pi k} \sin 2kt$$

and

$$\mathbf{V}_k = \frac{20}{\pi k}(-j1)$$

The impedance offered by the RL circuit at this frequency is

$$\mathbf{Z}_k = 4 + j(2k)2 = 4 + j4k$$

and thus the component of the forced response at this frequency is

$$\mathbf{I}_k = \frac{\mathbf{V}_k}{\mathbf{Z}_k} = \frac{-j5}{\pi k(1 + jk)}$$

Transforming to the time domain, we have

$$i_k(t) = \frac{5}{\pi k}\frac{1}{\sqrt{1+k^2}}\cos\,(2kt - 90^\circ - \tan^{-1} k)$$

$$= \frac{5}{\pi(1+k^2)}\left(\frac{\sin 2kt}{k} - \cos 2kt\right)$$

Since the response to the dc component is obviously 1.25 A, the forced response may be expressed as the summation

$$i_f(t) = 1.25 + \frac{5}{\pi}\sum_{\substack{k=1\\ \text{odd}}}^{\infty}\left[\frac{\sin 2kt}{k(1+k^2)} - \frac{\cos 2kt}{1+k^2}\right]$$

The familiar natural response of this simple circuit is the single exponential term (characterizing the single zero of the input impedance)

$$i_n(t) = Ae^{-2t}$$

The complete response is therefore the sum

$$i(t) = i_f(t) + i_n(t)$$

and, since $i(0) = 0$, it is necessary to select A so that

$$A = -i_f(0)$$

Letting $t = 0$, we find that $i_f(0)$ is given by

$$i_f(0) = 1.25 - \frac{5}{\pi}\sum_{\substack{k=1\\ \text{odd}}}^{\infty}\frac{1}{1+k^2}$$

Although we could express A in terms of this summation, it is more convenient to use the numerical value of the summation. The sum of the first 5 terms of $\Sigma 1/(1+k^2)$ is 0.671, the sum of the first 10 terms is 0.695, the sum of the first 20 terms is 0.708, and the exact sum[3] is 0.720 to three significant figures. Thus

$$A = -1.25 + \frac{5}{\pi}0.720 = -0.104$$

and

$$i(t) = -0.104e^{-2t} + 1.25 + \frac{5}{\pi}\sum_{\substack{k=1\\ \text{odd}}}^{\infty}\left[\frac{\sin 2kt}{k(1+k^2)} - \frac{\cos 2kt}{1+k^2}\right]$$

[3]The sum of this series is known in closed form: $\displaystyle\sum_{\substack{k=1\\ \text{odd}}}^{\infty}\frac{1}{1+k^2} = \frac{\pi}{4}\tanh\frac{\pi}{2}$.

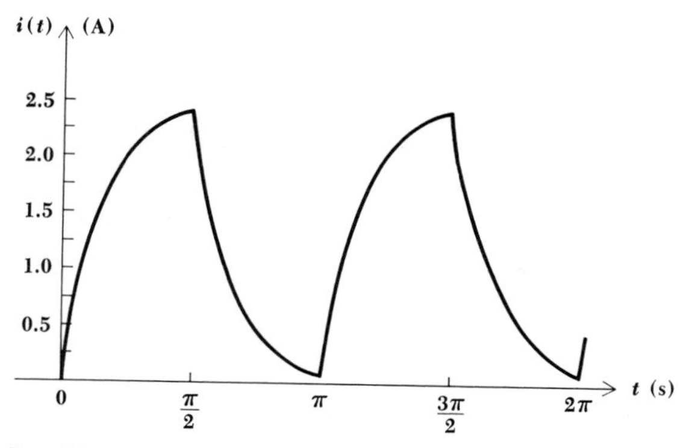

Fig. 17-7 The initial portion of the complete response of the circuit of Fig. 17-6*b* to the forcing function of Fig. 17-6*a*.

In obtaining this solution, we have had to use many of the most general concepts introduced in this and the preceding 16 chapters. Some we did not have to use because of the simple nature of this particular circuit, but their places in the general analysis were indicated above. In this sense, we may look upon the solution of this problem as a significant achievement in our introductory study of circuit analysis. In spite of this glorious feeling of accomplishment, however, it must be pointed out that the complete response, as obtained above in analytical form, is not of much value as it stands; it furnishes no clear picture of the nature of the response. What we really need is a sketch of $i(t)$ as a function of time. This may be obtained by a laborious calculation at a sufficient number of instants of time; an available digital computer can be of great assistance here. It may be approximated by the graphical addition of the natural response, the dc term, and the first few harmonics; this is an unrewarding task. When all is said and done, the most informative solution of this problem is probably obtained by making a repeated transient analysis. That is, the form of the response can certainly be calculated in the interval from $t = 0$ to $t = \pi/2$ s; it is an exponential rising toward 2.5 A. After determining the value at the end of this first interval, we have an initial condition for the next $\frac{1}{2}\pi$-s interval. The process is repeated until the response assumes a generally periodic nature. The method is eminently suitable to this example, for there is negligible change in the current waveform in the successive periods $\pi/2 < t < 3\pi/2$ and $3\pi/2 < t < 5\pi/2$. The complete current response is sketched in Fig. 17-7.

17-5 COMPLEX FORM OF THE FOURIER SERIES

In obtaining a frequency spectrum, we have seen that the amplitude of each frequency component depends on both a_k and b_k; that is, the sine term and the cosine term both contribute to the amplitude. The exact expression for this amplitude is $\sqrt{a_k^2 + b_k^2}$. It is possible to obtain this amplitude directly by using a form of Fourier series in which each term is a cosine function with a phase angle; the amplitude and phase angle are functions of $f(t)$ and k. An even more convenient and concise form of the Fourier series is obtained if the sines and cosines are expressed as exponential functions with complex multiplying constants.

Let us first take the trigonometric form of the Fourier series:

$$f(t) = a_0 + \sum_{n=1}^{\infty} (a_n \cos n\omega_0 t + b_n \sin n\omega_0 t)$$

and then substitute the exponential forms for the sine and cosine. After rearranging,

$$f(t) = a_0 + \sum_{n=1}^{\infty} \left(e^{jn\omega_0 t} \frac{a_n - jb_n}{2} + e^{-jn\omega_0 t} \frac{a_n + jb_n}{2} \right)$$

We now define a complex constant $\mathbf{c}_n$:

$$\mathbf{c}_n = \tfrac{1}{2}(a_n - jb_n) \tag{25}$$

The values of a_n, b_n, and $\mathbf{c}_n$ all depend on n and $f(t)$. Suppose we now replace n by $(-n)$; how do the values of the constants change? The coefficients a_n and b_n are defined by (10) and (11), and it is evident that

$$a_{-n} = a_n$$

but

$$b_{-n} = -b_n$$

From (25), then,

$$\mathbf{c}_{-n} = \tfrac{1}{2}(a_n + jb_n) \tag{26}$$

Also

$$\mathbf{c}_0 = a_0$$

We may therefore express $f(t)$ as

$$f(t) = \mathbf{c}_0 + \sum_{n=1}^{\infty} \mathbf{c}_n e^{jn\omega_0 t} + \sum_{n=1}^{\infty} \mathbf{c}_{-n} e^{-jn\omega_0 t}$$

or

$$f(t) = \sum_{n=0}^{\infty} \mathbf{c}_n e^{jn\omega_0 t} + \sum_{n=1}^{\infty} \mathbf{c}_{-n} e^{-jn\omega_0 t}$$

Finally, instead of summing the second series over the positive integers from 1 to ∞, let us sum over the negative integers, from -1 to $-\infty$:

$$f(t) = \sum_{n=0}^{\infty} \mathbf{c}_n e^{jn\omega_0 t} + \sum_{n=-1}^{-\infty} \mathbf{c}_n e^{jn\omega_0 t}$$

or

$$f(t) = \sum_{n=-\infty}^{n=\infty} \mathbf{c}_n e^{jn\omega_0 t} \tag{27}$$

Equation (27) is the complex form of the Fourier series for $f(t)$; its conciseness is one of the most important reasons for its use. In order to obtain the expression by which a particular complex coefficient $\mathbf{c}_k$ may be evaluated, we substitute (10) and (11) into (25):

$$\mathbf{c}_k = \frac{1}{T}\int_{-T/2}^{T/2} f(t)\cos k\omega_0 t\,dt - j\frac{1}{T}\int_{-T/2}^{T/2} f(t)\sin k\omega_0 t\,dt$$

use the exponential equivalents of the sine and cosine, and simplify:

$$\mathbf{c}_k = \frac{1}{T}\int_{-T/2}^{T/2} f(t)e^{-jk\omega_0 t}\,dt \tag{28}$$

Thus, a single concise equation serves to replace the two equations required for the trigonometric form of the Fourier series. Instead of evaluating two integrals to find the Fourier coefficients, only one integration is required; moreover, it is almost always a simpler integration. It should be noted that the integral (28) contains the multiplying factor $1/T$, whereas the integrals for a_n and b_n both contain the factor $2/T$.

Since the amplitude of the Fourier component at a frequency $k\omega_0$ is $\sqrt{a_k^2 + b_k^2}$, the amplitude, expressed in terms of $\mathbf{c}_k$, must be $2|\mathbf{c}_k|$. In other words, the magnitude of $\mathbf{c}_k$ is equal to one-half the amplitude of the component of the frequency spectrum at the frequency $k\omega_0$.

Let us consider a numerical example to illustrate the use of the complex form of the Fourier series. A train of rectangular pulses of amplitude V_0 and duration τ, recurring periodically every T s, is shown in Fig. 17-8*a*. The fundamental frequency is, therefore,

$$f_0 = \frac{1}{T}$$

In order to determine the line spectrum of $v(t)$, we shall determine the $\mathbf{c}_k$ first and then obtain the corresponding amplitudes of the frequency components by multiplying each $|\mathbf{c}_k|$ by 2. The value of a general complex coefficient is first found from (28):

$$\begin{aligned}
\mathbf{c}_k &= \frac{1}{T}\int_{-T/2}^{T/2} f(t)e^{-jk\omega_0 t}\,dt \\
&= \frac{V_0}{T}\int_{t_0}^{t_0+\tau} e^{-jk\omega_0 t}\,dt \\
&= \frac{V_0}{-jk\omega_0 T}\left(e^{-jk\omega_0(t_0+\tau)} - e^{-jk\omega_0 t_0}\right) \\
&= \frac{2V_0}{k\omega_0 T}e^{-jk\omega_0(t_0+\tau/2)}\sin\left(\tfrac{1}{2}k\omega_0\tau\right) \\
&= \frac{V_0\tau}{T}\,\frac{\sin\left(\tfrac{1}{2}k\omega_0\tau\right)}{\tfrac{1}{2}k\omega_0\tau}\,e^{-jk\omega_0(t_0+\tau/2)}
\end{aligned}$$

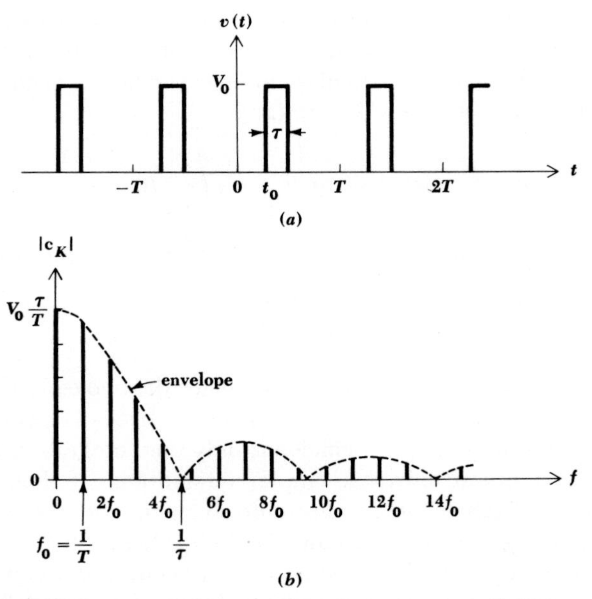

Fig. 17-8 (*a*) A periodic sequence of rectangular pulses. (*b*) The corresponding discrete line spectrum.

The magnitude of $\mathbf{c}_k$ is therefore

$$|\mathbf{c}_k| = \frac{V_0\tau}{T}\left|\frac{\sin\left(\frac{1}{2}k\omega_0\tau\right)}{\frac{1}{2}k\omega_0\tau}\right| \tag{29}$$

and the angle of $\mathbf{c}_k$ is

$$\text{ang } \mathbf{c}_k = -k\omega_0\left(t_0 + \frac{\tau}{2}\right) \qquad \text{(possibly plus } 180°\text{)} \tag{30}$$

The trigonometric factor in (29) occurs frequently in modern communication theory, and it is called the *sampling function*. Thus, we define

$$\text{Sa}(x) = \frac{\sin x}{x}$$

Because of the way in which it helps to determine the amplitude of the various frequency components in $v(t)$, it is worthwhile discovering the important characteristics of this function. First, we note that $\text{Sa}(x)$ is zero whenever x is an integral multiple of π; that is,

$$\text{Sa}(k\pi) = 0 \qquad k = 1, 2, 3, \ldots$$

When x is zero, the function is indeterminate, but it is easy to show that its value is unity:

$$\text{Sa}(0) = 1$$

The magnitude of $\text{Sa}(x)$ therefore decreases from unity at $x = 0$ to zero at $x = \pi$. As x increases from π to 2π, $|\text{Sa}(x)|$ increases from zero to a maximum less than unity, and then decreases to zero once again. As x continues to increase, the successive maxima continually become smaller because the numerator of $\text{Sa}(x)$ cannot exceed unity and the denominator is continually increasing.

Now let us return to the line spectrum of $v(t)$. We may express (29) in terms of the fundamental cyclic frequency f_0:

$$|\mathbf{c}_k| = \frac{V_0\tau}{T}\left|\frac{\sin(k\pi f_0\tau)}{k\pi f_0\tau}\right| \tag{31}$$

The amplitude of any harmonic (at a frequency kf_0) is obtained from (31) by using the known values τ and $T = 1/f_0$ and selecting the desired value of k. Instead of evaluating (31) at these discrete frequencies, let us sketch the *envelope* of $|\mathbf{c}_k|$ by considering the frequency kf_0 to be a continuous variable. That is, $f(= kf_0)$ can actually take on only the discrete values of the harmonic frequencies f_0, $2f_0$, $3f_0$, and so forth, but we may think of k for the moment as a continuous variable. When f is zero, $|\mathbf{c}_k|$ is evidently $V_0\tau/T$, and when f has increased to $1/\tau$, $|\mathbf{c}_k|$ is zero. The resultant envelope is sketched as the broken line of Fig. 17-8*b*. The line spectrum is then obtained by simply erecting vertical lines at each harmonic frequency, as shown in the sketch. The amplitudes shown are those of the c_k; except for the dc component, the amplitudes of the corresponding sinusoids are twice as great. The particular case sketched applies to the case where $\tau/T = 1/(1.5\pi) = 0.212$. In this example, it happens that there is no harmonic exactly at that frequency at which the envelope amplitude is zero; another choice of τ or T could produce such an occurrence, however.

There are several observations and conclusions which we may make about the line spectrum of a periodic sequence of rectangular pulses. With respect to the envelope of the discrete spectrum, it is evident that the "width" of the envelope depends upon τ, and not upon T. As a matter of fact, the shape of the envelope is not a function of T. It follows that the bandwidth of a filter which is designed to pass the periodic pulses is a function of the pulse width τ, but not of the pulse period T; an inspection of Fig. 17-8*b* indicates that the required bandwidth is about $1/\tau$ Hz. If the pulse period T is increased (or the pulse repetition frequency f_0 is decreased), the bandwidth $1/\tau$ does not change, but the number of spectral lines between zero frequency and $1/\tau$ Hz increases, albeit discontinuously; the amplitude of each line is inversely proportional to T. Finally, a shift in the time origin does not change the line spectrum; that is, $|\mathbf{c}_k|$ is not a function of t_0. The relative phases of the frequency components do change with the choice of t_0.

Drill Problem

17-5 Determine the general coefficient $\mathbf{c}_k$ in the complex Fourier series for each of the three waveforms shown in Fig. 17-3.

$$\frac{2}{k\pi}\sin\frac{k\pi}{2};\quad -j\frac{1}{k\pi}(1-\cos k\pi);\quad -j\frac{4}{k^2\pi^2}\sin\frac{k\pi}{2}$$

PROBLEMS

☐ **1** A voltage waveform is defined by: $v = 0,\ -T/2 < t < -d/2;\ v = 100,\ -d/2 < t < d/2;\ v = 0,\ d/2 < t < T/2$. Express d in terms of T so that the amplitude of the kth harmonic component of the Fourier series is the maximum possible value.

☐ **2** The half-wave rectifier output shown in Fig. 17-2*a* is replaced by the full-wave rectifier output, such that $f(t + 0.2) = f(t)$. Compare the discrete line spectrum with that shown in Fig. 17-2*b*.

☐ **3** (*a*) Show that the derivative of the triangular waveform of Fig. 17-3*c* is twice the value of the waveform of Fig. 17-3*b* for all t. (*b*) Show that the Fourier series for the two waveforms have a corresponding relationship.

☐ **4** (*a*) Obtain the Fourier series for the waveform displayed in Fig. 17-9. (*b*) Find the sum of the first four terms of the series at $t = 2$.

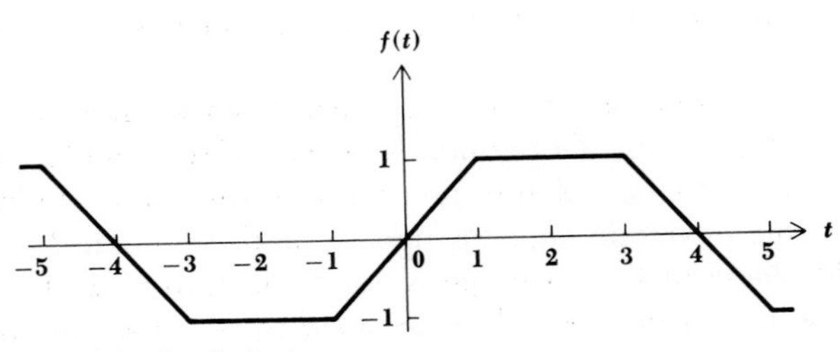

Fig. 17-9 See Prob. 4.

☐ **5** The ideal voltage source $v_s(t)$, where $v_s(t) = -5 + 10\cos(\pi t/3)$ for $-1 < t < 1$ and zero elsewhere in the period interval $-2 < t < 2$, is connected to the series combination of 100 Ω, 40 μF, and 2.5 H. Find the ratio of the amplitude of the 64th harmonic to that of the fundamental for: (*a*) the source voltage; (*b*) the source current.

□ **6** A rectangular current pulse of amplitude 20 mA, duration 5 ms, and period 50 ms begins at $t = 0$. (*a*) Find the average power supplied to a 100-Ω resistor by this current. (*b*) What fraction of the total power is represented by the dc component? (*c*) by the fundamental component? (*d*) by the fifth harmonic component?

□ **7** A square wave of voltage with a period of 4π ms has an amplitude of ±4 V. Draw the discrete frequency spectrum of the current produced in a 50-mH inductor by this voltage.

□ **8** (*a*) Find the Fourier series for the current waveform of Fig. 17-10. (*b*) If this waveform is shifted 4 ms later (to the right) in time, determine the Fourier series.

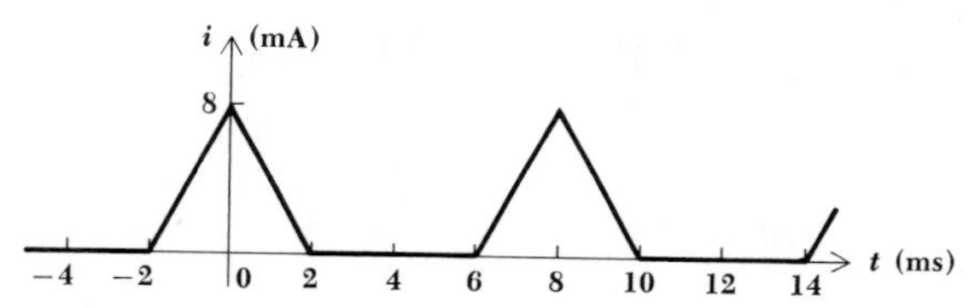

Fig. 17-10 See Prob. 8.

□ **9** In the interval from $t = 0$ to $t = 3$, $f(t)$ is defined by the sketch of Fig. 17-11. (*a*) Define $f(t)$ in the interval $3 < t < 6$ so that $f(t)$ is an odd function with $T = 6$. (*b*) Define $f(t)$ in the interval $3 < t < 6$ so that $f(t)$ is an even function with $T = 6$. (*c*) Define $f(t)$ in the interval $-9 < t < 0$ so that $f(t)$ is an odd function with half-wave symmetry with $T = 12$. Determine the Fourier series for the waveform of: (*d*) part (*a*); (*e*) part (*b*); (*f*) part (*c*).

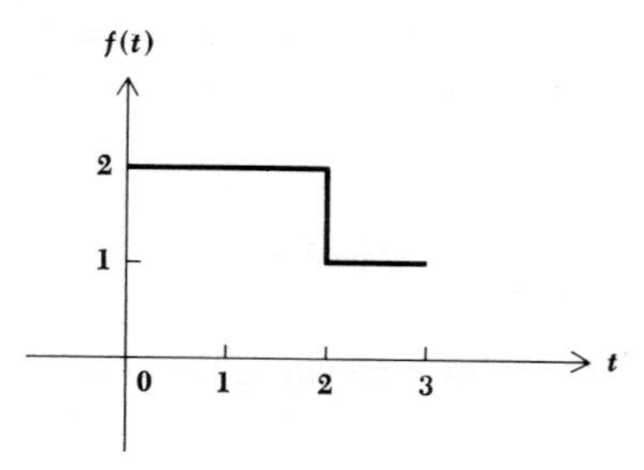

Fig. 17-11 See Prob. 9.

□ **10** Determine whether the given function is odd, even, or neither: (*a*) $t/(t^2 + 1)$; (*b*) te^{-2t}; (*c*) $\cos(\sin 2t)$; (*d*) $\sin(\cos 2t)$; (*e*) $t \sin|t|$; (*f*) $-0.5 + u(t)$; (*g*) $f(t) + f(-t) + 3$.

☐ **11** Taking advantage of the symmetry present, determine the Fourier series for: (*a*) $f(t) = \sin t$, $0 < t < \pi/4$; $f(t) = \cos t$, $\pi/4 < t < \pi/2$; $T = \pi/2$; (*b*) $f(t) = 0$ for $2 < t < 6$ and $10 < t < 14$; $f(t) = 100$, $-2 < t < 2$; $f(t) = -100$, $6 < t < 10$; $T = 16$.

☐ **12** Sketch an example of a waveform that has neither even, odd, nor half-wave symmetry, but whose Fourier series contains no cosine terms.

☐ **13** Determine the Fourier series for the waveforms shown in Fig. 17-12*a* and *b*, and sketch the line spectrum for each.

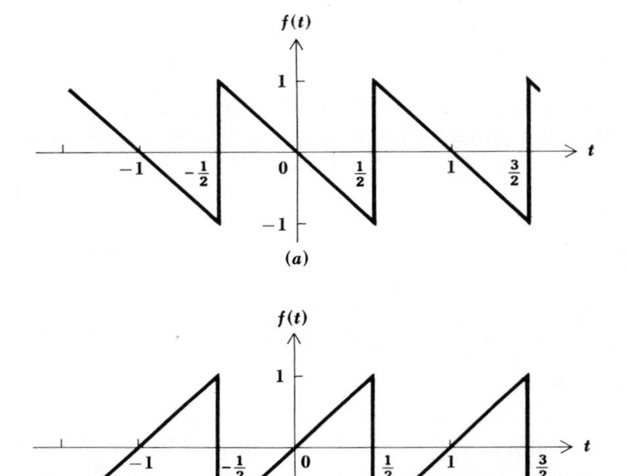

Fig. 17-12 See Prob. 13.

☐ **14** A Fourier series may be obtained for a waveform by numerical methods when an analytical expression for $f(t)$ is not known or is unwieldy. The integral for a_k

$$a_k = \frac{2}{T}\int_0^T f(t)\cos k\omega_0 t\, dt$$

may be expressed as an approximate sum by subdividing the interval of integration into M equal parts:

$$a_k \doteq \frac{2}{T}\sum_{m=1}^{M} f(t_m)(\cos k\omega_0 t_m)\frac{T}{M} = \frac{2}{M}\sum_{m=1}^{M} f(t_m)\cos k\omega_0 t_m$$

The value of $f(t_m)$ and $\cos k\omega_0 t_m$ may be determined at the midpoint of each interval. A similar summation is used to evaluate b_k. By dividing the period into

20 equal parts, find the approximate amplitude of the $1\frac{1}{2}$-Hz component of the waveform described by: $f(t) = 1/(1-t)$, $-1 < t < 0$, and $f(t) = 1/(1+t)$, $0 < t < 1$, with $T = 2$.

☐ **15** An ideal voltage source, an open switch, a 2-kΩ resistor, and a 50-μF capacitor are in series. The voltage waveform is a square wave similar to that shown in Fig. 17-3*a*, but with a peak-to-peak amplitude of 100 V and a period of 10 ms. The switch is closed at $t = 0^+$ and the resistor voltage is the desired response. (*a*) Work in the frequency domain of the *k*th harmonic to find the forced response to this component. Express the total forced response as a trigonometric Fourier series. (*b*) Specify the functional form of the natural response. (*c*) Determine the complete response.

☐ **16** Find the complete response of the circuit described in Prob. 15 by considering the input as a sequence of suitable unit-step forcing functions. The response is to be determined and plotted over an interval which represents two periods of the forcing function. Can you show that the maximum and minimum values of the resistor voltage (forced response) are 51.25 and −51.25 V, respectively?

☐ **17** A parallel *RLC* circuit is energized by a square-wave-current forcing function alternating between $\pm I_0$ with a period of 0.5 ms. Specify the resonant frequency and Q_0 of the *RLC* circuit if the current through the inductor is to be essentially a 150-kHz sinusoid. Assume that no other component of the inductor current may have an amplitude greater than 5 per cent of the desired component.

☐ **18** An AM (amplitude-modulated) broadcasting station operating at 920 kHz carries an exciting flute duet, with one flute playing A (440 Hz) and the other tootling C (512 Hz). For 40 per cent modulation by each signal, the voltage waveform applied to the antenna might be given by: $v = 500(1 + 0.4 \sin 880\pi t + 0.4 \sin 1024\pi t) \cos 1.84 \times 10^6 \pi t$ V. Draw the discrete frequency spectrum for the voltage waveform and determine its rms value.

☐ **19** A pulsed communication signal consists of a burst of 250-kHz energy every 1.25 ms that lasts 0.2 ms. Determine the frequency spectrum of this signal, noting the atmospheric pollution it produces.

☐ **20** Determine the complex Fourier series of the waveform illustrated in Fig. 17-13.

Fig. 17-13 See Prob. 20.

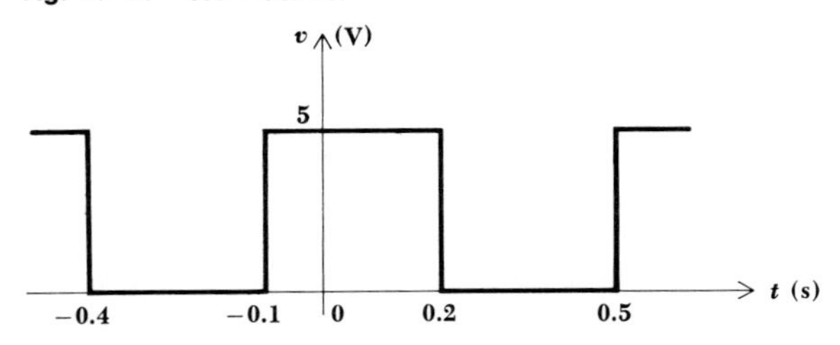

☐ **21** A periodic waveform is given by $i(t) = Ae^{-bt}$, $0 < t < 1$, with $T = 1$. By finding the complex Fourier series, show that the value of b that will maximize the amplitude of the 5-Hz component in the frequency spectrum is about 5.27.

☐ **22** A waveform having both odd and half-wave symmetry is specified by $f(t) = 2t$, $-1 < t < 1$. Let $T = 4$; find the complex Fourier series and sketch the discrete frequency spectrum.

Chapter Eighteen
Fourier Transforms

18-1 INTRODUCTION

In this chapter we shall begin to consider the use of transform methods in studying the behavior of linear circuits through a discussion of the Fourier transform. The Laplace transform is the subject of the next chapter. Before beginning a formal discussion, however, let us look back to Chap. 9.

At that time we recalled the facilitation of certain arithmetic operations (multiplication, division, extraction of roots, and raising numbers to powers) by the use of logarithms, which *transformed* the operations of multiplication and division, for example, to the simpler operations of addition and subtraction. Thus, it was often more useful to work with the logarithm of

14.9 than with the number itself. We might interpret 1.173 as the "logarithmic transform" of 14.9. That is, these two numbers make up a log-transform pair defined by

$$14.9 = 10^{1.173}$$

and

$$1.173 = \log 14.9$$

There is a unique one-to-one correspondence between the number and its logarithmic transform: given the number, we can find its logarithm, and, given the logarithm, we can calculate the corresponding number.

We applied this concept of transformation when we analyzed the behavior of circuits excited by steady-state sinusoidal forcing functions. If we had tried to work exclusively in the time domain, we would have been confronted with some rather staggering sets of integrodifferential equations that might have chilled less stalwart hearts than ours. The concept of the phasor transform was introduced to ease our computational labors. A forcing function, $v(t) = 10 \cos (3t - \pi/4)$, for example, was transformed into the phasor, $\mathbf{V} = 10e^{-j\pi/4}$,

$$10 \cos \left(3t - \frac{\pi}{4}\right) = \text{Re}(10e^{-j\pi/4}e^{j3t})$$

Again, there is a one-to-one correspondence, on this occasion between the time function and its phasor transform.

With these thoughts in mind, we are now prepared to discuss two *integral transforms,* the Fourier and Laplace transforms, which are extremely important in the study of many types of engineering systems, including linear electric circuits. Up to now we have not really needed them, because most of our circuits, along with their forcing and response functions, have been rather simple. The use of relatively powerful integral-transform techniques on them could be likened to using an electric locomotive to crack walnuts; we might lose sight of the meat. But now we have progressed to a point where our forcing functions, and perhaps our circuits, are getting too complicated for the tools we have developed so far. We want to look at forcing functions that are *not* periodic, and in later studies at (random) forcing functions or signals that are not expressible at all as time functions. This desire is often gratified through use of the integral-transform methods which now burst upon us.

18-2 DEFINITION OF THE FOURIER TRANSFORM

Let us proceed to define the Fourier transform by first recalling the spectrum of the periodic train of rectangular pulses we obtained at the end of Chap. 17. That was a *discrete* line spectrum, which is the type that we must always obtain for periodic functions of time. The spectrum was discrete in the sense that it was not a smooth or continuous function of frequency; instead, it had nonzero values only at specific frequencies.

There are many important forcing functions, however, that are not periodic functions of time, such as a single rectangular pulse, a step function, a ramp function, or a rather strange type of function called the *impulse function* to be defined later in this chapter. Frequency spectra may be obtained for such nonperiodic functions, but they will be *continuous* spectra in which some energy, in general, may be found in any frequency interval, no matter how small.

We shall develop this concept by beginning with a periodic function and then letting the period become infinite. Our experience with the periodic rectangular pulses at the end of Chap. 17 should indicate that the envelope will decrease in amplitude without otherwise changing shape, and that more and more frequency components will be found in any given frequency interval. In the limit we should expect an envelope of vanishingly small amplitude, filled with an infinite number of frequency components, separated by vanishingly small frequency intervals. The number of frequency components between, say, 0 and 100 Hz becomes infinite, but the amplitude of each one approaches zero. At first thought, a spectrum of zero amplitude is a puzzling concept. We know that the line spectrum of a periodic forcing function shows the amplitude of each frequency component. But what does the zero-amplitude continuous spectrum of a nonperiodic forcing function signify? That question will be answered in the following section; now we proceed to carry out the limiting procedure suggested above.

We begin with the exponential form of the Fourier series:

$$f(t) = \sum_{n=-\infty}^{n=\infty} \mathbf{c}_n e^{jn\omega_0 t} \tag{1}$$

where

$$\mathbf{c}_n = \frac{1}{T}\int_{-T/2}^{T/2} f(t)e^{-jn\omega_0 t}\,dt \tag{2}$$

and

$$\omega_0 = \frac{2\pi}{T} \tag{3}$$

We now let

$$T \to \infty$$

and thus, from (3), ω_0 must become vanishingly small. We represent this limit by a differential

$$\omega_0 \to d\omega$$

Thus

$$\frac{1}{T} = \frac{\omega_0}{2\pi} \rightarrow \frac{d\omega}{2\pi} \tag{4}$$

Finally, the frequency of any "harmonic" $n\omega_0$ must now correspond to the general frequency variable which describes the continuous spectrum. In other words, n must tend to infinity as ω_0 approaches zero, such that the product is finite:

$$n\omega_0 \rightarrow \omega \tag{5}$$

When these four limiting operations are applied to (2), we find that $\mathbf{c}_n$ must approach zero, as we had previously presumed. If we multiply each side of (2) by the period T and then undertake the limiting process, a nontrivial result is obtained:

$$\mathbf{c}_n T \rightarrow \int_{-\infty}^{\infty} f(t) e^{-j\omega t}\, dt$$

The right side of this expression is a function of ω (and *not* of t), and we represent it by $\mathbf{F}(j\omega)$:

$$\mathbf{F}(j\omega) = \int_{-\infty}^{\infty} f(t) e^{-j\omega t}\, dt \tag{6}$$

In order to apply the limiting process to (1), we multiply and divide the summation by T,

$$f(t) = \sum_{n=-\infty}^{n=\infty} \mathbf{c}_n T e^{jn\omega_0 t} \frac{1}{T}$$

and use (4), (5), and the new quantity $\mathbf{F}(j\omega)$. In the limit, the summation becomes an integral, and

$$f(t) = \frac{1}{2\pi} \int_{-\infty}^{\infty} \mathbf{F}(j\omega) e^{j\omega t}\, d\omega \tag{7}$$

Equations (6) and (7) are collectively called the *Fourier transform pair.* The function $\mathbf{F}(j\omega)$ is the *Fourier transform* of $f(t)$, and $f(t)$ is the *inverse* Fourier transform of $\mathbf{F}(j\omega)$.

This transform-pair relationship is most important! We should memorize it, draw arrows pointing to it, and mentally keep it on the conscious level henceforth and forevermore.[1] We emphasize the importance of these relations by repeating and blocking them off below:

$$\mathbf{F}(j\omega) = \int_{-\infty}^{\infty} e^{-j\omega t} f(t)\, dt \qquad (8a)$$

$$f(t) = \frac{1}{2\pi} \int_{-\infty}^{\infty} e^{j\omega t} \mathbf{F}(j\omega)\, d\omega \qquad (8b)$$

One question is appropriate to raise at this time. For the Fourier transform relationships given above, can we obtain the Fourier transform of *any* arbitrarily chosen $f(t)$? It turns out that the answer is affirmative for essentially any voltage or current that we can produce. A sufficient condition for the existence of $\mathbf{F}(j\omega)$ is that

$$\int_{-\infty}^{\infty} |f(t)|\, dt < \infty$$

This condition is not necessary, however, for some functions that do not meet it still have a Fourier transform; the step function is one such example. Furthermore, we shall see later that $f(t)$ does not even need to be nonperiodic in order to have a Fourier transform; the Fourier series representation for a periodic time function is just a special case of the more general Fourier transform representation.

As in our previous work with logarithms and phasor transforms, the Fourier transform-pair relationship is unique. For a given $f(t)$ there is one specific $\mathbf{F}(j\omega)$; and for a given $\mathbf{F}(j\omega)$ there is one specific $f(t)$.

Let us use the Fourier transform to obtain the continuous spectrum of a single rectangular pulse. We shall select that pulse in Fig. 17-8*a* (repeated as Fig. 18-1*a*) which occurs in the interval $t_0 < t < t_0 + \tau$. Thus

$$f(t) = \begin{cases} V_0 & t_0 < t < t_0 + \tau \\ 0 & t < t_0 \text{ and } t > t_0 + \tau \end{cases}$$

The Fourier transform of $f(t)$ is found from (6):

$$\mathbf{F}(j\omega) = \int_{t_0}^{t_0+\tau} V_0 e^{-j\omega t}\, dt$$

[1]Future used-car dealers and politicians may forget it.

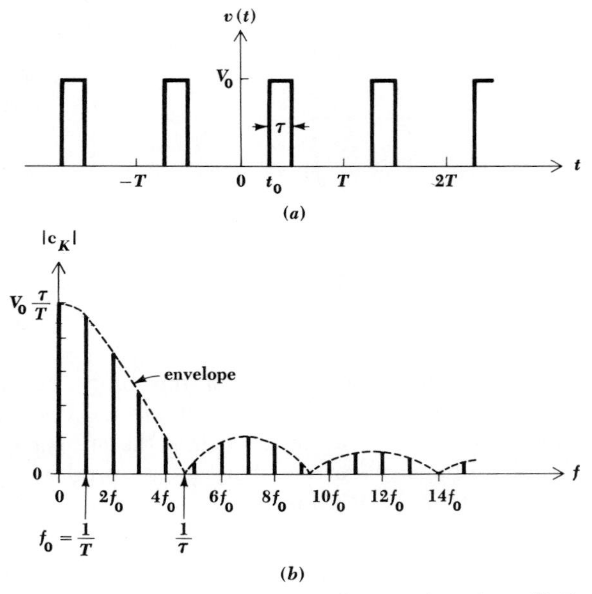

Fig. 18-1 (*a*) A periodic sequence of rectangular pulses. (*b*) The corresponding discrete line spectrum.

and this may be easily integrated and simplified:

$$\mathbf{F}(j\omega) = V_0\tau \frac{\sin \frac{1}{2}\omega\tau}{\frac{1}{2}\omega\tau} e^{-j\omega(t_0+\tau/2)}$$

The magnitude of $\mathbf{F}(j\omega)$ yields the continuous frequency spectrum, and it is obviously of the form of the sampling function. The value of $\mathbf{F}(0)$ is $V_0\tau$. The shape of the spectrum is identical with the broken-line curve of Fig. 18-1*b*. A plot of $|\mathbf{F}(j\omega)|$ as a function of ω does *not* indicate the magnitude of the voltage present at any given frequency. What is it, then? Examination of (7) will show that $\mathbf{F}(j\omega)$ is dimensionally "volts per unit frequency," a concept that may be strange to many of us. In order to understand this a little better, we next look into some of the properties of $\mathbf{F}(j\omega)$.

Drill Problems

18-1 For each of the time functions given, determine $\mathbf{F}(j\omega)$ and $\mathbf{F}(j5)$. Let $f(t) = 0$ outside the range of t specified. (*a*) $I_m \cos 5t$, $-0.1\pi < t < 0.1\pi$; (*b*) $I_m \cos 5t$, $-\pi < t < \pi$; (*c*) $I_m \cos 5t$, $-10\pi < t < 10\pi$.

$$Ans.\ -\frac{2\omega I_m}{25-\omega^2}\sin 10\pi\omega,\ 31.4I_m;\ \frac{2\omega I_m}{25-\omega^2}\sin \pi\omega,\ 3.14I_m;$$
$$\frac{10I_m}{25-\omega^2}\cos 0.1\pi\omega,\ 0.314I_m$$

18-2 For each of the $\mathbf{F}(j\omega)$ given, determine $f(t)$ and $f(0)$. Let $\mathbf{F}(j\omega) = 0$ outside the frequency range specified. (*a*) 1, $-0.5 < \omega < 0.5$; (*b*) 0.1, $-5 < \omega < 5$; (*c*) 10, $-0.05 < \omega < 0.05$.

$$Ans.\ \frac{0.1\sin 5t}{\pi t},\ \frac{1}{2\pi};\ \frac{\sin 0.5t}{\pi t},\ \frac{1}{2\pi};\ \frac{10\sin 0.05t}{\pi t},\ \frac{1}{2\pi}$$

18-3 SOME PROPERTIES OF THE FOURIER TRANSFORM

Our object in this section is to establish several of the mathematical properties of the Fourier transform and, even more important, to understand its physical significance. We begin by using Euler's identity to replace $e^{-j\omega t}$ in (8*a*),

$$\mathbf{F}(j\omega) = \int_{-\infty}^{\infty} f(t)\cos\omega t\,dt - j\int_{-\infty}^{\infty} f(t)\sin\omega t\,dt \tag{9}$$

Since $f(t)$, $\cos\omega t$, and $\sin\omega t$ are all real functions of time, both the integrals in (9) are real functions of ω. Thus, by letting

$$\mathbf{F}(j\omega) = A(\omega) + jB(\omega) = |\mathbf{F}(j\omega)|e^{j\theta(\omega)} \tag{10}$$

we have

$$A(\omega) = \int_{-\infty}^{\infty} f(t)\cos\omega t\,dt \tag{11}$$

$$B(\omega) = -\int_{-\infty}^{\infty} f(t)\sin\omega t\,dt \tag{12}$$

$$|\mathbf{F}(j\omega)| = \sqrt{A^2(\omega) + B^2(\omega)} \tag{13}$$

and

$$\theta(\omega) = \tan^{-1}\frac{B(\omega)}{A(\omega)} \tag{14}$$

Replacing ω by $-\omega$ shows that $A(\omega)$ and $|\mathbf{F}(j\omega)|$ are both even functions of ω, while $B(\omega)$ and $\theta(\omega)$ are both odd functions of ω.

Now, if $f(t)$ is an even function of t, then the integrand of (12) is an odd function of t, and the symmetrical limits force $B(\omega)$ to be zero; thus the Fourier transform $\mathbf{F}(j\omega)$ is a real, even function of ω, and the phase function $\theta(\omega)$ is zero or π for all ω. However, if $f(t)$ is an odd function of t, then $A(\omega) = 0$, $\mathbf{F}(j\omega)$ is both odd and a pure imaginary function of ω; $\theta(\omega)$ is $\pm\pi/2$.

Finally, we note that the replacement of ω by $-\omega$ in (9) forms the conjugate of $\mathbf{F}(j\omega)$. Thus,

$$\mathbf{F}(-j\omega) = A(\omega) - jB(\omega) = \mathbf{F}^*(j\omega)$$

and we have

$$\mathbf{F}(j\omega)\mathbf{F}(-j\omega) = \mathbf{F}(j\omega)\mathbf{F}^*(j\omega) = A^2(\omega) + B^2(\omega) = |\mathbf{F}(j\omega)|^2$$

With these basic mathematical properties of the Fourier transform in mind, we are now ready to consider its physical significance. Let us suppose that $f(t)$ is either the voltage across or the current through a 1-Ω resistor, so that $f^2(t)$ is the instantaneous power delivered to the 1-Ω resistor by $f(t)$. Integrating this power over all time, we obtain the total energy delivered by $f(t)$ to the 1-Ω resistor,

$$W_{1\Omega} = \int_{-\infty}^{\infty} f^2(t)\, dt \tag{15}$$

Now let us resort to a little trickery. Thinking of the integrand above as $f(t)$ times itself, we replace one of those functions by (8*b*),

$$W_{1\Omega} = \int_{-\infty}^{\infty} f(t) \left[\frac{1}{2\pi}\int_{-\infty}^{\infty} e^{j\omega t}\mathbf{F}(j\omega)\, d\omega\right] dt$$

Since $f(t)$ is not a function of the variable of integration ω, we move it inside the bracketed integral, and then interchange the order of integration,

$$W_{1\Omega} = \frac{1}{2\pi}\int_{-\infty}^{\infty} \left[\int_{-\infty}^{\infty} \mathbf{F}(j\omega)e^{j\omega t}f(t)\, dt\right] d\omega$$

Next we shift $\mathbf{F}(j\omega)$ outside the inner integral, causing that integral to become $\mathbf{F}(-j\omega)$,

$$W_{1\Omega} = \frac{1}{2\pi}\int_{-\infty}^{\infty} \mathbf{F}(j\omega)\mathbf{F}(-j\omega)\, d\omega = \frac{1}{2\pi}\int_{-\infty}^{\infty} |\mathbf{F}(j\omega)|^2\, d\omega$$

Collecting these results,

$$\int_{-\infty}^{\infty} f^2(t)\, dt = \frac{1}{2\pi}\int_{-\infty}^{\infty} |\mathbf{F}(j\omega)|^2\, d\omega \tag{16}$$

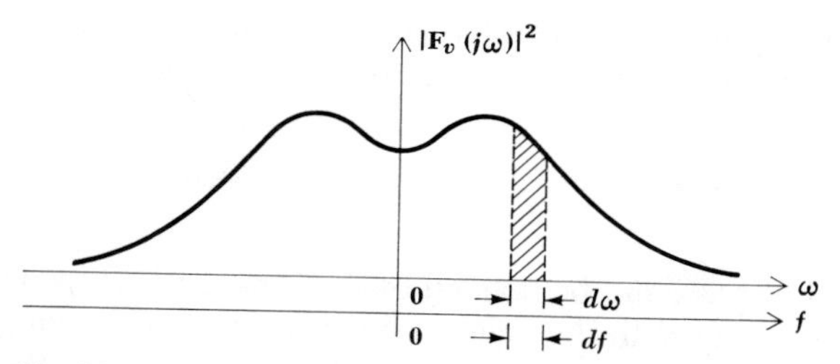

Fig. 18-2 The area of the slice $|\mathbf{F}_v(j\omega)|^2\, df$ is the 1-Ω energy associated with $v(t)$ lying in the bandwidth df.

Equation (16) is a very useful expression known as Parseval's theorem.[2] This theorem, along with (15), tells us that the energy associated with $f(t)$ can be obtained either from an integration over all time in the time domain or by ($1/2\pi$ times) an integration over all frequency in the frequency domain.

Parseval's theorem also leads us to a greater understanding and interpretation of the meaning of the Fourier transform. Consider a voltage $v(t)$ with Fourier transform $\mathbf{F}_v(j\omega)$ and 1-Ω energy W_{1v},

$$W_{1v} = \frac{1}{2\pi}\int_{-\infty}^{\infty} |\mathbf{F}_v(j\omega)|^2\, d\omega = \frac{1}{\pi}\int_{0}^{\infty} |\mathbf{F}_v(j\omega)|^2\, d\omega$$

where the right-most equality follows from the fact that $|\mathbf{F}_v(j\omega)|^2$ is an even function of ω. Then, since $\omega = 2\pi f$, we can write

$$W_{1v} = \int_{-\infty}^{\infty} |\mathbf{F}_v(j\omega)|^2\, df = 2\int_{0}^{\infty} |\mathbf{F}_v(j\omega)|^2\, df \tag{17}$$

Figure 18-2 illustrates a typical plot of $|\mathbf{F}_v(j\omega)|^2$ as a function of both ω and f. If we divide the frequency scale up into vanishingly small increments df, (17) shows us that the area of a differential slice under the $|\mathbf{F}_v(j\omega)|^2$ curve, having a width df, is equal to $|\mathbf{F}_v(j\omega)|^2\, df$. This area is shown shaded. The sum of all such areas, as f ranges from minus to plus infinity, is the total 1-Ω energy contained in $v(t)$. Thus, $|\mathbf{F}_v(j\omega)|^2$ is the (1-Ω) *energy density* or energy per unit bandwidth (J/Hz) of $v(t)$, and, by integrating $|\mathbf{F}_v(j\omega)|^2$ over the appropriate frequency interval, we should be able to calculate that portion of the total energy lying within the given interval.

As an example, let us assume that $v(t)$ is the voltage forcing function

[2] Marc-Antoine Parseval-Deschenes was a rather obscure French mathematician, geographer, and occasional poet who published these results in 1805, 17 years before Fourier published his theorem.

applied to the input of an ideal bandpass filter,[3] and that $v_0(t)$ is the output voltage. For $v(t)$ we select the one-sided [i.e., $v(t) = 0$ for $t < 0$] exponential pulse,

$$v(t) = 4e^{-3t}u(t)$$

and let the filter passband be defined by $1 < |f| < 2$. The energy in $v_o(t)$ will therefore be equal to the energy of that part of $v(t)$ having frequency components in the intervals, $1 < f < 2$ and $-2 < f < -1$. We determine the Fourier transform of $v(t)$,

$$\mathbf{F}_v(j\omega) = 4\int_{-\infty}^{\infty} e^{-j\omega t}e^{-3t}u(t)\,dt = 4\int_0^{\infty} e^{-(3+j\omega)t}\,dt$$
$$= \frac{4}{3 + j\omega}$$

and then we may calculate the total 1-Ω energy in the input signal either by

$$W_{1v} = \frac{1}{2\pi}\int_{-\infty}^{\infty} |\mathbf{F}_v(j\omega)|^2\,d\omega = \frac{8}{\pi}\int_{-\infty}^{\infty}\frac{d\omega}{9 + \omega^2}$$
$$= \frac{16}{\pi}\int_0^{\infty}\frac{d\omega}{9 + \omega^2} = \frac{8}{3}\quad \text{J}$$

or

$$W_{1v} = \int_{-\infty}^{\infty} v^2(t)\,dt = 16\int_0^{\infty} e^{-6t}\,dt = \frac{8}{3}\quad \text{J}$$

The total energy in $v_o(t)$, however, is smaller:

$$W_{1o} = \frac{1}{2\pi}\int_{-4\pi}^{-2\pi}\frac{16d\omega}{9 + \omega^2} + \frac{1}{2\pi}\int_{2\pi}^{4\pi}\frac{16d\omega}{9 + \omega^2} = \frac{16}{\pi}\int_{2\pi}^{4\pi}\frac{d\omega}{9 + \omega^2}$$
$$= \frac{16}{3\pi}\left(\tan^{-1}\frac{4\pi}{3} - \tan^{-1}\frac{2\pi}{3}\right) = 0.358\quad \text{J}$$

Thus we see that an ideal bandpass filter enables us to remove energy from prescribed frequency ranges while still retaining the energy contained in other frequency ranges. The Fourier transform helps us to describe the filtering action quantitatively without actually evaluating $v_o(t)$, although we shall see later that the Fourier transform can also be used to obtain the expression for $v_o(t)$ if we wish to do so.

[3] An ideal bandpass filter is a two-port network which allows all those frequency components of the input signal for which $\omega_1 < |\omega| < \omega_2$ to pass unattenuated from the input terminals to the output terminals; all components having frequencies outside this so-called *passband* are completely attenuated.

Drill Problems

18-3 Given $i(t) = 2e^{-3|t|}$ A, find: (*a*) $A_i(2)$; (*b*) $B_i(1)$; (*c*) $|\mathbf{F}_i(j0)|$.

Ans. 0; 0.923; 1.333

18-4 Find the 1-Ω energy associated with the current, $i(t) = 2e^{-3|t|}$ A, in the frequency interval: (*a*) $-1.5 < \omega < 1.5$; (*b*) $-3 < \omega < 3$; (*c*) $-6 < \omega < 6$ rad/s.

Ans. 0.734; 1.092; 1.278 J

18-4 THE UNIT-IMPULSE FUNCTION

Before continuing our discussion of the Fourier transform, we need to pause briefly to define a new singularity function called the *unit-impulse* or *delta function.* We shall see that the unit-impulse function enables us to confirm our previous statement that periodic, as well as nonperiodic, time functions possess Fourier transforms. In addition, our analysis of general *RLC* circuits can be enhanced by utilization of the unit impulse. We have been avoiding the possibility that the voltage across a capacitor or the current through an inductor might change by a finite amount in zero time, for in these cases the capacitor current or the inductor voltage would have had to assume infinite values. Although this is not physically possible, it is mathematically possible. Thus, if a step voltage $V_0u(t)$ is applied directly across an uncharged capacitor C, the time constant is zero, since R_{eq} is zero and C is finite. This means that a charge CV_0 must be established across the capacitor in a vanishingly small time. We might say, with more insight than rigor, that "infinite current flowing for zero time produces a finite charge" on the capacitor. This type of phenomenon can be described through use of the unit-impulse function.

We shall define the unit impulse as a function of time which is zero when its argument, generally $(t - t_0)$, is less than zero; which is also zero when its argument is greater than zero; which is infinite when its argument is zero; and which has unit area. Mathematically, the defining statements are

$$\delta(t - t_0) = 0 \qquad t \neq t_0 \tag{18}$$

and

$$\int_{-\infty}^{\infty} \delta(t - t_0)\, dt = 1 \tag{19}$$

where the symbol δ ("curly" delta) is used to represent the unit impulse. In view of the functional values expressed by (18), it is apparent that the limits on the integral in (19) may be any values which are less than t_0 and

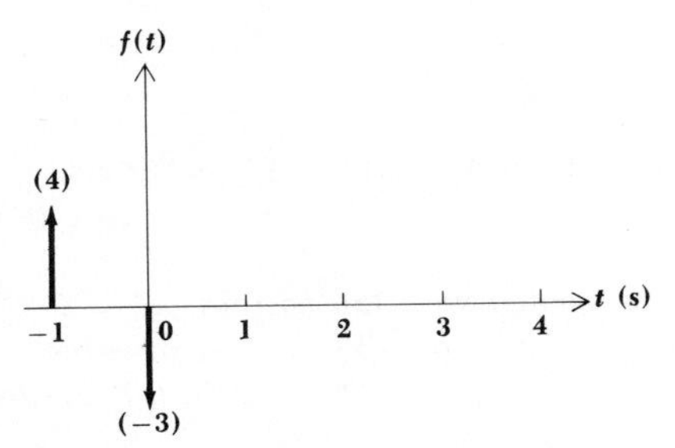

Fig. 18-3 A positive and negative impulse are plotted graphically as functions of time. The strengths of the impulses are 4 and -3, respectively, and thus $f(t) = 4\delta(t+1) - 3\delta(t)$.

greater than t_0. In particular, we may let t_0^- and t_0^+ represent values of time which are arbitrarily close to t_0 and then express (19) as

$$\int_{t_0-}^{t_0+} \delta(t - t_0)\,dt = 1 \tag{20}$$

For the most part, we shall concern ourselves with circuits having only a single discontinuity, and we shall select our time scale so that the switching operation occurs at $t = 0$. For this special case, the defining equations are

$$\delta(t) = 0 \qquad t \neq 0 \tag{21}$$

and

$$\int_{-\infty}^{\infty} \delta(t)\,dt = 1 \tag{22}$$

or

$$\int_{0-}^{0+} \delta(t)\,dt = 1 \tag{23}$$

The unit impulse may be multiplied by a constant also; this cannot, of course, affect (18) or (21), because the value must still be zero when the argument is not zero. However, multiplication of any of the integral expressions by a constant shows that the area under the impulse is now equal to the constant multiplying factor; this area is called the *strength* of the impulse. Thus, the impulse $5\delta(t)$ has a strength of 5, and the impulse $-10\delta(t-2)$ has a strength of -10. If the unit impulse is multiplied by a function of time, then the strength of the impulse must be the value of that function at the time for which the impulse *argument* is zero. In other words, the strength of the impulse $e^{-t/2}\delta(t-2)$ is 0.368, and the strength of the impulse $\sin(5\pi t + \pi/4)\,\delta(t)$ is 0.707. It is therefore possible to write the following integrals[4] which make the same statement in mathematical form:

$$\int_{-\infty}^{\infty} f(t)\,\delta(t)\,dt = f(0) \tag{24}$$

or

$$\int_{-\infty}^{\infty} f(t)\,\delta(t - t_0)\,dt = f(t_0) \tag{25}$$

The graphical symbol for an impulse is shown in Fig. 18-3, where $f(t) = 4\delta(t + 1) - 3\delta(t)$ is plotted as a function of time. It is customary to indicate the strength of the impulse in parentheses adjacent to the impulse. Note that no attempt should be made to indicate the strength of an impulse by adjusting its amplitude; each spike has infinite amplitude, and all impulses should be drawn with the same convenient amplitude. Positive and negative impulses should be drawn above and below the time axis, respectively. In order to avoid confusion with the ordinate, the lines and arrows used to form the impulses are drawn thicker than are the axes.

We shall find it convenient to be familiar with several other interpretations of the unit impulse. We seek graphical forms which do not have infinite amplitudes, but which will approximate an impulse as the amplitude increases. Let us first consider a rectangular pulse, such as that shown in Fig. 18-4. The pulse width is selected as Δ and its amplitude as $1/\Delta$, thus forcing the area of the pulse to be unity, regardless of the magnitude of Δ. As Δ decreases, the amplitude $1/\Delta$ increases, and the rectangular pulse becomes a better approximation to a unit impulse. The response of a circuit element to a unit impulse may be determined by finding its response to this rectangular pulse and then letting Δ approach zero. However, since the impulse response is easily found, we should also realize that this response may in itself be an acceptable approximation to the response produced by a short rectangular pulse.

[4]They are called *sifting integrals* because the integral sifts out a particular value of $f(t)$.

Fig. 18-4 A rectangular pulse of unit area which approaches a unit impulse as $\Delta \to 0$.

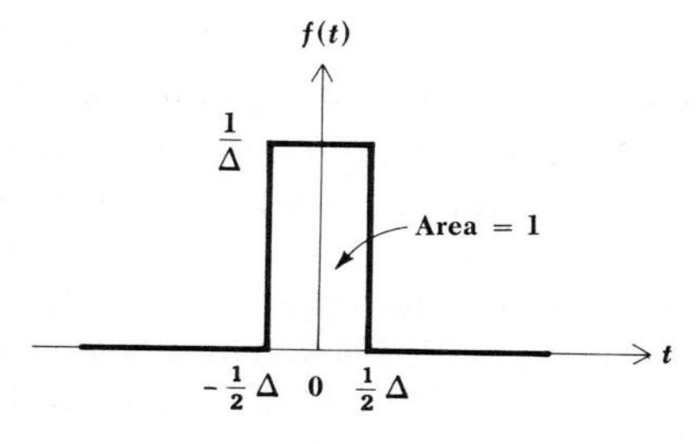

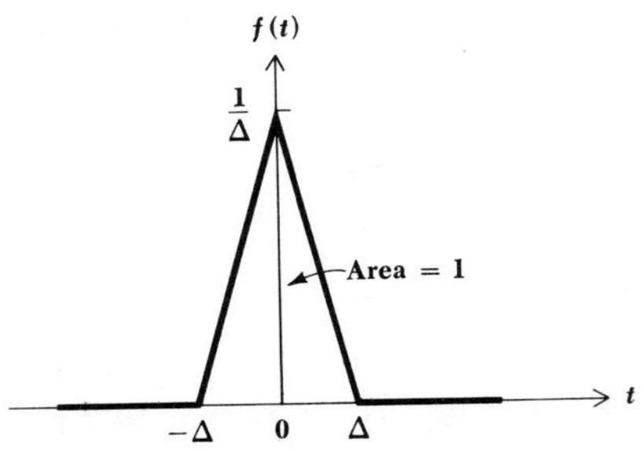

Fig. 18-5 A triangular pulse of unit area which approaches a unit impulse as $\Delta \to 0$.

A triangular pulse, sketched in Fig. 18-5, may also be used to approximate the unit impulse. Since we again prefer unit area, a pulse of amplitude $1/\Delta$ must possess an overall width of 2Δ. As Δ approaches zero, the triangular pulse approaches the unit impulse. There are many other pulse shapes which, in the limit, approach the unit impulse, but we shall take the negative exponential as our last limiting form. We first construct such a waveform with unit area by finding the area under the general exponential,

$$f(t) = \begin{cases} 0 & t < 0 \\ Ae^{-t/\tau} & t > 0 \end{cases}$$

or

$$f(t) = Ae^{-t/\tau}u(t)$$

Thus

$$\text{Area} = \int_0^\infty Ae^{-t/\tau}\,dt = -\tau Ae^{-t/\tau}\Big|_0^\infty = \tau A$$

and we thus must set $A = 1/\tau$. The time constant will be very short, and we shall again use Δ to represent this short time. Thus, the exponential function

$$f(t) = \frac{1}{\Delta}e^{-t/\Delta}u(t)$$

approaches the unit impulse as $\Delta \to 0$. This representation of the unit impulse indicates that the exponential decay of a current or voltage in a circuit approaches an impulse (but not necessarily a unit impulse) as the time constant is reduced.

For our final interpretation of the unit impulse, let us try to establish a relationship with the unit-step function. The function shown in Fig. 18-6*a* is almost a unit step; however, Δ s is required for it to complete the linear change from zero to unit amplitude. Beneath this modified unit-step

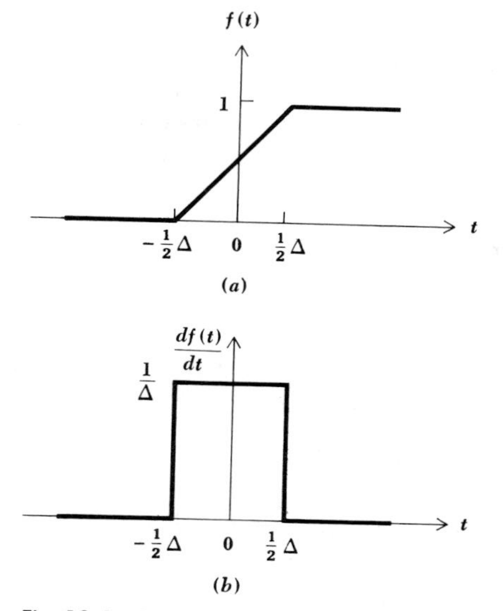

Fig. 18-6 (*a*) A modified unit-step function; the transition from zero to unity is linear over a Δ-s time interval. (*b*) The derivative of the modified unit step. As $\Delta \to 0$, (*a*) becomes the unit step and (*b*) approaches the unit impulse.

function, Fig. 18-6*b* shows its derivative; since the linear portion of the modified step rises at the rate of 1 unit every Δ s, the derivative must be a rectangular pulse of amplitude $1/\Delta$ and width Δ. This, however, was the first function we considered as an approximation to the unit impulse, and we know that it approaches the unit impulse as Δ approaches zero. But the modified unit step approaches the unit step itself as Δ approaches zero, and we conclude that the unit impulse may be regarded as the time derivative of the unit-step function.[5] Mathematically,

$$\delta(t) = \frac{du(t)}{dt} \tag{26}$$

and conversely,

$$u(t) = \int_{0^-}^{t} \delta(t)\, dt \qquad t > 0 \tag{27}$$

where the lower limit may in general be any value of t less than zero. Either (26) or (27) could be used as the definition of the unit impulse if we wished.

[5]The fact that the derivative of the unit step does not exist at the point of the discontinuity creates quite a skeptical attitude in many mathematicians; nevertheless, the impulse is a useful analytical function.

We thus see another method suggesting itself for the determination of the response to a unit impulse. If we can find the unit-step response, then the linear nature of our circuits requires the response to a unit impulse to be the derivative of the response to the unit step. From the opposite point of view, if the response to a unit impulse is known, then the integral of this response must be the unit-step response.

Drill Problems

18-5 Find the strength of the impulse defined by: (*a*) df/dt, where $f(t) = 2u(t - \pi)$; (*b*) the limit as $a \to 0$ of $(1/a)e^{-2|t|/a}$; (*c*) the limit as $b \to 0$ of $b/(x^2 + b^2)$; $b > 0$.

Ans. 1; 2; π

18-6 Evaluate: (*a*) $\int_{-1}^{1} 3e^{-2x^2}\,\delta(x)\,dx$; (*b*) $\int_{0}^{1} 4(\sin^2 \pi t)\,\delta(t - 0.25)\,dt$; (*c*) $\int_{-0.35}^{0.25} 5\tau\,\delta(1 - \sin 5\pi\tau)\,d\tau$.

Ans. -1; 2; 3

18-5 FOURIER TRANSFORM PAIRS FOR SOME SIMPLE TIME FUNCTIONS

We now seek the Fourier transform of the unit impulse $\delta(t - t_0)$. That is, we are interested in the spectral properties or frequency-domain description of this singularity function. If we use the notation $\mathfrak{F}\{\ \}$ to symbolize "Fourier transform of $\{\ \}$," then

$$\mathfrak{F}\{\delta(t - t_0)\} = \int_{-\infty}^{\infty} e^{-j\omega t}\,\delta(t - t_0)\,dt$$

From our earlier discussion of this type of integral, and by (25) in particular, we have

$$\mathfrak{F}\{\delta(t - t_0)\} = e^{-j\omega t_0} = \cos \omega t_0 - j \sin \omega t_0 \tag{28}$$

This complex function of ω leads to the 1-Ω energy density function,

$$|\mathfrak{F}\{\delta(t - t_0)\}|^2 = \cos^2 \omega t_0 + \sin^2 \omega t_0 = 1$$

This remarkable result says that the (1-Ω) energy per unit bandwidth is unity *at all frequencies,* and that the total energy in the unit impulse is infinitely large.[6] No wonder, then, that we must conclude that the unit impulse is "impractical" in the sense that it cannot be generated in the laboratory.

Since there is a unique one-to-one correspondence between a time function and its Fourier transform, we can say that the inverse Fourier transform of $e^{-j\omega t_0}$ is $\delta(t - t_0)$. Utilizing the symbol $\mathfrak{F}^{-1}\{\quad\}$ for the inverse transform, we have

$$\mathfrak{F}^{-1}\{e^{-j\omega t_0}\} = \delta(t - t_0)$$

Thus, we now know that

$$\frac{1}{2\pi}\int_{-\infty}^{\infty} e^{+j\omega t}e^{-j\omega t_0}\,d\omega = \delta(t - t_0)$$

even though we would fail in an attempt at the direct evaluation of this improper integral. Symbolically, we may write

$$\delta(t - t_0) \Leftrightarrow e^{-j\omega t_0} \tag{29}$$

where $\Leftrightarrow$ indicates that the two functions constitute a Fourier transform pair.

Continuing with our consideration of the unit-impulse function, suppose that a certain time function $f(t)$ is known to have a Fourier transform given by

$$\mathbf{F}(j\omega) = \delta(\omega - \omega_0)$$

which is a unit impulse *in the frequency domain* located at $\omega = \omega_0$. Then $f(t)$ must be

$$f(t) = \mathfrak{F}^{-1}\{\mathbf{F}(j\omega)\} = \frac{1}{2\pi}\int_{-\infty}^{\infty} e^{j\omega t}\,\delta(\omega - \omega_0)\,d\omega$$

Again using the sifting property of the unit impulse, we obtain

$$f(t) = \frac{1}{2\pi}e^{j\omega_0 t}$$

[6]Note, for example, from Fig. 18-4 that the total energy in the unit impulse is

$$\lim_{\Delta\to 0}\int_{\Delta/2}^{\Delta/2}\left(\frac{1}{\Delta}\right)^2 dt = \lim_{\Delta\to 0}\left(\frac{1}{\Delta}\right) = \infty$$

Thus we may now write

$$\frac{1}{2\pi} e^{j\omega_0 t} \Leftrightarrow \delta(\omega - \omega_0)$$

or

$$e^{j\omega_0 t} \Leftrightarrow 2\pi\delta(\omega - \omega_0) \tag{30}$$

Also, by a simple sign change we obtain

$$e^{-j\omega_0 t} \Leftrightarrow 2\pi\delta(\omega + \omega_0) \tag{31}$$

Clearly, the time function is complex in both (30) and (31), and does not exist in the real world of the laboratory. Time functions like $\cos \omega_0 t$, for example, can be produced with laboratory equipment, but a function like $e^{j\omega_0 t}$ cannot.

However, we know that

$$\cos \omega_0 t = \tfrac{1}{2} e^{j\omega_0 t} + \tfrac{1}{2} e^{-j\omega_0 t}$$

and it is easily seen from the definition of the Fourier transform that

$$\mathfrak{F}\{f_1(t)\} + \mathfrak{F}\{f_2(t)\} = \mathfrak{F}\{f_1(t) + f_2(t)\} \tag{32}$$

Therefore,

$$\begin{aligned} \mathfrak{F}\{\cos \omega_0 t\} &= \mathfrak{F}\{\tfrac{1}{2} e^{j\omega_0 t}\} + \mathfrak{F}\{\tfrac{1}{2} e^{-j\omega_0 t}\} \\ &= \pi\delta(\omega - \omega_0) + \pi\delta(\omega + \omega_0) \end{aligned}$$

which indicates that the frequency-domain description of $\cos \omega_0 t$ shows a *pair* of impulses, located at $\omega = \pm\omega_0$. This should not be a great surprise, for in our discussion of the complex-frequency plane in Sec. 13-2, we noted that a sinusoidal function of time was always represented by a pair of imaginary frequencies located at $\mathbf{s} = \pm j\omega_0$. We have, therefore,

$$\cos \omega_0 t \Leftrightarrow \pi[\delta(\omega + \omega_0) + \delta(\omega - \omega_0)] \tag{33}$$

Before establishing the Fourier transforms of any more time functions, we should understand just where we are heading, and why. Thus far we have determined several Fourier transform pairs. As we build up our knowledge of such pairs, they, in turn, can be used to obtain more pairs and eventually we will have a catalog of most of the familiar time functions encountered in circuit analysis, along with the corresponding Fourier transforms. Thus we will have not only the time-domain descriptions of such functions, but their frequency-domain descriptions as well. Then, just as the use of phasor transforms simplified the determination of the steady-state sinusoidal response, we shall find that the use of the Fourier transforms of various forcing functions can simplify the determination of the complete response, both the natural and the forced components. When we extend our thinking to the use of the Laplace transform in the following chapter, we shall even be able to account for the troublesome initial conditions that have plagued us in the past. With these thoughts in mind, let us look at just a few more transform pairs with the goal of listing our findings in a form that will be useful for quick reference later.

The first forcing function that we considered many chapters ago was a dc voltage or current. To find the Fourier transform of a constant function of time, $f(t) = K$, our first inclination might be to substitute this constant into the defining equation for the Fourier transform and evaluate the resulting integral. If we did, we would find ourselves with an indeterminate expression on our hands. Fortunately, however, we have already solved this problem, for from (31),

$$e^{-j\omega_0 t} \Leftrightarrow 2\pi\delta(\omega + \omega_0)$$

We see that, if we simply let $\omega_0 = 0$, then the resulting transform pair is

$$1 \Leftrightarrow 2\pi\delta(\omega) \tag{34}$$

from which it follows that

$$K \Leftrightarrow 2\pi K\,\delta(\omega) \tag{35}$$

and our problem is solved. The frequency spectrum of a constant function of time consists only of a component at $\omega = 0$, which we knew all along.

As another example, let us obtain the Fourier transform of a singularity function known as the *signum function*, sgn (t), defined by

$$\text{sgn}\,(t) = \begin{cases} -1 & t < 0 \\ 1 & t > 0 \end{cases} \tag{36}$$

Again, if we should try to substitute this time function into the defining equation for the Fourier transform, we would face an indeterminate expression upon substitution of the limits of integration. This same problem will arise every time we attempt to obtain the Fourier transform of a time function that does not approach zero as $|t|$ approaches infinity. Eventually, we will avoid this situation by defining a new kind of integral transform called the *Laplace transform*, which contains a built-in convergence factor that will cure many of the inconvenient ills associated with the evaluation of certain Fourier transforms.

The signum function under consideration can be written as

$$\operatorname{sgn}(t) = \lim_{a\to 0}\,[e^{-at}u(t) - e^{at}u(-t)]$$

Notice that the expression within the brackets *does* approach zero as $|t|$ gets very large. Using the definition of the Fourier transform, we obtain

$$\mathfrak{F}\{\operatorname{sgn}(t)\} = \lim_{a\to 0}\left[\int_0^{\infty} e^{-j\omega t}e^{-at}\,dt - \int_{-\infty}^{0} e^{-j\omega t}e^{at}\,dt\right]$$

$$= \lim_{a\to 0}\frac{-j2\omega}{\omega^2 + a^2} = \frac{2}{j\omega}$$

The real component is zero since $\operatorname{sgn}(t)$ is an odd function of t. Thus,

$$\operatorname{sgn}(t) \Leftrightarrow \frac{2}{j\omega} \tag{37}$$

As a final example in this section, let us look at the familiar unit-step function, $u(t)$. Making use of our work on the signum function above, we represent the unit step by

$$u(t) = \tfrac{1}{2} + \tfrac{1}{2}\operatorname{sgn}(t)$$

and obtain the Fourier transform pair

$$u(t) \Leftrightarrow \left[\pi\delta(\omega) + \frac{1}{j\omega}\right] \tag{38}$$

Table 18-1 presents the conclusions drawn from the examples discussed in this section, along with a few others that have not been detailed here.

Drill Problems

18-7 Find $\mathbf{F}(j2)$ for each function of time given: (*a*) $5u(t)$; (*b*) $2u(t + 0.8) - 2u(t - 0.8)$; (*c*) $e^{-0.2t}(\cos t)u(t)$.

Ans. $0.655\underline{/-80.6°}$; 2.00; $2.50\underline{/-90°}$

18-8 Evaluate $f(t)$ at $t = 0.1$ if $\mathfrak{F}\{f(t)\} =$: (*a*) $10/(3 + j\omega)$; (*b*) $15\delta(\omega - 4) + 15\delta(\omega + 4)$; (*c*) $\dfrac{\sin 0.2\omega}{0.1\omega} - j\dfrac{6}{\omega}$.

Ans. 4.40; 7.41; 8.00

Table 18-1 Some familiar Fourier transform pairs

$f(t)$	$f(t)$	$\mathfrak{F}\{f(t)\} = \mathbf{F}(j\omega)$	$\lvert\mathbf{F}(j\omega)\rvert$
	$\delta(t - t_0)$	$e^{-j\omega t_0}$	
	$e^{j\omega_0 t}$	$2\pi\delta(\omega - \omega_0)$	
	$\cos \omega_0 t$	$\pi[\delta(\omega + \omega_0) + \delta(\omega - \omega_0)]$	
	1	$2\pi\,\delta(\omega)$	
	$\operatorname{sgn}(t)$	$\dfrac{2}{j\omega}$	
	$u(t)$	$\pi\,\delta(\omega) + \dfrac{1}{j\omega}$	
	$e^{-\alpha t}u(t)$	$\dfrac{1}{\alpha + j\omega}$	
	$e^{-\alpha t} \cos \omega_d t \cdot u(t)$	$\dfrac{\alpha + j\omega}{(\alpha + j\omega)^2 + {\omega_d}^2}$	
	$u(t + \frac{1}{2}T) - u(t - \frac{1}{2}T)$	$T\dfrac{\sin \frac{\omega T}{2}}{\frac{\omega T}{2}}$	

18-6 THE FOURIER TRANSFORM OF A GENERAL PERIODIC TIME FUNCTION

In Sec. 18-2 we remarked that we would be able to show that even periodic time functions possess Fourier transforms. Since a promise made is a debt unpaid, let us now establish this fact on a rigorous basis. Consider a periodic time function $f(t)$ with period T and Fourier series expansion, as outlined by (1), (2), and (3),

$$f(t) = \sum_{n=-\infty}^{\infty} \mathbf{c}_n e^{jn\omega_0 t} \tag{1}$$

where

$$\mathbf{c}_n = \frac{1}{T}\int_{-T/2}^{T/2} f(t)e^{-jn\omega_0 t}\,dt \tag{2}$$

and

$$\omega_0 = \frac{2\pi}{T} \tag{3}$$

Bearing in mind that the Fourier transform of a sum is just the sum of the transforms of the terms in the sum, and that $\mathbf{c}_n$ is not a function of time, we can write

$$\mathfrak{F}\{f(t)\} = \mathfrak{F}\left\{\sum_{n=-\infty}^{\infty} \mathbf{c}_n e^{jn\omega_0 t}\right\} = \sum_{n=-\infty}^{\infty} \mathbf{c}_n \mathfrak{F}\{e^{jn\omega_0 t}\}$$

Substituting the transform of $e^{jn\omega_0 t}$ given in (30), we have

$$f(t) \Leftrightarrow 2\pi \sum_{n=-\infty}^{\infty} \mathbf{c}_n \delta(\omega - n\omega_0) \tag{39}$$

This shows that $f(t)$ has a discrete spectrum consisting of impulses located at points on the ω axis given by $\omega = n\omega_0$, $n = \ldots, -2, -1, 0, 1, \ldots$. The strength of each impulse is 2π times the value of the corresponding Fourier coefficient appearing in the complex form of the Fourier series expansion for $f(t)$.

Now, if the expression for the transform pair shown in (39) is valid, then the inverse Fourier transform of the right side should be $f(t)$. This inverse transform can be written as

$$\mathfrak{F}^{-1}\{\mathbf{F}(j\omega)\} = \frac{1}{2\pi}\int_{-\infty}^{\infty} e^{j\omega t}\left[2\pi \sum_{n=-\infty}^{\infty} \mathbf{c}_n \delta(\omega - n\omega_0)\right] d\omega \stackrel{?}{=} f(t)$$

Since the exponential term does not contain the index of summation n, we can interchange the order of the integration and summation operations,

$$\mathfrak{F}^{-1}\{\mathbf{F}(j\omega)\} = \sum_{n=-\infty}^{\infty} \int_{-\infty}^{\infty} \mathbf{c}_n e^{j\omega t} \delta(\omega - n\omega_0)\, d\omega \stackrel{?}{=} f(t)$$

Because it is not a function of the variable of integration ω, $\mathbf{c}_n$ can be treated as a constant. Then, using the sifting property of the impulse, we obtain

$$\mathfrak{F}^{-1}\{\mathbf{F}(j\omega)\} = \sum_{n=-\infty}^{\infty} \mathbf{c}_n e^{jn\omega_0 t} \stackrel{?}{=} f(t)$$

which is exactly the same as (1), the complex Fourier series expansion for $f(t)$. The question marks in the preceding equations can now be removed, and the existence of the Fourier transform for a periodic time function is established. This should come as no great surprise, however. In the last section we evaluated the Fourier transform of a cosine function, which is certainly periodic, although we made no direct reference to its periodicity. However, we did use a backhanded approach in getting the transform. But now we have a mathematical tool by which the transform can be obtained more directly. To demonstrate this procedure, consider $f(t) = \cos \omega_0 t$ once more. First we evaluate the Fourier coefficients $\mathbf{c}_n$,

$$\mathbf{c}_n = \frac{1}{T} \int_{-T/2}^{T/2} \cos \omega_0 t \; e^{-jn\omega_0 t}\, dt = \begin{cases} \frac{1}{2} & n = \pm 1 \\ 0 & \text{otherwise} \end{cases}$$

Then

$$\mathfrak{F}\{f(t)\} = 2\pi \sum_{n=-\infty}^{\infty} \mathbf{c}_n \delta(\omega - n\omega_0)$$

This expression only has values that are nonzero when $n = \pm 1$, and it follows, therefore, that the entire summation reduces to

$$\mathfrak{F}\{\cos \omega_0 t\} = \pi[\delta(\omega - \omega_0) + \delta(\omega + \omega_0)]$$

which is precisely the expression that we obtained before. What a relief!

Drill Problem

18-9 Determine the Fourier transform for: (*a*) $10 \cos 5t$; (*b*) $10 \sin 5t$; (*c*) $10 \cos (5t + \pi/3)$.

Ans. $10\pi[\delta(\omega + 5) + \delta(\omega - 5)]$; $j10\pi[\delta(\omega + 5) - \delta(\omega - 5)]$; $10\pi[e^{-j\pi/3}\delta(\omega + 5) + e^{j\pi/3}\delta(\omega - 5)]$

18-7 CONVOLUTION AND CIRCUIT RESPONSE IN THE TIME DOMAIN

Before continuing our discussion of the Fourier transform, let us note again where all this is leading. Our goal is a technique for simplifying problems in linear circuit analysis which involve the determination of explicit expressions for response functions caused by the application of one or more forcing functions. We accomplish this by utilizing a mathematical quantity called the *system function* or *transfer function* of the circuit. It turns out that this system function is the Fourier transform of the unit-impulse response of the circuit. The specific analytical technique that will be used requires the evaluation of the Fourier transform of the forcing function, the multiplication of this transform by the system function to obtain the transform of the response function, and then the inverse-transform operation to obtain the response function. By these means some relatively complicated integral expressions will be reduced to simple functions of ω, and the mathematical operations of integration and differentiation will be replaced by the simpler operations of algebraic multiplication and division. With these remarks in mind, let us now proceed to examine the unit-impulse response of a circuit and eventually establish its relation to the system function. Then we can look at some specific analysis problems.

Consider an electric network N without initial stored energy to which a forcing function $x(t)$ is applied. At some point in this circuit, a response function $y(t)$ is present. We show this in block diagram form in Fig. 18-7*a* along with general sketches of typical time functions. The forcing function is arbitrarily shown to exist only over the interval $a < t < b$. Thus, $y(t)$ can exist only for $t > a$. The question that we now wish to answer is this: if we know the form of $x(t)$, then how is $y(t)$ described? To answer this question, it is obvious that we need to know something about N. Suppose, therefore, that our knowledge of N is the way it responds when the forcing function is a unit impulse. That is, we are assuming that we know $h(t)$, the response function resulting from a unit impulse being supplied as the forcing function at $t = 0$, as shown in Fig. 18-7*b*. The function $h(t)$ is commonly called the unit-impulse response function or the *impulse response*. This is a very important descriptive property for an electric circuit. Instead of applying the unit impulse at time $t = 0$, suppose that it were applied at time $t = \lambda$. It is evident that the only change in the output would be a time delay. Thus, the output becomes $h(t - \lambda)$ when the input is $\delta(t - \lambda)$, as shown in Fig. 18-7*c*. Next, suppose that the input impulse were to have some strength other than unity. Specifically, let the strength of the impulse be numerically equal to the value of $x(t)$ when $t = \lambda$. This value $x(\lambda)$ is a constant; we know that the multiplication of a single forcing function in a linear circuit by a constant simply causes the response to change proportionately. Thus, if the input is changed to $x(\lambda)\delta(t - \lambda)$, then the response becomes $x(\lambda)h(t - \lambda)$, as shown in Fig. 18-7*d*. Now let us

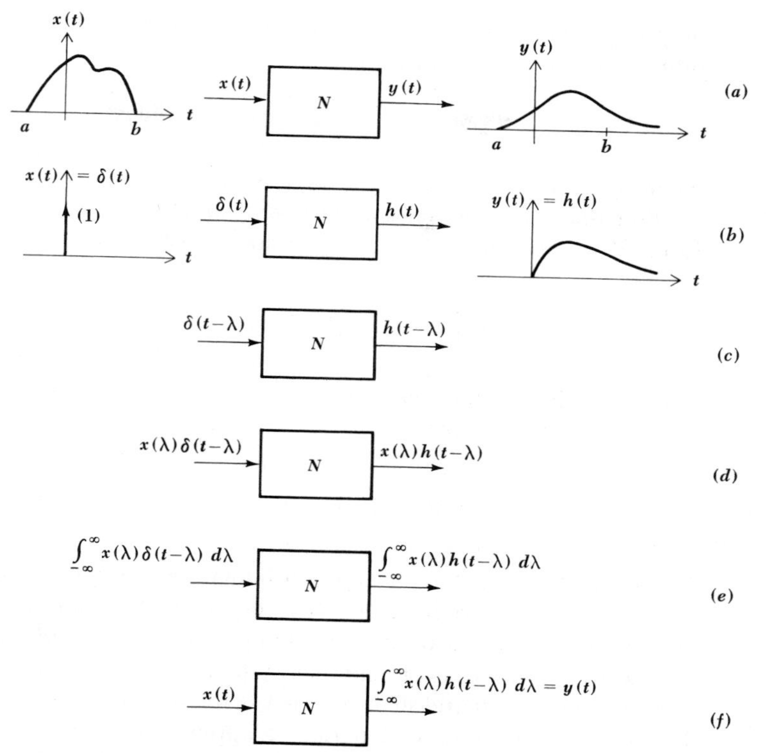

Fig. 18-7 A conceptual development of the convolution integral, $y(t) = \int_{-\infty}^{\infty} x(\lambda)h(t - \lambda)\, d\lambda$.

sum this latest input over all possible values of λ and use the result as a forcing function for N. Linearity decrees that the output must be equal to the sum of the responses resulting from the use of all possible values of λ. Loosely speaking, the integral of the input produces the integral of the output, as shown in Fig. 18-7*e*. But what is the input now? Using the sifting property of the unit impulse, we see that the input is simply $x(t)$, the original input.

Our question is now answered. When $x(t)$, the input to N, is known, and when $h(t)$, the impulse response of N, is known, then $y(t)$, the output or response function, is expressed by

$$y(t) = \int_{-\infty}^{\infty} x(\lambda)h(t - \lambda)\, d\lambda \qquad (40)$$

as shown in Fig. 18-7*f*. This important relationship is known far and wide as the *convolution integral*. In words, this last equation states that "the output is equal to the input convolved with the impulse response." It is often abbreviated by means of

$$y(t) = x(t) * h(t) \tag{41}$$

where the asterisk is read "convolved with."

Equation (40) sometimes appears in a slightly different but equivalent form. If we let $z = t - \lambda$, the expression for $y(t)$ becomes

$$y(t) = \int_{\infty}^{-\infty} -x(t-z)h(z)\,dz = \int_{-\infty}^{\infty} x(t-z)h(z)\,dz$$

and, since the symbol that we use for the variable of integration is unimportant, we can write

$$y(t) = x(t) * h(t) = \int_{-\infty}^{\infty} x(z)h(t-z)\,dz = \int_{-\infty}^{\infty} x(t-z)h(z)\,dz \tag{42}$$

These two forms of the convolution integral are worth memorizing.

The result that we have in (42) is very general. It applies to any linear system. However, we are interested in *physically realizable* systems, those that *do* exist or *could* exist, and such systems have a property that modifies the convolution integral slightly. That is, the response of the system cannot begin before the forcing function is applied. In particular, $h(t)$ is the response of the system resulting from the application of a unit impulse at $t = 0$. Therefore, $h(t)$ cannot exist for $t < 0$. It follows that, in the second integral of (42), the integrand is zero when $z < 0$; in the first integral, the integrand is zero when $t - z$ is negative or when $z > t$. Therefore, for realizable systems the limits of integration change in the convolution integrals:

$$y(t) = x(t) * h(t) = \int_{-\infty}^{t} x(z)h(t-z)\,dz = \int_{0}^{\infty} x(t-z)h(z)\,dz \tag{43}$$

Equations (42) and (43) are both valid, but the latter is more specific when we are speaking of *realizable* linear systems.

Before discussing the significance of the impulse response of a circuit any further, let us consider a numerical example which will give us some insight into just how the convolution integral can be evaluated. Although the expression itself is simple enough, the evaluation is sometimes troublesome, especially with regard to the values used as the limits of integration.

Suppose that the input is a rectangular voltage pulse that starts at $t = 0$, has a duration of 1 s, and is 1 V in amplitude,

$$x(t) = v_i(t) = u(t) - u(t-1)$$

Suppose also that the impulse response of this circuit is known to be an exponential function of the form:[7]

$$h(t) = 2e^{-t}u(t)$$

We wish to evaluate the output voltage $v_o(t)$, and we can write the answer immediately in integral from,

$$y(t) = v_o(t) = v_i(t) * h(t) = \int_0^\infty v_i(t - z)h(z)\,dz$$

$$= \int_0^\infty [u(t - z) - u(t - z - 1)][2e^{-z}u(z)]\,dz$$

Obtaining this expression for $v_o(t)$ was simple enough, but the presence of the many unit-step functions tends to make its evaluation confusing. Careful attention must be paid to determine those portions of the range of integration in which the integrand is zero. Let us first use some graphical assistance to help us understand what it says. We first draw some horizontal z axes lined up one above the other, as shown in Fig. 18-8. We know what $v_i(t)$ looks like, so we know what $v_i(z)$ looks like also; this is plotted as Fig. 18-8*a*. The function $v_i(-z)$ is simply $v_i(z)$ run backwards with respect to z, or rotated about the ordinate axis; it is shown in Fig. 18-8*b*. Next we wish to represent $v_i(t - z)$, which is $v_i(-z)$ after it is shifted to the right by an amount $z = t$ as shown in Fig. 18-8*c*. On the next z axis in Fig. 18-8*d*, the hypothetical impulse response is plotted. Finally, we multiply the two functions $v_i(t - z)$ and $h(z)$. The result is shown in Fig. 18-8*e*. Since $h(z)$ does not exist prior to $t = 0$ and $v_i(t - z)$ does not exist for $z > t$, notice that the product of these two functions has nonzero values only in the interval $0 < z < t$ for the case shown where $t < 1$; when $t > 1$, then the nonzero values for the product are obtained in the interval, $(t - 1) < z < t$. The *area* under the product curve (shown shaded in the figure) is numerically equal to the value of v_o corresponding to the specific value of t selected in Fig. 18-8*c*. As t increases from zero to unity, the area under the product curve continues to rise, and thus $v_o(t)$ continues to rise. But as t increases beyond $t = 1$, the area under the product curve, which is equal to $v_o(t)$, starts decreasing and approaches zero. For $t < 0$, the curves representing $v_i(t - z)$ and $h(z)$ do not overlap at all, so the area under the product curve is obviously zero. Now let us use these graphical concepts to obtain an explicit expression for $v_o(t)$.

For values of t that lie between zero and unity, we must integrate from

[7] A description of one possible circuit to which this impulse response might apply is developed in Prob. 27.

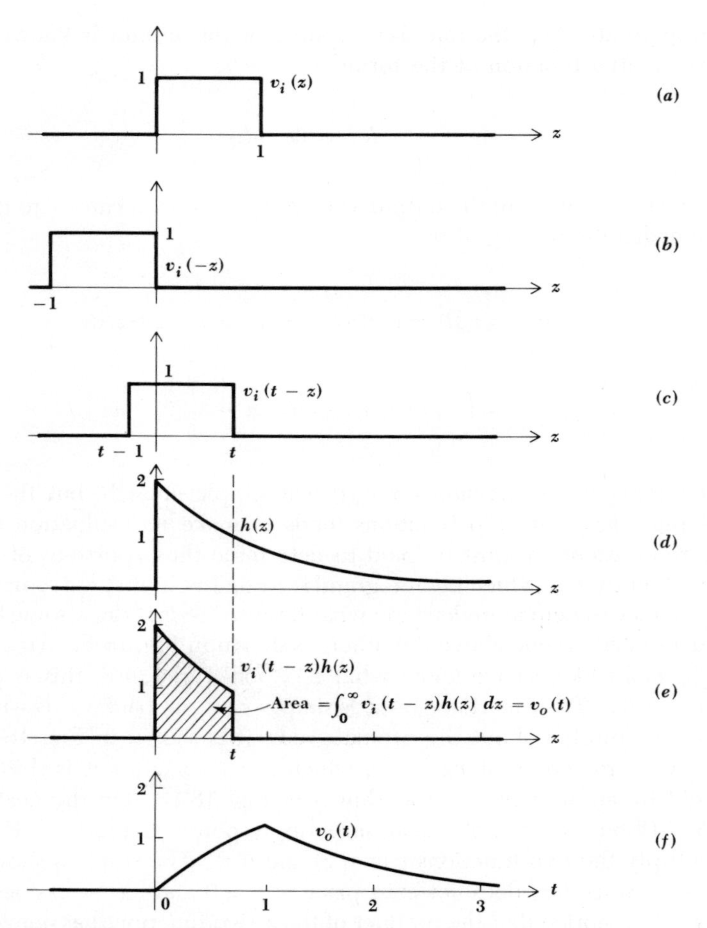

Fig. 18-8 Graphical concepts in evaluating a convolution integral.

$z = 0$ to $z = t$; for values of t that exceed unity, the range of integration is $t - 1 < z < t$. Thus, we may write

$$v_o(t) = \begin{cases} 0 & t < 0 \\ \displaystyle\int_0^t 2e^{-z}\,dz = 2(1 - e^{-t}) & 0 < t < 1 \\ \displaystyle\int_{t-1}^t 2e^{-z}\,dz = 2(e - 1)e^{-t} & t > 1 \end{cases}$$

This function is shown plotted versus the time variable t in Fig. 18-8f, and our problem is completed. There is a great deal of information and technique wrapped up in this one example problem. The drill problems below offer an opportunity to make sure that the procedure is understood sufficiently well to make it worthwhile passing on to new material.

Drill Problems

18-10 Given the impulse response for a network N, $h(t) = u(t) - u(t-1)$, and the input signal, $v_i = 2e^{-t}u(t)$ V, determine $v_o(t)$ at $t =$: (*a*) -0.5; (*b*) 0.5; (*c*) 1.5 s.

Ans. *0; 0.767; 0.787* V

18-11 When the current source, $i_s = \delta(t)$ A, is applied to a certain network, the output voltage is $5[u(t) - u(t-4)]$ V. If the source, $i_s = 2t[u(t) - u(t-3)]$ A, is applied, find the output voltage at $t =$: (*a*) 2; (*b*) 3.62; (*c*) 5 s.

Ans. *20; 40; 45* V

18-8 THE SYSTEM FUNCTION AND RESPONSE IN THE FREQUENCY DOMAIN

In the previous section the problem of determining the output of a physical system in terms of the input and the impulse response was solved by using the convolution integral and working entirely in the time domain. The input, the output, and the impulse response are all *time* functions. Now let us see whether some analytical simplification can be wrought by working with frequency-domain descriptions of these three functions.

To do this we examine the Fourier transform of the system output, utilizing the basic definition of the Fourier transform and the output expressed by the convolution integral (42),

$$\mathfrak{F}\{v_o(t)\} = \mathbf{F}_o(j\omega) = \int_{-\infty}^{\infty} e^{-j\omega t}\left[\int_{-\infty}^{\infty} v_i(t-z)h(z)\,dz\right] dt$$

where we again assume no initial energy storage. At first glance this expression may seem rather formidable, but it can be reduced to a result that is surprisingly simple. We may move the exponential term inside the inner integral because it does not contain the variable of integration z. Next we reverse the order of integration, obtaining

$$\mathbf{F}_o(j\omega) = \int_{-\infty}^{\infty}\int_{-\infty}^{\infty} e^{-j\omega t} v_i(t-z)h(z)\,dt\,dz$$

Since it is not a function of t, we can extract $h(z)$ from the inner integral and simplify the integration with respect to t by a change of variable, $t - z = x$,

$$\mathbf{F}_o(j\omega) = \int_{-\infty}^{\infty} h(z) \int_{-\infty}^{\infty} e^{-j\omega(x+z)} v_i(x)\, dx\, dz$$

$$= \int_{-\infty}^{\infty} e^{-j\omega z} h(z) \int_{-\infty}^{\infty} e^{-j\omega x} v_i(x)\, dx\, dz$$

But now the sum is starting to break through, for the inner integral is merely the Fourier transform of $v_i(t)$. Furthermore, it contains no z terms and can be treated as a constant in any integration involving z. Thus, we can move this transform, $\mathbf{F}_i(j\omega)$, completely outside all the integral signs,

$$\mathbf{F}_o(j\omega) = \mathbf{F}_i(j\omega) \int_{-\infty}^{\infty} e^{-j\omega z} h(z)\, dz$$

Finally, the remaining integral exhibits our old friend once more, another Fourier transform! This one is the Fourier transform of the impulse response, which we shall designate by the notation $\mathbf{H}(j\omega)$. Therefore, all our work has boiled down to the simple result:

$$\mathbf{F}_o(j\omega) = \mathbf{F}_i(j\omega)\mathbf{H}(j\omega) = \mathbf{F}_i(j\omega)\mathfrak{F}\{h(t)\} \tag{44}$$

This is another important result; it defines for us the meaning of the *system function* $\mathbf{H}(j\omega)$. The system function is the ratio of the Fourier transform of the response function to the Fourier transform of the forcing function. Moreover, the system function and the impulse response comprise a Fourier transform pair:

$$h(t) \Leftrightarrow \mathbf{H}(j\omega) \tag{45}$$

The development in the preceding paragraph also serves to prove the general statement that the Fourier transform of the convolution of two time functions is the product of their Fourier transforms,

$$\mathfrak{F}\{f(t) * g(t)\} = \mathbf{F}_f(j\omega)\mathbf{F}_g(j\omega) \tag{46}$$

To recapitulate, if we know the Fourier transforms of the forcing function and the impulse response, then the Fourier transform of the response function can be obtained as their product. The result is a description of the response function in the frequency domain; if we wish to do so, we can obtain the time-domain description of the response function by taking the inverse Fourier transform. Thus we see that the process of convolution in the time domain has been reduced to the relatively simple operation of multiplication in the frequency domain. This is one fact that makes the use of integral transforms so attractive.

The foregoing comments might make us wonder why we would ever choose to work in the time domain, but we must always remember that we seldom get something for nothing. A poet once said, "Our sincerest laughter, with some pain is fraught."[8] The pain herein is the occasional difficulty in obtaining the inverse Fourier transform of a response function, for reasons of mathematical complexity. On the other hand, a modern digital computer can convolve two time functions with magnificent celerity. For that matter, it can also obtain an FFT (fast Fourier transform)[9] quite rapidly. Consequently there is no clear-cut advantage between working in the time domain or the frequency domain. A decision must be made each time a new problem arises; it should be based on the given information available and on the computational facilities at hand.

Now let us attempt a frequency-domain analysis of the problem that we worked in the preceding section with the convolution integral. Recalling that we had a forcing function of the form

$$v_i(t) = u(t) - u(t - 1)$$

and a unit-impulse response defined by

$$h(t) = 2e^{-t}u(t)$$

we first obtain the corresponding Fourier transforms. The forcing function is the difference between two unit-step functions. These two functions are identical, except that one is initiated 1 s after the other. We shall evaluate the response due to $u(t)$, and the response due to $u(t - 1)$ will be the same, but delayed in time by 1 s. The difference between these two partial responses will be the total response due to $v_i(t)$.

The Fourier transform of $u(t)$ was obtained in Sec. 18-5,

$$\mathfrak{F}\{u(t)\} = \pi\delta(\omega) + \frac{1}{j\omega}$$

The system function is obtained by taking the Fourier transform of $h(t)$, listed in Table 18-1,

$$\mathfrak{F}\{h(t)\} = \mathbf{H}(j\omega) = \mathfrak{F}\{2e^{-t}u(t)\} = \frac{2}{1 + j\omega}$$

[8]A cultural message from P. B. Shelley, "To a Skylark," 1821.

[9]The fast Fourier transform is a type of *discrete* Fourier transform, which is a numerical approximation to the (continuous) Fourier transform we have been considering. Many references concerning it are given in G. D. Bergland, A Guided Tour of the Fast Fourier Transform, *IEEE Spectrum*, vol. 6, no. 7, pp. 41–52, July 1969.

The inverse transform of the product of these two functions yields that component of $v_o(t)$ caused by $u(t)$,

$$v_{o1}(t) = \mathfrak{F}^{-1}\left\{\frac{2\pi\delta(\omega)}{1 + j\omega} + \frac{2}{j\omega(1 + j\omega)}\right\}$$

Using the sifting property of the unit impulse, the inverse transform of the first term is just a constant equal to unity. After breaking the second term up into partial fractions,

$$v_{o1}(t) = 1 + \mathfrak{F}^{-1}\left\{\frac{2}{j\omega} - \frac{2}{1 + j\omega}\right\}$$

the inverse transforms are taken from Table 18-1 once more:

$$\begin{aligned} v_{o1}(t) &= 1 + \operatorname{sgn}(t) - 2e^{-t}u(t) \\ &= 2u(t) - 2e^{-t}u(t) \\ &= 2(1 - e^{-t})u(t) \end{aligned}$$

It follows that $v_{o2}(t)$, the component of $v_o(t)$ produced by $u(t - 1)$, is

$$v_{o2}(t) = 2(1 - e^{-(t-1)})u(t - 1)$$

Therefore,

$$\begin{aligned} v_o(t) &= v_{o1}(t) + v_{o2}(t) \\ &= 2(1 - e^{-t})u(t) - 2(1 - e^{-t+1})u(t - 1) \end{aligned}$$

The discontinuities at $t = 0$ and $t = 1$ dictate a separation into three time intervals:

$$v_o(t) = \begin{cases} 0 & t < 0 \\ 2(1 - e^{-t}) & 0 < t < 1 \\ 2(e - 1)e^{-t} & t > 1 \end{cases}$$

This is the same result we obtained by working this problem with the convolution integral in the time domain.

It appears that some ease of solution is realized by working this particular problem in the frequency domain rather than in the time domain. We are able to trade a relatively complicated time-domain integration for a simple multiplication of two Fourier transforms in the frequency domain, and (this is a most noteworthy point) we did not have to pay the price of having a difficult Fourier transform to invert in determining $v_o(t)$.

Drill Problems

18-12 If $h(t) = 2e^{-3t}u(t)$ and $v_i(t) = 3e^{-2t}u(t)$, first find $\mathbf{H}(j\omega)$, $\mathbf{V}_i(j\omega)$, and $\mathbf{V}_o(j\omega) = \mathbf{V}_i(j\omega)\mathbf{H}(j\omega)$, and then evaluate $v_o(t)$ at $t =$: (*a*) 0.5; (*b*) 1.0; (*c*) 1.5 s.

Ans. *0.232; 0.513; 0.869* V

18-13 Use frequency-domain methods to find the time-domain response of a network having a system function, $j2\omega/(1 + j2\omega)$, if the input is: (*a*) a unit impulse; (*b*) a unit-step function; (*c*) $\cos t$.

Ans. $e^{-t/2}u(t)$; $\delta(t) - 0.5e^{-t/2}u(t)$; $0.8 \cos t - 0.4 \sin t$

18-9 THE PHYSICAL SIGNIFICANCE OF THE SYSTEM FUNCTION

In this section we shall try to connect several aspects of the Fourier transform with work we have completed in earlier chapters.

Given a general linear two-port network N without any initial energy storage, we assume sinusoidal forcing and response functions, arbitrarily assumed to be voltages, as shown in Fig. 18-9. We let the input voltage be simply $A \cos(\omega_x t + \theta)$, and the output can be described in general terms as $v_o(t) = B \cos(\omega_x t + \phi)$, where the amplitude B and phase angle ϕ are functions of ω_x. In phasor form, we can write the forcing and response functions as $\mathbf{V}_i = Ae^{j\theta}$ and $\mathbf{V}_o = Be^{j\phi}$. The ratio of the phasor response to the phasor forcing function is a complex number that is a function of ω_x,

$$\frac{\mathbf{V}_o}{\mathbf{V}_i} = \mathbf{G}(\omega_x) = \frac{B}{A}e^{j(\phi-\theta)}$$

where B/A is the amplitude of $\mathbf{G}$ and $\phi - \theta$ is its phase angle. This function $\mathbf{G}(\omega_x)$ is the frequency-response function for N, and it could be obtained in the laboratory by varying ω_x over a large range of values, and measuring the amplitude B/A and phase $\phi - \theta$ for each value of ω_x. If we would then plot each of these parameters as a function of frequency, the resultant pair of curves would completely describe the frequency-response function.

Now let us hold these comments in the backs of our minds for a moment as we consider a slightly different aspect of the same analysis problem.

Fig. 18-9 Sinusoidal analysis can be used to determine the transfer function, $\mathbf{H}(j\omega_x) = (B/A)e^{j(\phi-\theta)}$, where B and ϕ are functions of ω_x.

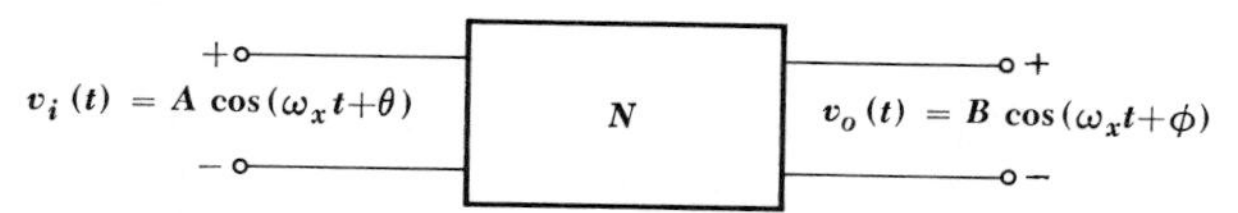

For the circuit with sinusoidal input and output shown in Fig. 18-9, what is the system function $\mathbf{H}(j\omega)$? To answer this question, we begin with the definition of $\mathbf{H}(j\omega)$ as the ratio of the Fourier transforms of the output and the input. Both of these time functions involve the functional form, $\cos(\omega_x t + \beta)$, whose Fourier transform we have not evaluated as yet, although we can handle $\cos \omega_x t$. The transform we need is

$$\mathfrak{F}\{\cos(\omega_x t + \beta)\} = \int_{-\infty}^{\infty} e^{-j\omega t} \cos(\omega_x t + \beta)\, dt$$

If we make the substitution, $\omega_x t + \beta = \omega_x \tau$, then

$$\begin{aligned}\mathfrak{F}\{\cos(\omega_x t + \beta)\} &= \int_{-\infty}^{\infty} e^{-j\omega\tau + j\omega\beta/\omega_x} \cos \omega_x \tau\, d\tau \\ &= e^{j\omega\beta/\omega_x}\, \mathfrak{F}\{\cos \omega_x t\} \\ &= \pi e^{j\omega\beta/\omega_x}[\delta(\omega - \omega_x) + \delta(\omega + \omega_x)]\end{aligned}$$

This is a new Fourier transform pair,

$$\cos(\omega_x t + \beta) \Leftrightarrow \pi e^{j\omega\beta/\omega_x}[\delta(\omega - \omega_x) + \delta(\omega + \omega_x)] \tag{47}$$

which we can now use to evaluate the desired system function,

$$\begin{aligned}\mathbf{H}(j\omega) &= \frac{\mathfrak{F}\{B\cos(\omega_x t + \phi)\}}{\mathfrak{F}\{A\cos(\omega_x t + \theta)\}} \\ &= \frac{\pi B e^{j\omega\phi/\omega_x}[\delta(\omega - \omega_x) + \delta(\omega + \omega_x)]}{\pi A e^{j\omega\theta/\omega_x}[\delta(\omega - \omega_x) + \delta(\omega + \omega_x)]} \\ &= \frac{B}{A} e^{j\omega(\phi - \theta)/\omega_x}\end{aligned}$$

Now we recall the expression for $\mathbf{G}(\omega_x)$,

$$\mathbf{G}(\omega_x) = \frac{B}{A} e^{j(\phi - \theta)}$$

where B and ϕ were evaluated at $\omega = \omega_x$, and we see that evaluating $\mathbf{H}(j\omega)$ at $\omega = \omega_x$ gives

$$\mathbf{H}(j\omega_x) = \mathbf{G}(\omega_x) = \frac{B}{A} e^{j(\phi - \theta)}$$

Since there is nothing special about the x subscript, we conclude that the system function and the frequency-response function are identical:

$$\mathbf{H}(j\omega) = \mathbf{G}(\omega) \tag{48}$$

The fact that one argument is ω while the other is indicated by $j\omega$ is immaterial and arbitrary; the j merely makes possible a more direct comparison between the Fourier and Laplace transforms.

Equation (48) represents a direct connection between Fourier transform techniques and sinusoidal steady-state analysis. Our previous work on steady-state sinusoidal analysis using phasors was but a special case of the more general techniques of Fourier transform analysis. It was "special" in the sense that the inputs and outputs were sinusoids, whereas the use of Fourier transforms and system functions enables us to handle non-sinusoidal forcing functions and responses.

As an example of the application of these profound generalities, let us look at the simple RL series circuit shown in Fig. 18-10a. We seek the voltage across the inductor when the input voltage is a simple exponentially decaying pulse. We need the system function; but it is not necessary to apply an impulse, find the impulse response, and then determine its inverse transform. Instead we use (48) to obtain the system function $\mathbf{H}(j\omega)$ by assuming that the input and output voltages are both sinusoids described by their corresponding phasors, as shown in Fig. 18-10b. Using voltage division, we have

$$\mathbf{H}(j\omega) = \frac{\mathbf{V}_o}{\mathbf{V}_i} = \frac{j2\omega}{4 + j2\omega}$$

The transform of the forcing function is

$$\mathcal{F}\{v_i(t)\} = \frac{5}{3 + j\omega}$$

Fig. 18-10 (a) The response $v_o(t)$ caused by $v_i(t)$ is desired. (b) The system function $\mathbf{H}(j\omega)$ may be determined by sinusoidal steady-state analysis: $\mathbf{H}(j\omega) = \mathbf{V}_o/\mathbf{V}_i$.

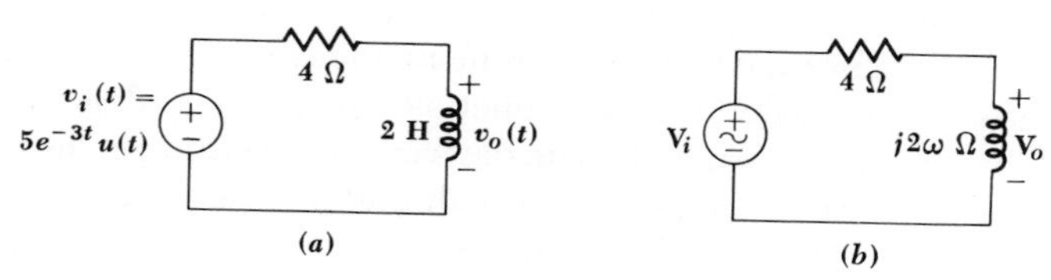

and thus the transform of $v_o(t)$ is given as

$$\begin{aligned}
\mathfrak{F}\{v_o(t)\} &= \mathbf{H}(j\omega)\mathfrak{F}\{v_i(t)\} \\
&= \frac{j2\omega}{4 + j2\omega}\,\frac{5}{3 + j\omega} \\
&= \frac{15}{3 + j\omega} - \frac{10}{2 + j\omega}
\end{aligned}$$

where the partial fractions appearing in the last step help to determine the inverse Fourier transform

$$\begin{aligned}
v_o(t) &= \mathfrak{F}^{-1}\left\{\frac{15}{3 + j\omega} - \frac{10}{2 + j\omega}\right\} \\
&= 15e^{-3t}u(t) - 10e^{-2t}u(t) \\
&= 5(3e^{-3t} - 2e^{-2t})u(t)
\end{aligned}$$

Our problem is completed without fuss, convolution, or differential equations.

Returning again to (48), the identity between the system function $\mathbf{H}(j\omega)$ and the sinusoidal steady-state frequency-response function $\mathbf{G}(\omega)$, we may now consider the system function as the ratio of the output phasor to the input phasor. Suppose that we hold the input-phasor amplitude at unity and the phase angle at zero. Then the output phasor is $\mathbf{H}(j\omega)$. Under these conditions, if we record the output amplitude and phase as functions of ω, for all ω, we have recorded the system function $\mathbf{H}(j\omega)$ as a function of ω, for all ω. We thus have examined the system response under the condition that an infinite number of sinusoids, all with unity amplitude and zero phase, were successively applied at the input. Now suppose that our input is a single unit impulse and look at the impulse response $h(t)$. Is the information we examine really any different from that we just obtained? The Fourier transform of the unit impulse is a constant equal to unity, indicating that all frequency components are present, all with the same magnitude, and all with zero phase. Our system response is the sum of the responses to all these components. The result might be viewed at the output on a cathode-ray oscilloscope. It is evident that the system function and the impulse-response function contain equivalent information regarding the response of the system.

We therefore have two different methods of describing the response of a system to a general forcing function; one is a time-domain description, and the other a frequency-domain description. Working in the time domain, we convolve the forcing function with the impulse response of the system to obtain the response function. As we saw when we first considered convolution, this procedure may be interpreted by thinking of the input as a continuum of impulses of different strengths and times of application; the output which results is a continuum of impulse responses.

In the frequency domain, however, we determine the response by multiplying the Fourier transform of the forcing function by the system function. In this case we interpret the transform of the forcing function as a frequency spectrum, or a continuum of sinusoids. Multiplying this by the system function, we obtain the response function, also as a continuum of sinusoids.

Whether we choose to think of the output as a continuum of impulse responses or as a continuum of sinusoidal responses, the linearity of the network and the superposition principle enable us to determine the total output as a time function by summing over all frequencies (the inverse Fourier transform), or as a frequency function by summing over all time (the Fourier transform).

Unfortunately, both of these techniques have some difficulties or limitations associated with their use. In using convolution, the integral itself can often be rather difficult to evaluate when complicated forcing functions or impulse response functions are present. Furthermore, from the experimental point of view, we cannot really measure the impulse response of a system because we cannot actually generate an impulse. Even if we approximate the impulse by a short high-amplitude pulse, we should probably drive our system into saturation and out of its linear operating range.

With regard to the frequency domain, we encounter one absolute limitation in that we may easily hypothesize forcing functions that we would like to apply theoretically that do not possess Fourier transforms. Moreover, if we wish to find the time-domain description of the response function, we must evaluate an inverse Fourier transform, and some of these inversions can be extremely difficult.

Finally, neither of these techniques offers a very convenient method of handling initial conditions.

The greatest benefits derived from the use of the Fourier transform arise through the abundance of useful information it provides about the spectral properties of a signal, particularly the energy or power per unit bandwidth.

Most of the difficulties and limitations associated with the Fourier transform are overcome with the use of the Laplace transform. We shall see that it is defined in such a way that there is a built-in convergence factor that enables transforms to be determined for a much wider range of input time functions than the Fourier transform can accommodate. Moreover, we shall find that initial conditions may be handled in a manner that may make us wonder why this marvelous technique was not introduced before the last chapter. Finally, the spectral information that is so readily available in the Fourier transform can also be obtained from the Laplace transform, at least for most of the time functions that arise in real engineering problems.

Well, why *has* this all been withheld until now? The best answer is probably because these powerful techniques can overcomplicate the solution of simple problems and tend to obscure the physical interpretation of the performance of the simpler networks. For example, if we are interested only in the forced response, then there is little point in using the Laplace transform and obtaining both the forced and natural response after laboring through a difficult inverse transform operation.

So much for generalities; on to Laplace.

Drill Problem

18-14 Using Fourier transform methods, find v_C in the circuit of Fig. 18-11 at $t = 0.5$ s if $i_s =$: (*a*) $\delta(t)$; (*b*) $u(t)$; (*c*) $u(t) - u(t - 0.4)$.

Ans. 148; 214; 215 V

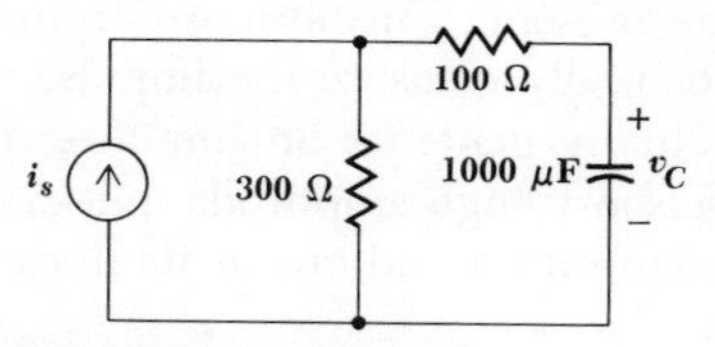

Fig. 18-11 See Drill Prob. 18-14.

PROBLEMS

☐ **1** Find the Fourier transforms for the waveforms illustrated in Fig. 18-12.

Fig. 18-12 See Prob. 1.

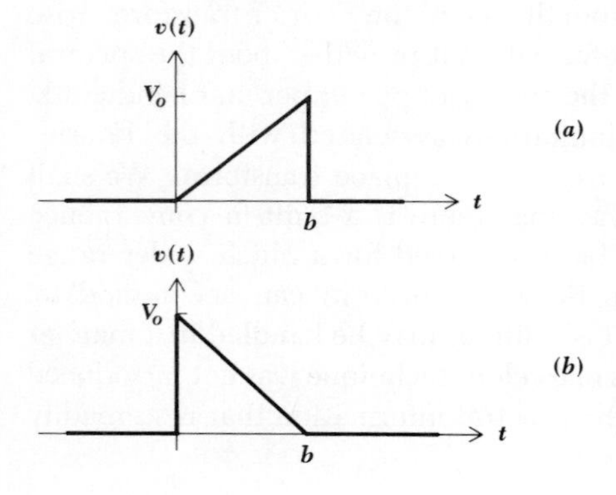

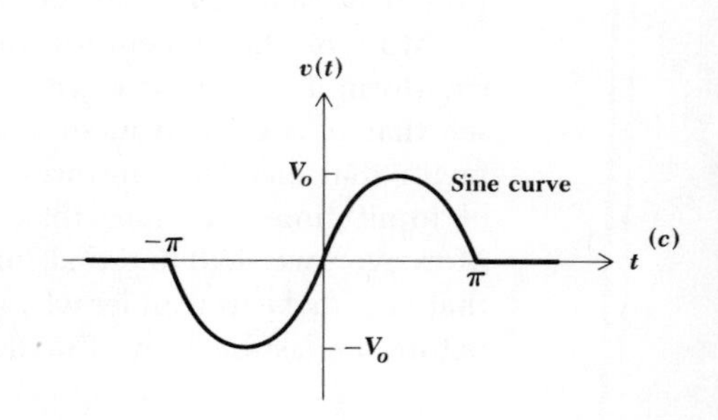

☐ **2** Given the function, $f(t) = \sin 2\pi t$, $0 < t < \frac{1}{4}$, and $f(t) = 0$ for $t > \frac{1}{4}$ and $t < -\frac{1}{4}$, define $f(t)$ for $-\frac{1}{4} < t < 0$ so that: (*a*) $f(t)$ is an odd function and determine $\mathbf{F}(j\omega)$; (*b*) $f(t)$ is an even function and determine $\mathbf{F}(j\omega)$.

☐ **3** Determine the Fourier transform of the waveform shown in Fig. 18-13.

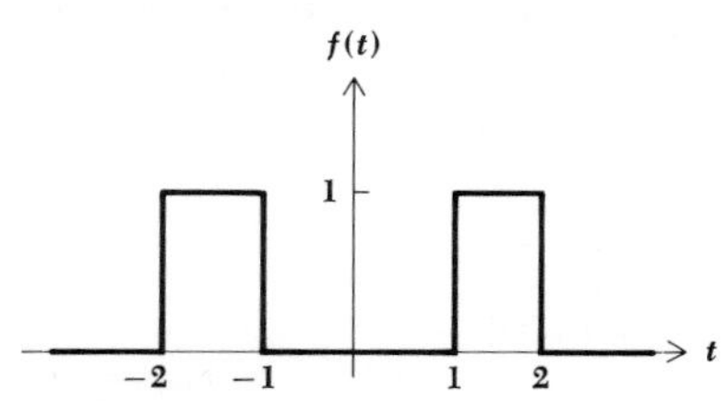

Fig. 18-13 See Prob. 3.

☐ **4** For each continuous frequency spectrum given, determine $f(t)$: (*a*) $\mathbf{F}(j\omega) = e^{-|\omega|}$; (*b*) $\mathbf{F}(j\omega) = e^{-|\omega|}u(|\omega| - 1)$; (*c*) $\mathbf{F}(j\omega) = e^{-|\omega|}u(1 - |\omega|)$; (*d*) $\mathbf{F}(j\omega) = e^{-\omega}u(\omega)$.

☐ **5** If $\mathbf{F}(j\omega)$ is both a real and an even function of ω: (*a*) show that $f(t) = (1/\pi)\int_0^\infty \mathbf{F}(j\omega)\cos\omega t\, d\omega$. If $\mathbf{F}(j\omega) = 1/|\omega|$ for $1 < |\omega| < 2$ and is zero elsewhere: (*b*) evaluate $f(0)$; (*c*) use graphical or numerical methods to approximate $f(\pi/2)$.

☐ **6** Find the total 1-Ω energy associated with the voltage: (*a*) $Ae^{-bt}u(t)$; (*b*) $Ae^{-b|t|}$.

☐ **7** Given the current pulse, $i(t) = te^{-bt}u(t)$: (*a*) find the total 1-Ω energy associated with this waveform by two different methods. (*b*) What fraction of this energy is present in the frequency band from $-b$ to b rad/s?

☐ **8** (*a*) Let $\alpha > 0$ and show that the Fourier transform of $f(t) = e^{-\alpha t}(\cos\omega_d t)u(t)$ is $(\alpha + j\omega)/[(\alpha + j\omega)^2 + \omega_d^2]$. Find the total 1-Ω energy associated with $e^{-t}(\cos t)u(t)$ by using a: (*b*) time-domain integration; (*c*) frequency-domain integration. $\left\{\textit{Hint: } \int_0^\infty [(x^2 + 1)/(x^4 + 4)]\, dx = 3\pi/8.\right\}$

☐ **9** An ideal source, $i_s = \delta(t)$ mA, is applied to a 2-μF capacitor and a 500-Ω resistor in parallel, as illustrated in Fig. 18-14*a*. Find: (*a*) $i_R(t)$; (*b*) $i_C(t)$; (*c*) $v(t)$. Determine the total energy: (*d*) dissipated in the resistor; (*e*) provided by the source.

Fig. 18-14 (*a*) See Prob. 9. (*b*) See Prob. 10.

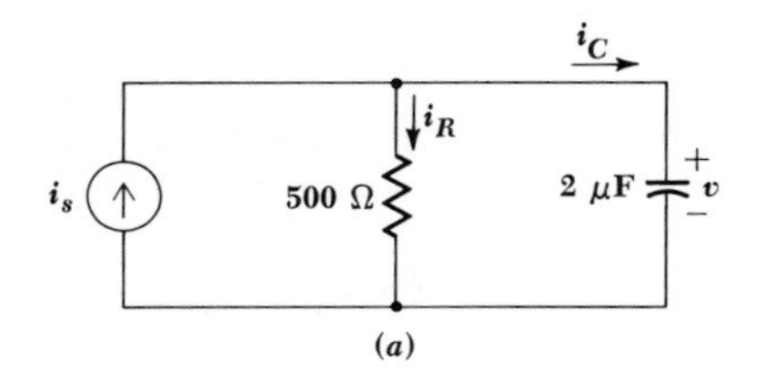

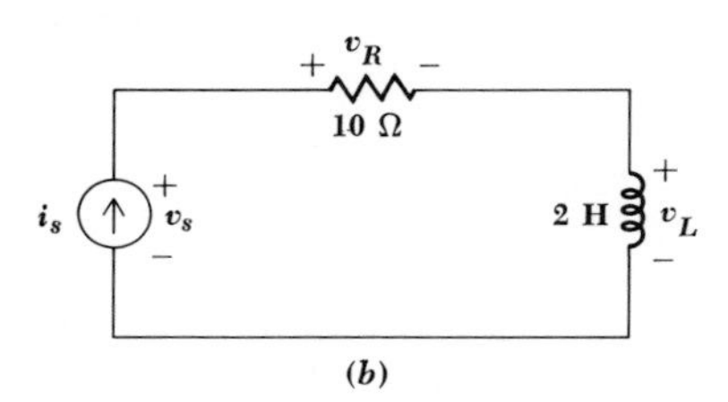

☐ **10** With reference to the circuit shown in Fig. 18-14*b*, let $i_s = 0.1u(t)$ A and find: (*a*) $v_R(t)$; (*b*) $v_L(t)$; (*c*) $v_s(t)$.

☐ **11** (*a*) In Fig. 18-15, let $v_1(t) = \delta(t)$ V, and find $v_2(t)$. (*b*) Let $v_1(t) = u(t)$ V, and find $v_2(t)$. (*c*) Show that the forcing function in part (*a*) is the derivative of the forcing function in (*b*), and that the response in (*a*) is the derivative of the response in (*b*).

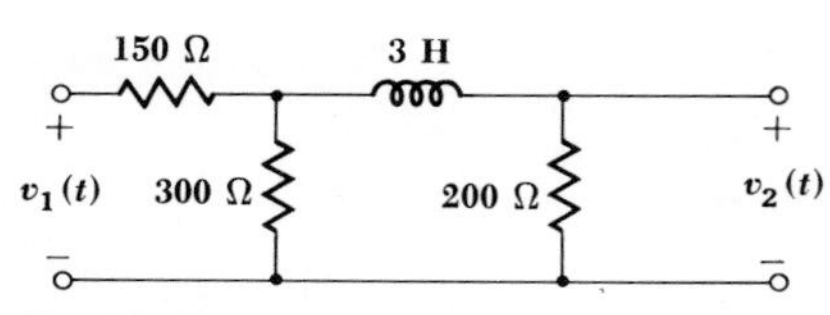

Fig. 18-15 See Prob. 11.

☐ **12** A rectangular voltage pulse, $v(t) = 10k\left[u(t) - u\left(t - \frac{1}{k}\right)\right]$ V, is applied in series with a 100-μF capacitor and a 20-kΩ resistor. Find the capacitor voltage at $t = 1$ if: (*a*) $k = 1$; (*b*) $k = 10$; (*c*) $k = 100$; (*d*) in the limit as $k \to \infty$.

☐ **13** Find the Fourier transform of each of the following functions of time: (*a*) $e^{-(1+t^2)}\delta(t)$; (*b*) $[\text{sgn}\,(t)][u(t+1) - u(t-1)]$; (*c*) $10\cos^2 4t$.

☐ **14** Use the definition of the Fourier transform to prove the following results: (*a*) $\mathfrak{F}\{f(t-t_0)\} = e^{-j\omega t_0}\mathfrak{F}\{f(t)\}$; (*b*) $\mathfrak{F}\{df(t)/dt\} = j\omega\mathfrak{F}\{f(t)\}$; (*c*) $\mathfrak{F}\{f(kt)\} = (1/|k|)\mathbf{F}(j\omega/k)$, where $\mathfrak{F}\{f(t)\} = \mathbf{F}(j\omega)$; (*d*) $\mathfrak{F}\{f(-t)\} = \mathbf{F}(-j\omega)$.

☐ **15** Prove the following Fourier transform pairs: (*a*) $e^t u(-t) \Leftrightarrow 1/(1 - j\omega)$; (*b*) $e^{-|t|} \Leftrightarrow 2/(1+\omega^2)$; (*c*) $e^{-\alpha t}(\sin \omega_d t)u(t) \Leftrightarrow \omega_d/[(\alpha + j\omega)^2 + \omega_d^2]$.

☐ **16** Find and sketch as a function of time the inverse Fourier transform corresponding to $\mathbf{F}(j\omega) =$: (*a*) $j\omega/(1+\omega^2)$; (*b*) $1/[(1+j\omega)(2+j\omega)]$.

☐ **17** Find $f(t)$ if $\mathfrak{F}\{f(t)\} =$: (*a*) $1/(1 - \omega^2 + j\omega)$; (*b*) $1/[(1+j\omega)(2+j\omega)(3+j\omega)]$.

☐ **18** (*a*) Given $f(t) \Leftrightarrow \mathbf{F}(j\omega)$, show that $tf(t) \Leftrightarrow j\,d\mathbf{F}(j\omega)/d\omega$. (*b*) Find $\mathfrak{F}^{-1}\{1/(1+j\omega)^2\}$.

☐ **19** Find the Fourier transform of the periodic waveform shown in Fig.: (*a*) 18-16*a*; (*b*) 18-16*b*.

☐ **20** Given $\mathbf{F}(j\omega) = 10\pi\delta(\omega) - j10\sum_{n=1}^{\infty}(1/n)[\delta(\omega + 20\pi n) - \delta(\omega - 20\pi n)]$, find $f(0.025)$.

☐ **21** Convolve $x(t)$ with $h(t)$ for these pairs of functions: (*a*) $x(t) = h(t) = u(t) - u(t-1)$; (*b*) $x(t) = 2e^t u(t)$, $h(t) = (\cos t)u(t)$.

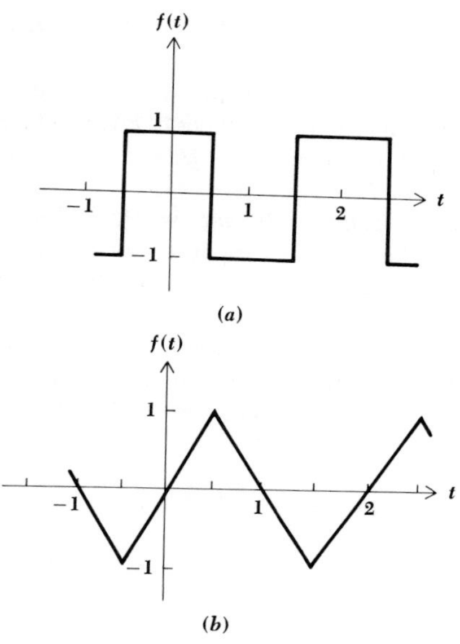

Fig. 18-16 See Prob. 19.

☐ **22** (*a*) Sketch the waveshape versus *t* which results when the two functions shown in Fig. 18-17*a* are convolved. (*b*) Repeat for Fig. 18-17*b*.

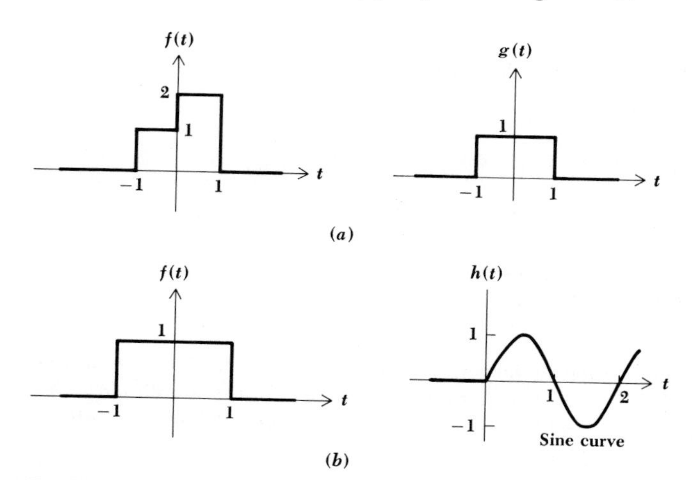

Fig. 18-17 See Probs. 22 and 23.

☐ **23** Show that both forms of the convolution integral given by (42) in Sec. 18-7 lead to the same result for the time functions shown in Fig. 18-17*a*.

☐ **24** If $y(t) = f(t) * g(t)$, find $y(t)$ for $\mathbf{F}(j\omega) = 1/(1 + \omega^2)$ and $\mathbf{G}(j\omega) = 2 + \delta(\omega)$ by convolution in the: (*a*) frequency domain; (*b*) time domain.

☐ **25** If $\mathbf{Y}(j\omega) = \mathbf{F}(j\omega)\mathbf{H}(j\omega)$, find $y(t)$ for $\mathbf{F}(j\omega) = \mathbf{H}(j\omega) = 1/(1 + \omega^2)$.

☐ **26** Find the system function, $\mathbf{H}(j\omega) = \mathbf{V}_i(j\omega)/\mathbf{I}_i(j\omega)$, and impulse response of a one-port network if the input voltage, $v_i(t) = 100 \cos \omega_0 t$ V, produces the input current, $\mathbf{I}_i(j\omega) = 100\pi[\delta(\omega + \omega_0) + \delta(\omega - \omega_0)](1 - j/2\omega)$.

☐ **27** In Sec. 18-7 the impulse response, $h(t) = 2e^{-t}u(t)$, was hypothesized. In order to develop one network which has such a response: (*a*) determine $\mathbf{H}(j\omega) = \mathbf{V}_o(j\omega)/\mathbf{V}_i(j\omega)$. (*b*) Either by inspecting $h(t)$ or $\mathbf{H}(j\omega)$, note that the network has a single energy storage element. Arbitrarily selecting an *RC* circuit with $R = 1\ \Omega$, $C = 1$ F, to provide the necessary time constant, determine the form of the circuit to give $\frac{1}{2}h(t)$ or $\frac{1}{2}\mathbf{H}(j\omega)$. (*c*) Place an ideal-voltage amplifier in cascade with the network to provide the proper multiplicative constant. What is the gain of the amplifier?

☐ **28** Find $v_o(t)$ in the *LC* circuit shown in Fig. 18-18 if $v_i(t) = \delta(t)$. (*Hint:* See Prob. 15*c*.)

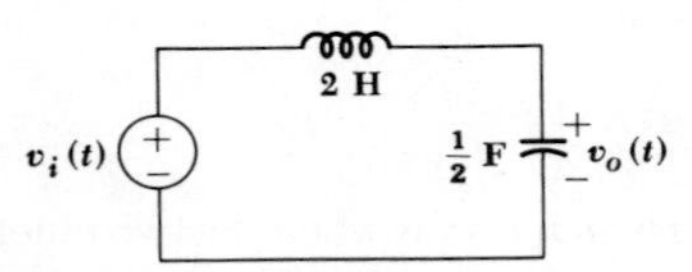

Fig. 18-18 See Prob. 28.

☐ **29** Find $i_o(t)$ in the circuit shown in Fig. 18-19 if $v_i(t) =$: (*a*) $\delta(t)$ V; (*b*) $u(t)$ V; (*c*) $e^{-t}u(t)$ V.

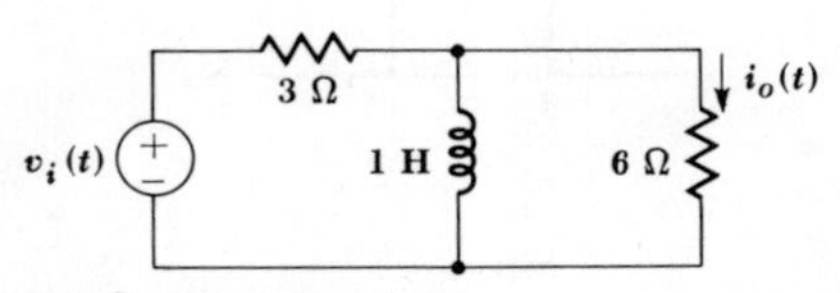

Fig. 18-19 See Prob. 29.

☐ **30** The source in the circuit shown in Fig. 18-20 is $v_i(t) = 100[u(t) - u(t - 0.001)]$ V. Sketch curves as a function of ω of the amplitude of $\mathbf{V}_i(j\omega)$, $\mathbf{H}(j\omega) = \mathbf{V}_o(j\omega)/\mathbf{V}_i(j\omega)$, and $\mathbf{V}_o(j\omega)$. In what way does the *RC* network affect the spectrum of the input signal?

Fig. 18-20 See Prob. 30.

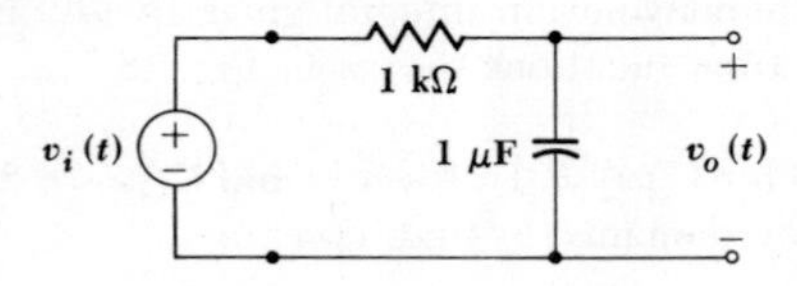

Chapter Nineteen
Laplace Transform Techniques

19-1 INTRODUCTION

As we begin this last chapter on the Laplace transform, it may be beneficial to pause long enough to review our progress to date, first because we have covered a considerable amount of ground since page 1 and some tying together is advisable, and secondly because we need to see how the Laplace transform fits into our hierarchy of analytical methods.

Our goal has constantly been one of analysis: given some forcing function at one point in a linear circuit, determine the response at some other point. For the first several chapters, we played only with dc forcing functions and responses of the form V_0e^0. However, after the introduction of inductance and capacitance, the dc excitation of simple RL and RC circuits produced responses varying exponentially with time $V_0e^{\sigma t}$. When we considered the RLC circuit, the responses took on the form of the exponentially varying sinusoid $V_0e^{\sigma t}\cos(\omega t + \theta)$. All of this work was accomplished in the time domain, and the dc forcing function was the only one we considered.

As we advanced to the use of the sinusoidal forcing function, the tedium and complexity of solving the integrodifferential equations caused us to begin casting about for an easier way to work problems. The phasor transform was the result, and we might remember that we were led to it through consideration of a complex forcing function of the form $V_0e^{j\theta}e^{j\omega t}$. As soon as we concluded that we did not need the factor containing t, we were left with the phasor $V_0e^{j\theta}$; we had arrived at the frequency domain.

After gaining some facility in sinusoidal steady-state analysis, a modicum of pleasant cerebration led us to apply a forcing function of the form $V_0e^{j\theta}e^{(\sigma+j\omega)t}$, and we thereby invented the complex frequency $\mathbf{s}$, relegating all our previous functional forms to special cases: dc ($\mathbf{s} = 0$), exponential ($\mathbf{s} = \sigma$), sinusoidal ($\mathbf{s} = j\omega$), and exponential sinusoid ($\mathbf{s} = \sigma + j\omega$). Such was our status at the end of Chap. 16.

Continuing to strive for mastery over an ever greater number of different types of forcing functions, we turned next to nonsinusoidal periodic functions. Here the infinite series developed by Fourier, $a_0 + \Sigma(a_n \cos n\omega_0 t + b_n \sin n\omega_0 t)$, was found to be capable of representing almost any periodic function in which we might be interested. Thanks to linearity and superposition, the response could then be found as the sum of the responses to the individual sinusoidal terms.

Remembering our success with the complex exponential function, we next developed the complex form of the Fourier series, $\Sigma\mathbf{c}_n e^{jn\omega_0 t}$, suitable again only for periodic functions. However, by letting the period of a periodic sequence of pulses increase without limit, we arrived at a form applicable to a single pulse, the inverse Fourier transform:

$$v(t) = \frac{1}{2\pi}\int_{-\infty}^{\infty} e^{j\omega t}\mathbf{V}(j\omega)\, d\omega$$

This then is our current status. What more could we possibly need? It turns out that there are several things. First, there are a few time functions for which a Fourier transform does not exist, such as the increasing exponential, many random signals, and other time functions that are not absolutely integrable. Second, we have not yet permitted initial-energy storage in the networks whose transient or complete response we desired. Both of these objections are overcome with the Laplace transform, and in addition we shall note a simpler nomenclature and ease of manipulation.

19-2 DEFINITION OF THE LAPLACE TRANSFORM

The step that we must now take to develop the Laplace transform is suggested by using the inverse Fourier transform to interpret $v(t)$ as the sum (integral) of an infinite number of terms, each having the form

$$\left[\frac{\mathbf{V}(j\omega)\,d\omega}{2\pi}\right]e^{j\omega t}$$

Upon comparing this with that of the complex forcing function,

$$[V_0 e^{j\theta}]e^{j\omega t}$$

which led us to the phasor $V_0 e^{j\theta}$, we should note that both bracketed terms are complex quantities in general, and, therefore, $(1/2\pi)\mathbf{V}(j\omega)\,d\omega$ can also be interpreted as some kind of a phasor. Of course, the frequency differential leads to a vanishingly small amplitude, but we add an infinite number of such terms together when we perform the integration.

Our final step, therefore, parallels that which we took when we introduced phasors in the complex-frequency domain: we now let the time variation have the form, $e^{(\sigma+j\omega)t}$.

To do this, let us consider the Fourier transform of $e^{-\sigma t}v(t)$, rather than $v(t)$ itself. For $v(t)$ we have

$$\mathbf{V}(j\omega) = \int_{-\infty}^{\infty} e^{-j\omega t} v(t)\,dt \tag{1}$$

and

$$v(t) = \frac{1}{2\pi}\int_{-\infty}^{\infty} e^{j\omega t}\mathbf{V}(j\omega)\,d\omega \tag{2}$$

Letting

$$g(t) = e^{-\sigma t} v(t) \tag{3}$$

we see that

$$\mathbf{G}(j\omega) = \int_{-\infty}^{\infty} e^{-j\omega t} e^{-\sigma t} v(t)\,dt$$

$$= \int_{-\infty}^{\infty} e^{-(\sigma+j\omega)t} v(t)\,dt$$

or

$$\mathbf{G}(j\omega) = \mathbf{V}(\sigma + j\omega) = \int_{-\infty}^{\infty} e^{-(\sigma+j\omega)t} v(t)\,dt \tag{4}$$

through comparison with (1). Taking the inverse Fourier transform, we obtain

$$g(t) = \frac{1}{2\pi}\int_{-\infty}^{\infty} e^{j\omega t}\mathbf{G}(j\omega)\,d\omega$$
$$= \frac{1}{2\pi}\int_{-\infty}^{\infty} e^{j\omega t}\mathbf{V}(\sigma + j\omega)\,d\omega$$

With the help of (3) we have

$$e^{-\sigma t}v(t) = \frac{1}{2\pi}\int_{-\infty}^{\infty} e^{j\omega t}\mathbf{V}(\sigma + j\omega)\,d\omega$$

or, upon transferring $e^{-\sigma t}$ inside the integral,

$$v(t) = \frac{1}{2\pi}\int_{-\infty}^{\infty} e^{(\sigma+j\omega)t}\mathbf{V}(\sigma + j\omega)\,d\omega$$

We now replace $\sigma + j\omega$ by the single complex variable s, with $ds = j\,d\omega$, and have

$$v(t) = \frac{1}{2\pi j}\int_{\sigma-j\infty}^{\sigma+j\infty} e^{st}\mathbf{V}(\mathbf{s})\,ds \tag{5}$$

where the limits have become $\sigma - j\infty$ and $\sigma + j\infty$. In terms of s, (4) may be written

$$\mathbf{V}(\mathbf{s}) = \int_{-\infty}^{\infty} e^{-st}v(t)\,dt \tag{6}$$

Equation (6) defines the *two-sided* or *bilateral Laplace transform* of $v(t)$. The terms "two-sided" or "bilateral" are used to emphasize the fact that both positive and negative times are included in the range of integration. Equation (5) is the inverse Laplace transform, and the two equations constitute the two-sided Laplace transform pair.

If we continue the process we began earlier in this section, we may note that $v(t)$ is now represented as the sum (integral) of terms of the form

$$\left[\frac{\mathbf{V}(\mathbf{s})\,ds}{2\pi j}\right]e^{st} = \left[\frac{\mathbf{V}(\mathbf{s})\,d\omega}{2\pi}\right]e^{st}$$

Such terms have the same form as those we encountered when we used phasors to represent exponentially varying sinusoids,

$$[\mathbf{V}_0 e^{j\theta}]e^{st}$$

where the bracketed term was a function of $\mathbf{s}$. The two-sided Laplace transform may therefore be interpreted as expressing $v(t)$ as the sum (integral) of an infinite number of vanishingly small terms with complex frequency, $\mathbf{s} = \sigma + j\omega$. The variable is $\mathbf{s}$ or ω, and σ should be thought of as governing the convergence factor $e^{-\sigma t}$. That is, by including this exponential term in (6), more positive values of σ ensure that the function $e^{-\sigma t}v(t)\,u(t)$ is absolutely integrable for almost any $v(t)$ that we might meet. The more negative values of σ are needed when $t < 0$. Thus, the two-sided Laplace transform exists for a wider class of functions $v(t)$ than does the Fourier transform. The exact conditions required for the existence of the (one-sided) Laplace transform are given in the following section.

In many of our circuit analysis problems, the forcing and response functions do not exist forever in time, but rather they are initiated at some specific instant that we usually select as $t = 0$. Thus, for time functions that do not exist for $t < 0$, or for those time functions whose behavior for $t < 0$ is of no interest, the time-domain description can be thought of as $v(t)u(t)$. The corresponding Laplace transform is then

$$\mathbf{V}(\mathbf{s}) = \int_{-\infty}^{\infty} e^{-\mathbf{s}t}v(t)u(t)\,dt = \int_{0}^{\infty} e^{-\mathbf{s}t}v(t)\,dt$$

This defines the *one-sided Laplace transform* of $v(t)$, or simply the *Laplace transform* of $v(t)$, one-sided being understood. The inverse transform expression remains unchanged, but when evaluated, it is understood to be valid only for $t > 0$. Here then is the definition of the Laplace transform pair that will be used from now on:

$$\mathbf{V}(\mathbf{s}) = \int_{0}^{\infty} e^{-\mathbf{s}t}v(t)\,dt \tag{7}$$

$$v(t) = \frac{1}{2\pi j}\int_{\sigma-j\infty}^{\sigma+j\infty} e^{\mathbf{s}t}\mathbf{V}(\mathbf{s})\,d\mathbf{s} \tag{8}$$

$$v(t) \Leftrightarrow \mathbf{V}(\mathbf{s})$$

These are memorable expressions.

The script $\mathfrak{L}$ may also be used to indicate the direct or inverse Laplace transform operation:

$$\mathbf{V}(\mathbf{s}) = \mathfrak{L}\{v(t)\}$$

and

$$v(t) = \mathfrak{L}^{-1}\{\mathbf{V}(\mathbf{s})\}$$

Drill Problems

19-1 If $v(t) = 10$ for $-2 < t < 1$ and is zero elsewhere, determine its: (*a*) Fourier transform; (*b*) two-sided Laplace transform; (*c*) Laplace transform.

Ans. $\frac{-10}{\mathbf{s}}(e^{-\mathbf{s}} - 1)$; $\frac{-10}{\mathbf{s}}(e^{-\mathbf{s}} - e^{2\mathbf{s}})$; $\frac{-10}{j\omega}(e^{-j\omega} - e^{j2\omega})$

19-2 If $\sigma = 3$, determine the Laplace transform of: (*a*) $e^{-4t}u(t)$; (*b*) $e^{t}u(t)$; (*c*) $e^{4t}u(t)$.

Ans. $1/(2 + j\omega)$; $1/(7 + j\omega)$; *does not exist*

19-3 Find the Laplace transform of: (*a*) $2u(t)$; (*b*) $e^{-3t}u(t)$; (*c*) $(2 + e^{-3t})u(t)$. In each case state the range of σ for which the transform exists.

Ans. $2/\mathbf{s}$, $\sigma > 0$; $1/(\mathbf{s} + 3)$, $\sigma > -3$; $(2/\mathbf{s}) + 1/(\mathbf{s} + 3)$, $\sigma > 0$

19-3 LAPLACE TRANSFORMS OF SOME SIMPLE TIME FUNCTIONS

In this section we shall begin to build up a catalog of Laplace transforms for those time functions most frequently encountered in circuit analysis. This will be done, at least initially, by utilizing the definition,

$$\mathbf{V}(\mathbf{s}) = \int_0^\infty e^{-\mathbf{s}t} v(t)\, dt = \mathcal{L}\{v(t)\}$$

which, along with the expression for the inverse transform,

$$v(t) = \frac{1}{2\pi j} \int_{\sigma - j\infty}^{\sigma + j\infty} e^{\mathbf{s}t} \mathbf{V}(\mathbf{s})\, d\mathbf{s} = \mathcal{L}^{-1}\{\mathbf{V}(\mathbf{s})\}$$

establishes a one-to-one correspondence between $v(t)$ and $\mathbf{V}(\mathbf{s})$. That is, for every $v(t)$ for which $\mathbf{V}(\mathbf{s})$ exists, there is a unique $\mathbf{V}(\mathbf{s})$. At this point, we may be looking with some trepidation at the rather ominous form given for the inverse transform. Fear not! As we shall see shortly, an introductory study of Laplace transform theory does not require actual evaluation of this integral. By going from the time domain to the frequency domain and taking advantage of the uniqueness mentioned above, we shall be able to generate a catalog of transform pairs such that nearly every transform that we wish to invert will already have the corresponding time function listed in our catalog.

Before we start evaluating some transforms, however, we must pause to consider whether there is any chance that the transform may not even exist for some $v(t)$ that concerns us. In our study of Fourier transforms, we were forced to take a kind of back-door approach in finding several transforms. This was true with time functions that were not absolutely integrable, such as the unit step function. In the case of the Laplace transform, the use of realizable linear circuits and real-world forcing functions almost never leads to troublesome time functions. Technically, a set of conditions sufficient to ensure the absolute convergence of the Laplace integral for Re (s) $> k$ are:

1. The function $v(t)$ is integrable in every finite interval $t_1 < t < t_2$, where $0 \leq t_1 < t_2 < \infty$.
2. The limit, $\lim_{t \to \infty} e^{-kt}|v(t)|$, exists for some value of k.

Time functions that do not satisfy these conditions are seldom encountered by the circuit analyst.[1]

In considering the Laplace transform of time functions having an impulse or higher-order singularity function at $t = 0$, the defining integral for the Laplace transform must be taken with the lower limit at $t = 0^-$ in order to include the effect of the discontinuity.[2]

Now let us look at some specific transforms. In a somewhat vengeful mood, we first examine the Laplace transform of the unit-step function $u(t)$ which caused us some earlier trouble. From the defining equation, we may write

$$\mathcal{L}\{u(t)\} = \int_0^\infty e^{-\mathbf{s}t} u(t)\, dt = \int_0^\infty e^{-\mathbf{s}t}\, dt$$

$$= \left(-\frac{1}{\mathbf{s}} e^{-\mathbf{s}t} \right)_0^\infty = \frac{1}{\mathbf{s}}$$

Thus,

$$u(t) \Leftrightarrow \frac{1}{\mathbf{s}} \tag{9}$$

and our first Laplace transform pair has been established with great ease.

Another singularity function whose transform is of considerable interest is the unit impulse function $\delta(t - t_0)$, where $t_0 > 0^-$:

$$\mathcal{L}\{\delta(t - t_0)\} = \int_{0^-}^\infty e^{-\mathbf{s}t}\, \delta(t - t_0)\, dt = e^{-\mathbf{s}t_0}$$

$$\delta(t - t_0) \Leftrightarrow e^{-\mathbf{s}t_0} \tag{10}$$

[1] Examples of such functions are e^{t^2} and e^{e^t}, but not t^n or n^t.

[2] For a further discussion on the selection of $t = 0^-$ as the lower limit, refer to George R. Cooper and Clare D. McGillem, "Methods of Signal and System Analysis," Chaps. 6 and 7, Holt, Rinehart, and Winston, New York, 1967. This text presents a somewhat more detailed discussion of the Laplace transform and its applications.

In particular, note that we obtain

$$\delta(t) \Leftrightarrow 1 \tag{11}$$

for $t_0 = 0$.

Recalling our past interest in the exponential function, we examine its transform,

$$\mathcal{L}\{e^{-\alpha t}u(t)\} = \int_0^\infty e^{-\alpha t}e^{-\mathbf{s}t}\,dt$$

$$= \left[-\frac{1}{\mathbf{s}+\alpha}e^{-(\mathbf{s}+\alpha)t}\right]_0^\infty$$

$$= \frac{1}{\mathbf{s}+\alpha}$$

and

$$e^{-\alpha t}u(t) \Leftrightarrow \frac{1}{\mathbf{s}+\alpha} \tag{12}$$

If α is negative, it is understood that Re $(\mathbf{s}) > -\alpha$.

As a final example, for the moment, let us consider the ramp function $tu(t)$. We obtain,

$$\mathcal{L}\{tu(t)\} = \int_0^\infty te^{-\mathbf{s}t}\,dt = \frac{1}{\mathbf{s}^2}$$

$$tu(t) \Leftrightarrow \frac{1}{\mathbf{s}^2} \tag{13}$$

either by a straightforward integration by parts or by use of a table of definite integrals.

In order to accelerate the process of deriving more Laplace transform pairs, we shall pause to develop several useful theorems in the following section.

Drill Problems

19-4 Determine $\mathbf{F}(\mathbf{s})$ if $f(t) =$: (*a*) $5u(t)$; (*b*) $5u(t-5)$; (*c*) $5u(t+5)$.
Ans. $5/\mathbf{s}$; $5/\mathbf{s}$; $5e^{-5\mathbf{s}}/\mathbf{s}$

19-5 Find $\mathcal{L}^{-1}\{\mathbf{V}(\mathbf{s})\}$ for the following $\mathbf{V}(\mathbf{s})$: (*a*) $2/(\mathbf{s}+2)$; (*b*) $2/\mathbf{s}$; (*c*) $2/\mathbf{s}^2$.
Ans. $2u(t)$; $2t\,u(t)$; $2e^{-2t}u(t)$

19-4 SEVERAL BASIC THEOREMS FOR THE LAPLACE TRANSFORM

The further evaluation of Laplace transforms is facilitated by applying several basic theorems. One of the simplest and most obvious is the linearity theorem: the Laplace transform of the sum of two or more time functions is equal to the sum of the transforms of the individual time functions. For two time functions we have

$$\mathcal{L}\{f_1(t) + f_2(t)\} = \int_0^\infty e^{-\mathbf{s}t}[f_1(t) + f_2(t)]\,dt$$

$$= \int_0^\infty e^{-\mathbf{s}t} f_1(t)\,dt + \int_0^\infty e^{-\mathbf{s}t} f_2(t)\,dt$$

$$= \mathbf{F}_1(\mathbf{s}) + \mathbf{F}_2(\mathbf{s})$$

As an example of the use of this theorem, suppose that we have a Laplace transform $\mathbf{V}(\mathbf{s})$ and want to know the corresponding time function $v(t)$. It will often be possible to decompose $\mathbf{V}(\mathbf{s})$ into the sum of two or more functions, say $\mathbf{V}_1(\mathbf{s})$ and $\mathbf{V}_2(\mathbf{s})$, whose inverse transforms, $v_1(t)$ and $v_2(t)$, are already tabulated. It then becomes a simple matter to apply the linearity theorem and write

$$v(t) = \mathcal{L}^{-1}\{\mathbf{V}(\mathbf{s})\} = \mathcal{L}^{-1}\{\mathbf{V}_1(\mathbf{s}) + \mathbf{V}_2(\mathbf{s})\}$$

$$= \mathcal{L}^{-1}\{\mathbf{V}_1(\mathbf{s})\} + \mathcal{L}^{-1}\{\mathbf{V}_2(\mathbf{s})\} = v_1(t) + v_2(t)$$

As a specific example, let us determine the inverse Laplace transform of

$$\mathbf{V}(\mathbf{s}) = \frac{1}{(\mathbf{s} + \alpha)(\mathbf{s} + \beta)}$$

Although it is possible to substitute this expression into the defining equation for the inverse transform, it is much easier to utilize the linearity theorem. Making use of the partial fraction expansion that we all should have seen in our introductory calculus courses, we can split the given transform into the sum of two simpler transforms,

$$\mathbf{V}(\mathbf{s}) = \frac{1/(\beta - \alpha)}{\mathbf{s} + \alpha} + \frac{1/(\alpha - \beta)}{\mathbf{s} + \beta}$$

But we have already evaluated inverse transforms of the form shown on the right, and thus

$$v(t) = \frac{1}{\beta - \alpha} e^{-\alpha t} u(t) + \frac{1}{\alpha - \beta} e^{-\beta t} u(t)$$

$$= \frac{1}{\beta - \alpha} (e^{-\alpha t} - e^{-\beta t}) u(t)$$

If we wished we could now include this as a new entry in our catalog of Laplace pairs,

$$\frac{1}{\beta - \alpha} (e^{-\alpha t} - e^{-\beta t}) u(t) \Leftrightarrow \frac{1}{(\mathbf{s} + \alpha)(\mathbf{s} + \beta)} \tag{14}$$

It is noteworthy that a transcendental function of t in the time domain transforms into a simpler rational function of $\mathbf{s}$ in the frequency domain. Simplifications of this kind are of paramount importance in transform theory. In this example also note that we made use of the fact that

$$kv(t) \Leftrightarrow k\mathbf{V}(\mathbf{s}) \tag{15}$$

where k is a constant of proportionality. This result is obviously a direct result of the definition of the Laplace transform.

We are now able to consider two theorems that might be considered collectively as the raison d'être for Laplace transforms in circuit analysis—the time differentiation and integration theorems. These will help us transform the derivatives and integrals appearing in the time-domain circuit equations.

Let us look at time differentiation first by considering a time function $v(t)$ whose Laplace transform $\mathbf{V}(\mathbf{s})$ is known to exist. We want the transform of the first derivative of $v(t)$,

$$\mathcal{L}\left\{\frac{dv}{dt}\right\} = \int_0^\infty e^{-\mathbf{s}t} \frac{dv}{dt}\, dt$$

This can be integrated by parts,

$$U = e^{-\mathbf{s}t} \qquad dV = \frac{dv}{dt}\, dt$$

with the result,

$$\mathcal{L}\left\{\frac{dv}{dt}\right\} = [v(t)e^{-\mathbf{s}t}]_0^\infty + \mathbf{s}\int_0^\infty e^{-\mathbf{s}t} v(t)\, dt$$

The first term on the right must approach zero as t increases without limit; otherwise $\mathbf{V}(\mathbf{s})$ would not exist. Hence,

$$\mathcal{L}\left\{\frac{dv}{dt}\right\} = 0 - v(0) + \mathbf{s}\mathbf{V}(\mathbf{s})$$

and

$$\frac{dv}{dt} \Leftrightarrow \mathbf{s}\mathbf{V}(\mathbf{s}) - v(0) \tag{16}$$

When there is a discontinuity at $t = 0$, $v(0)$ must be interpreted as $v(0^-)$.

Similar relationships may be developed for higher-order derivatives,

$$\frac{d^2v}{dt^2} \Leftrightarrow \mathbf{s}^2\mathbf{V}(\mathbf{s}) - \mathbf{s}v(0) - v'(0) \tag{17}$$

$$\frac{d^3v}{dt^3} \Leftrightarrow \mathbf{s}^3\mathbf{V}(\mathbf{s}) - \mathbf{s}^2v(0) - \mathbf{s}v'(0) - v''(0) \tag{18}$$

where $v'(0)$ is the value of the first derivative of $v(t)$ evaluated at $t = 0$ (or at $t = 0^-$ if there is a discontinuity at $t = 0$), $v''(0)$ is the initial value of the second derivative of $v(t)$, and so forth. When all initial conditions are zero, we see that differentiating once with respect to t in the time domain corresponds to one multiplication by $\mathbf{s}$ in the frequency domain; differentiating twice in the time domain corresponds to multiplication by $\mathbf{s}^2$ in the frequency domain, and so on. Thus, differentiation in the time domain is equivalent to multiplication in the frequency domain. This is a substantial simplification! We should also begin to see that, when the initial conditions are not zero, their presence is still accounted for. A simple example will serve to demonstrate this.

Suppose that we have the series RL circuit shown in Fig. 19-1. The network is driven by a unit-step voltage, and we assume an initial value of the current (at $t = 0^-$) of 5 A.[3] Using Kirchhoff's voltage law to write the single loop equation in the time domain, we have

$$2\frac{di}{dt} + 4i = 3u(t)$$

[3]We could have established this current by letting the source be $20u(-t) + 3u(t)$ V, or $20 - 17u(t)$ V, or other expressions of this nature.

Fig. 19-1 A circuit which is analyzed by transforming the differential equation $2\,di/dt + 4i = 3u(t)$ into $2[\mathbf{s}\mathbf{I}(\mathbf{s}) - i(0^-)] + 4\mathbf{I}(\mathbf{s}) = 3/\mathbf{s}$.

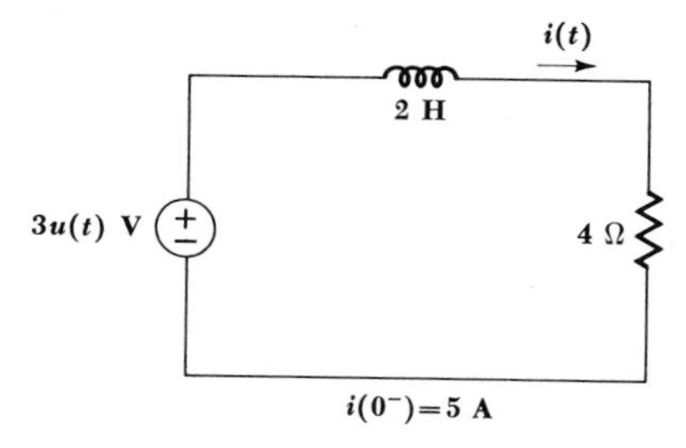

Instead of solving this differential equation as we have done previously, we first transform to the frequency domain by taking the Laplace transform of each term,

$$2[s\mathbf{I}(s) - i(0^-)] + 4\mathbf{I}(s) = \frac{3}{s}$$

We next solve for $\mathbf{I}(s)$, substituting $i(0^-) = 5$,

$$(2s + 4)\mathbf{I}(s) = \frac{3}{s} + 10$$

and

$$\mathbf{I}(s) = \frac{1.5}{s(s+2)} + \frac{5}{s+2}$$

$$= \frac{0.75}{s} - \frac{0.75}{s+2} + \frac{5}{s+2}$$

$$= \frac{0.75}{s} + \frac{4.25}{s+2}$$

and then use our previously determined transform pairs to invert:

$$i(t) = 0.75u(t) + 4.25e^{-2t}u(t)$$
$$= (0.75 + 4.25e^{-2t})u(t)$$

Our solution for $i(t)$ is complete. Both the forced response $0.75u(t)$ and the natural response $4.25e^{-2t}u(t)$ are present, and the initial condition was automatically incorporated into the solution. The method illustrates a very painless way of obtaining the complete solution of a differential equation.

The same kind of simplification can be accomplished when we meet the operation of integration with respect to time in our circuit equations. Let us determine the Laplace transform of the time function described by $\int_0^t v(x)\,dx$,

$$\mathcal{L}\left\{\int_0^t v(x)\,dx\right\} = \int_0^\infty e^{-st}\left[\int_0^t v(x)\,dx\right]dt$$

Integrating by parts, we let

$$u = \int_0^t v(x)\,dx \qquad dv = e^{-st}\,dt$$

$$du = v(t)\,dt \qquad v = -\frac{1}{s}e^{-st}$$

Then

$$\mathcal{L}\left\{\int_0^t v(x)\,dx\right\} = \left\{\left[\int_0^t v(x)\,dx\right]\left[-\frac{1}{\mathbf{s}}e^{-\mathbf{s}t}\right]\right\}_{t=0}^{t=\infty} - \int_0^\infty -\frac{1}{\mathbf{s}}e^{-\mathbf{s}t}v(t)\,dt$$

$$= \left[-\frac{1}{\mathbf{s}}e^{-\mathbf{s}t}\int_0^t v(x)\,dx\right]_0^\infty + \frac{1}{\mathbf{s}}\mathbf{V}(\mathbf{s})$$

But, since $e^{-\mathbf{s}t} \rightarrow 0$ as $t \rightarrow \infty$, the first term on the right vanishes at the upper limit, and when $t \rightarrow 0$, the integral in this term likewise vanishes. This leaves only the $\mathbf{V}(\mathbf{s})/\mathbf{s}$ term, so

$$\int_0^t v(x)\,dx \Leftrightarrow \frac{\mathbf{V}(\mathbf{s})}{\mathbf{s}} \tag{19}$$

and thus integration in the time domain corresponds to division by $\mathbf{s}$ in the frequency domain. Once more, a relatively complicated calculus operation in the time domain simplifies to an algebraic operation in the frequency domain.

As an example of how this helps us in circuit analysis, we shall determine $i(t)$ and $v(t)$ for $t > 0$ in the series RC circuit shown in Fig. 19-2. It is assumed that there was some initial energy stored in the capacitor prior to $t = 0$ so that $v(0) = 9$ V. We first write the single loop equation,

$$u(t) = 4i(t) + 16\int_{-\infty}^t i(t)\,dt$$

In order to apply the time-integration theorem, we must arrange for the lower limit of integration to be zero. Thus, we set

$$16\int_{-\infty}^t i(t)\,dt = 16\int_{-\infty}^0 i(t)\,dt + 16\int_0^t i(t)\,dt = v(0) + 16\int_0^t i(t)\,dt$$

Therefore,

$$u(t) = 4i(t) + v(0) + 16\int_0^t i(t)\,dt$$

Fig. 19-2 A circuit illustrating the use of the Laplace transform pair $\int_0^t i(t)\,dt \Leftrightarrow (1/\mathbf{s})\mathbf{I}(\mathbf{s})$.

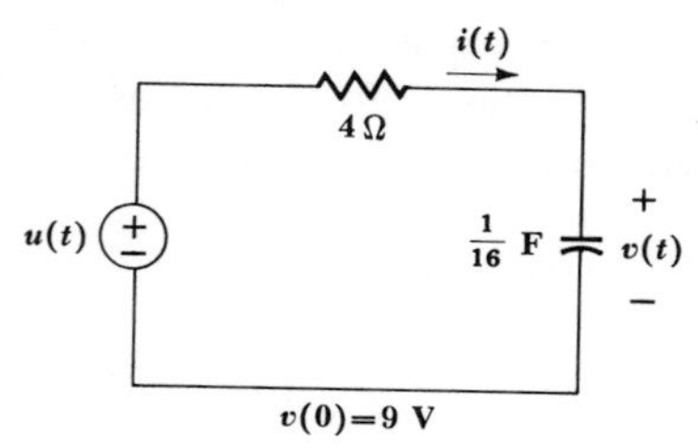

We next take the Laplace transform of both sides of this equation. Since we are utilizing the single-ended transform, $\mathcal{L}\{v(0)\}$ is simply $\mathcal{L}\{v(0)u(t)\}$, and thus

$$\frac{1}{\mathbf{s}} = 4\mathbf{I}(\mathbf{s}) + \frac{9}{\mathbf{s}} + \frac{16}{\mathbf{s}}\mathbf{I}(\mathbf{s})$$

and solving for $\mathbf{I}(\mathbf{s})$,

$$\mathbf{I}(\mathbf{s}) = \frac{-2}{\mathbf{s} + 4}$$

the desired result is immediately obtained,

$$i(t) = -2e^{-4t}u(t)$$

If $v(t)$ had been our desired response, we would simply have written a single nodal equation,

$$\frac{v(t) - u(t)}{4} + \frac{1}{16}\frac{dv}{dt} = 0$$

Taking the Laplace transform, we obtain

$$\frac{\mathbf{V}(\mathbf{s})}{4} - \frac{1}{4\mathbf{s}} + \frac{1}{16}\mathbf{s}\mathbf{V}(\mathbf{s}) - \frac{v(0)}{16} = 0$$

or

$$\mathbf{V}(\mathbf{s})\left(1 + \frac{\mathbf{s}}{4}\right) = \frac{1}{\mathbf{s}} + \frac{9}{4}$$

Thus,

$$\mathbf{V}(\mathbf{s}) = \frac{4}{\mathbf{s}(\mathbf{s} + 4)} + \frac{9}{\mathbf{s} + 4} = \frac{1}{\mathbf{s}} - \frac{1}{\mathbf{s} + 4} + \frac{9}{\mathbf{s} + 4}$$

$$= \frac{1}{\mathbf{s}} + \frac{8}{\mathbf{s} + 4}$$

and taking the inverse transform,

$$v(t) = (1 + 8e^{-4t})u(t)$$

we have quickly obtained the desired capacitor voltage without recourse to the usual differential equation solution. To check this result, we note that $\frac{1}{16}\, dv/dt$ should yield the previous expression for $i(t)$. For $t > 0$,

$$\frac{1}{16}\frac{dv}{dt} = \frac{1}{16}(-32)e^{-4t} = -2e^{-4t}$$

which is correct.

To illustrate the use of both the linearity theorem and the time-differentiation theorem, not to mention the addition of a most important pair to our forthcoming Laplace transform table, let us establish the Laplace transform of $\sin \omega t\, u(t)$. We could use the defining integral expression with integration by parts, but this is needlessly difficult. Instead, we use the relationship,

$$\sin \omega t = \frac{1}{2j}(e^{j\omega t} - e^{-j\omega t})$$

The transform of the sum of these two terms is just the sum of the transforms, and each term is an exponential function for which we already have the transform. We may immediately write

$$\mathcal{L}\{\sin \omega t\, u(t)\} = \frac{1}{2j}\left(\frac{1}{s - j\omega} - \frac{1}{s + j\omega}\right) = \frac{\omega}{s^2 + \omega^2}$$

$$\sin \omega t\, u(t) \Leftrightarrow \frac{\omega}{s^2 + \omega^2} \qquad (20)$$

We next use the time-differentiation theorem to obtain the transform of $\cos \omega t\, u(t)$, which is proportional to the derivative of $\sin \omega t$. That is,

$$\mathcal{L}\{\cos \omega t\, u(t)\} = \mathcal{L}\left\{\frac{1}{\omega}\frac{d}{dt}[\sin \omega t\, u(t)]\right\}$$

$$= \frac{1}{\omega} s \frac{\omega}{s^2 + \omega^2}$$

and

$$\cos \omega t\, u(t) \Leftrightarrow \frac{s}{s^2 + \omega^2} \qquad (21)$$

Drill Problems

19-6 By expansion into partial fractions, find the inverse Laplace transform of: (*a*) $1/[s(s + 3)]$; (*b*) $1/[s^2(s + 3)]$; (*c*) $1/[s(s + 1)(s + 3)]$

Ans. $\frac{1}{3}(1 - e^{-3t})u(t)$; $(\frac{1}{3} - \frac{1}{2}e^{-t} + \frac{1}{6}e^{-3t})u(t)$; $(-\frac{1}{9} + \frac{1}{3}t + \frac{1}{9}e^{-3t})u(t)$

19-7 Use transform techniques to find $i(t)$ for $t > 0$ if: (*a*) $2\, di/dt + 5i = 3u(t)$, $i(0) = 4$; (*b*) $d^2i/dt^2 + 4\, di/dt + 3i = -14\delta(t)$, $i'(0) = 2$, $i(0) = 3$; (*c*) $i(t)$ is the current indicated in Fig. 19-3*a*.

Ans. $5(1 + e^{-2.5t})u(t)$; $1.5(3e^{-3t} - e^{-t})u(t)$; $(0.6 + 3.4e^{-2.5t})u(t)$ A

Fig. 19-3 (*a*) See Drill Prob. 19-7. (*b*) See Drill Prob. 19-8.

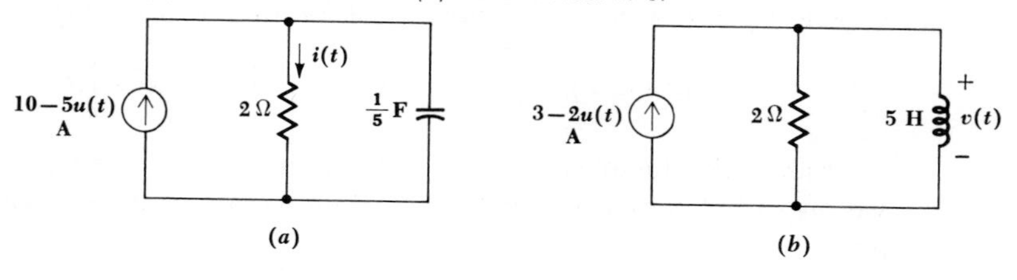

19-8 Determine $v(t)$ for $t > 0$ if: (*a*) $2\int_0^t v(t)\,dt + 5v = 8$; (*b*) $2\int_{-\infty}^t v(t)\,dt + 8\,dv/dt = 8$, $\int_{-\infty}^0 v(t)\,dt = 4$, $v(0) = 1$; (*c*) $v(t)$ is the indicated voltage in Fig. 19-3*b*.
Ans. $\cos(t/2)u(t)$; $-4e^{-0.4t}u(t)$; $1.6e^{-0.4t}u(t)$ V

19-5 CONVOLUTION AGAIN

In our study of the Fourier transform, we discovered that the Fourier transform of $f_1(t) * f_2(t)$, the convolution of two time functions, was simply the product of the transforms of the individual functions. This then led to the very useful concept of the system function, defined as the ratio of the transform of the system output to the transform of the system input. Exactly the same fortunate circumstances exist when we work with the Laplace transform, as we shall now see.

Let $\mathbf{F}_1(\mathbf{s})$ and $\mathbf{F}_2(\mathbf{s})$ be the Laplace transforms of $f_1(t)$ and $f_2(t)$, and consider the Laplace transform of $f_1(t) * f_2(t)$,

$$\mathcal{L}\{f_1(t) * f_2(t)\} = \mathcal{L}\left\{\int_{-\infty}^{\infty} f_1(\lambda) f_2(t-\lambda)\,d\lambda\right\}$$

As we discovered in our earlier look at convolution, one of these time functions will often be the forcing function that is applied at the input terminals of a linear circuit, and the other will be the unit-impulse response of the circuit. That is, the response of a linear circuit is just the convolution of the input and the impulse response.

Since we are now dealing with time functions that do not exist prior to $t = 0$ (the definition of the Laplace transform forces us to assume this), the lower limit of integration can be changed to zero. Then, using the definition of the Laplace transform, we get

$$\mathcal{L}\{f_1(t) * f_2(t)\} = \int_0^{\infty} e^{-\mathbf{s}t}\left[\int_0^{\infty} f_1(\lambda) f_2(t-\lambda)\,d\lambda\right] dt$$

Since $e^{-\mathbf{s}t}$ does not depend upon λ, we can move this factor inside the inner integral. If we do this and also reverse the order of integration, the result is

$$\mathcal{L}\{f_1(t) * f_2(t)\} = \int_0^{\infty}\left[\int_0^{\infty} e^{-\mathbf{s}t} f_1(\lambda) f_2(t-\lambda)\,dt\right] d\lambda$$

Continuing with the same type of trickery, we note that $f_1(\lambda)$ does not depend upon t, so it can be moved outside the inner integral,

$$\mathcal{L}\{f_1(t) * f_2(t)\} = \int_0^{\infty} f_1(\lambda)\left[\int_0^{\infty} e^{-\mathbf{s}t} f_2(t-\lambda)\,dt\right] d\lambda$$

We then make the substitution, $x = t - \lambda$, in the bracketed integral (where we may treat λ as a constant) and, while we are at it, remove the factor $e^{-\mathbf{s}\lambda}$,

$$\mathcal{L}\{f_1(t) * f_2(t)\} = \int_0^\infty e^{-\mathbf{s}\lambda} f_1(\lambda) \left[\int_0^\infty e^{-\mathbf{s}x} f_2(x)\, dx \right] d\lambda$$

The bracketed term is $\mathbf{F}_2(\mathbf{s})$, which is not a function of t and can therefore be moved outside both integral signs. What remains is the Laplace transform $\mathbf{F}_1(\mathbf{s})$, and we have

$$f_1(t) * f_2(t) \Leftrightarrow \mathbf{F}_1(\mathbf{s})\mathbf{F}_2(\mathbf{s}) \tag{22}$$

which is the desired result. Stated slightly differently, we may conclude that the inverse transform of the product of two transforms is the convolution of the individual inverse transforms, a result that is sometimes useful in obtaining inverse transforms.

As an example of the use of the convolution theorem, let us reconsider the first example of Sec. 19-4, in which we were given the transform,

$$\mathbf{V}(\mathbf{s}) = \frac{1}{(\mathbf{s} + \alpha)(\mathbf{s} + \beta)}$$

and obtained the inverse transform by a partial fraction expansion. We now identify $\mathbf{V}(\mathbf{s})$ as the product of two transforms,

$$\mathbf{V}_1(\mathbf{s}) = \frac{1}{\mathbf{s} + \alpha}$$

$$\mathbf{V}_2(\mathbf{s}) = \frac{1}{\mathbf{s} + \beta}$$

where

$$v_1(t) = e^{-\alpha t} u(t)$$

and

$$v_2(t) = e^{-\beta t} u(t)$$

The desired $v(t)$ can be immediately expressed as

$$v(t) = \mathcal{L}^{-1}\{\mathbf{V}_1(\mathbf{s})\mathbf{V}_2(\mathbf{s})\} = v_1(t) * v_2(t)$$

$$= \int_0^\infty v_1(\lambda) v_2(t - \lambda)\, d\lambda$$

$$= \int_0^\infty e^{-\alpha\lambda} u(\lambda) e^{-\beta(t-\lambda)} u(t - \lambda)\, d\lambda$$

$$= \int_0^t e^{-\alpha\lambda} e^{-\beta t} e^{\beta\lambda}\, d\lambda$$

$$= e^{-\beta t} \int_0^t e^{(\beta-\alpha)\lambda}\, d\lambda$$

$$= e^{-\beta t} \frac{e^{(\beta-\alpha)t} - 1}{\beta - \alpha} u(t)$$

and finally,

$$v(t) = \frac{1}{\beta - \alpha}(e^{-\alpha t} - e^{-\beta t})u(t)$$

which is the same result that we obtained before using the partial fraction expansion. Note that it is necessary to insert the unit step $u(t)$ in the result because all Laplace transforms are valid only for nonnegative time.

Was the result easier to obtain by this method? Not unless one is in love with convolution integrals. The partial-fraction-expansion method is usually simpler, assuming that the expansion itself is not too cumbersome.

As we have noted several times before, the output $v_o(t)$ at some point in a linear circuit can be obtained by convolving the input $v_i(t)$ with the unit-impulse response $h(t)$. However, we must remember that the impulse response results from the application of a unit impulse at $t = 0$ *with all initial conditions zero.* Under these conditions, the Laplace transform of the output is

$$\mathcal{L}\{v_o(t)\} = \mathbf{V}_o(\mathbf{s}) = \mathcal{L}\{v_i(t) * h(t)\} = \mathbf{V}_i(\mathbf{s})[\mathcal{L}\{h(t)\}]$$

Thus, the ratio $\mathbf{V}_o(\mathbf{s})/\mathbf{V}_i(\mathbf{s})$ is equal to the transform of the impulse response, which we shall denote by $\mathbf{H}(\mathbf{s})$,

$$\mathcal{L}\{h(t)\} = \mathbf{H}(\mathbf{s}) = \frac{\mathbf{V}_o(\mathbf{s})}{\mathbf{V}_i(\mathbf{s})} \tag{23}$$

By definition, this is known as the *transfer function,* the ratio of the Laplace transform of the output (or response) to the Laplace transform of the input (or forcing function) when all initial conditions are zero. The impulse response and the transfer function make up a Laplace transform pair. This is an important fact that we shall utilize later to analyze the behavior of some circuits that would previously have baffled us.

Drill Problems

19-9 Use the convolution integral to find the inverse Laplace transform of $\mathbf{F}_1(\mathbf{s})\mathbf{F}_2(\mathbf{s}) =$: (*a*) $(1/\mathbf{s})(1/\mathbf{s})$; (*b*) $(1/\mathbf{s})(1/\mathbf{s})^2$; (*c*) $(1/\mathbf{s})[1/(\mathbf{s}+1)]$.

Ans. $tu(t)$; $(1 - e^{-t})u(t)$; $\frac{1}{2}t^2u(t)$

19-10 In the circuit of Fig. 19-4, let the input $i_s(t) = \delta(t)$, and determine $h(t)$ and $\mathbf{H}(\mathbf{s})$ if the output is taken as: (*a*) $v_C(t)$; (*b*) $i_C(t)$; (*c*) $i_R(t)$.

Ans. $50e^{-50t}u(t)$, $50/(\mathbf{s}+50)$ A; $\delta(t) - 50e^{-50t}u(t)$, $\mathbf{s}/(\mathbf{s}+50)$ A; $0.5 \times 10^6 e^{-50t}u(t)$, $0.5 \times 10^6/(\mathbf{s}+50)$ V

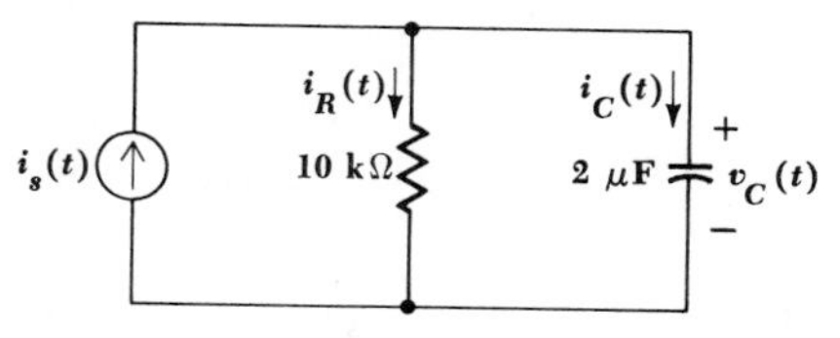

Fig. 19-4 See Drill Prob. 19-10.

19-6 TIME-SHIFT AND PERIODIC FUNCTIONS

By now we have obtained a number of entries for the catalog of Laplace transform pairs that we agreed to construct earlier. Included are the transforms of the impulse function, the step function, the exponential function, the ramp function, the sinusoidal function, and the sum of two exponentials. In addition, we have noted the consequences in the s-domain of the time-domain operations of differentiation, integration, and convolution; and we have used the Laplace transform to define what we mean by a transfer function. These results, plus several others, are collected together for quick reference in Tables 19-1 and 19-2. Some of the tabulated relationships are unfamiliar, however, and we shall now develop a few additional useful theorems to help us obtain them.

As we have seen in some of our earlier transient problems, not all forcing functions begin at $t = 0$. What happens to the transform of a time function if that function is simply shifted in time by some known amount? In particular, if the transform of $f(t)u(t)$ is the known function $\mathbf{F}(\mathbf{s})$, then what is the transform of $f(t-a)u(t-a)$, the original time function delayed by a seconds (and not existing for $t < a$)? Working directly from the definition of the Laplace transform, we get

$$\mathcal{L}\{f(t-a)u(t-a)\} = \int_0^\infty e^{-\mathbf{s}t} f(t-a)u(t-a)\,dt = \int_a^\infty e^{-\mathbf{s}t} f(t-a)\,dt$$

for $t \geq a$. Choosing a new variable of integration, $\tau = t - a$, we obtain

$$\mathcal{L}\{f(t-a)u(t-a)\} = \int_0^\infty e^{-\mathbf{s}(\tau+a)} f(\tau)\,d\tau = e^{-a\mathbf{s}}\mathbf{F}(\mathbf{s}) \tag{24}$$

This result is known as the *time-shift theorem*, and it simply states that, if a time function is delayed by a time a in the time domain, the result in the frequency domain is a multiplication by $e^{-a\mathbf{s}}$.

As an example of the application of this theorem, consider the rectangular pulse described by

$$v(t) = u(t-2) - u(t-5)$$

Table 19-1 Laplace transform pairs

$f(t) = \mathcal{L}^{-1}\{\mathbf{F}(s)\}$	$\mathbf{F}(s) = \mathcal{L}\{f(t)\}$
$\delta(t)$	1
$u(t)$	$\frac{1}{s}$
$tu(t)$	$\frac{1}{s^2}$
$\frac{t^{n-1}}{(n-1)!}u(t),\ n = 1, 2, \ldots$	$\frac{1}{s^n}$
$e^{-\alpha t}u(t)$	$\frac{1}{s+\alpha}$
$te^{-\alpha t}u(t)$	$\frac{1}{(s+\alpha)^2}$
$\frac{t^{n-1}}{(n-1)!}e^{-\alpha t}u(t),\ n = 1, 2, \ldots$	$\frac{1}{(s+\alpha)^n}$
$\frac{1}{\beta-\alpha}(e^{-\alpha t} - e^{-\beta t})u(t)$	$\frac{1}{(s+\alpha)(s+\beta)}$
$\sin \omega t\ u(t)$	$\frac{\omega}{s^2+\omega^2}$
$\cos \omega t\ u(t)$	$\frac{s}{s^2+\omega^2}$
$\sin(\omega t + \theta)u(t)$	$\frac{s \sin\theta + \omega\cos\theta}{s^2+\omega^2}$
$\cos(\omega t + \theta)u(t)$	$\frac{s \cos\theta - \omega\sin\theta}{s^2+\omega^2}$
$e^{-\alpha t} \sin \omega t\ u(t)$	$\frac{\omega}{(s+\alpha)^2+\omega^2}$
$e^{-\alpha t} \cos \omega t\ u(t)$	$\frac{s+\alpha}{(s+\alpha)^2+\omega^2}$

which has unit value for the time interval, $2 < t < 5$, and has zero value elsewhere. We know that the transform of $u(t)$ is just $1/\mathbf{s}$, and since $u(t-2)$ is simply $u(t)$ delayed by 2 s, the transform of this delayed function is $e^{-2\mathbf{s}}/\mathbf{s}$. Similarly, the transform of $u(t-5)$ is $e^{-5\mathbf{s}}/\mathbf{s}$. It follows, then, that the desired transform is

$$\mathbf{V}(\mathbf{s}) = \frac{e^{-2\mathbf{s}} - e^{-5\mathbf{s}}}{\mathbf{s}}$$

It was not necessary to revert to the definition of the Laplace transform in order to determine $\mathbf{V}(\mathbf{s})$.

The time-shift theorem is also useful in evaluating the transform of periodic time functions. Suppose that $f(t)$ is periodic with a period T for positive values of t. The behavior of $f(t)$ for $t < 0$ has no effect on the (one-sided) Laplace transform, as we know. Thus, $f(t)$ can be written as

$$f(t) = f(t - nT) \qquad n = 0, 1, 2, \ldots$$

If we now define a new time function which is nonzero only in the first period of $f(t)$,

$$f_1(t) = [u(t) - u(t - T)]f(t)$$

Table 19-2 Laplace transform operations

Operation	$f(t)$	$F(s)$
Addition	$f_1(t) \pm f_2(t)$	$\mathbf{F}_1(\mathbf{s}) \pm \mathbf{F}_2(\mathbf{s})$
Scalar multiplication	$kf(t)$	$k\mathbf{F}(\mathbf{s})$
Time differentiation	$\dfrac{df}{dt}$	$\mathbf{sF}(\mathbf{s}) - f(0)$
	$\dfrac{d^2f}{dt^2}$	$\mathbf{s}^2\mathbf{F}(\mathbf{s}) - \mathbf{s}f(0) - f'(0)$
	$\dfrac{d^3f}{dt^3}$	$\mathbf{s}^3\mathbf{F}(\mathbf{s}) - \mathbf{s}^2f(0) - \mathbf{s}f'(0) - f''(0)$
Time integration	$\int_0^t f(t)\,dt$	$\dfrac{1}{\mathbf{s}}\mathbf{F}(\mathbf{s})$
	$\int_{-\infty}^t f(t)\,dt$	$\dfrac{1}{\mathbf{s}}\mathbf{F}(\mathbf{s}) + \dfrac{1}{\mathbf{s}}\int_{-\infty}^0 f(t)\,dt$
Convolution	$f_1(t) * f_2(t)$	$\mathbf{F}_1(\mathbf{s})\mathbf{F}_2(\mathbf{s})$
Time shift	$f(t - a)u(t - a),\ a \geq 0$	$e^{-a\mathbf{s}}\mathbf{F}(\mathbf{s})$
Frequency shift	$f(t)e^{-at}$	$\mathbf{F}(\mathbf{s} + a)$
Frequency differentiation	$-tf(t)$	$\dfrac{d\mathbf{F}(\mathbf{s})}{d\mathbf{s}}$
Frequency integration	$\dfrac{f(t)}{t}$	$\int_{\mathbf{s}}^{\infty} \mathbf{F}(\mathbf{s})\,d\mathbf{s}$
Scaling	$f(at),\ a \geq 0$	$\dfrac{1}{a}\mathbf{F}\left(\dfrac{\mathbf{s}}{a}\right)$
Initial value	$f(0^+)$	$\lim_{\mathbf{s}\to\infty} \mathbf{sF}(\mathbf{s})$
Final value	$f(\infty)$	$\lim_{\mathbf{s}\to 0} \mathbf{sF}(\mathbf{s})$, all poles of $\mathbf{sF}(\mathbf{s})$ in LHP
Time periodicity	$f(t) = f(t + nT),\ n = 1, 2, \ldots$	$\dfrac{1}{1 - e^{-T\mathbf{s}}}\mathbf{F}_1(\mathbf{s})$, where $\mathbf{F}_1(\mathbf{s}) = \int_0^T f(t)e^{-\mathbf{s}t}\,dt$

then the original $f(t)$ can be represented as the sum of an infinite number of such functions, delayed by integral multiples of T. That is,

$$f(t) = [u(t) - u(t-T)]f(t) + [u(t-T) - u(t-2T)]f(t) + [u(t-2T) - u(t-3T)]f(t) + \cdots$$
$$= f_1(t) + f_1(t-T) + f_1(t-2T) + \cdots$$

or

$$f(t) = \sum_{n=0}^{\infty} f_1(t - nT)$$

The Laplace transform of this sum is just the sum of the transforms,

$$\mathbf{F}(s) = \sum_{n=0}^{\infty} \mathcal{L}\{f_1(t - nT)\}$$

so that the time-shift theorem leads to

$$\mathbf{F}(s) = \sum_{n=0}^{\infty} e^{-nTs}\mathbf{F}_1(s)$$

where

$$\mathbf{F}_1(s) = \mathcal{L}\{f_1(t)\} = \int_0^T e^{-st} f(t)\, dt$$

Since $\mathbf{F}_1(s)$ is not a function of n, it can be removed from the summation, and $\mathbf{F}(s)$ becomes

$$\mathbf{F}(s) = \mathbf{F}_1(s)[1 + e^{-Ts} + e^{-2Ts} + \cdots]$$

When we apply the binomial theorem to the bracketed expression, it simplifies to $1/(1 - e^{-Ts})$. Thus, we conclude that the periodic function $f(t)$, with period T, has a Laplace transform expressed by

$$\mathbf{F}(s) = \frac{\mathbf{F}_1(s)}{1 - e^{-Ts}} \tag{25}$$

where

$$\mathbf{F}_1(s) = \mathcal{L}\{[u(t) - u(t-T)]f(t)\} \tag{26}$$

is the transform of the first period of the time function.

To illustrate the use of this transform theorem for periodic functions, let us apply it to the familiar rectangular pulse train, Fig. 19-5. We may describe this periodic function analytically:

$$v(t) = \sum_{n=0}^{\infty} V_0[u(t - nT) - u(t - nT - \tau)] \qquad t > 0$$

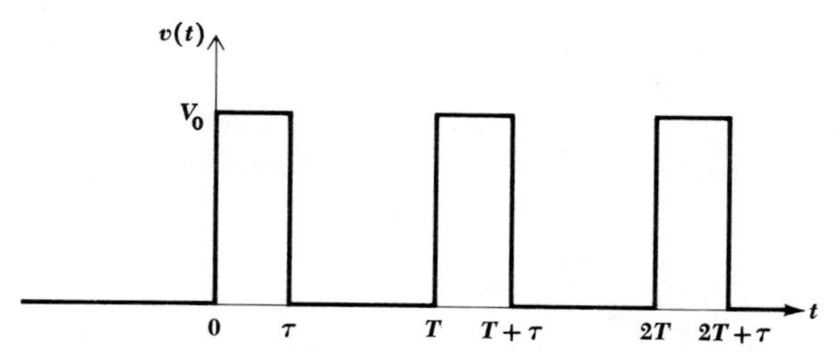

Fig. 19-5 A periodic train of rectangular pulses for which $\mathbf{F}(\mathbf{s}) = (V_0/\mathbf{s})(1 - e^{-\mathbf{s}\tau})/(1 - e^{-\mathbf{s}T})$.

The function $\mathbf{V}_1(\mathbf{s})$ is simple to calculate,

$$\mathbf{V}_1(\mathbf{s}) = V_0 \int_0^{\tau} e^{-\mathbf{s}t}\, dt = \frac{V_0}{\mathbf{s}}(1 - e^{-\mathbf{s}\tau})$$

Now, to obtain the desired transform, we just divide by $(1 - e^{-\mathbf{s}T})$,

$$\mathbf{V}(\mathbf{s}) = \frac{V_0}{\mathbf{s}} \frac{1 - e^{-\mathbf{s}\tau}}{1 - e^{-\mathbf{s}T}} \tag{27}$$

We should note how the two theorems described in this section show up in the transform (27). The $(1 - e^{-\mathbf{s}T})$ factor in the denominator accounts for the periodicity of the function, the $e^{-\mathbf{s}\tau}$ term in the numerator arises from the time delay of the negative square wave that turns off the pulse, and the $V_0/\mathbf{s}$ factor is, of course, the transform of the step functions involved in $v(t)$.

Drill Problems

19-11 Determine the Laplace transform of the functions of time depicted in Fig. 19-6.

Ans. $1.5(1 - e^{-2\mathbf{s}})/\mathbf{s}^2$; $1.5e^{-\mathbf{s}}(1 - e^{-2\mathbf{s}})/\mathbf{s}^2$; $1.5(e^{-\mathbf{s}} - e^{-3\mathbf{s}})(1 - e^{-2\mathbf{s}})/\mathbf{s}^2$

Fig. 19-6 See Drill Prob. 19-11.

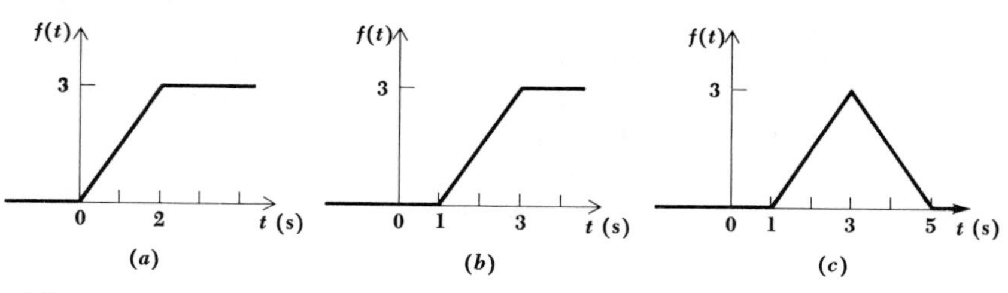

19-12 Find the Laplace transforms of the periodic functions illustrated in Fig. 19-7.

Ans. $10/(1 - e^{-s/5})$; $(50/s^2) - 10e^{-0.2s}/(s - se^{-0.2s})$; $(10/s)(e^{-s/10} - e^{-s/5})/(1 - e^{-0.3s})$

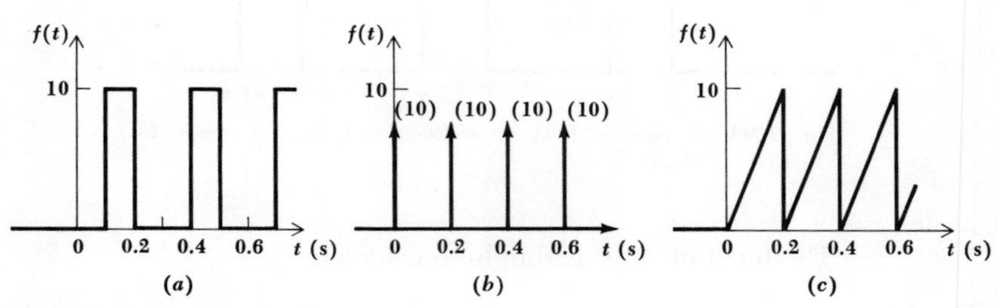

Fig. 19-7 See Drill Prob. 19-12.

19-7 SHIFTING, DIFFERENTIATION, INTEGRATION, AND SCALING IN THE FREQUENCY DOMAIN

Other theorems appearing in Table 19-2 specify the results in the time domain of simple operations on s in the frequency domain. We shall obtain several easily in this section and then see how they may be applied to derive additional transform pairs.

The first new theorem establishes a relationship between $\mathbf{F}(\mathbf{s}) = \mathcal{L}\{f(t)\}$ and $\mathbf{F}(\mathbf{s} + a)$. We consider the Laplace transform of $e^{-at}f(t)$,

$$\mathcal{L}\{e^{-at}f(t)\} = \int_0^\infty e^{-\mathbf{s}t}e^{-at}f(t)\,dt = \int_0^\infty e^{-(\mathbf{s}+a)t}f(t)\,dt$$

Looking carefully at this result, we note that the integral on the right is identical to that defining $\mathbf{F}(\mathbf{s})$ with one exception: $(\mathbf{s} + a)$ appears in place of $\mathbf{s}$. Thus,

$$e^{-at}f(t) \Leftrightarrow \mathbf{F}(\mathbf{s} + a) \tag{28}$$

Thus, we conclude that replacing $\mathbf{s}$ by $(\mathbf{s} + a)$ in the frequency domain corresponds to multiplication by e^{-at} in the time domain. This is known as the *frequency-shift* theorem. It can be put to immediate use in evaluating the transform of the exponentially damped cosine function that we used extensively in previous work. Beginning with the known transform of the cosine function,

$$\mathcal{L}\{\cos \omega_0 t\} = \mathbf{F}(\mathbf{s}) = \frac{\mathbf{s}}{\mathbf{s}^2 + \omega_0^2}$$

then the transform of $e^{-at} \cos \omega_0 t$ must be $\mathbf{F}(\mathbf{s} + a)$,

$$\mathcal{L}\{e^{-at} \cos \omega_0 t\} = \mathbf{F}(\mathbf{s} + a) = \frac{(\mathbf{s} + a)}{(\mathbf{s} + a)^2 + \omega_0^2} \tag{29}$$

Next let us examine the consequences of differentiating $\mathbf{F}(\mathbf{s})$ with respect to $\mathbf{s}$. The result is

$$\begin{aligned} \frac{d}{d\mathbf{s}} \mathbf{F}(\mathbf{s}) &= \frac{d}{d\mathbf{s}} \int_0^\infty e^{-\mathbf{s}t} f(t)\, dt \\ &= \int_0^\infty -te^{-\mathbf{s}t} f(t)\, dt \\ &= \int_0^\infty e^{-\mathbf{s}t} [-tf(t)]\, dt \end{aligned}$$

which is clearly the Laplace transform of $[-tf(t)]$. We therefore conclude that differentiation with respect to $\mathbf{s}$ in the frequency domain results in multiplication by $-t$ in the time domain, or

$$-tf(t) \Leftrightarrow \frac{d}{d\mathbf{s}} \mathbf{F}(\mathbf{s}) \tag{30}$$

Suppose now that $f(t)$ is the unit-ramp function $tu(t)$ whose transform we know is $1/\mathbf{s}^2$. We can use our newly acquired frequency-differentiation theorem to determine the inverse transform of $1/\mathbf{s}^3$ as follows:

$$\frac{d}{d\mathbf{s}} \left(\frac{1}{\mathbf{s}^2} \right) = -\frac{2}{\mathbf{s}^3} \Leftrightarrow -t\mathcal{L}^{-1} \left\{ \frac{1}{\mathbf{s}^2} \right\} = -t^2$$

and

$$\frac{t^2 u(t)}{2} \Leftrightarrow \frac{1}{\mathbf{s}^3} \tag{31}$$

Continuing with the same procedure, we find

$$\frac{t^3}{3!} u(t) \Leftrightarrow \frac{1}{\mathbf{s}^4} \tag{32}$$

and in general

$$\frac{t^{n-1}}{(n-1)!} u(t) \Leftrightarrow \frac{1}{\mathbf{s}^n} \tag{33}$$

The effect on $f(t)$ of integrating $\mathbf{F}(\mathbf{s})$ with respect to $\mathbf{s}$ may be shown by beginning with the definition once more,

$$\mathbf{F}(\mathbf{s}) = \int_0^\infty e^{-\mathbf{s}t} f(t)\, dt$$

performing the frequency integration from $\mathbf{s}$ to ∞,

$$\int_{\mathbf{s}}^\infty \mathbf{F}(\mathbf{s})\, d\mathbf{s} = \int_{\mathbf{s}}^\infty \left[\int_0^\infty e^{-\mathbf{s}t} f(t)\, dt\right] d\mathbf{s}$$

interchanging the order of integration,

$$\int_{\mathbf{s}}^\infty \mathbf{F}(\mathbf{s})\, d\mathbf{s} = \int_0^\infty \left[\int_{\mathbf{s}}^\infty e^{-\mathbf{s}t}\, d\mathbf{s}\right] f(t)\, dt$$

and performing the inner integration,

$$\int_{\mathbf{s}}^\infty \mathbf{F}(\mathbf{s})\, d\mathbf{s} = \int_0^\infty \left[-\frac{1}{t} e^{-\mathbf{s}t}\right]_{\mathbf{s}}^\infty f(t)\, dt = \int_0^\infty \frac{f(t)}{t} e^{-\mathbf{s}t}\, dt$$

Thus,

$$\frac{f(t)}{t} \Leftrightarrow \int_{\mathbf{s}}^\infty \mathbf{F}(\mathbf{s})\, d\mathbf{s} \tag{34}$$

For example, we have already established the transform pair,

$$\sin \omega_0 t\, u(t) \Leftrightarrow \frac{\omega_0}{\mathbf{s}^2 + \omega_0^2}$$

Therefore,

$$\mathcal{L}^{-1}\left\{\frac{\sin \omega_0 t\, u(t)}{t}\right\} = \int_{\mathbf{s}}^\infty \frac{\omega_0\, d\mathbf{s}}{\mathbf{s}^2 + \omega_0^2} = \tan^{-1} \frac{\mathbf{s}}{\omega_0} \Bigg|_{\mathbf{s}}^\infty$$

and we have

$$\frac{\sin \omega_0 t\, u(t)}{t} \Leftrightarrow \frac{\pi}{2} - \tan^{-1} \frac{\mathbf{s}}{\omega_0} \tag{35}$$

We next develop the time-scaling theorem of Laplace transform theory by evaluating the transform of $f(at)$, assuming that $\mathcal{L}\{f(t)\}$ is known. The procedure is very simple:

$$\mathcal{L}\{f(at)\} = \int_0^\infty e^{-\mathbf{s}t} f(at)\, dt = \frac{1}{a} \int_0^\infty e^{-(\mathbf{s}/a)\lambda} f(\lambda)\, d\lambda$$

where the change of variable, $at = \lambda$, has been employed. The last integral is recognizable as $1/a$ times the Laplace transform of $f(t)$, except that s is replaced by s/a in the transform. It follows that

$$f(at) \Leftrightarrow \frac{1}{a}\mathbf{F}(s/a) \tag{36}$$

As an elementary example of the use of this time-scaling theorem, consider the determination of the transform of a 1-kHz cosine wave. Assuming we know the transform of a 1-rad/s cosine wave,

$$\cos t\, u(t) \Leftrightarrow \frac{s}{s^2 + 1}$$

the result is

$$\mathcal{L}\{\cos 2000\pi t\, u(t)\} = \frac{1}{2000\pi}\frac{s/2000\pi}{(s/2000\pi)^2 + 1} = \frac{s}{s^2 + (2000\pi)^2}$$

which is correct.

The time-scaling theorem offers us some computational advantages, for it enables us to work initially in a slowed-down world, where time functions may extend for several seconds and periodic functions possess periods of the order of magnitude of a second. Practical engineering work, however, usually involves time functions that vary much more rapidly with time than our examples might have indicated. The time slowdown is used only to simplify the computational aspects of our problems. These results may then be translated easily to the real world through the use of the time-scaling theorem.

Drill Problems

19-13 Find: (*a*) $\mathcal{L}^{-1}\{10/[(s + 3)^2 + 4]\}$; (*b*) $\mathcal{L}^{-1}\{e^{-3s-3}/(s + 1)\}$; (*c*) $\mathcal{L}^{-1}\{2/[(s + 3)^2]\}$.

Ans. $e^{-t}u(t - 3)$; $2e^{-3t}t\, u(t)$; $5e^{-3t} \sin 2t\, u(t)$

19-14 Find the Laplace transform of: (*a*) $te^{-2t}u(t)$; (*b*) $t^2e^{-2t}u(t)$; (*c*) $t^3e^{-2t}u(t)$.

Ans. $1/(s + 2)^2$; $2/(s + 2)^3$; $6/(s + 2)^4$

19-15 If $\mathcal{L}^{-1}\{s^{-5/2}\} = (4t^{3/2}/3\sqrt{\pi})u(t)$, find: (*a*) $\mathcal{L}^{-1}\{s^{-3/2}\}$; (*b*) $\mathcal{L}^{-1}\{s^{-1/2}\}$; (*c*) $\mathcal{L}^{-1}\{s^{-7/2}\}$.

Ans. $(t^{-1/2}/\sqrt{\pi})u(t)$; $(2t^{1/2}/\sqrt{\pi})u(t)$; $(8t^{5/2}/15\sqrt{\pi})u(t)$

19-16 If $\mathcal{L}\{f(t)\} = (s^2 + 1)/(s^5 + s^2 + 1)$, determine the Laplace transform of: (*a*) $f(2t)$; (*b*) $f(t/2)$; (*c*) $tf(t)$.

Ans. $\dfrac{1}{4}\dfrac{s^2 + \frac{1}{4}}{s^5 + s^2/8 + \frac{1}{32}}$; $4\dfrac{s^2 + 4}{s^5 + 8s^2 + 32}$; $3\dfrac{s^4(s^2 + \frac{5}{3})}{(s^5 + s^2 + 1)^2}$

19-8 THE INITIAL-VALUE AND FINAL-VALUE THEOREMS

The last two fundamental theorems that we shall discuss are known as the initial-value and final-value theorems. They will enable us to evaluate $f(0^+)$ and $f(\infty)$ by examining the limiting values of $\mathbf{F}(\mathbf{s})$.

To derive the initial-value theorem, we consider the Laplace transform of the derivative once again,

$$\mathcal{L}\left\{\frac{df}{dt}\right\} = \mathbf{s}\mathbf{F}(\mathbf{s}) - f(0^-) = \int_{0^-}^{\infty} e^{-\mathbf{s}t}\frac{df}{dt}\,dt$$

We now let s approach infinity. By breaking the integral into two parts,

$$\lim_{\mathbf{s}\to\infty}[\mathbf{s}\mathbf{F}(\mathbf{s}) - f(0^-)] = \lim_{\mathbf{s}\to\infty}\left(\int_{0^-}^{0^+} e^0\frac{df}{dt}\,dt + \int_{0^+}^{\infty} e^{-\mathbf{s}t}\frac{df}{dt}\,dt\right)$$

we see that the second integral must approach zero in the limit since the integrand itself approaches zero. Also, $f(0^-)$ is not a function of **s**, and it may be removed from the left limit,

$$-f(0^-) + \lim_{\mathbf{s}\to\infty}[\mathbf{s}\mathbf{F}(\mathbf{s})] = \lim_{\mathbf{s}\to\infty}\int_{0^-}^{0^+} df = \lim_{\mathbf{s}\to\infty}[f(0^+) - f(0^-)]$$

$$= f(0^+) - f(0^-)$$

and finally

$$f(0^+) = \lim_{\mathbf{s}\to\infty}[\mathbf{s}\mathbf{F}(\mathbf{s})]$$

or

$$\lim_{t\to 0^+} f(t) = \lim_{\mathbf{s}\to\infty}[\mathbf{s}\mathbf{F}(\mathbf{s})] \qquad (37)$$

This is the mathematical statement of the *initial-value theorem*. It states that the initial value of the time function $f(t)$ can be obtained from its Laplace transform $\mathbf{F}(\mathbf{s})$ by first multiplying the transform by s and then letting s approach infinity. Note that the initial value of $f(t)$ that is obtained is the limit from the right.

The initial-value theorem, along with the final-value theorem that we shall consider in a moment, is useful in checking the results of a transformation or an inverse transformation. For example, when we first calculated the transform of $\cos\omega_0 t\, u(t)$, we obtained $\mathbf{s}/(\mathbf{s}^2 + \omega_0^2)$. After noting that $f(0^+) = 1$, a partial check on the validity of this result is achieved by applying the initial-value theorem,

$$\lim_{\mathbf{s}\to\infty}\left(\mathbf{s}\frac{\mathbf{s}}{\mathbf{s}^2 + \omega_0^2}\right) = 1$$

and the check is accomplished.

The final-value theorem is not quite as useful as the initial-value theorem, for it can be used only with a certain class of transforms, those whose poles lie entirely within the left half of the s plane, except for a simple pole at $\mathbf{s} = 0$. We again consider the Laplace transform of df/dt,

$$\int_{0^-}^{\infty} e^{-\mathbf{s}t} \frac{df}{dt}\, dt = \mathbf{sF(s)} - f(0^-)$$

this time in the limit as s approaches zero,

$$\lim_{\mathbf{s}\to 0} \int_{0^-}^{\infty} e^{-\mathbf{s}t} \frac{df}{dt}\, dt = \lim_{\mathbf{s}\to 0} [\mathbf{sF(s)} - f(0^-)] = \int_{0^-}^{\infty} \frac{df}{dt}\, dt$$

We assume that both $f(t)$ and its first derivative are transformable. Now, the last term of this equation is readily expressed as a limit,

$$\int_{0^-}^{\infty} \frac{df}{dt}\, dt = \lim_{t\to\infty} \int_{0^-}^{t} \frac{df}{dt}\, dt = \lim_{t\to\infty} [f(t) - f(0^-)]$$

By recognizing that $f(0^-)$ is a constant, a comparison of the last two equations shows us that

$$\lim_{t\to\infty} f(t) = \lim_{\mathbf{s}\to 0} [\mathbf{sF(s)}] \tag{38}$$

which is the *final-value theorem.* In applying this theorem, it is necessary to know that $f(\infty)$, the limit of $f(t)$ as t becomes infinite, exists, or, what amounts to the same thing, that the poles of $\mathbf{sF(s)}$ all lie *within* the left half of the s plane; they may not even lie on the $j\omega$ axis.

As a straightforward example of the application of this theorem, let us consider the function $f(t) = (1 - e^{-at})u(t)$, where $a > 0$. We see immediately that $f(\infty) = 1$. The transform of $f(t)$ is

$$\mathbf{F(s)} = \frac{1}{\mathbf{s}} - \frac{1}{\mathbf{s} + a} = \frac{a}{\mathbf{s}(\mathbf{s} + a)}$$

Multiplying by s and letting s approach zero, we obtain

$$\lim_{\mathbf{s}\to 0} [\mathbf{sF(s)}] = \lim_{\mathbf{s}\to 0} \frac{a}{\mathbf{s} + a} = 1$$

which agrees with $f(\infty)$.

If $f(t)$ is a sinusoid, however, so that $\mathbf{F(s)}$ has poles on the $j\omega$ axis, then a blind use of the final-value theorem might lead us to conclude that the final value is zero. We know, however, that the final value of either $\sin \omega_0 t$ or $\cos \omega_0 t$ is indeterminate. So, beware of $j\omega$-axis poles.

We now have all the tools to apply the Laplace transform to the solution of problems that we either could not solve previously, or that we solved with considerable stress and strain.

Drill Problem

19-17 Find $f(0^+)$ and $f(\infty)$ for each of the following transforms, without determining $f(t)$ first: (*a*) $(2 - e^{-3s})/[s(s^2 + s + 4)]$; (*b*) $2s/(s^2 + s + 4)$; (*c*) $2/(1 - e^{-s})$.

Ans. 0, 0.25; 2, 0; ∞, unknown

19-9 THE TRANSFER FUNCTION H(s)

Earlier in this chapter we defined the transfer function $\mathbf{H}(\mathbf{s})$ as the Laplace transform of the impulse response $h(t)$, initial energy being zero throughout the circuit. Before we can make the most effective use of the transfer function, we need to show that it may be obtained very simply for any linear circuit by frequency-domain analysis. We do this by an argument similar to that presented for the Fourier transform and $\mathbf{H}(j\omega)$ in Sec. 18-9. But we shall be brief.

We work in the complex-frequency domain and apply an input, $v_i(t) = Ae^{\sigma_x t}\cos(\omega_x t + \theta)$, which is $\mathbf{V}_i(\mathbf{s}_x) = Ae^{j\theta}$ in phasor form. Both A and θ are functions of $\mathbf{s}_x = \sigma_x + j\omega_x$, the complex frequency of the excitation. The response is $v_o(t) = Be^{\sigma_x t}\cos(\omega_x t + \phi)$, $\mathbf{V}_o(\mathbf{s}_x) = Be^{j\phi}$. Thus,

$$\frac{\mathbf{V}_o(\mathbf{s}_x)}{\mathbf{V}_i(\mathbf{s}_x)} = \mathbf{G}(\mathbf{s}_x) = \frac{B}{A}e^{j(\phi-\theta)} \tag{39}$$

To determine $\mathbf{H}(\mathbf{s})$, we need to find the ratio of the transform of the output to the transform of the input,

$$\mathbf{H}(\mathbf{s}) = \frac{\mathcal{L}\{Be^{\sigma_x t}\cos(\omega_x t + \phi)\}}{\mathcal{L}\{Ae^{\sigma_x t}\cos(\omega_x t + \theta)\}}$$

The required transform is obtained from Table 19-1 by replacing s by $\mathbf{s} - \sigma_x$ in the transform for $\cos(\omega t + \theta)$. We have

$$\mathbf{H}(\mathbf{s}) = \frac{B\left\{\dfrac{[(\mathbf{s} - \sigma_x)\cos\phi - \omega_x \sin\phi]}{[(\mathbf{s} - \sigma_x)^2 + \omega_x^2]}\right\}}{A\left\{\dfrac{[(\mathbf{s} - \sigma_x)\cos\theta - \omega_x \sin\theta]}{[(\mathbf{s} - \sigma_x)^2 + \omega_x^2]}\right\}}$$

At $\mathbf{s} = \mathbf{s}_x$, this simplifies to

$$\mathbf{H}(\mathbf{s}_x) = \frac{B}{A}\frac{j\omega_x \cos\phi - \omega_x \sin\phi}{j\omega_x \cos\theta - \omega_x \sin\theta}$$

$$= \frac{B}{A}\frac{\cos\phi + j\sin\phi}{\cos\theta + j\sin\theta} = \frac{B}{A}e^{j(\phi-\theta)}$$

which is identical to (39). Since there is no special significance to the x subscript, it follows that

$$\mathbf{H}(\mathbf{s}) = \mathbf{G}(\mathbf{s}) \tag{40}$$

Thus, we may find $\mathbf{H}(\mathbf{s})$ by using normal frequency-domain methods with all elements expressed in terms of their impedances at a complex frequency $\mathbf{s}$.

Let us illustrate how this technique may be used to find both $h(t)$ and an output voltage for the circuit shown in Fig. 19-8*a*. At this time we assume that there is no initial energy storage in the network. We first construct the frequency-domain circuit, shown in Fig. 19-8*b*. The ratio of $\mathbf{V}_o(\mathbf{s})/\mathbf{V}_i(\mathbf{s})$ may be found by determining $\mathbf{Z}_i(\mathbf{s})$, the impedance of the three parallel branches at the right,

$$\mathbf{Z}_i(\mathbf{s}) = \frac{1}{\mathbf{s}/24 + \frac{1}{30} + 1/(24 + 48/\mathbf{s})} = \frac{120(\mathbf{s} + 2)}{5\mathbf{s}^2 + 19\mathbf{s} + 8}$$

Fig. 19-8 (*a*) An example in which the transfer function $\mathbf{H}(\mathbf{s}) = \mathbf{V}_o(\mathbf{s})/\mathbf{V}_i(\mathbf{s})$ is to be obtained by frequency-domain analysis. Initial conditions are all zero. (*b*) The frequency-domain circuit.

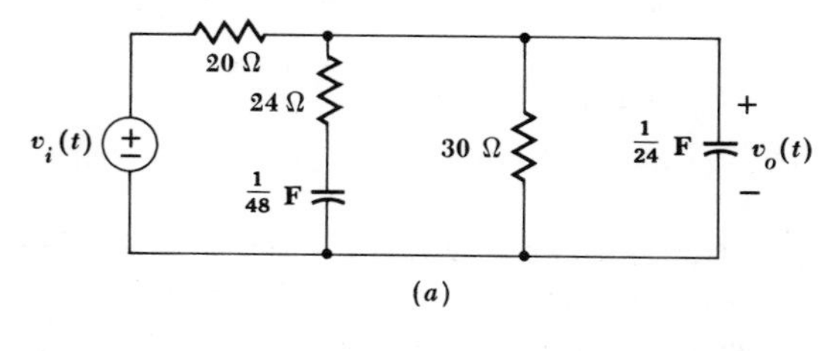

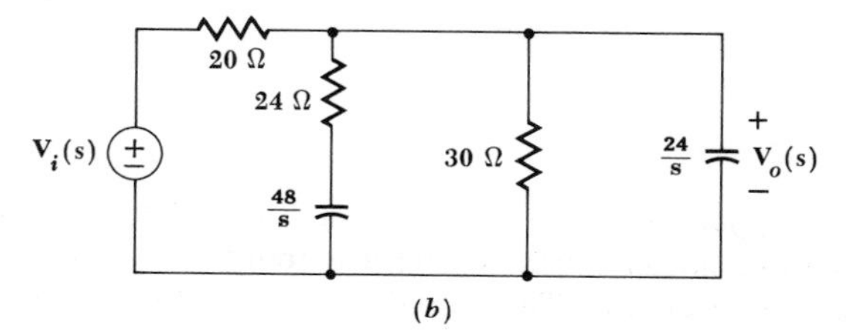

and then using voltage division,

$$\frac{\mathbf{V}_o(\mathbf{s})}{\mathbf{V}_i(\mathbf{s})} = \frac{\mathbf{Z}_i(\mathbf{s})}{20 + \mathbf{Z}_i(\mathbf{s})} = \frac{6(\mathbf{s} + 2)}{5\mathbf{s}^2 + 25\mathbf{s} + 20}$$

Thus,

$$\mathbf{H}(\mathbf{s}) = \frac{1.2(\mathbf{s} + 2)}{(\mathbf{s} + 1)(\mathbf{s} + 4)}$$

To find $h(t)$, we need $\mathcal{L}^{-1}\{\mathbf{H}(\mathbf{s})\}$:

$$\mathcal{L}^{-1}\{\mathbf{H}(\mathbf{s})\} = \mathcal{L}^{-1}\left\{\frac{0.4}{\mathbf{s} + 1} + \frac{0.8}{\mathbf{s} + 4}\right\}$$

and

$$h(t) = (0.4e^{-t} + 0.8e^{-4t})u(t)$$

Thus, if $v_i(t) = \delta(t)$, $v_o(t) = h(t) = (0.4e^{-t} + 0.8e^{-4t})u(t)$.

For a different input, say $v_i(t) = 50 \cos 2t\, u(t)$ V, we may make use of the transfer function concept,

$$\mathbf{V}_o(\mathbf{s}) = \mathbf{H}(\mathbf{s})\mathbf{V}_i(\mathbf{s})$$

where

$$\mathbf{V}_i(\mathbf{s}) = \mathcal{L}\{50 \cos 2t\, u(t)\} = \frac{50\mathbf{s}}{\mathbf{s}^2 + 4}$$

and

$$\mathbf{V}_o(\mathbf{s}) = \frac{1.2(\mathbf{s} + 2)}{(\mathbf{s} + 1)(\mathbf{s} + 4)} \frac{50\mathbf{s}}{\mathbf{s}^2 + 4}$$

Expanding in partial fractions,

$$\mathbf{V}_o(\mathbf{s}) = \frac{-4}{\mathbf{s} + 1} + \frac{-8}{\mathbf{s} + 4} + \frac{6 + j6}{\mathbf{s} + j2} + \frac{6 - j6}{\mathbf{s} - j2}$$

$$= \frac{-4}{\mathbf{s} + 1} + \frac{-8}{\mathbf{s} + 4} + \frac{12\mathbf{s} + 24}{\mathbf{s}^2 + 4}$$

and

$$v_o(t) = (-4e^{-t} - 8e^{-4t} + 12 \cos 2t + 12 \sin 2t)u(t)$$

The solution is straightforward, and we should feel confident of being able to find the response for any input that is Laplace-transformable. We consider the presence of initial-energy storage in the next (and final) section.

Drill Problem

19-18 Find the impulse response of the circuit of Fig. 19-9 by constructing the frequency-domain circuit, finding the appropriate transfer function, and evaluating the inverse Laplace transform. The output is: (*a*) $v_A(t)$; (*b*) $v_B(t)$; (*c*) $v_C(t)$.

Ans. $(6e^{-4t} - 24te^{-4t})u(t)$; $\delta(t) - (6e^{-4t} - 8te^{-4t})u(t)$; $(6e^{-4t} - 8te^{-4t})u(t)$ V

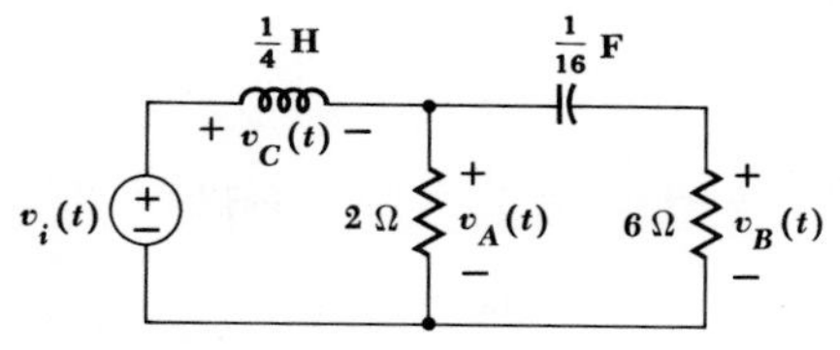

Fig. 19-9 See Drill Prob. 19-18.

19-10 THE COMPLETE RESPONSE

When initial energy is present in a circuit, the Laplace transform method may be used to obtain the complete response by any of several different methods. We shall consider two of them.

The first is the more fundamental, for it involves writing the differential equations for the network and then taking the Laplace transform of those equations. The initial conditions appear when a derivative or integral is transformed. The second technique requires each initial capacitor voltage or inductor current to be replaced by an equivalent dc source, often called an *initial-condition generator*. The elements then carry no initial energy, and the transfer function procedure of the preceding section can be followed.

Let us illustrate the differential equation approach by considering the same circuit we just analyzed, but with nonzero initial conditions this time, as shown in Fig. 19-10. We let $v_1(0^-) = 10$ V and $v_2(0^-) = 25$ V.

The differential equations for this circuit may be obtained by writing nodal equations in terms of v_1 and v_2. At the v_1 node,

$$\frac{v_1 - v_2}{24} + \frac{1}{48} v_1' = 0$$

or

$$2v_2 = 2v_1 + v_1' \tag{41}$$

Fig. 19-10 The response $v_2(t)$ is obtained for this network with the initial conditions, $v_1(0^-) = 10$ V, $v_2(0^-) = 25$ V.

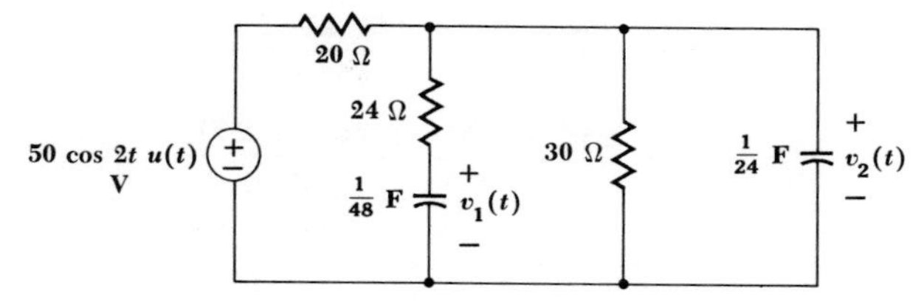

while at the v_2 node,

$$\frac{v_2 - 50\cos 2t\,u(t)}{20} + \frac{v_2 - v_1}{24} + \frac{v_2}{30} + \frac{1}{24}v_2' = 0$$

or

$$v_1 = v_2' + 3v_2 - 60\cos 2t\,u(t) \tag{42}$$

Identifying v_2 as the desired response, we eliminate v_1 and v_1' by taking the derivative of (42), remembering that $du(t)/dt = \delta(t)$,

$$v_1' = v_2'' + 3v_2' + 120\sin 2t\,u(t) - 60\delta(t) \tag{43}$$

and substituting (42) and (43) into (41),

$$2v_2 = 2[v_2' + 3v_2 - 60\cos 2t\,u(t)] + [v_2'' + 3v_2' + 120\sin 2t\,u(t) - 60\delta(t)]$$

or

$$v_2'' + 5v_2' + 4v_2 = (120\cos 2t - 120\sin 2t)u(t) + 60\delta(t)$$

We now take the Laplace transform,

$$\mathbf{s}^2\mathbf{V}_2(\mathbf{s}) - \mathbf{s}v_2(0^-) - v_2'(0^-) + 5\mathbf{s}\mathbf{V}_2(\mathbf{s}) - 5v_2(0^-) + 4\mathbf{V}_2(\mathbf{s}) = \frac{120\mathbf{s} - 240}{\mathbf{s}^2 + 4} + 60$$

collect terms,

$$(\mathbf{s}^2 + 5\mathbf{s} + 4)\mathbf{V}_2(\mathbf{s}) = \mathbf{s}v_2(0^-) + v_2'(0^-) + 5v_2(0^-) + \frac{120\mathbf{s} - 240}{\mathbf{s}^2 + 4} + 60$$

let $v_2(0^-) = 25$,

$$(\mathbf{s}^2 + 5\mathbf{s} + 4)\mathbf{V}_2(\mathbf{s}) = 25\mathbf{s} + 125 + v_2'(0^-) + \frac{120\mathbf{s} - 240}{\mathbf{s}^2 + 4} + 60$$

and need a value for $v_2'(0^-)$. This we may obtain from the two circuit equations (41) and (42) by evaluating each term at $t = 0^-$. Actually, we need use only (42) in this problem:

$$v_1(0^-) = v_2'(0^-) + 3v_2(0^-) - 0$$

and

$$v_2'(0^-) = -65$$

Thus,

$$\begin{aligned} \mathbf{V}_2(\mathbf{s}) &= \frac{25\mathbf{s} + 120 + 120[(\mathbf{s} - 2)/(\mathbf{s}^2 + 4)]}{(\mathbf{s} + 1)(\mathbf{s} + 4)} \\ &= \frac{25\mathbf{s}^3 + 120\mathbf{s}^2 + 220\mathbf{s} + 240}{(\mathbf{s} + 1)(\mathbf{s} + 4)(\mathbf{s}^2 + 4)} \qquad (44) \\ &= \frac{23/3}{\mathbf{s} + 1} + \frac{16/3}{\mathbf{s} + 4} + \frac{12\mathbf{s} + 24}{\mathbf{s}^2 + 4} \end{aligned}$$

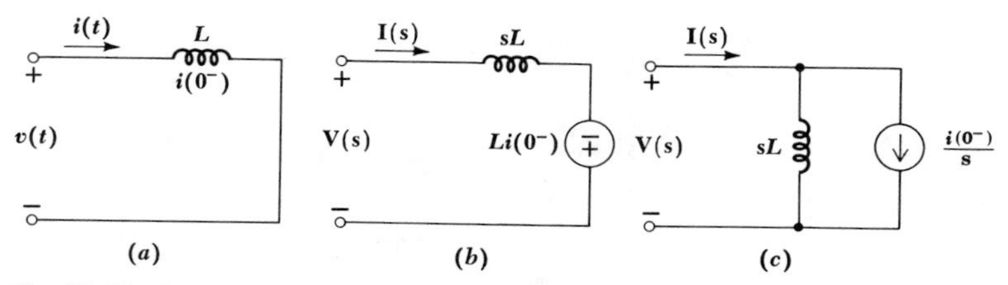

Fig. 19-11 (a) An inductor L with initial current $i(0^-)$ is shown in the time domain. (b) and (c) Frequency-domain networks that are equivalent to (a) for Laplace-transform analysis.

from which the time-domain response is obtained,

$$v_2(t) = (^{23}\!/_3 e^{-t} + {}^{16}\!/_3 e^{-4t} + 12 \cos 2t + 12 \sin 2t)u(t)$$

Now let us determine the frequency-domain equivalent of an inductor L with initial current $i(0^-)$. The time-domain network of Fig. 19-11a is described by

$$v(t) = Li'$$

and, therefore,

$$\mathbf{V}(\mathbf{s}) = \mathbf{s}L\mathbf{I}(\mathbf{s}) - Li(0^-) \tag{45}$$

or

$$\mathbf{I}(\mathbf{s}) = \frac{\mathbf{V}(\mathbf{s})}{\mathbf{s}L} + \frac{i(0^-)}{\mathbf{s}} \tag{46}$$

The frequency-domain equivalents can be read directly from (45) and (46), and they are shown in Figs. 19-11b and c, respectively. It may be helpful to note that the voltage source in Fig. 19-11b is the transform of an impulse, while the current source in c is the transform of a step.

Equivalent networks for an initially charged capacitor are obtained by a similar procedure; the results are shown in Fig. 19-12.

Fig. 19-12 (a) A capacitor C with initial voltage $v(0^-)$ is shown in the time domain. (b) and (c) Frequency-domain networks that are equivalent to (a) for Laplace-transform analysis.

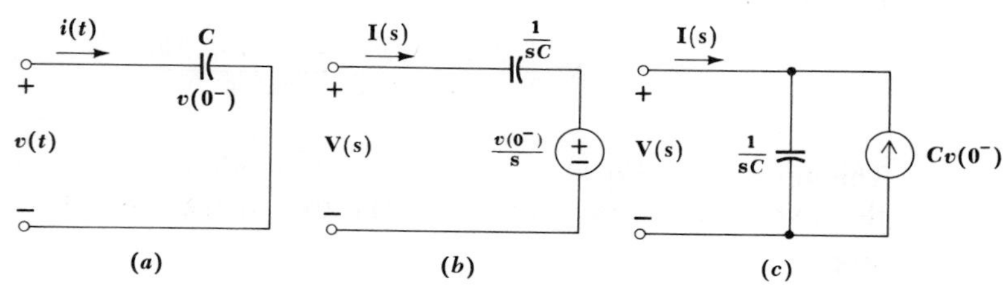

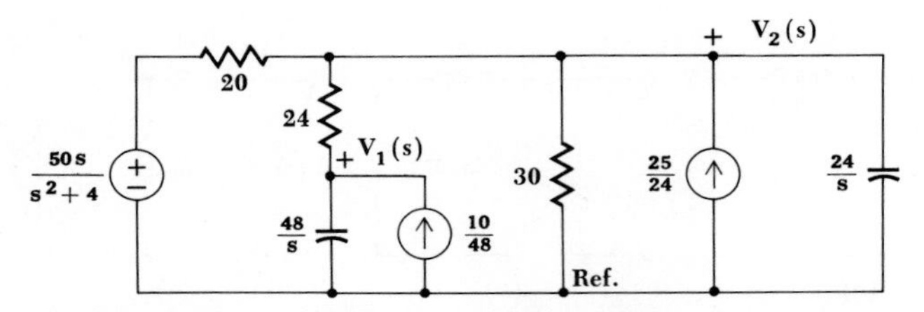

Fig. 19-13 The frequency-domain equivalent of the circuit of Fig. 19-10. The current sources $^{10}/_{48}$ and $^{25}/_{24}$ provide initial time-domain voltages of 10 and 25 V across the $^{1}/_{48}$-F and $^{1}/_{24}$-F capacitors, respectively.

We may now use these results to construct a frequency-domain equivalent of the circuit shown in Fig. 19-10, including the effect of the initial conditions. The result is shown in Fig. 19-13. Current sources are used for the initial conditions to expedite the writing of nodal equations.

We must now use both the superposition principle and the concept of the transfer function to see that $\mathbf{V}_2(\mathbf{s})$ is composed of the sum of three terms, one due to each source acting alone. Moreover, each of these sources has a transfer function to $\mathbf{V}_2(\mathbf{s})$. Each transfer function could be obtained by applying the standard frequency-domain analysis methods. We need not solve three little problems, however, for we can accomplish the frequency-domain analysis with all three sources operating. We do this by writing two nodal equations:

$$\frac{\mathbf{V}_1(\mathbf{s}) - \mathbf{V}_2(\mathbf{s})}{24} + \frac{\mathbf{s}\mathbf{V}_1(\mathbf{s})}{48} - \frac{10}{48} = 0$$

or

$$(\mathbf{s} + 2)\mathbf{V}_1 - 2\mathbf{V}_2 = 10 \tag{47}$$

and

$$\frac{\mathbf{V}_2(\mathbf{s}) - 50\mathbf{s}/(\mathbf{s}^2 + 4)}{20} + \frac{\mathbf{V}_2(\mathbf{s}) - \mathbf{V}_1(\mathbf{s})}{24} + \frac{\mathbf{V}_2(\mathbf{s})}{30} - \frac{25}{24} + \frac{\mathbf{s}\mathbf{V}_2(\mathbf{s})}{24} = 0$$

or

$$(\mathbf{s} + 3)\mathbf{V}_2 - \mathbf{V}_1 = \frac{60\mathbf{s}}{\mathbf{s}^2 + 4} + 25 \tag{48}$$

Using (47) to eliminate $\mathbf{V}_1$ in the above equation, we have

$$\mathbf{V}_1 = \frac{10 + 2\mathbf{V}_2}{\mathbf{s} + 2}$$

and then

$$\mathbf{V}_2 = \frac{25\mathbf{s}^3 + 120\mathbf{s}^2 + 220\mathbf{s} + 240}{(\mathbf{s} + 1)(\mathbf{s} + 4)(\mathbf{s}^2 + 4)}$$

This agrees with Eq. (44) of the preceding section, and we need not repeat the inverse transform operation. The two methods check.

In comparing their use, it is probably true that the transfer function method with initial condition generators is a little faster. This becomes a safer statement to make as the complexity of the network increases. However, we must also bear in mind the fact that the closer we stay to fundamentals, the less apt we are to become confused with special techniques and procedures. Certainly the differential equation approach is the more basic. The use of the Laplace transform may then be thought of simply as a handy method of solving linear differential equations.

In finishing our discussion of the Laplace transform, it is helpful to compare its uses with those of the Fourier transform. The latter, of course, does not exist for as wide a variety of time functions; there are also more j's scattered about in the transform expressions, and the Fourier transform is more difficult to apply to circuits containing initial-energy storage. Thus, transient problems are more easily handled with the Laplace transform. However, when spectral information about a signal is desired, such as the distribution of energy across the frequency band, the Fourier transform is the more convenient.

And now we approach the last few paragraphs of the last section of the last chapter. Looking back at six hundred pages of linear circuit analysis, we should ask ourselves what we have accomplished. Are we prepared to tackle a really practical problem, or have we just been tilting at quixotic windmills? Can we analyze a multistage active filter, a complicated telemetry receiver, or a large interconnected power grid? Before we confess that we cannot, or at least admit that there are other things we would rather do, let us consider how far we could go in such a complex problem.

We certainly have developed substantial skills in writing accurate sets of equations to describe the behavior of increasingly complicated linear circuits. This in itself is no mean accomplishment. Our only real limitation is our inability to produce numerical results as the complexity of the circuit increases. For that reason we concentrated on those examples which suited computational abilities that were only human. However, we did develop an increased understanding of the most important analysis techniques in the process.

In the last several chapters we have been developing those techniques which facilitate the process of obtaining specific numerical answers, transform and frequency-domain methods. We could now write a set of time-domain or frequency-domain equations for any large-scale circuit we chose, but the numerical solution would strain our accuracy and stretch our patience. If our discussion were to proceed any further, we would have to resort to inhuman means to solve these equations. This is the point at which we turn to the computer. Its patience is virtually unlimited, its accuracy is astonishing, and its speed is almost beyond comprehension. The detailed procedure by which the computer arrives at its numerical results, however, is generated by us, and if we make an error in the program, the computer will proceed to generate erroneous results, also with speed, accuracy, and patience.

In summary, then, we can now produce sets of accurate descriptive equations that characterize almost any given linear circuit configuration, and we have some modest abilities for solving the simpler sets of these equations. Our next step, if we were to continue, would be to begin writing our equations in those forms most compatible to a digital computer. But we must stop sometime, so let it be now.

Drill Problems

19-19 By writing an appropriate differential equation and then taking the Laplace transform of both sides, find $v_2(t)$ in the circuit of Fig. 19-14 if $v_s(t) =$: (*a*) $12u(t)$; (*b*) $12 \cos t\, u(t)$; (*c*) $12e^{-t}u(t)$.

Ans. $4e^{-t/2}u(t)$; $(-12e^{-t/2} + 16e^{-t})u(t)$; $(-2.4e^{-t/2} + 6.4 \cos t - 3.2 \sin t)u(t)$ V

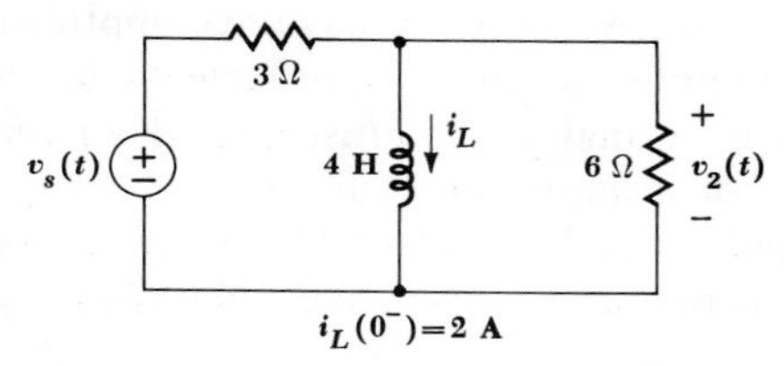

Fig. 19-14 See Drill Probs. 19-19 and 19-20.

19-20 Rework Drill Prob. 19-19 by installing a suitable initial-condition generator in the circuit and then working completely in the frequency domain.

Ans. Same

PROBLEMS

☐ **1** Find the two-sided Laplace transform of each of the following functions and state the range of σ for which each exists: (*a*) $2e^{-3|t|}$; (*b*) $2e^{-3t}[u(t+3) - u(t-3)]$; (*c*) $2e^{3t}[u(t+3) - u(t-3)]$.

☐ **2** Given $\mathbf{V}(\mathbf{s}) = \mathcal{L}\{v(t)\}$, under what conditions is $\mathcal{F}\{v(t)\} = \mathbf{V}(\mathbf{s} = j\omega)$?

☐ **3** Repeat Prob. 1 for the (one-sided) Laplace transform.

☐ **4** Determine $\mathbf{F}(\mathbf{s})$ if $f(t) =$: (*a*) $2e^{\sin t}\,\delta(t)$; (*b*) $tu(t-2)$; (*c*) $\delta(t-2)$.

☐ **5** Let $f_1(t) = \delta(t) + \delta(t + 1)$ and $f_2(t) = \delta(t)$. (*a*) Sketch $f_1(t)$ and $f_2(t)$. (*b*) Show that $\mathbf{F}_1(\mathbf{s}) = \mathbf{F}_2(\mathbf{s})$. (*c*) Reconcile the answers to (*a*) and (*b*) with the statement in Sec. 19-3 that there is a one-to-one correspondence between $f(t)$ and $\mathbf{F}(\mathbf{s})$.

☐ **6** Find $f(10)$ if $\mathbf{F}(\mathbf{s}) =$ (*a*) $5/(10\mathbf{s} + 1)$; (*b*) $1/(5\mathbf{s} - 1)$.

☐ **7** Find the partial fraction expansion of each of the following functions: (*a*) $2\mathbf{s}/(\mathbf{s}^2 + 5\mathbf{s} + 4)$; (*b*) $2\mathbf{s}^2/(\mathbf{s}^2 + 5\mathbf{s} + 4)$; (*c*) $2/(\mathbf{s}^2 + 2\mathbf{s} + 2)$; (*d*) $\mathbf{s}^3/(\mathbf{s}^2 + 2\mathbf{s} + 1)$; (*e*) $(4\mathbf{s} + 1)/(\mathbf{s}^4 + 2\mathbf{s}^3 + \mathbf{s}^2)$.

☐ **8** Find the inverse Laplace transform for: (*a*) $(3\mathbf{s} + 1)/(\mathbf{s}^3 + 5\mathbf{s}^2 + 6\mathbf{s})$; (*b*) $\mathbf{s}^2/(\mathbf{s}^3 + 6\mathbf{s}^2 + 11\mathbf{s} + 6)$; (*c*) $(\mathbf{s}^2 + 2\mathbf{s} + 6)/(\mathbf{s}^3 + \mathbf{s}^2 + 4\mathbf{s} + 4)$.

☐ **9** Determine $\mathcal{L}^{-1}\{\mathbf{F}(\mathbf{s})\}$ if $\mathbf{F}(\mathbf{s}) =$: (*a*) $(2\mathbf{s} + 3)/(\mathbf{s}^2 + 1)$; (*b*) $\mathbf{s}^2/(\mathbf{s}^4 + 3\mathbf{s}^2 + 2)$; (*c*) $(\mathbf{s}^2 + 3\mathbf{s} + 2)/\mathbf{s}^2$.

☐ **10** Use Laplace transform techniques to solve each of the following differential equations, subject to the initial conditions specified: (*a*) $y'' + 4y = 0$, $y(0) = 4$, $y'(0) = 6$; (*b*) $y' + 2y = 8t\,u(t), y(0) = 1$; (*c*) $f^{(3)} + 4f' = 8u(x), f(0) = 1, f'(0) = 2$, $f''(0) = -4$.

☐ **11** In the circuit shown in Fig. 19-15: (*a*) let $v_s(t) = 3t\,u(t)$ and find $i(t)$; (*b*) let $v_s(t) = 3\delta(t)$ and find $v(t)$.

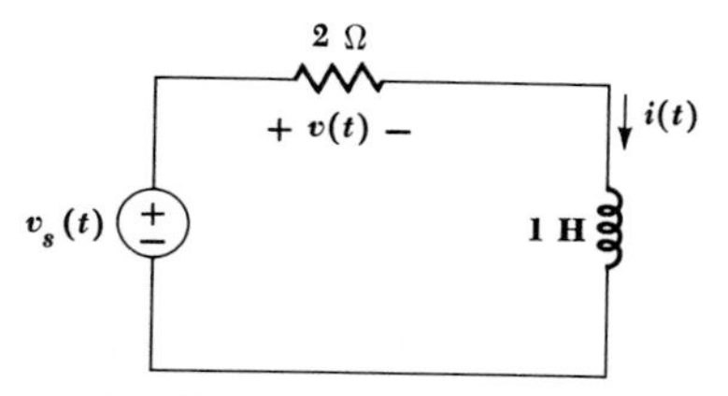

Fig. 19-15 See Prob. 11.

☐ **12** Find $v_C(t)$ for the circuit of Fig. 19-16 if $v_s(t) =$: (*a*) $\delta(t)$; (*b*) $2 \cos 500t\, u(t)$.

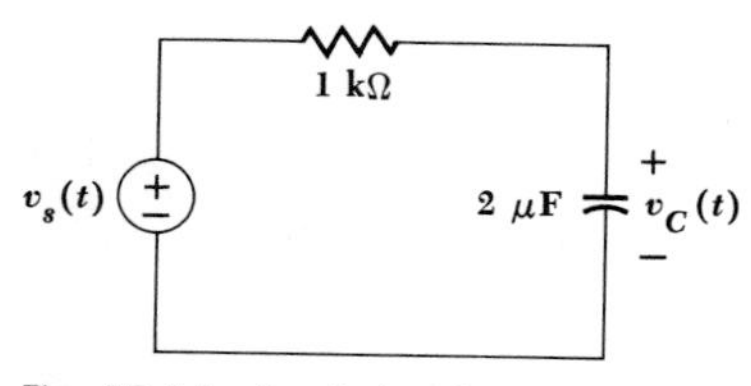

Fig. 19-16 See Prob. 12.

☐ **13** The voltage pulse described by $10(e^{-800t} - e^{-2000t})u(t)$ V is applied to a series *RL* circuit in which $L = 4$ H and $R = 20$ kΩ. Find the current.

□ **14** Repeat Prob. 13 if a $\frac{1}{16}$-μF capacitor is inserted in series with the resistor and inductor.

□ **15** Find $f_1(t) * f_2(t)$ if $f_1(t) = 2e^{-4t}u(t)$ and $f_2(t) = 5 \cos 3t\, u(t)$.

□ **16** The impulse response of a certain network is $0.1 \sin 60t\, u(t)$. What is the output produced by an input $0.1 \sin 120t\, u(t)$?

□ **17** (*a*) Use convolution to find the inverse Laplace transform of $e^{-as}/\mathbf{s}$. (*b*) Use convolution to express the inverse transform of $\mathbf{F}_1(\mathbf{s})e^{-as}$ in terms of $f_1(t) = \mathcal{L}^{-1}\{\mathbf{F}_1(\mathbf{s})\}$.

□ **18** By using convolution, evaluate: (*a*) $\mathcal{L}^{-1}\{(1/\mathbf{s}^2)(1/\mathbf{s}^2)\}$; (*b*) $\mathcal{L}^{-1}\{(1/\mathbf{s}^2)(1/\mathbf{s}^4)\}$.

□ **19** Determine the Laplace transform of the time functions shown in: (*a*) Fig. 19-17*a*; (*b*) Fig. 19-17*b*.

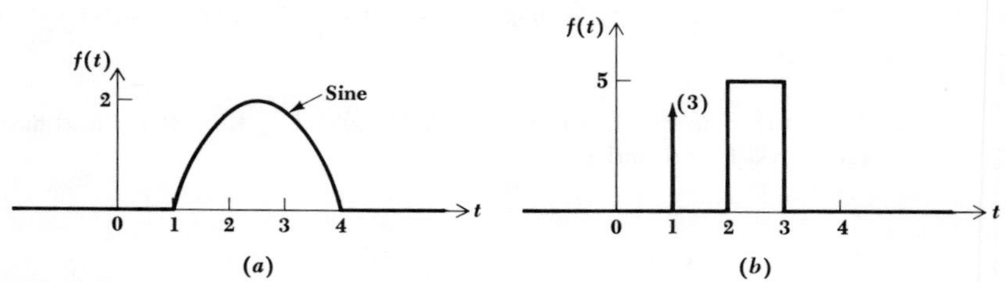

Fig. 19-17 See Prob. 19.

□ **20** Use the time-shift theorem to obtain the Laplace transform of the: (*a*) nonperiodic staircase function shown in Fig. 19-18*a*; (*b*) nonperiodic sawtooth function of Fig. 19-18*b*.

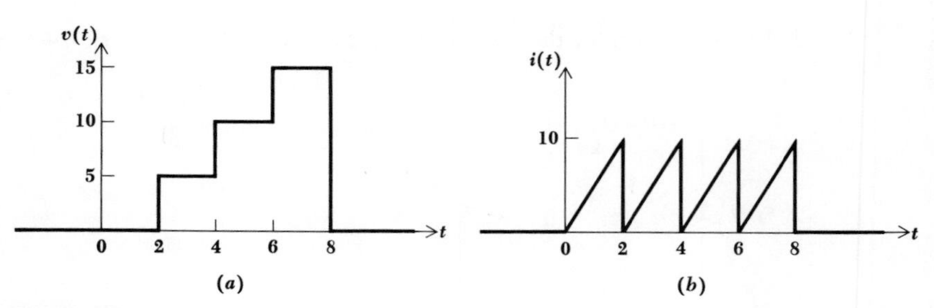

Fig. 19-18 See Prob. 20.

□ **21** Find and sketch as functions of t: (*a*) $\mathcal{L}^{-1}\{4e^{-5\mathbf{s}}[(9\mathbf{s} + 2)/(3\mathbf{s}^2 + \mathbf{s})]\}$; (*b*) $\mathcal{L}^{-1}\{(1 - e^{-2\mathbf{s}})^2\}$; (*c*) $\mathcal{L}^{-1}\{4e^{-3\mathbf{s}} \cosh 2\mathbf{s}\}$; (*d*) $\mathcal{L}^{-1}\{4e^{-3\mathbf{s}} \sinh 2\mathbf{s}\}$.

☐ **22** Find the Laplace transform of the following periodic functions: (*a*) full-wave rectifier output, shown in Fig. 19-19*a*; (*b*) half-wave rectifier output, shown in Fig. 19-19*b*.

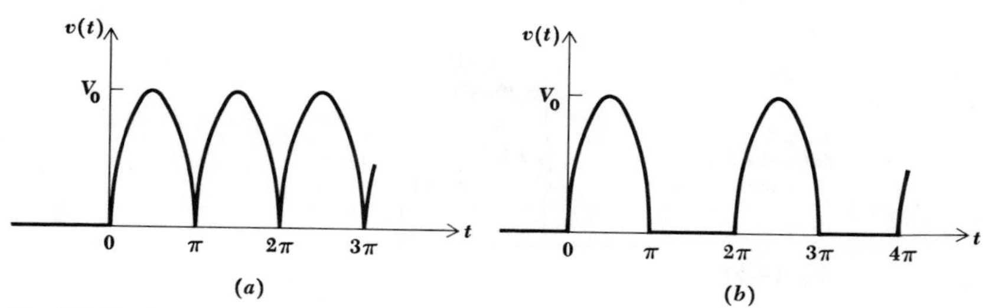

Fig. 19-19 See Prob. 22.

☐ **23** Find $\mathcal{L}\{i(t)\}$ for the periodic current waveform shown in: (*a*) Fig. 19-20*a*; (*b*) Fig. 19-20*b*.

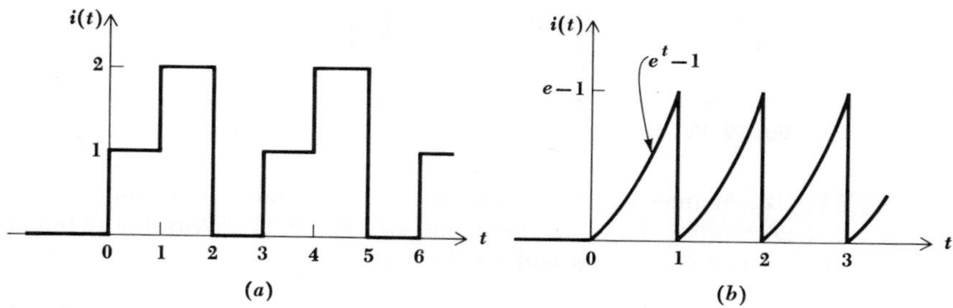

Fig. 19-20 See Prob. 23.

☐ **24** Determine $\mathcal{L}\{u(\sin \pi t)u(t)\}$.

☐ **25** Find $\mathcal{L}\{f(t)\}$ for $f(t) =$: (*a*) $2te^{-3t} \sin 4t\, u(t)$; (*b*) $\sin^3 4t\, u(t)$; (*c*) $t^2e^{-3t}u(t-1)$; (*d*) $e^{-3t}u(t)\int_0^t t \sin 4t\, dt$.

☐ **26** Find $\mathcal{L}^{-1}\{\mathbf{F}(\mathbf{s})\}$ if $\mathbf{F}(\mathbf{s}) =$: (*a*) $(\mathbf{s}e^{-2\mathbf{s}})/(\mathbf{s}+3)$; (*b*) $\ln[(\mathbf{s}+1)/(\mathbf{s}+3)]$; (*c*) $\mathbf{s}\ln[(\mathbf{s}+1)/(\mathbf{s}+3)]$.

☐ **27** Evaluate both sides of the final value theorem for each of the following transforms and its inverse: (*a*) $2\mathbf{s}/(\mathbf{s}^2+2\mathbf{s}+2)$; (*b*) $2/(\mathbf{s}^3+2\mathbf{s}^2+2\mathbf{s})$; (*c*) $1/(\mathbf{s}^4+4)$.

☐ **28** Construct a function of **s** as the ratio of two polynomials in **s** such that both $\lim_{\mathbf{s}\to\infty}[\mathbf{sF}(\mathbf{s})] = 7$ and $\lim_{\mathbf{s}\to 0}[\mathbf{sF}(\mathbf{s})] = 11$. Find the inverse transform and then investigate $\lim_{t\to 0} f(t)$ and $\lim_{t\to\infty} f(t)$.

☐ **29** For each of the following $\mathbf{V}(s)$, find $v(0^+)$: (*a*) $(2s^3 + 1)/(s^4 + 10^4)$; (*b*) $[(2 + 3s^2)(3 + 4s)]/[s^3(5 + 2s)]$; (*c*) $(1/s)[(1 - e^{-6s})/(1 - e^{-2s})]$.

☐ **30** Use Laplace transform techniques to find the complete response $v(t)$ of the circuit shown in Fig. 19-21.

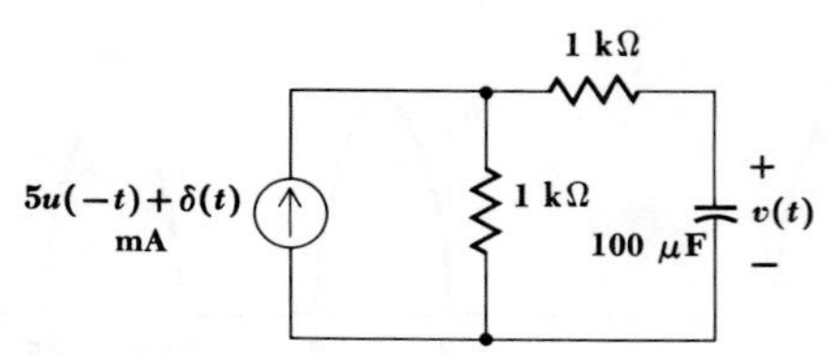

Fig. 19-21 See Prob. 30.

☐ **31** (*a*) Draw the frequency-domain circuit that is valid for $t > 0$ for the circuit shown in Fig. 19-22. (*b*) Find $i(t)$ for $t > 0$.

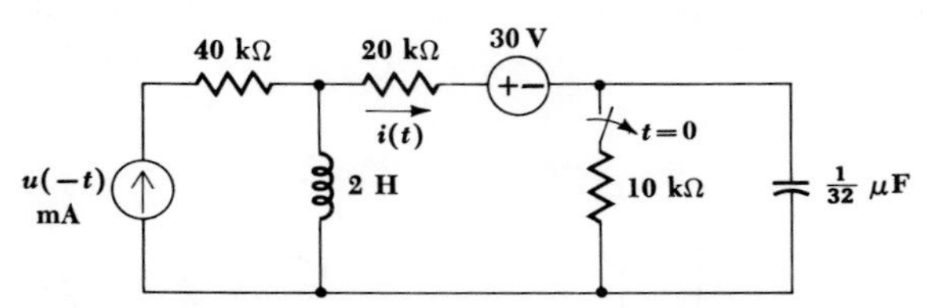

Fig. 19-22 See Prob. 31.

☐ **32** Sources and switches not shown have produced the initial conditions, $v(0^-) =$ 20 V, $i(0^-) = 2$ A, in the circuit of Fig. 19-23. Write the differential equations for this circuit and find $v(t)$ for $t > 0$.

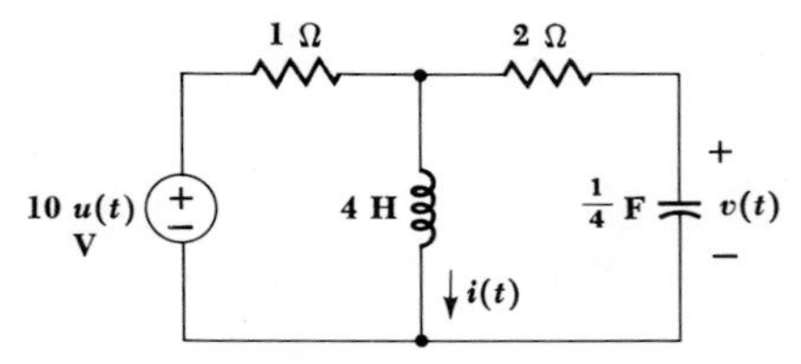

Fig. 19-23 See Prob. 32.

☐ **33** A voltage pulse rises linearly from 0 at $t = 0$ to 20 V at $t = 2$ s, at which time it falls to zero and remains at zero thereafter. This voltage is applied to the series combination of 50 kΩ and 40 μF. Find and sketch the capacitor voltage for $t > 0$.

☐ **34** A single sinusoidal current pulse, $i_s(t) = 5 \sin 2t[u(t) - u(t - \pi/2)]$ A, is applied to the parallel combination of a $\frac{1}{8}$-H inductor and a $\frac{1}{2}$-F capacitor. Find the capacitor voltage $v(t)$ for $t > 0$.

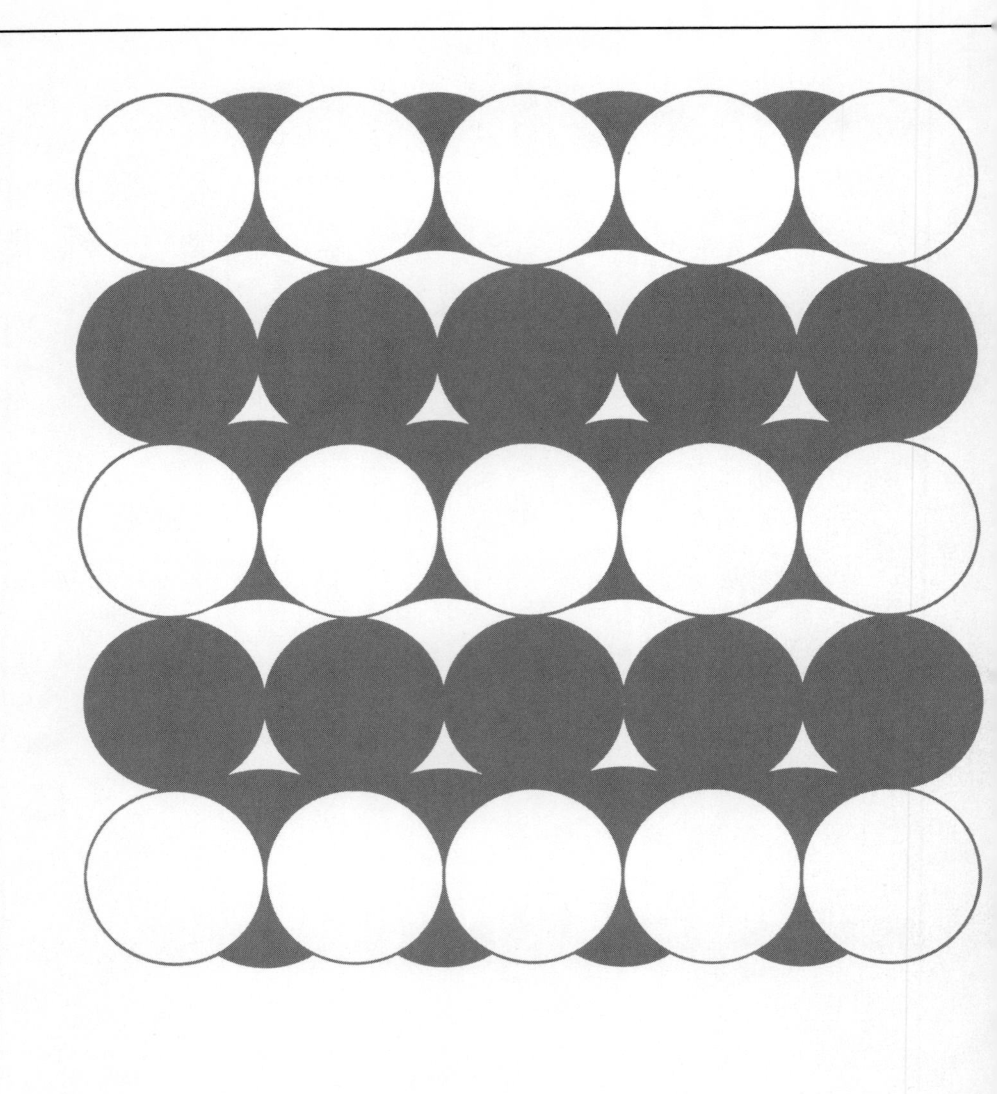

Part Seven
APPENDIXES

Appendix One Determinants

In Sec. 3-2 of Chap. 3 we obtained a system of three equations for the four-node circuit of Fig. 3-2:

$$7v_1 - 3v_2 - 4v_3 = -11 \tag{3}$$

$$-3v_1 + 6v_2 - 2v_3 = 3 \tag{4}$$

$$-4v_1 - 2v_2 + 11v_3 = 25 \tag{5}$$

This set of equations could have been solved by a systematic elimination of the variables. This procedure is lengthy, however, and may never yield answers if done unsystematically for a greater number of simultaneous equations. A much more orderly method involves using determinants and Cramer's rule, as discussed in most courses in college algebra. The use of determinants has the additional advantages that it leads naturally into the expression of the circuit elements in terms of matrices, and it establishes a method of analyzing a general circuit which will be helpful in proving general theorems. It should be pointed out that the number of arithmetic steps required to solve a large set of simultaneous equations by determinants is excessive; a digital computer would be programmed to use another method. This appendix consists of a brief review of the determinant method and nomenclature.

Consider (3), (4), and (5). The array of the constant coefficients of the equations is called a *matrix*,

$$\mathbf{G} = \begin{bmatrix} 7 & -3 & -4 \\ -3 & 6 & -2 \\ -4 & -2 & 11 \end{bmatrix}$$

where the symbol **G** has been selected since each element of the matrix is a conductance value. A matrix has no "value"; it is merely an ordered array of elements. We use boldface type to represent a matrix, and we enclose the array itself by square brackets.

The *determinant* of a square matrix has a value, however. To be precise, we should say that the determinant of a matrix *is* a value, but common usage enables us to speak of both the array itself and its value as the determinant. We shall symbolize a determinant by Δ, and employ a suitable subscript to denote the matrix to which the determinant refers. Thus,

$$\Delta_G = \begin{vmatrix} 7 & -3 & -4 \\ -3 & 6 & -2 \\ -4 & -2 & 11 \end{vmatrix}$$

Note that simple vertical lines are used to enclose the determinant.

The value of any determinant is obtained by expanding it in terms of its minors. To do this, we select any row j or any column k, multiply each element in that row or column by its minor and by $(-1)^{j+k}$, and then add these products. The *minor* of the element appearing in both row j and column k is the determinant which is obtained when row j and column k are removed; it is indicated by Δ_{jk}.

As an example, let us expand the determinant Δ_G along column 3. We first multiply the (-4) at the top of this column by $(-1)^{3+1} = 1$ and then by its minor,

$$(-4)(-1)^{3+1}\begin{vmatrix} -3 & 6 \\ -4 & -2 \end{vmatrix}$$

and then repeat for the other two elements in column 3, adding the results,

$$\Delta_G = (-4)\begin{vmatrix} -3 & 6 \\ -4 & -2 \end{vmatrix} - (-2)\begin{vmatrix} 7 & -3 \\ -4 & -2 \end{vmatrix} + 11\begin{vmatrix} 7 & -3 \\ -3 & 6 \end{vmatrix}$$

The minors now contain only two rows and columns. They are of *order* two, and their values are easily determined by expanding in terms of minors again, here a trivial operation. Thus, for the first determinant, we expand along the first column by multiplying (-3) by $(-1)^{1+1}$ and its minor, which is merely the element (-2), and then multiplying (-4) by (-1) and by 6. Thus,

$$\begin{vmatrix} -3 & 6 \\ -4 & -2 \end{vmatrix} = (-3)(-2) - 6(-4) = 30$$

It is usually easier to remember the result for a second-order determinant as "upper left times lower right minus upper right times lower left." Finally,

$$\begin{aligned} \Delta_G &= -4[(-3)(-2) - 6(-4)] + 2[7(-2) - (-3)(-4)] + 11[7(6) - (-3)(-3)] \\ &= -4(30) + 2(-26) + 11(33) \\ &= 191 \end{aligned}$$

For practice, let us expand this same determinant along the first row,

$$\begin{aligned} \Delta_G &= 7\begin{vmatrix} 6 & -2 \\ -2 & 11 \end{vmatrix} - (-3)\begin{vmatrix} -3 & -2 \\ -4 & 11 \end{vmatrix} + (-4)\begin{vmatrix} -3 & 6 \\ -4 & -2 \end{vmatrix} \\ &= 7(62) + 3(-41) - 4(30) \\ &= 191 \end{aligned}$$

The expansion by minors is valid for a determinant of any order.

Repeating these rules for evaluating a determinant in more general terms, we would say, given a matrix $\mathbf{A}$,

$$\mathbf{A} = \begin{bmatrix} a_{11} & a_{12} & \cdots & a_{1N} \\ a_{21} & a_{22} & \cdots & a_{2N} \\ \cdots & \cdots & \cdots & \cdots \\ a_{N1} & a_{N2} & \cdots & a_{NN} \end{bmatrix}$$

then Δ_A may be obtained by expansion in terms of minors along any row j:

$$\begin{aligned}\Delta_A &= a_{j1}(-1)^{j+1}\,\Delta_{j1} + a_{j2}(-1)^{j+2}\,\Delta_{j2} + \cdots + a_{jN}(-1)^{j+N}\,\Delta_{jN}\\ &= \sum_{n=1}^{N} a_{jn}(-1)^{j+n}\,\Delta_{jn}\end{aligned}$$

or along any column k:

$$\begin{aligned}\Delta_A &= a_{1k}(-1)^{1+k}\,\Delta_{1k} + a_{2k}(-1)^{2+k}\,\Delta_{2k} + \cdots + a_{Nk}(-1)^{N+k}\,\Delta_{Nk}\\ &= \sum_{n=1}^{N} a_{nk}(-1)^{n+k}\,\Delta_{nk}\end{aligned}$$

The *cofactor* C_{jk} of the element appearing in both row j and column k is simply $(-1)^{j+k}$ times the minor Δ_{jk}. Thus, $C_{11} = \Delta_{11}$, but $C_{12} = -\Delta_{12}$ We may now write

$$\Delta_A = \sum_{n=1}^{N} a_{jn}C_{jn} = \sum_{n=1}^{N} a_{nk}C_{nk}$$

As an example, let us consider this third-order determinant:

$$\Delta = \begin{vmatrix} 4 & -1 & -2 \\ -1 & 6 & -3 \\ -2 & -3 & 10 \end{vmatrix}$$

We find

$$\Delta_{11} = \begin{vmatrix} 6 & -3 \\ -3 & 10 \end{vmatrix} = 51 \qquad \Delta_{12} = \begin{vmatrix} -1 & -3 \\ -2 & 10 \end{vmatrix} = -16$$

and

$$C_{11} = 51 \qquad C_{12} = 16$$

We next consider Cramer's rule, which enables us to find the values of the unknown variables. Let us again consider (3), (4), and (5); we define the determinant Δ_1 as that determinant which is obtained when the first column of Δ_G is replaced by the three constants on the right sides of the three equations. Thus,

$$\Delta_1 = \begin{vmatrix} -11 & -3 & -4 \\ 3 & 6 & -2 \\ 25 & -2 & 11 \end{vmatrix}$$

In order to take advantage of minors identical to those we just evaluated above, we expand along the first column:

$$\Delta_1 = -11\begin{vmatrix} 6 & -2 \\ -2 & 11 \end{vmatrix} - 3\begin{vmatrix} -3 & -4 \\ -2 & 11 \end{vmatrix} + 25\begin{vmatrix} -3 & -4 \\ 6 & -2 \end{vmatrix}$$

$$= -682 + 123 + 750 = 191$$

Cramer's rule then states that

$$v_1 = \frac{\Delta_1}{\Delta_G} = \frac{191}{191} = 1 \quad \text{V}$$

and $$v_2 = \frac{\Delta_2}{\Delta_G} = \frac{\begin{vmatrix} 7 & -11 & -4 \\ -3 & 3 & -2 \\ -4 & 25 & 11 \end{vmatrix}}{191} = \frac{581 - 63 - 136}{191} = 2 \quad \text{V}$$

and finally,

$$v_3 = \frac{\Delta_3}{\Delta_G} = \frac{\begin{vmatrix} 7 & -3 & -11 \\ -3 & 6 & 3 \\ -4 & -2 & 25 \end{vmatrix}}{191} = \frac{1092 - 291 - 228}{191} = 3 \quad \text{V}$$

Cramer's rule is applicable to a system of N simultaneous linear equations in N unknowns; for the ith variable v_i:

$$v_i = \frac{\Delta_i}{\Delta_G}$$

Appendix Two
A Proof of Thévenin's Theorem

We shall prove Thévenin's theorem in the same form in which it is stated in Sec. 3-6 of Chap. 3, repeated here for reference:

> Given any linear circuit, rearrange it in the form of two networks A and B that are connected together by two resistanceless conductors. Define a voltage v_{oc} as the open-circuit voltage which would appear across the terminals of A if B were disconnected so that no current is drawn from A. Then all the currents and voltages in B will remain unchanged if A is killed (all independent voltage sources and independent current sources in A replaced

by short circuits and open circuits, respectively) and an independent voltage source v_{oc} is connected, with proper polarity, in series with the dead (inactive) A network.

We shall effect our proof by showing that the original A network and the Thévenin equivalent of the A network both cause the same current to flow into the terminals of the B network. If the currents are the same, then the voltages must be the same; in other words, if we apply a certain current, which we might think of as a current source, to the B network, then the current source and the B network constitute a circuit which has a specific input voltage as a response. Thus, the current determines the voltage. Alternatively we could, if we wished, show that the terminal voltage at B is unchanged, because the voltage also determines the current uniquely. If the input voltage and current to the B network are unchanged, then it follows that the currents and voltages *throughout* the B network are also unchanged.

Let us first prove the theorem for the case where the B network is inactive (no independent sources). After this step has been accomplished, we may then use the superposition principle to extend the theorem to include B networks which contain independent sources.

The current i, flowing in the upper conductor from the A network to the B network in Fig. A2-1*a*, is therefore caused entirely by the independent sources present in the A network. Suppose now that we add an additional voltage source v_x, which we shall call the Thévenin source, in the conductor in which i is measured, as shown in Fig. A2-1*b*, and then adjust the magnitude and time variation of v_x until the current is reduced to zero. By our definition of v_{oc}, then, the voltage across the terminals of A must be v_{oc} since $i = 0$. B contains no independent sources, and no current is entering its terminals; therefore, there is no voltage across the terminals of the B network, and by Kirchhoff's voltage law the voltage of the Thévenin source is v_{oc} volts, $v_x = v_{oc}$. Moreover, since the Thévenin source and the A network jointly deliver no current to B, and since the A network by itself delivers a current i, then superposition requires that the Thévenin source acting by itself deliver $-i$ A to B. The source acting alone in a reversed direction, as shown in Fig. A2-1*c*, therefore produces a current in the upper lead i. This situation, however, is the same as the conclusion reached by Thévenin's theorem: the Thévenin source v_{oc} acting in series with the inactive A network is equivalent to the given network.

Now let us consider the case where the B network may be an active network. We now think of the current i, flowing from the A network to the B network in the upper conductor, as being composed of two parts i_A and i_B, where i_A is the current produced by A acting alone and the current i_B is due to B acting alone. Our ability to divide the current into these

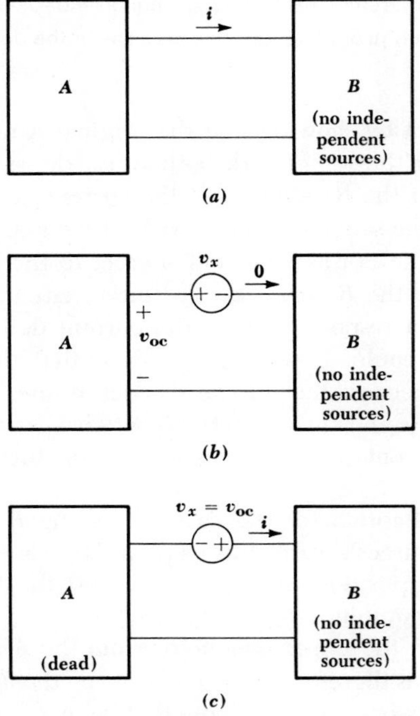

Fig. A2-1 (*a*) A general linear network A and a network B that contains no independent sources. (*b*) The Thévenin source is inserted in the circuit and adjusted until $i = 0$. No voltage appears across network B and thus $v_x = v_{oc}$. The Thévenin source thus produces $-i$ A while network A provides $+i$ A. (*c*) The Thévenin source is reversed and network A is killed. The current is therefore i.

two components is a direct consequence of the applicability of the superposition principle to these two linear networks; the complete response and the two partial responses are indicated by the diagrams of Fig. A2-2.

The partial response i_A has already been considered; if network B is inactive, we know that network A may be replaced by the Thévenin source and the inactive A network. In other words, of the three sources which we must keep in mind, those in A, those in B, and the Thévenin source, the partial response i_A occurs when A and B are dead and the Thévenin source is active. Preparing for the use of superposition, we now let A remain inactive, but turn on B and turn off the Thévenin source; by definition, the partial response i_B is obtained. Superimposing the results, the response when A is dead and both the Thévenin source and B are active is $i_A + i_B$. This sum is the original current i, and the situation wherein the Thévenin source and B are active, but A is dead, is the desired Thévenin

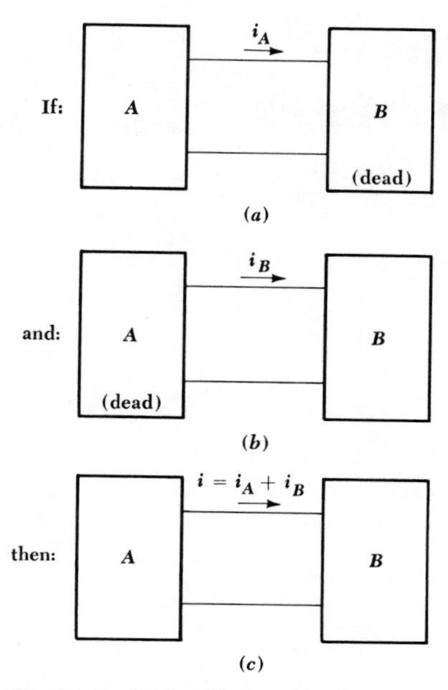

Fig. A2-2 Superposition enables the current i to be considered as the sum of two partial responses.

equivalent circuit. Thus the active network A may be replaced by its Thévenin source, the open-circuit voltage, in series with the inactive A network, regardless of the status of the B network; it may be either active or inactive.

Appendix Three Complex Numbers

This Appendix includes sections covering the definition of a complex number, the basic arithmetic operations for complex numbers, Euler's identity, and the exponential and polar forms of the complex number. We first introduce the concept of a complex number.

A3-1 THE COMPLEX NUMBER

Our early training in mathematics dealt exclusively with real numbers, such as 4, $-2/7$, and π. Soon, however, we began to encounter algebraic equations, such as $x^2 = -3$, which could not be satisfied by any real number.

Such an equation can be solved only through the introduction of the *imaginary unit* or the *imaginary operator,* which we shall designate[1] by the symbol j. By definition, $j^2 = -1$, and thus $j = \sqrt{-1}$, $j^3 = -j$, $j^4 = 1$, and so forth. The product of a real number times the imaginary operator is called an *imaginary number,* and the sum of a real number and an imaginary number is called a *complex number.* Thus, a number having the form $a + jb$, where a and b are real numbers, is a complex number.[2]

We shall designate a complex number by means of a special single symbol; thus, $\mathbf{A} = a + jb$. The complex nature of the number is indicated by the use of bold-face type; in hand-written material, a dot or a bar over the letter is customary. The complex number **A** above is described as having a *real component* or *real part* a and an *imaginary component* or *imaginary part* b. This is also expressed as

$$\text{Re}\,[\mathbf{A}] = a \qquad \text{Im}\,[\mathbf{A}] = b$$

The imaginary component of **A** is *not* jb. By definition, the imaginary component is a real number.

It should be noted that all real numbers may be regarded as complex numbers having imaginary parts equal to zero. The real numbers are therefore included in the system of complex numbers, and we may now consider them as a special case. When we define the fundamental arithmetic operations for complex numbers, we should therefore expect them to reduce to the corresponding definitions for real numbers if the imaginary part of every complex number is set equal to zero.

Since any complex number is completely characterized by a pair of real numbers, such as a and b in the example above, we can obtain some visual assistance by representing a complex number graphically on a rectangular, or cartesian, coordinate system. By providing ourselves with a real axis and an imaginary axis, as shown in Fig. A3-1, we form a *complex plane* or *Argand diagram* on which any complex number can be represented as a single point. The complex numbers $\mathbf{M} = 3 + j1$ and $\mathbf{N} = 2 - j2$ are indicated. It is important to understand that this complex plane is only a visual aid; it is not at all essential to the mathematical statements which follow.

We shall define two complex numbers as being equal if, and only if, their real parts are equal and their imaginary parts are equal. Graphically, then, to each point in the complex plane there corresponds only one complex number, and conversely, to each complex number there corre-

[1]The mathematicians designate the imaginary operator by the symbol i, but it is customary to use j in electrical engineering in order to avoid confusion with the symbol for current.
[2]The choice of the words imaginary and complex is unfortunate. They are used here and in the mathematical literature as technical terms to designate a class of numbers. To interpret imaginary as "not pertaining to the physical world" or complex as "complicated" is neither justified nor intended.

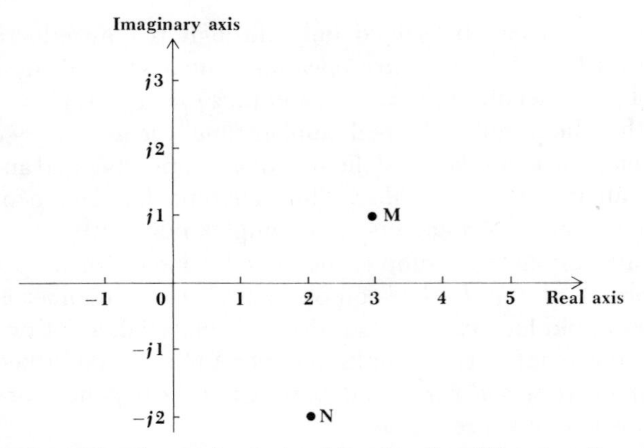

Fig. A3-1 The complex numbers $\mathbf{M} = 3 + j1$ and $\mathbf{N} = 2 - j2$ are shown on the complex plane.

sponds only one point in the complex plane. Thus, given the two complex numbers.

$$\mathbf{A} = a + jb \qquad \text{and} \qquad \mathbf{B} = c + jd$$

then, if

$$\mathbf{A} = \mathbf{B}$$

it is necessary that

$$a = c \qquad \text{and} \qquad b = d$$

A complex number expressed as the sum of a real number and an imaginary number, such as $\mathbf{A} = a + jb$, is said to be in *rectangular* or *cartesian* form. Other forms for a complex number will appear shortly.

Let us now define the fundamental operations of addition, subtraction, multiplication, and division for complex numbers. The sum of two complex numbers is defined as the complex number whose real part is the sum of the real parts of the two complex numbers and whose imaginary part is the sum of the imaginary parts of the two complex numbers. Thus,

$$(a + jb) + (c + jd) = (a + c) + j(b + d)$$

For example,

$$(3 + j4) + (4 - j2) = 7 + j2$$

The difference of two complex numbers is taken in a similar manner; for example,

$$(3 + j4) - (4 - j2) = -1 + j6$$

Addition and subtraction of complex numbers may also be accomplished graphically on the complex plane. Each complex number is represented as a vector, or directed line segment, and the sum is obtained by completing the parallelogram, illustrated by Fig. A3-2*a*, or by connecting the vectors in a head-to-tail manner, as shown in Fig. A3-2*b*. A graphical sketch is often useful as a check for a more exact numerical solution.

The product of two complex numbers is defined by

$$(a + jb)(c + jd) = (ac - bd) + j(bc + ad)$$

This result may be easily obtained by the direct multiplication of the two binomial terms, using the rules of the algebra of real numbers, and then simplifying the result by letting $j^2 = -1$. For example,

$$\begin{aligned}(3 + j4)(4 - j2) &= 12 - j6 + j16 - 8j^2 \\ &= 12 + j10 + 8 \\ &= 20 + j10\end{aligned}$$

It is easier to multiply the complex numbers by this method, particularly if we immediately replace j^2 by -1, than it is to substitute into the general formula which defines the multiplication.

Before defining the operation of division for complex numbers, we

Fig. A3-2 (*a*) The sum of the complex numbers $\mathbf{M} = 3 + j1$ and $\mathbf{N} = 2 - j2$ is obtained by constructing a parallelogram. (*b*) The sum of the same two complex numbers is found by a head-to-tail combination.

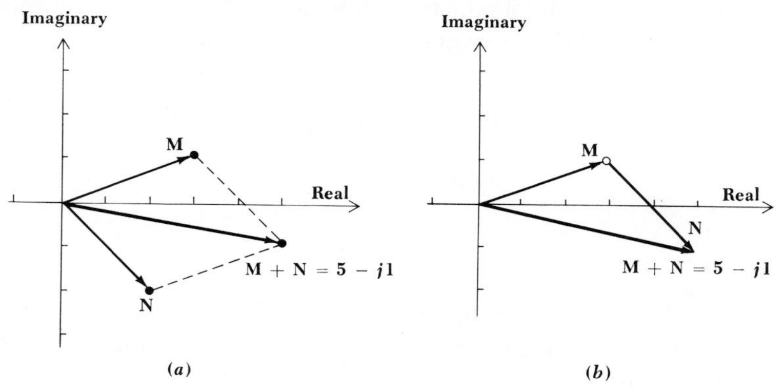

should define the conjugate of a complex number. The *conjugate* of the complex number $\mathbf{A} = a + jb$ is $a - jb$ and is represented as $\mathbf{A}^*$. The conjugate of any complex number is therfore easily obtained by merely changing the sign of the imaginary part of the complex number. Thus, if

$$\mathbf{A} = 5 + j3$$

then

$$\mathbf{A}^* = 5 - j3$$

It is evident that the conjugate of any complicated complex expression may be found by replacing every complex term in the expression by its conjugate.

The definitions of addition, subtraction, and multiplication show that the following statements are true: the sum of a complex number and its conjugate is a real number; the difference of a complex number and its conjugate is an imaginary number; and the product of a complex number and its conjugate is a real number. It is also evident that if $\mathbf{A}^*$ is the conjugate of $\mathbf{A}$, then $\mathbf{A}$ is the conjugate of $\mathbf{A}^*$; in other words, $\mathbf{A} = (\mathbf{A}^*)^*$. A complex number and its conjugate are said to form a *conjugate complex pair* of numbers.

We now define the quotient of two complex numbers:

$$\frac{\mathbf{A}}{\mathbf{B}} = \frac{(\mathbf{A})(\mathbf{B}^*)}{(\mathbf{B})(\mathbf{B}^*)}$$

and thus

$$\frac{a + jb}{c + jd} = \frac{(ac + bd) + j(bc - ad)}{c^2 + d^2}$$

We multiply numerator and denominator by the conjugate of the denominator in order to obtain a denominator which is real; this process is called *rationalizing the denominator.* As a numerical example,

$$\frac{3 + j4}{4 - j2} = \frac{(3 + j4)(4 + j2)}{(4 - j2)(4 + j2)}$$

$$= \frac{4 + j22}{16 + 4}$$

$$= 0.2 + j1.1$$

The addition or subtraction of two complex numbers which are each expressed in rectangular form is a relatively simple operation; multiplication or division of two complex numbers in rectangular form, however, is a rather unwieldy process. These latter two operations will be found to be much simpler when the complex numbers are given in either exponential or polar form. These forms will be introduced in Secs. A3-3 and A3-4.

Drill Problems

A3-1 Given $\mathbf{A} = 2 + j4$, $\mathbf{B} = -1 + j3$, $\mathbf{C} = 3 - j2$, find: (*a*) $\mathbf{A} - \mathbf{B}$; (*b*) $3\mathbf{C} + 2\mathbf{A}$; (*c*) $-2\mathbf{B} + \mathbf{A} - 2\mathbf{C}$.

Ans. $-2 + j2$; $3 + j1$; $13 + j2$

A3-2 Given $\mathbf{A} = 2 + j4$, $\mathbf{B} = -1 + j3$, $\mathbf{C} = 3 - j2$, find: (*a*) $\mathbf{AB}$; (*b*) $(\mathbf{A} + \mathbf{B})(\mathbf{A} - \mathbf{B})$; (*c*) $j(\mathbf{A}^2 + 2\mathbf{C})(\mathbf{A} - \mathbf{C})$.

Ans. $-14 + j2$; $-4 + j22$; $48 - j66$

A3-3 Given $\mathbf{A} = 2 + j4$, $\mathbf{B} = -1 + j3$, $\mathbf{C} = 3 - j2$, find: (*a*) $\mathbf{A}^*\mathbf{B}$; (*b*) $(\mathbf{A}^*\mathbf{B}^*)^*$; (*c*) $(\mathbf{A} - \mathbf{A}^*)(\mathbf{B} + \mathbf{B}^*)\mathbf{CC}^*$.

Ans. $-14 + j2$; $-j208$; $10 + j10$

A3-4 Given $\mathbf{A} = 2 + j4$, $\mathbf{B} = -1 + j3$, $\mathbf{C} = 3 - j2$, find: (*a*) $(\mathbf{A} + \mathbf{C})/\mathbf{B}$; (*b*) $\mathbf{AC}/(\mathbf{A} + \mathbf{B})$; (*c*) $(2\mathbf{A} - 3\mathbf{B}^*)/(\mathbf{A} - \mathbf{C}^*)$.

Ans. $0.1 - j1.7$; $1.4 - j1.8$; $5.4 - j6.2$

A3-2 EULER'S IDENTITY

In Chap. 9 we begin to encounter functions of time which contain complex numbers, and we are concerned with the differentiation and integration of these functions. In most cases we must differentiate and integrate these functions with respect to the real variable t. In only a few isolated cases is it necessary to differentiate or integrate with respect to a complex variable. Neither of these cases requires any special treatment.

The operations of differentiation and integration with respect to complex variables are defined using the same limiting processes that are employed for real variables, and it can be shown in a straightforward manner that complex variables and complex constants can be treated just as though they were real variables or real constants when performing the operations of differentiation or integration. In other words, if $\mathbf{f}(t)$ is a complex function of time, such as

$$\mathbf{f}(t) = a \cos ct + jb \sin ct$$

then

$$\frac{d\mathbf{f}(t)}{dt} = -ac \sin ct + jbc \cos ct$$

and $$\int \mathbf{f}(t)\,dt = \frac{a}{c}\sin ct - j\frac{b}{c}\cos ct + \mathbf{C}$$

where the constant of integration **C** is a complex number in general.

At this time we must make use of a very important fundamental relationship known as Euler's identity (pronounced "oilers"). We shall prove this identity, for it is extremely useful in representing a complex number in a form other than rectangular form. We begin by forming the complex quantity **B**,

$$\mathbf{B} = \cos\theta + j\sin\theta \tag{1}$$

where θ is a dimensionless real number expressed in radians. If we differentiate **B** with respect to θ, we obtain

$$\frac{d\mathbf{B}}{d\theta} = -\sin\theta + j\cos\theta = j(\cos\theta + j\sin\theta)$$

or $$\frac{d\mathbf{B}}{d\theta} = j\mathbf{B}$$

and $$\frac{d\mathbf{B}}{\mathbf{B}} = j\,d\theta \tag{2}$$

Now let us integrate each side of (2):

$$\ln\mathbf{B} = j\theta + \mathbf{C}$$

where **C** is a complex constant of integration. We evaluate **C** by returning to (1) and letting $\theta = 0$; thus, $\mathbf{B} = 1 + j0$ when $\theta = 0$. Hence, $\mathbf{C} = 0$, and

$$\ln\mathbf{B} = j\theta \qquad \text{or} \qquad \mathbf{B} = e^{j\theta}$$

and thus we obtain Euler's identity,

$$e^{j\theta} = \cos\theta + j\sin\theta \tag{3}$$

If we had begun with the conjugate of (1), we should have obtained an alternative form of Euler's identity,

$$e^{-j\theta} = \cos\theta - j\sin\theta \tag{4}$$

By adding and subtracting (3) and (4), we obtain the two expressions which we used without proof in our study of the underdamped natural response of the parallel and series *RLC* circuits,

$$\cos\theta = \tfrac{1}{2}(e^{j\theta} + e^{-j\theta}) \tag{5}$$

$$\sin\theta = -j\tfrac{1}{2}(e^{j\theta} - e^{-j\theta}) \tag{6}$$

Another method whereby Euler's identity may be developed is through a comparison of the infinite power-series representations of $e^{j\theta}$, $\cos\theta$, and $\sin\theta$.

Drill Problem

A3-5 Make use of Euler's identity to evaluate: (*a*) $e^{j0.5}$; (*b*) $e^{1+j0.5}$; (*c*) $\cos(j0.5)$.
Ans. *1.128; 0.878 + j0.479; 2.39 + j1.30*

A3-3 THE EXPONENTIAL FORM

Let us now take Euler's identity

$$e^{j\theta} = \cos\theta + j\sin\theta$$

and multiply each side by the real number C,

$$Ce^{j\theta} = C\cos\theta + jC\sin\theta \tag{7}$$

The right side of (7) consists of the sum of a real number and an imaginary number and thus represents a complex number in rectangular form; let us call this complex number **A**, where $\mathbf{A} = a + jb$. By equating the real parts

$$a = C\cos\theta \tag{8}$$

and the imaginary parts

$$b = C\sin\theta \tag{9}$$

squaring and adding (8) and (9),

$$a^2 + b^2 = C^2$$

or

$$C = \sqrt{a^2 + b^2} \tag{10}$$

and dividing (9) by (8),

$$\frac{b}{a} = \tan\theta$$

or

$$\theta = \tan^{-1}\frac{b}{a} \tag{11}$$

we obtain the relationships (10) and (11), which enable us to determine C and θ from a knowledge of a and b. For example, if $\mathbf{A} = 4 + j2$, then we identify a as 4 and b as 2 and find C and θ:

$$C = \sqrt{4^2 + 2^2} = 4.47$$

$$\theta = \tan^{-1} 2/4 = 26.6^\circ$$

We could use this new information to write $\mathbf{A}$ in the form

$$\mathbf{A} = 4.47 \cos 26.6^\circ + j4.47 \sin 26.6^\circ$$

but it is the form of the left side of (7) which will prove to be the more useful:

$$\mathbf{A} = Ce^{j\theta} = 4.47e^{j26.6^\circ}$$

A complex number expressed in this manner is said to be in *exponential form.* The real multiplying factor C is known as the *magnitude* or *amplitude* and the real quantity θ appearing in the exponent is called the *argument* or *angle.* A mathematician would always express θ in radians and would write

$$\mathbf{A} = 4.47e^{j0.464}$$

but electrical engineers customarily work in terms of degrees. The use of the degree symbol (°) in the exponent should make confusion impossible.

To recapitulate, if we have a complex number which is given in rectangular form,

$$\mathbf{A} = a + jb$$

and wish to express it in exponential form,

$$\mathbf{A} = Ce^{j\theta}$$

we may find C and θ by (10) and (11). If we are given the complex number in exponential form, then we may find a and b by (8) and (9).

One question will be found to arise in the determination of the angle θ by using the arc-tangent relationship (11). This function is multivalued, and an appropriate angle must be selected from various possibilities. One method by which the choice may be made is to select an angle for which the sine and cosine have the proper signs to produce the required values of a and b from (8) and (9). For example, let us convert

$$\mathbf{V} = 4 - j3$$

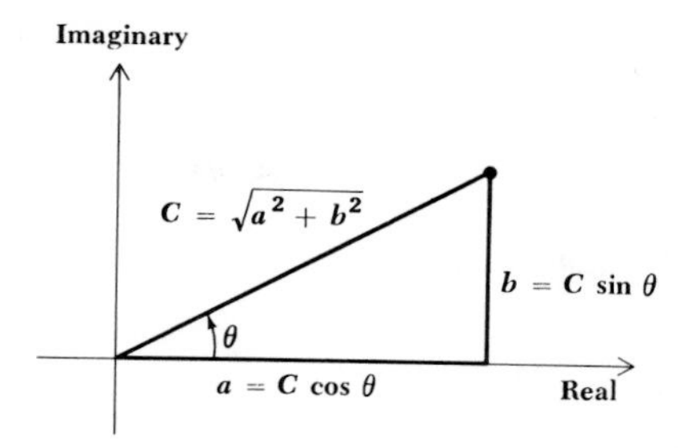

Fig. A3-3 A complex number may be represented by a point in the complex plane through choosing the correct real and imaginary parts from the rectangular form, or by selecting the magnitude and angle from the exponential form.

to exponential form. The amplitude is

$$C = \sqrt{4^2 + (-3)^2} = 5$$

and the angle is

$$\theta = \tan^{-1} \frac{-3}{4} \tag{12}$$

A value of θ must be selected which leads to a positive value for $\cos \theta$, since $4 = 5 \cos \theta$, and a negative value for $\sin \theta$, since $-3 = 5 \sin \theta$. We therefore obtain $\theta = -36.9°$, $323.1°$, $-396.9°$, and so forth. Any of these angles is correct, and we usually select that one which is the simplest, here $-36.9°$. We should note that the solution of (12), $\theta = 143.1°$, is not correct because $\cos \theta$ is negative and $\sin \theta$ is positive.

A simpler method of selecting the correct angle is available if we represent the complex number graphically in the complex plane. Let us first select a complex number, given in rectangular form, $\mathbf{A} = a + jb$, which lies in the first quadrant of the complex plane, as illustrated in Fig. A3-3. If we draw a line from the origin to the point which represents the complex number, we shall have constructed a right triangle whose hypotenuse is evidently the magnitude of the exponential representation of the complex number. In other words, $C = \sqrt{a^2 + b^2}$. Moreover, the counterclockwise angle which the line makes with the positive real axis is seen to be the angle θ of the exponential representation, because $a = C \cos \theta$ and $b = C \sin \theta$. Now if we are given the rectangular form of a complex number which lies in another quadrant, such as $\mathbf{V} = 4 - j3$, which is depicted in Fig. A3-4, the correct angle is graphically evident, either $-36.9°$ or $323.1°$ for this example. The sketch may often be visualized and need not be drawn.

If the rectangular form of the complex number has a negative real part,

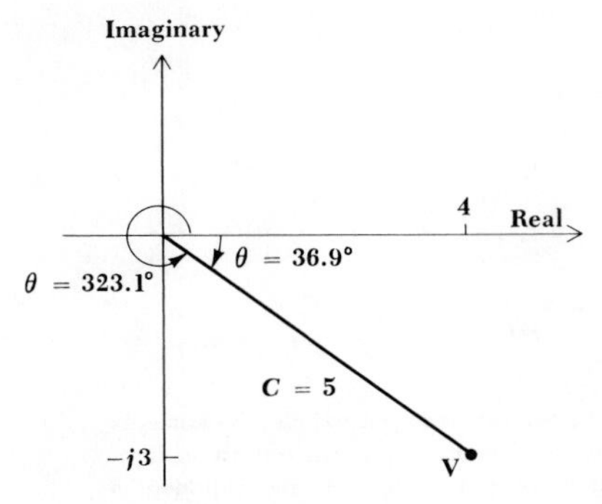

Fig. A3-4 The complex number $\mathbf{V} = 4 - j3 = 5e^{-j36.9°}$ is represented in the complex plane.

it is often easier to work with the negative of the complex number, thus avoiding angles greater than 90° in magnitude. For example, given

$$\mathbf{I} = -5 + j2$$

we write

$$\mathbf{I} = -(5 - j2)$$

and then transform to exponential form,

$$\mathbf{I} = -Ce^{j\theta}$$

where

$$C = \sqrt{29} = 5.39 \qquad \text{and} \qquad \theta = \tan^{-1}\frac{-2}{5} = -21.8°$$

Thus, the exponential form of $\mathbf{I}$ is

$$\mathbf{I} = -5.39e^{-j21.8°}$$

The negative sign may be removed from the complex number by increasing or decreasing the angle by 180°, as shown by reference to a sketch in the complex plane. Thus, we may write

$$\mathbf{I} = 5.39e^{j158.2°} \qquad \text{or} \qquad \mathbf{I} = 5.39e^{-j201.8°}$$

Any of these three forms is acceptable.

One last remark about the exponential representation of a complex number should be made. Two complex numbers, both written in exponential form with positive amplitudes, are equal if, and only if, their magnitudes are equal and their angles are equivalent. Equivalent angles are those which differ by multiples of 360°. For example, if $\mathbf{A} = Ce^{j\theta}$ and $\mathbf{B} = De^{j\phi}$, then if $\mathbf{A} = \mathbf{B}$, it is necessary that $C = D$ and $\theta = \phi \pm 360°n$, where $n = 0$, 1, 2, 3, ...

Drill Problems

A3-6 Express each of the following numbers in exponential form, using an angle less than 180° in magnitude: (*a*) $2 - j3$; (*b*) $-4 - j3$; (*c*) $0.2 - j3$.

Ans. $-5e^{j36.9°}$; $3.01e^{-j86.18°}$; $3.61e^{-j56.3°}$

A3-7 Express each of the following numbers in rectangular form: (*a*) $7e^{-j28°}$; (*b*) $5e^{j200°}$; (*c*) $6e^{-j162°}$.

Ans. $-5.71 - j1.854$; $-4.70 - j1.710$; $6.18 - j3.29$

A3-4 THE POLAR FORM

The third (and last) form in which we may represent a complex number is essentially the same as the exponential form, except for a slight difference in symbolism. We use an angle sign ($\angle$) to replace the combination (e^j). Thus, the exponential representation of a complex number $\mathbf{A}$,

$$\mathbf{A} = Ce^{j\theta}$$

may be written somewhat more concisely as

$$\mathbf{A} = C\underline{/\theta}$$

The complex number is now said to be expressed in *polar* form, a name which suggests the representation of a point in a (complex) plane through the use of polar coordinates.

It is apparent that the transformation from rectangular to polar form or from polar form to rectangular form is basically the same as transformations between rectangular and exponential form. The same relationships exist between C, θ, a, and b.

The complex number

$$\mathbf{A} = 10 + j10$$

is thus written in exponential form as

$$\mathbf{A} = 14.14e^{j45°}$$

and in polar form as

$$\mathbf{A} = 14.14\underline{/45^\circ}$$

In order to appreciate the utility of the exponential and polar forms, let us consider the multiplication and division of two complex numbers represented in exponential or polar form. If we are given

$$\mathbf{A} = 5\underline{/53.1^\circ} \quad \text{and} \quad \mathbf{B} = 15\underline{/-36.9^\circ}$$

then the expression of these two complex numbers in exponential form

$$\mathbf{A} = 5e^{j53.1^\circ} \qquad \mathbf{B} = 15e^{-j36.9^\circ}$$

enables us to write the product as a complex number in exponential form whose amplitude is the product of the amplitudes and whose angle is the algebraic sum of the angles, in accordance with the normal rules for multiplying two exponential quantities,

$$(\mathbf{A})(\mathbf{B}) = (5)(15)e^{j(53.1^\circ - 36.9^\circ)}$$

or

$$\mathbf{AB} = 75e^{j16.2^\circ} = 75\underline{/16.2^\circ}$$

From the definition of the polar form, it is evident that

$$\frac{\mathbf{A}}{\mathbf{B}} = 0.333\underline{/90^\circ}$$

Addition and subtraction of complex numbers are accomplished most easily by operating on complex numbers in rectangular form, and the addition or subtraction of two complex numbers given in exponential or polar form should begin with the conversion of the two complex numbers to rectangular form. The reverse situation applies to multiplication and division; two numbers given in rectangular form should be transformed to polar form, unless the numbers happen to be small integers. For example, if we wish to multiply $(1 - j3)$ by $(2 + j1)$, it is easier to multiply them directly as they stand and obtain $(5 - j5)$. If the numbers can be multiplied mentally, then time is wasted in transforming them to polar form.

We should now endeavor to become familiar with the three different forms in which complex numbers may be expressed and with the rapid conversion from one form to another. The relationships between the three forms seem almost endless, and the following lengthy equation summarizes the various interrelationships:

$$\mathbf{A} = a + jb = \text{Re}\,[\mathbf{A}] + j\,\text{Im}\,[\mathbf{A}] = Ce^{j\theta} = \sqrt{a^2 + b^2}\, e^{j \tan^{-1}(b/a)}$$

$$= \sqrt{a^2 + b^2}\,\underline{/\tan^{-1}(b/a)} = \frac{b}{\sin\,[\tan^{-1}(b/a)]}\,\underline{/\tan^{-1}(b/a)}$$

$$= C \cos\theta + jC \sin\theta = C\underline{/\theta}$$

Most of the conversions from one form to another will be done with the help of a slide rule. The transformation which we shall find most important to be able to achieve accurately and simply is that from rectangular to polar or exponential form. Although this may be done by finding the square root of the sum of the squares of the rectangular components, it is easier for most complex number manipulators to find the magnitude as

$$|a + jb| = \frac{b}{\sin\,[\tan^{-1}(b/a)]}$$

Each student should refer to his own slide-rule manual for the easiest technique. In case of confusion, help should be solicited from an instructor, a fellow classmate, or a friendly high school student.

We shall find that complex numbers are a convenient mathematical artifice which facilitates the analysis of real physical situations. Inevitably in a physical problem a complex number is somehow accompanied by its conjugate.

Drill Problems

A3-8 Evaluate each of the following, expressing the answer in polar form: (*a*) $(2 + 3\underline{/60^\circ})(3\underline{/150^\circ} + 3\underline{/30^\circ})$; (*b*) $1/(1.82\underline{/-68.5^\circ} - 0.717\underline{/103.2^\circ})$; (*c*) $[-j17 + (4/j) + 5\underline{/90^\circ}]/(2.50\underline{/45^\circ} + 2.10\underline{/-30^\circ})$.

Ans. $0.396\underline{/70.8^\circ}$; $4.37\underline{/-101.3^\circ}$; $13.08\underline{/126.6^\circ}$

A3-9 Solve for **A**: (*a*) $\mathbf{A}^2 + j2\mathbf{A} - 1 = 0$; (*b*) $\mathbf{A} - e^{2\underline{/30^\circ}} = 0$; (*c*) $e^{\mathbf{A}} = -1$.

Ans. $\pm j3.14$; $3.05 + j4.76$; $-j1$

Index

Standard Abbreviations

alternating current	ac
ampere	A
coulomb	C
cycle per second	cps (avoid)
decibel	dB
degree Celsius	°C
degree Kelvin	°K
direct current	dc
electronvolt	eV
farad	F
foot	ft
gram	g
henry	H
hertz	Hz
hour	h
inch	in
joule	J
kilogram	kg
meter	m
meter-kilogram-second	mks

mho	℧
minute	min
neper	Np
newton	N
newton-meter	N-m
ohm	Ω
pound-force	lbf
power factor	PF
radian	rad
resistance-inductance-capacitance	RLC
revolutions per second	rps
root-mean-square	rms
second	s
vacuum-tube voltmeter	VTVM
volt	V
voltampere	VA
watt	W
watthour	Wh

[Note: Standard prefixes of the decimal system are tabulated in Sec. 1–2.]